Methods in Enzymology

Volume 289
SOLID-PHASE PEPTIDE SYNTHESIS

METHODS IN ENZYMOLOGY

EDITORS-IN-CHIEF

John N. Abelson   Melvin I. Simon

DIVISION OF BIOLOGY
CALIFORNIA INSTITUTE OF TECHNOLOGY
PASADENA, CALIFORNIA

FOUNDING EDITORS

Sidney P. Colowick and Nathan O. Kaplan

Methods in Enzymology

Volume 289

Solid-Phase Peptide Synthesis

EDITED BY

Gregg B. Fields

DEPARTMENT OF LABORATORY MEDICINE AND PATHOLOGY
BIOMEDICAL ENGINEERING CENTER
UNIVERSITY OF MINNESOTA
MINNEAPOLIS, MINNESOTA

ACADEMIC PRESS

San Diego   London   Boston   New York   Sydney   Tokyo   Toronto

Academic Press
15 East 26th Street, 15th Floor, New York, New York 10010, USA
http://www.apnet.com

Academic Press Limited
24-28 Oval Road, London NW1 7DX, UK
http://www.hbuk.co.uk/ap/

International Standard Book Number: 0-12-182190-0

PRINTED IN THE UNITED STATES OF AMERICA
97 98 99 00 01 02 MM 9 8 7 6 5 4 3 2 1

Table of Contents

Section I. Methods for Solid-Phase Assembly of Peptides

Section II. Analytical Techniques

Section III. Specialized Applications

Contributors to Volume 289

Article numbers are in parentheses following the names of contributors.
Affiliations listed are current.

FERNANDO ALBERICIO (7, 15), *Department of Organic Chemistry, University of Barcelona, E-08028 Barcelona, Spain*

DIANNE ALEWOOD (2), *Centre for Drug Design and Development, The University of Queensland, Q 4072 Australia*

PAUL ALEWOOD (2), *Centre for Drug Design and Development, The University of Queensland, Q 4072 Australia*

RUTH HOGUE ANGELETTI (32), *Department of Developmental and Molecular Biology, Albert Einstein College of Medicine, Bronx, New York 10461*

IOANA ANNIS (10), *Department of Chemistry, University of Minnesota, Minneapolis, Minnesota 55455*

ERIC ATHERTON (4), *Zeneca-Cambridge Research Biochemicals, Northwich, Cheshire CW9 7RA, United Kingdom*

GEORGE BARANY (8, 10, 27), *Departments of Chemistry and Laboratory Medicine and Pathology, University of Minnesota, Minneapolis, Minnesota 55455*

ELISAR BARBAR (27), *Department of Biochemistry, University of Minnesota, St. Paul, Minnesota 55108*

ŠÁRKA BERANOVÁ-GIORGIANNI (21), *Department of Physiology, The Charles B. Stout Neuroscience Mass Spectrometry Laboratory, The University of Tennessee, Memphis, Tennessee 38163*

CHRISTOPHER BLACKBURN (9), *PerSeptive Biosystems, Inc., Framingham, Massachusetts 01701*

LYNDA F. BONEWALD (32), *Departments of Medicine and Biochemistry, University of Texas Health Science Center, San Antonio, Texas 78284*

ANDREW C. BRAISTED (14), *Department of Protein Engineering, Genentech, Inc., South San Francisco, California 94080*

DANIEL J. BURDICK (22), *Bio-Organic Chemistry Department, Genentech, Inc., South San Francisco, California 94080*

PAUL J. CACHIA (19), *The Canadian Bacterial Diseases Network of Centres of Excellence, University of Alberta, Edmonton, Alberta T6G 2H7 Canada*

LOUIS A. CARPINO (7), *Department of Chemistry, University of Massachusetts, Amherst, Massachusetts 01003*

MARK W. CRANKSHAW (17), *Department of Molecular Biology and Pharmacology, Washington University School of Medicine, St. Louis, Missouri 63110*

T. A. CROSS (31), *Center for Interdisciplinary Magnetic Resonance in the National High Magnetic Field Laboratory, Institute of Molecular Biophysics and Department of Chemistry, Florida State University, Talahassee, Florida 32306-4005*

PHILIP E. DAWSON (13, 24), *The Scripps Research Institute, La Jolla, California 92037*

DOMINIC M. DESIDERIO (21), *Departments of Neurology and Biochemistry, The Charles B. Stout Neuroscience Mass Spectrometry Laboratory, The University of Tennessee, Memphis, Tennessee 38163*

YOAV DORI (26), *Department of Chemical Engineering and Materials Science, Biomedical Engineering Center, University of Minnesota, Minneapolis, Minnesota 55455-0392*

MIKAEL ELOFSSON (11), *Organic Chemistry 2, Center for Chemistry and Chemical Engineering, The Lund Institute of Technology, Lund University, S-22100 Lund, Sweden*

GREGG B. FIELDS (5, 25, 26, 32), *Department of Laboratory Medicine and Pathology, Biomedical Engineering Center, University of Minnesota, Minneapolis, Minnesota 55455-0392*

MICHAEL C. FITZGERALD (24), *The Scripps Research Institute, La Jolla, California 92037*

ERNEST GIRALT (15), *Department of Organic Chemistry, University of Barcelona, E-08028 Barcelona, Spain*

JOHN GORKA (17), *Department of Molecular Biology and Pharmacology, Washington University School of Medicine, St. Louis, Missouri 63110*

GREGORY A. GRANT (17), *Departments of Medicine and Molecular Biology and Pharmacology, Washington University School of Medicine, St. Louis, Missouri 63110*

CHRISTOPHER M. GROSS (27), *Department of Biochemistry, University of Minnesota, St. Paul, Minnesota 55108; Department of Chemistry, University of Minnesota, Minneapolis, Minnesota 55455*

CYNTHIA A. GUY (5), *Department of Laboratory Medicine and Pathology, Biomedical Engineering Center, University of Minnesota, Minneapolis, Minnesota 55455-0392*

BALAZS HARGITTAI (10), *Departments of Chemistry and Laboratory Medicine and Pathology, University of Minnesota, Minneapolis, Minnesota 55455*

ROBERT S. HODGES (19), *Department of Biochemistry and the Medical Research Council, Group in Protein Structure and Function, University of Alberta, Edmonton, Alberta T6G 2H7, Canada*

J. KEVIN JUDICE (14), *Department of Bioorganic Chemistry, Genentech, Inc., South San Francisco, CA 94080*

STEVEN A. KATES (9), *PerSeptive Biosystems, Inc., Framingham, Massachusetts 01701*

STEPHEN B. H. KENT (13, 24), *Gryphon Sciences, South San Francisco, California 94080*

JAN KIHLBERG (11), *Organic Chemistry 2, Umeå University, S-90187 Umeå, Sweden*

LESLIE H. KONDEJEWSKI (19), *The Protein Engineering Network of Centres of Excellence, University of Alberta, Edmonton, Alberta T6G 2H7 Canada*

VIKTOR KRCHNÁK (16), *Robotics and Synthesis Automation, Trega Biosciences, Inc., San Diego, California 92121*

JANELLE L. LAUER (25), *Department of Laboratory Medicine and Pathology, Biomedical Engineering Center, University of Minnesota, Minneapolis, Minnesota 55455-0392*

MICHAL LEBL (16), *Robotics and Synthesis Automation, Trega Biosciences, Inc., San Diego, California 92121*

PAUL LLOYD-WILLIAMS (15), *Department of Organic Chemistry, University of Barcelona, E-08028 Barcelona, Spain*

STEPHEN LOVE (2), *Centre for Drug Design and Development, The University of Queensland, Q 4072 Australia*

COLIN T. MANT (19), *Department of Biochemistry and the Medical Research Council, Group in Protein Structure and Function, University of Alberta, Edmonton, Alberta T6G 2H7, Canada*

KEVIN H. MAYO (30), *Department of Biochemistry, Biomedical Engineering Center, University of Minnesota, Minneapolis, Minnesota 55455*

JAMES B. MCCARTHY (26), *Department of Laboratory Medicine and Pathology, Biomedical Engineering Center, University of Minnesota, Minneapolis, Minnesota 55455-0392*

MORTEN MELDAL (6), *Carlsberg Laboratory, Department of Chemistry, DK-2500 Valby, Denmark*

BRUCE MERRIFIELD (1), *The Rockefeller University, New York, New York 10021*

WIM MEUTERMANS (2), *Centre for Drug Design and Development, The University of Queensland, Q 4072 Australia*

LES MIRANDA (2), *Centre for Drug Design and Development, The University of Queensland, Q 4072 Austalia*

OSCAR D. MONERA (19), *The Protein Engineering Network of Centres of Excellence, University of Alberta, Edmonton, Alberta T6G 2H7 Canada*

WILLIAM T. MOORE (23), *Protein Chemistry Laboratory, Department of Pathology and Laboratory Medicine, The School of Medicine, University of Pennsylvania, Philadelphia, Pennsylvania 19104*

TOM W. MUIR (13, 24), *Laboratory of Synthetic Protein Chemistry, The Rockefeller University, New York, New York 10021*

TEIKA PAKALNS (26), *Department of Chemical Engineering and Materials Science, Biomedical Engineering Center, University of Minnesota, Minneapolis, Minnesota 55455-0392*

JOHN WILLIAM PERICH (12), *School of Chemistry, The University of Melbourne, Parkville, Victoria 3032, Australia*

LOURDES A. SALVADOR (11), *Organic Chemistry 2, Center for Chemistry and Chemical Engineering, The Lund Institute of Technology, Lund University, S-22100 Lund, Sweden*

AGUSTIN SANCHEZ (20), *Beckman Center, Stanford University, Stanford, California 94305-5425*

ALAN J. SMITH (18, 20), *Beckman Center, Stanford University, Stanford, California 94305-5425*

MICHAEL F. SONGSTER (8), *Solid Phase Sciences, Biosearch Technologies, Inc., San Rafael, California 94903*

JANE C. SPETZLER (28), *Department of Microbiology and Immunology, Vanderbilt University, Nashville, Tennessee 37232*

JOHN M. STEWART (3), *Department of Biochemistry, University of Colorado Medical School, Denver, Colorado 80262*

JOHN T. STULTS (22), *Protein Chemistry Department, Genentech, Inc., South San Francisco, California 94080*

JAMES P. TAM (28), *Department of Microbiology and Immunology, Vanderbilt University Medical School, Nashville, Tennessee 37232*

MATTHEW TIRRELL (26), *Department of Chemical Engineering and Materials Science, Biomedical Engineering Center, University of Minnesota, Minneapolis, Minnesota 55455-0392*

GEOFFREY W. TREGEAR (29), *Howard Florey Institute, University of Melbourne, Parkville, Victoria 3052, Australia*

JOHN D. WADE (29), *Howard Florey Institute, University of Melbourne, Parkville, Victoria 3052, Australia*

DONALD A. WELLINGS (4), *Zeneca-Cambridge Research Biochemicals, Northwich, Cheshire CW9 7RA, United Kingdom*

JAMES A. WELLS (14), *Department of Protein Engineering, Genentech, Inc., South San Francisco, CA 94080*

DAVID WILSON (2), *Centre for Drug Design and Development, The University of Queensland, Q 4072 Australia*

CLARE WOODWARD (27), *Department of Biochemistry, University of Minnesota, St. Paul, Minnesota 55108*

YING-CHING YU (26), *Department of Laboratory Medicine and Pathology and Chemical Engineering and Materials Science, Biomedical Engineering Center, University of Minnesota, Minneapolis, Minnesota 55455-0392*

Preface

It has been nearly forty years since the advent of the solid-phase method for the assembly of peptides. Prior volumes of *Methods in Enzymology* have included certain aspects of the solid-phase technique, such as the synthesis of bioactive peptides and combinatorial libraries, or analytical techniques used to characterize synthetic peptides. However, this is the first volume of the series dedicated entirely to the practice of solid-phase peptide synthesis, including the principles and methods for peptide construction and analysis. Both the *tert*-butyloxycarbonyl (Boc)- and 9-fluorenylmethoxycarbonyl (Fmoc)-based synthetic chemistries are described. A variety of analytical techniques, ranging from the more traditional approaches (amino acid analysis, Edman degradation sequence analysis, high-performance liquid chromatography) to the most current innovations (mass spectrometry, capillary electrophoresis), are presented. A section has been devoted to specialized techniques for creating modified peptides and small proteins, which represent some of the current and exciting peptide-related research areas. Included in this section are the construction of phosphorylated, glycosylated, and cyclic peptides, and assembly of proteins using peptide fragments. The volume concludes with specific applications of synthetic peptides for the study of structural and biological problems of interest. These applications include utilization of several "designed" peptides, whereby site-specific isotopic labels have been incorporated for structural studies or lipophilic compounds have been added to promote the formation of distinct molecular architecture.

The collected works presented in this volume represent the practicality and power of solid-phase peptide synthesis for attacking scientific problems in biological, biophysical, and biochemical disciplines.

GREGG B. FIELDS

METHODS IN ENZYMOLOGY

VOLUME 284. Lipases (Part A: Biotechnology)
Edited by BYRON RUBIN AND EDWARD A. DENNIS

VOLUME 285. Cumulative Subject Index Volumes 263, 264, 266–289

VOLUME 286. Lipases (Part B: Enzyme Characterization and Utilization)
Edited by BYRON RUBIN AND EDWARD A. DENNIS

VOLUME 287. Chemokines
Edited by RICHARD HORUK

VOLUME 288. Chemokine Receptors
Edited by RICHARD HORUK

VOLUME 289. Solid Phase Peptide Synthesis
Edited by GREGG B. FIELDS

VOLUME 290. Molecular Chaperones (in preparation)
Edited by GEORGE H. LORIMER AND THOMAS O. BALDWIN

Section I

Methods for Solid-Phase Assembly of Peptides

[1] Concept and Early Development of Solid-Phase Peptide Synthesis

By Bruce Merrifield

Introduction

A book on peptide chemistry cannot begin without expressing due respect for Theodore Curtius[1] who achieved the first syntheses of a protected peptide and Emil Fischer[2] who synthesized the first free dipeptide in 1901. Fischer is properly considered to be the founder of the field of peptide chemistry and originator of the term peptide. His achievements were remarkable and his vision was even greater. Already, at the turn of the twentieth century he could clearly foresee the day when a protein would be synthesized. In spite of the lack of detailed knowledge about the composition and structure of the proteins, he made an attack on the problem himself and set the stage for those who were to follow. Subsequent to this brilliant beginning, progress was slow for the next 40 years. The towering exception was the invention of the carbobenzoxy group by Bergmann and Zervas[3] in 1932. The discovery of a readily reversible protecting group for the α-amine of the amino acids had eluded all the early chemists until that time. Now it could be expected that the situation would change rapidly. With the exception of the Bergmann laboratory and a few others, however, progress continued to be slow for several more years, and much fundamental work remained to be done.

During the 1950s peptide synthesis was rapidly developing into a mature science, and methods were devised that would allow the chemical synthesis of almost any small peptide.[4,5] With the end of World War II and new peptide laboratories being established by first rate chemists the field began to blossom. The older acid azide coupling method of Curtius and the acid chloride method of Fischer were being supplemented with the introduction of mixed anhydrides by Wieland[6] and also by Boissonnas, Vaughan, Kenner,

[1] T. Curtius, *J. Pract. Chem.* **24**, 239 (1881).

[2] E. Fischer and E. Forneau, *Ber. Dtsch. Chem. Ges.* **34**, 2868 (1901).

[3] M. Bergmann and L. Zervas, *Ber. Dtsch. Chem. Ges.* **65**, 1192 (1932).

[4] E. Schröder and K. Lubke, "Methods of Peptide Synthesis," Vol. 1. Academic Press, New York, 1965.

[5] M. Bodanszky and M. A. Ondetti, "Peptide Synthesis." Wiley (Interscience), New York, 1966.

[6] T. Wieland and H. Bernhard, *Liebigs Ann. Chem.* **572**, 190 (1951).

Anderson, and others. At about the same time active esters were demonstrated to be good coupling reagents. They included thiophenyl esters,[7] cyanomethyl esters,[8] and nitrophenyl esters,[9] and later various halogenated phenyl esters.[10] The application of carbodiimides to the formation of the peptide bond by Sheehan's group[11] was also a major advance.

Eventually, the benzyloxycarbonyl function was supplemented with the more acid-labile *tert*-butyloxycarbonyl group introduced by Carpino[12] and developed by McKay and Albertson[13] and Anderson and McGregor[14] and with very acid-labile groups such as the nitrophenyl sulfenyl group from the Zervas laboratory[15] and Sieber and Iselin[16] biphenylisopropyloxycarbonyl derivatives. Side-chain protection for the trifunctional amino acids was also markedly advanced during this period, and at the same time methods for the selective removal of the protecting groups were being studied and improved.

The question of partial racemization was of major concern and was examined in great detail. Fortunately it was found that N^α protection of amino acids by urethane derivatives virtually eliminated the racemization problem for stepwise synthesis, and later new techniques were developed to replace the largely racemization-free azide method for fragment coupling.

Now the time had come when peptide chemists could aim their sights at the synthesis of bioactive peptides, the hormones, antibiotics, growth factors, toxins, etc. The landmark achievement was by Vincent du Vigneaud and his group in 1953, when they successfully accomplished the first synthesis of a peptide hormone, oxytocin.[17] Very soon other impressive results were forthcoming from several laboratories, for example, on the synthesis of vasopressin, gramicidin S, α-melanocytic stimulating hormone (MSH), β-MSH, renin substrate, and fragments of adrenocorticotropin (ACTH).

While all this activity was proceeding at a rapid pace I entered the field as a newcomer in 1954, with the aim of preparing some penta- and heptapeptide growth factors. The methods were largely in place to permit

[7] T. Wieland, W. Schäfer, and E. Bokelmann, *Liebigs Ann. Chem.* **573,** 99 (1951).
[8] R. Schwyzer, M. Feurer, and B. Iselin, *Helv. Chim. Acta* **38,** 83 (1955).
[9] M. Bodanszky, *Nature (London)* **175,** 685 (1955).
[10] J. Kovács, L. Kisfaludy, and M. Q. Ceprini, *J. Am. Chem. Soc.* **89,** 183 (1967).
[11] J. C. Sheehan and G. P. Hess, *J. Am. Chem. Soc.* **77,** 1067 (1955).
[12] L. A. Carpino, *J. Am. Chem. Soc.* **79,** 4427 (1957).
[13] F. C. McKay and N. F. Albertson, *J. Am. Chem. Soc.* **79,** 4686 (1957).
[14] G. W. Anderson and A. C. McGregor, *J. Am. Chem. Soc.* **79,** 6180 (1957).
[15] L. Zervas, D. Borovas, and E. Gazis, *J. Am. Chem. Soc.* **85,** 3660 (1963).
[16] P. Sieber and B. Iselin, *Helv. Chim. Acta* **51,** 622 (1968).
[17] V. du Vigneaud, C. Ressler, J. M. Swan, C. W. Roberts, P. G. Katsoyannis, and S. Gordon, *J. Am. Chem. Soc.* **75,** 4879 (1953).

the synthesis of this group of small peptides, and I could obtain pure products of the correct structure to study their growth promoting activity toward certain strains of bacteria,[18] and even to prepare a good competitive antagonist. However, this was a very labor-intensive process that took months to achieve, and the overall yields were low. I also prepared some peptide polymers and copolymers by N-carboxy anhydride procedures, which were necessarily of ill-defined molecular weight and composition.

By 1959, I had reached the conclusion that a new approach to peptide synthesis was needed. It was then that I had the idea for solid-phase peptide synthesis, which I believed could be of great help in simplifying and accelerating the synthetic process. I could imagine that small peptides and even those of perhaps 10 residues could be prepared if the idea could be developed. I even believed that automation could be possible. What I could not foresee was the possibility of synthesizing molecules of the size and composition of a protein or enzyme. Nor could I envision the discovery of the vast number of biologically active peptides that were to be found over the next decades and the role that solid-phase peptide synthesis was to have in their chemical synthesis and the study of their mechanisms of action. The application to multiple simultaneous syntheses and to peptide libraries was, of course, far out of sight.

This volume presents in considerable detail the main features of solid-phase synthesis as they have been developed over the years by many people in many laboratories. For this short introduction, I plan only to outline the general principle and main features of the approach and to describe some of the early experiments[19] leading to the development of a working method.

Concept of Solid-Phase Peptide Synthesis

The initial idea of solid-phase peptide synthesis is outlined in Fig. 1. Conceptually the unique feature of this approach was the use of a polymeric support to assist in the synthesis of another polymeric compound. By the late 1950s, a vast amount of information was known about the chemistry of polymers, and many chemical transformations of polymers had been carried out to change their properties. However, polymers had not previously been used to prepare another class of compounds that could later be removed from the polymer support and isolated in the free state.

The process required a solid, insoluble support to hold the first amino acid residue and the subsequent stages of the peptide as it was lengthened.

---

[18] R. B. Merrifield and D. W. Woolley, *J. Am. Chem. Soc.* **80,** 6635 (1958).
[19] R. B. Merrifield, "Life during a Golden Age of Peptide Chemistry." (J. I. Seeman, Series ed.), Profiles, Pathways and Dreams, p. 1. American Chemical Society, Washington, DC, 1993.

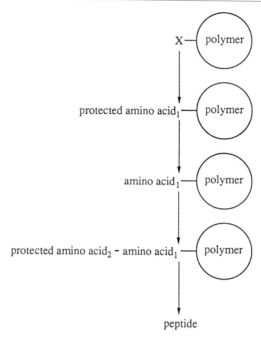

Fig. 1. Basic idea for solid-phase peptide synthesis.

A preference for a covalent attachment at the carboxyl end was clear from the beginning. Therefore, the N terminus needed to be protected and deprotected at each cycle of the stepwise synthesis, and the resulting amino group had to be converted to a free NH_2 after each deprotection. It was then necessary to activate the next N-protected amino acid for coupling to the growing peptide chain. These three steps, deprotection, neutralization, and coupling, would be repeated until the desired sequence was assembled. Finally, the covalent anchoring bond would be cleaved and the peptide liberated into solution. These were the simple basic steps necessary to carry out the new approach. Because the growing peptide was in the solid phase throughout the synthesis, the process was given the name solid-phase peptide synthesis.

The potential advantages of the proposed synthetic scheme appeared to be speed, simplicity, and yield. Once the support containing the first amino acid is placed in a suitable reaction vessel or a packed column it would not have to be transferred or removed until the end of the synthesis. If the solid support were completely insoluble in all solvents used, the peptide would also be insoluble, and it could be readily filtered and washed without transfer to other containers and the usual physical losses would be avoided. The coupling reaction could be driven to completion by a molar excess and a high concentration of the protected amino acid. Thorough washing by large volumes of solvents would efficiently remove excess reagents and soluble by-products and thereby effect a rapid partial purification at each step without the need for difficult and time-consuming isolation and crystallization procedures after each synthetic cycle. Incomplete reactions and the gradual buildup of insoluble side reaction by-products were recognized as obvious potential disadvantages.

Development of Solid-Phase Peptide Synthesis

It was soon apparent that each of the variables, the support, the solvents, the anchoring bond, the N^α protection and deprotection, the activation and coupling steps, and the cleavage step had to be studied in detail and optimized. However, because they were not independent, optimization of all steps had to come together at the same time. A good coupling reaction in the wrong solvent or on the wrong support would not do, or poor chemistry applied to a potentially useful support would be ineffective. It is mainly for this reason that it took about 3 years to devise a useful, compatible system.

The key to the entire solid-phase process is the support. The first attempts to develop the method were with cellulose because it was already known that cellulose columns were effective for the purification of peptides and proteins. It was reasoned that the synthetic peptide and the solid cellulose would be compatible and that the reactants could enter the matrix and the peptide product could exit freely. In practice, efforts to esterify cellulose or regenerated cellulose with N-protected or free amino acids did not appear to be very satisfactory, although a small degree of substitution was achieved, and a dipeptide was assembled. Therefore, this material was abandoned, although years later, after the chemistry had changed, it was possible to use LH-Sephadex as the support, and others have reported the successful application of cellulose paper and cotton thread.

Materials such as polyvinyl alcohol could be highly substituted but were too soluble and were only examined briefly. Commercial ion-exchange resins such as IRC-50 were examined in more detail and gave the first successful dipeptide by the solid-phase approach. Following saponification of the copolymer of methyl methacrylate and divinylbenzene, the carboxyl groups of the resin were converted to the acid chloride with thionyl chloride. Carbobenzoxyglycine was esterified with p-nitrophenol, reduced to the amino phenyl ester, and coupled to the resin acid chloride. After deprotection with 30% (v/v) HBr in acetic acid, the second protected amino acid was coupled as the p-nitrophenyl ester. Much better results were obtained when p-aminobenzyl bromide was used for esterification of Z-Phe. Acid deprotection and coupling with Z-Gly-ONp gave the dipeptide–resin. Cleavage with sodium in liquid ammonia gave the free dipeptide, Gly-Phe, with an equimolar ratio of amino acids. Disappointingly, extension of the chain did not continue with equimolar ratios of succeeding residues, and the search for a better resin was resumed. Part of the problem was believed to have derived from the high concentration of carboxyl groups and ionic repulsion or from the fact that the functional groups were present in the monomers before polymerization and were not all accessible. More likely, the chemistry was simply not adequate. HBr was too strong an acid and nitrophenyl esters were not reactive enough. This again points out the need for all components of the system to be near optimum at the same time.

Copolymers of styrene and divinylbenzene, either prederivatized or derivatized after polymerization, were examined next. The ion-exchange resin Dowex 50, containing sulfonic acid groups, was available and therefore selected for the new experiments. Here again the highly substituted, highly charged resin was not easily derivatized. The sulfonyl chloride resin, although only 4% cross-linked, gave only 0.05 mmol Phe/g after reaction with Z-Phe p-aminobenzyl resin (theory ~5.5 mmol/g). The excess sulfonic acid groups were avoided by sulfonating a copolymer of styrene and divinylbenzene under controlled conditions, but in these experiments too high a cross-linked resin was used. However, the first solid-phase synthesis of a peptide amide was achieved on this support.

The critical change was to turn to a controlled level of chloromethyl groups on a styrene-divinylbenzene resin bead containing only 2% of cross-linker. By this time the requirement for better swelling of the resin in organic solvents was being appreciated. The low cross-linking and use of good swelling solvents such as dichloromethane or dimethylformamide were effective in this regard. It was found years later that this copolymer can become highly solvated and swollen, and reach a volume 50-fold greater than the dry bead. These experiments also showed that a large amount of peptide could be accommodated without filling the resin. The attachment of the first carbobenzoxy amino acid was by a benzyl ester via the chloromethyl resin. Unfortunately, the chemistry was still not right, and the differential stability of the benzyl urethane and the benzyl ester to HBr was not adequate. Use of 10% HBr in acetic acid was a barely usable compromise. Ring bromination or nitration of the resin greatly increased the stability of the benzyl ester so that a satisfactory synthesis of a model tetrapeptide could be achieved. This was the first really acceptable solid-phase synthesis and led to the first report on the method in 1962 and the first full publication[20] in 1963. The overall yield was good, the analytical data were satisfactory, and lack of significant racemization was demonstrated. The critical question of the consequences of incomplete reactions was discussed, and the value of capping low levels of unreacted peptide chains was recognized.

The replacement of the carbobenzoxy group by the *tert*-butyloxycarbonyl (Boc) group and the replacement of nitrophenyl esters by the dicyclohexylcarbodiimide coupling reagent were major subsequent improvements, and reduction of the cross-linking to 1% divinylenzene was an important further change. The Boc group could be removed in 4 N HCl in dioxane or by 50% trifluoroacetic acid in dichloromethane without serious losses of peptide from the resin support. The diimide reaction was much faster than the active ester one and was best in dichloromethane, but it could also be done in dimethylformamide or in N-methylpyrrolidinone. The final cleavage of the anchoring benzyl ester was in 30% acetic acid or in hydrogen fluoride containing 10% of a carbonium ion scavenger. The process was also automated within the next 3 years, as envisioned at the beginning. This was the early recommended general procedure for solid-phase peptide synthesis (Fig. 2), and it served well for a number of years.

Further Improvements in Solid-Phase Peptide Synthesis

Changes and improvements were gradually introduced by our laboratory and by many others, and they have been reviewed a number of times (e.g., see Refs. 21–25). They were designed to give faster, more complete reactions and to avoid various side reactions. Improvements have also dealt with methods for attachment and removal of the peptide chain and of orthogonal combinations of bonds at the α-amine, α-carboxyl, and side-chain groups that allow very selective removal in any order desired. These are based on acidic, basic, and neutral deprotection methods. For this purpose several new protecting groups were adapted to solid-phase synthesis, especially the important base-labile fluorenylmethyloxycarbonyl group

[20] R. B. Merrifield, *J. Am. Chem. Soc.* **85,** 2149 (1963).

[21] R. B. Merrifield, *Adv. Enzymol.* **32,** 221 (1969).

[22] J. M. Stewart and J. D. Young, "Solid Phase Peptide Synthesis." Freeman, San Francisco, 1969.

[23] B. W. Erickson and R. B. Merrifield, *in* "The Proteins" (H. Neurath and R. L. Hill, eds.), 3rd Ed., Vol. 2, p. 255. Academic Press, New York, 1976.

[24] G. B. Fields, Z. Tian, and G. Barany, *in* "Synthetic Peptides: A User's Guide" (G. A. Grant, ed.), p. 77, Freeman, New York, 1992.

[25] R. B. Merrifield, *in* "Peptides: Synthesis, Structures and Applications" (B. Gutte, ed.), p. 93. Academic Press, San Diego, 1995.

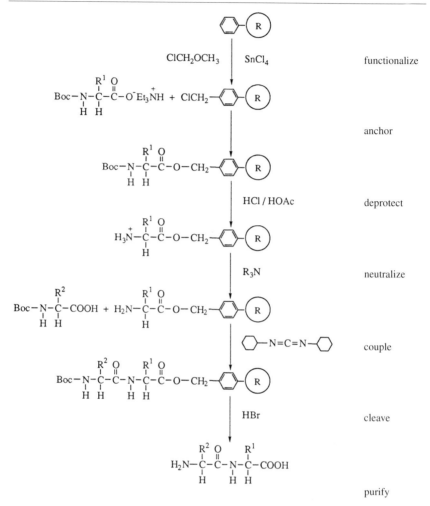

Fig. 2. An early scheme for solid-phase peptide synthesis. ® represents a copoly(styrene-1% divinylbenzene) resin bead. Only one of the approximately 10^{14} aromatic rings is shown.

of Carpino.[12] Amino acid activation and coupling methods have received much attention, and some important new reagents have been forthcoming, including phosphonium and uronium salts, acid fluorides, and urethane protected N-carboxy anhydrides.

In addition, many alternate supports have been developed, such as macroreticular polymers, polyethylene glycol, polyacrylamides, polyoxyethylene grafted on to polystyrene, controlled pore glass, carbohydrates, and others. Some of these supports are highly effective, while others are not.

Automation has continued and a number of synthesizers have been developed, including several commercial ones. They are based both on the original batch method in which the resin is placed in a reaction vessel and suspended in solvent by shaking or stirring, followed by filtration and introduction of new solvent and reagents, or on the continuous flow approach in which the swollen resin is packed in a column and solvents and reagents are pumped through. This was, in fact the first procedure examined, but both poor physical properties and poor chemistry prevented the achievement of satisfactory results at that time.

The general problem of yield and purity of products has been a concern from the first introduction of the solid-phase principle for peptide synthesis. The requirement for purification procedures was obvious, but the fact that only simple washing procedures could be used at intermediate stages and classic methods including crystallization could not be applied until the last step after cleavage from the support had occurred caused some critics to believe that it would not be possible to prepare peptides of adequate purity to satisfy their needs. In 1976, I commented,[23]

> The pessimist says that a pure product cannot be isolated and that, even if it were, its purity cannot be demonstrated. Since this attitude produces no progress, we prefer the pragmatic approach of being aware of the problem of purity and simply using the best methods currently available during the synthesis, isolation, and characterization of the product. Improvements in separation methods are appearing regularly; what cannot be achieved today may seem simple tomorrow. In the meantime, much useful information can be gained using peptides purified by the available procedures.

It has come to pass that many highly effective preparative and analytical methods have been developed. Preparative high-performance liquid chromatography (HPLC) has been remarkably effective for the fractionation and isolation of purified products, and ion-exchange methods for this purpose have also improved. Affinity purification has been especially valuable for selective isolation of biologically active peptides and proteins. Analytical HPLC and capillary electrophoresis now allow sensitive detection of impurities, and nuclear magnetic resonance (NMR) spectroscopy and mass spectrometry are extremely effective for identifying impurities and for establishing the composition of the purified synthetic peptides.

Detailed studies on the physical and chemical properties of the solid-phase system were begun early and have continued to the present. The important properties are good physical and chemical stability, easily controlled derivatization, good solvation and swelling of both resin and peptide, ready permeability to reagents, rapid and complete chemical reactions, and minimization of peptide aggregation during the synthesis.

In the very early days of solid-phase synthesis it was believed that the solid support effectively isolated the peptide sites, leading to reactions that resembled those at infinite dilutions and thereby avoiding intermolecular reactions between separate resin-bound sites, which in turn would promote intramolecular reactions. It was soon found, however, that there is very significant polymer chain motion even on the cross-linked resins. This was demonstrated by electron spin resonance (ESR) and NMR measurements. The motional rates for the aromatic side chains and aliphatic backbone atoms of 1% cross-linked polystyrene in CH_2Cl_2 are high (10^8/sec) and equivalent to those of soluble polystyrene, which indicates that the polymers are highly solvated. For pendant peptides the rates were as high as 10^{10}/sec. Chemical experiments also showed significant polymer chain motion of 200–300 Å.

It was demonstrated early on by autoradiography that the peptide chains were uniformly distributed throughout the matrix of lightly cross-linked copoly(styrene-divinylbenzene) beads. The coupling reactions proceeded by rapid second order rates (99% reaction within 10 to 100 sec), and the mass transfer was at least an order of magnitude faster than the initial coupling rate. In the cases studied, the coupling rate was independent of the peptide length (up to 60 residues), and the extent or completion of the reaction, as measured by a very sensitive quantitation of deletion peptides, was greater than 99.9% and was also independent of peptide length or loading of the resin (up to at least 0.95 mmol/g). More recent work using mass spectrometric measurement of deletion and addition peptides showed an average extent of coupling of a model decapeptide to be 99.82%, and after 20 residues the measured average yield was 99.93%. These are excellent results and if they could be achieved in all syntheses, little more developmental work would be necessary.

There are, however, so-called difficult sequences in which less complete reactions are encountered. The problem is generally believed to be due to peptide self-aggregation, which is largely a result of formation of hydrogen-bonded structures that reduce peptide solvation and reactivity. Experiments show that the aggregated clusters do not readily dissociate even when the amount of unreacted peptide–resin decreases during a coupling reaction. In my view peptide aggregation is dependent on the structure of the peptide and its environment and not on interactions between the peptide and the resin support. This is the important remaining problem in solid-phase peptide synthesis. The effect is minimized by appropriate selection of the support and the reaction solvent, by decreasing the peptide loading, by chaotropic agents and other additives, by elevated temperature, and by suitable derivatization of the amide bond. Together, these measures are near to the elimination of this last major obstacle.

Summary

There are several reasons for the success of the solid-phase approach to peptide synthesis. The first is the ease of the procedure, the acceleration of the overall process, and the ability to achieve good yields of purified products. The second was the unanticipated discovery of many new biologically active peptides and the expanded need for synthetic peptides to help solve problems in virtually all disciplines of biology. In many cases, the solid-phase technique has been the method of choice.

This approach, of course, does not replace the classic solution synthesis methods, but rather supplements them. The choice of techniques depends on the objectives of the synthesis. When carefully worked out, the solution methods can give high yields of highly purified products in large quantities. Many superb syntheses of active peptides have been achieved in this way. The solid-phase method has also yielded many large active peptides. It is particularly useful when large numbers of analogs, in relatively small quantities, are required as in structure–function studies on hormones, growth factors, antibiotics, and other biologically active peptides or for determining the antigenic epitopes of proteins. In addition, it has on occasion been scaled up for production of kilogram quantities.

One of the unique uses of solid-phase synthesis has been the synthesis of peptide libraries. Most of the work on this new field in which thousands or millions of peptides are prepared simultaneously has been by solid-phase methods. This new technique is proving to be of great practical importance in rapid drug discovery of peptide, peptide mimetic, and nonpeptide compounds. Developments in screening methods now allow the examination of large numbers of compounds, and active products with structures unpredictable from natural product sequences are being found in this way.

The properties of the solid-phase system, the changes in the chemistry, and the applications of the technique to biological problems are discussed in detail in subsequent articles of this volume.

[2] Rapid *in Situ* Neutralization Protocols for Boc and Fmoc Solid-Phase Chemistries

By Paul Alewood, Dianne Alewood, Les Miranda, Stephen Love, Wim Meutermans, and David Wilson

Introduction

The concept of solid-phase peptide synthesis (SPPS) was elucidated by Bruce Merrifield in the late 1950s and early 1960s.[1] The initial chemical strategies employed *tert*-butyloxycarbonyl (Boc)/benzyl protection, carbodiimide-based coupling, and stepwise chain assembly on a cross-linked polystyrene resin. By the mid-1980s, after a period of chemical optimization and successful practice of this novel approach, manual SPPS had been established in many laboratories. A typical synthesis schedule of that period is illustrated in Fig. 1.

Since that period, a better understanding of the physicochemical nature of the peptide–resin[2] combined with elimination of chronic side reactions[3,4] and significant improvements in resins[5] has underpinned the development of more efficient solid-phase synthesis. The chemical synthesis of peptides and proteins, using both manual and machine-assisted protocols, has been under continuous development in our laboratory.[6–8] In the early 1990s we reported simple manual protocols for Boc and fluorenylmethyloxycarbonyl (Fmoc) chemistries that allowed the rapid generation of high-quality peptides.[6] We chose to develop robust, efficient chemistry and simple laboratory procedures based on our understanding of the nature of the growing peptide–resin in relation to SPPS.[8,9] These protocols were significantly en-

---

[1] R. B. Merrifield, *J. Am. Chem. Soc.* **85,** 2149 (1963).

[2] S. B. H. Kent, *Annu. Rev. Biochem.* **57,** 957 (1988).

[3] For example, S. B. H. Kent, A. R. Mitchell, M. Englehard, and R. B. Merrifield, *Proc. Natl. Acad. Sci. USA* **76,** 2180 (1979).

[4] S. B. H. Kent and R. B. Merrifield, *Int. J. Pept. Protein Res.* **22,** 57 (1983).

[5] A. R. Mitchell, S. B. H. Kent, M. Englehard, and R. B. Merrifield, *J. Org. Chem.* **43,** 2845 (1978).

[6] P. F. Alewood, M. Croft, M. Schnolzer, and S. B. H. Kent, *in* "Peptides 1990" (E. Giralt and D. Andreu, eds.), p. 174. ESCOM, Leiden, The Netherlands, 1991.

[7] M. Schnolzer, P. Alewood, A. Jones, D. Alewood, and S. B. H. Kent, *Int. J. Pept. Protein Res.* **40,** 180 (1992).

[8] S. B. H. Kent, D. Alewood, P. Alewood, M. Baca, A. Jones, and M. Schnolzer, *in* "Innovation and Perspectives in Solid Phase Synthesis" (R. Epton, ed.), p. 1. Intercept Limited, Andover, UK, 1992.

[9] V. K. Sarin, S. B. H. Kent, and R. B. Merrifield, *J. Am. Chem. Soc.* **102,** 5463 (1980).

0076-6879/97 $25.00

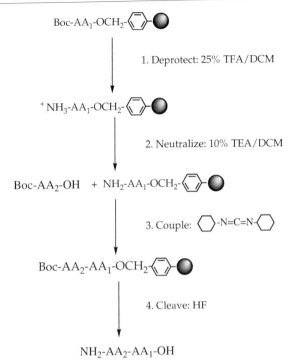

Boc-AA₁-OCH₂-

1. Deprotect: 25% TFA/DCM

⁺NH₃-AA₁-OCH₂-

2. Neutralize: 10% TEA/DCM

Boc-AA₂-OH   +   NH₂-AA₁-OCH₂-

3. Couple: ◯-N=C=N-◯

Boc-AA₂-AA₁-OCH₂-

4. Cleave: HF

NH₂-AA₂-AA₁-OH

Fɪɢ. 1. Early SPPS scheme. Typically the precise protocols required repeated resin washes with dichloromethane, and other polar solvents such as 2-propanol. Cycle times were generally 2–3 hrs. TFA, Trifluoroacetic acid; TEA, triethylamine; AA, amino acid.

hanced[10] through the incorporation of the conceptually and in practice simple modification of *in situ* neutralization.[7,10–15]

The protocols for Boc chemistry are characterized by the use of a single and inexpensive solvent dimethylformamide (DMF), to maintain maximum solvation of the peptide–resin. This ensures swelling of the peptide–resin

[10] M. Schnolzer, P. F. Alewood, A. Jones, and S. B. H. Kent, *in* "Peptides: Chemistry and Biology" (J. Smith and J. Rivier, eds.), p. 623. ESCOM, Leiden, The Netherlands, 1992.

[11] D. Le-Nguyen, A. Heitz, and B. Castro, *J. Chem. Soc. Perkin Trans 1*, 1915 (1987).

[12] J. P. Briand, J. Coste, A. Van Dorsselaer, B. Raboy, J. Neimark, B. Castro, and S. Muller, *in* "Peptides 1990" (E. Giralt and D. Andreau, eds.), p. 80. ESCOM, Leiden, The Netherlands, 1990.

[13] Y. Kiso, T. Kimura, Y. Fujiwara, H. Sakikawa, and K. Akaji, *Chem. Pharm. Bull.* **38**, 270 (1990).

[14] Y. Kiso, Y. Fujiwara, T. Kimura, A. Nishitani, and K. Akaji, *Int. J. Pept. Protein Res.* **40**, 308 (1992).

[15] J. Jezek and R. A. Houghten, *Collect. Czech. Chem. Commun.* **59**, 691 (1993).

while potentially aiding the disaggregation of intermolecularly H-bonded aggregates,[16] a major contributor to "difficult sequences." Neat trifluoroacetic acid (TFA) is used to achieve fast and efficient removal of the N^α-Boc group. Rapid flow washes of the swollen peptide–resin, combined with high concentrations (0.3–0.5 M) of activated amino acids and *in situ* neutralization, give significantly improved coupling efficiencies. Furthermore, we are able to routinely achieve three coupling cycles per hour in either the manual or machine-assisted mode. Importantly, each coupling step is monitored by the quantitative ninhydrin reactions.[17] This precise recording of each synthesis pinpoints "difficult and/or slow couplings" and gives a permanent record of each synthesis.

Together with modern analytical techniques for peptide analysis such as mass spectrometry (MS) and MS combined with liquid chromatography (LC–MS), it is now feasible to observe and determine the nature of the various side products in SPPS.[18] Such analyses are critical for future improvements in peptide synthesis.

Manual *in Situ* Neutralization

In 1991, we described practical *in situ* neutralization Boc chemistry protocols[10] that held significant promise for fast, efficient SPPS. The simple chemistry (Fig. 2) required the activation of Boc-amino acids in a polar solvent plus the addition of diisopropylethylamine (DIEA). Although early efforts employed carbodiimides with 1-hydroxybenzotriazole (HOBt) for activation, these were readily displaced by use of the more convenient and effective uronium salts[19] in the presence of base as the primary form of activation chemistry. In particular, N-[1H-(benzotriazol-1-yl)(dimethylamino)methylene]-N-methylmethanaminimum hexafluorophosphate N-oxide (HBTU) in combination with DIEA is used routinely in our laboratory to efficiently activate all protected amino acids (Table I). The *in situ* nature of the neutralization was particularly convenient as we could add an amount of DIEA sufficient to activate the protected amino acid with concurrent neutralization of the $TFA^- \cdot {}^+NH_3$-peptide–resin. These changes resulted in fewer steps in the protocol, a reduction of the synthesis cycle time from 20–30 min to around 15 min, and significantly cleaner products.

---

[16] For a discussion see Refs. 2 and 8.
[17] V. K. Sarin, S. B. H. Kent, J. P. Tam, and R. B. Merrifield, *Anal. Biochem.* **117**, 147 (1981).
[18] M. Schnolzer, A. Jones, P. F. Alewood, and S. B. H. Kent, *Anal. Biochem.* **204**, 335 (1992).
[19] R. Knorr, A. Trzeciak, W. Bannwarth, and D. Gillessen, *Tetrahedron Lett.* **30**, 1927 (1989).

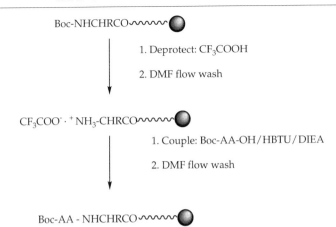

FIG. 2. *In situ* neutralization SPPS. The N^α-Boc group was removed by treatment with 100% TFA, and the peptide–resin salt was rinsed with a single flow wash with DMF. The activated Boc-amino acid was added with sufficient excess of DIEA to neutralize the peptide–resin salt. Coupling was performed for 10 min, followed by a single flow wash with DMF prior to the next cycle.

Synthesis of Enterotoxin STh(1–19), Aba[10,18]

Enterotoxigenic *Escherichia coli* produces two homologous heat-stable enterotoxins, STh and STp, which are responsible for acute diarrhea in infants and domestic animals.[20] These toxins are peptides of 19 and 18

TABLE I
In Situ NEUTRALIZATION PROTOCOL FOR MANUAL SYNTHESIS

Synthesis cycle	Component	Time and mode
Deprotect	100% TFA	Two 1-min shakes
Wash	DMF	1-min flow wash
Couple	Activated Boc-AA[a]	10-min shake
		Take resin sample for ninhydrin test
Wash	DMF	1-min flow wash

[a] HBTU/DIEA: Mix 2 mmol Boc-AA, 4 ml of 0.5 M HBTU (2 mmol) in DMF, and 2.5 mmol DIEA (460 μl). Activate for 2 min and add to protonated resin.

[20] S. Aiomoto, T. Takao, Y. Shimonishi, S. Hara, T. Takeda, S. Takeda, and T. Miwatani, *Eur. J. Biochem.* **129,** 257 (1982).

STh 1-19   NSSNYCCELCCNPACTGCY

Fig. 3. Amino acid sequence and disulfide linkages of STh 1–19. The enterotoxigenic region is underlined.

amino acid residues, respectively, and contain three intramolecular disulfide bridges, as shown for STh (Fig. 3). The physiological response of ST is initiated by binding to its receptor protein on the membrane epithelial cells.[20,21] This leads ultimately to stimulation of fluid secretion.

An analog of STh (STh 1–19, Aba[10,18]), missing one of the disulfide pairs, is synthesized using single coupled *in situ* protocols with L-α-amino-*n*-butyric acid (Aba) as an isosteric replacement for Cys. The remaining Cys side chains are orthogonally protected using either MeBzl (4-methylbenzyl) or Acm (acetamidomethyl) groups for later off-resin selective disulfide formation. Chain assembly is successfully performed on a Boc-Tyr(Cl-Z)-OCH$_2$-Pam (phenylacetamido) polystyrene resin (Fig. 4). The overall yield of correct peptide sequence from the synthesis is approximately 98%, and the average coupling yield is 99.9%.

After HF cleavage the resultant peptide is characterized by high-performance liquid chromatography (HPLC) and MS analysis (Fig. 5). Analysis by MS confirms that the major product contains the correct molecular weight (M_r 2155). A minor product (M_r 71) is also present, a result of premature removal of the Acm protecting group during HF cleavage.

Disulfide bond formation is carried out in two stages (Fig. 6). The first disulfide bond is formed via aerial oxidation in ammonium bicarbonate (pH 8.0, 24 hr). The oxidized, di-Acm product is purified by reversed-phase (RP)-HPLC, and then the second disulfide bond is formed by removal of Acm with concurrent oxidation [I$_2$ (10 equivalents) in 80/20 (v/v) acetic acid/water]. On completion of the oxidation the desired product is purified by RP-HPLC and analyzed by electrospray mass spectrometry (ESMS) (data not shown).

Evaluation of Coupling Reagents

The ideal peptide coupling reagent for peptide and protein synthesis has been sought by peptide chemists since the early 1900s. A coupling reagent that is effective through difficult synthetic sequences, minimizes side reactions during chain assembly, and reduces coupling times is still actively pursued in many laboratories.

[21] C. L. Gyles, *Can. J. Microbiol.* **38,** 734 (1992).

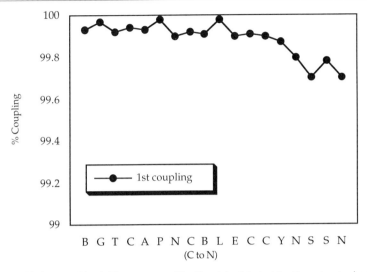

FIG. 4. Chain assembly yields as measured by the ninhydrin test for the enterotoxin analog STh 1–19, Aba[10,18]. B is L-α-amino-*n*-butyric acid.

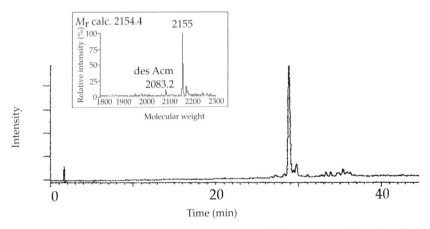

FIG. 5. Analysis by HPLC and MS (inset) of the crude HF cleavage product of synthetic STh 1–19, Aba[10,18]. Chromatography was carried out using a 1%/min gradient at 1 ml/min from 0 to 60% acetonitrile on a Brownlee C_8 analytical cartridge, with UV detection at 214 nm. The MS data were collected from 300 to 2250 mass units at OR 80 V on a PE-SCIEX API-III mass spectrometer.

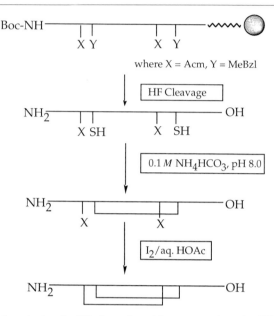

FIG. 6. Schematic for selective disulfide formation of the enterotoxin analog STh 1–19, Aba[10,18].

Although inefficient coupling reagents and lengthy cycle times may be tolerable for the synthesis of simple short peptides (<15 residues), in longer syntheses (15–60 residues) poor couplings often lead to failure, that is, where the target peptide is not observed or is a minor part of a melange. With the success of chemoselective ligation strategies[22] in the construction of small proteins, the burden now lies with peptide chemists to deliver highly homogeneous peptides containing up to 60 residues. Consequently, the development of rapid coupling protocols using a superior coupling reagent is highly desirable and an ongoing focus of our laboratory.

Carpino[23] has demonstrated the effectiveness of N-[(dimethylamino)-1H-1,2,3-triazolo[4,5-b]pyridino-1-ylmethylene]-N-methylmethanaminium hexafluorophosphate N-oxide (HATU) activation in difficult couplings. To compare the suitability and relative performance of HATU with that of its benzotriazole analog HBTU, we have synthesized the well-known difficult sequence from the acyl carrier protein [ACP(65–74)] using both reagents and compared their relative coupling efficiencies. Single couplings were used for both syntheses, and identical protocols were used except where

[22] M. Schnolzer and S. B. H. Kent, *Science* **256**, 221 (1992).
[23] L. A. Carpino, *J. Am. Chem. Soc.* **115**, 4397 (1993).

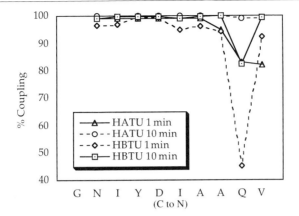

FIG. 7. Comparative chain assembly yields for the synthesis of ACP(65–74) using HBTU and HATU activation chemistry.

HATU was substituted for HBTU. Coupling yields were determined by the quantitative ninhydrin test at 1-, 5-, and 10-min intervals (Fig. 7). Both peptides were cleaved from the resin with HF and analyzed by RP-HPLC and ESMS (Fig. 8).

As expected, the substitution of HATU for HBTU had a dramatic impact. Using HATU with coupling times of 1 min, the first six residues were assembled with stepwise yields of approximately 99.9%. In the difficult region (residues 7–10 from the C terminus), good couplings (>99.5%) could be attained after 10 min. By comparison, HBTU was effective only if 10-

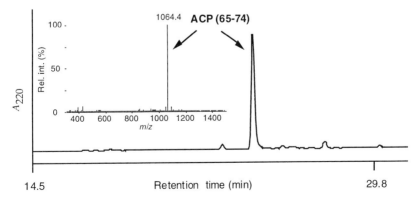

FIG. 8. HPLC and MS (inset) of the crude cleavage product from the synthesis of ACP(65–74) using HATU activation chemistry.

min couplings were maintained, with the exception of coupling Gln to Ala where double coupling was clearly necessary.

Machine-Assisted *in Situ* Neutralization Peptide Synthesis

The manual *in situ* neutralization synthesis protocols have been adapted for machine-assisted synthesis on the Applied Biosystems (ABI, Foster City, CA) 430A peptide synthesizer from peptide synthesis cycles previously reported.[7] This involves some plumbing modifications to the instrument and associated programming changes, in addition to those previously described.[7]

Two three-way switching valves are inserted between the reaction vessel and the instrument valve blocks. These replace the reaction vessel in-line filters (which have a tendency to block and reduce the solvent flow) and bypass the valve block. This results in a more controlled flow of solvent both into and out of the reaction vessel.

The TFA bottle is changed to a larger 2-liter vessel (Plasticote) replacing one of the dichloromethane (DCM) bottles (which is now largely redundant given the small volume of DCM used), and larger diameter lines are installed for both TFA and DMF delivery to the reaction vessel. This results in faster and more consistent flow rates, thereby increasing the efficiency of the Boc deprotection and the DMF flow washes. The transfer line from the activator block, via the concentrator valve block, to the reaction vessel is replaced by a single larger diameter line from the activator valve block directly to the reaction vessel. This totally bypasses the activator and concentrator vessels. Thus, the protected amino acid, which is dissolved and activated within the normal cartridge, can be transferred directly to the TFA deprotected peptide–resin in the reaction vessel.

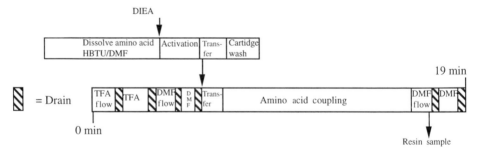

FIG. 9. Cycle diagram for machine-assisted *in situ* SPPS on a ABI 430A peptide synthesizer for a 20-ml reaction vessel. The amino acid coupling time, drain times, and DMF washes vary depending on reaction vessel size. The amino acid activation mixture contains 2 mmol of amino acid, 0.5 *M* HBTU/DMF (4 ml), and DIEA (1 ml). These cycles essentially parallel the manual cycles.

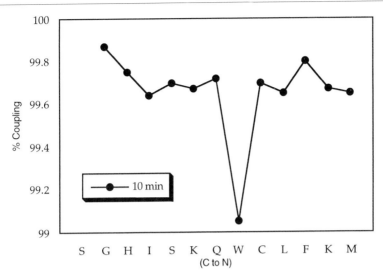

Fɪɢ. 10. Chain assembly yields for the machine-assisted synthesis of PsENOD40.

With programming changes to incorporate these modifications the instrument is able to carry out *in situ* Boc amino acid chemistry with 12-min synthesis cycles per amino acid (7- to 20-ml reaction vessel).[24] However, in the work described here we chose amino acid coupling times of 10 min with HBTU activation to parallel the *in situ* manual cycles. This extends the total cycle time to 20 min (Fig. 9).

Synthesis of PsENDO40 and Urocortin

To illustrate the efficacy of these changes we have synthesized the plant hormone PsENDO40, Met-Lys-Phe-Leu-Cys-Trp-Gln-Lys-Ser-Ile-His-Gly-Ser[25] (chosen for its amino acid diversity), using single coupling cycles. The average coupling yield is 99.7% (Fig. 10) by quantitative ninhydrin determination of residual free amine.[17] The lowest coupling yield observed is that of Trp (99.05%); this is a direct consequence of the DCM wash before and after the addition of TFA. This wash step is included to prevent pyroglutamate formation of the Gln-peptide–resin.[8] The crude cleaved

[24] S. B. H. Kent and J. J. Kent, unpublished results (1992).
[25] K. Van de Sande, K. Pawlowski, I. Czaja, U. Wieneke, J. Schell, J. Schmidt, R. Walden, M. Matvienko, J. Wellink, A. van Kammen, H. Franssen, and T. Bisseling, *Science* **273**, 370 (1996).

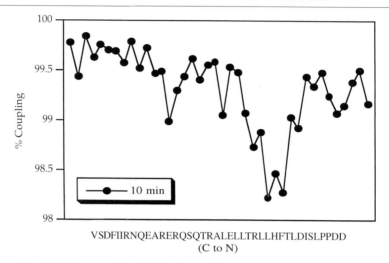

VSDFIIRNQEARERQSQTRALELLTRLLHFTLDISLPPDD
(C to N)

FIG. 11. Chain assembly yields for the machine-assisted synthesis of urocortin.

PsENOD40 is homogeneous by both HPLC and mass spectral analysis (data not shown).

The synthesis of the 40-residue C-terminally amidated peptide urocortin[26] is more challenging. Urocortin is a newly discovered corticotropin releasing hormone-like peptide that has a marked effect on the dietary control of mice.[26] The average coupling yield for the single-couple cycles on a p-methylbenzhydrylamine (MBHA) resin is 99.32% (Fig. 11). The poor couplings observed from residues 24 to 31 may well reflect the nature of the resin. In our laboratory we have demonstrated significant batch-to-batch variance in the quality of commercially available MBHA resins. Subsequent HPLC and mass spectral analysis of the peptide again show the high purity of the crude product (Fig. 12).

The synthetic protocols continue to undergo revision with further improvements to both the speed and the efficiency of the cycles. We anticipate that substitution of the coupling agent HBTU with HATU will significantly reduce the coupling time and allow routine use of efficient single-coupled machine cycles.

[26] J. Vaughan, C. Donaldson, J. Bittencourt, M. H. Perrin, K. Lewis, S. Sutton, R. Chan, A. V. Turnbull, D. Lovejoy, C. Rivier, J. Rivier, P. E. Sawchenko, and W. Vale, *Nature* (*London*) **378,** 287 (1995).

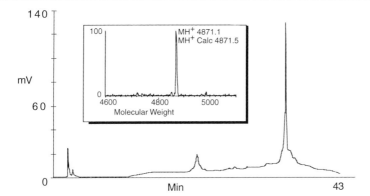

FIG. 12. Analysis by HPLC (C_{18} Vydac column, 2%/min gradient from 0 to 60% acetonitrile/ 0.1% TFA; $\lambda = 214$ nm) of crude urocortin and reconstructed mass spectrum (MH^+ calculated, 4871.5; found, 4871.1).

In Situ Neutralization Protocols for Fmoc Solid-Phase Peptide Synthesis

Although the benefits of *in situ* neutralization over a separate neutralization step are well established in Boc chemistry they have been sparingly employed in Fmoc SPPS, as the deprotected terminal amine is never protonated during standard coupling cycles. *In situ* neutralization is thus not applicable to the standard Fmoc chain assembly unless we introduce a protonation step in the coupling cycle. Beyerman and Bienert,[27] for instance, found that when a TFA wash (neat TFA) is performed prior to *in situ* neutralization coupling, the yields from Fmoc-based assembly of $(Val)_n$ or $(Ala)_n$ on MBHA resin improved significantly. Beyerman *et al.* ascribe the effect to the disaggregation properties of strong acids such as TFA,[28] rather than to simple protonation/*in situ* neutralization of the terminal amine.

Because this strong acid chemistry is incompatible with Fmoc linker and side-chain protecting group stabilities, we considered protonating the resin-bound amine with the incoming Fmoc-protected amino acid prior to its activation during "*in situ*" coupling. To examine the effect of this protonation on coupling yield, we have carried out a comparative study

---

[27] M. Beyermann and M. Bienert, *Tetrahedron Lett.* **33**, 3745 (1992).
[28] M. Beyermann, A. Klose, H. Wenschuh, and M. Bienert, *in* "Peptides 1992" (C. H. Schneider and A. N. Eberle, eds.), p. 263. ESCOM, Leiden, The Netherlands. 1993.

where the difficult sequence STAT91(699–709)[29] is assembled using either neutral (i.e., standard Fmoc SPPS) or *in situ* coupling protocols. Both protocols use the same activating reagent (HBTU) in the same concentration (0.5 M) and excess (5 equivalents) with the same coupling time. Protocol 2 differs from 1 only in that the amino acid is added to the resin prior to its activation. Both chain assemblies are performed on the same resin (Boc-Val-OCH$_2$-Pam) with coupling yields being determined by the quantitative ninydrin test.[17]

1. Neutral Protocol: Five equivalents of Fmoc-amino acid are preactivated in DMF using HBTU and DIEA (6 equivalents) as described earlier. The activated amino acid (0.5 M) is then added to the resin-bound amine.

2. *In Situ* Protocol: A 0.5 M solution (5 equivalents) of Fmoc-amino acid/HBTU (1/1) in DMF is added to the resin-bound amine. After 5 min, DIEA (6 equivalents) is added to initiate the activation of the amino acid and the *in situ* coupling step.

Figure 13 shows the coupling yields obtained after 10 min and 1 hr. The results clearly indicate that all couplings in this difficult sequence are substantially improved when carried out on protonated resin-bound amines using *in situ* neutralization protocols.

The modified Fmoc protocols allow successful synthesis of STAT91(699–709) using 1-hr couplings for residues 7–10. After HF-induced deprotection and cleavage, the peptide is obtained in good yield and purity with no detectable HBTU-related side products (data not shown). This is in contrast with the results obtained from our standard neutral protocol, where forcing conditions (up to 15-hr coupling at 30°) are required for an effective synthesis. Protonation thus significantly increases accessibility of the terminal amine in the solid-phase synthesis of STAT91(699–709).

To examine the generality of this approach we have carried out a similar comparison with the solid-phase synthesis of ACP(65–74). Protocols 1 and 2 are employed with 10-min coupling times; trityl (Trt) protection is maintained on the Asn and Gln side chains. Although the coupling yields for protocol 2 are generally higher when compared to protocol 1, HPLC–ESMS analysis of the resulting crude product (after HF cleavage) revealed the presence of a new side product for the *in situ* assembly. At the difficult Gln^{66} coupling (Fig. 14a,b) resin-bound amine (Ala^{67}) reacts with HBTU (during the protonation step) to form a blocked guanidine adduct. We postulate that the sterically hindered amino acid Fmoc-Gln(Trt)-OH only partially protonates this aggregated resin-bound amine. Some support for

[29] W. Meutermans and P. F. Alewood, *Tetrahedron Lett.* **37,** 4765 (1996).

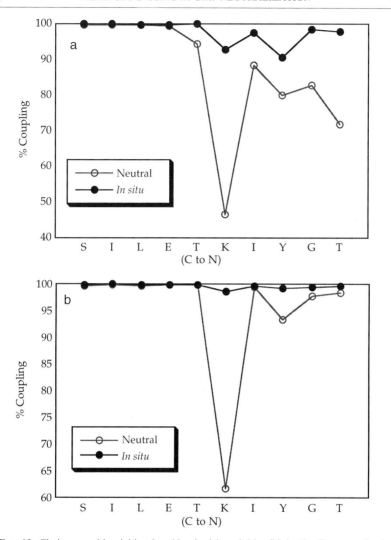

FIG. 13. Chain assembly yields after 10 min (a) and 1 hr (b) in the Fmoc synthesis of STAT91(699–709), TGYIKTELISV, using protocols 1 (Neutral) and 2 (*In situ*).

this arose when, in the *in situ* assembly of a different easy sequence, Fmoc-Gln(Trt)-OH was readily incorporated with no observable HBTU-derived adduct (data not shown).

Further evidence for this partial protonation was found in the following comparative experiment. In one case, ACP(65–74) assembly involves a precoupling treatment of 2 equivalents of Fmoc-Gln(Trt)-OH in DMF,

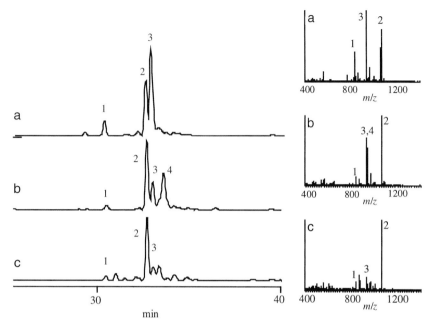

FIG. 14. Analysis by HPLC (Vydac C_{18} column, 1%/min gradient from 0 to 60% acetonitrile/ 0.1% TFA; λ = 214 nm) and MS of crude ACP(65–74) after (a) protocol 1, (b) protocol 2, and (c) TFA protonation prior to adding activated amino acid. Peak 1 is ACP(65–74) desVal-65, desGln-66 (MH$^+$ 836); peak 2 is ACP(65–74) (MH$^+$ 1064); peak 3 is ACP(65–74) desGln-66 (MH$^+$ 935); and peak 4 is the HBTU adduct of ACP(67–74) (MH$^+$ 934).

followed by a 10-min *in situ* coupling with preactivated (by HBTU) Fmoc-Gln(Trt)-OH. In the second case, ACP(67–74) is assembled and then treated with 2 equivalents of TFA in DMSO prior to a 10-min *in situ* coupling with preactivated Fmoc-Gln(Trt)-OH. After further chain assembly and HF cleavage, the crude products are analyzed by RP-HPLC and ESMS. Whereas the amino acid pretreatment had little or no effect when compared to protocol 1 (data not shown), protonation of the aminopeptide–resin with 2 equivalents of TFA considerably enhanced the yield and purity of the target peptide (Fig. 14c).

In summary, a mild protonation step to facilitate *in situ* coupling can be achieved by simply changing the order of addition of the reagents (protocol 2). This provides a useful alternative in the difficult couplings region when standard Fmoc SPPS fails (protocol 1). Where Fmoc-Gln(Trt)-OH acylation of highly aggregated resin-bound amine is required [and possibly Fmoc-Asn(Trt)-OH], protonation should be accomplished with dilute TFA, prior to *in situ* coupling with preactivated amino acid and excess base.

Acknowledgments

This work was supported (in part) by the Australian Research Grants Council. We thank Stephen Kent for support and encouragement in the continuing development of more effective SPPS chemistry.

[3] Cleavage Methods Following Boc-Based Solid-Phase Peptide Synthesis

By JOHN M. STEWART

Introduction

In solid-phase peptide synthesis (SPPS) using Boc (*tert*-butyloxycarbonyl) amino acids, peptide chains are linked to the resin supports as benzyl esters or by bonds of similar chemical stability, and side-chain functional groups are protected by groups of similar reactivity. When the desired product is a peptide with a C-terminal carboxyl group synthesized on Merrifield-type resin or a peptide amide on a 4-methylbenzhydrylamine (MBHA) resin, simultaneous deprotection of the side-chain groups and cleavage of the peptide from the resin is most commonly done by treatment with a hard acid reagent, such as hydrogen fluoride (HF). Although anhydrous HF is an inconvenient (it dissolves glass, is volatile) and toxic reagent, it has been used successfully in SPPS since the mid-1960s, and is still the cleavage reagent of choice in most cases. Once adequate equipment is at hand and suitable protocols have been established, use of HF is safe, rapid, and effective. Methods have been developed to overcome most of the limitations of HF cleavage, such as unwanted side reactions with certain amino acids. If equipment for handling HF is not available, peptide–resins may be cleaved by using trifluoromethanesulfonic acid (TFMSA) or trifluoromethanesulfonic acid trimethylsilyl ester (TMSOTf), although these reagents may cause more serious side reactions with some peptides.

Much of the versatility of the Merrifield system of Boc SPPS lies in the fact that different reagents can be used to cleave the finished peptide from the resin, thus yielding a variety of final products.[1] For example, in addition to production of free peptides by acidolysis or hydrogenolysis, some options are (1) ammonolysis to produce a peptide amide, (2) aminolysis to produce substituted peptide amides, (3) hydrazinolysis to produce peptide hydraz-

---

[1] J. M. Stewart and J. D. Young, "Solid Phase Peptide Synthesis." Pierce, Rockford, Illinois, 1984.

                                        0076-6879/97 $25.00

ides for further azide couplings in solution, (4) reduction to yield peptide alcohols, and (5) transesterification to yield peptide esters. A limitation of these methods is that they are not generally applicable to peptides containing Asp or Glu residues, because the side-chain carboxyl groups will be similarly modified.

Types of Cleavage Reactions

Acidolysis

The most popular reagent for cleavage of peptides from Merrifield, phenylacetamidomethyl (PAM), or MBHA resins at the end of the synthesis is anhydrous liquid HF. Of all the cleavage procedures used until now, this appears to be the most versatile and least harmful to a wide variety of peptides. HF is a toxic, corrosive gas (boiling point 19°), and it must always be used in an adequate fume hood. Because it attacks glass very rapidly, with an exothermic reaction, all equipment for handling HF must be made exclusively of plastic or noncorrosive metal. The commercially available fluorocarbon vacuum line for handling HF allows all of the operations necessary for successful SPPS cleavage reactions to be carried out without any hazard to operators. It is important that all persons using the equipment understand its operation thoroughly and follow explicit instructions carefully. This commercial apparatus also allows transfer of HF under vacuum; this is particularly important for removal of HF at the end of the cleavage reaction. Several side reactions have been shown to be temperature dependent and are much worse at elevated temperatures. Simpler equipment for handling HF has been constructed of heavy-walled fluorocarbon bottles, tubing, and valves; however, these generally do not withstand vacuum, and the operator must use a stream of nitrogen for removing HF. A consequence is that the finished peptide is exposed to HF at elevated temperature for an extended time, which can promote derivatization of Asp and Glu side chains by acylation of scavengers and cause increased N → O shifts of Ser and Thr residues.

HF cleavage is generally done at 0° for 45–60 min; these conditions will generally cleave the peptide effectively from the resin and remove most side-chain blocking groups. Because many of the side-chain blocking groups are removed by HF at temperatures significantly lower than 0°, some investigators prefer to do cleavage at a lower temperature, or at least a preliminary cleavage at a lower temperature followed by cleavage at 0°. This may give a significant improvement in results, particularly if the peptide is large and contains many residues with susceptible side chains. Some chemists have treated the peptide–resin with HF at −10 for 30 min and then at 0° for

30 min. The $-10°$ cleavage does not remove cyclohexyl esters of Asp and Glu.

When side-chain blocking groups are cleaved by acidolysis, reactive carbonium or acylium ions are formed that can attack easily alkylatable residues in the peptide. Benzyl and *tert*-butyl carbonium ions, for example, can readily alkylate Met, Cys, Tyr, and Trp residues. These destructive alkylations can be largely prevented if a large excess of a suitable nucleophilic scavenger is included in the HF reaction mixture. Inclusion of 10% anisole in the HF has been most frequently used for this purpose, but other nucleophiles may be even more effective. Resorcinol, *m*-cresol, thioanisole, dimethyl sulfide (DMS), ethyl methyl sulfide, methionine, 1,2-dithioethane, and indole have all been used for this purpose, with favorable results. We recommend *m*-cresol instead of the more usual *p*-cresol, as it is a liquid and is conveniently measured by pipetting. For peptides containing Cys(Meb), where Meb is 4-methylbenzyl, a mixture of *m*-cresol and *p*-thiocresol is effective and also helps keep the Cys reduced. Met sulfoxide can be reduced if 2-mercaptopyridine is used as a scavenger. Many investigators use a mixture of several scavengers simultaneously, with apparently improved results in the case of very sensitive peptides.

Many of the side reactions encountered with "problem" peptides in standard HF cleavage can be overcome if the "low–high" HF procedure is used.[2] If the cleavage mixture consists of an equimolar mixture of HF and DMS (1 : 3 by volume), the cleavage mechanism changes from the usual S_N1 (where carbonium and acylium ions are produced) to S_N2, where the high concentration of DMS causes it to attack the initial protonated intermediate before separation of a discrete carbonium ion can occur. This procedure prevents alkylation of Tyr by *O*-benzyl or *O*-dichlorobenzyl groups, formation of succinimide peptides from Asp-Gly sequences, and acylation of scavenger molecules by Glu side chains. It reduces Met sulfoxide residues to Met. *m*-Cresol is recommended as a scavenger in this mixture. If *p*-thiocresol is added to the cleavage mixture, formyl groups are cleaved from Trp residues. This "low HF" procedure does not cleave toluenesulfonyl (Tos) or nitro (NO_2) groups from arginine, cyclohexyl esters (OcHx) from Asp and Glu, methylbenzyl (Meb) from Cys, or dinitrophenyl (Dnp) from His, and it may not cleave peptides from the amide-yielding MBHA resins if the C-terminal residue is sterically hindered. A thorough study of the chemistry of peptide–resin acidolysis has been published.[3]

If the low HF procedure is followed by a standard HF cleavage, remain-

[2] J. P. Tam, W. F. Heath, and R. B. Merrifield, *J. Am. Chem. Soc.* **105**, 6442 (1983).
[3] J. P. Tam and R. B. Merrifield, *in* "The Peptides" (C. W. Smith, ed.), Vol. 8, p. 185. Academic Press, New York, 1987.

ing standard blocking groups will be cleaved, and peptides will be more effectively cleaved from MBHA resins. His(Dnp) is not cleaved by acidolysis, and the Dnp group should be removed from the peptide–resin first by treatment with thiophenol. Cyclohexyl esters of Asp and Glu are now commonly used because of their greater stability to cleavage than benzyl esters. The slower cleavage of these esters means that acylium ions produced by cleavage of peptide side-chain esters will have less time to attack anisole or other scavenger molecules to produce undesirable ketone by-products.

If an HF apparatus is not available, the original cleavage reagent introduced by Merrifield, HBr in trifluoroacetic acid (TFA), can be used. Because HBr is not very soluble in TFA, the cleavage is performed by bubbling a constant stream of HBr through a suspension of the resin in TFA. The commercially available HBr in glacial acetic acid should not be used for peptides containing Ser and Thr residues, because it causes acetylation of hydroxyl groups. Trifluoroacetylation has not been shown to occur with HBr–TFA. The major drawback is that Arg blocking groups are not cleaved and must be removed later in a separate reaction if Arg is present.

Another effective cleavage reagent is TFMSA ("triflic acid"). Like HBr, this reagent does not remove nitro or tosyl blocking groups from Arg. Moreover, it is a more powerful Friedel–Crafts catalyst than is HF, and greater problems with side reactions may be anticipated. A mixture of very effective scavengers must generally be used with this cleavage reagent. The trimethylsilyl ester of TFMSA, TMSOTf ("TMS triflate"), is also a powerful reagent for peptide–resin cleavage. These reagents are used in TFA solution. A related reagent is trimethylsilyl bromide (TMS-Br), which is less vigorous than the two TFMSA reagents above. Some of the advantages of the low–high HF procedure can be obtained by doing an initial treatment with TMS-Br, followed by TFMSA or TMSOTf.[4] Boron tribromide has also been used for SPPS cleavage, either in dichloromethane (DCM) or TFA solution. In TFA solution the actual reagent is $B(TFA)_3$.[5]

A complex of HF with pyridine is commercially available and has had some application as an SPPS cleavage reagent. One problem this reagent has in common with TFMSA is lack of sufficient volatility so that the reagent may be removed conveniently by evaporation under vacuum. When these reagents are used for cleavage, the peptide must be precipitated from solution with a dry solvent such as ether, or else the acid can be diluted with a large volume of ice water and the crude peptide immediately separated by chromatography.

[4] M. Nomizu, Y. Inagaki, T. Yamashita, A. Ohkubo, A. Otaka, N. Fujii, P. P. Roller, and H. Yajima, *Int. J. Peptide Protein Res.* **37**, 145 (1991).
[5] J. Pless and W. Bauer, *Angew. Chem., Int. Ed. Engl.* **12**, 147 (1973).

Ammonolysis and Aminolysis

The benzyl ester linking the peptide to the resin in the classic Merrifield SPPS system can be cleaved by ammonia and amines to yield peptide amides. These reactions are quite slow, and must be carried out in solvents which swell the resin. Among others, dimethylformamide (DMF) has been used for this purpose, although it may be partially cleaved under the conditions of the reaction and lead to formation of peptide dimethylamides. Ammonolysis in liquid ammonia is very slow, as this solvent does not swell the resin. A small amount of tertiary amine may improve the rate by catalyzing the reaction.

Reductive Cleavage to Alcohol

Peptide alcohols (in which the C-terminal carboxyl group is reduced to an alcohol) can be synthesized by borohydride reduction of the C-terminal resin ester in peptides lacking Asp and Glu residues.[6] Reductive cleavage proceeds rapidly when the peptide–resin is treated with $LiBH_4$ in tetrahydrofuran (THF). If the peptide contains Glu and Asp residues protected as esters, these side chains will also be reduced to the alcohols.

Other Cleavage Reactions

Peptide esters can be formed by transesterification of the peptide from the Merrifield resin, using a tertiary amine as catalyst. Reasonable rates are attained only if the alcohol is simple. Side-chain blocking groups are not removed in this procedure and must be cleaved by subsequent treatment with an appropriate reagent, such as HF. Glu and Asp side-chain esters will also be converted to the same ester as the C-terminal residue. Dimethylaminoethanol has been reported to give satisfactory transesterification of peptide–resins, and it yields the dimethylaminoethyl ester. This ester can by hydrolyzed very rapidly by treatment with mild aqueous base to yield the peptide acid.

Hydrazinolysis of peptide–resins yields blocked peptide hydrazides, which can be used for further synthesis in solution by the azide condensation method. Here again, Glu and Asp residues will also react.

If the peptide does not contain residues that will cause problems under reduction conditions (Cys, Trp), catalytic transfer hydrogenolysis is an additional alternative for cleavage of peptide acids from Merrifield peptide–resins.[7] The peptide–resin is soaked in a DMF solution of a palladium salt,

[6] J. M. Stewart and D. H. Morris, U.S. Patent 4,254,023 (1981).
[7] M. K. Anwar, A. F. Spatola, C. D. Bossinger, E. Flanigan, R. C. Liu, D. B. Olsen, and D. Stevenson, *J. Org. Chem.* **8,** 50 (1983).

and then ammonium formate is added to serve as hydrogen donor. One reported side reaction is saturation of the indole ring of Trp residues.

Procedures for Peptide–Resin Cleavage

Preparation of Peptide–Resin for Cleavage

In preparation for cleavage, the Boc group should be removed by deprotection after coupling the last amino acid to minimize *tert*-butylation of susceptible residues (Tyr, Trp, Cys) in the product. The final wash of the peptide–resin should be with DCM, which swells the resin. The DCM should be thoroughly removed without drawing air through the resin; condensation of atmospheric moisture can cause problems and delay thorough drying of the peptide–resin.

Removal of Dinitrophenyl Groups from Histidine Residues

If His(Dnp) residues are present, the Dnp group should be removed before cleavage of the peptide–resin. Suspend the peptide–resin in a round-bottom flask with a magnetic stirring bar in the minimum amount of purified DMF needed to slurry the resin (about 5 ml/g resin). Add 20 mol of thiophenol (0.102 ml/mmol) per mole of His(Dnp) present. Mix at room temperature for 1 hr. Thiolysis is rapid and is probably complete in 15 min. As the cleavage proceeds, the mixture becomes intensely yellow due to the color of Dnp–thiophenol. Wash the resin thoroughly with DMF, water, ethanol, and DCM and dry it thoroughly. A small amount of the yellow Dnp–thiophenol usually will remain adsorbed to the resin and will contaminate the crude peptide. It is easily removed during purification of the product.

Cleavage by Anhydrous Hydrogen Fluoride

Anhydrous HF is an extremely toxic, corrosive, and volatile (bp 19°) liquid. It dissolves glass in an extremely rapid, exothermic reaction. However, it is very easily handled in a fluorocarbon (Teflon–Kel-F) vacuum line. Do not use makeshift vacuum lines constructed of polyethylene or polypropylene; these plastics become extremely brittle at low temperatures and may shatter suddenly with serious consequences. Hydrogen fluoride reacts vigorously and exothermically with water.

Preparation for HF cleavage. Figure 1 diagrams a suitable fluorocarbon HF vacuum line. These are commercially available (Protein Research Foundation, Osaka, Japan; Peptides International, Louisville, KY; Beltech, Bel-

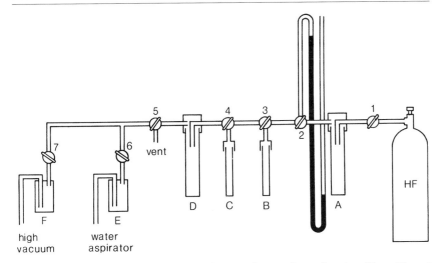

FIG. 1. Diagram of vacuum line for HF cleavage. See text for explanation. (From Stewart and Young,[1] with permission.)

mont, CA). Lines of this type have been in continual use for many years, and they will apparently last indefinitely if given care. Stopcocks and O rings will need to be replaced as they wear. Stopcocks 1, 6, and 7 (Fig. 1) are two-way; stopcocks 2, 3, 4, and 5 are three-way. The manometer connected to stopcock 2 is filled with mercury, and the Hg is capped with a layer of fluorocarbon oil (suitable oil is used for infrared samples). A is the HF drying and storage reservoir, B and C are reaction vessels, D is the trap for waste, E is a plastic safety trap for the water pump, and F is a KOH pellet-filled plastic safety trap for the high vacuum pump. If local restrictions do not allow discharge of HF into the sewer, a calcium oxide trap is added to the left side of stopcock 5 to absorb the HF. The O rings in the connections are ethylene–propylene elastomer (Parker Seals) and will need to be replaced occasionally. After a new set of O rings has been exposed to HF and "set," they should not be moved or stressed; leaks will develop, and the rings will have to be replaced. The traps between the HF line and the vacuum pumps must be made of plastic; they can be fabricated from 250-ml polypropylene graduated cylinders, polyethylene 1/4-inch tubing, and rubber stoppers. Stopcocks 6 and 7 can be commercially available polypropylene or polyethylene units, connected with heavy-wall rubber tubing. Rubber glazes on exposure to HF; rubber–plastic connections should not be moved after exposure to HF, or they will leak. The entire vacuum line is mounted in a good fume hood. Mark each reaction vessel to indicate the height of 5 and 10 ml contents.

If, through carelessness, peptide, resin, or CoF_3 is allowed to bump badly and occlude a stopcock or connecting tube, it will be necessary to open a vessel containing condensed HF. Working entirely in the hood, chill the vessel to be opened in dry ice–ethanol, unscrew the vessel carefully, and stand it in the back of the hood. From a distance direct a stream of water from a wash bottle into the vessel containing the HF. When the vigorous reaction has subsided, remove the vessel and wash out the contents. Carefully clean stopcocks and tubing contaminated by solids. Solids in a stopcock will cause leaks and abrade the soft fluorocarbon surface.

No operation should be performed that can possibly lead to contact with HF liquid or vapors. The entire HF operation must be carried out in a good hood; **breathing HF can cause death.** The commonest HF burns result from accidental contact with HF after which the HF is not completely washed from under the fingernails; several hours later, a painful burn is noticed. The pain can be alleviated by applying calcium gluconate (saturated solution in glycerol). After contact with HF, wash well with water and soak in water, then apply calcium gluconate solution.

The cleavage should be started early enough in the day to ensure at least 4 hr of uninterrupted time. Do not operate the line alone. Wear safety glasses and gloves at all times. Keep a record on the HF tank of the HF removed from it, the date, and the amount remaining. The entire HF line should be checked regularly for leaks: evacuate the line, close stopcock 5 (Fig. 1), and observe that the Hg in the manometer remains at a suitable high level.

For ensuring anhydrous HF, place a few grams of CoF_3 in the reservoir, add a stirring bar, and screw the reservoir into the opening snugly finger-tight. Evacuate the apparatus with the water aspirator until the vacuum reads about (atmospheric pressure minus vapor pressure of water)/2. This may be about 360 torr at sea level. Close the line (stopcock 5, Fig. 1) and chill the reservoir in a dry ice–ethanol Dewar flask for about 5–10 min. Do not use dry ice–acetone, because acetone is much more flammable. Fluorocarbons (such as Kel-F) have a low thermal conductivity and require some time to cool down; once cold they require some time to warm up. Open the HF cylinder valve and carefully open stopcock 1 (Fig. 1), so that HF distills into the reservoir. When a good rate of HF distillation is obtained, the manometer will usually show a vacuum of about 20 torr. If a good transfer rate is not obtained, close stopcock 1 and reevacuate the line. Collect enough HF in the reservoir for three or four cleavages; this predried HF will be ready for immediate use. The reservoir can be replenished from the cylinder during the cleavage reaction.

When sufficient HF has collected, close both the tank valve and stopcock 1 (Fig. 1). During HF distillation the approximate level of HF in the cylinder is indicated by a cold zone on the cylinder wall. The level of HF in the reservoir can be observed by shining a flashlight behind the vessel and gently tapping the vessel to observe the liquid motion. It is important to keep the dry ice–solvent level well up to the top of the vessel while distilling HF into the reservoir except for the time it takes to observe the liquid level.

Performing Cleavage Reaction. An adequate volume of HF and scavengers must be used to avoid excessive side reactions. For 0.2 mmol of peptide–resin (about 0.5 g) use 9 ml of HF and 1.0 ml of anisole (or other scavengers). Place the peptide–resin, a small Teflon-covered stirring bar, and scavengers into a small reaction vessel (B, Fig. 1). Insert the baffle in the top of the reaction vessel. Make sure there are no resin grains on the rim of the reaction vessel. Screw the vessel into place, snugly finger-tight. Place a 1-liter beaker containing a large magnetic stirring bar and a little water under the HF reservoir; start the stirrer. Evacuate the reaction vessel with the water aspirator until the manometer shows good vacuum. Do not evacuate too long or the anisole may distill off. Close stopcock 5 (Fig. 1) and be sure the vacuum holds; now is the time to correct any leaks. Cool the reaction vessel in a dry ice—ethanol bath for about 5 min. Again, be sure that the dry ice is well up to the top of the reaction vessel.

With the reservoir stirrer going, very slowly open the manometer stopcock 2 (Fig. 1), watching that the Hg level in the manometer lowers gradually and that the HF in the reservoir does not boil over. Be careful to avoid bumping of CoF_3 from the reservoir. When stable distillation is achieved, add room temperature water to the reservoir bath if necessary to accelerate distillation. Distill the desired amount of HF into the reaction vessel (10 ml is the maximum amount recommended for the small reaction vessel) and then turn stopcock 2 (Fig. 1) so that the reaction vessel is totally closed.

Immerse the reaction vessel in a water bath containing crushed ice and stir magnetically for the desired length of time. After the anisole melts, make sure the resin is mixing, well suspended in the liquid HF. The resin color is brick red in the presence of anisole; without anisole it is a brilliant blue color. The usual conditions of cleavage are 45–60 min at 0°. This allows 15 min to warm the reaction vessel contents to 0°.

At the end of the reaction time, turn on the water pump, evacuate the line, and very slowly open the reaction vessel stopcock 3 (Fig. 1) to the water vacuum, watching that the HF does not boil too high or fast. Maintain magnetic stirring until no liquid is left. Continue to evacuate the system with the water aspirator until the resin is cream colored, which occurs when the HF is gone. The time required is usually about 45 min. Chill trap D (Fig. 1) in a dry ice–ethanol Dewar; open stopcock 5 to vent the aspirator, and turn off the water aspirator. When trap D is cold (about 10 min), connect the reaction vessel to the mechanical vacuum pump and pump about 1 hr.

Workup of Cleavage Mixture. Use ethyl acetate to transfer the resin–peptide mixture to a fritted funnel. If the peptide is very hydrophobic and may be soluble in ethyl acetate, use ethyl ether to transfer the mixture to the filter funnel. Scavengers and related by-products will be more thoroughly removed by ethyl acetate, yielding a cleaner crude peptide. Use three organic washes, allowing time for the solvent to penetrate the resin particles. If water will be used to extract the peptide, air-dry the resin. Suction may be used, as condensation of atmospheric moisture is not a problem at this stage.

Extract the crude peptide from the resin with an appropriate solvent. For water-soluble peptides, 10% acetic acid is useful. For more hydrophobic peptides, glacial acetic acid is preferred. Some chemists use both these solvents in succession. Trifluoroacetic acid is a very effective solvent for extraction of hydrophobic peptides. It is volatile (bp 70°) and evaporates easily under reduced pressure. Alternatively, it can be poured into a large volume of cold, dry ether to precipitate the peptide. Use three extractions, allowing the solvent time to diffuse into the resin thoroughly before removing it by suction. Filter the peptide solution directly into the round-bottom flask that will be used for rotary evaporation or directly into the lyophilizer flask if you plan to omit the concentration step. Save both the resin and the organic washes until you are sure you have recovered the peptide in the expected extract.

The solution of crude peptide may be reduced to near dryness by rotary evaporation under reduced pressure. If foaming is a problem during evaporation, omit the concentration step and lyophilize the solution directly. Dissolve the residue in glacial acetic acid and lyophilize. Alternatively, acetic acid extracts may be lyophilized directly. Peptide solutions that are less than 10% or more than 90% acetic acid can usually be lyophilized satisfactorily. More nearly equal mixtures of water and acetic acid form eutectic mixtures and will usually not stay frozen.

To reverse any N → O shifts in Ser or Thr peptides, dissolve the lyophilized peptide in water, neutralize the solution with NH_4OH, and relyophilize.

Low HF Cleavage

It is important to begin the cleavage procedure[2] early in the day. In the standard small HF cleavage vessel mix peptide–resin (0.2 mmol peptide), 1.0 ml *m*-cresol, and 6.5 ml DMS. If the peptide contains formyl Trp, the scavenger mixture should be 0.75 ml *m*-cresol, 0.25 ml *p*-thiocresol, and

6.5 ml DMS. Attach the cleavage vessel to the vacuum line, evacuate, and distill in 2.5 ml of HF. Do not prolong the evacuation, or the volatile DMS will be lost. Cleave for 2 hr at 0°. Using the water aspirator, distill off the HF and DMS. The *m*-cresol and *p*-thiocresol will remain in the cleavage vessel to serve as scavengers in the following standard HF cleavage. Distill 10 ml of HF into the cleavage vessel and perform a standard high HF cleavage, as above.

Cleavage with HF–Pyridine

Place the peptide–resin (0.2 mmol peptide) and a magnetic stirring bar in a fluorocarbon or polyethylene vessel that can be tightly closed and will withstand vacuum. Add the appropriate scavenger (see above) and 10 ml of HF–pyridine (Aldrich, Milwaukee, WI; 70% HF). Close the vessel and stir 1 hr at room temperature. Continue to stir while evaporating the reagent with a high vacuum pump (evacuate very carefully at first!), using a CaO trap to collect the HF and a cold trap and a KOH trap to protect the pump. Work up as described above.

Cleavage by HBr–Trifluoroacetic Acid

Cleavage of the peptide from the resin using HBr–TFA[1] can be carried out in a manual synthesis vessel or a round-bottom flask fitted with a bubbling tube for introduction of HBr and an exit tube fitted with a $CaCl_2$ drying tube. The resin must be deprotected for removal of Boc groups before cleavage. Suspend the peptide–resin in TFA (10 ml per g resin) in the cleavage vessel. If the peptide contains Cys, Met, or Tyr, in addition to benzyl, tosyl, nitro, or carbobenzoxy groups, dissolve 50 mol of anisole or methyl ethyl sulfide in the TFA for each mole of sensitive amino acid; 15 mol of Met per mole of sensitive residues has also been found to give satisfactory protection.

Bubble a slow stream of anhydrous HBr for 90 min through the resin suspension and out through a $CaCl_2$ drying tube. A shorter cleavage time may be adequate. If the peptide contains Tyr or Trp, bubble the HBr first through a scrubber tube containing a solution of 2 g of anisole or resorcinol in TFA. Use a trap between the HBr cylinder and the cleavage vessel or scrubber to prevent drawing back of solvent into the cylinder. The HBr flow must be adjusted occasionally, because the needle valve often tends to shut itself off, and, if it does, the solvent will be drawn back into the tubing. Use a fairly rapid stream of bubbles for a few minutes to sweep out the vessels and saturate the solvents, then use a slow stream of bubbles just sufficient to suspend the resin and keep it dispersed. After 90 min, close the HBr tank and close the HBr inlet tube (stopcock). Remove the HBr valve after each use, wash with methanol, and dry thoroughly. Wash the resin three times with TFA (5 ml per g resin each time), allowing the TFA to extract the resin about 1 min with each wash, and then evaporate the peptide solution to dryness under reduced pressure without heating. Dissolve the peptide several times in a suitable solvent, acetic acid–water (3:1, v/v) or methanol–water (1:1, v/v), and evaporate the solvent under reduced pressure to remove excess HBr (the peptide is always quite acidic at this stage). If the cleavage solution contained anisole, extract the crude product thoroughly with ether (Sephadex chromatography has also been used for removal of anisole, Met, and ethyl methyl sulfide) and then dry the peptide under high vacuum. The crude peptide from HBr–TFA cleavage usually contains much nonpeptide material, especially if the resin was not initially extracted with TFA to remove linear polystyrene, so the weight at this point is not a true index of the yield. Yields of crude peptide can be based on amino acid analysis of hydrolyzates.

Caution: The entire cleavage process should be done in a good fume hood. TFA causes serious burns; wash off any spillage on the skin immediately. **Do not breathe the TFA vapor.**

Standard Cleavage by Trifluoromethanesulfonic Acid in Trifluoroacetic Acid: "High" TFMSA Cleavage

The peptide–resin should be dry and without Boc or His(Dnp) groups.[8] Working in a good hood, place the peptide–resin (0.5 g, 0.2 mmol peptide) in a round-bottom flask with a magnetic stirring bar. Add 1.5 ml of a scavenger mixture of thioanisole:ethanedithiol (2:1, by volume) and allow to soak a few minutes. Add 10 ml of TFA, chill in an ice bath, and stir while slowly adding 1.0 ml of TFMSA. Allow to warm to room temperature and stir for 30–90 min. Use the minimum length of reaction time found to carry out complete cleavage; the longer times will be needed for resistant blocking groups and hindered C-terminal residues. Arg(Tos) and Cys(Meb) groups and benzylhydrylamine (BHA) resin links are not cleaved; use Arg(Mts) or Arg(Pmc) and Cys(Mob) (where Mts is mesitylenesulfonyl, Pmc is 2,2,5,7,8-pentamethylchroman-6-sulfonyl, and Mob is 4-methoxybenzyl).

Filter the reaction mixture slowly on a fritted glass funnel directly into 200 ml of cold, stirred ether. More complete precipitation will be obtained if the ether is chilled in dry ice–ethanol. Wash the resin three times with small aliquots of TFA. Add 1.0 ml of dry pyridine dropwise to the cold

---

[8] J. P. Tam, W. F. Heath, and R. B. Merrifield, *J. Am. Chem. Soc.* **108**, 5242 (1986).

stirred ether suspension to complete peptide precipitation. Collect the crude peptide on a fine-pore fritted funnel, wash with cold ether, and continue product workup by standard procedures.

"Low" TFMSA Cleavage

To avoid serious side reactions in TFMSA cleavage, use the following procedure,[8] which is analogous to the low-HF procedure. Use the high TFMSA conditions with the following changes. The cleavage reagent mixture is TFMSA–TFA–DMS–m-cresol (10:50:30:10, by volume), and it is made by adding the TFMSA to the chilled mixture of the other components. If the peptide contains Trp(For), the reagent is TFMSA–TFA–DMS–m-cresol–1,2-ethanedithiol (EDT) (10:50:30:8:2, by volume). Add the cold reagent to the peptide–resin and stir for 4 hr at 0°. Work up as for the "high TFMSA" procedure.

"High" Cleavage by Trifluoromethanesulfonic Acid Trimethylsilyl Ester

To 0.5 g peptide–resin in a round-bottom flask add 20 ml of ice-cold TFA containing 1.0 M TMSOTf (molecular weight 222, d = 1.08; 4.1 ml), 1.0 M thioanisole (molecular weight 124, d = 1.06; 2.4 ml), and 0.5 ml m-cresol (add 0.5 ml ethanedithiol for Trp peptides) and stir the mixture for 2 hr at 0°.[4] Work up the reaction as for the high TFMSA procedure. Removal of Arg(Tos) groups will require longer reaction time.

"Low" Cleavage by Trimethylsilyl Bromide

The following procedure is analogous to low HF cleavage for removal of troublesome protecting groups prior to cleavage of the peptide from the resin with high TMSOTf conditions. Treat peptide–resin (0.5 g) in a round-bottom flask with 20 ml of cold (0°) TFA containing 1.0 M TMSBr (molecular weight 153; d = 1.16; 2.65 ml), 1.0 M thioanisole (molecular weight 124, d = 1.06; 2.4 ml), 0.5 ml m-cresol, and 0.5 ml ethanedithiol (for Trp peptides) and stir for 1 hr at 0°. Filter the resin on a fritted funnel and wash three times with TFA. The resin can be transferred back to the flask for high TMSOTf cleavage or dried for HF cleavage.

Cleavage by Boron Tribromide in Trifluoroacetic Acid: Boron tris(trifluoroacetate)

Boron *tris*(trifluoroacetate) [B(TFA)$_3$; BTFA][5] can be purchased as a 1 M solution in TFA (Aldrich; Fluka, Ronkonkoma, NY). Alternatively, it can be prepared from BBr$_3$ and TFA by adding 3 equivalents of TFA (molecular weight 114; d = 1.54) to 1 equivalent of chilled (0°) 1 M BBr$_3$

solution (Aldrich, Fluka) and evaporating to dryness to remove the by-product HBr. Dissolve the solid BTFA in TFA to reconstitute a $1.0 M$ solution, chill to $0°$, and add this solution to a mixture of 0.2 mmol of peptide–resin and 1.5 ml of thioanisole as well as any other scavengers needed for the particular peptide (e.g., ethanedithiol for Trp peptides). Usually about 5–8 equivalents of BTFA reagent is used per equivalent of cleavable functional groups on the peptide–resin. After stirring for 1 hr at $0°$, filter off the resin and wash three times with TFA. Rotary evaporate the solvent under reduced pressure. Apply the crude peptide to a gel column for separation from reagents. Alternatively, boron may be removed by repeatedly dissolving the reaction residue in methanol and evaporating under reduced pressure. In this procedure benzyl and Z groups are cleaved, as are Merrifield resin links, Arg(Tos), and Cys(Mob). Do not use anisole as a scavenger with BTFA.

Cleavage from Merrifield Resins by Catalytic Transfer Hydrogenolysis

Add 1.0 g of peptide–resin to a solution of 1.0 g palladium(II) acetate in 14 ml of DMF and stir for 2 hr.[7] Add a solution of 1.5 g ammonium formate in 1.5 ml of water; stir 2 hr. Filter on a fritted funnel, washing twice each with acetic acid and water. Concentrate the peptide solution under vacuum and purify conventionally.

This procedure may not completely cleave benzyl (Bzl) ethers. If Ser(Bzl) or Thr(Bzl) is present in the peptide, after the initial 2-hr hydrogenolysis add 20 ml acetic acid and 0.5 g of palladium on charcoal. Stir for an additional 2 hr. Filter and work up as described above.

Cleavage by Ammonolysis

Working in a good fume hood, suspend the peptide–resin in anhydrous methanol (20 ml/g resin) in a heavy-wall pressure bottle and add a magnetic stirring bar. A round-bottom flask can be used for small runs if it is in perfect condition without any scratches, chips, or nicks. Chill the suspension to $-20°$ in an ice–CaCl$_2$ bath and saturate it with anhydrous NH$_3$, excluding moisture (use the KOH vent tube). Use a safety trap and watch the operation continually to be sure that solvent is not drawn back into the cylinder. Wire or tape a tight rubber stopper in place and stir the suspension at room temperature for 2–4 days. Chill the bottle, remove the stopper, allow the mixture to come to room temperature in a good hood, and remove the resin by filtration. Wash the resin three times with 5 ml methanol and evaporate the combined filtrate and washings in a rotary evaporator. Save the resin for additional treatment if the yield is low.

If the C-terminal residue of the peptide is sterically hindered, conversion of the initially formed methyl ester to amide may not be complete, and the ester will contaminate the product. If 2-propanol or trifluoroethanol is used as solvent for ammonolysis no esters are formed.

In some cases, particularly if the peptide product is not soluble in methanol, yields have been improved by using methanol : dioxane or methanol : DMF as solvent for the ammonolysis; these solvents swell the resin better than methanol. If the yield is unsatisfactory, it may be possible to improve it by first removing the peptide from the resin as the methyl ester (see below) and then converting it to the amide in solution.

Cleavage by Hydrazinolysis

Suspend the peptide–resin in purified DMF (5 ml per g resin) and add anhydrous hydrazine (30 equivalents per equivalent of peptide). Stir the mixture for 2 days at room temperature. Remove the resin by filtration and wash with DMF. Evaporate the combined filtrate and washings under high vacuum, and purify the peptide hydrazide by a suitable procedure. Hydrazine boils at 113°; DMF boils at 153°.

Because Boc and formyl amine protecting groups and *tert*-butyl esters remain intact during the hydrazinolysis, hydrazinolysis can be used to produce protected peptide hydrazides suitable for further coupling in solution. Methyl and benzyl esters on side chains will be converted to hydrazides. The tosyl group is cleaved from His, and Arg(NO$_2$) is destroyed.

The procedure of Honda *et al.*[9] may be used to prepare anhydrous hydrazine from the hydrate. Add 1 kg of hydrazine hydrate during 3 hr from a dropping funnel to a stirred, gently refluxing mixture of 2 liters of toluene and 2 kg of CaO. Continue refluxing for 10 hr, and then distill the toluene–hydrazine azeotrope into a Dean–Stark receiver. Remove the hydrazine, which collects as the lower layer, and return the toluene to the flask. The yield, which is about 75%, can be improved on successive runs by reusing the same toluene.

Caution: Hydrazine is an eye and skin irritant; use care in handling it. Hydrazine is flammable and can explode during distillation if oxygen is present.

Cleavage by Transesterification

Methyl Esters. Suspend the peptide–resin in anhydrous methanol (40 ml per g resin), and add triethylamine (TEA; 50 equivalents per equivalent of peptide). Stir the mixture at room temperature for 20 hr. Remove the resin by filtration, evaporate the solvent, and purify the peptide ester by a suitable technique.

---

[9] I. Honda, Y. Shimonishi, and S. Sakakibara, *Bull. Chem. Soc. Jpn.* **40,** 2415 (1967).

If the C-terminal residue is sterically hindered and the above procedure does not give a reasonable yield, suspend the peptide–resin (1 g) in 25 ml DMF, 25 ml methanol, and 12 ml TEA. Stir for 24 hr at 45°. Remove the resin by filtration and again repeat the cleavage in the same way. Peptide esters thus obtained have been converted to amides in difficult cases by dissolving the methyl ester in a mixture of 50 ml DMF, 40 ml ethylene glycol, and 25 ml water, saturating with anhydrous ammonia at 0°, and stirring in a closed pressure vessel for 2 days at room temperature. Asp and Glu benzyl esters will be transesterified, but cyclohexyl esters should remain largely intact.

Ethyl Esters. Stir a suspension of peptide–resin in 10% TEA in ethanol at 45° for 90 hr. A fair yield of peptide benzyl ester has been obtained by reaction at 80° for 40 hr. Other primary alcohols have been similarly used.

Cleavage by Reduction: Synthesis of Peptide Alcohols

Swell 1 g (0.4 mmol) dry peptide–resin in 30 ml of anhydrous, peroxide-free THF in a round-bottom flask with magnetic stirring and exclusion of moisture.[1,6] Add 2 mmol (58 mg) of $LiBH_4$ in portions and stir 1 hr at room temperature. Add 1 ml glacial acetic acid to destroy excess hydride, filter on a fritted glass funnel, and wash the resin with several small portions of glacial acetic acid. Evaporate the combined filtrate and washings in a rotary evaporator and lyophilize the residue from glacial acetic acid. Purify the peptide alcohol by a suitable procedure. This procedure will convert all esters in the peptide (Glu and Asp) to alcohols, and it will deprotect bromobenzyloxycarbonyl (Tyr) residues.

[4] Standard Fmoc Protocols

By Donald A. Wellings and Eric Atherton

Introduction

The application of 9-fluorenylmethoxycarbonyl (Fmoc)[1] protection to solid-phase peptide synthesis has grown in prominence since two research

---

[1] E. Atherton and R. C. Sheppard, *in* "The Peptides: Analysis, Synthesis, Biology" (S. Undenfriend and J. Meienhofer, eds.), Vol. 9. p. 1. Academic Press, New York, 1987.

METHODS IN ENZYMOLOGY, VOL. 289   0076-6879/97 $25.00

groups in the late 1970s published independently on the utility of this group for α-amino protection in the solid phase.[2,3] The Fmoc group had remained relatively dormant in the literature since the initial publication by Carpino in 1970.[4] However, two interim publications emerged, one from the Merrifield laboratory using a sulfonic acid derivative of the Fmoc group as an adjunct to purification after solid-phase assembly and cleavage[5] and the other using an isolated Fmoc-amino acid at the amino terminus of an otherwise conventional *tert*-butoxycarbonyl/benzyl (Boc/Bzl) assembly process.[6]

In an initial paper by Carpino the Fmoc group was removed by refluxing ammonia, and hence its utility was considered to be operationally limited. The realization that the group was rapidly cleaved with secondary base, particularly piperidine diluted with N,N-dimethylformamide (DMF), and the fact that careful examination established the absence of side reactions with other potentially sensitive amino acid derivatives, dramatically increased its utility.[7]

The method that emerged is operationally simple and chemically less complex than the Boc procedure. It is a mild procedure, and because of the Fmoc base lability and orthogonal nature relative to the previously accepted acid-labile protecting groups it has allowed an element of chemical versatility in solid-phase strategies that previously did not exist. Indeed, it is the method of choice for the solid-phase synthesis of most modified peptide species including phosphorylated, sulfated, and glycosylated peptides (covered later in this volume). Its use in combination with groups such as Boc, allyl, and Dde (see below) has allowed access to a new dimension of peptide species previously not attainable from solid-phase synthesis procedures.

This article deals with standard protocols for Fmoc solid-phase peptide synthesis. It covers Fmoc-amino acid synthesis, compatible side-chain protection, solid phases (and the derivatization of), linkage agents, assembly methods, cleavage from the solid support, and, briefly, workup and purification protocols. Other more specialized procedures using the Fmoc group are dealt with in Chapters 5, 7, and 11.

[2] C. D. Chang and J. Meienhofer, *Int. J. Pept. Protein Res.* **11**, 246 (1978).
[3] E. Atherton, H. Fox, D. Harkiss, C. J. Logan, R. C. Sheppard, and B. J. Williams, *J. Chem. Soc. Chem. Commun.,* 537 (1978).
[4] L. A. Carpino and G. Y. Han, *J. Am. Chem. Soc.* **92**, 5748 (1970).
[5] R. B. Merrifield and A. E. Bach, *J. Org. Chem.* **43**, 4808 (1978).
[6] S. S. Wang, *J. Am. Chem. Soc.* **95**, 1328 (1973).
[7] For a comprehensive review of Fmoc solid-phase synthesis, see E. Atherton and R. C. Sheppard, "Solid Phase Synthesis: A Practical Approach." IRL Press at Oxford Univ. Press, Oxford, 1989.

Solid Supports

Since the original concept of solid-phase peptide synthesis was reported by Merrifield in 1962, the design of the polymer support has sought varied attention. Work carried out by Letsinger and Kornet almost simultaneously with Merrifield's work was published in 1963.[9] While Merrifield worked using microporous polystyrene cross-linked with low levels of divinylbenzene, Letsinger and Kornet used a more highly cross-linked macroreticular polystyrene matrix.

Merrifield had the foresight in 1963[8] to point out that technical difficulties arising from solubility problems might hinder the synthesis of long-chain polypeptides. Solvents that solvate the polystyrene matrix sometimes bring about aggregation of peptide chains containing certain sequences of amino acids. The consequent aggregation can result in steric inaccessibility of reagents to reactive sites. Several strategies have been employed that use support matrices with solvation properties more comparable to those of the growing peptide chain. Inman and Dintzis[10] suggested the use of polyacrylamide itself as a support for solid-phase peptide synthesis. However, owing to the internal hydrogen bonding, the resin could only be permeated by highly polar solvents such as water.

The pioneering work carried out by Sheppard and co-workers established that poly(N,N-dimethylacrylamide) cross-linked by incorporation of bisacryloylethylenediamine had favorable swelling properties in N,N-dimethylformamide and other solvents used in peptide synthesis.[11] These polyacrylamide solid phases have been modified to increase the functionality up to five times (5 mmol/g) the capacity of normally loaded supports and have been shown to be of practicable use.[12] Although this type of support, in its various forms and particularly the original microporous polystyrene resins, still dominate solid-phase peptide synthesis, a number of new supports have gained recognition.

The introduction from the Sheppard group[11] of low-pressure continuous flow methodology for solid-phase synthesis, allowing the peptide to be assembled in an enclosed column, prompted the preparation of compatible solid supports. The original support used was polydimethylacrylamide polymerized within the pores of a kieselguhr matrix (Pepsyn K).[13] Polyhipe,

---

[8] R. B. Merrifield, *J. Am. Chem. Soc.* **85**, 1249 (1963).

[9] R. L. Letsinger and M. J. Kornet, *J. Am. Chem. Soc.* **85**, 3045 (1963).

[10] J. K. Inman and H. M. Dintzis, *Biochemistry* **8**, 4074 (1969).

[11] E. Atherton, D. L. J. Clive, and R. C. Sheppard, *J. Am. Chem. Soc.* **97**, 6584 (1975).

[12] R. Epton, G. Marr, B. J. McGinn, P. W. Small, D. A. Wellings, and A. Williams, *Int. J. Biol. Macromol.* **7**, 289 (1985).

[13] A. Dryland and R. C. Sheppard, *J. Chem. Soc. Perkin Trans. I*, 125 (1986).

another development of this principle, incorporates polyacrylamide into a novel macroreticular polystyrene matrix.[14] Following on from this several different supports emerged that can be used both in a batchwise and continuous flow manner. Mutter and co-workers[15] developed the concept of polyethylene glycol (PEG) grafted onto a rigid polystyrene core. Further optimization of this support by Bayer and Rapp led to Tentagel[16] and the related PEG–polystyrene developed by Barany et al.[17] These latter two have more recently gained in importance for the solid-phase synthesis of peptides. Other supports that may attain prominence in the future are polyethylene glycol–polyacrylamide (PEGA) resins developed by Meldal[18] and the cross-linked ethoxylate acrylate resin (CLEAR) introduced by Kempe and Barany.[19]

The choice of commercially available supports for solid-phase peptide synthesis is now exhaustive, ranging from supports containing methyl ester or amine functional groups through to resins with linkage agents and the first amino acids already attached. The protocols described here cover the conversion of methyl ester-based supports to an amine followed by attachment of an internal reference amino acid,[11] linkage agent,[11] and first amino acid, and then the repetitive cycle for peptide chain elongation.

Preparation of Fmoc-Amino Acids

Besides the advantages that they confer to solid-phase peptide synthesis, Fmoc-amino acids are usually easy to prepare in high yield. They exist in a crystalline state and are stable as the free acid when stored in the cold, in a dry form. They also solubilize relatively freely in solvents of choice for solid-phase syntheses.

Many reagents have been suggested for the introduction of the Fmoc group, but two dominate. They are fluorenylmethyl chloroformate (Fmoc-Cl)[20] and fluorenylmethylsuccinimidyl carbonate (Fmoc-ONSu).[21] Both give high yields of the amino acid derivatives. However, with certain amino acids, particularly those with less hindered side chains (Gly, Ala), Fmoc-Cl

[14] P. W. Small and D. C. Sherrington, J. Chem. Commun., 1589 (1989).
[15] H. Hellerman, H.-W. Lucas, J. Maul, V. N. Rajasekharan Pillai, and M. Mutter, Macromol. Chem. 184, 2603 (1983).
[16] E. Bayer and W. Rapp, in "Poly(ethylene) Glycol Chemistry" (J. M. Harris, ed.), p. 325, 1992.
[17] G. Barany, N. A. Solé, R. J. van Abel, F. Albericio, and M. Selsted, in "Innovation and Perspectives in Solid Phase Synthesis" (R. Epton, ed.), p. 29. Intercept, Andover, UK, 1992.
[18] M. Meldal, Tetrahedron Lett. 33, 3077 (1992).
[19] M. Kempe and G. Barany, J. Am. Chem. Soc. 118, 7083 (1996).
[20] L. A. Carpino and G. Y. Han, J. Org. Chem. 38, 4218 (1973).
[21] G. F. Sigler, W. D. Fuller, N. C. Chuturverdi, M. Goodman, and M. Verlander, Biopolymers 22, 2157 (1983); L. Lapatsani, G. Milias, K. Proussios, and M. Kolovos, Synthesis, 671 (1983).

can give small but detectable amounts of dipeptides and even tripeptides. Fmoc-ONSu is the reagent of choice for eliminating this oligomer formation. Other procedures have been proposed that eliminate the possibility of this side reaction, such as prior bistrimethylsilylation,[22] but the simple protocols employing Fmoc-Cl and Fmoc-ONSu remain those of choice.

Three illustrative procedures (see later) for the introduction of the Fmoc group are described below. One uses Fmoc-ONSu for the preparation of Fmoc-Ile-OH, and the other two use Fmoc-Cl for the preparation of Fmoc-His(Boc)-OH and Fmoc-Cys(Acm)-OH, where Acm is the acetamidomethyl group.

Side-Chain Protection

Because of the base lability of the Fmoc group, acid-labile side-chain protecting groups are normally employed. Where appropriate these are based on the *tert*-butyl moiety: *tert*-butyl ethers for Ser, Thr, and Tyr, *tert*-butyl esters for Asp and Glu, and the Boc group for His and Lys. The indole group of Trp can be incorporated unprotected, but when the side chain is considered to be particularly susceptible to side reactions the Boc group can be employed.[23] The trityl group has also been extensively utilized for the protection of Cys[24] and the amide bearing side-chain groups of both Asn and Gln.[25] Also for Cys, the Acm group is extensively used when a protecting group on the sulfur needs to be maintained after cleavage of the peptide from solid support. A host of other protecting groups are applicable for Cys when differential disulfide bond formation is required (see Chapter 10).

The guanidino group of Arg is commonly protected by the 4-methoxy-2,3,6-trimethylbenzenesulphonyl (Mtr)[26] or the more acid-labile 2,2,5,7,8-pentamethylchroman-6-sulfonyl (Pmc) group.[27] The commercial availability of the 2,2,4,6,7-pentamethyldihydrobenzofuran-5-sulfonyl (Pbf)[28] group could increase its popularity particularly in conjunction with the Trp-containing peptides.[29]

[22] D. R. Bolin, I.-I. Sytwa, F. Humiec, and J. Meienhofer, *Int. J. Pept. Protein Res.* **33**, 353 (1989).
[23] P. White, in "Peptides: Chemistry and Biology" (J. A. Smith and J. Rivier, eds.), p. 1537. Escom Science Publishers, Leiden, The Netherlands, 1992.
[24] S. N. McCurdy, *Pept. Res.* **2**, 147 (1989).
[25] P. Sieber and B. Riniker, *Tetrahedron Lett.* **32**, 739 (1991).
[26] E. Atherton, R. C. Sheppard, and J. D. Wade, *J. Chem. Soc. Chem. Commun.*, 1060 (1983).
[27] J. Green, O. M. Ogunjobi, R. Ramage, and A. J. S. Stewart, *Tetrahedron Lett.* **29**, 4341 (1988).
[28] L. A. Carpino, H. Shroff, S. A. Trido, E. M. E. Mansour, H. Wenschuh, and F. Albericio, *Tetrahedron Lett.* **34**, 7829 (1993).
[29] C. G. Fields and G. B. Fields, *Tetrahedron Lett.* **34**, 6661 (1993).

Other compatible protecting groups that require different chemical methods for removal have gained in prominence. In conjunction with the Fmoc group they have allowed the syntheses of relatively complex peptide structures. Groups such as allyl, allyloxycarbonyl (Alloc), 1-(4,4-dimethyl-2,6-dioxocyclohex-1-ylidene) (Dde), and the common *tert*-butyl moiety allow greater chemical versatility when used in conjunction with Fmoc for α-amino group protection.

The utility of these various side-chain protecting groups and combinations of such will be illustrated in other parts of this volume. Here the standard Fmoc methods are described in combination with acid-labile (trifluoroacetic acid) side-chain protecting groups. All of the Fmoc-amino acid derivatives mentioned are commercially available.

Linkage Agents

The concept of adding individual reversible linkage agents to suitably amino modified supports was first introduced in the mid-1970s,[11] and a host of linkage agents are now commonly used in Fmoc solid-phase synthesis. For the synthesis of fully deprotected peptides terminating in a carboxyl group the linkage agent 4-hydroxymethylphenoxyacetic acid[7] is most often employed. It can be introduced onto an amino support using standard coupling agents. However, care must be taken to avoid the reaction mixture becoming basic, so as to prevent multiple additions.

For the synthesis of peptide amides the initial linkage used with Fmoc was 4-hydroxymethylbenzoic acid,[7,30] which, with the same precautions, can be attached as above, but particular care should be exercised to prevent basic conditions from being generated as the hydroxyl function is particularly susceptible to acylation. This linkage agent is completely resistant to acids, and consequently acid-labile protecting groups can be removed prior to cleavage. To produce peptide amides the deprotected peptide–resin is subjected to treatment with ammonia. Other bases/nucleophiles can be used to liberate modified C-terminal moieties.

The most popular linkage for the preparation of peptide amides is the Rink linker, 4-(Fmoc-amino-2′,4′-dimethoxybenzyl) phenoxyacetic acid.[7,31] This protected linkage agent can readily be attached to an amino support using common activation methods. This linkage can be cleaved with trifluoroacetic acid, simultaneously removing acid-labile side-chain protecting groups, to liberate fully deprotected peptide amides. Many other linkage agents, such as the PAL [5-(4-aminomethyl-3,5-dimethoxyphenoxy)valeric

[30] E. Brown, R. C. Sheppard, and B. J. Williams, *J. Chem. Soc. Perkin Trans. 1,* 1161 (1983).
[31] H. Rink, *Tetrahedron Lett.* **28,** 3787 (1987).

acid] linker,[32] have been utilized for producing peptide amides and peptides with modified amide groups.

For the synthesis of fully protected peptides, to be further used in fragment condensation strategies, the linkage agent 3-methoxy-4-hydroxymethylphenoxyacetic acid was first employed.[7,33] This linker can be introduced onto amino supports in the usual manner, and it is cleaved by dilute trifluoroacetic acid (1–2%, v/v) to produce protected peptides. Other linkage agents conferring increased acid lability are the 4-hydroxymethyl-3,5-dimethoxyphenoxyvaleric acid[34] and the 2-chlorotrityl group.[35] The latter linkage agent is susceptible to cleavage by dilute acetic acid solutions. The allyl group (utilized for preparing glycosylated peptides) can also be used for preparing fully protected peptides.[36] The allyl group is cleaved by Pd(0).

Peptide Assembly Methods

The procedures described follow on from derivatization of the solid support and addition of the linkage agent. However, it is useful to add an internal reference amino acid between the linkage agent and the resin support.[7,11] This aids analytical evaluation of in-process amino acid analyses, allowing confirmation of efficient loading of the first amino acid attached to the support and a check that the peptide is fully cleaved postassembly. The reference amino acid is a normal amino acid addition and is described in the following protocols.

Addition of C-Terminal Amino Acid. After addition of the linker, the peptide chain can then be assembled. If the first amino acid is to be coupled to an amino function on the linker, then standard coupling protocols can be employed (see below). However, if ester bond formation is required precaution should be taken to reduce racemization of the C-terminal amino acid during the acylation reaction. Coupling with the carboxylate salt of the first amino acid to the halogenated linker is popularly employed, and many other methods have also been described to reduce enantiomer formation.

The procedure described in these protocols, utilizing diisopropylcarbodiimide (DIPCDI) for activation and a catalytic amount of 4-dimethylamino-

[32] F. Albericio, N. Knieb-Cordonier, S. Biancalana, L. Gera, I. R. Masada, D. Hudson, and G. Barany, *J. Org. Chem.* **55**, 3730 (1990).

[33] E. Atherton, L. R. Cameron, L. E. Cammish, A. Dryland, P. Goddard, G. P. Priestly, J. D. Richards, R. C. Sheppard, J. D. Wade, and B. J. Williams, *in* "Innovation and Perspectives in Solid Phase Synthesis" (R. Epton, ed.), p. 11. Intercept, Andover, UK, 1992.

[34] F. Albericio and G. Barany, *Tetrahedron Lett.* **32**, 1015 (1991).

[35] K. Barlos, D. Gates, S. Kapolos, G. Papahotion, W. Schafer, and Y. Wenquin, *Tetrahedron Lett.* **30**, 3947 (1989).

[36] H. Kunz and B. Dombo, *Angew. Chem., Int. Ed. Engl.* **27**, 711 (1988).

pyridine (DMAP) for ester bond formation, keeps racemization down to amounts that are not detectable if the conditions described are strictly adhered to.

Standard Coupling Procedures. For normal coupling reactions traditional reagents such as DIPCDI and active esters such as pentafluorophenyl esters[37] have been effective in amide bond formation on the solid-phase. The phosphonium- and uronium-based coupling agents have gained in prominence. These reagents activate Fmoc-amino acids in an efficient manner, lead to rapid amide bond formation, are easy to use, and are devoid of side reactions if the activation is carried out properly. Typical of the species are benzotriazol-1-yloxytris(dimethylamine)phosphonium hexafluorophosphate (BOP) developed by Castro *et al.,*[38] and the benzotriazoletetramethyluronium salts *O*-benzotriazole-1-yl-*N,N,N',N'*-tetramethyluronium hexafluorophosphate (HBTU) and the corresponding tetrafluoroborate (TBTU) introduced by Knorr *et al.*[39] Complementary to these latter two is the coupling agent 1-hydroxy-1-azabenzotriazoleuronium salt (HATU) developed by Carpino,[40] which is particularly effective with hindered couplings.

Many other inventive coupling procedures have been proposed and have been shown to be effective with Fmoc-amino acids. The utility of these is also illustrated in this volume.[40a] Here detail is given to the commonly employed methods.

Fmoc Group Removal. Many different methods have been proposed for removing the Fmoc group from the growing peptide chain.[7,41] Here the standard method utilizing piperidine in DMF (20%, v/v) is described.

Monitoring Acylation Reaction. Peptide bond formation has been monitored by many qualitative and quantitative techniques, but the most routinely applied is the Kaiser test.[42] This is a qualitative test done in real time and uses a ninhydrin-based system. The test is convenient for primary amino groups, but when acylations are performed onto a secondary amino function such as Pro N-terminating peptides, the isatin test[43] can be used. Both procedures are described. In addition, amino acid analysis following

---

[37] E. Atherton, L. R. Cameron, and R. C. Sheppard, *Tetrahedron* **44**, 843 (1988).
[38] B. Castro, J. R. Domoy, G. Evin, and C. Selve, *Tetrahedron Lett.,* 1214 (1975).
[39] R. Knorr, A. Trzeciak, W. Bannworth, and D. Gilleson, *Tetrahedron Lett.* **30**, 1927 (1989).
[40] L. A. Carpino, *J. Am. Chem. Soc.* **115**, 4397 (1993).
[40a] F. Alberico and L. A. Carpino, *Methods Enzymol.* **289**, [7], 1997 (this volume).
[41] S. A. Kates, N. A. Solé, M. Bayermann, G. Barany, and F. Alberico, *Pept. Res.* **9**, 106 (1996).
[42] E. Kaiser, R. L. Colescott, C. D. Bossinger, and P. J. Cook, *Anal. Biochem.* **84**, 595 (1970).
[43] E. Kaiser, C. D. Bossinger, R. L. Colescott, and D. D. Olser, *Anal. Chem. Acta* **118**, 149 (1976).

hydrolysis of peptide–resins gives useful information, albeit following on from assembly procedures. This protocol is also described.

Cleavage from Solid Support

The procedure normally employed to remove the peptide from the support to produce both peptide acids and amides is to treat the resin with a trifluoroacetic acid–scavenger mix. This treatment releases the peptide from the resin while simultaneously removing the side-chain protecting groups. The scavenger mixture is very much dependent on the constituent amino acids and the order in which they appear in the peptide chain. For instance, Met and Trp at the C terminus of a peptide have the tendency to add back intramolecularly to resin-bound carbocations, Met through its thioether moiety and Trp through its indole nucleus. The proximity of amino acids one to another can also be problematic, and it has been shown that Mtr side-chain-protected Arg next to Trp causes problems with the alkylation of the Trp side chain.

When such problems are foreseen it is advantageous to preadd the scavengers to the peptide–resin and thoroughly mix prior to the addition of the trifluoroacetic acid. The procedures described in this section are currently employed. More detailed cleavage conditions are described in Ref. 44.

Chromatographic Purification of Peptides

Synthetic peptides can be purified by preparative countercurrent, ion-exchange, normal-phase, and reversed-phase chromatography. The most versatile and efficacious approach utilizes the high-pressure reversed-phase mode of chromatography. This procedure is described here due to its more general applicability.

Protocols for Fmoc / tert-Butyl Solid-Phase Peptide Synthesis

The general methods described below are for polydimethylacrylamide resin but are equally applicable to most polymer supports. In the case of polystyrene-based matrices dichloromethane is often added as a cosolvent to the N,N-dimethylformamide (DMF) commonly used for solid-phase peptide synthesis. Dichloromethane, added in varying proportions up to 50%, improves swelling of the polystyrene and related supports, thus allowing improved diffusion of reagents to active sites.

---

[44] C. A. Guy and G. B. Fields, *Methods Enzymol.* **289,** [5], 1997 (this volume).

Many automated instruments are commercially available for continuous flow and batchwise solid-phase peptide synthesis. These instruments are supplied with appropriate operating procedures and manuals. The procedures described below relate to careful manual operation using batchwise methodology.

Although reagents and solvents can be specially purified for peptide synthesis, the quality of Fmoc-amino acids from most manufacturers is satisfactory. Analytical reagent (AR) grade solvents are adequate for most operations.

Operation of Nitrogen-Stirred Solid (Gel)-Phase Peptide Synthesis Reactor

The general manually operated reactor consists of a sintered glass column (number 3 porosity sinter), a three-way tap, and a three-necked round-bottom flask (Fig. 1). This system facilitates the washing and agitation of bead form gel resins. When the tap is in position A, nitrogen (at 1–2 psi) can be vented up through the glass sinter to agitate the reaction solution. With the tap in position B, a vacuum can be applied to draw off solvents and reaction solutions.

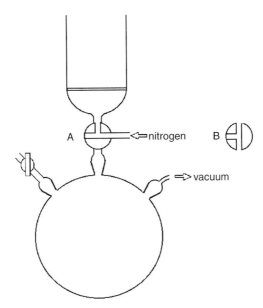

FIG. 1. Diagram of manually operated nitrogen-stirred solid-phase peptide synthesis reactor.

If the resin is dry it is generally given 10 1-min washes with DMF (10 ml/g resin) unless otherwise stated in the relevant operating procedure. If the resin requires shrinkage, a gradient of solvents is used to minimize physical damage to the resin that can be caused by rapid collapse. In this circumstance, wash the resin with DMF (five times, 10 ml/g resin), dichloromethane (three times, 10 ml/g resin), and diethyl ether (three times, 10 ml/g resin). Pass nitrogen through the resin for 10 min to evaporate off residual diethyl ether.

Analytical Procedures

Color Tests: Kaiser Test

The Kaiser test is a qualitative test for primary amines; hence, it cannot be reliably applied to the detection of Pro and other N-substituted amino acids.

Reagents

Ninhydrin
Phenol
Potassium cyanide
Ethanol
Pyridine

Test Solutions

A. Ninhydrin (5%, w/v) in ethanol (or benzyl alcohol)
B. Phenol (4:1, w/v) in ethanol
C. Potassium cyanide (2%, v/v, of a 1 mmol/liter aqueous solution) in pyridine

Procedure. The test is carried out by adding 4 drops of A, 2 drops of B, and 2 drops of C to the test sample (usually 4–5 mg of peptide–resin) contained in a small glass vial and heating at 100° for 5 min. A blue coloration to the solution is a positive result, indicating that the coupling reaction is incomplete. It is important to note that on some occasions certain amino acid residues can give unusual colorations ranging from red to blue (Asn, Cys, Ser, and Thr in particular).

Isatin Test

The isatin test is a qualitative test for peptides terminating in Pro.

Reagents

Isatin
Benzyl alcohol
Boc-Phe-OH

Test Solution. Add isatin (2 g) to benzyl alcohol (60 ml) and stir the mixture for 2 hr at ambient temperature. Filter to remove insoluble isatin and add Boc-Phe-OH (2.5 g) to the filtrate.

Procedure. The test is carried out by adding 2–3 drops of the above isatin solution to the test sample (usually 4–5 mg of peptide–resin) contained in a small glass vial and heating at 100° for 5 min. If a blue color appears on the beads, then the isatin test is positive, indicating that the coupling reaction is incomplete. The blue color on the beads is more readily observed when the supernatant liquid is removed from the vial by decantation and the beads repeatedly washed by decantation with acetone.

Amino Acid Analysis

Reagents. Constant boiling hydrochloric acid containing 0.1% (w/v) phenol.

Apparatus

Pico-tag work station (Waters, Marlborough, MA) or similar
Pico-tag tube (Waters) or similar
Pico-tag vial (Waters) or similar
Vacuum pump and dry ice trap

Procedures for Preparing Samples for Amino Acid Analysis. Place the sample (approximately 1–2 mg) to be hydrolyzed in a Pico-tag tube. Place tube in a Pico-tag vial and add constant boiling HCl (0.2 ml) to the vial.

Fill trap of Pico-tag work station with dry ice. Check that heating block is set at required temperature (120° ± 1° for an 18-hr hydrolysis and 150° ± 1° for a 1-hr hydrolysis). Ensure that the vacuum knob is closed (i.e., screwed in); then turn on the pump. Screw the top onto the vial tightly and also ensure that the tap is open. Attach the vial to the Pico-tag vacuum port and open up the vacuum. Allow the acid to degas until the bubbles stop. Do not allow the vacuum gauge to go below 1 torr. Close the vial and then shut off the vacuum. Remove the vial from the Pico-tag vacuum port and place it in the heating block. Open up the vacuum and switch off the pump.

After the required time remove the vial from the block and allow to cool. Release the vacuum on the vial. Remove the tube from the vial and rinse the vial out to remove the excess acid. Return the tube to the vial, put the top on the vial, and make sure the top is open. Fill the dry ice trap on the vacuum pump and apply a vacuum to the Pico-tag vial. Allow the sample to dry until there is no evidence of any acid in the tube. The sample can now be subjected to amino acid analysis (see Chapter 18).

Note: When polystyrene-based supports are used, propionic acid is often added to assist in swelling the peptide–resin during hydrolysis.

Preparation of Fmoc-Amino Acids

N-Fmoc-L-isoleucine. L-Ile (66 g, 0.50 mol) and sodium carbonate (53 g) are dissolved in a mixture of water (650 ml) and acetone (650 ml). Fmoc-ONSu (165 g, 0.49 mol) is added over a period of 60 min to the briskly stirred solution while the pH is kept between pH 9 and pH 10 by the addition of aqueous sodium carbonate (1 mol/liter). After stirring overnight, ethyl acetate (2 liters) is added and the mixture acidified with aqueous hydrochloric acid (6 mol/liter). The ethyl acetate layer is separated and washed with water (four times, 1.5 liters), dried over anhydrous magnesium sulfate, and evaporated to approximately 500 ml. The product crystallizes on the addition of petroleum ether, yielding 162 g (93%), mp 143°, $[\alpha]_D$ −12.1° (c = 1 in DMF).

Bis-Fmoc-L-histidine. L-His (4.56 g, 30 mmol) is vigorously stirred in water (120 ml) and the solution is adjusted to pH 8.2 with aqueous sodium carbonate (10%, w/v). The solution is then cooled in an ice bath, and Fmoc-Cl (19.41 g, 75 mmol) dissolved in dioxane (20 ml) is added dropwise over 20–30 min with vigorous stirring. The solution is maintained at pH 8.2 by further addition of sodium carbonate (10%, w/v). The reaction mixture is stirred for 3 hr at room temperature, during which time a white solid precipitates, and is then added to water (1.5 liters). The aqueous phase is extracted with ether (three times), then with ethyl acetate, and acidified with dilute hydrochloric acid. The precipitated solid is collected, washed thoroughly with water, and dried under high vacuum over phosphorus pentoxide. The yield is 16.4 g (91%), mp 161–163°, $[\alpha]_D$ −6.8° (c = 1 in DMF). A small sample is recrystallized from dimethylformamide/ether giving mp 163–165°, $[\alpha]_D$ −6.7° (c = 1 in DMF). (Found: C, 72.18; H, 4.92; N, 7.34; $C_{36}H_{29}N_3O_6$ requires C, 72.11; H, 4.88; N, 7.01%.)

This preparation is difficult and gives somewhat variable results, probably because of the sparing solubility of the sodium salt in the reaction mixture.

N$^\alpha$-Fmoc-N^{im}-tert-butoxycarbonyl-L-histidine and Its Cyclohexylamine Salt. Bis-Fmoc-L-His (10 g, 16.6 mmol) is suspended in DMF (100 ml), and diisopropylethylamine (5.38 g, 42 mmol) is added with stirring. Di-*tert*-butyl dicarbonate (9.08 g, 42 mmol) is added and stirring continued until the reaction is complete as determined by high-performance liquid chromatography (HPLC) (5.2 hr). The solvent is removed at 0.1 mm Hg and 30°, and the residual syrup dissolved in ethyl acetate, washed with citric acid solution (10%, w/v, 200 and 100 ml), and then washed with water (four times, 50 ml). The organic layer is dried with anhydrous sodium sulfate, filtered, and evaporated. The residual oil is dissolved in diethyl ether (75 ml) and added dropwise with stirring to petroleum ether (bp 80–100°, 300 ml). The precipitate is filtered, washed with petroleum ether, and dried under vacuum to give a crude solid (7.6 g). This is dissolved in ethyl acetate (5 ml) and warm diethyl ether (75 ml) and is added dropwise to a stirred petroleum ether mixture (bp 80–100°, 300 ml). The product (5.6 g) is collected. Fmoc-His(Boc)-OH shows mp 95–100°, $[\alpha]_D$ −5.7° (c = 1 in DMF) and is demonstrated by HPLC to be greater than 97% pure. The cyclohexylammonium salt prepared by addition of cyclohexylamine to an ethereal solution of the free acid and recrystallization from methanol shows mp 149–151°, $[\alpha]_D$ +15.2° (c = 1 in DMF). (Found: C, 66.71; H, 7.16; N, 9.65; $C_{32}H_{40}O_6N_4$ requires C, 66.64; H, 6.99; N, 9.72%.)

Caution should be exercised in deviating from this procedure, because other preparations have been found to contain significant racemate.

N-Fmoc-S-acetamidomethyl-L-cysteine. S-Acetamidomethyl-L-cys monohydrate (3.15 g, 15 mmol) is dissolved in a mixture of aqueous sodium carbonate (10%, w/v, 45 ml) and dioxane (20 ml). The solution is stirred and cooled to ice temperature, and a solution of Fmoc-Cl (94.41 g, 17.0 mmol) in dioxane (15 ml) is added dropwise over 30 min. The ice bath is removed, and the mixture is stirred for 1.5 hr at room temperature, at which time thin-layer chromatography (TLC) indicates almost complete reaction. After a further 1 hr the reaction mixture is poured into water (400 ml), extracted with ether (three times, 100 ml), and acidified with aqueous citric acid (10%, w/v) to pH 3. The resultant white crystalline solid is collected and freed from citric acid by partitioning between ethyl acetate and water, and the organic phase is washed several times with water, dried over anhydrous sodium sulfate, and evaporated. The product is dissolved in 2-propanol, and then water is added to cloud point. The mixture is cooled, and the title product crystallizes (4.5 g, 75%), mp 150–154°, $[\alpha]_D$ −27.5° (c = 1 in ethyl acetate). (Found: C, 60.25; H, 5.4; N, 6.9; $C_{21}H_{22}N_2O_5S$ requires C, 60.86; H, 5.34; N, 7.76%.)

In other preparations the compound failed to crystallize in a pure form and was purified on a Lobar C silica column (Merck) using chloroform–methanol–acetic acid (370 : 20 : 10, v/v/v) as elutant prior to recrystallization.

Functionalization of Polyamide Resins

The following protocols describe synthesis on 1 mmol of resin.

Reagents

Pepsyn gel resin
N,N-Dimethylformamide (DMF)
Dichloromethane (CH_2Cl_2)
Diethyl ether
Ethylenediamine
Fmoc-L-norleucine (Fmoc-L-Nle)
Piperidine
Diisopropylethylamine (DIEA)
Linkage agent (see below)
Diisopropylcarbodiimide (DIPCDI)
O-Benzotriazol-1-yl-*N,N,N',N'*-tetramethyluronium tetrafluorobor-
ate (TBTU)

Linkage Agents

2-(4-Hydroxymethylphenoxy)acetic acid (HMPA)
4-Hydroxymethylbenzoic acid (HMBA)
2-(4-Hydroxymethyl-3-methoxyphenoxy)acetic acid (HALLA)
4-[(*N*-9-Fluorenylmethoxycarbonyl)-2,4-dimethoxyphenylamino-
methyl]phenoxyacetic acid (Rink linker)

Apparatus

50-ml Conical flask
Solid (gel)-phase nitrogen-stirred reactor (Fig. 1) with number 3 porous
glass sinter

Procedure

AMINOETHYLAMIDATION. Place Pepsyn gel resin (1 g) in a conical flask
(50 ml), add ethylenediamine (30 ml), stir gently to mix the contents, and
leave the mixture to stand overnight. Transfer to a solid-phase reactor,
filter the ethylenediamine off the resin, and wash the resin with DMF (ten
times, 10 ml each), DIEA solution (three times, 10 ml, 10%, v/v, in DMF),
DMF (five times, 10 ml), and DMF (five times, 10 ml). Carry out a Kaiser
test on the filtrate of the last wash (1 drop) to confirm that all the ethylenediamine has been washed from the resin. If the test is positive repeat the last
five DMF washes and check the Kaiser test again. Repeat until the color
test is negative.

COUPLING OF INTERNAL REFERENCE STANDARD USING TBTU. Place the
Fmoc-Nle (2.5 equivalents) and HBTU (2.38 equivalents) as dry powders
in a beaker followed by DMF (~10 ml). Add DIEA (5 equivalents) and

stir the mixture gently for 2–3 min. Transfer the solution to the resin and add further DMF to make the resin just mobile to nitrogen agitation if necessary. Stir the mixture gently at intervals to ensure thorough mixing. After 1 hr remove a sample of resin (4–5 mg) and place this in a vial. Wash the sample (by decantation or filtration) using DMF (twice, 1 ml), CH_2Cl_2 (1 ml), and diethyl ether (twice, 1 ml). Submit this sample to the Kaiser test. A negative test result indicates that the coupling is complete. The reaction can be left overnight if required. When the coupling is complete, draw off the reaction solution and wash the resin with crude DMF (ten times, 10 ml).

Fmoc REMOVAL. Add piperidine/DMF (10 ml, 20%, v/v) to the resin, mix well, and allow the mixture to stand for 10 min. Draw off the piperidine/DMF and repeat the treatment with fresh piperidine/DMF (10 ml, 20%, v/v). Draw off the piperidine/DMF and wash the resin with DMF (ten times, 10 ml).

ATTACHMENT OF LINKAGE AGENT. Place the linkage agent (3 equivalents) and HOBt (6 equivalents) as dry powders in the reactor followed by the minimum amount of DMF required to make the resin just mobile to nitrogen agitation. Add DIPCDI (4 equivalents) and stir the mixture gently at intervals to ensure thorough mixing. After 1 hr remove a sample of resin (4–5 mg) and place it in a vial. Wash the sample (by decantation or filtration) using DMF (twice, 1 ml), CH_2Cl_2 (1 ml), and diethyl ether (twice, ml). Subunit the sample to the Kaiser test. A negative test result indicates that the coupling is complete. If the reaction is not complete, add additional DIPCDI (1 equivalents), stir the mixture, and then allow the reaction to proceed overnight.

When the coupling is complete, draw off the reaction solution and wash the resin with DMF (ten times, 10 ml), CH_2Cl_2 (three times, 10 ml), CH_2Cl_2/diethyl ether (two times, 10 ml, 1 : 1, v/v), and diethyl ether (three times, 10 ml). Pass nitrogen through the resin for 10 min to evaporate off residual diethyl ether. Cover sinter with filter paper and leave the resin to air-dry overnight.

If the correct amount of resin is present in the reactor for the scale of synthesis required, then the assembly can be continued. Otherwise the amount of resin of a particular loading related to the length, amounts, and purity requirements are listed below. This is of course a rough guide and is dependent on the amino acid sequence.

Recommended Scales for Routine Peptide Syntheses

   1. For production of 50–100 mg of >80% purity material:

Length of peptide (number of residues)	Amount of resin (g of 1 mmol/g capacity)
5–15	1
15–20	1
20–30	1
30–40	0.5

2. For production of 10 mg of >95% purity material:

Length of peptide (number of residues)	Amount of resin (g of 1 mmol/g capacity)
5–15	1
15–20	1
20–30	1
30–40	0.5

3. For production of 25–75 mg of >95% purity material:

Length of peptide (number of residues)	Amount of resin (g of 1 mmol/g capacity)
5–15	2
15–20	1
20–30	1
30–40	0.5

For assemblies requiring up to 1 g of resin use a 30 mm diameter reactor. For assemblies requiring 2 g of resin, a 60 mm diameter reactor will be needed (see Fig. 1).

Loading of C-Terminal Amino Acids to Hydroxy-Functionalized Resins (Manually Operated Nitrogen-Stirred Reactor)

Reagents

Fmoc-amino acid
N,*N*-Dimethylformamide (DMF)
Diisopropylcarbodiimide (DIPCDI)
4-Dimethylaminopyridine (DMAP)

Dichloromethane (CH$_2$Cl$_2$)
Diethyl ether
Acetic anhydride

Apparatus

Solid (gel)-phase nitrogen-stirred reactor (Fig. 1)

Procedure. Place the appropriate hydroxy-functionalized resin in a reactor of sufficient capacity for the synthesis, followed by the Fmoc-amino acid to be coupled (3 equivalents). Add sufficient DMF to make the resin just mobile to nitrogen agitation. DIPCDI (4 equivalents) can then be added followed by the dropwise addition of a solution of DMAP dissolved in DMF (0.1 equivalents, ~50 mmol/liter). Allow the reaction to proceed for 60 min, draw off the reaction solution, and wash the resin using DMF (two times, 10 ml/g). Recouple the amino acid using the above procedure.

Wash the resin with DMF (five times, 10 ml/g) and acetylate any remaining hydroxy functions (before Fmoc removal). Add sufficient DMF to make the swollen resin just mobile to nitrogen agitation. Acetic anhydride (6 equivalents) can then be added followed by a solution of DMAP in DMF (0.1 equivalents, 50 mmol/liter). After 1 hr draw off the reaction solution, wash the resin with DMF (ten times, 10 ml/g), and then proceed with peptide chain elongation.

The loading efficiency can be determined at this point by hydrolysis of a sample after Fmoc group removal of a small portion (10 mg) as stipulated. Hydrolysis followed by amino acid analysis will determine the C-terminal amino acid : Nle ratio.

Removal of 9-Fluorenylmethoxycarbonyl (Fmoc) Amino-Terminal Protecting Group

Reagents

N,N-Dimethylformamide (DMF)
Piperidine

Apparatus

Solid (gel)-phase nitrogen-stirred reactor

Procedure. Add piperidine/DMF (10 ml/g, 20%, v/v) to the resin contained in the reactor and agitate for 3 min. Draw off the solvents and repeat this procedure for a further 7 min. Draw off the solvents and wash the resin with DMF (ten times, 10 ml/g). Submit a portion of the resin (4–5 mg) to the Kaiser test (or isatin test in the case of Pro). A strong positive

Kaiser test (or isatin test for Pro) should be obtained at this stage. If a strong positive test is not observed, the piperidine/DMF treatment may have to be extended in this and subsequent deprotections.

During an assembly peptide–resin can be stored with the Fmoc group removed after washing according to the shrink cycle (see below). However, this is to be avoided at the dipeptide stage with benzyl ester-based linkage agents because of the potential formation of diketopiperazines resulting in cleavage of the peptide from the support.

Peptide Bond Formation

Coupling with O-Benzotriazol-1-yl-N,N,N′,N′-tetramethyluronium Tetrafluoroborate

Reagents

N,N-Dimethylformamide (DMF)
Fmoc-amino acid
O-Benzotriazol-1-yl-$N,N,N′,N′$-tetramethyluronium tetrafluoroborate (TBTU)
Diisopropylethylamine (DIEA)

Apparatus

Solid (gel)-phase nitrogen-stirred reactor
Procedure. Place the Fmoc-amino acid (2.5 equivalents) and TBTU (2.38 equivalents) in a beaker. Add DMF (minimum to obtain solution) and DIEA (5 equivalents), and allow the mixture to react for 2–3 min. Add this preactivated amino acid solution to the swollen resin and add further DMF if necessary until the resin is just mobile to nitrogen agitation. Use a spatula to gently stir the mixture if necessary and allow the reaction to proceed with monitoring by the color tests.

The coupling should be monitored after approximately 15 min, then at regular intervals (~15-min intervals) using the following procedure. Withdraw a small sample (4–5 mg) of resin from the reactor and place it in a vial. Wash the sample (by decantation or filtration) using DMF (twice, 1 ml), CH_2Cl_2 (1 ml), and diethyl ether (twice, 1 ml). Submit this sample to the Kaiser test (or isatin test in the case of Pro-terminating peptides). A negative test should be obtained. When the reaction is complete draw off the reaction solution and wash the resin with DMF (ten times, 10 ml/g resin). This deprotection and acylation procedure constitutes the repetitive cyclic procedure used to assemble the fully protected peptide.

The cycle is summarized as follows.
Washes: 10 times with DMF, 1 min each
Deprotection: 20% piperidine in DMF, 3 min and 7 min
Washes: 10 times with DMF, 1 min each
Acylation: 2.5 equivalents of the Fmoc-amino acid preactivated with
    2.38 equivalents of TBTU in the presence of 5 equivalents of DIEA
    in the minimum volume of DMF

Cleavage of Peptides from Polyamide Resins

Reagents

Peptide–resin (Pepsyn gel)
Trifluoroacetic acid (TFA)
1,2-Ethanedithiol (EDT)
Phenol
Anisole
Dichloromethane (CH_2Cl_2)
Diethyl ether

Apparatus

Round-bottom flasks and stoppers
Filtration equipment with number 3 porous glass sinter
Measuring cylinder
Recommended Scavengers for Trifluoroacetic Acid Cleavages. The scavenger mixture is dependent on the sequence of the peptide to be cleaved. The following tabulation shows the amino acids that require addition of a particular scavenger during cleavage.

Amino acid	Scavenger
Arg	Phenol
Cys	EDT
Met	EDT
Trp	Anisole and EDT
Tyr	Phenol

Note: If the peptide is on the Rink linker resin, then EDT should always be added as a scavenger.

Proportion of Scavenger to Trifluoroacetic Acid

Number of scavengers	% Scavenger	% TFA
0	(5.0% water)	95%
1	5.0% scavenger	95%
2	2.5% of each	95%
3	2.0% of each	94%

Quantities of Reagents for Trifluoroacetic Acid Cleavages. For 1 g of peptide–resin use 20 ml of cleavage mixture and 50 ml of diethyl ether per trituration. The quantities should be rounded to the nearest 10 ml for ease of handling.

Cleavage Times for Trifluoroacetic Acid Cleavages. Peptides not containing Arg are cleaved for 90 min. Peptides containing Arg (protected by Mtr) are cleaved overnight (18 hr). If a peptide was assembled on the Rink linker and requires an overnight cleavage, then it is best to leave the resin in contact with the TFA for only 90 min, then filter off the resin, and leave the filtrate overnight.

Procedure. Prior to cleavage the Fmoc group should be removed and the peptide–resin should have been passed through a standard shrink cycle to remove any traces of DMF. Weigh the peptide–resin into a round-bottom flask. Prepare the cleavage mixture in a measuring cylinder, using TFA and the required scavengers (see above). Purge the solution with nitrogen for 5 min. Add the cleavage mixture to the resin and mix gently. There should be enough TFA solution to thoroughly wet the peptide–resin.

When the cleavage time is complete, filter into a preweighed round-bottom flask and wash the resin with TFA (two times, using twice the bed volume). Do not fill the round-bottom flask more than half full with TFA solution. Remove the TFA by rotary evaporation under reduced pressure until a thick oil is left. Add diethyl ether (50 ml/g peptide–resin) to precipitate the peptide and extract the scavengers. Allow the peptide to settle and then remove the diethyl ether by decantation or filtration. Repeat the trituration (or wash) with two further lots of diethyl ether. If triturating, after the final wash leave the flask on its side to allow the final traces of diethyl ether to evaporate off. Wash the residual resin with CH_2Cl_2 (three times, using approximately twice the resin bed volume) and diethyl ether (three times, approximately twice the resin bed volume). Allow the resin to air-dry. The residual resin can be hydrolyzed and subjected to amino acid analysis to determine the amount of peptide cleaved based on the ratio of peptide to the internal reference amino acid.

Purification of Peptides by Reversed-Phase High-Performance
Liquid Chromatography

The instrumentation employed for purification by HPLC must be used
in accordance with the manufacturer's operators manual.

Reagents

Acetonitrile (CH_3CN), far-UV HPLC grade
Water, far-UV HPLC grade
Trifluoroacetic acid (TFA)
Ammonium bicarbonate
Methanol

Buffer Systems

System I: Component A is 0.1% TFA/water (v/v), B is 0.1% TFA/
CH_3CN (v/v), and C is methanol.
System II: Component A is 50 mM ammonium bicarbonate, and B is
40% A in CH_3CN (v/v)

Procedure. There are four decisions to be made prior to starting a purification.

BUFFER SYSTEM. The buffer system chosen for purification is dependent
on the nature of the peptide. In the majority of cases buffer system I above
will be suitable. System II has some limitations due to instability of some
peptides at high pH.

QUANTITY TO BE PURIFIED. The following tabulation should be used as
a general guide to the minimum amount of crude peptide to be purified
and the column required to do the purification in one attempt.

Required yield (mg)	Purity (%)	Minimum amount of crude peptide to be purified (g)	Column diameter (mm)
50	>80	0.5	20–25
100	>80	1.0	40–50
10	>95%	0.3	20–25
50	>95%	1.0	40–50
100	>95%	2.0	40–50

COLUMN SIZE. For amounts of peptide between 300 and 600 mg use a
20–25 mm diameter purification column. For amounts over 600 mg and up
to 3 g of peptide use a 40–50 mm diameter purification column.

PURIFICATION GRADIENT. Select a suitable gradient by using the following equation: $Y = (\frac{2}{3} X) - 10$, where X is the concentration of solvent B at which the peptide is eluted on an equivalent analytical HPLC column and Y is the starting point of purification gradient. The gradient is then 15% (A to B) over 45 min from this starting point. For example, a peptide that elutes at 30% on an analytical HPLC column would be run on a gradient of 10 to 25% B over 45 min.

The following points should be considered: (a) The comparison shown in the above calculation is consistent with C_{18} analytical columns and equivalent C_{18} purification columns. (b) If a peptide elutes below a concentration of 15% B on the analytical system start the gradient at 0% B.

Chromatography. First, prepare both sample loading systems. Leave the injection loop in line with the column during the column washing stage. Wash the injection port with water (5 ml), methanol (5 ml), and the starting buffer for the gradient to be used (5 ml). The injection port should be washed during an initial methanol wash of the column. Wash the line to the pump heads with water (10 ml), methanol (10 ml), and the starting buffer for the gradient to be used (10 ml). This line should be washed prior to the initial column washing.

Next, the column needs to be prepared. The following gradient will ensure that the column is free from most peptide-related contaminants.

Time	% A	% B	% C
Initial	0	0	100
5	0	0	100
10	0	100	0
15	0	100	0
20	100	0	0
25	100	0	0
30		Initial conditions of purification gradient	

When changing from one solvent to another on a purification system (or analytical system) a gradient should always be used to prolong the life of the column. Recommended flow rates are 10 ml/min for a 20–25 mm diameter purification column and 35 ml/min for a 40–50 mm diameter column. At the end of the above column cleaning gradient, the system should be left initializing the new gradient for at least 10 min before loading the sample.

The ideal solvent for dissolution of peptides prior to application to the purification column is the starting buffer for the gradient chosen (this includes 0.1% TFA unless acid needs to be avoided for reasons of chemical instability). The minimum amount of solvent should be used to dissolve the peptide. The sample should be filtered prior to application to the purification column. If the sample volume is greater than the loop size, multiple injections can be made or the sample can often be loaded through the pump heads.

Having loaded the sample, start the gradient and collect fractions across the eluted peaks at suitable intervals (0.5 to 1 min). If the peptide has not eluted by the end of the gradient, lengthen the gradient by doing a further 5% change over 15 min, and continue extending the gradient by 5% over 15 min until the peptide has eluted. While the gradient is running, the sample loading system through which the peptide was loaded should be washed to prevent blockage. The fractions are then analyzed by HPLC. Having decided which fractions are to be combined, a coelution is done by removing a drop from each fraction and injecting a portion of the resulting solution on HPLC. If the coelution is of the required purity the fractions are combined in a round-bottom flask and lyophilized.

[5] Trifluoroacetic Acid Cleavage and Deprotection of Resin-Bound Peptides Following Synthesis by Fmoc Chemistry

By CYNTHIA A. GUY and GREGG B. FIELDS

Introduction

The successful assembly of a peptide sequence represents only half the challenge of solid-phase peptide synthesis (SPPS). Following assembly on the resin, the peptide is side-chain deprotected and removed from the support. Typically, the side-chain protecting groups and linkers designed for fluorenylmethyloxycarbonyl (Fmoc) protocols are labile to trifluoroacetic acid (TFA) (Fig. 1). Although the stepwise addition of amino acids using Fmoc chemistry has been accepted as a more mild alternative to *tert*-butyloxycarbonyl (Boc) chemistry, the final step involving simultaneous removal of side-chain protecting groups and peptide from the solid support with TFA still results in the exposure of vulnerable residues to a reactive pool of carbocations. Certain chemicals, known as scavengers, can be added to the TFA to mediate irreversible modification of the peptide by reacting with the cations as they are released.

The procedures described herein are suitable for the various linkers cleaved by TFA,[1-3] whether producing C-terminal acids or amides. Com-

---

[1] G. B. Fields and R. L. Noble, *Int. J. Pept. Protein Res.* **35**, 161–214 (1990).

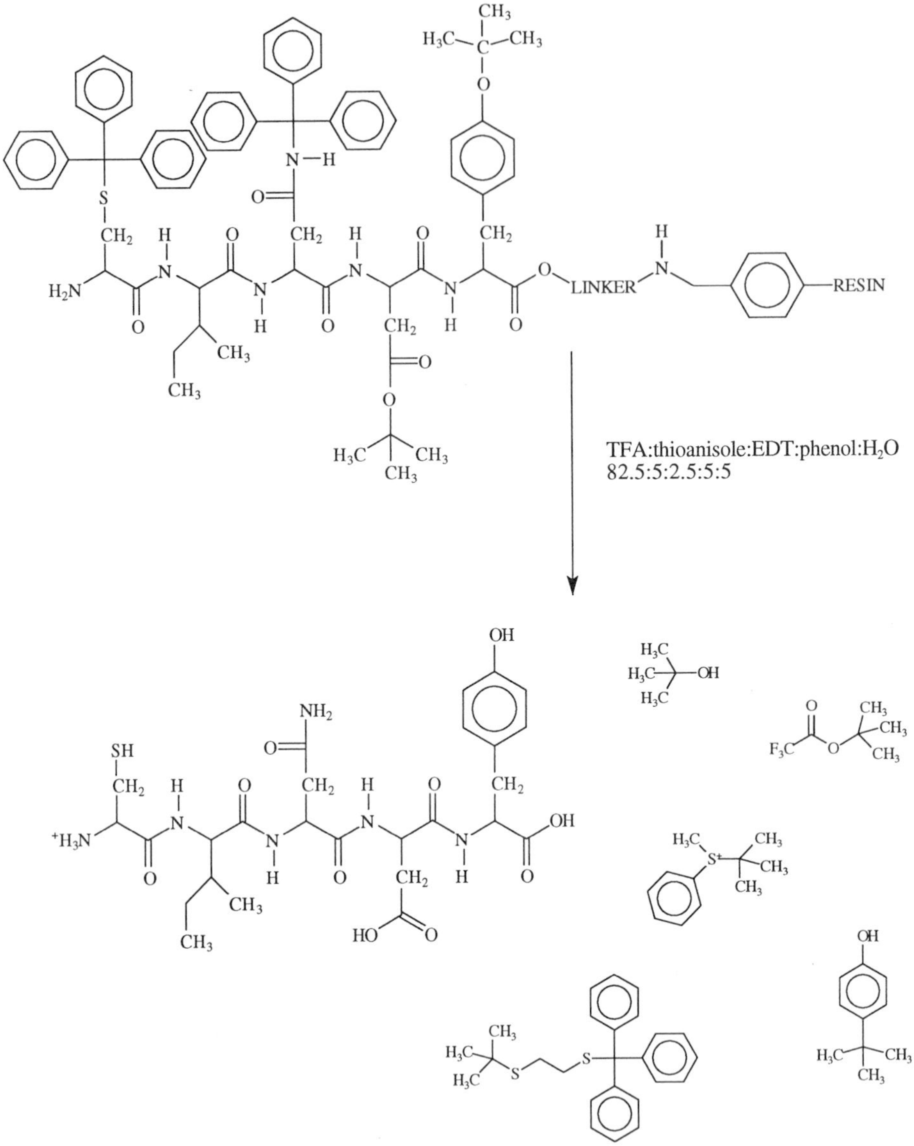

FIG. 1. Reaction of TFA and scavengers with a fully protected peptide–resin. In addition to the fully deprotected peptide, some possible scavenger-protecting group adducts are shown.

mercially available and suggested linker–resins for generating peptide acids are 4-alkoxybenzyl alcohol (Wang), 4-hydroxymethylphenoxyacetic acid (HMPA/PAB), and 2-chlorotrityl chloride. For producing peptide amides, the following linkers can be used: 4-(2',4'-dimethoxyphenylaminomethyl) phenoxymethyl (Rink amide), 4-(2',4'-dimethoxyphenylaminomethyl) phenoxymethylbenzhydrylamine (Rink amide MBHA), 2,4-dimethoxy-4'-(carboxymethyloxy)benzhydrylamine, 4-methoxy-4'-(carboxypropyloxy) benzhydrylamine, 5-amino-10,11-dihydro-5H-dibenzo[a,d]cycloheptenyl-2-oxyacetyl-Nle-4-methylbenzhydrylamine, 4-succinylamino-2,2',4'-trimethoxybenzhydrylamine (SAMBHA), and 5-(4-aminomethyl-3,5-dimethoxyphenoxy)valeric acid (PAL). Some linkers, including 4-(2',4'-dimethoxyphenylhydroxymethyl)phenoxymethyl (Rink acid), 4-(4-hydroxymethyl-3-methoxyphenoxy)butyric acid (HMPB), 2-methoxy-4-alkoxybenzyl alcohol (SASRIN), and 9-aminoxanthen-3-yloxy, are designed so that peptides can be liberated from the resin with dilute acid while keeping the side-chain protecting groups intact. Cleavage of such desired products is addressed elsewhere in this volume.[4]

Under certain circumstances, TFA can cleave both at the linker–peptide bond and at the attachment point of the linker to the resin. Consequently, two events may occur. The free carbonium ion of the linker may alkylate Trp, with the resulting peptide exhibiting a more hydrophobic high-performance liquid chromatography (HPLC) elution behavior and a higher mass than the desired peptide.[5–10] In addition, the cleavage yield may be noticeably reduced due to a reaction between a Trp or Cys residue and the linker ion while still attached to the resin.[5,7,10,11] These reactions can be suppressed and are addressed later in this chapter when discussing the use of scavengers.

---

[2] G. B. Fields, Z. Tian, and G. Barany, *in* "Synthetic Peptides: A User's Guide" (G. A. Grant, ed.), pp. 77–183. W. H. Freeman & Co., New York, 1992.

[3] M. F. Songster and G. Barany, *Methods Enzymol.* **289**, [8], 1997 (this volume).

[4] F. Albericio, P. Lloyd-Williams, and E. Giralt, *Methods Enzymol.* **289**, [15], 1997 (this volume).

[5] B. Riniker and B. Kamber, *in* "Peptides 1988" (G. Jung and E. Bayer, eds.), p. 115. de Gruyter, Berlin, 1989.

[6] C. G. Fields, V. L. VanDrisse, and G. B. Fields, *Pept. Res.* **6**, 39 (1993).

[7] F. Albericio, N. Kneib-Cordonier, S. Biancalana, L. Gera, R. I. Masada, D. Hudson, and G. Barany, *J. Org. Chem.* **55**, 3730 (1990).

[8] P. D. Gesellchen, R. B. Rothenberger, D. E. Dorman, J. W. Paschal, T. K. Elzey, and C. S. Campbell, *in* "Peptides: Chemistry, Structure and Biology" (J. E. Rivier and G. R. Marshall, eds.), p. 957. ESCOM, Leiden, The Netherlands, 1990.

[9] P. Sieber and B. Riniker, *in* "Innovation and Perspectives in Solid Phase Synthesis" (R. Epton, ed.), p. 577. Solid Phase Conference Coordination, Birmingham, UK, 1990.

[10] E. Atherton, L. R. Cameron, and R. C. Sheppard, *Tetrahedron* **44**, 843 (1988).

[11] B. Riniker, A. Flörsheimer, H. Fretz, P. Sieber, and B. Kamber, *Tetrahedron* **49**, 9307 (1993).

Alternatively, some linkers have been found not to be as labile as reported and will require longer cleavage times or a stronger acid. Riniker et al.[11] have attributed this reduced lability to a neutralization of the TFA by amide bonds. Riniker et al. recommend a cycle of treating the peptide–resin several times with TFA solution, filtering, and adding fresh TFA, to attain an ionic strength high enough for complete cleavage from the support. Alternatively, Kiso and co-workers[12] describe a procedure using tetrafluoroboric acid (HBF_4) to increase cleavage yields over those obtained from standard TFA protocols, without sacrificing peptide integrity. Prudent practice would suggest performing a series of pilot or minicleavages to optimize deprotection conditions for a particular linker–resin being used for the first time. However, the use of linkers stable to TFA permits the removal of side-chain protecting groups prior to peptide–resin cleavage. The advantage of using this method to prevent Trp modification by *tert*-butyl groups was demonstrated by Sheppard and co-workers.[13]

Although there is only one linker per peptide, there are often a plethora of side-chain protecting groups with which to contend. The most commonly used acid-labile side-chain protecting groups for Fmoc chemistry are *tert*-butyl (*t*Bu) for Asp, Glu, Ser, Thr, and Tyr, triphenylmethyl (Trt) for Cys, His, Asn, and Gln (or 2,4,6-trimethoxybenzyl, Tmob, for Asn and Gln), *tert*-butyloxycarbonyl (Boc) for Lys and Trp, and 2,2,5,7,8-pentamethylchroman-6-sulfonyl (Pmc) or 2,2,4,6,7-pentamethyldihydrobenzofuran-5-sulfonyl (Pbf) for Arg. On side-chain deprotection with TFA, highly reactive carbocations[14–16] and sulfonyl species[11,17–22] are generated from these protecting groups that can alkylate or sulfonate susceptible amino acid residues.

[12] K. Akaji, Y. Nakagawa, Y. Fujiwara, K. Fujino, and Y. Kiso, *Chem. Pharm. Bull.* **41**, 1244–1248 (1993).

[13] T. Johnson and R. C. Sheppard, *J. Chem. Soc., Chem. Commun.*, 1653 (1991).

[14] B. F. Lundt, N. L. Johansen, A. Volund, and J. Markussen, *Int. J. Pept. Protein Res.* **12**, 258 (1978).

[15] J. P. Tam, in "Macromolecular Sequencing and Synthesis" (D. H. Schlesinger, ed.), p. 153. Alan R. Liss, New York, 1988.

[16] S. F. Brady, R. Hirschmann, and D. F. Veber, *J. Org. Chem.* **42**, 143 (1977).

[17] R. Ramage, J. Green, and A. J. Blake, *Tetrahedron* **47**, 6353 (1991).

[18] J. Green, O. M. Ogunjobi, R. Ramage, A. S. J. Stewart, S. McCurdy, and R. Noble, *Tetrahedron Lett.* **29**, 4341 (1988).

[19] E. Jaeger, H. A. Remmer, G. Jung, J. Metzger, W. Oberthür, K. P. Rücknagel, W. Schäfer, J. Sonnenbichler, and I. Zetl, *Biol. Chem. Hoppe-Seyler* **374**, 349 (1993).

[20] B. Riniker and A. Hartmann, in "Peptides: Chemistry, Structure and Biology" (J. E. Rivier and G. R. Marshall, eds.), p. 950. ESCOM, Leiden, The Netherlands, 1990.

[21] P. Sieber, *Tetrahedron Lett.* **28**, 1637 (1987).

[22] D. S. King, C. G. Fields, and G. B. Fields, *Int. J. Pept. Protein Res.* **36**, 255 (1990).

TFA itself can react with tBu cations to form tBu trifluoroacetate[14] or form an adduct with the scavenger ethanedithiol (EDT).[21] The amino acids most vulnerable to the consequences of TFA deprotection are Met, Tyr, and Trp. Met is susceptible to oxidation[23,24] or $tert$-butylation.[14,25–27] Tyr may also be $tert$-butylated[26,28] or sulfonated by a Pmc group.[20] Trp, without Boc side-chain protection, is especially vulnerable to ozonolysis[29] or modification by the linker,[5,10] side-chain protecting groups,[9,13,14,18,20,21,26,30–39] or scavengers themselves[21] (see Fig. 2).[40] In a study of the correlation between the proximity of Trp to Arg in a sequence and extent of modification of Trp by Pmc, Lebl and co-workers[41] found that this side reaction was greatest when Trp and Arg were separated by one residue. When the vulnerable sequence Trp-Xxx-Arg was located at the

[23] K. Hofmann, W. Haas, M. J. Smithers, R. D. Wells, Y. Wolman, N. Yanalhara, and G. Zanetti, *J. Am. Chem. Soc.* **87**, 620 (1965).

[24] K. Norris, J. Halstrom, and K. Brunfeldt, *Acta Chem. Scand.* **25**, 945 (1971).

[25] R. L. Noble, D. Yamashiro, and C. H. Li, *J. Am. Chem. Soc.* **98**, 2324 (1976).

[26] P. Sieber, *in* "Peptides 1968" (E. Bricas, ed.), p. 236. North-Holland, Amsterdam, 1968.

[27] P. Sieber, B. Riniker, M. Brugger, B. Kamber, and W. Rittel, *Helv. Chim. Acta* **53**, 2135 (1970).

[28] B. F. Lundt, N. L. Johansen, and J. Markussen, *Int. J. Pept. Protein Res.* **14**, 344 (1979).

[29] E. Scoffone, A. Previero, C. A. Benassi, and P. Pajetta, *in* "Peptides 1963" (L. Zervas, ed.), p. 183. Pergamon, Oxford, 1966.

[30] M. Löw, L. Kisfaludy, E. Jaeger, P. Thamm, S. Knof, and E. Wünsch, *Hoppe-Seyler's Z. Physiol. Chem.* **359**, 1637 (1978).

[31] A. Fontana and C. Toniolo, *in* "Progress in the Chemistry of Organic Natural Products" (W. Herz, H. Grieseback, and G. W. Kirby, eds.), Vol. 33, p. 309. Springer-Verlag, Berlin, 1976.

[32] M. Löw, L. Kisfaludy, and P. Sohár, *Hoppe-Seyler's Z. Physiol. Chem.* **359**, 1643 (1978).

[33] J. L. Harrison, G. M. Petrie, R. L. Noble, H. S. Beilan, S. N. McCurdy, and A. R. Culwell, *in* "Techniques in Protein Chemistry" (T. E. Hugli, ed.), p. 506. Academic Press, San Diego, 1989.

[34] E. Jaeger, P. Thamm, S. Knof, and E. Wünsch, *Hoppe-Seyler's Z. Physiol. Chem.* **359**, 1629 (1978).

[35] E. Jaeger, P. Thamm, S. Knof, E. Wünsch, M. Löw, and L. Kisfaludy, *Hoppe-Seyler's Z. Physiol. Chem.* **359**, 1617 (1978).

[36] H. Gausepohl, M. Kraft, and R. W. Frank, *Int. J. Pept. Protein Res.* **34**, 287 (1989).

[37] E. Wünsch, E. Jaeger, L. Kisfaludy, and M. Low, *Angew. Chem., Int. Ed. Eng.* **16**, 317 (1977).

[38] S. Sakakibara, *in* "Peptides: Proceedings of the 5th American Peptide Symposium" (M. Goodman and J. Meienhofer, eds.), p. 436. Halsted Press, New York, 1977.

[39] Y. B. Alakhov, A. A. Kirushkin, V. M. Lipkin, and G. W. A. Milne, *J. Chem. Soc. Sect. D: Chem. Commun.*, 406 (1970).

[40] C. G. Fields and G. B. Fields, *in* "Innovation and Perspectives in Solid Phase Synthesis: Peptides, Proteins and Nucleic Acids" (R. Epton, ed.), p. 251. Mayflower Worldwide Limited, Birmingham, UK, 1994.

[41] A. Stierandová, N. F. Sepetov, G. V. Nikiforovich, and M. Lebl, *Int. J. Pept. Protein Res.* **43**, 31 (1994).

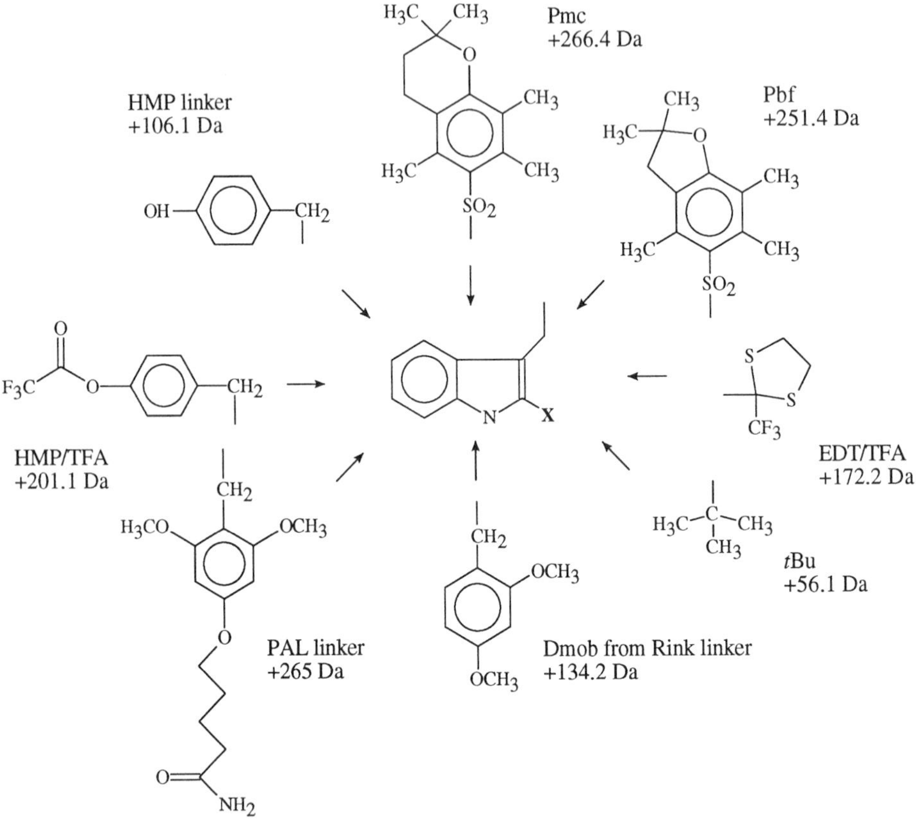

FIG. 2. Possible modifications of the indole moiety of Trp. Although modifications can occur at any position on the indole ring, the 2 position as shown is most frequent site of modification.

C terminus rather than at the N terminus, the extent of modification was increased. Several investigators have demonstrated the advantage of Boc protection for the Trp indole nitrogen to inhibit electrophilic alkylation of the indole moiety.[11,42-44] Purified yields of a synthetic fibronectin fragment peptide, Trp-Gln-Pro-Pro-Arg-Ala-Arg-Ile-Tyr (Fig. 3) and its scrambled version, Arg-Pro-Gln-Ile-Pro-Trp-Ala-Arg-Tyr (Fig. 4), synthesized without side-chain protection on the Trp, were routinely 60%. After the protocol

[42] P. White, in "Peptides: Proceedings of the Twelfth American Peptide Symposium" (J. A. Smith and J. E. Rivier, eds.), p. 537. ESCOM, Leiden, The Netherlands, 1992.

[43] H. Choi and J. V. Aldrich, Int. J. Pept. Protein Res. 42, 58 (1993).

[44] C. G. Fields and G. B. Fields, Tetrahedron Lett. 34, 6661 (1993).

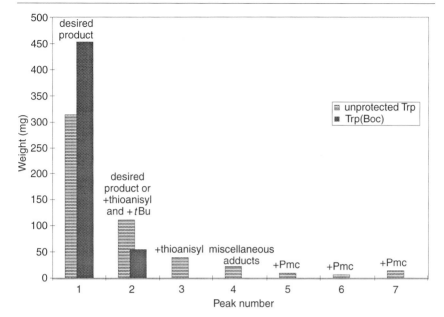

FIG. 3. Identity (bar labels) and amount (y axis) of products (x axis) obtained from the reversed-phase HPLC separation of crude peptide V, Trp-Gln-Pro-Pro-Arg-Ala-Arg-Ile-Tyr.

was changed to include Boc protection of the Trp indole, typical yields increased to 90% (Figs. 3 and 4). Preparative chromatograms in Fig. 5 demonstrate the extent of heterogeneity before (Fig. 5A,C) and after (Fig. 5B,D) use of Trp(Boc).

When using linkers designed to be cleaved with a low concentration of TFA, it is important to realize that concomitant deprotection of the peptide may not be complete. Two-stage TFA treatments have been recommended by Barany and co-workers[7,45,46] and Rink and others,[47–50] whereby the peptide is released from the resin by a dilute TFA solution [2–5% in

[45] F. Albericio and G. Barany, Int. J. Pept. Protein Res. **30,** 206 (1987).
[46] Y. Han, S. L. Bontems, P. Hegyes, M. C. Munson, C. A. Minor, S. A. Kates, F. Albericio, and G. Barany, J. Org. Chem. **61,** 6326 (1996).
[47] H. Rink and P. Sieber, in "Peptides 1988" (G. Jung and E. Bayer, eds.), p. 139. de Gruyter, Berlin, 1989.
[48] K. M. Otteson, J. L. Harrison, A. Ligutom, and P. Ashcroft, in "Poster Presentations at the Eleventh American Peptide Symposium," p. 34. Applied Biosystems, Foster City, CA, 1989.
[49] H. Rink, in "Peptide Chemistry 1987" (T. Shiba and S. Sakakibara, eds.), p. 279. Protein Research Foundation, Osaka, 1987.
[50] H. Rink, Tetrahedron Lett. **28,** 3787 (1987).

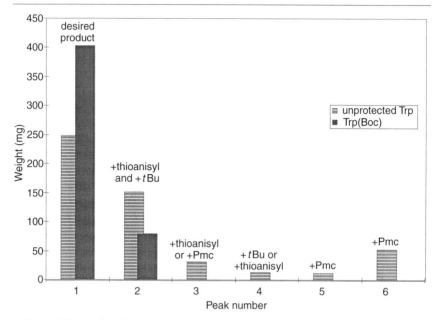

FIG. 4. Identity (bar labels) and amount (y axis) of products (x axis) obtained from the reversed-phase HPLC separation of crude scrambled peptide V, Arg-Pro-Gln-Ile-Pro-Trp-Ala-Arg-Tyr.

dichloromethane (DCM)], filtered, and then allowed to stir in an increased concentration of TFA (50–95%) to fully deprotect the side chains. An additional problem during removal of side-chain protecting groups is their occasional lack of lability.[51] Depending on the location and number in a sequence, tBu,[17] Trt,[11] and especially Pmc groups may require assistance for their complete removal. Increasing cleavage time from 1 to 2 hr completely eliminated peaks from a chromatogram attributed to peptides with intact tBu groups on Glu and/or Asp.[52] Friede and co-workers found that Trt on an N-terminal Asn is quite stable to TFA, regardless of scavenger composition, requiring 12 hr of treatment for complete removal.[51] Sometimes suitable cleavage conditions cannot be found, in which case the peptide must be resynthesized using an alternative protecting group.

Many of the problems mentioned above, including irreversible modifications, oxidations, and removal of stubborn protecting groups, can be circumvented by the addition of nucleophilic scavengers to the TFA. The

[51] M. Friede, S. Denery, J. Neimark, S. Keiffer, H. Gausepohl, and J. P. Briand, *Pept. Res.* **5**, 145 (1992).

[52] J. A. Borgia, C. G. Fields, G. B. Fields, and T. R. Oegema, Jr., unpublished results (1996).

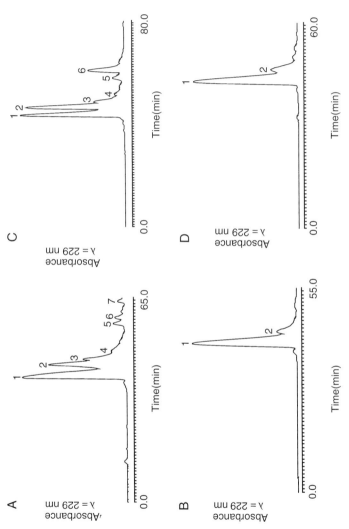

FIG. 5. Chromatograms of (A) crude peptide V synthesized with no side-chain protection on Trp and cleaved with reagent R, (B) crude scrambled peptide V synthesized with no side-chain protection on Trp and cleaved with reagent R, (C) crude peptide V synthesized with Trp(Boc) and cleaved with reagent K (the cleavage mixture described in the text), and (D) crude scrambled peptide V synthesized with Trp(Boc) and cleaved with reagent K. Analytical chromatograms (not shown) of the two peptides synthesized with Trp(Boc) and cleaved with reagent R were identical to those cleaved with reagent K. All peptides were purified on a Vydac C$_{18}$ column (22 mm × 25 cm, 10 μm, 300 Å) at a flow rate of 5 ml/min, with a gradient of 1% B/min where A is 0.1% TFA/water and B is 0.1% TFA/acetonitrile, on a Rainin (Woburn, MA) Autoprep HPLC system. Eluent was monitored at λ = 229 nm by a Rainin UV-1 detector.

cleavage mixture described in this article, reagent K, evolved from a search for a combination of scavengers that would be most effective at suppressing side reactions in a broad range of peptide sequences, especially in those containing Trp.[22] There are a variety of cleavage cocktails described in the literature that have been developed for specific sequences, and which may also be applicable to a wide range of peptides. Reagent R, developed by Barany and co-workers,[7] has performed similarly[53] to or not as well as[54,55] reagent K in instances where direct comparisons have been made. Barany and co-workers also developed reagent B, which substitutes triisopropylsilane for sulfurous compounds.[53] These alternative mixtures could be tested and compared to reagent K using the minicleavages described below.

Although there have been many different nucleophilic species suggested as useful scavengers in the literature, only the ones recommended in this protocol are described. One of the four, 1,2-ethanedithiol (EDT), competes favorably with Pmc, linker cations,[5] and tBu and tBu trifluoroacetate,[14] preventing attachment to Trp, Tyr, or Met. EDT will also prevent the reattachment of Trt to Cys.[56–59] The addition of EDT inhibits the aforementioned reaction of Trp or Cys with the linker still bound to the resin[5,11] and suppresses the oxidation of Trp by acid-catalyzed ozonolysis.[60,61] As mentioned earlier, EDT can form an adduct with TFA and modify Trp. Water, the second scavenger added to a cocktail containing EDT, will inhibit the formation of this adduct. Added as 5% of the total TFA mixture, water will also quench the reactivity of tBu[22,26,62] and Pmc ions[18,20,22,63] and eliminate a side reaction of Pmc removal, namely, sulfation of Ser or Thr.[19] Thioanisole, the third scavenger, will both prevent oxidation[64] and tert-

[53] N. A. Solé and G. Barany, *J. Org. Chem.* **57,** 5399 (1992).

[54] R. H. Angeletti, L. F. Bonewald, and G. B. Fields, *Methods Enzymol.* **289,** [32], 1997 (this volume).

[55] R. J. Van Abel, Y.-Q. Tang, V. S. V. Rao, C. H. Dobbs, D. Tran, G. Barany, and M. E. Selsted, *Int. J. Pept. Protein Res.* **45,** 401 (1994).

[56] S. N. McCurdy, *Pept. Res.* **2,** 147 (1989).

[57] I. Photaki, J. Taylor-Papadimitriou, C. Sakarellos, P. Mazarakis, and L. Zervas, *J. Chem. Soc. C,* 2683 (1970).

[58] E. E. Büllesbach, W. Danho, H.-J. Helbig, and H. Zahn, *in* "Peptides 1978" (I. Z. Siemion and G. Kupryszweski, eds.), p. 643. Wroclaw University Press, Wroclaw, 1970.

[59] D. A. Pearson, M. Blanchette, M. L. Baker, and C. A. Guindon, *Tetrahedron Lett.* **30,** 2739 (1989).

[60] S. S. Wang and R. B. Merrifield, *Int. J. Protein Res.* **1,** 235 (1969).

[61] J. J. Sharp, A. B. Robinson, and M. D. Kamen, *J. Am. Chem. Soc.* **95,** 6097 (1973).

[62] A. Scarso, J. Brison, J. P. Durieux, and A. Loffet, *in* "Peptides 1980" (K. Brunfeldt, ed.), p. 321. Scriptor, Copenhagen, 1981.

[63] R. Ramage, J. Green, and O. M. Ogunjobi, *Tetrahedron Lett.* **30,** 2149 (1989).

[64] H. Yajima, J. Kanaki, M. Kitajima, and S. Funakoshi, *Chem. Pharm. Bull.* **28,** 1214 (1980).

butylation of Met and accelerate the removal of aryl sulfonyls[33,65] by the push–pull mechanism.[66] Although chemicals such as dimethyl sulfide and ethyl methyl sulfide are structurally analogous to Met, thioanisole is preferred because of its dual functionality. Last, phenol was added to further assist in scavenging Pmc cations,[22] an observation verified by Solé and Barany.[53] The inclusion of all four scavengers ensures a high molar composition sufficient to scavenge all cations generated during cleavage of even the longest peptides. Furthermore, as some studies have demonstrated the shortcomings of using some of these scavengers in solo,[7,18,22,37,53,67–77] this medley apparently works in concert to effectively prevent the side reactions.

Very specialized advances have been made in peptide chemistry. Incorporation of sulfated Tyr and phosphorylated[78] or glycosylated[79] Tyr, Thr, and Ser have become commonplace. Researchers have carefully monitored and optimized cleavage conditions to ensure the preservation of their modified residue. Yagami and co-workers[80] reported that cleavage of peptides containing $Tyr(SO_3Na)$ must be performed at $4°$ in the absence of thiols. In the case of a sulfated octapeptide, Barany and co-workers optimized cleavage conditions to maximize deprotection and minimize loss of sulfate, recommending a 15-min treatment at room temperature with TFA–DCM–2-mercaptoethanol–anisole (50 : 45 : 3 : 2).[46] Peptides containing

[65] M. Fujino, M. Wakimasu, and C. Kitada, *Chem. Pharm. Bull.* **29**, 2825 (1981).

[66] Y. Kiso, M. Satomi, K. Ukawa, and T. Akita, *J. Chem. Soc., Chem. Commun.*, 1063 (1980).

[67] Biosearch Technical Bulletin 9000-02, Milligen/Biosearch Division of Millipore, Burlington, MA, and Novato, CA (1988).

[68] Y. Masui, N. Chino, and S. Sakakibara, *Bull. Chem. Soc. Jpn.* **53**, 464 (1980).

[69] C.-D. Chang, A. M. Felix, M. H. Jiminez, and J. Meienhofer, *Int. J. Pept. Protein Res.* **15**, 485 (1980).

[70] G. Breipohl, J. Knolle, and W. Stüber, *Tetrahedron Lett.* **28**, 5651 (1987).

[71] J. D. Wade, *in* "Workshop Manual on Fmoc-Polyamide Peptide Synthesis." Howard Florey Institute of Experimental Physiology and Medicine, University of Melbourne, Parkville, Victoria, Australia, 1987.

[72] A. N. Eberle, E. Atherton, A. Dryland, and R. C. Sheppard, *J. Chem. Soc., Perkin Trans.* **1**, 361 (1986).

[73] D. Hudson, *in* "Peptides 1988" (G. Jung and E. Bayer, eds.), p. 211. de Gruyter, Berlin, 1989.

[74] A. Grandas, E. Pedroso, E. Giralt, C. Granier, and J. Van Rietschoten, *Tetrahedron* **42**, 6703 (1986).

[75] A. Dryland and R. C. Sheppard, *J. Chem. Soc., Perkin Trans. 1*, 125 (1986).

[76] M. Bodanszky and A. Bodanszky, *in* "Peptides 1984" (U. Ragnarsson, ed.), p. 101. Almqvist and Wiksell International, Stockholm, 1984.

[77] E. Atherton and R. C. Sheppard, *in* "Solid Phase Peptide Synthesis: A Practical Approach." IRL Press, Oxford, 1989.

[78] J. W. Perich, *Methods Enzymol.* **289**, [12], 1997 (this volume).

[79] J. Kihlberg, M. Elofsson, and L. A. Salvador, *Methods Enzymol.* **289**, [11], 1977 (this volume).

[80] T. Yagami, S. Shiwa, S. Futaki, and K. Kitagawa, *Chem. Pharm. Bull.* **41**, 376 (1993).

phosphorylated Thr or Tyr, however, do sustain cleavage by traditional means.[81]

The method presented here may be thought of as universal, and it is applicable to a wide variety of peptide sequences. This procedure has been used routinely in our laboratory to successfully cleave more than 1000 peptides. In many cases, following this protocol will result in a satisfactory product, completely deprotected and untainted. With some peptides, especially those of great length or containing Trp, multiple Arg, or Trt, deprotection may not be so simplistic. In these cases, it is suggested that several small minicleavages be performed with various scavenger mixtures and durations to determine the optimal cleavage conditions. The procedure for these small-scale (5–10 mg) cleavages follows at the end of this article.

Laboratory-Scale Cleavages (50 mg–2 g Peptide–Resin)

Before beginning the procedure, be familiar with the Material Safety Data Sheet for each chemical. This procedure must be performed in a certified fume hood. Latex gloves and protective eyeware and clothing should be worn at all times!

Equipment and Supplies Needed. Needed equipment includes a fume hood, 50- and 100-ml round-bottom glass flasks with glass stoppers, cork supports, disposable glass test tubes, P200 and P1000 pipettes, P5000 pipette or 5- and 10-ml graduated cylinders, Pasteur pipettes, vortex mixer such as the Fisher (Pittsburgh, PA) Vortex Genie equipped with recessed platform and elastic rings for flasks or a platform or wrist action shaker which has flask attachments, fritted glass disk funnels of fine porosity, glass filtering flasks with side arm, flask adapters, vacuum line with tubing for attachment to filtering flasks, 50 ml conical centrifuge tubes with plug seals, access to −20° freezer or 4° refrigerator approved for flammable storage or an ice bath, refrigerated tabletop centrifuge (with the ability to cool down to ~4°, accommodate 50-ml conical tubes, and have a rotor speed of 3000 rpm), lyophilizer equipped with at least one cooling trap at −50° or lower and preferably a second cooling trap at −110°, and a noncorrosive vacuum pump.

Chemicals Needed. Trifluoroacetic acid (TFA; Biotechnical or HPLC grade), EDT, thioanisole, phenol, water (HPLC grade), bleach, methyl *tert*-butyl ether (HPLC grade), dimethylformamide (DMF), dimethyl sulfoxide (DMSO), aqueous ammonia or trimethylamine, guanidine or urea, and 2-propanol (depending on the peptide sequence) are required.

[81] J. W. Perich, *Int. J. Pept. Protein Res.* **40,** 134 (1992).

Protocol

1. Determine which cleavage reagent to use and the duration of cleavage. Cleave resin with 5% water/TFA for 1 hr unless otherwise indicated. If peptide contains

    a. *t*Bu or Trt groups, start with 2 hr cleavage time;

    b. More than one Asn(Trt), use an additional hour per each;

    c. Met or unprotected Trp, use only 1-hr cleavage time, or do a time course, and use reagent K;

    d. More than one Arg(Pmc), use an additional hour per each and use reagent K;

    e. Cys(Trt), use reagent K;

    f. Cys(Acm), avoid using reagent K as the thiols may remove the acetamidomethyl (Acm) groups.[82,83]

These criteria are only a guideline. Conditions may be optimized first using the minicleavage procedure. The goal is to maximize the yield of desired peptide. Longer cleavage times may increase the yield of peptide released from resin and ensure complete removal of side-chain protecting groups but may induce modifications to Met or Trp. There are some cases in which using reagent K is undesirable, for example, when the peptide is not to be purified or does not precipitate in ether. For some long peptides (>40 residues), we have seen the value of adding a second aliquot of reagent K after 2 hr where several hours of cleavage are required. Refreshing the TFA is apparently necessary to offset the neutralizing effects of the peptide backbone, as described by Riniker *et al.*[11]

2. Prepare 1.5 ml of total cocktail for every 100 mg of peptide–resin.

3. Weigh peptide–resin into a round-bottom flask. For cocktail volumes of 15 ml or less, use a 50-ml flask; for 15–30 ml, use a 100-ml flask.

4. Locate TFA, scavengers, Pipetman, and pipette tips, and fill a test tube with bleach.

5. Prepare the cleavage cocktail as described below.

6. Eject the pipette tips that were used on the sulfurous scavengers into a test tube containing bleach.

Preparation of 5% water in TFA: Pipette water (5% of total volume) into flask containing dry resin (this amount of water will barely be noticeable), then measure TFA (95% of total volume) and pour or pipette into flask containing resin and water.

Preparation of reagent K: Weigh phenol (5% by weight or 75 mg/ml total cocktail volume) into a clean, dry test tube. Be very careful

[82] R. Eritja, J. P. Ziehler-Martin, P. A. Walker, T. D. Lee, K. Legesse, F. Albericio, and B. E. Kaplan, *Tetrahedron* **43,** 2675 (1987).

[83] E. Atherton, R. C. Sheppard, and P. Ward, *J. Chem. Soc., Perkin Trans. 1,* 2065 (1985).

not to get this on your skin! Pipette water, thioanisole, and ethanedithiol (5, 5, and 2.5% of total volume, respectively) into the test tube containing phenol. Pipette the contents of the test tube into a flask containing the resin. Measure TFA (82.5% of total cocktail volume) and pour or pipette into the flask containing resin and scavengers.

7. Place the stoppered flask on the vortexer or shaker. Shake for the amount of time determined in step 1.

8. Shortly before the cleavage time is up, locate Pasteur pipettes and bulbs, cool methyl *tert*-butyl ether in an ice bath or refrigerator or freezer approved for flammable storage, set temperature of centrifuge to about 0°, and attach a small side-arm filtering flask (50–125 ml) to a vacuum source via rubber tubing, preferably with an in-line base trap for neutralizing TFA vapors.

9. After the cleavage time, pipette the resin solution onto the filter. Wash the round-bottom flask with several pipette volumes of TFA, and add the washes to the solution in the filter. Allow the vacuum to remain on to reduce the volume of the peptide/TFA solution in the flask. Reduce to a volume that just covers the bottom of the flask. The peptide may precipitate out.

10. Fill a centrifuge tube about two-thirds full with cold methyl *tert*-butyl ether. Squirt the peptide/TFA solution into the ether using a Pasteur pipette.

11. Centrifuge at approximately 0°, speed setting of 3000–3500 rpm, for 10 min.

12. Decant the ether to waste. If the peptide does not precipitate with ether, place the capped tube in −20° freezer overnight. If the peptide precipitates, decant the ether to waste. If no precipitate has formed, remove the ether under rotary evaporation.

13. Add 1 ml TFA to the cleavage flask to rinse and/or dissolve any remaining peptide. Add this solution to the pellet in the centrifuge tube. Shake or vortex vigorously to homogenize the pellet. Add methyl *tert*-butyl ether (wash 2) to about two-thirds the total centrifuge tube volume and centrifuge. Peptides cleaved with 5% water/TFA need only two washes with ether. For reagent K, three washes are needed.

14. Decant the ether to waste. Add more ether (wash 3, if required), shake, and centrifuge. If ether is still discolored after the washes, continue to wash until clear.

15. Rinse all glassware, pipettes, and tubes that came into contact with scavengers with bleach before taking them out of the hood.

16. After the final ether extraction, allow the pellet to partially dry under the hood for 1 hr or so. Allowing remaining ether to evaporate

should make dissolving easier, but if the pellet is too dry, it may be difficult to dissolve.

17. Add 1 ml 0.1% TFA/water or HPLC aqueous phase. Shake on the vortexer for up to 5 min.

18. Look at the sequence. Count the number of hydrophobic amino acids, negative charges, positive charges, and polar groups.

Hydrophobic	Negative charge	Positive charge	Polar
Alanine (A)	Aspartic acid (D)	Histidine (H)	Asparagine (N)
Cysteine (C)	Glutamic acid (E)	Lysine (K)	Glutamine (Q)
Methionine (M)	Free acid terminus	Arginine (R)	Threonine (T)
Proline (P)		Free amino terminus	Tyrosine (Y)
Tryptophan (W)			Serine (S)
Phenylalanine (F)			
Leucine (L)			
Isoleucine (I)			
Valine (V)			

19. If the net charge is negative, add trimethylamine or aqueous ammonia dropwise until dissolved. If more than one-third of the residues in the peptide are polar and/or hydrophobic, add 1 ml of 0.1% (v/v) acetonitrile or HPLC organic phase. Keep alternating with aqueous solution used in step 17 until dissolved. Ideally the peptide should be dissolved in as low a percentage of organic solvent as possible. If the peptide is difficult to dissolve, add 1 ml 2-propanol, 1 ml dimethylformamide, and/or 1 ml dimethyl sulfoxide. Shake for at least 5 min between additions.

20. If the peptide is still not dissolved after adding 10 ml of solvent, add a small amount of guanidine or urea (small spatulaful).

21. If peptide will still not dissolve, dilute to no more than 20% organic solvent and lyophilize. Start the procedure from step 17 again, once the peptide is dry.

Minicleavages (5–10 mg Peptide–Resin)

Before beginning the procedure, be familiar with the Material Safety Data Sheet for each chemical. This procedure must be performed in a certified fume hood. Latex gloves and protective eyewear and clothing should be worn at all times!

Equipment and Supplies Needed. Necessary equipment includes a fume hood, disposable glass test tubes, test tube rack, P20 and P200 pipettes,

Pasteur pipette, vortex mixer or platform shaker which will continuously shake test tubes, and cotton or glass wool.

Chemicals Needed. Required reagents include TFA (Biotechnical or HPLC grade), EDT, thioanisole, phenol, water (HPLC grade), bleach, methl *tert*-butyl ether (HPLC grade), acetic acid, aqueous ammonia, and 2-propanol (depending on peptide sequence).

Protocol

1. Using the cleavage criteria given above, select the most obvious cleavage conditions. Usually the amount of time for cleavage is the desired variable, although performing test cleavages with and without reagent K can be informative.

2. To quantitatively evaluate the cleavage by analytical HPLC, weigh the same amount of resin (5.0–10.0 mg) into a test tube for each condition.

3. Locate TFA, scavengers, Pipetman, and pipette tips, and fill a test tube with bleach.

4. Use 150 μl total cleavage cocktail/10 mg peptide–resin. Prepare enough cocktail for all test samples in a single, separate test tube.

5. Prepare the cleavage cocktail as described below; eject the pipette tips that were used on the sulfurous scavengers into the test tube of bleach.

Preparation of 5% water in TFA: Prepare enough for all samples. Pipette water (5% of total volume) and TFA (95% of total volume) into test tube. Pipette appropriate aliquots into each test tube containing resin.

Preparation of reagent K: Prepare enough cocktail for all samples. Weigh phenol (5% by weight or 75 mg/ml total cocktail volume) into a clean, dry test tube. Be very careful not to get this on your skin! Pipette water, thioanisole, and EDT, and TFA (5, 5, 2.5, and 82.5% of total volume, respectively) into the test tube containing phenol. Pipette appropriate aliquots into each test tube containing resin.

6. Place the test tubes on the vortexer. Shake for the desired amount of time.

7. To quench the reaction, fill test tube about one-third full, keeping volumes constant, with water, 0.1% TFA/water, or 30% acetic acid. Add methyl *tert*-butyl ether until the test tube is two-thirds full.

8. Mix the layers using a Pasteur pipette and bulb, filling and releasing a few times.

9. Allow the layers to separate. Carefully remove the ether layer with the Pasteur pipette either manually with the bulb or by vacuum with the pipette attached to a side-arm flask via tubing.

10. Wash the aqueous layer once or twice more with ether if reagent K was used.

11. Allow the aqueous peptide–resin solution to vortex for a few minutes to evaporate residual ether. If the solution is cloudy, add 2-propanol, acetic acid, or aqueous ammonia to dissolve peptide, using the above guidelines and noting the volumes.

12. Take a small piece of cotton or glass wool and insert it into a clean Pasteur pipette to act as a filter. Place this pipette in a clean test tube and pipette the aqueous peptide–resin solution through the cotton filter. Allow the pipette to drain by gravity.

13. Rinse all glassware, pipettes, and tubes that came into contact with scavengers with bleach before taking out of the hood.

14. The peptide solution is now ready for analysis. The composition of each peptide can be quantitatively compared among test samples by HPLC if constant volumes were maintained. The presence of aromatic side-chain protecting groups can be detected by UV absorbance in the range of 254–280 nm. Adducts can be confirmed by mass spectral analysis. The test tubes are suitable for concentration in a Speed Vac, or solutions can be transferred for lyophilization.

[6] Properties of Solid Supports

By Morten Meldal

Solid Supports in Synthesis, Applications, and Limitations

This article is not intended to give a comprehensive account of previous work in the field of peptide synthesis resins; rather, it attempts to give an overview and an understanding of the factors that determine the properties of the solid supports, in particular those of the polymer matrix itself.

The solid support has a large influence on the outcome of solid-phase synthesis in particular when the synthesis carried out becomes more difficult or chemically demanding.[1-3] Even though this was realized many years ago,

---

[1] H. Gausepohl, W. Rapp, E. Bayer, and R. W. Frank, *in* "Innovation and Perspectives in Solid Phase Synthesis" (R. Epton, ed.), p. 381. Intercept Limited, Andover, UK, 1992.

[2] J. Bedford, T. Johnson, W. Jun, and R. C. Sheppard, *in* "Innovation and Perspectives in Solid Phase Synthesis" (R. Epton, ed.), p. 213. Intercept Limited, Andover, UK, 1992.

[3] M. Meldal, *in* "Peptides 1992, Proceedings of the Twenty-Third European Peptide Symposium" (C. H. Schneider and A. N. Eberle, eds.), p. 61. ESCOM, Leiden, The Netherlands, 1993.

0076-6879/97 $25.00

surprisingly little has been done to improve and tailor the solid supports for special purposes. Originally the Merrifield resin[4] was 2% cross-linked chloromethylated divinylbenzene–polystyrene resin, and this was later modified to have only 1% cross-linking due to problems observed in synthesis of difficult peptides.[5,6] With some minor modifications this resin together with Sheppard's polyamide resins[7] have been used extensively and only during the last decade have new, improved resins based on composite polymers appeared, prompted by the demands of resins for general organic synthesis and library techniques. There is no complete systematic study of the quality of different kinds of supports for peptide synthesis, and the data presented in the literature can rarely be compared in a meaningful way. In general, the resins used for organic synthesis should be compatible with strongly acidic and basic conditions, with radical, carbene, carbanion, and carbenium ion chemistry, with reducing or oxidizing conditions, or with conditions of nitrations and halogenations.[8–10] The process of generating organic molecules often requires use of a wide range of different reaction conditions, and the support should furthermore be mechanically robust and stable to variation in temperature. In the field of libraries[9] and solid-phase assays,[11–13] it is furthermore required that the resins are biocompatible and swell in aqueous buffers, show little nonspecific binding to biomolecules, and allow nonhindered access of, for example, enzymes to the interior of the resin. All these requirements impose tremendous restrictions to the design of suitable resins, and originality is required to solve future problems.

The methods of polymer analysis for linear polymers involve determination of chain lengths and chain-length distributions by gel-permeation chromatography (GCP) and light scattering as well as determination of viscosity and dynamic properties of polymer melts and solutions.[14] Three-dimen-

---

[4] B. Merrifield, *J. Am. Chem. Soc.* **85**, 2149 (1963).

[5] V. K. Sarin, S. B. H. Kent, A. R. Mitchell, and R. B. Merrifield, *J. Am. Chem. Soc.* **106**, 7845 (1984).

[6] K. C. Pugh, E. J. York, and J. M. Stewart, *Int. J. Pept. Protein Res.* **40**, 208 (1992).

[7] E. Atherton, D. L. J. Clive, and R. C. Sheppard, *J. Am. Chem. Soc.* **97**, 6584 (1975).

[8] M. S. Deshpande, *Tetrahedron Lett.* **35**, 5613 (1994).

[9] M. Rinnová and M. Lebl, *Collect. Czech, Chem. Commun.* **61**, 171 (1996).

[10] P. Hodge, *in* "Innovation and Perspectives in Solid Phase Synthesis" (R. Epton, ed.), p. 273. SPCC, Birmingham, UK, 1990.

[11] M. Meldal, I. Svendsen, K. Breddam, and F. I. Auzanneau, *Proc. Natl. Acad. Sci. U.S.A.* **91**, 3314 (1994).

[12] M. Meldal and I. Svendsen, *J. Chem. Soc., Perkin Trans. 1,* 1591 (1995).

[13] M. Meldal, *in* "Methods in Molecular Biology. Combinatorial Peptide Libraries" (C. Shmuel, ed.), in press. Humana, Totowa, New Jersey.

[14] J.-M. Guenet and M. Klein, *Macromol. Chem. Macromol. Symp.* **39**, 85 (1996).

sional cross-linked polymers can be analyzed by gel-phase[15,16] or solid-state magic angle spinning[17-19] (MAS) nuclear magnetic resonance (NMR) spectroscopy. The NMR techniques can give information about the composition (integration), homogeneity (chemical shifts), dynamic behavior of the resin (T_1 values and linewidth),[16,20-22] and the presence of functional groups. The T_1 relaxation times found for different parts of a polymer are extremely important parameters because they give a direct measure of the mobility of different kinds of atoms in the polymer network.[16,20,23] The presence of functional groups can also be determined by infrared (IR) spectroscopy, in particular by Fourier transform (FT) techniques on flattened beads,[24] and both IR and Raman spectroscopy can be used to determine the distribution and population of local polymer conformations.[25,26] Beaded resins are characterized by bead size and size distribution.[27,28] The degrees of swelling in various solvents give information on the overall polar or hydrophobic character of the resin and of the amphipathic nature of the polymer chains.[6,16] The swelling of the resin[29] or the conductance of the resin suspension[30,31] can be monitored continuously to provide information on the reaction kinetics. The permeability of macromolecules determined by GCP[16] gives a measure of the porosity of the resin and the performance of instantaneous chemical reactions; for example, use of colored or fluorescent reagents can give a measure of the rate of mass transport and homogeneity

[15] J.-C. Alfred, J.-L. Aubagnac, M. Calmes, J. Daunis, B. Elamrani, R. Jacquire, and G. Nkusi, *Tetrahedron* **44**, 4407 (1988).

[16] F. I. Auzanneau, M. Meldal, and K. Bock, *J. Pept. Sci.* **1**, 31 (1995).

[17] J. C. Castro-Palomino and R. R. Schmidt, *Tetrahedron Lett.* **36**, 5343 (1995).

[18] R. C. Anderson, M. A. Jarema, M. J. Shapiro, J. P. Stokes, and M. Ziliox, *J. Org. Chem.* **60**, 2650 (1995).

[19] W. L. Fitch, G. Detre, and C. P. Holmes, *J. Org. Chem.* **59**, 7955 (1994).

[20] R. Roy, F. D. Tropper, A. J. Williams, and J.-R. Brisson, *Can. J. Chem.* **71**, 1995 (1993).

[21] H. Rockelmann and H. Sillescu, *Z. Physik. Chem. Neue Folge* **92**, 263 (1974).

[22] H. Sillescu and R. Brüssau, *Chem. Phys. Lett.* **5**, 525 (1970).

[23] F. Albericio, M. Pons, E. Pedroso, and E. Giralt, *J. Org. Chem.* **54**, 360 (1989).

[24] B. Yan and G. Kumaravel, *Tetrahedron* **52**, 843 (1996).

[25] M. Kobayashi and T. Kozasa, *Appl. Spectrosc.* **47**, 1417 (1993).

[26] S. Masatoki, M. Takamura, H. Matsuura, K. Kamogawa, and T. Kitagawa, *Chem. Lett.*, 991 (1995).

[27] R. Arshady and A. Ledwith, *Reactive Polymers* **1**, 159 (1983).

[28] R. Arshady, *J. Chromatogr.* **586**, 199 (1991).

[29] I. L. Rodionov, M. B. Baru, and V. T. Ivanov, *Reactive Polymers* **16**, 311 (1992).

[30] J. Fox, R. Newton, P. Heegard, and C. Schafer-Nielsen, *in* "Innovation and Perspectives in Solid Phase Synthesis" (R. Epton, ed.), p. 141. SPCC, Birmingham, UK, 1990.

[31] N. McFerran and B. Walker, *in* "Innovation and Perspectives in Solid Phase Synthesis" (R. Epton, ed.), p. 261. SPCC, Birmingham, UK, 1990.

of the resin (phased reactions).[32,33] The comparison of synthesis of aggregating peptides[1-3,32] can afford information on the ability of the resin to assist solvation and break aggregation of the peptide by direct amphipathic interaction between the polymer and the peptide chains.[3] By synthesis of different types of difficult peptides the nature of the polymer–peptide interaction can be revealed. The solid-phase kinetics can be monitored with a solid-phase spectrophotometer[34,35] using bromphenol blue or 3-hydroxy-4-oxo-1,2,3-benzotriazine as an amino group reporter, and the release of fluorenylmethyloxycarbonyl (Fmoc) can be monitored by recording the UV absorption of the reagent solution.[1] The incorporation of spin labels on functional groups allows measurements by electron spin resonance (ESR) spectroscopy of functional group spacing.[36]

The preparation of beaded resins and some of their properties have previously been reviewed.[27,28,37] The supports that have been used in solid-phase synthesis can be divided into four physically different groups.

Gel Type Supports. The most used supports are of the gel type in Fig. 1, due to the optimal behavior of gels for high yield synthesis. The equal distribution of functional groups throughout a highly solvated and inert polymer network is ideal for the assembly of large molecules such as peptides.[28] The capacity can be adjusted to the desired assembly usually to afford a problem-free synthesis and a high yield per volume of resin. The polymer network is flexible, and the resin can expand or exclude solvent to accommodate the growing molecule within the gel. Four types of gel resins have been developed for peptide synthesis. (1) The hydrophobic polystyrene resins are of the 1% cross-linked type produced from inexpensive styrene and divinylbenzene, and they are substituted with various forms of functionalities mostly starting with chloromethylation or aminomethylation. Because the chloromethylation reaction has the disadvantage of introducing additional cross-linking and of incomplete conversion to other functional groups, aminomethylation was developed. The additional chemistry required to convert the polystyrene beads into a functional resin increase the cost of production considerably. (2) Polyacrylamide resins were devel-

[32] M. Meldal and R. C. Sheppard, *in* "Peptides 1986, Proceedings of the Nineteenth European Peptide Symposium" (D. Theodoropoulos, ed.), p. 131. de Gruyter, Berlin, 1987.

[33] M. Meldal, *Tetrahedron Lett.* **33**, 3077 (1992).

[34] L. Cameron, M. Meldal, and R. C. Sheppard, *J. Chem. Soc., Chem. Commun.*, 270 (1987).

[35] L. R. Cameron, J. L. Holder, M. Meldal, and R. C. Sheppard, *J. Chem. Soc., Perkin Trans. 1*, 2895 (1988).

[36] E. M. Cilli, R. Marchetto, S. Schreier, and C. R. Nakaie, *in* "Peptides 1995, Proceedings of the Fourteenth American Peptide Symposium" (P. T. P. Kaumaya and R. S. Hodges, eds.), p. 103. Mayflower Scientific, Kingswinford, UK, 1996.

[37] R. Arshady, *J. Chromatogr.* **586**, 181 (1991).

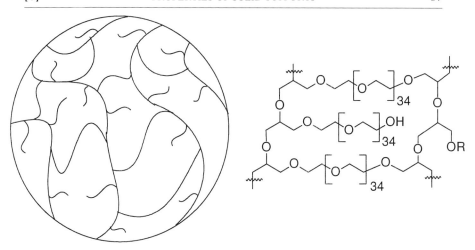

FIG. 1. Principal structure of a three-dimensional polymer network typical for the gel type of resin, for example, polystyrene or poly-*N,N*-dimethylacrylamide, shown with the example of a PEG-cross-linked polyoxypropylene polymer used in solid-phase synthesis.

oped as a hydrophilic alternative to the polystyrene resins and are obtained by cross-linking poly-*N,N*-dimethylacrylamide with 5% bis-*N,N'*-acryloyl-ethylenediamine and as functional groups *N*-acryloylsarcosine methyl ester.[38] The ester is converted to a functional group after polymerization. (3) Polyethylene glycol (PEG) grafted resins were developed as a mechanically more stable resin and to space the site of synthesis from the polymer backbone. They are formed either by graft polymerization on polystyrene beads (Tentagels, Rapp Polymere, Tübingen, Germany)[39] or by reaction of preformed oligooxyethylenes with aminomethylated polystyrene beads.[40] It was found that the grafted resins also improved the performance in the synthesis of difficult peptides. (4) The PEG-based resins[33,41-43] are composed either exclusively of a PEG/polypropylene glycol (PPG) network or of a combination of PEG with a small amount of polyamide or polystyrene. The resins have been obtained by partial derivatization of PEG with epi-

[38] P. Kanda, R. C. Kennedy, and J. T. Sparrow, *Int. J. Pept. Protein Res.* **38**, 385 (1991).

[39] W. Rapp, L. Zhang, R. Häbish, and E. Bayer, *in* "Peptides 1988, Proceedings of the Twentieth European Peptide Symposium" (G. Jung and E. Bayer, eds.), p. 199. de Gruyter, Berlin, 1989.

[40] D. D. Ho, A. V. Neumann, A. S. Perelson, W. Chen, J. M. Leonard, and M. Markowitz, *Nature* (*London*) **373**, 123 (1995).

[41] M. Renil and M. Meldal, *Tetrahedron Lett.*, **37**, 6185 (1996).

[42] M. Renil and M. Meldal, *Tetrahedron Lett.* **36**, 4647 (1995).

[43] M. Kempe and G. Barany, *J. Am. Chem. Soc.* **118**, 7083 (1996).

chlorohydrin, chloromethylstyrene, or acryloyl chloride (using bisamino-propyl-PEG, Jeffamines, Fluka, Buchs, Switzerland). The resulting mixture of non-, mono-, and difunctionalized PEG can be mixed with other mono-mers to vary the polymer composition, and polymerization gives a highly cross-linked polymer with long cross-linkers of PEG.

Surface Type Supports. Several materials have been used for surface functionalization for solid-phase synthesis. Among these are beads made from sintered polyethylene (Fig. 2), cellulose fibers (cotton, paper, Sepha-rose, and LH-20), porous highly cross-linked polystyrene or polymethacry-late, controlled pore glass, and silicas. Furthermore, protein or chitin precip-itates have been employed in organic solvents.

Composites and Supported Gels. When continuous flow synthesis was introduced it soon became apparent that the polystyrene and polyamide gels used in batch synthesis were not stable to the flow conditions. To increase the mechanical stability the gel type of polymers were supported by rigid matrices as presented in Fig. 3. The first example used kieselguhr as matrix support by polymerizing the polyamide monomer solution after absorption into granules of the inorganic matrix. Later the irregular kiesel-

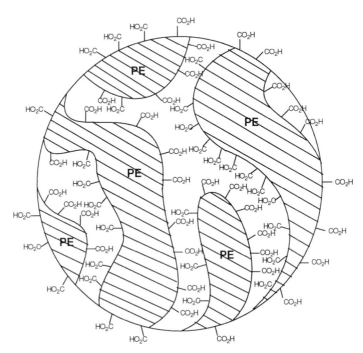

FIG. 2. Physical structure of a functionalized surface particle illustrated by chromium-trioxide oxidized porous polyethylene (PE).

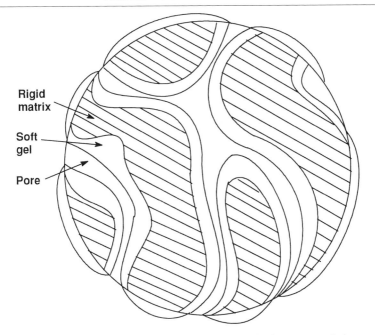

FIG. 3. Principal structure of a physically supported soft gel polymer network, for example, poly-N,N-dimethylacrylamide supported by kieselguhr or highly cross-linked porous polystyrene.

guhr was substituted by a highly cross-linked polystyrene sponge, to the interior of which the polyamide was grafted by covalent attachment. Teflon membranes have also been modified by polymethacrylate as a flow-stable gel for peptide synthesis.

Brush Polymers. A special type of polymer is the brush polymer presented in Fig. 4, where a linear component, for example, polystyrene is grafted onto, for example, a polyethylene film or tubing. The polymer chains extend from the rigid surface in a brushlike fashion. Functionalization can be achieved along the extending chains, and the functional groups can be further modified with functionalized spacers for synthesis.

Composition and Properties of Gel Type Polymers

The monomer components of beaded gel type polymers are mixtures of a functional group component, a cross-linker, and a filler component, all containing activated double bonds for radical suspension polymerization. To form a homogeneous polymer the monomer components must have a comparative reaction rate in the radical chain polymerization reaction.

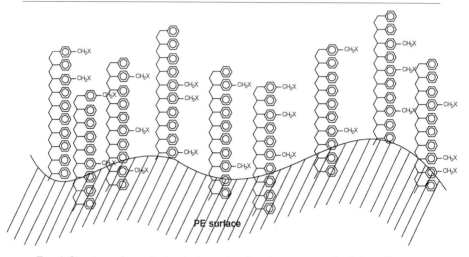

FIG. 4. Structure of a grafted polystyrene brush polymer on a polyethylene film surface. The polystyrene can be directly used as a support for synthesis, or X can be a second graft polymer, for example, poly-N,N-dimethylacrylamide or PEG.

Cross-linked polystyrene is obtained by suspension polymerization[27,44,45] of styrene and divinylbenzene, and the product is a hydrophobic swelling bead, which shows best swelling properties in nonpolar solvents such as toluene and dichloromethane.[6,27] The control of bead size is quite good,[28,46] and virtually monosized beads have been obtained.[44,45] Resins with average bead sizes from 10 to 200 μm are available; however, with seeding procedures beads as large as 600 μm for grafting and library synthesis have been prepared.[45,47] The diffusion of reagents in these large beads is not as fast as in the smaller beads. The polymer chains are not formed in an ordered manner, and particularly the regions around cross-linking sites are rigid.[48,49] These regions of the polystyrene gel may show a semicrystalline arrangement of helical chains with regularly arranged aromatic phenyl groups[25]

[44] S. Wang and X. Zhang, Polym. Adv. Techn. 2, 93 (1991).
[45] F. Svec and J. M. J. Frechet, Science 273, 205 (1996).
[46] M. Eggenweiler, N. Clausen, H. Fritz, L. Zhang, and E. Bayer, in "Peptides 1994, Proceedings of the Twenty-Third European Peptide Symposium" (H. L. S. Maia, ed.), p. 275. ESCOM, Leiden, The Netherlands, 1995.
[47] W. Rapp, M. Maier, G. Schlotterbeck, M. Pirsch, K. Albert, and E. Bayer, in "Peptides 1995, Proceedings of the Fourteenth American Peptide Symposium" (T. P. Kaumaya and R. S. Hodges, eds.), p. 313. Mayflower Scientific, Kingswinford, UK, 1996.
[48] S. Amelar, C. E. Eastman, R. L. Morris, M. A. Smeltzly, T. P. Lodge, and E. D. von Meerwall, Macromolecules 24, 3505 (1991).
[49] N. Lehsaini, N. Boudenne, J. G. Zilliox, and D. Sarazin, Macromolecules 27, 4353 (1994).

and solvent (e.g., $CHCl_3$ or benzene)-filled cavities.[14,50] The cross-linker distribution and primary structure effects in polystyrenes can lead to differences in reactivity of several orders of magnitude for functional groups in the resin.[28] As expected, polystyrene gels formed from linear polymers at high concentration show increased polymer chain entanglement when compared to gels formed from more dilute solutions.[14,51,52]

Different gel resins (polystyrene and polyamide) were labeled with difluoromethylene in the functional groups of the resins. ^{19}F NMR T_1 relaxation experiments on these resins demonstrated the functional groups of 1% cross-linked polystyrene to have an overall greater mobility than an unspecified polyacrylamide resin, in dimethylformamide (DMF) and in particular in chloroform.[23] The mobility correlated well with the rate of a peptide fragment coupling. However, the mobilities of polymer chains in the polyamide resins are very dependent on monomer composition. The exact choice of monomer, for example, methacrylate versus acrylate, may have major influence on the properties and homogeneity of the resulting polymer.[53]

The solvation of pure polystyrene–divinylbenzene beads is not very efficient,[14,22,50] and when peptide synthesis is carried out the swelling may change dramatically as a result of solvation of the growing peptide chains. Depending on the peptide careful selection of solvent mixtures may be required to match the solubility parameters of both peptide and resin.[54] This is particularly the case when synthesis is carried out in polar solvents. The change in swelling and the fragile character of the resin toward the end of synthesis make the resin less useful for continuous flow synthesis. The rigid character of the divinylbenzene cross-linker has been improved by substitution with a tetraethylene glycol diacrylate cross-linker.[55,56] The functional groups are introduced into polystyrene by chloro-, hydroxy-, or aminomethylation, and the high density of phenyl groups allows a high degree of functionalization. However, consideration should be given to the additional cross-linking obtained, and the practical loading for synthesis should be kept at less than 0.5 mmol/g resin to ensure a successful outcome

[50] F. Müller-Plathe, *Chem. Phys. Lett.* **252,** 419 (1996).
[51] B. Chu, K. Kubota, and K. M. Abbey, *Polym. Preprints* (*Am. Chem. Soc., Div. Polym. Chem.*) **22,** 70 (1981).
[52] L. Fang, W. Brown, and C. Konák, *Polymer* **31,** 1960 (1990).
[53] L. Winther, K. Almdal, W. B. Pedersen, J. Kops, and R. H. Berg, *in* "Peptides 1993, Proceedings of the Thirteenth American Peptide Symposium" (R. S. Hodges and J. A. Smith, eds.), p. 872. ESCOM, Leiden, The Netherlands, 1994.
[54] G. B. Fields and C. G. Fields, *J. Am. Chem. Soc.* **113,** 4202 (1996).
[55] M. Renil, R. Nagaraj, and V. N. R. Pillai, *Tetrahedron* **50,** 6681 (1994).
[56] M. Renil and V. N. R. Pillai, *Proc. Indian Acad. Sci. Chem. Sci.* **106,** 1123 (1994).

of the synthesis. In the assembly of long and difficult peptides a maximum loading of 1–2 mmol/g corresponding to approximately 0.01 mmol/ml of the final peptide is preferred. Acetoxystyrene has been used as a direct means to introduce phenolic functional groups for peptide synthesis.[57]

The properties of the polystyrene have been modified by grafting of polyethylene glycol, often used by itself as a soluble support for synthesis,[58,59] to the functional groups of the polystyrene. Polymerization grafting of oxyethylene onto the functional hydroxymethyl groups affords a PEG grafting with a chain length of 7–10 ethylene glycol residues or more where the PEG chain acts as a spacer for the peptide attachment.[39,60] The PEG chains may also be grafted by attachment of long polyethylene glycol chains through an oxycarbonylglycine bond to a functional aminomethyl groups of the polystyrene.[61,62] The weight ratio of PEG to polystyrene is often about 1 : 1 in these resins, and the presence of PEG results in a complete change in the swelling behavior. The beads swell in both nonpolar and polar solvents except water. More important, they assume a state of complete swelling from the start of the synthesis, and swelling does not change significantly during the synthesis. The high degree of swelling gives the beads a more firm and flow-stable character.[39] The improved synthesis obtained with these resins have been ascribed to an environmental effect of the PEG chains.[63–65] Furthermore, the use of monodisperse small beads increases the efficiency of synthesis.[46]

The hydrophobic polymer network of polystyrene gel beads does not allow biomolecules in aqueous solution to penetrate into the interior of

[57] R. Arshady, G. W. Kenner, and A. Ledwith, *J. Polym. Sci. Polym. Chem. Ed.* **12,** 2017 (1974).

[58] H. Gehrhardt and M. Mutter, *Polym. Bull.* **18,** 487 (1987).

[59] M. Mutter, R. Uhmann, and E. Bayer, *Liebigs Ann. Chem.,* 901 (1975).

[60] W. Rapp, L. Zhang, and E. Bayer, *in* "Innovation and Perspectives in Solid Phase Synthesis" (R. Epton, ed.), p. 205. SPCC, Birmingham, UK, 1990.

[61] S. Zalipsky, F. Albericio, and G. Barany, *in* "Peptides 1985, Proceedings of the Ninth American Peptide Symposium" (C. M. Deber, V. J. Hruby, and K. D. Kopple, eds.), p. 257. Pierce, Rockford, Illinois, 1986.

[62] S. Zalipsky, J. L. Chang, F. Albericio, and G. Barany, *React. Polym.* **22,** 243 (1994).

[63] F. Albericio, J. Bacardit, G. Barany, J. M. Coull, M. Egholm, E. Giralt, G. W. Griffin, S. A. Kates, E. Nicolas, and N. A. Sole, *in* "Peptides 1994, Proceedings of the Twenty-Third European Peptide Symposium" (H. L. S. Maia, ed.), p. 271. ESCOM, Leiden, The Netherlands, 1995.

[64] G. Barany, N. A. Sole, R. J. van Abel, F. Albericio, and M. E. Selsted, *in* "Innovation and Perspectives in Solid Phase Synthesis" (R. Epton, ed.), p. 29. Intercept Limited, Andover, UK, 1992.

[65] M. Beyermann, H. Wenschuh, P. Henklein, and M. Bienert, *in* "Innovation and Perspectives in Solid Phase Synthesis" (R. Epton, ed.), p. 349. Intercept Limited, Andover, UK, 1992.

the beads.[66] This is true even when the polystyrene is grafted with polyethylene glycol.

Polar resins are obtained in a beaded form by inverse suspension radical polymerization[7,37,38] in tetrachloromethane/heptane of monomers dissolved in water droplets. In contrast to polystyrene, polyacrylamide or poly-N,N-dimethylacrylamide is more flexible[16,67] and has an open coiled structure with high mobility of the polymer chains, allowing a much higher degree of cross-linking. The dependence of polyamide gels on cross-linking has been examined,[68,69] and the shear modulus of the gel was found to vary linearly with the amount of cross-linker used. According to diffusion studies polyamide gels show some inhomogeneities probably in the cross-linked regions.[70] The high flexibility of the polyacrylamides allows proteins to diffuse inside, and they are the preferred polymer gels for electrophoresis.[71] However, the solvation of the polymer chains is weak[72] and gives even highly cross-linked poly-N,N-dimethylacrylamide beads a fragile and weak character. The unsupported gel beads[7,73] are not stable enough for continuous flow synthesis, where the gel tends to block column filters and collapse.[74] This problem is partially solved by using a step of fluidizing the resin by inverse flow for each coupling reaction using a high capacity polyacrylamide resin with phenol as the functional group.[75] The polydimethylacrylamide resins swell best in polar solvents, for example, DMF, and have improved properties over polystyrene for synthesis under polar conditions.[76] Less swelling is expected in toluene or dichloromethane; however, they do show enough swelling in aqueous buffers to be useful for solid-phase enzyme assays.[11] A hydrolyzable acetal cross-linker can be incorporated in the

[66] J. Vagner, V. Krchnak, N. F. Sepetov, P. Strop, K. S. Lam, G. Barany, and M. Lebl, *in* "Innovation and Perspectives in Solid Phase Synthesis" (R. Epton, ed.), p. 347. Mayflower Worldwide, Kingswinford, UK, 1994.

[67] I. Soutar, L. Swanson, F. G. Thorpe, and C. Zhu, *Macromolecules* **29**, 918 (1996).

[68] R. Nossal, *Macromolecules* **18**, 49 (1985).

[69] S. K. Patel, F. Rodriguez, and C. Cohen, *Polym. Mater. Sci. Eng.* **58**, 1120 (1988).

[70] T.-P. Hsu, D. S. Ma, and C. Cohen, *Polymer* **24**, 1273 (1983).

[71] T. Zewert and M. Harrington, *Electrophoresis* **13**, 817 (1992).

[72] T. Schwartz, J. Francois, and G. Weill, *Polymer* **21**, 247 (1980).

[73] R. Arshady, E. Atherton, D. L. J. Clive, and R. C. Sheppard, *J. Chem. Soc., Perkin Trans. 1*, 529 (1981).

[74] E. Atherton, E. Brown, and R. C. Sheppard, *J. Chem. Soc., Chem. Commun.*, 1151 (1981).

[75] P. A. Baker, A. F. Coffey, and R. Epton, *in* "Innovation and Perspectives in Solid Phase Synthesis" (R. Epton, ed.), p. 435. SPCC, Birmingham, UK, 1990.

[76] J. T. Sparrow, N. G. Kneib-Cordonier, P. Kanda, N. U. Obeyesekere, and J. S. McMurray, *in* "Peptides 1994, Proceedings of the Twenty-Third European Peptide Symposium" (H. L. S. Maia, ed.), p. 281. ESCOM, Leiden, The Netherlands, 1995.

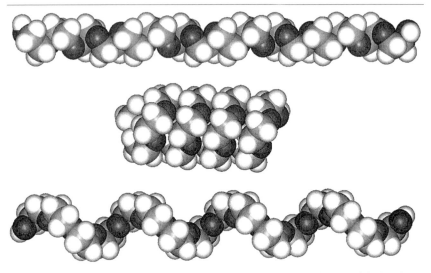

FIG. 5. Three different regular structures of the PEG helix in which the vicinal carbon–oxygen bonds maintain the most preferred gauche conformation. The upper extended helix is the more polar, displaying the oxygen atoms on the surface of the helix.

polyacrylamide polymer to yield on cleavage the soluble linear polymer, which can be used directly, for example, for immunization.[77]

Optimal support properties can be achieved by using polyethylene glycol as the major component of the resin.[33] The reason is the so-called environmental effect previously observed with PEG-grafted polystyrene.[64] This effect is caused by two properties of the PEG molecules. Because of the vicinal arrangements of carbon–oxygen bonds throughout the chain, PEG assumes helical structures with gauche interaction between the polarized bonds. This effect is well-known, for example, in carbohydrate chemistry, and has been demonstrated to be realized for the majority of the ethylene glycol residues of the polymer.[26] As presented in Fig. 5 there are three helical arrangements of PEG that allow the low energy gauche conformations of vicinal oxygen–carbon bonds. One is largely hydrophobic with the oxygen atoms in the interior of the helix, the second is of intermediate polarity, and the third is quite hydrophilic, exposing the oxygen atoms to the environment. The PEG molecule therefore has an amphipathic nature and is solvated well both by polar and nonpolar solvents. Furthermore, the organized structure of polymer fragments affords a large and highly

[77] P. Goddard, J. S. McMurray, R. C. Sheppard, and P. Emson, *J. Chem. Soc., Chem. Commun.*, 1025 (1988).

organized solvent shell,[14,78] and this in turn leads to a high tension of solvation in a cross-linked polymer. At the same time the mobility and reorganization of the PEG molecules create a highly dynamic process. The result is extremely fast mass transfer, and nonhindered peptide coupling reactions are usually complete within a few minutes.[33] However, PEG molecules in block polymers have been shown to have a glass transition at $-45°$,[79] and, as with any other polymer matrices, reactions at very low temperature could be problematic in PEG-based resins. The amphipathic nature of the polyethylene glycol also helps to unzip aggregating molecules by interacting and wrapping around the molecule. It has therefore been used for protein conjugation and to induce refolding and stabilization of proteins.[80] There is a similar effect of the PEG on aggregating peptides during synthesis, where the PEG molecules facilitate slow deaggregation and acylation of otherwise aggregated and nonreactive peptide chains.[3]

There are four different types of cross-linked PEG-based polymers available for synthesis. Partial reaction of polyethylene glycol with sodium hydride and epichlorohydrin yield a mixture of PEG mono- and bismethyloxirane. This can be bulk polymerized at high temperature in the presence of $(CH_3)_3COK$ to give an inert polymer with only ether bonds and hydroxyl groups as the functional groups.[41] The hydroxyl groups have been alkylated or acylated with handles for synthesis. This resin has been shown to be much superior to the amide-containing resins for solid-phase glycosylation reactions in which reactive carbenium ions are generated.

Similarly, partial alkylation of PEG with chloromethyl styrene gives a mixture of non-, mono-, and bisalkylated PEG that can be polymerized in a beaded form by inverse suspension polymerization or just by bulk polymerization followed by swelling and granulation.[41] This resin contains only around 10% polystyrene and behaves differently than the grafted resins described above. The resin shows good swelling in water as well as in very nonpolar media such as toluene.

Partial acryloylation of bis-2-aminopropyl PEG with acryloyl chloride followed by radical initiated inverse suspension polymerization of the resulting macromonomer mixture yields resins that are ideally suited for continuous flow peptide synthesis.[16,33,81] These PEGA resins contain only polyether backbones and polyacrylamide backbones linked together by amide bonds that are inert and stable under all peptide synthesis conditions

---

[78] J. J. Point, C. Coutelier, and D. Villers, *J. Polym. Sci. Phys. Ed.* **23**, 231 (1985).

[79] S. Fakirov, K. Goranov, E. Bosvelieva, and A. D. Chesne, *Macromol. Chem.* **193**, 2391 (1992).

[80] N. V. Katre, *Adv. Drug Delivery Rev.* **10**, 91 (1996).

[81] M. Meldal, F. I. Auzanneau, and K. Bock, in "Innovation and Perspectives in Solid Phase Synthesis" (R. Epton, ed.), p. 259. Mayflower Worldwide, Kingswinford, UK, 1994.

including cleavage with strong acids or aqueous base. The family of PEGA resins are flow stable and beaded, and they can be used as solid supports in split synthesis of libraries. These resins can easily be modified by addition of small amounts of monomers such as N,N-dimethylacrylamide, acrylamide, acrylates, or acrylonitrile. The capacity of the resins can be increased to 0.8 mmol/g by addition of small functional group monomers.[42] Furthermore, the chain length of the polymer and the size distribution has a large influence on the porosity and permeability of the resin to large molecules.[13] For highly cross-linked resins the size distribution of the PEG molecules has a large effect on the swelling capacity. This is due to the use of excess of cross-linkers because the shorter PEG chain cross-linkers stretch out first during swelling and determine the degree of swelling of the polymer.[81] Alternatively, in a similar polymer the PEG and the polyacrylic backbones are linked by acrylic ester bonds.[71,82]

Additional cross-linking can be incorporated by use of branched PEG molecules with acrylic ester residues for formation of a PEG cross-linked ethoxylate acrylate resin polymer named CLEAR. The ester bonds resulted in a resin with properties similar to the PEGA resins except for the stability toward nucleophiles. An increased rate of mass transport was obtained by use of porogens in the beading process.[43,82]

The different properties of the polymers are reflected in the behavior in bioassays in aqueous systems.[11–13,16,83,84] The more hydrophobic polystyrene resins do not swell at all in water, and proteins and cells often show a large degree of nonspecific binding to the polystyrene (PS) surface. The PEG-grafted polystyrenes do swell to some extent in aqueous buffer, and there is little or no nonspecific binding of proteins. However, the proteins do not penetrate into the PEG–PS polymer, which can be advantageous in studies of protein binding. The kieselguhr-supported poly-N,N-dimethylacrylamide supports are porous and flexible enough to allow the 27-kDa subtilisin into the interior to give a complete substrate cleavage.[11] The PEGA polymers containing polyamide swell depending on the length and size distribution of PEG cross-linkers 10–60 times the weight of the polymer, allowing even large proteins of 30–70 kDa to diffuse into the polymer.[16] The diffusion of subtilisin is complete within 1 hr, and the diffusion can be followed using fluorogenic quenched peptide substrates. The activities of enzymes are reduced 3- to 20-fold inside the resin as compared to solution; however,

[82] M. Kempe and G. Barany, in "Peptides 1995, Proceedings of the Fourteenth American Peptide Symposium" (P. T. P. Kaumaya and R. S. Hodges, eds.), p. 865. Mayflower Scientific, Kingswinford, UK, 1996.

[83] M. Meldal, F.-I. Auzanneau, O. Hindsgaul, and M. M. Palcic, *J. Chem. Soc., Chem. Commun.*, 1849 (1994).

[84] M. Meldal, *Methods (San Diego)* 6, 417 (1994).

the activity is maintained over long periods owing to the stabilizing effect of the PEG polymer. This was exploited in a quantitative transfer of galactose to a resin-bound glycopeptide using a glycosyltransferase.[83]

Functionalized Surfaces

There are some problems associated with synthesis directly on a functionalized surface. The surface area has to be very large to give useful loadings for preparative synthesis and product analysis. At the same time the topology of the surface has to be regular and contain uniform cavities and pores. This has been realized by application of both synthetic and natural porous or fibrous materials. A second problem is the proper solvation of the synthesized molecules directly linked to a rigid surface. Preferentially the surface should not interact with the molecule synthesized, and the surface should be properly solvated in the solvent used for synthesis. Furthermore, the distance between the attachment sites should be large to allow nonhindered access of reagents and to prevent aggregation on the surface.

Sintered polyethylene particles[85] or polyethylene filters[86] can be functionalized by chromium trioxide and ozone oxidation to generate carboxylic acid groups. This process also increases the porosity and surface area by creation of new cracks and cavities in the polyethylene to yield a loading of 80–90 μmol/g resin.[85] Synthesis performance has been promising using this new resin. The property of the surface may also be changed by derivatization with a polar carbohydrate or PEG spacers for direct biomolecular solid-phase assays.[86] Surface functionalized rigid porous polystyrene[45] has been used for peptide synthesis. Thus, amino functionalized porous beads have been used for synthesis of a pentapeptide in 26% yield. Similarly, a macroporous highly cross-linked methacrylate polymer[87] with phenolic functional groups (0.25 mmol/g) has been obtained by radical polymerization that allowed an oxytocin analog to be synthesized with high yield and purity.[88]

[85] R. M. Cook and D. Hudson, in "Peptides 1995, Proceedings of the Thirteenth American Peptide Symposium" (P. T. P. Kaumaya and R. S. Hodges, eds.), p. 39. Mayflower Scientific, Birmingham, UK, 1996.

[86] J. A. Buettner, D. Hudson, C. R. Johnson, M. J. Ross, and K. Shoemaker, in "Innovation and Perspectives in Solid Phase Synthesis" (R. Epton, ed.), p. 169. Mayflower Worldwide, Birmingham, UK, 1994.

[87] K. J. Shea, G. J. Stoddard, D. M. Shavelle, F. Wakui, and R. M. Choate, Macromolecules 23, 4497 (1990).

[88] L. Andersson and M. Lindquist, in "Peptides 1992, Proceedings of the Twenty-Second European Peptide Symposium" (C. H. Schneider and A. N. Eberle, eds.), p. 265. ESCOM, Leiden, The Netherlands, 1993.

Cellulose fibers from filter paper[63,89-91] or cotton[92-94] are inexpensive sources of large surface polymers and can be modified by O-acylation with protected amino acids. Derivatization of cellulose can be achieved by direct attachment to hydroxyls on the cotton[92] or paper[89,91] to form amino acid esters, or the amino groups can be obtained by cyanoethylation followed by borane reduction to provide amino groups. On the microscopic level, the synthesis is then performed on the surface of the large cellulose fibers. The large surface area of the rigid fibers provides a reasonable yield of functional groups, and synthesis of a large variety of peptides has compared favorably with other support materials. The fibrous cellulose sheets or woven cotton pieces are easy to manipulate and are well suited for multiple or combinatorial synthesis.[95] When fibrous materials are used, the topography of the fiber surface may have a large influence on the performance of synthesis. The natural cellulose fibers from cotton have a smooth surface and can therefore behave differently from paper cellulose fibers obtained by wood processing, and different types of cellulose paper may also behave differently. Peloza (Tessek, Prague) beads of non-cross-linked cellulose fibers have been used with cyanoethylation and reduction to provide a 0.7 mmol/g loading. The synthesis performance on these beads was satisfactory for angiotensins and luteinizing hormone-releasing hormone (LH-RH).[96]

With chitin particles as a solid support[97] the amino function of the 2-amino-2-deoxy-β-D-glucose units can be used as functional groups. The result of synthesis has not been satisfactory compared with results usually obtained on gel-type supports, and high temperature and low loading are required to afford acceptable coupling yields.

The cellulose (and chitin) type of supports cannot be used with strong acids, for example, HF, owing to the cleavage of glycosidic bonds under such conditions. However, they are surprisingly stable to trifluoroacetic acid (TFA) for shorter times, although some insignificant loss (5–10%) of

[89] B. Blankemeyer-Menge and R. Frank, in "Innovation and Perspectives in Solid Phase Synthesis" (R. Epton, ed.), p. 465. SPCC, Birmingham, UK, 1990.
[90] B. Blankemeyer-Menge and R. Frank, Tetrahedron Lett. 29, 5871 (1988).
[91] R. Frank and R. Doering, Tetrahedron 44, 6031 (1988).
[92] J. Eichler, A. Beinert, A. Stierandova, and M. Lebl, Pept. Res. 4, 296 (1991).
[93] J. Eichler, M. Bienert, N. F. Sepetov, P. Stolba, V. Krchnák, O. Smékal, V. Gut, and M. Lebl, in "Innovation and Perspectives in Solid Phase Synthesis" (R. Epton, ed.), p. 337. SPPC, Kingswinford, UK, 1990.
[94] M. Rinnová, J. Jezek, P. Malon, and M. Lebl, Pept. Res. 6, 88 (1993).
[95] R. Frank, in "Innovation and Perspectives in Solid Phase Synthesis" (R. Epton, ed.), p. 509, Mayflower Worldwide, Kingswinford, UK, 1994.
[96] D. R. Englebretsen and D. R. K. Harding, Int. J. Pept. Protein Res. 40, 487 (1992).
[97] W. Neugebauer, R. E. Williams, J.-R. Barbier, and G. Willick, Int. J. Pept. Protein Res. 47, 269 (1996).

peptide is observed because of the ester linkage on prolonged treatment with piperidine or TFA. Sepharose and LH-20[98,99] have been used as supports for solid-phase peptide synthesis. Both are synthetic dextran polymer beads also composed of fibers in a porous network. The fragile nature and acid lability of the glycosidic bonds of the dextran are serious limitations in the use of these materials as solid supports. However, they are perfectly compatible with aqueous buffers and can be used in combined chemo/enzymatic synthesis.[99a]

Protein precipitates of bovine serum albumin (BSA) have been used for synthesis with reasonable results for smaller peptides. The method could not be recommended for general peptide synthesis but is practical for immunological studies where the product can be used directly in immunizations.

Silica gels can be modified with aminopropylsilane to yield a satisfactory loading and synthetic performance for smaller peptides. Furthermore, this support can be used in high yield (60–70%) conversions with enzymes and has been used with glycosyltransferases to afford glycopeptides.[100]

Controlled pore glass (CPG) with a pore diameter of 500 Å has been used in synthesis.[1] The coupling rates during peptide assembly indicated absence of aggregation because the peptides were fixed on the solid surface and unable to rearrange. However, the loadings on CPG were too low to be of practical use, and problems associated with synthesis of long peptides on the rigid surface have not been evaluated.

Supported Gels

When continuous flow synthesis was introduced it was found that the polystyrene and polydimethylacrylamide gels were too fragile for the mechanical wear of the flow pressure, and composites were prepared where a rigid matrix supported the soft gels. Kieselguhr, a porous, inert, inorganic, but highly heterogeneous material prepared as granules from natural sediments, was used as a matrix for dimethylacrylamide polymerization to afford a resin that was stable to flow and had excellent properties for synthesis.[74] However, problems caused by aggregation were often observed during difficult peptide synthesis, and reaction rates were slow compared

[98] G. P. Vlasov, A. Y. Bilibin, N. N. Skvortsova, U. Kalejs, N. Y. Kozhevnikova, and G. Aukone, *in* "Peptides 1994, Proceedings of the Twenty-Third European Peptide Symposium" (H. L. S. Maia, ed.), p. 273. ESCOM, Leiden, The Netherlands, 1995.

[99] A. Orlowska, E. Holodowics, and S. Drabarek, *Pol. J. Chem.* **55,** 2349 (1981).

[99a] O. Blixt and T. Norberg, *J. Carbohydr. Chem.* **16,** 143 (1997).

[100] M. Schuster, P. Wang, J. C. Paulson, and C.-H. Wong, *J. Am. Chem. Soc.* **116,** 1135 (1994).

to the unsupported gel polymers.[32] Today many macroporous materials with uniform sizes and properties have been described.[45,101]

Improved flow properties and homogeneity was obtained by substituting the kieselguhr with highly cross-linked (50%) macroporous polystyrene granules as the rigid matrix.[102–104] In addition to the better flow properties, the poly-N,N-dimethylacrylamide was also grafted to the surface by introduction of aminomethyl groups on the polystyrene and acryloylation.[103] The acryloyl groups form a part of the polydimethylacrylamide to afford a covalently attached gel with less tendency to form fines during mechanical manipulations.[102] In resins with a rigid porous matrix there is a minor problem of having a large polymer surface in contact with the small reactive chemicals (e.g., piperidine), which only slowly are released once they have penetrated into the rigid polymer. To reduce this diffusion of reagents the supporting polymer must be highly cross-linked.[101]

Polypropylene shaped into arrays of pins or small caps attachable to pins[105] has been modified by grafting via oxidation and amination with acrylamide, which has then been included in radical polymerization of a covalently bound polydimethylacrylamide coating. This supported gel on an array of rigid pins is then used in a multiple synthesis of up to 96 peptides simultaneously. The supported gel is flexible and has essentially the same properties as beaded polydimethylacrylamide resins.

Membranes have also been used for supporting gels. Polypropylene membranes have been coated with cross-linked hydroxypropylacrylate for peptide synthesis and compared with conventional gel type supports for a standard synthesis. It is difficult to compare gels with membranes on the basis of the results because the synthesis performed very well in all cases tested.[106] A polytetrafluoroethylene (PTFE) membrane (2000 Å pores) was coated with a 50-Å layer of a swelling polydimethylacrylamide cross-linked with methylenebisacrylamide and aminopropylmethacrylamide.[107] This resin was used for DNA synthesis and compared favorably to aminopropylated CPG (500 Å pores), affording fewer deletion oligomers and increased yield.

[101] M. H. B. Skovby and J. Kops, *J. Appl. Polym. Sci.* **39**, 169 (1990).

[102] D. P. Gregory, N. Bhaskar, R. C. Sheppard and S. Singleton, *in* "Innovation and Perspectives in Solid Phase Synthesis" (R. Epton, ed.), p. 391. Intercept Limited, Andover, UK, 1992.

[103] D. C. Sherrington, *in* "Innovation and Perspectives in Solid Phase Synthesis" (R. Epton, ed.), p. 71. SPCC, Birmingham, UK, 1990.

[104] P. W. Small and D. C. Sherrington, *J. Chem. Soc., Chem. Commun.*, 1589 (1989).

[105] A. M. Bray, N. J. Maeji, and H. M. Geysen, *Tetrahedron Lett.* **31**, 5811 (1990).

[106] S. B. Daniels, M. S. Bernatowicz, J. M. Coull, and H. Köster, *Tetrahedron Lett.* **30**, 4345 (1989).

[107] R. Fitzpatrick, P. Goddard, R. Stankowski, and J. Coull, *in* "Innovation and Perspectives in Solid Phase Synthesis" (R. Epton, ed.), p. 157. Mayflower Worldwide, Birmingham, UK, 1994.

Brush or Film Polymers

Brush polymers are polymers of linear chains grafted onto a rigid surface. They are often composed of polystyrene on a surface of polyethylene, and long chains can be obtained by γ-ray initiated graft polymerization of 440% linear polystyrene in methanol on non-cross-linked polyethylene.[108] The γ rays result in some cross-linking of the polyethylene, thus increasing the rigid character of the surface. By careful functionalization by amidomethylation every tenth phenyl group has been modified to give a loading of 1 mmol/g and excellent properties for synthesis of smaller peptides. The brush polymers have been modified by a second grafting using polyethylene glycol or polyamide attached to the functional groups,[109] affording a polymer that wets more readily in polar solvents. The rigid polyethylene surface imposes a restriction on the swelling of the polystyrene chains to be one-dimensional. Thus, depending on the number of grafts per surface area, there is an inherent problem with the geometry of a brush polymer in the case of synthesis of large and bulky molecules. A certain amount of heterogeneity is also suspected in the region of the polyethylene–polystyrene interface. However, mass transport is rapid because the polystyrene is linear and flexible.

Methods

Suspension Polymerization of Phenolic Polystyrene Resin.[57] Suspension polymerizations are performed in a 7 cm wide by 18 cm high cylindrical polymerization apparatus with four intrusions of 1 cm along the length of the cylinder. A central guided glass stirrer with two sets of 30° tilted parallel stirrer blades provides a even stirring motion throughout the biphasic reaction mixture, and the top has three side arms for addition of reagents and for argon purging. Polymerizations are carried out at 70–80°. Deionized water (400 ml) and suspension agent (1.16 g comprising 400 mg polyvinylpyrrolidine, 680 mg calcium sulfate, and 80 mg calcium orthophosphate) are added to the apparatus, purged with nitrogen, and heated on a water bath at a stirring speed of 400–600 rpm. At temperature equilibrium a mixture of monomers (20 ml comprising 10 mol% 4-acetoxystyrene, 1.3 mol% divinylbenzene, and 88.7 mol% styrene) containing azabisisobutyronitrile (400 mg) is added in one portion and stirring is continued at 80° for 12 hr. The resin is rinsed by agitating with water (3 liters) followed by

---

[108] R. Berg, K. Amdal, W. B. Pedersen, A. Holm, J. P. Tam, and R. B. Merrifield, *J. Am. Chem. Soc.* **111**, 8024 (1989).
[109] A. Capperucci, A. Degl'Innocenti, M. Funicello, G. Mauriello, P. Scafato, and P. Spagnolo, *J. Org. Chem.* **60**, 2254 (1995).

decantation of fines. This is repeated until no more fines can be observed. Decantation is continued with portions (3 liters) of water–methanol (5:1 and 1:1, v/v) and of water–methanol–acetone (7:2:1 and 0:1:1, by volume). The product is washed with methanol–acetone and dried at 50°, affording an 80–90% yield of polymer.

Synthesis of Macromonomer of Partially Acryloylated Acr$_{0.77}$ PEG 1900. Acr$_{0.77}$ PEG 1900 is prepared to be used directly in a 850-ml polymerization apparatus. A solution of acryloyl chloride (1.8 ml, 0.77 equivalents) in CH$_2$Cl$_2$ (30 ml) is added dropwise to a solution of PEG 1900 (58 g, 29 mmol) in CH$_2$Cl$_2$ (40 ml) stirred at 0°. After 1 hr at 20°, concentration of the reaction mixture gives crude Acr$_{0.77}$ PEG 1900 as an opalescent, colorless thick oil.

Inverse Suspension Polymerization Procedure. A mixture n-heptane–carbon tetrachloride (6:4, v/v, 470 ml) is purged with argon for 5 min in the polymerization reactor described above. The solution is warmed to 70° and the stirring speed adjusted to 650 rpm. During this period a mixture of the Acr$_{0.77}$ PEG 1900 monomers (60 g) in water (95 ml) is purged with argon. N,N-Dimethylacrylamide (5 or 10 g) is added, and after a further 5 min purging, solutions of sorbitan monolaurate (500 mg) in DMF (2.5 ml) and ammonium peroxodisulfate (705 mg) in water (2.5 ml) are added to the mixture of monomers, which is then rapidly poured into the polymerization reactor. After 2 min, TEMED (N,N,N',N'-tetramethylethylenediamine, 2 ml) is added to the mixture, and the sticky point is reached within 30 sec. The reaction is allowed to proceed under 650 rpm stirring at 70–75° for 2–3 hr. The cooled resin is filtered off and washed twice with 2 volumes of ethanol then with several volumes of water, and the wet resin is passed through a steel net (1 mm^2 holes). It is transferred back to the filter and washed three times with 2 volumes of ethanol. It is washed with 2 volumes of CH$_2$Cl$_2$, aspirated dry, and dried successively under low vacuum (water pump) and high vacuum (freeze-dryer) for a period of 2 days to afford 36 g of dry resin (90% of the theoretical yield).

Determination of Swelling. Swelling of the resin is most conveniently and reproducibly performed in a 2-ml disposable plastic syringe with a flat polypropylene piston and fitted with a 2-mm Teflon filter. Dry resin (~100 mg) is measured into the syringe, and the syringe is dried overnight on a lyophilizer. The dry resin is weighed and swelled in the appropriate solvent. After 30 min the piston is inserted, and a force of about 50 psi is supplied to the piston with a 10-lb weight. The volume of the compressed resin is measured and the piston released by withdrawing the weight until it lifts from the resin bed. The volume of the uncompressed resin is measured in milliliters per gram of dry resin.

Determination of Loading. Swelling of the resin is most conveniently and reproducibly performed in a 2-ml disposable plastic syringe fitted with a 2-mm Teflon filter. Dry resin ($\sim$10 mg) is measured into the syringe, and the syringe is dried overnight on a lyophilizer. The dry resin is weighed and swelled in DMF. The resin is reacted with Fmoc-Gly-OPfp ($\sim$5 equivalents) for 2 hr, and the excess reagent washed out carefully with 20 volumes of DMF. The resin is aspirated dry, and the Fmoc group is cleaved with 20% piperidine in DMF (1.0 ml for 10 min). The cleavage mixture is eluted and washed out with DMF. The cleavage mixture and washings are diluted to 25 ml with DMF, and the absorbance is measured and compared with the absorption of a standard solution. The absorbance is proportional to the concentration of dibenzfulvene–piperidine adduct formed in the deprotection reaction.

Summary

Many supports including composite materials and functionalized surfaces are available for solid-phase synthesis. In the process of selecting the proper support it is important to consider the optimal performance during solid-phase synthesis. For most purposes the mechanically stable beaded gel resins are preferred. These resins are homogeneous, and the loading and physical and chemical properties can easily be varied. Optimal properties have been obtained by radical polymerization of end group acryloylated long-chain polyethylene glycols. However, polystyrene resins or amide bond free PEG-based resins[41] may be more suited for general organic synthesis where reactivity of radicals, carbenes, carbanions, carbenium ions, or strong Lewis acids have to be considered.[8–10,110,111] Loading of the resins can have a dramatic effect on the outcome of a synthesis and has to be considered separately for each synthesis. Synthesis of long peptides with 50–100 amino acids imposes completely different requirements on the performance, swelling, and loading than a large-scale synthesis of, for example, the pentapeptide enkephalin.

Automated multiple synthesizers constructed for columns of beaded gel or composite supports are available from many suppliers. It is therefore expected that the optimization of support properties will continue in order to meet new synthetic challenges.

In the synthesis for solid-phase screening of binding of biomolecules to ligands directly on the resin beads, it is an advantage if the resin is not

[110] L. Yan, C. M. Taylor, J. Goodnow, and D. Kahne, *J. Am. Chem. Soc.* **116,** 6953 (1994).
[111] H. Paulsen, A. Schleyer, N. Mathieux, M. Meldal, and K. Bock, *J. Chem. Soc., Perkin Trans. 1,* **281** (1997).

permeable to the biomolecule so unbound molecules can easily be removed by washing. This is the case with polystyrene-based resins, but they do, however, often show nonspecific adhesion of proteins owing to the hydrophobic character of the polystyrene. Modification of the functional groups of polystyrene with polyethylene glycol as spacers for synthesis of the binding ligands can increase the available ligand concentration on the bead surface and eliminate most of the nonspecific adhesion. In contrast to binding studies, solid-phase assays of enzymes require beads that are permeable to the enzyme, as the progress of reaction can be followed and the product of reaction analyzed. The available amount on the surface of the polystyrene-based beads ($\sim$0.3%) is not enough for product analysis. Therefore, in the case of enzyme assays, highly swelling permeable PEG-based gel resins or functionalized surfaces of a polar and porous matrix are preferred.

[7] Coupling Reagents and Activation

By Fernando Albericio and Louis A. Carpino

The controlled formation of a peptide bond (the so-called coupling reaction) between two amino acids requires activation of the carboxyl group of one for facile reaction with the amino group of the other. The process of activation is the aspect of peptide synthesis that has been most extensively developed in recent years. An essential feature of all coupling methods is that, in addition to giving the peptide bond in good yield, the configurational integrity of the carboxylic component must be maintained. This duality, good yield and absence of racemization, is often difficult to achieve, because the best methods usually involve conversion of the acid to a derivative bearing a good leaving group. Such leaving groups tend to increase the acidity of the α-proton and favor formation of an oxazolone (**4**). Loss of configuration is especially prominent if oxazolone formation occurs, but it also can occur at the stage of the activated carboxyl derivative (Fig. 1).

In stepwise solid-phase peptide synthesis (SPPS)[1,2] the problem of racemization is less dramatic than for other strategies. The N^α-protecting group of the amino acid to be coupled (the one with the carboxylic function activated) is normally a urethane function such as 9-fluorenylmethoxycar-

---

[1] R. B. Merrifield, *Angew. Chem. Int. Ed. Engl.* **24**, 799 (1985).
[2] G. B. Fields, Z. Tian, and G. Barany, *in* "Synthetic Peptides: A User's Guide" (G. A. Grant, ed.), p. 77. Freeman, New York, 1992.

                    0076-6879/97 $25.00

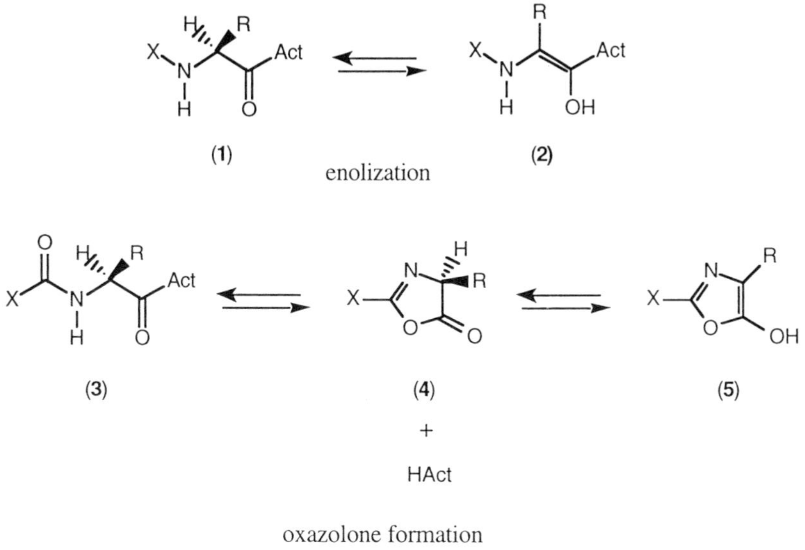

(1)   enolization   (2)

(3)   (4)   (5)

+

HAct

oxazolone formation

FIG. 1. Mechanisms of racemization.

bonyl (Fmoc) or *tert*-butoxycarbonyl (Boc). The presence of an alkoxy group (**1–5**, X = *O*-alkyl) in these carbamates reduces the tendency to give an oxazolone, and if formed such oxazolones are less sensitive to racemization.[3] Furthermore, an important feature of the solid-phase approach is the general use of large excesses of reagents. The coupling reactions are therefore generally faster than in solution, thus minimizing loss of configuration. Although it is widely accepted that for stepwise SPPS the risk of racemization is practically nil, this is not completely true, and the possibility of loss of configuration should always be kept in mind, especially for certain sensitive amino acids such as Cys or His.

In this article we discuss the coupling methods currently most commonly used in SPPS. The reader should also consult the bibliographies found in the references cited; a review of other methods such as solution techniques are outside the scope of this article. The two main classes of coupling technique involve (1) those that require *in situ* activation of the carboxylic acid and (2) those that depend on an activated species that has previously been prepared, isolated, purified, and characterized.

---

[3] D. S. Kemp, *in* "The Peptides: Analysis, Synthesis, Biology" (E. Gross and J. Meienhofer, eds.), Vol. 1, p. 315. Academic Press, New York, 1979.

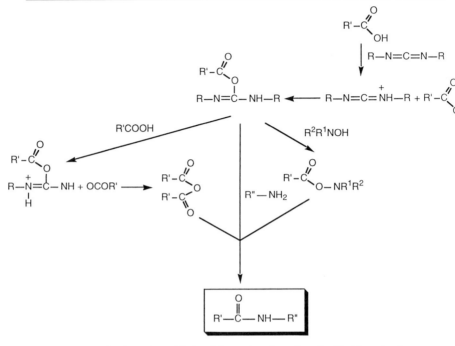

Fig. 2. Mechanism of peptide bond formation through carbodiimide activation.

Coupling Reagents

The most widely used coupling reagents are carbodiimides, on the one hand, and phosphonium and aminium salts, on the other hand.

Carbodiimides

Reaction of a protected carboxylic acid with a carbodiimide[4,5] is believed to involve a labile *O*-acylisourea, which reacts with the amino component to give the corresponding amide (Fig. 2). If 2 equivalents of carboxylic acid are used, the intermediate *O*-acylisourea reacts with the second equivalent of acid to give the corresponding symmetric anhydride. If the activation

[4] J. C. Sheehan and G. P. Hess, *J. Am. Chem. Soc.* **77,** 1067 (1955).
[5] D. H. Rich and J. Singh, *in* "The Peptides: Analysis, Synthesis, Biology" (E. Gross and J. Meienhofer, eds.), Vol. 1, p. 241. Academic Press, New York, 1979.

process is carried out in the presence of a hydroxylamine derivative [1-hydroxybenzotriazole (HOBt, **6**)[6] or 1-oxo-2-hydroxydihydrobenzotriazine (HODhbt, **7**)[7] (R_2R_1NOH in Fig. 2)] an active ester is obtained. Any one of the three active species, *O*-acylisourea, symmetric anhydride, or active ester, is an excellent acylating reagent. The main advantage of using *N*-hydroxy compounds as additives is to reduce loss of configuration at the carboxylic acid residue. In addition, by carrying out the coupling of side-chain unprotected Gln and Asn with carbodiimides in the presence of HOBt, no dehydration of the carboxamide residue occurs. The preparation of ODhbt esters is accompanied by formation of a by-product, 3-(2-azido-benzoloxy)-4-oxo-3,4-dihydro-1,2,3-benzotriazine, which itself can react with the amino group to terminate chain growth.[7]

7-Aza-1-hydroxybenzotriazole (HOAt, **8**)[8] has been described as being superior to HOBt as an additive for both solution and solid-phase syntheses. HOAt enhances coupling rates and reduces the risk of racemization,[9,10] possibly because it incorporates into the HOBt structure a nitrogen atom strategically placed at position 7 of the aromatic system. Incorporation of a nitrogen atom in the benzene ring has two consequences. First, the electron-withdrawing influence of a nitrogen atom (regardless of its position) effects stabilization of the leaving group, leading to greater reactivity. Second, placement specifically at the 7 position makes feasible a classic neighboring group effect (Fig. 3, **10**),[8] which can both speed up the reactivity and reduce loss of configuration. The corresponding 4-isomer (**9**), lacking the ability to take part in such a neighboring group effect, has no influence on the extent of stereomutation during the segment coupling reaction relative to HOBt.

For carrying out SPPS by the Boc/benzyl (Bzl) strategy, among carbodiimides (Fig. 4) the *N,N'*-dicyclohexyl derivative (DCC, **11**) is the most widely used, as the by-product *N,N'*-dicyclohexylurea (DCU) can be eliminated from the reaction vessel owing to its solubility during removal of the Boc group with trifluoroacetic acid (TFA). The high solubility in *N,N*-dimethylformamide (DMF) of the urea derived from *N,N'*-diisopropylcar-bodiimide (DIPCDI, **12**) makes it the carbodiimide of choice when the Fmoc/*tert*-butyl (*t*Bu) strategy is used. In fact, use of DCC is not possible

[6] W. König and R. Geiger, *Chem. Ber.* **103,** 788 (1970).
[7] W. König and R. Geiger, *Chem. Ber.* **103,** 2034 (1970).
[8] L. A. Carpino, *J. Am. Chem. Soc.* **115,** 4397 (1993).
[9] L. A. Carpino, A. El-Fahan, C. A. Minor, and F. Albericio, *J. Chem. Soc., Chem. Commun.,* 201 (1994).
[10] L. Carpino, A. El-Faham, and F. Albericio, *Tetrahedron Lett.* **35,** 2279 (1994).

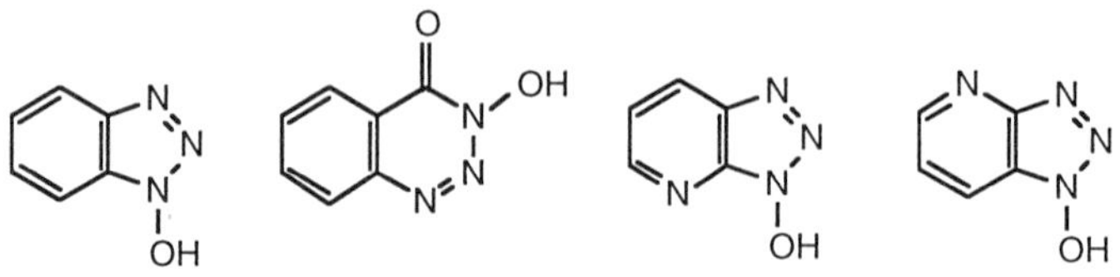

HOBt (6)   HODhbt (7)   HOAt (8)   4-HOAt (9)

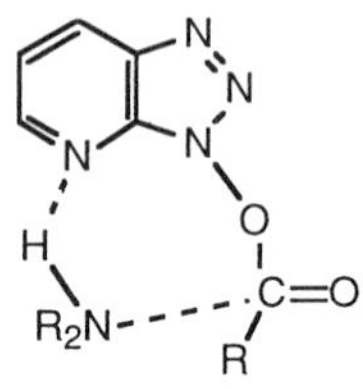

(10)

FIG. 3. Structures (6–9) of additives for SPPS. Assisted basic catalysis during the coupling of oxy-7-azobenzotriazole (OAt) esters (10).

in this case, because DCU would plug the frit of either the reaction vessel or the column in the case of batch or continuous-flow syntheses, respectively. Finally, 1-ethyl-3-(3'-dimethylaminopropyl)carbodiimide hydrochloride (EDC, 13), the urea by-product of which is soluble in aqueous solvent mixtures, is mainly used for syntheses carried out in solution.

Preformed symmetric anhydrides are more often used for Boc/Bzl-based procedures than for Fmoc/tBu-based syntheses. These highly reactive intermediates are most efficiently prepared from the protected amino acid (2 equivalents) and carbodiimide (1 equivalent) in neat dichloromethane

DCC (11)   DIPCDI (12)

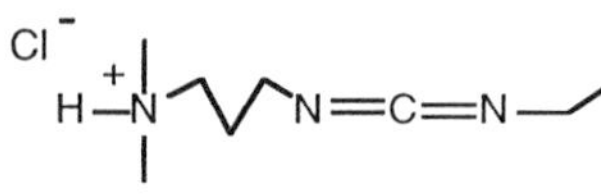

EDC (13)

FIG. 4. Structures (11–13) of common carbodiimides.

(DCM).[5,11] For Fmoc-amino acids that are not totally soluble in DCM, mixtures with DMF may be used.[2]

Carbodiimide-mediated couplings are usually carried out with preactivation of the protected amino acid (see below), at either 25° or 4°. If desired, after filtration of DCU and evaporation of DCM, DMF may be used as the coupling medium. The activation process is slower in DMF. For large-scale syntheses (~10 mmol of peptide), it is mandatory to preactivate at 4°, because the exothermic nature of the reaction increases the risk of racemization, even when urethane-type protecting groups are used.

Carbodiimides as well as other coupling reagents should be treated with care, as they are acute skin irritants for susceptible individuals. Thus, it is recommended that they be manipulated in a well-ventilated hood, using glasses, gloves, and, if possible, a face mask. DCC, because of its low melting point, can be handled as a liquid by gentle warming of the reagent container.[12] HOBt normally crystallizes with one molecule of water. Use of the hydrated form is perfectly satisfactory, but if the anhydrous material is needed the dehydration should be carried out with care. Heating of HOBt or HOAt above 180° can cause rapid exothermic decomposition.[13]

Phosphonium and Aminium Salts

Although Kenner and co-workers[14] were the first to describe the use of acylphosphonium salts as coupling reagents, these species (Fig. 5) have only been widely adopted after the extensive studies of Castro and Coste,[15,16] who described the applicability of (benzotriazol-1-yloxy)tris(dimethylamino)phosphonium hexafluorophosphate (BOP, **14**) and (benzotriazol-1-yloxy)tris(pyrrolidino)phosphonium hexafluorophosphate (PyBOP, **15**). These coupling reagents incorporate in their structure an equivalent of HOBt, and the final reactive species are the corresponding oxybenzotriazole (OBt) esters. Formation of the OBt ester is achieved in the presence of an equivalent of a tertiary base such as *N,N*-diisopropylethylamine (DIEA) or *N*-methylmorpholine (NMN).[17] The presence of an extra equivalent of HOBt can accelerate the coupling process.[18] The precise nature of the

---

[11] R. B. Merrifield, L. D. Vizioli, and H. G. Boman, *Biochemistry* **21**, 5020 (1982).
[12] J. S. Albert and A. D. Hamilton, *in* "Encyclopedia of Reagents for Organic Synthesis" (L. A. Paquette, ed.), Vol. 3, p. 1751. Wiley, Chichester, UK, 1995.
[13] S. A. Kates, F. Albericio, and L. A. Carpino, *in* "Encyclopedia of Reagents for Organic Synthesis" (L. A. Paquette, ed.), Vol. 4, p. 2784. Wiley, Chichester, UK, 1995.
[14] G. Gawne, G. Kenner, and R. C. Sheppard, *J. Am. Chem. Soc.* **91**, 5669 (1969).
[15] B. Castro, J. R. Dormoy, G. Evin, and C. Selve, *Tetrahedron Lett.*, 1219 (1975).
[16] J. Coste, D. Le-Nguyen, and B. Castro, *Tetrahedron Lett.* **31**, 205 (1990).
[17] B. Castro, J.-R. Dormoy, G. Evin, and C. Selve, *J. Chem. Res. (S)* 182 (1977).
[18] D. Hudson, *J. Org. Chem.* **53**, 617 (1988).

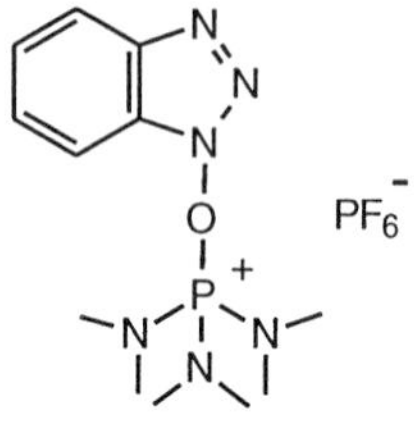

BOP (14) 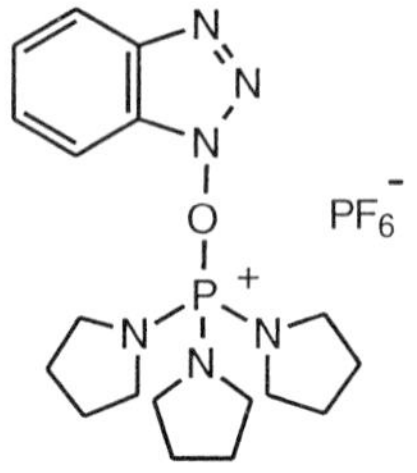   PyBOP (15)

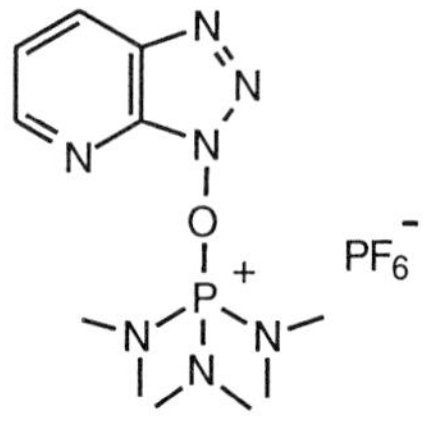

AOP (16) 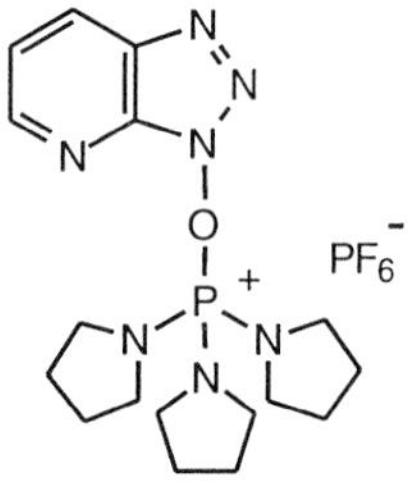   PyAOP (17)

Fig. 5. Structures (14–17) of phosphonium-type coupling reagents.

intermediates involved in the use of the BOP reagent is not known. Some controversy has arisen regarding the possible intermediacy of an acylphosphonium salt and its lifetime. Kim and Patel[19] reported that such intermediates can exist at $-20°$ in the absence of excess HOBt. However, Coste and Campagne[20] suggested that this species is very unstable and even at low temperature undergoes conversion to the active ester. Fmoc derivatives of Tyr and Thr have been incorporated through a BOP-based protocol without side-chain protection.[21,22]

Phosphonium salts derived from HOAt such as (7-azabenzotriazol-1-yloxy)tris(dimethylamino)phosphonium hexafluorophosphate (AOP, 16) and (7-azabenzotriazol-1-yloxy)tris(pyrrolidino)phosphonium hexafluorophosphate (PyAOP, 17) have also been prepared and are generally more efficient than BOP and PyBOP.[9,23] The pyrrolidino derivatives (PyBOP

[19] M. H. Kim and D. V. Patel, *Tetrahedron Lett.* **35**, 5603 (1994).
[20] J. Coste and J.-M. Campagne, *Tetrahedron Lett.* **36**, 4253 (1995).
[21] A. Fournier, C.-T. Wang, and A. M. Felix, *Int. J. Pept. Protein Res.* **31**, 86 (1988).
[22] A. Fournier, W. Danho, and A. M. Felix, *Int. J. Pept. Protein Res.* **33**, 133 (1989).
[23] S. A. Kates, E. Diekmann, A. El-Faham, L. W. Herman, D. Ionescu, B. F. McGuinness, S. A. Triolo, F. Albericio, and L. A. Carpino, *in* "Techniques in Protein Chemistry VII" (D. R. Marshak, ed.), p. 515. Academic Press, New York, 1996.

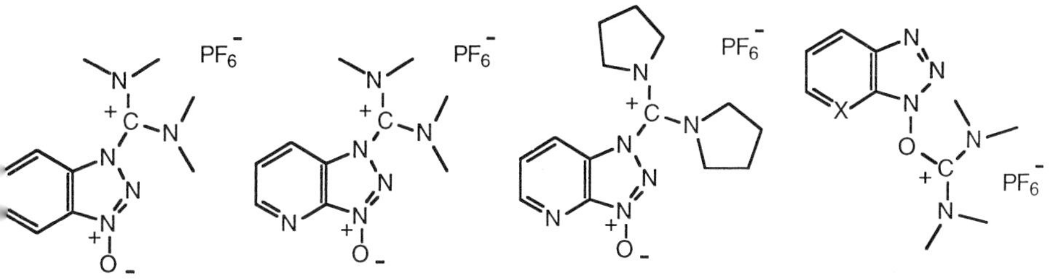

HBTU (18)   HATU (19)   HAPyU (20)   (21), X = CH, N

FIG. 6. Structures (**18–21**) of aminium-type coupling reagents.

and PyAOP) are slightly more reactive than the dimethylamino derivatives (BOP and AOP) and in the activation step do not liberate, as do the latter two reagents, hexamethylphosphoric triamide (HMPA), a compound that has been classified as a potential human carcinogen.[24]

The preparation and use in peptide synthesis of analogs of phosphonium salts (Fig. 6) such as N-[(1H-benzotriazol-1-yl)(dimethylamino) methylene]-N-methylmethanaminium hexafluorophosphate N-oxide (HBTU) (**18**) and N-[(dimethylamino)-1H-1,2,3-triazolo[4,5-b]pyridin-1-yl-methylene]-N-methylmethanaminium hexafluorophosphate N-oxide (HATU) (**19**) bearing a positive carbon atom in place of the phosphonium residue have also been reported. Although initially assigned a uronium-type structure,[8,9,25,26] presumably by analogy with the corresponding phosphonium salts, more recently it has been determined by X-ray analysis that HBTU (**18**), HATU (**19**), and 1-(1-pyrrolidinyl-1H-1,2,3-triazolo [4,5-b]pyridin-1-ylmethylene)pyrrolidinium hexafluorophosphate N-oxide (HAPyU, **20**) crystallize as aminium salts (guanidinium N-oxides) rather than the corresponding uronium salts (**21**).[27,28] Nuclear magnetic resonance (NMR) studies in the case of HAPyU show that the same structure is found in solution.[28] A possible rationale for the greater stability of the N- over the O-forms can be found in the concept of Y-aromaticity.[29] TBTU and

[24] R. R. Dykstra, *in* "Encyclopedia of Reagents for Organic Synthesis" (L. A. Paquette, ed.), Vol. 4, p. 2668. Wiley, Chichester, UK, 1995.

[25] V. Dourtoglou, J.-C. Ziegler, and B. Gross, *Tetrahedron Lett.,* 1269 (1978).

[26] R. Knorr, A. Trzeciak, W. Bannwarth, and D. Gillessen, *Tetrahedron Lett.* **30,** 1927 (1989).

[27] I. Abdelmoty, F. Albericio, L. A. Carpino, B. M. Foxman, and S. A. Kates, *Lett. Pept. Sci.* **1,** 52 (1994).

[28] P. Henklein, B. Costisella, V. Wray, T. Domke, L. A. Carpino, A. El-Faham, S. A. Kates, A. Abdelmoty, and B. M. Foxman, *in* "Peptides 1996, Proceedings of the Twenty-fourth European Peptide Symposium" (R. Epton, ed.), in press. Mayflower Worldwide, Birmingham, UK, 1997.

[29] P. Gund, *J. Chem. Ed.* **49,** 100 (1972).

TATU, the tetrafluoroborate salts related to **18** and **19**, have also been prepared and used in SPPS. These derivatives are more soluble in DMF than the hexafluorophosphate salts.

The superiority of the 7-aza-based coupling reagents (HOAt, PyAOP, HATU, and HApyU) has been demonstrated by SPPS of the decapeptide model ACP(65–74) (fragment of acyl carrier protein, H-Val-Gln-Ala-Ala-Ile-Asp-Tyr-Ile-Asn-Gly-NH$_2$), which exhibits difficult couplings at Ile-72, Ile-69, Val-75, and Asn-73. For testing purposes, peptide elongation was carried out using deliberately reduced coupling times and excesses of reagents to magnify differences among various coupling techniques.[9] It was found that HOAt/carbodiimide, PyAOP, HATU, and HApyU are clearly superior to HOBt/carbodiimide, PyBOP, and HBTU; and that phosphonium and aminium salts are preferred to carbodiimide/active ester methods; and, finally, that addition of HOXt to HXTU (X = A, B) couplings does not significantly improve the coupling yields, with the exception of the coupling of Asn(Trt) to C-terminal Gly (Trt is the trityl or triphenylmethyl group). HOAt-derived coupling reagents are also very favorable for the coupling of hindered amino acids as shown by excellent syntheses of peptides containing consecutive *N*-methylamino acids or dimethylglycine (Aib) or diethylglycine (Deg) units.[9,30,31] Examples include [MeLeu^1]cyclosporin A (CsA)[32] and alamethicinamide.[31] In the latter, results obtained with HATU are only slightly inferior to those obtained via isolated acid fluorides, which represent the reagents of choice for such highly hindered amino acids.[33]

A special case involves deprotection of the second residue and the coupling of the third when hydroxymethylbenzyl resins are used. These resins are prone to lead to substantial amounts of diketopiperazine (DKP) formation during deprotection of the second amino acid of the sequence.[34,35] This intramolecular cyclization will be favored by the presence of Pro or Gly in the first two positions of the sequence. For Boc chemistry, the formation of DKP can be diminished using coupling methods in which neutralization of the dipeptide–resin is carried out *in situ* during coupling

[30] Y. M. Angell, C. García-Echeverría, and D. H. Rich, *Tetrahedron Lett.* **35**, 5981 (1994).
[31] F. Albericio, I. Abdelmoty, J. M. Bofill, L. A. Carpino, A. El-Faham, B. M. Foxman, M. Gairí, E. Giralt, G. W. Griffin, S. A. Kates, P. Lloyd-Williams, C. A. Minor, L. M. Scarmoutzos, H. N. Shroff, S. Triolo, and H. Wenschuh, *in* "Peptides 1994. Proceedings of the Twenty-Third European Peptide Symposium" (H. L. S. Maia, ed.), p. 209. ESCOM, Leiden, The Netherlands, 1995.
[32] Y. Angell, T. L. Thomas, G. R. Flentke, and D. H. Rich, *J. Am. Chem. Soc.* **117**, 7279 (1995).
[33] H. Wenschuh, M. Beyermann, H. Haber, J. K. Seydel, E. Krause, M. Bienert, L. A. Carpino, A. El-Faham, and F. Albericio, *J. Org. Chem.* **60**, 405 (1995).
[34] B. F. Gisin and R. B. Merrifield, *J. Am. Chem. Soc.* **94**, 3102 (1972).
[35] E. Giralt, R. Eritja, and E. Pedroso, *Tetrahedron Lett.* **22**, 3779 (1981).

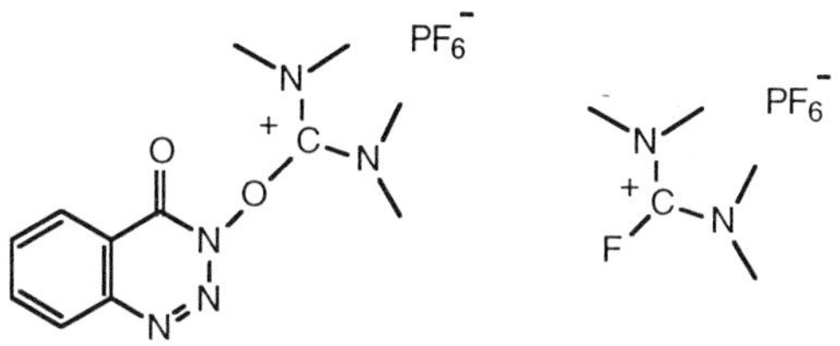

HDTU (22)   TFFH (23)

FIG. 7. Structure of HDTU (22) and TFFH (23).

of the third amino acid, rather than by neutralization in a separate wash step by means of a tertiary amine. A convenient protocol involves removing the Boc group with TFA–DCM and carrying out coupling of the next amino acid via PyBOP or PyAOP in the presence of DIEA.[36] For the Fmoc/tBu strategy, the second amino acid has to be introduced using N^α-Trt protection, since the trityl group can be selectively removed with a very dilute acid solution (0.2–1% TFA in DCM) in the presence of tBu type protecting groups.[37] Although the formation of DKP is also catalyzed by acids, the extent of this side reaction is more severe in the presence of bases. Therefore, more care must be taken when using the Fmoc/tBu strategy, which involves treatment with a secondary amine.

O-(3,4-Dihydro-4-oxo-1,2,3-benzotriazin-3-yl)-1,1,3,3-tetramethyl-uronium hexafluorophosphate (HDTU, 22, Fig. 7), a uronium salt derived from HODhbt, has also been prepared.[26,38] Although for solution syntheses HDTU has given results comparable to HATU, in the solid-phase mode the performance of HDTU is inferior to that of HATU or HAPyU. Furthermore, the appearance of extra peaks in the high-performance liquid chromatography (HPLC) traces is indicative of the formation of triazine side products.[38]

Among the most reactive of the common coupling reagents are the preformed Fmoc amino acid fluorides (see below). Rather than use an isolated acid fluoride, a more convenient method of making use of these efficient reagents is to generate them *in situ* via the aminium reagent tetramethylfluoroformamidinium hexafluorophosphate (TFFH, 23).[39,40] With TFFH even His and Arg, which are the only two proteinogenic amino acids that cannot be converted to shelf-stable Fmoc-protected amino acid

---

[36] M. Gairí, P. Lloyd-Williams, F. Albericio, and E. Giralt, *Tetrahedron Lett.* **31,** 7363 (1990).
[37] J. Alsina, E. Giralt, and F. Albericio, *Tetrahedron Lett.* **37,** 4195 (1996).
[38] L. A. Carpino, A. El-Faham, and F. Albericio, *J. Org. Chem.* **60,** 3561 (1995).
[39] L. A. Carpino and A. El-Faham, *J. Am. Chem. Soc.* **117,** 5401 (1995).
[40] L. A. Carpino, M. Beyermann, H. Wenschuh, and M. Bienert, *Acc. Chem. Res.* **29,** 268 (1996).

fluorides, can be routinely coupled in this form. In some cases for these two amino acids as well as for Asn better results are obtained if 1 equivalent of HOAt is present during the coupling process.[9,39]

The use of these various onium salts requires careful attention to the tertiary base used and the preactivation time. Although in the case of some automatic synthesizers the preactivation time comes dictated by the instrument, in others and for manual syntheses it can be modulated. For onium salts incorporating HOAt, the activation of ordinary amino acids gives the corresponding OAt esters almost instantly. Thus, in such cases, the preactivation time should be kept to a minimum, because on standing alone the activated species can give rise to several side reactions including racemization and formation of δ-lactam (Arg) or cyano derivatives (Asn or Gln) or α-aminocrotonic acid (Thr) (see below). The same consideration applies to coupling reagents that incorporate HOBt. For TFFH longer preactivation times may be needed depending on the specific amino acid and activating base.

Regarding the use of a base during the coupling process, for those coupling reactions that involve amino acids which are not likely to lose their configuration (all except Cys and His), the reactions are carried out in the presence of 1.5–2 equivalents of a tertiary amine such as DIEA or NMM. For the coupling of protected peptides, where resistance to conversion to oxazolone does not apply, and for the coupling of Cys and His, the use of only 1 equivalent of a weaker or more hindered base is to be recommended; for such systems 2,4,6-trimethylpyridine (TMP) or collidine and the more basic 2,6-di-*tert*-butyl-4-(dimethylamino)pyridine [DB(DMAP)] are very promising.[41,42]

Related reagents such as 2-(benzotriazol-1-yl)oxy-1,3-dimethylimidazolidinium hexafluorophosphate (BOI)[43] and 2-[2-oxo-1(2H)-pyridyl]-1,1,3,3-bispentamethyleneuronium tetrafluoroborate (TOPPipU)[44] have also been used successfully in SPPS. However, other reagents such as bromotris(pyrrolidino)phosphonium hexafluorophosphate (PyBroP)[45] and N,N'-bis(2-oxo-3-oxazolidinyl)phosphinic acid chloride (BOP-Cl),[46] which perform well in solution, are not suitable for SPPS.

[41] L. A. Carpino and A. El-Faham, *J. Org. Chem.* **59**, 685 (1994).

[42] L. A. Carpino, D. Ionescu, and A. El-Faham, *J. Org. Chem.* **61**, 2460 (1996).

[43] Y. Kiso, Y. Fujiwara, T. Kimura, A. Nishitani, and K. Akaji, *Int. J. Pept. Protein Res.* **40**, 308 (1992).

[44] P. Henklein, M. Beyermann, M. Bienert, and R. Knorr, in "Proceedings of the Twenty-first European Peptide Symposium" (E. Giralt and D. Andreu, eds.), p. 67. ESCOM Leiden, The Netherlands, 1991.

[45] J. Coste, E. Frérot, and P. Jouin, *J. Org. Chem.* **59**, 2437 (1994).

[46] J. Diago-Meseguer, A. L. Palomo-Coll, J. R. Fernandez-Lizarbe, and A. Zugaza-Bilbao, *Synthesis*, 547 (1980).

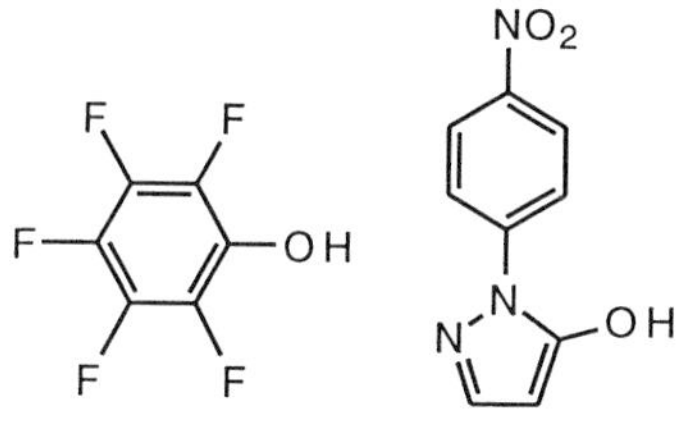

$$NO_2$$

HOPfp (**24**)   Hpp (**25**)

FIG. 8. Precursors (**24** and **25**) of commonly used active esters.

Active Species

The major difficulty in the assembly of peptides via isolated intermediates is the necessity of having on hand the full range of such activated reagents. However, only protected amino acids themselves are required when coupling reagents are used. The main advantages of using activated species are two: (1) the high reactivity of the isolated intermediates and (2) the simplicity of the by-products released (e.g., carbon dioxide and fluoride ion). The most often used active species are active esters, urethane-protected α-amino acid N-carboxyanhydrides (UNCAs), and acid fluorides.

The commonly used active esters (Fig. 8) are those derived from pentafluorophenol (HOPfp, **24**) and HODhbt (**7**), available by reaction of the protected amino acids, the hydroxy compound, and DCC.[7,47] Pentafluorophenyl esters can also be prepared from pentafluorophenyl trifluoroacetate.[48] During the preparation of HODhbt esters it is possible to remove the triazine side product; therefore, the use of such preformed esters is safe and free of undesired side reactions, in contrast to the situation involving HODhbt as additive. The OBt and OAt esters are less stable than the OPfp and ODhbt esters and are not generally isolated. On the other hand, the OBt esters of N-trityl amino acids have been used in the isolated form.[49] Pentafluorophenyl esters react somewhat slowly, although the addition of HOAt speeds up their reactivity.[8] Use of pentafluorophenyl esters allows incorporation of Asn without protection of the side-chain amide function.[50] Fmoc derivatives of Tyr and Ser have also been incorporated as preformed active esters without side-chain protection.[51,52] Owing to their solubility,

[47] L. Kisfaludy and I. Schön, *Synthesis,* 325 (1983).
[48] M. Green and J. Berman, *Tetrahedron Lett.* **31**, 5851 (1990).
[49] K. Barlos, D. Papaioannou, and D. Theodoropoulos, *Int. J. Pept. Protein Res.* **23**, 300 (1984).
[50] H. Gausepohl, M. Kraft, and R. W. Frank, *Int. J. Pept. Protein Res.* **34**, 287 (1989).
[51] L. Otvos, I. Elekes, and V. M.-Y. Lee, *Int. J. Pept. Protein Res.* **34**, 129 (1989).
[52] C. G. Fields, G. B. Fields, R. L. Noble, and T. A. Cross, *Int. J. Pept. Protein Res.* **33**, 298 (1989).

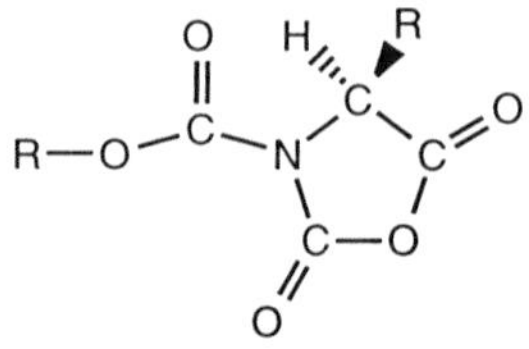

UNCA (26)

Fig. 9. Structure of UNCA (26).

esters derived from 1-(4′-nitrophenyl)-pyrazolin-5-one (Hpp, **25**)[53] have been used successfully for the solid-phase incorporation of poorly soluble biomolecules such as biotin and fluorescein.[54]

The main advantage of UNCAs (**26**, Fig. 9)[55] relative to classic N-carboxyanhydrides or Leuchs' anhydrides[56] is their greater stability and lack of sensitivity toward polymerization. UNCAs are conveniently prepared by first phosgenation of bistrimethylsilyl amino acids and then reaction with Fmoc-Cl or BocON in the presence of NMM or pyridine, respectively.[57] UNCAs are very soluble in most organic solvents, and they are most reactive in DMF. Their reactivity is comparable to BOP- and HBTU-style reagents. The fact that only carbon dioxide is released on coupling is a distinct advantage relative to other coupling methods.[57,58]

The practical utility of protected amino acid chlorides[40] is restricted to bifunctional amino acids with Fmoc protection for the N^α-amino function or trifunctional amino acids whose side chains are benzyl-protected because of the incompatibility of Boc and *tert*-butyl-based protecting groups with the conditions used to prepare these derivatives. Such Fmoc-amino acid chlorides can be prepared from the Fmoc-amino acids and thionyl chloride and are storable indefinitely in a dry atmosphere.[59] Fmoc derivatives of Tyr, Thr, and Ser with *tert*-butyl protected side chains can be prepared,

[53] D. Hudson, *Pept. Res.* **3**, 51 (1990).

[54] J. Kremsky, M. Pluskal, S. Casey, H. Perry-O'Keefe, S. A. Kates, and N. D. Sinha, *Tetrahedron Lett.* **37**, 4313 (1996).

[55] W. D. Fuller, M. P. Cohen, M. Shabankareh, R. K. Blair, and M. Goodman, *J. Am. Chem. Soc.* **112**, 7414 (1990).

[56] H. Leuchs, *Ber. Dtsch. Chem. Ges.* **39**, 857 (1906).

[57] W. D. Fuller, M. Goodman, F. Naider, and Y.-F. Zhu, *Biopolymers (Pept. Sci.)* **40**, 183 (1996).

[58] J.-A. Fehrentz, C. Genu-Dellac, H. Amblard, F. Winternitz, A. Loffet, and J. Martinez, *J. Pept. Sci.* **1**, 124 (1995).

[59] L. A. Carpino, B. J. Cohen, K. E. Stephens, S. Y. Sadat-Aalee, J.-H. Tien, and D. C. Langridge, *J. Org. Chem.* **51**, 3734 (1986).

but loss of *t*Bu was observed on storage.[60] The coupling of these derivatives requires the presence of a hydrogen chloride acceptor. In the presence of a tertiary amine such as DIEA or NMM the corresponding oxazolone is formed, which, although also a coupling reagent, is less reactive than the acyl chloride.[60,61] If a hindered base such as 2,6-di-*tert*-butylpyridine is used as hydrogen chloride scavenger, only a small amount of oxazolone is formed.[60] In addition, these derivatives can be coupled in the presence of a 1:1 mixture of an amine and HOBt. The corresponding OBt ester is formed initially.[60] Fmoc-amino acid chlorides have been successfully applied in solid-phase syntheses,[60] as well as for the incorporation of the first amino acid onto hindered hydroxyl resins.[62,63] All these reactions occur without significant racemization.

More widely useful than the acid chlorides, the protected amino acid fluorides[40] can be easily prepared from the amino acids with cyanuric fluoride in the presence of pyridine,[64–66] or via (diethylamino)sulfur trifluoride (DAST) in the absence of base,[67] or via TFFH in the presence of DIEA.[39] Advantages of fluorides relative to the chloro derivatives are their total compatibility with *tert*-butyl-based protecting groups, their greater stability, even in the presence of moisture, and their lack of conversion to oxazolone in the presence of tertiary amines. They are also very soluble in organic solvents (>1 M in DMF), thus facilitating acylation through the effect of high concentration. The acid fluorides are stable for extended periods (3 days or longer) when dissolved in DMF and therefore are suitable for use in multiple peptide synthesizers.[68] A key difference between the chlorides and fluorides is the ability of the latter to effect acylation in the total absence of base, thereby reducing the risk of racemization.[69] Acid fluorides have been successfully used for the SPPS of peptides containing difficult

[60] L. A. Carpino, H. G. Chao, M. Beyermann, and M. Bienert, *J. Org. Chem.* **56,** 2635 (1991).
[61] N. L. Benoiton, *Biopolymers* (*Pept. Sci.*) **40,** 245 (1996).
[62] K. Akaji, Y. Kiso, and L. A. Carpino, *J. Chem. Soc., Chem. Commun.,* 584 (1990).
[63] K. Akaji, H. Tanaka, H. Itoh, J. Imai, Y. Fujiwara, T. Kimura, and Y. Kiso, *Chem. Pharm. Bull.* **38,** 3471 (1990).
[64] G. A. Olah, M. Nojima, and I. Kerekes, *Synthesis,* 487 (1973).
[65] J.-N. Bertho, A. Loffet, C. Pinel, F. Reuther, and G. Sennyey, *Tetrahedron Lett.* **32,** 1303 (1991).
[66] L. A. Carpino, E. M. E. Mansour, and D. Sadat-Aalee, *J. Org. Chem.* **56,** 2611 (1991).
[67] C. Kaduk, H. Wenschuh, H. C. Beyermann, K. Forner, L. A. Carpino, and M. Bienert, *Lett. Pept. Sci.* **2,** 285 (1995).
[68] H. Wenschuh, M. Beyermann, S. Rothemund, L. A. Carpino, and M. Bienert, *Tetrahedron Lett.* **36,** 1247 (1995).
[69] H. Wenschuh, M. Beyermann, A. El-Faham, S. Ghassemi, L. A. Carpino, and M. Bienert, *J. Chem. Soc., Chem. Commun.,* 669 (1995).

sequences,[70] and most importantly for the synthesis of peptides containing very hindered units such as those which incorporate two or more consecutive Aib residues or N-substituted amino acids. Comparison studies showed that among three methods which are said to be useful for hindered systems in solution, namely, isolated acid fluorides, UNCAs, and PyBroP, only the first successfully allowed the incorporation of four consecutive Aib residues into segment 36–45 of corticotropin-releasing hormone (CRH).[70] The Fmoc-amino acid fluorides made possible the first solid-phase syntheses of the petaibols, naturally occurring peptides containing up to 60% Aib residues.[33]

Difficult Sequences, Solvents, and Temperature

Incomplete acylations are often observed during the coupling of hindered amino acids, such as Ile, Val, and Thr, or those having β-branched side chains,[71,72] an effect that is increased when several of these amino acids follow each other in the sequence. If revealed by the ninhydrin test, these incomplete couplings can be improved by recoupling. In such cases, use of a different coupling method and/or solvent is advisable for the second coupling. More intractable coupling difficulties are due to intrachain or interchain hydrogen bonding.[73,74] Intrachain interactions occur most frequently at reverse turns, whereas interchain aggregations occur when ordered structures such as helices or β sheets are formed within the matrix. The development of predictive methods for interchain interaction has been attempted,[72] but correlation of the secondary structure of protected peptides on a resin support is unreliable and often differs from related effects seen in solution.[74] Interchain interactions often become evident in the initial stages of the synthesis (5–15 residues) and then disappear.[71] This effect has been rationalized, for a polystyrene resin, by the chain clustering that takes place when the peptide content reaches nearly 50% of the polystyrene, becoming predominantly a polyamide resin.[74]

To overcome coupling difficulties due to interchain interaction several strategies can be used.

[70] H. Wenschuh, M. Beyermann, E. Krause, M. Brudel, R. Winter, M. Schümann, L. A. Carpino, and M. Bienert, *J. Org. Chem.* **59,** 3275 (1994).

[71] S. M. Meister and S. B. H. Kent, *in* "Peptides: Structure and Function, Proceedings of the Eight American Peptide Symposium" (V. J. Hruby and D. H. Rich, eds.), p. 103. Pierce, Rockford, Illinois, 1983.

[72] R. C. de L. Milton, S. C. F. Milton, and P. A. Adams, *J. Am. Chem. Soc.* **112,** 6039 (1990).

[73] D. H. Live and S. B. H. Kent, *in* "Peptides: Structure and Function, Proceedings of the Eighth American Peptide Symposium" (V. J. Hruby and D. H. Rich, eds.), p. 65. Pierce, Rockford, Illinois, 1983.

[74] J. P. Tam and Y.-A. Lu, *J. Am. Chem. Soc.* **117,** 12058 (1995).

1. Use of resins containing functionalized sites evenly distributed and with low loadings (0.1–0.2 mmol/g).[74,75]

2. Use of elevated temperatures. Although all steps in SPPS are usually carried out at 25°, several examples have been described where yields of difficult couplings could be increased by carrying out the reactions at 50°.[76–79] The routine use of high temperatures (50–60°) for peptides containing Asn, Gln, and Glu should be avoided because dehydration is possible in the case of Asn and Gln,[77] and pGlu can be formed from Glu.[78]

3. Use of polar solvents such as DMF, N-methylpyrrolidone (NMP), or dimethyl sulfoxide (DMSO)[71,73,80–82] (the use of the latter can cause the oxidation of methionine to the sulfoxide[83]) or even mixtures of 2,2,2-trifluoroethanol (TFE)[84] and 1,1,1,3,3,3-hexafluoro-2-propanol (HFIP)[85] in DCM, which can disrupt aggregation. TFE is preferred to HFIP, as the latter can consume the activated species.[86]

4. Use of denaturants or chaotropic agents. Salt additives such as LiCl, LiBr, KSCN, LiClO$_4$, and NaClO$_4$ can increase resin swelling and improve coupling yields by disrupting β sheets.[87,88] Perchlorate salts do not oxidize Trp, and KSCN does not modify free amino groups on the peptide–resin, although the latter can catalyze rearrangement of the active O-acylisourea to the inactive N-acylurea. Preincubation of the peptide–resin with 0.8 M KSCN and preactivation of the protected amino acid either via DCC or

[75] K. C. Pugh, E. J. York, and J. M. Stewart, *Int. J. Pept. Protein Res.* **40**, 208 (1992).

[76] J. P. Tam, *Int. J. Pept. Protein Res.* **29**, 421 (1987).

[77] D. H. Lloyd, G. M. Petrie, R. L. Noble, and J. P. Tam, *in* "Peptides: Structure and Function, Proceedings of the Eleventh American Peptide Symposium" (J. E. Rivier and G. R. Marshall, eds.), p. 909, ESCOM, Leiden, The Netherlands, 1990.

[78] A. K. Rabinovich and J. E. Rivier, *in* "Peptides: Structure and Function, Proceedings of the Thirteenth American Peptide Symposium" (R. S. Hodges and J. A. Smith, eds.), p. 71. ESCOM, Leiden, The Netherlands, 1994.

[79] S. A. Kates, N. A. Solé, M. Beyermann, G. Barany, and F. Albericio, *Pept. Res.* **9**, 106 (1996).

[80] S. B. H. Kent, *in* "Peptides: Structure and Function, Proceedings of the Ninth American Peptide Symposium" (C. M. Deber, V. J. Hruby, and K. D. Kopple, eds.), p. 407. Pierce, Rockford, Illinois, 1985.

[81] S. B. H. Kent, *Annu. Rev. Biochem.* **57**, 957 (1988).

[82] G. B. Fields and C. G. Fields, *J. Am. Chem. Soc.* **113**, 4202 (1991).

[83] S. A. Kates, S. B. Daniels, and F. Albericio, *Anal. Biochem.* **212**, 303 (1993).

[84] D. Yamashiro, J. Blake, and C. H. Li, *Tetrahedron Lett.*, 1469 (1976).

[85] S. C. F. Milton and R. C. d. L. Milton, *Int. J. Pept. Protein Res.* **36**, 193 (1990).

[86] H. Kuroda, Y.-N. Chen, T. Kimura, and S. Sakakibara, *Int. J. Pept. Protein Res.* **40**, 294 (1992).

[87] J. M. Stewart and W. A. Klis, *in* "Innovation and Perspectives in Solid-Phase Synthesis and Related Technologies: Peptides, Polypeptides and Oligonucleotides; Macro-organic Reagents and Catalysts" (R. Epton, ed.), p. 1. SPCC, Birmingham, UK, 1990.

[88] A. Thaler, D. Seebach, and F. Cardinaux, *Helv. Chim. Acta* **74**, 628 (1991).

DCC/HOXt was found to be particularly effective.[87] In some cases, it has been reported that although the use of these chaotropic salts inhibits interchain aggregation and helps to dissolve protected peptides, the efficiency of coupling is lower compared with unmodified solvents.[89,90] In the absence of additives the formation of N-acylureas can occur during carbodiimide-mediated coupling reactions. Rearrangement is not important in DCM, but it can be extensive when activation is carried out in DMF.[91] If DMF is needed to effect efficient coupling, a solvent change can be made following preactivation in DCM.

5. Replacement of the secondary amide bond by an alkyl residue which can subsequently be removed. An example involves the 2-hydroxy-4-methoxybenzyl (Hmb) group, which is cleavable by TFA in the presence of scavengers and therefore compatible only with the Fmoc/tBu strategy. This methodology works by avoidance of interchain hydrogen bond formation.[92]

Another type of aggregation can occur in regions containing apolar side-chain protecting groups.[93] In this case the use of polar mixtures during coupling is not enough to overcome the problem. In the Fmoc/tBu strategy the lack of polar side-chain protecting groups interferes with proper solvation of the peptide–resin. In this case, the use of solvent mixtures containing both a polar and nonpolar component, such as tetrahydrofuran (THF)–NMP (7 : 13, v/v) or TFE–DCM (1 : 4, v/v) can be useful.[82]

Monitoring and Capping

Procedure for the Ninhydrin Test

The most convenient method for quick monitoring of the coupling process is the ninhydrin test.[94,95] Ninhydrin reacts with primary amines to give the dye known as Ruhemann's purple.

Two solutions are required and can be made as follows. For solution A, mix 40 g of phenol with 10 ml of absolute ethanol. Warm until dissolved, Stir with 4 g of Amberlite mixed-bed resin MB-3 (Aldrich, Milwaukee,

[89] J. C. Hendrix, K. J. Halverson, J. T. Jarrett, and P. T. Lansbury, *J. Org. Chem.* **55,** 4517 (1990).
[90] B. Riniker, A. Flörsheimer, H. Fretz, P. Sieber, and B. Kamber, *Tetrahedron* **49,** 9307 (1993).
[91] G. Barany and R. B. Merrifield, *in* "The Peptides: Analysis, Synthesis, Biology" (E. Gross and J. Meienhofer, eds.), Vol. 2, p. 1. Academic Press, New York, 1979.
[92] T. Johnson, M. Quibell, and R. C. Sheppard, *J. Pept. Sci.* **1,** 11 (1995).
[93] E. Atherton, V. Woolley and R. C. Sheppard, *J. Chem. Soc., Chem. Commun.,* 970 (1980).
[94] E. Kaiser, R. L. Colescott, C. D. Bossinger, and P. I. Cook, *Anal. Biochem.* **34,** 595 (1970).
[95] V. K. Sarin, S. B. H. Kent, J. P. Tam, and R. B. Merrifield, *Anal. Biochem.* **117,** 147 (1981).

WI) for 45 min and filter. In a separate vessel, dissolve 65 mg of KCN in 100 ml of water. Dilute 2 ml of this solution to 100 ml with pyridine (freshly distilled from ninhydrin). Stir with 4 g of Amberlite mixed-bed resin MB-3 for 45 min and filter. The solutions are mixed. For solution B, dissolve 2.5 g of ninhydrin in 50 ml of absolute ethanol. Store in the dark under nitrogen.

The test procedure is as follows. (1) Transfer a sample containing approximately 1–3 mg of resin to a test tube. (2) Add 9 drops of solution A and 3 drops of solution B. (3) Mix well and place the tube in a heating block preadjusted to 110° for 3 min. (4) Place the tube in cold water. If free amino groups are present, a purple color is formed in the solution or on the resin beads. The sensitivity for detection is in the range of 1 μmol/g resin (99.5% coupling for resins having a functionalization level of 0.2–0.5 mmol/g).

Procedure for the Chloranil Test

For Pro and other secondary amines, the chloranil test can be used.[96] Chloranil (2,3,5,6-tetrachloro-1,4-benzoquinone) reacts with secondary and primary amines, in the presence of acetone or acetaldehyde, respectively, to give a green-blue benzoquinone derivative.

The procedure is as follows. (1) Transfer a sample containing approximately 1 mg of resin to a test tube. (2) Add 200 μl of acetone for secondary amines or acetaldehyde for primary amines. (3) Add 50 μl of a saturated solution of chloranil in toluene. (4) Shake the test tube occasionally over 5 min. If free amino groups are present, a green or blue color is formed on the resin beads. The sensitivity for detection of Pro is in the range of 2–5 μmol/g resin (97–99% coupling for resins having a level of functionalization of 0.2–0.5 mmol/g), and for primary amines 5–8 μmol/g resin, when approximately 1 mg of peptide–resin is assayed.

Procedure for Bromophenol Blue Test

The coupling end point can also be monitored by a noninvasive test based on the use of the acid–base indicator bromophenol blue,[97] which in the presence of free amino groups gives a deep blue color that turns to greenish yellow in the absence of free amino groups. This monitoring can be performed both continuously, with bromophenol blue simply being added to the acylating agent, or discontinuously, with a sample of the peptide resin being removed from the reactor and treated with the reagent.

[96] T. Christensen, *Acta Chem. Scand.* **33B,** 763 (1979).
[97] V. Krchnák, J. Vágner, P. Safar, and M. Lebl, *Collect. Czech. Chem. Commun.* **53,** 2542 (1988).

For the test, 3 drops of a 1% (w/v) solution of bromophenol blue in dimethylacetamide is added to the reactor containing the acylating reagents, causing the suspension to turn dark blue. After the suspension turns greenish yellow, the next step of the synthesis is carried out. Alternatively, (1) transfer a sample containing approximately 1–3 mg of resin to a test tube; (2) add 3 drops of a 1% solution of bromphenol blue in dimethylacetamide; and (3) shake the test tube occasionally over 1 min. If free amino groups are present, a blue-green color is formed on the resin beads. The sensitivity is in the range of 3 μmol/g resin (99.5% of coupling for resins having a level of functionalization of 0.7 mmol/g).

Comments

The sensitivity of these tests decreases with the length of the peptide attached to the resin. Thus, for medium to small peptides (up to 15 amino acid residues) the accuracy is good, but for longer peptides the results are often misleading.

Some protocols recommend a capping reaction for unreacted chains following each coupling. Others call for the introduction of an extra step that could, however, cause side reactions (see below). The capping step is usually carried out by acetylation with acetic anhydride–DIEA (1:1, v/v, 30 equivalents) in DMF. In some automatic synthesizers acetic anhydride is replaced by *N*-acetylimidazole, which as a solid is easier to handle.[2]

Side Reactions

Although the most important side reaction is the loss of configuration at the C-terminal carboxyl residue, there are other problems that should be considered. These include modification of the structure of the C-activated residue, its depletion with a consequent decrease in coupling yield, double insertion, or termination of the growing peptide chain.

In SPPS, all coupling strategies described above are reported to proceed with only negligible levels of racemization for all proteinogenic amino acids except Cys. For this residue, several reports have appeared regarding the incursion of substantial stereomutation (up to 40%) when some of the common coupling methods are applied.[98,99] Thus, there is a loss of configuration without regard to the side-chain protecting group used [acetamidomethyl (Acm); Trt; 2,4,6-trimethoxybenzyl (Tmob); and 9*H*-xanthen-9-yl (Xan) have been tested for Fmoc chemistry]. The stereomutation is

[98] Y. Han, F. Albericio, and G. Barany, *J. Org. Chem.* **62,** 3841 (1977).
[99] E. T. Kaiser, G. J. Nicholson, H. J. Kohlbau, and W. Voelter, *Tetrahedron Lett.* **37,** 1187 (1996).

more important in DMF than in DMF–DCM mixtures. For phosphonium or aminium coupling reagents, the loss of configuration increases notably with preactivation time, whereas for DIPCDI-based coupling it is lower after 5 min of preactivation. Substitution of TMP for DIEA or NMM leads to cleaner products. In addition it is preferable to use only 1 equivalent of base rather than 2. Optimized conditions are as follows: (1) DIPCDI–HOXt (X = A or B) in DMF–DCM (1 : 1, v/v), with 5 min of preactivation; (2) symmetric anhydride prepared in DCM–DMF (9 : 1, v/v) and coupled in DMF–DCM (1 : 1, v/v); (3) Pfp esters in either DMF or DMF–DCM (1 : 1, v/v); and (4) phosphonium and aminium salts in the presence of HOXt with 1 equivalent of TMP and without preactivation.[98]

Asn and Gln should be coupled with their side chains protected or in the form of an active ester; otherwise, the carboxamide group is converted to the corresponding cyano derivative.[50] Likewise, Thr can undergo a β-elimination giving α-aminocrotonic acid.[100]

During the activation step, Arg can cyclize giving the corresponding δ-lactam. Although this reaction is more pronounced when unprotected Arg is used, it is often encountered with ω-tosyl-based protection of Arg.[91] For reasons still not clear, the coupling of Fmoc-Asn(Trt)-OH via phosphonium or aminium reagents is subject to some deficiencies. Thus, the incorporation of these residues is more satisfactory if 1 equivalent of HOAt is present.[9,39] These two side reactions do not provoke any undesirable modification of the peptide but, by lowering the concentration of the activated species, favor the formation of deletion peptides that lack Arg or Asn in the final product. For Fmoc chemistry, if side-chain urethane-type protecting groups are used for Arg, acylation of the guanidino residue occurs along with subsequent partial conversion of Arg to Orn.[101]

Intramolecular rearrangement of symmetric anhydrides gives bis-protected dipeptide derivatives, which can again be activated, thus leading to the incorporation of an extra residue. This side reaction is most favored when relatively unhindered amino acids such as Gly and Ala are incorporated.[102]

An important side reaction can occur during the synthesis of His-containing peptides using the Boc/Bzl strategy if the imidazole ring of His has been protected with the Tos residue and HOBt is used subsequent to the coupling step. Detosylation can occur via HOBt,[103] and the resulting free imidazole moiety can be acetylated or otherwise acylated; then, during the

[100] S. Sakakibara, *Biopolymers* (*Pept. Sci.*) **37**, 17 (1995).
[101] H. Rink, P. Sieber, and F. Raschdorf, *Tetrahedron Lett.* **25**, 621 (1984).
[102] R. B. Merrifield, A. R. Mitchell, and J. E. Clarke, *J. Org. Chem.* **39**, 660 (1974).
[103] T. Fujii and S. Sakakibara, *Bull. Chem. Soc. Jpn.* **47**, 3146 (1974).

neutralization step, a N^{im} to N^{α} transfer of the acetyl or acyl moiety can occur, giving a terminated peptide (acetylation) or insertion of an extra residue (acylation).[104–106] To avoid these side reactions, it is recommended that the imidazole residue be protected by the dinitrophenyl (Dnp) group.[107]

The use of aminium salts should be carried out with caution, because such salts can react with the amino component leading to a guanidino derivative, a process which terminates the peptide chain.[108,109] This side reaction is not important during the coupling of single protected amino acids, because activation is fast and the aminium salt is rapidly consumed. However, during the much slower activation of hindered amino acids or protected peptide segments, the aminium salt can react with the amino component.

Examples of Coupling Reactions

Couplings can be carried out either automatically in a synthesizer or manually. Manual synthesis can be performed in a polypropylene syringe fitted with a polyethylene disk and a stopcock, with occasional stirring, or in a mechanically shaken silanized screw-cap reaction vessel with a Teflon-lined cap, a sintered glass frit, and a stopcock. In this section, the discussion will be limited to protocols for manual synthesis, because those for automatic synthesizers are either dictated by the manufacturer or easily adapted from the manual procedure. The concentration of the activated species should be maintained at maximum (0.6–1 M when the solubility of reagents allows). NMP or other convenient solvents as discussed above can be used instead of DMF. Coupling times can also be modified, but usually 15, 30, and 60 min are used for fast, regular, and extended cycles.

Boc-Amino Acids via Symmetric Anhydrides. Boc-amino acid (8 equivalents) is dissolved in DCM at 4°, then DCC (4 equivalent) is added and the mixture left for 5–15 min at 4°.[110] The solution is filtered and the filtrate diluted with DMF and added to the resin carrying the free amino group. After 30 min the resin is filtered and washed with DMF.

[104] T. Ishiguro and C. Eguchi, *Chem. Pharm. Bull.* **37**, 506 (1989).
[105] M. Kusunoki, S. Nakagawa, K. Seo, T. Hamara, and T. Fukuda, *Int. J. Pept. Protein Res.* **36**, 381 (1990).
[106] C. Celma, F. Albericio, E. Pedroso, and E. Giralt, *Pept. Res.* **5**, 62 (1992).
[107] F. Chillemi and R. B. Merrifield, *Biochemistry,* **8**, 4344 (1969).
[108] H. Gausepohl, U. Pieles, and R. W. Frank, *in* "Peptides: Chemistry and Biology, Proceedings of the Twelfth American Peptide Symposium" (J. A. Smith and J. E. Rivier, eds.), p. 523. ESCOM, Leiden, The Netherlands, 1992.
[109] S. C. Story and J. V. Aldrich, *Int. J. Pept. Protein Res.* **43**, 292 (1994).
[110] G. B. Fields, K. M. Otteson, C. G. Fields, and R. L. Noble, *in* "Innovation and Perspectives in Solid Phase Synthesis" (R. Epton, ed.), p. 241. SPCC, Birmingham, UK, 1991.

Fmoc-Amino Acids via Symmetric Anhydrides. The same method is used although the Fmoc-amino acid is dissolved in DCM containing the minimum amount of DMF needed to dissolve the Fmoc-amino acid.[2] DIPCDI is used instead of DCC, and after preactivation the solution is not filtered because the diisopropylurea does not precipitate. This method can be used for all protected amino acids, except Boc- and Fmoc-Asn/Gln, side-chain protected Boc- and Fmoc-Arg, and Boc-His(Dnp).

Boc- and Fmoc-Amino Acids via HOXt Esters (X = B, A). The method is the same as described except that only 4 equivalent of protected amino acid is dissolved in DCM or DCM–DMF and then HOXt (4 equivalent) in DMF is added.[9,110]

Boc- and Fmoc-Amino Acids via Phosphonium and Aminium Salts. Protected amino acid (4 equivalent) and the phosphonium or aminium salt (4 equivalent) are dissolved in DMF or NMP, the mixture added to the resin bearing the free amino group, and finally the base (4–8 equivalent) added.[9,18,23,38,39,110–112] After 30 min the resin is filtered and washed with DMF or NMP. For Fmoc-Asn(Trt) coupling, HOXt (4 equivalent) should be added. Some researchers prefer to carry out a preactivation step involving the protected amino acid, the coupling reagent, the base, and HOXt for 10 min. This protocol is preferred in the case of aminium salt couplings where chain termination via formation of the guanidino species can intervene. Alternatively, it is possible to reverse the addition of base and the other reagents, in order to avoid contact of the aminium salt with the free amino group, as the activation step is faster than formation of the chain-terminating guanidino derivative.

Boc-Amino Acids via Phosphonium and Aminium Salts with in Situ Neutralization.[36,37,113,114] The previously described procedure is used except that 8–12 equivalent of base is used along with preactivation. When this procedure is followed in order to avoid DKP formation, the use of 8 equivalent of base is preferred.

Boc- and Fmoc-Amino Acids via Preformed Active Esters. The preformed ester of the amino acid (4 equivalent) and HOXt (4 equivalent) are dissolved in DMF, and the solution is added to the resin bearing the

[111] C. G. Fields, D. H. Lloyd, R. L. Macdonald, K. M. Otteson, and R. L. Noble, *Pept. Res.* **4,** 95 (1991).
[112] S. A. Kates, S. A. Triolo, G. W. Griffin, L. W. Herman, G. Tarr, N. A. Solé, E. Diekmann, A. El-Faham, D. Ionescu, F. Albericio, and L. A. Carpino, *in* "Fourth International Symposium on Solid-Phase Synthesis and Combinatorial Libraries" (R. Epton, ed.), pp. 41, 1997. Mayflower Worldwide, Birmingham, UK, 1997.
[113] M. Schnölzer, P. Alewood, A. Jones, D. Alewood, and S. B. H. Kent, *Int. J. Pept. Protein Res.* **40,** 180 (1992).
[114] J. Sueiras-Diaz and J. Horton, *Tetrahedron Lett.* **33,** 2721 (1992).

free amino group.[9,115,116] After 30 min the resin is filtered and washed with DMF.

Boc- and Fmoc-Amino Acids via N-Carboxyanhydrides. The UNCA (4 equivalent) is dissolved in DMF, and the solution is added to the resin bearing the free amino group.[57] The addition of catalytic amounts of base (up to 1 equivalent) is optional, although recommended if a second coupling step is performed. After 30 min the resin is filtered and washed with DMF.

Fmoc-Amino Acid Chlorides. Fmoc-amino acid chloride (4 equivalent) and either the base (4 equivalent) or a mixture of HOXt–base (1 : 1, v/v, 4 equivalent) are dissolved in DMF and added to the resin bearing the free amino group.[59,60] After 30 min the resin is filtered and washed with DMF.

Boc- and Fmoc-Amino Acid Fluorides. The Fmoc-amino acid fluoride (4 equivalent) is dissolved in DMF and added to the resin bearing the free amino group.[33,69] The addition of base (4 equivalent) is optional. After 30 min the resin is filtered and washed with DMF.

Acknowledgments

Work in the authors' laboratories is supported by funds from CICYT (PB95-1131), Generalitat de Catalunya [Grup Consolidat (1995SGR 494) i Centre de Referència en Biotecnología], the National Institutes of Health (GM-09708), and the National Science Foundation (CHE-9314038).

[115] E. Atherton, L. R. Cameron, and R. C. Sheppard, *Tetrahedron* **44**, 843 (1988).
[116] E. Atherton, J. L. Holder, M. Meldal, R. C. Sheppard, and R. M. Valerio, *J. Chem. Soc., Perkin Trans 1,* 2887 (1988).

[8] Handles for Solid-Phase Peptide Synthesis

By MICHAEL F. SONGSTER* and GEORGE BARANY

Introduction

One important aspect to achieving milder and/or more versatile chemical methods for solid-phase peptide synthesis[1–6] is to specify the mode

* Current address: Solid Phase Sciences, Inc., 40 Mark Drive, San Rafael, CA 94903.
[1] R. B. Merrifield, *J. Am. Chem. Soc.* **85**, 2149 (1963).
[2] G. Barany and R. B. Merrifield, *in* "The Peptides" (E. Gross and J. Meienhofer, eds.), Vol. 2, p. 1. Academic Press, New York, 1979.
[3] R. B. Merrifield, *Science* **232**, 341 (1986).
[4] G. Barany, N. Kneib-Cordonier, and D. G. Mullen, *Int. J. Pept. Protein Res.* **30**, 705 (1987).

of attachment (*anchoring*) of the starting (usually terminal) residue to a polymeric support. The classic Merrifield scheme relies on the trifluoroacetic acid (TFA)-labile *tert*-butyloxycarbonyl (Boc) group for N^α-amino protection and TFA-stable groups, for example, suitable "fine-tuned" derivatives of benzyl and cyclohexyl alcohol, for side-chain protection. The synthetic scheme begins with esterification of the protected C-terminal amino acid residue onto a chloromethyl-functionalized polystyrene resin support. After completion of chain assembly, the resultant protected peptide *p*-alkylbenzyl ester-resin is treated with liquid hydrogen fluoride (HF) or equivalent strong acids to achieve cleavage from the resin and full deprotection of the peptide. Alternatively, the peptide can be released from the support by treatment with strong bases or nucleophiles. Because strong cleavage reagents can promote partial destruction of sensitive structures found in peptides, a number of research teams have sought strategies conducive to *milder* reaction conditions. The concept of *orthogonal* protection,[2,7] which involves the use of two or more independent classes of groups that can be removed through differing chemical mechanisms in any order and in the presence of each other, is widely accepted as one desirable approach toward achieving this goal. The most popular orthogonal scheme involves Carpino's N^α-9-fluorenylmethyloxycarbonyl (Fmoc) group,[8,9] which is removed through a base-catalyzed β-elimination mechanism,[10,11] along with acid-labile *tert*-butyl (*t*Bu) and other derivatives for side-chain protection and base-stable anchoring linkages such as Wang's *p*-alkoxybenzyl ester system[12,13] and relatives.[14–18] In this case, final

---

[5] E. Atherton and R. C. Sheppard, "Solid Phase Peptide Synthesis: A Practical Approach." IRL, Oxford, 1989.

[6] G. B. Fields, Z. Tian, and G. Barany, *in* "Synthetic Peptides: A User's Guide" (G. A. Grant, ed.), p. 259. Freeman, New York, 1992.

[7] G. Barany and R. B. Merrifield, *J. Am. Chem. Soc.* **99**, 7363 (1977).

[8] L. A. Carpino and G. Y. Han, *J. Org. Chem.* **37**, 3404 (1972).

[9] G. B. Fields and R. L. Noble, *Int. J. Pept. Protein Res.* **35**, 161 (1990).

[10] E. Atherton, H. Fox, D. Harkiss, C. J. Logan, R. C. Sheppard, and B. J. Williams, *J. Chem. Soc., Chem. Commun.*, 537 (1978).

[11] C.-D. Chang, A. M. Felix, M. H. Jimenez, and J. Meienhofer, *Int. J. Pept. Protein Res.* **15**, 485 (1980).

[12] S.-S. Wang, *J. Am. Chem. Soc.* **95**, 1328 (1973).

[13] S.-S. Wang, J. P. Tam, B. S. H. Wang, and R. B. Merrifield, *Int. J. Pept. Protein Res.* **18**, 459 (1981).

[14] E. Atherton, C. J. Logan, and R. C. Sheppard, *J. Chem. Soc., Perkin Trans. 1*, 538 (1981).

[15] R. C. Sheppard and B. J. Williams, *Int. J. Pept. Protein Res.* **20**, 451 (1982).

[16] F. Albericio and G. Barany, *Int. J. Pept. Protein Res.* **23**, 342 (1984).

[17] F. Albericio and G. Barany, *Int. J. Pept. Protein Res.* **26**, 92 (1985).

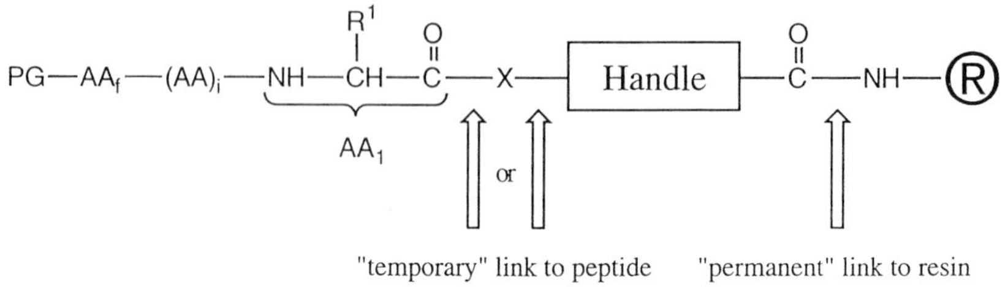

$$PG—AA_f—(AA)_i—NH—CH—C—X—\boxed{Handle}—C—NH—®$$

$$\underbrace{}_{AA_1}$$

"temporary" link to peptide   "permanent" link to resin

Fig. 1. Handle concept for solid-phase peptide synthesis. PG, protecting group, usually Boc or Fmoc; AA, amino acid residue, numbered from C-terminal (1) to N-terminal (f); X, usually O (generally gives peptide acid on cleavage) or NH (gives peptide amide on cleavage), although other functional groups are possible. Prior to formation of the temporary link, X may be derived from either AA_1 or from the handle; similarly, cleavage of the temporary link can occur on either side of X depending on the precise experimental design. For convenience, the permanent link is drawn here as an amide (easily established quantitatively), although other linkages are possible. Often, an "internal reference" amino acid (IRAA) is inserted between the handle and the functional group on the resin.

cleavage/deprotection is carried out with TFA in the presence of suitable scavengers.

Anchoring to the resin, and the reciprocal cleavage step, is usually accomplished through the use of bifunctional spacer molecules known as *handles* or *linkers* (Fig. 1). Such handles become attached *permanently* to a functionalized resin at one end, often through a stable amide bond, and are linked *temporarily* to a growing peptide chain at the other end. Handles provide considerable control and generality to the synthetic plan, because two discrete chemical steps, each of which can be optimized, are used to create the just-designated temporary and permanent links. Cleavage of the temporary peptide–handle bond results in release of the completed peptide from the resin. The design and implementation of handles is focused on (1) the lability and/or stability of the peptide–handle bond to various conditions and (2) the C-terminal functionality, for example, acid, amide, or other, desired for the final peptide. The former factor dictates the mode and conditions of the final cleavage step, as well as which protecting groups can be used compatibly during the synthesis; the latter factor is a function of the precise chemistry selected for attachment and/or cleavage. In comparison to the traditional anchoring strategy in which functionalization of a support provides directly a resin-bound analog to a C^α-carboxyl-protecting reagent, the handle approach has a number of advantages which include

---

[18] M. S. Bernatowicz, T. Kearney, R. S. Neves, and H. Köster, *Tetrahedron Lett.* **30,** 4341 (1989).

the following: (1) substitution levels (i.e., *loading*) can be controlled better by using handles; (2) *any* resin support or material that has been previously functionalized can be used as a parent for synthesis; (3) handle structures can be fine-tuned for optimal yields of both anchoring and cleavage steps; (4) side reactions, including racemization, may be minimized; and (5) handles are used readily in conjunction with "internal reference" amino acids (IRAAs)[5,19–22] which facilitate monitoring of yields of the various steps.

Often, the handle is connected to the support first (creation of permanent linkage of Fig. 1), before attachment of the initial amino acid (temporary linkage). The more involved *preformed* variation reverses the order of steps, making it possible to first purify the handle intermediate and remove possible side products that arise during creation of the temporary linkage. Subsequently, one can achieve *quantitative* attachments to the support (permanent linkage), and thereby circumvent problems associated with extraneous polymer-bound functional groups. Specific examples are described later.

This chapter surveys handles, both standard and preformed, that meet the aforementioned criteria and have proved their usefulness over the years. We emphasize, with experimental details, some handles developed in our laboratory (Table I) that illustrate a gamut of possibilities for the C terminus produced, for the applied cleavage mechanisms (e.g., acidolysis, fluoridolysis, or photolysis), and for compatibility with N^α-amino and side-chain protecting groups. The latter point is significant, because segment condensation/convergent strategies for peptide synthesis require cleavage from the support, either by graduated lability or by an orthogonal mode, while retaining protecting groups on all functions except the ones to be coupled.[23–27] For more extensive discussions or other perspectives concerning handle chemistry and/or the applications of handles to peptide synthesis (including detailed descriptions of various esterification protocols), the interested reader is referred to additional review articles and the original literature cited here and elsewhere.[2,4–6,9,26,27]

[19] E. Atherton, D. L. J. Clive, and R. C. Sheppard, *J. Am. Chem. Soc.* **97**, 6584 (1975).
[20] G. R. Matsueda and E. Haber, *Anal. Biochem.* **104**, 215 (1980).
[21] F. Albericio, N. Kneib-Cordonier, S. Biancalana, L. Gera, R. I. Masada, D. Hudson, and G. Barany, *J. Org. Chem.* **55**, 3730 (1990), and references cited therein.
[22] F. Albericio and G. Barany, *Int. J. Pept. Protein Res.* **41**, 307 (1993).
[23] F. Albericio, P. Lloyd-Williams, and E. Giralt, *Methods Enzymol.* **289**, [15], 1997 (this volume).
[24] E. T. Kaiser, *Acc. Chem. Res.* **22**, 47 (1989).
[25] B. Riniker, A. Flörsheimer, H. Fretz, P. Sieber, and B. Kamber, *Tetrahedron* **49**, 9307 (1993).
[26] P. Lloyd-Williams, F. Albericio, and E. Giralt, *Tetrahedron* **49**, 11065 (1993).
[27] H. Benz, *Synthesis*, 337 (1994).

TABLE I
HANDLES FOR SOLID-PHASE PEPTIDE SYNTHESIS DEVELOPED IN OUR LABORATORY[a]

Handle	Cleavage conditions	Resulting C terminus
3-(4-Hydroxymethylphenoxy)propionic acid (PAB)[b]	TFA	Acid
$HOCH_2$—C$_6$H$_4$—$O(CH_2)_2C(=O)$—OH		
5-(4-Hydroxymethyl-3,5-dimethoxyphenoxy)valeric acid (HAL)[c]	Dilute TFA	Acid
CH_3O, $HOCH_2$—C$_6$H$_2$(CH_3O)—$O(CH_2)_4C(=O)$—OH		
3-Nitro-4-hydroxymethylbenzoic acid (ONb)[d]	$h\nu$ (350 nm)	Acid
$HOCH_2$—C$_6$H$_3$(O_2N)—$C(=O)$—OH		
(3 or 4)-[[[(4-Hydroxymethyl)phenoxy-*tert*-butylphenyl]silyl]phenyl]pentanedioic acid, monoamide (Pbs)[e]	$(n\text{-Bu})_4N^+F^-$	Acid
$HOCH_2$—C$_6$H$_4$—O—Si(tBu)(Ph)—C$_6$H$_4$—NH—$C(=O)$—$(CH_2)_3$$C(=O)$—OH		
5-(4-Aminomethyl-3,5-dimethoxyphenoxy)valeric acid (PAL)[f]	TFA	Amide
CH_3O, H_2NCH_2—C$_6$H$_2$(CH_3O)—$O(CH_2)_4C(=O)$—OH		

TABLE I (continued)

Handle	Cleavage conditions	Resulting C terminus
5-(9-Aminoxanthen-2 or 3-oxy)valeric acid (XAL)[g]	Dilute TFA	Amide
3-Nitro-4-aminomethylbenzoic acid (Nonb)[h]	hν (350 nm)	Amide

[a] Unless indicated otherwise, handle structures are drawn to match the orientation of Fig. 1, that is, the "temporary" link will be established on the far left-hand side, and the "permanent" link on the far right-hand side. For the first four handles, the temporary link is an ester between the carboxyl of the first amino acid and the hydroxyl of the handle, and cleavage under the conditions shown gives an acid as the resulting C terminus. For the remaining handles, the amino group is suitably protected. After deblocking, the handle amine is coupled to the carboxyl of the first amino acid to create the temporary link, and cleavage under the conditions shown gives an amide as the resulting C terminus. In all cases, the permanent link is an amide between the handle carboxyl (activated as appropriate) and an amino-functionalized support. Further details on these various points are outlined in later footnotes.

[b] Handle carboxyl protected/activated as 2,4,5-trichlorophenyl ester. Handle hydroxyl esterified with carboxyl of first amino acid, in preformed handle variation (for that step, the Tcp ester serves the dual function as a protecting group). For details, see F. Albericio and G. Barany, *Int. J. Pept. Protein Res.* **26**, 92 (1985), and references cited therein.

[c] Handle carboxyl free (*in situ*) or activated as 2,4,5-trichlorophenyl ester. Handle cleavage possible while retaining side-chain protecting groups. For details, see this article (Schemes 1 and 2, text and experimental) as well as G. Barany and F. Albericio, *in* "Peptides 1990" (E. Giralt and D. Andreu, eds.), p. 139. Escom, Leiden, The Netherlands, 1991; F. Albericio and G. Barany, *Tetrahedron Lett.* **32**, 1015 (1991); and M. F. Songster, "Design, Synthesis, and Implementation of Handles for Solid-Phase Peptide Synthesis," Ph.D. Thesis, University of Minnesota, Minneapolis (1996).

[d] Handle carboxyl free or activated as 2,4,5-trichlorophenyl ester. Preformed handle variations have been described. Handle cleavage possible while retaining side-chain protecting groups. For details, see G. Barany and F. Albericio, *J. Am. Chem. Soc.* **107**, 4936 (1985); N. Kneib-Cordonier, F. Albericio, and G. Barany, *Int. J. Pept. Protein Res.* **35**, 527 (1990).

[e] Handle carboxyl activated as 2,4,5-trichlorophenyl ester. Handle hydroxyl esterified with carboxyl of first amino acid, in preformed handle variation. Handle cleavage possible while retaining side-chain protecting groups. For details, see D. G. Mullen and G. Barany, *J. Org. Chem.* **53**, 5240 (1988).

[f] Handle amine protected with Fmoc (occasionally with Trt), handle carboxyl free. Standard as well as reductive amination routes to PAL and its relatives are discussed in the text. For details, see this article (Scheme 5, text and experimental) as well as F. Albericio and G. Barany, *Int. J. Pept. Protein Res.* **30**, 206 (1987); F. Albericio, N. Kneib-Cordonier, S. Biancalana, L. Gera, R. I. Masada,

(continued)

TABLE I (*continued*)

D. Hudson, and G. Barany, *J. Org. Chem.* **55,** 3730 (1990), and references cited therein; S. K. Sharma, M. F. Songster, T. L. Colpitts, P. Hegyes, G. Barany, and F. J. Castellino, *J. Org. Chem.* **58,** 4993 (1993); and M. F. Songster and G. Barany, *in* "Innovation and Perspectives in Solid Phase Synthesis: Peptides, Proteins and Nucleic Acids: Biological and Biomedical Applications, 1994" (R. Epton, ed.), p. 685. Mayflower Worldwide, Birmingham, UK, 1994.

g Handle amine (at top of structure as drawn) protected with Fmoc, handle carboxyl free. Handle cleavage possible while retaining side-chain protecting groups. For details, see this article (Scheme 6, text and experimental) as well as R. J. Bontems, P. Hegyes, S. L. Bontems, F. Albericio, and G. Barany, *in* "Peptides: Chemistry and Biology" (J. A. Smith and J. E. Rivier, eds.), p. 601. Escom, Leiden, The Netherlands, 1992; Y. Han and G. Barany, *in* "Innovation and Perspectives in Solid Phase Synthesis: Peptides, Proteins and Nucleic Acids: Biological and Biomedical Applications, 1994" (R. Epton, ed.), p. 525. Mayflower Worldwide, Birmingham, U.K., 1994; and Y. Han, S. L. Bontems, P. Hegyes, M. C. Munson, C. A. Minor, S. A. Kates, F. Albericio, and G. Barany, *J. Org. Chem.* **61,** 6326 (1996).

h Handle amine protected with Boc, Fmoc, Dts, or Tfa; handle carboxyl free. Handle cleavage possible while retaining side-chain protecting groups. For details, see R. P. Hammer, F. Albericio, L. Gera, and G. Barany, *Int. J. Pept. Protein Res.* **36,** 31 (1990).

Handles for Synthesis of Peptide Acids

The stepwise solid-phase synthesis of peptides is invariably carried out in the C → N direction, and consequently the C-terminal residue is generally attached to the support via its α-carboxyl group. If the anchoring linkage is an ester, cleavage often gives cleanly a peptide *acid,* the terminus of which is desirable for many biological and chemical applications. Standard benzyl esters are most usefully cleaved by acid, and sometimes in an orthogonal manner by catalytic transfer hydrogenolysis. Benzyl esters, as well as phenacyl and oxime ester linkages which are covered later (Table II), can also be cleaved by base. However, chemistry involving base treatments and/or nucleophilic displacements is less attractive because of the risk of racemization and other side reactions (discussed further in subsequent sections of this article).

Important themes for handles are exemplified in the 4-hydroxymethylphenylacetic acid (PAM, Fig. 2) system developed for Boc chemistry by Mitchell, Kent, and Tam while they were associated with the Merrifield laboratory.[28–30] First, the PAM structure is fine-tuned with a weakly electron-withdrawing *p*-acetamidomethyl function to create an ester anchoring

[28] A. R. Mitchell, B. W. Erickson, M. N. Ryabtsev, R. S. Hodges, and R. B. Merrifield, *J. Am. Chem. Soc.* **98,** 7357 (1976).

[29] A. R. Mitchell, S. B. H. Kent, M. Engelhard, and R. B. Merrifield, *J. Org. Chem.* **43,** 2845 (1978).

[30] J. P. Tam, S. B. H. Kent, T. W. Wong, and R. B. Merrifield, *Synthesis,* 955 (1979).

TABLE II

ORTHOGONALLY CLEAVABLE HANDLES AND RESINS FOR SOLID-PHASE SYNTHESIS OF C-TERMINAL PEPTIDE ACIDS[a]

Handle	Descriptor	Cleavage conditions	Refs.
Bromacetamido resin 	℞ = Polyacrylamide	$^-$OH	b
	℞ = Polystyrene	Li$^+$ $^-$SCH$_2$CH$_2$OH	c
Br—CH$_2$—C(=O)—NH—CH$_2$—℞			
2-(4-Carboxyphenylsulfonyl)ethanol [CASET(2)] 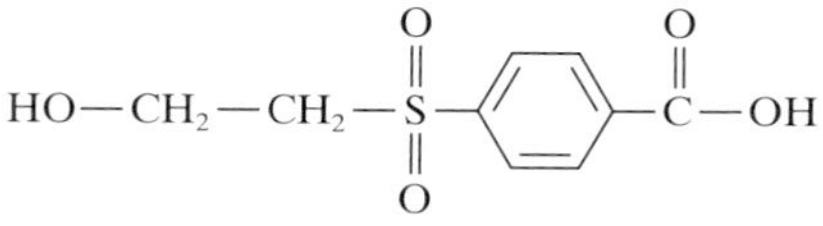		$^-$OH	d
HO—CH$_2$—CH$_2$—S(=O)(=O)—C$_6$H$_4$—C(=O)—OH			
2-Hydroxyethylsulfonylacetic acid 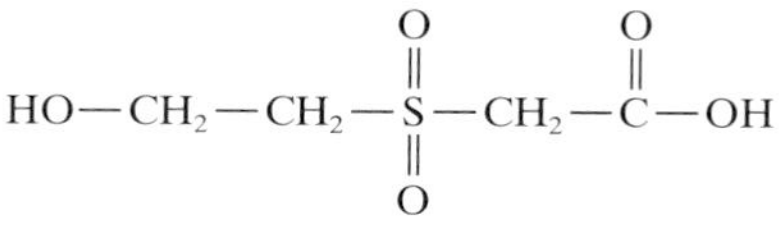		$^-$OH	e
HO—CH$_2$—CH$_2$—S(=O)(=O)—CH$_2$—C(=O)—OH			
(4-Bromocrotonyl)aminomethyl resin (HYCRAM) 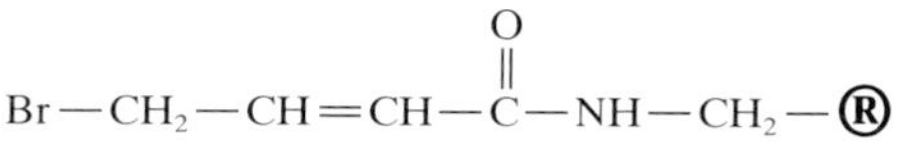		Pd(0)/morpholine	f
Br—CH$_2$—CH=CH—C(=O)—NH—CH$_2$—℞			

(continued)

TABLE II (*continued*)

Handle	Descriptor	Cleavage conditions	Refs.
(4-Hydroxy-*cis*-2-butene-1-oxy)alkanoic acid HO⎯⎯⎯O(CH₂)ₙC(=O)—OH	$n = 5$ $n = 1$	Pd(0)/HOBt (i) Pd(II)/Bu₃SnH (ii) H⁺	g h
2-Bromopropionyl resin Ⓡ⎯⎯ C(=O)⎯CH(Br)⎯CH₃		$h\nu$ (350 nm)	i
[4-(2-Bromopropionyl)phenoxy]acetic acid (PPOA) ⎯OCH₂C(=O)—OH, C(=O)⎯CH(Br)⎯CH₃		$h\nu$ (350 nm)	j
o-Nitro-(α-methyl)bromobenzyl resin Ⓡ⎯⎯ (O₂N, Br)⎯CH⎯CH₃		$h\nu$ (350 nm)	k

Label	Handle		Cleavage conditions
l	4-[4-(1-Hydroxyethyl)-2-methoxy-5-nitrophenoxy]butyric acid		$h\nu$ (350 nm)
m	p-Nitrobenzophenone oxime resin (Kaiser oxime resin)		Nucleophiles
	9-(Hydroxymethyl)fluorenyl handles		
n	(HOFmCOOH)	$R^1 = CO_2H$ $R^2 = H$	Piperidine/DMF
o	(HMFA)	$R^1 = H$ $R^2 = CH_2CO_2H$	Piperidine/DMF
p	(HMFS)	$R^1 = H$ $R^2 = NH(C=O)CH_2CH_2CO_2H$	Piperidine/DMF
q	3-Nitro-4-(2-hydroxyethyl))benzoic acid (NPE)		Piperidine/DMF or DBU/dioxane

(continued)

TABLE II (*continued*)

Handle	Descriptor	Cleavage conditions	Refs.
3-(4-Hydroxymethylphenyl)-3-trimethylsilylpropionic acid		$(n\text{Bu})_4\text{N}^+\text{F}^-$	*r*
4-[1-Hydroxy-2-(trimethylsilyl)ethyl]benzoic acid (SAC)		$(n\text{Bu})_4\text{N}^+\text{F}^-$	*s*
cis-Cobalt(III) amine complex		Dithiothreitol/DIEA/DMF	*t*
2-Azidomethyl-4-oxy-6,*N*-dimethyl-*N*-hydroxymethyl-benzamide resin		*n*Bu$_3$P/imidazole/DMF	*u*

2-Hydroxypropyldithio-2'-isobutyric acid (HPDI)

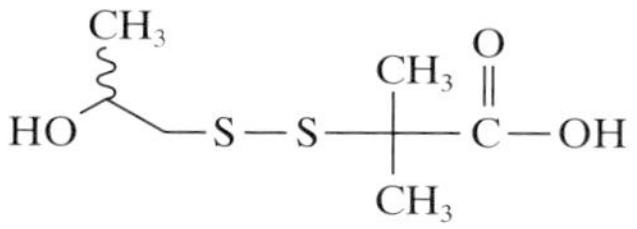

(i) $^-$CN, (ii) $^-$OH or
(i) TCEP^x, (ii) pH 9
or $^-$OH

v

4-Hydrazidylphenylsulfonyl resin

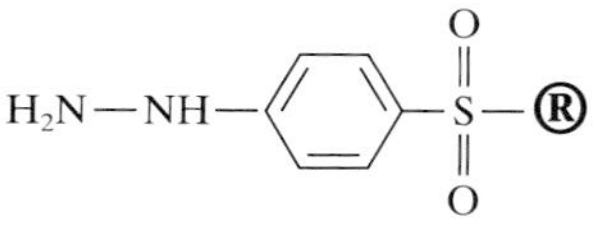

(i) O$_2$/Cu(Py)$_n$
(ii) H$_2$O

w

---

[a] Handle and resin structures are drawn to match the orientation of Fig. 1, with further subtleties described in the corresponding footnote of Table I. The third and fourth entries of Table I, namely, ONb and Pbs, would also fit here and are not repeated. In every case shown, the handle is attached to the resin prior to formation of the temporary ester linkage, which is created either by alkylation of a carboxylate with the handle/resin halide drawn on the far left of the structure, or by acylation (mediated by suitable condensing agents in the presence of appropriate bases and/or additives) of a carboxyl with the handle/resin hydroxyl drawn on the far left of the structure. Because the handles/resins shown are designed for orthogonal cleavage, they can in principle be used in ways that allow release of peptides from the support while retaining acid-labile side-chain protecting groups.

[b] F. Baleux, J. Daunis, R. Jacquier, and B. Calas, *Tetrahedron Lett.* **25**, 5893 (1984); B. Calas, J. Mery, J. Parello, and A. Cave, *Tetrahedron* **41**, 5331 (1985); and F. Baleux, B. Calas, and J. Mery, *Int. J. Pept. Protein Res.* **28**, 22 (1986).

[c] M. S. Shekhani, G. Grübler, H. Echner, and W. Voelter, *Tetrahedron Lett.* **31**, 339 (1990).

[d] R. Schwyzer, E. Felder, and P. Failli, *Helv. Chim. Acta* **67**, 1316 (1984).

[e] S. B. Katti, P. K. Misra, W. Haq, and K. B. Mathur, *J. Chem. Soc., Chem. Commun.*, 843 (1992).

[f] H. Kunz and B. Dombo, *Angew. Chem., Int. Ed. Engl.* **27**, 711 (1988); *Angew. Chem.* **100**, 732 (1988).

[g] B. Blankemeyer-Menge and R. Frank, *Tetrahedron Lett.* **29**, 5871 (1988).

[h] F. Guibé, O. Dangles, G. Balavoine, and A. Loffet, *Tetrahedron Lett.* **30**, 2641 (1989).

(continued)

TABLE II (continued)

[i] S.-S. Wang, J. Org. Chem. 41, 3258 (1976).

[j] D. Bellof and M. Mutter, Chimia 39, 317 (1985).

[k] A. Ajayaghosh and V. N. R. Pillai, Tetrahedron 44, 6661 (1988).

[l] C. P. Holmes, D. G. Jones, B. T. Frederick, and L.-C. Dong, Presented at the 14th American Peptide Symposium, Columbus, Ohio, 1995, Poster P004.

[m] W. F. DeGrado and E. T. Kaiser, J. Org. Chem. 45, 1295 (1980): S. H. Nakagawa and E. T. Kaiser, J. Org. Chem. 48, 678 (1983); E. T. Kaiser, Acc. Chem. Res. 22, 47 (1989); and P. T. Lansbury, Jr., J. C. Hendrix, and A. I. Coffman, Tetrahedron Lett. 30, 4915 (1989).

[n] M. Mutter and D. Bellof, Helv. Chim. Acta 67, 2009 (1984).

[o] Y.-Z. Liu, S.-H. Ding, J.-Y. Chu, and A. M. Felix, Int. J. Pept. Protein Res. 35, 95 (1990).

[p] F. Rabanal, E. Giralt, and F. Albericio, Tetrahedron Lett. 33, 1775 (1992); and F. Rabanal, E. Giralt, and F. Albericio, Tetrahedron 51, 1449 (1995).

[q] F. Albericio, J. Robles, D. Fernandez-Forner, Y. Palom, C. Celma, E. Pedroso, E. Giralt, and R. Eritja, in "Peptides 1990" (E. Giralt and D. Andreu, eds.) p. 134. Escom, Leiden, The Netherlands, 1991; and R. Eritja, J. Robles, D. Fernandez-Forner, F. Albericio, E. Giralt, and E. Pedroso, Tetrahedron Lett. 32, 1511 (1991).

[r] R. Ramage, C. A. Barron, S. Bielecki, and D. W. Thomas, Tetrahedron Lett. 28, 4105 (1987); and R. Ramage, C. A. Barron, S. Bielecki, R. Holden, and D. W. Thomas, Tetrahedron 48, 499 (1992).

[s] H.-G. Chao, M. S. Bernatowicz, P. D. Reiss, C. E. Klimas, and G. R. Matsueda, J. Am. Chem. Soc. 116, 1746 (1994).

[t] N. Mensi and S. S. Isied, J. Am. Chem. Soc. 109, 7882 (1987); and B. E. Arbo and S. S. Isied, Int. J. Pept. Protein Res. 42, 138 (1993).

[u] N. J. Osborn and J. A. Robinson, Tetrahedron 49, 2873 (1993).

[v] J. Brugidou and J. Méry, Pept. Res. 7, 40 (1994).

[w] A. N. Semenov and K. Y. Gordeev, Int. J. Pept. Protein Res. 45, 303 (1995).

[x] TCEP, tris(2-carboxyethyl)phosphine [P(CH$_2$CH$_2$CO$_2$H)$_3$].

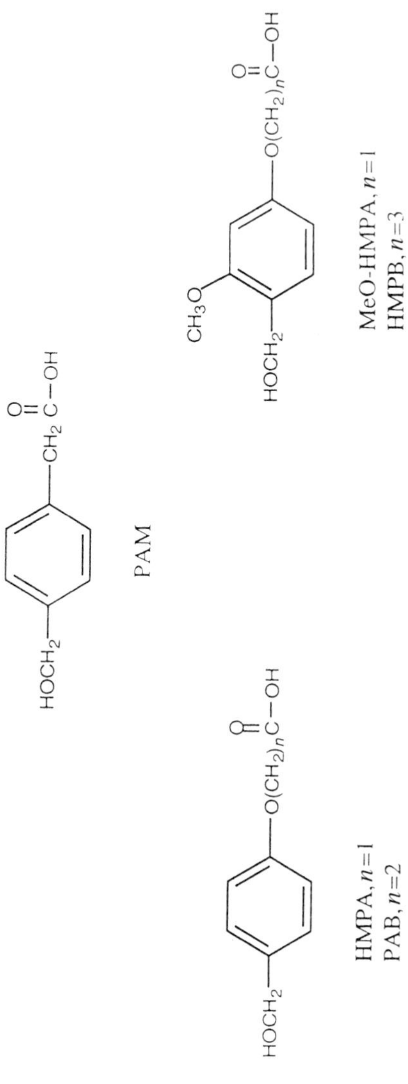

FIG. 2. Acid-labile handles for solid-phase synthesis of C-terminal peptide acids.

linkage that is somewhat more stable than the standard Merrifield p-alkyl-benzyl ester, and hence less prone to premature cleavage during repetitive Boc removal. Nonetheless, the p-(carbamoylmethyl)benzyl (PAM) ester is cleaved in high yield during the final strong acid step. Second, PAM is best established by forming in solution the temporary Boc-amino acid–handle bond (the handle carboxyl is protected during the esterification step, and deprotected in a separate step); this strategy gives a preformed handle that is purified and characterized, and then coupled quantitatively to amino-methylpolystyrene. Third, Boc solid-phase peptide syntheses have given demonstrably better results, and fewer side reactions have been observed, when PAM anchoring was used.

The previous example notwithstanding, much research on handles has focused on increasing their acid lability. This can be accomplished by fine-tuning the structural and electronic features of the handle. In general, acidolytic cleavage is driven by the ease of formation of the resultant carbonium ion, thus, a more stable carbonium ion is reflected by faster rates and/or use of weaker acids in the required cleavage. Structural stabilization generally follows the expected order $(CH_3^+ < RCH_2^+ < R_2CH^+ \cong C_6H_5CH_2^+ < R_3C^+)$, with electron-donating substituents such as alkyl or alkoxy on phenyl providing further stabilization. In accord with these pre-dictions, Wang et $al.$ designed the 4-alkoxybenzyl alcohol resin (handle connected onto chloromethyl–resin),[12,13] which evolved to the 4-hydroxy-methylphenoxyacetic acid (HMPA, Fig. 2) handle from Sheppard and co-workers[14,15] and the 3-(4-hydroxymethylphenoxy)propionic acid (PAB, Fig. 2) handle from us.[16,17] The simple change of a p-alkyl substituent in PAM to a p-alkoxy substituent in Wang/HMPA/PAB makes the resultant anchoring linkages cleavable by TFA rather than HF or trifluoromethanesulfonic acid (TFMSA). The trend toward acid lability continues further by introduction of an additional electron-donating group at one of the $ortho$ positions [i.e., 3-methoxy-4-hydroxymethylphenoxyacetic acid (MeO-HMPA, Fig. 2)[15] or 4-(4-hydroxymethyl-3-methoxyphenoxy)butyric acid (HMPB, Fig. 2);[31] note that the popular 2-methoxy-4-benzyloxybenzyl alcohol resin (SASRIN, Fig. 3)[32,33] bears the same relation to MeO-HMPA as Wang resin does to HMPA], that is, PAM < HMPA < MeO-HMPA/HMPB. In addi-tion, it has been shown that the length of the aliphatic spacer separating the electron-donating oxygen from the carboxamide link to the resin affects,

[31] A. Flörsheimer and B. Riniker, in "Peptides 1990" (E. Giralt and D. Andreu, eds.), p. 131. Escom, Leiden, The Netherlands, 1991.
[32] M. Mergler, R. Tanner, J. Gosteli, and P. Grogg, $Tetrahedron$ $Lett.$ **29**, 4005 (1988).
[33] M. Mergler, R. Nyfeler, R. Tanner, J. Gosteli, and P. Grogg, $Tetrahedron$ $Lett.$ **29**, 4009 (1988).

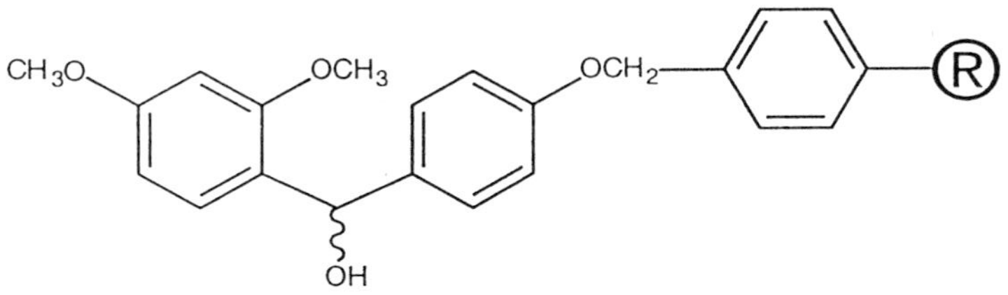

4-(2,4-Dimethoxyphenylhydroxymethyl)phenoxymethyl resin (Rink acid resin)

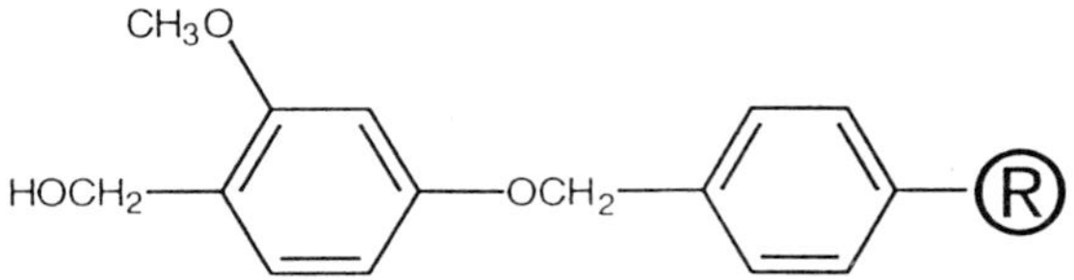

2-Methoxy-4-benzyloxybenzyl alcohol resin (SASRIN®) 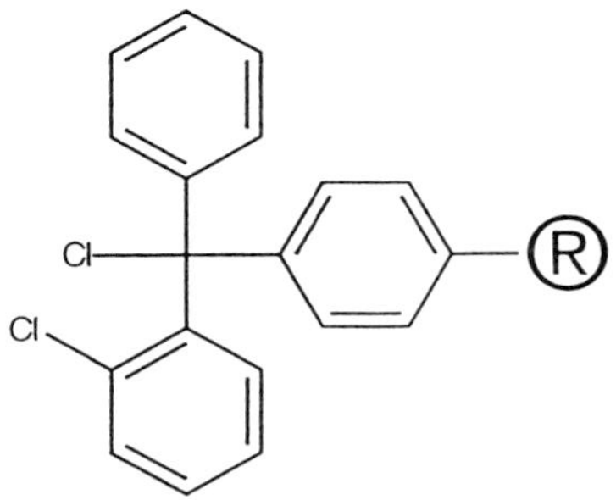   2-Chlorotrityl chloride resin (Barlos resin)

FIG. 3. Resins that provide highly acid-labile anchors for solid-phase synthesis of C-terminal peptide acids.

albeit to a somewhat lesser extent, the acid sensitivity of the handle (i.e., HMPA < PAB[17] and Meo-HMPA < HMPB[31]). The theme of adding electron-donating substituents to increase acid sensitivity of a handle culminates with our tris(alkoxy)benzyl ester (HAL)[34,35] system (Table I, second entry). The HAL anchor can be cleaved rapidly (~5 min) with very dilute acid [<0.1% (v/v) TFA in dichloromethane (DCM)] and is also somewhat labile to acetic acid–DCM (1:9, v/v, 78% cleavage after 24 hr). This means that HAL cleavage occurs without any premature loss of tBu side-chain protecting groups, and in this regard properties of HAL are better than those of Meo-HMPA, HMPB, and SASRIN, which require higher concentrations of TFA (about 1%) in DCM. However, unlike the esters derived from the Rink tris(alkoxy)benzhydryl alcohol[36] or the Barlos chlorotrityl[37,38] resins (Fig. 3), HAL was shown to be stable (<5% loss of peptide after 24 hr) to both 0.1 M 1-hydroxybenzotriazole (HOBt) in N,N-dimethylformamide (DMF) and 0.1 M Boc-amino acids in DMF. Therefore, HAL does not suffer the risk of premature release of growing peptides catalyzed by the free carboxyl group of the incoming amino acid, during each successive coupling cycle.[34]

The HAL linkage is established by esterification of an Fmoc-amino acid to a HAL-modified support,[34,39] which in turn is accessed through either active ester handle reagent **1**, a stable compound (Scheme 1) that is attached onto amino-functionalized supports,[34] or through a simplified method whereby the HAL linker **3** formed *in situ* is attached directly to the resin (Scheme 2).[39]

Preparation of 2,4,5-Trichlorophenyl-5-(4-hydroxymethyl-3,5-dimethoxyphenoxy)valerate (HAL-OTcp, 1)

Solid NaBH$_4$ (1.4 g, 37 mmol) is added over 15 min to a suspension of 5-(4-formyl-3,5-dimethoxyphenoxy)valeric acid (**2**,[21] 7.0 g, 25 mmol) in absolute ethanol (50 ml), and the reaction mixture is stirred under N$_2$ at 25° for 1 hr. The now homogeneous solution is poured into a slurry of ice in acetic acid–water (99:1, v/v, 500 ml) and extracted quickly with ethyl acetate (three times, 40 ml each time). The organic extract, which contains intermediate **3**, is washed with brine (50 ml), dried briefly (Na$_2$SO$_4$), and

[34] F. Albericio and G. Barany, *Tetrahedron Lett.* **32,** 1015 (1991).
[35] G. Barany and F. Albericio, *in* "Peptides 1990" (E. Giralt and D. Andreu, eds.), p. 139. Escom, Leiden, The Netherlands, 1991.
[36] H. Rink, *Tetrahedron Lett.* **28,** 3787 (1987).
[37] K. Barlos, D. Gatos, J. Kallitsis, G. Papaphotiu, P. Sotiriu, Y. Wenqing, and W. Schäfer, *Tetrahedron Lett.* **30,** 3943 (1989).
[38] K. Barlos, O. Chatzi, D. Gatos, and G. Stavropoulos, *Int. J. Pept. Protein Res.* **37,** 513 (1991).
[39] M. F. Songster, "Design, Synthesis, and Implementation of Handles for Solid-Phase Peptide Synthesis," Ph.D. Thesis, University of Minnesota, Minneapolis (1996).

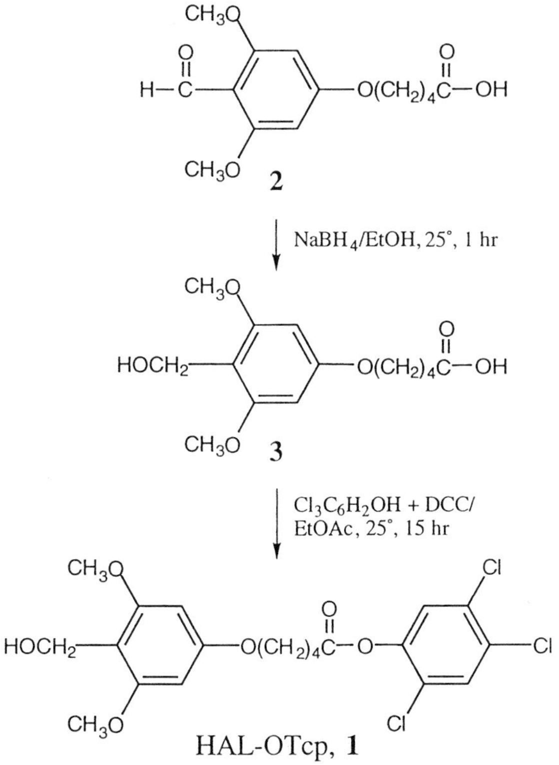

SCHEME 1. An optimized synthesis of 2,4,5-trichlorophenyl-5-(4-hydroxymethyl-3,5-dimethoxyphenoxy)valerate (HAL-OTcp, **1**),[39] modified from that published by Albericio and Barany.[34] Compound **1** is then coupled onto amino-functionalized supports to create HAL-® (an alternative route to the same intermediate is provided in Scheme 2). See text for all corresponding experimental procedures.

combined with *N,N'*-dicyclohexylcarbodiimide (DCC, 5.2 g, 25 mmol) and 2,4,5-trichlorophenol (5.0 g, 25 mmol). This solution is stirred under N_2 at 25° for 15 hr, whereupon the reaction is filtered through a Celite plug (Aldrich) and concentrated *in vacuo*. The oily residue is purified by flash

$$\mathbf{2} \xrightarrow[\text{25°, 30 min}]{\text{NaBH}_4\text{/EtOH}} \mathbf{3} \xrightarrow[\text{DMF, 25°, 1 hr}]{\text{H}_2\text{N-PEG-PS, BOP/HOBt/NMM}} \text{HAL-PEG-PS}$$

SCHEME 2. Simplified preparation of HAL–PEG–PS.[39] Structures **2** and **3** are in Scheme 1. Further relevant comments and experimental procedures are in the legend to Scheme 1 and in the text.

chromatography on oven-dried silica, eluting with ethyl acetate–hexane (1:1, v/v) to yield 8.7 g (76% based on **2**) of a thin-layer chromatography (TLC) (R_f 0.23, ethyl acetate–hexane, 1:1, v/v) and nuclear magnetic resonance (NMR) pure white solid.

Preparation of HAL-Resin

Method A. The coupling of handle reagent **1** to amino-functionalized resin is detailed here with polyethylene glycol–*graft*-polystyrene (PEG–PS). HCl·PEG–PS (5 g, 0.21 mmol/g) is washed with DCM (five times, 2 min each time), neutralized with diisopropylethylamine (DIEA)–DCM (1:19, v/v, five times, 2 min), washed again with DCM (five times, 2 min), followed by DMF (five times, 2 min), and coupled with a preactivated (10 min) solution of HAL-OTcp (**1**, 1.3 g, 3.2 mmol) and HOBt (0.43 g, 3.2 mmol) in DMF (10 ml). The reaction mixture is agitated for 15 hr, filtered, washed with DMF (five times, 2 min) and DCM (five times, 2 min), and found to be ninhydrin negative.

Method B. The abbreviated method of Scheme 2, with coupling of intermediate **3** to amino-functionalized resin, is detailed here with H-Ile-PEG–PS. HCl·PEG–PS (0.5 g, 0.21 mmol/g) is washed with DCM (five times, 2 min), neutralized with DIEA–DCM (1:19, v/v, five times, 2 min), washed again with DCM (five times, 2 min), followed by DMF (five times, 2 min), and coupled with Fmoc-Ile-OH [0.15 g, 0.42 mmol, preactivated (10 min) with benzotriazol-1-yl-oxytris-(dimethylamino)phosphonium hexafluorophosphate (BOP, 0.19 g, 0.42 mmol) and HOBt (64 mg, 0.42 mmol)] and *N*-methylmorpholine (NMM, 93 μl, 0.84 mmol) in DMF (2 ml) for 1 hr at 25°. After filtering, the resin is washed with DMF (five times, 2 min), followed by DCM (five times, 2 min), and, although ninhydrin negative, capped with acetic anhydride (Ac$_2$O, 79 μl, 0.84 mmol) in DCM (2 ml) for 30 min. The resin is next washed with DCM (five times, 2 min) and DMF (five times, 2 min), and Fmoc removal is achieved with piperidine–DMF (1:4, v/v, once for 2 min, twice for 10 min, twice for 2 min). Finally, washing with DMF (five times, 2 min) is carried out in preparation for immediate coupling with **3** as described below.

NaBH$_4$ (24 mg, 0.63 mmol) is added to a suspension of 5-(4-formyl-3,5-dimethoxyphenoxy)valeric acid (**2**,[21] 0.12 g, 0.42 mmol) in absolute ethanol (2.5 ml), and the reaction is stirred under N$_2$ at 25° for 30 min. The now homogeneous solution is concentrated under reduced pressure and allowed to stand for 30 min *in vacuo* to ensure total dryness. The residue is next taken up in DMF (2 ml), preactivated (10 min) with BOP (0.19 g, 0.42 mmol) and HOBt (64 mg, 0.42 mmol), and added along with NMM (93 μl, 0.84 mmol) to the H-Ile-PEG–PS prepared as detailed above. After

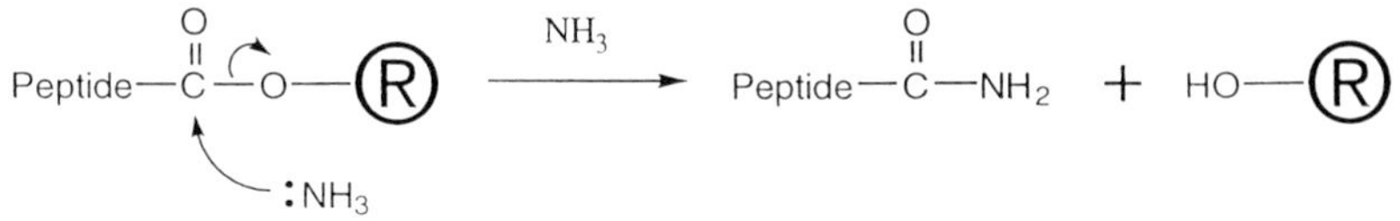

SCHEME 3. Preparation of C-terminal peptide amides by solid-phase synthesis involving ammonolytic cleavage of ester.

coupling at 25° for 1 hr, the resin (ninhydrin negative) is filtered, washed with DMF (five times, 2 min) and DCM (five times, 2 min), and dried *in vacuo* to provide HAL-Ile-PEG–PS (final loading 0.13 mmol/g, based on amino acid analysis of Ile).

The focus of this section so far is on substituted benzyl esters that are cleaved by acidolysis to provide peptide acids. For completeness, we point out that numerous handles, as well as functionalized resins, similarly give rise to peptide acids but involve ultimate cleavage by an orthogonal mechanism (Table II).

Handles for Synthesis of Peptide Amides

Although peptides are often required with C-terminal carboxylic acid end groups, many naturally occurring peptides are isolated as C-terminal amide derivatives and also represent valid targets for synthesis.[21] These include biologically important peptides such as oxytocin, secretin, apamin, calcitonin, thymosin, gastrins, and several releasing hormones from the brain.[40,41] Further, peptide amide fragments often exhibit increased biological activity relative to the corresponding acid derivatives,[41,42] in some cases because they are less prone to digestion by carboxypeptidases and therefore have increased life spans.[43,44] Finally, peptide amides are not ionized at physiological pH, and hence this net charge-preserving end group is preferred over carboxyls for immunological screening of partial protein sequences.

Solid-phase methods for the production of peptide amides involve either (1) anchoring through an ester bond which is cleaved by ammonolysis (Scheme 3) or (2) the use of suitable amino group-containing supports or

---

[40] "The Peptides: Analysis, Synthesis, Biology" (E. Gross and J. Meienhofer, eds.), Vols. 1–9. Academic Press, New York, 1980.
[41] N. Ling, F. Zeytin, P. Böhlen, F. Esch, P. Brazeau, W. B. Wehrenberg, A. Baird, and R. Guillemin, *Annu. Rev. Biochem.* **54,** 403 (1985).
[42] J. S. Morley, H. J. Tracey, and R. A. Gregory, *Nature (London)* **207,** 1356 (1965).
[43] P. G. Pietta, P. F. Cavallo, K. Takahashi, and G. Marshall, *J. Org. Chem.* **39,** 44 (1974).
[44] G. R. Matsueda and J. M. Stewart, *Peptides* **2,** 45 (1981).

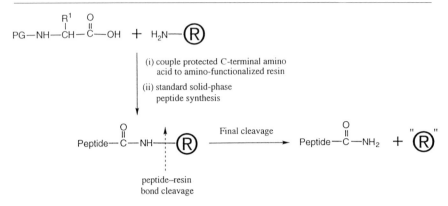

SCHEME 4. Preparation of C-terminal peptide amides by solid-phase synthesis involving appropriate amino-functionalized supports. When the amino group is part of a handle (compare with Fig. 1), then this provides a universal anchoring strategy insofar as a *single* reagent is used irrespective of the anchored residue.

handles that anchor the peptide through an amide bond and are designed to create an amide end group on cleavage (Scheme 4). The former approach was described first by Bodanszky and Sheehan in the mid-1960s,[45,46] following solution precedents.[47,48] Typically, a standard Merrifield peptide–resin (*p*-alkylbenzyl ester) is treated with a saturated solution of ammonia in methanol, or in another alcoholic solvent, for 2 to 4 days at 25°.[49,50] This method is simple and general, providing 60–80% yields, but its utility is severely limited by slow reaction times and side reactions such as racemization at the C terminus or undesired ammonolysis or rearrangement of sensitive side-chain functionalities [e.g., Asp(OBzl), Glu(OBzl), where Bzl is benzyl].[3,51–53] In addition, in some cases, particularly with systems where the C-terminal residue is sterically bulky and the solvent is methanol or ethanol, the product can be contaminated with significant quantities of peptide ester. In these cases, it is believed that actual cleavage is effected

[45] M. Bodanszky and J. T. Sheehan, *Chem. Ind. (London)*, 1423 (1964).
[46] M. Bodanszky and J. T. Sheehan, *Chem. Ind. (London)*, 1597 (1966).
[47] C. Ressler and V. du Vigneaud, *J. Am. Chem. Soc.* **76**, 3107 (1954).
[48] M. Zaoral and J. Rudinger, *Collect. Czech. Chem. Commun.* **20**, 1183 (1955).
[49] J. M. Stewart and J. D. Young, "Solid Phase Peptide Synthesis" 2nd Ed. Pierce, Rockford, Illinois, 1984.
[50] M. Manning, *J. Am. Chem. Soc.* **90**, 1348 (1968).
[51] M. Bodanszky and J. Martinez, *Synthesis,* 333 (1981).
[52] M. Christensen, O. Schou, and V. S. Pedersen, *Acta Chem. Scand., Ser. B* **35**, 573 (1981).
[53] P. Sieber, *Tetrahedron Lett.* **28**, 2107 (1987).

through base-catalyzed transesterification with the alcohol solvent, and the resultant ester is then incompletely converted to the amide in solution.[46,54]

In a general and likely preferred approach to peptide amides, the C-terminal amino acid residue, protected at the α-amine and side chains, is coupled via its α-carboxyl group onto an appropriate amino-functionalized support; following standard solid-phase synthesis, final cleavage of the completed peptide one atom over from the point of attachment transfers the amino function from the support to the released peptide (Scheme 4). This "one-step" strategy benefits by avoiding postsynthetic solution-phase transformations as well as problematic nucleophilic cleavages. The pioneering studies in this regard were reported in 1970 by Pietta and Marshall[55] with an unsubstituted benzhydrylamine (BHA) resin. These BHA resins have been used successfully,[43,56–59] although low cleavage yields for certain peptides (e.g., those with C-terminal Phe, Val, or Tyr)[44] prompted researchers to develop substituted benzhydrylamine resins with better cleavage characteristics (Table III). For example, the simple methyl substitution to p-methylbenzhydrylamine (MBHA) resins[44,52] is nearly ideal when final cleavage is planned by use of HF or TFMSA.[60–62] Substitution with sufficient electron-donating groups gives rise to resins and handles that are cleavable by TFA and thus compatible with Fmoc chemistry (Table III). The BHA system can also be modified in ways that allow cleavage by hydrogenolysis, photolysis, or in two stages (safety-catch mode, reduction followed by acidolysis) (Table III). Finally, a wide variety of non-BHA systems are also available (Table IV) that follow the general principle (Scheme 4) and use acidolysis, photolysis, or other chemical principles for final cleavage.

Handles for solid-phase synthesis of peptide amides under mild conditions, developed in our laboratory (Table I, last three entries), include the

[54] H. C. Beyerman, H. Hindriks, and E. W. B. de Leer, *Chem. Commun.*, 1668 (1968).

[55] P. G. Pietta and G. R. Marshall, *Chem. Commun.*, 650 (1970).

[56] G. W. Tregear, H. D. Niall, J. T. Potts, S. E. Leeman, and M. M. Chang, *Nature New Biol.* **232**, 87 (1971).

[57] J. Rivier, W. Vale, R. Burgus, N. Ling, M. Amoss, R. Blackwell, and R. Guillemin, *J. Med. Chem.* **16**, 545 (1973).

[58] C. Peña, J. M. Stewart, A. C. Paladini, J. M. Dellacha, and J. A. Santomé, *in* "Peptides: Chemistry, Structure and Biology" (R. Walter and J. Meienhofer, eds.), p. 523. Ann Arbor Science, Ann Arbor, Michigan, 1976.

[59] D. H. Coy and J. Gardner, *Int. J. Pept. Protein Res.* **15**, 73 (1980).

[60] C. G. Unson, D. Andreu, E. M. Gurzenda, and R. B. Merrifield, *Proc. Natl. Acad. Sci. U.S.A.* **84**, 4083 (1987).

[61] T. Inui, K. Hagiwara, K. Nakajima, T. Kimura, T. Nakajima, and S. Sakakibara, *Pept. Res.* **5**, 140 (1992).

[62] P. E. Thompson, H. H. Keah, P. T. Gomme, P. G. Stanton, and M. T. W. Hearn, *Int. J. Pept. Protein Res.* **46**, 174 (1995).

TABLE III
BENZHYDRYLAMINE-BASED HANDLES AND RESINS FOR SYNTHESIS OF PEPTIDE AMIDES[a]

Handle or resin	Cleavage conditions	Refs.

Benzhydrylamine resin (BHA; R^1, R^2 = H)	HF	b
4-Methylbenzhydrylamine resin (MBHA; R^1 = CH_3, R^2 = H)	HF	c
4-Methoxybenzhydrylamine resin (MOBHA; R^1 = CH_3O, R^2 = H)	HF	d
2,4-Dimethoxybenzhydrylamine resin (2,4-DMBHA; R^1, R^2 = CH_3O)	TFA	e
p-Nitrobenzhydrylamine resin (R^1 = NO_2, R^2 = H)	Pd/cyclohexa-1,4-diene	f
o-Nitrobenzhydrylamine resin (NBHA; R^1 = H, R^2 = NO_2)	$h\nu$ (350 nm)	g

4-(α-Aminobenzyl)phenoxyacetic acid (R^1, R^2 = H, n = 1)	HF	h
4-(α-Amino-4'-methoxybenzyl)phenoxybutyric acid (R^1 = CH_3O, R^2 = H, n = 3)	TFA	i
4-(α-Amino-4'-methoxybenzyl)-2-methylphenoxyacetic acid (R^1 = CH_3O, R^2 = CH_3, n = 1)	TFA	j

3-(α-Amino-4'-methoxybenzyl)-4-methoxyphenylpropionic acid (R^1, R^2 = CH_3O, R^3 = H)	TFA	k
3-(α-Amino-4'-methoxybenzyl)-4,6-dimethoxyphenylpropionic acid (R^1, R^2, R^3 = CH_3O)	Dilute TFA	l

TABLE III (*continued*)

Handle or resin	Cleavage conditions	Refs.
4-(α-Amino-2′,4′-dimethoxybenzyl)phenoxymethyl resin (Rink resin) 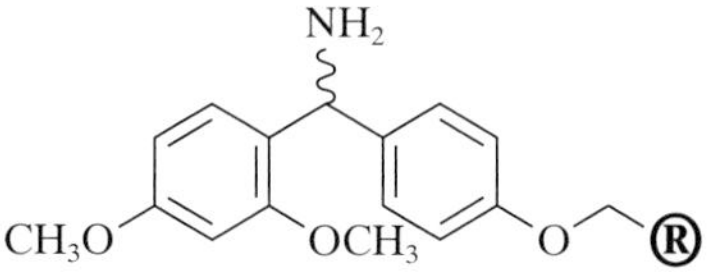	TFA	*m*
4-Succinylamino-2,2′,4′-trimethoxybenzhydrylamine (SAMBHA) 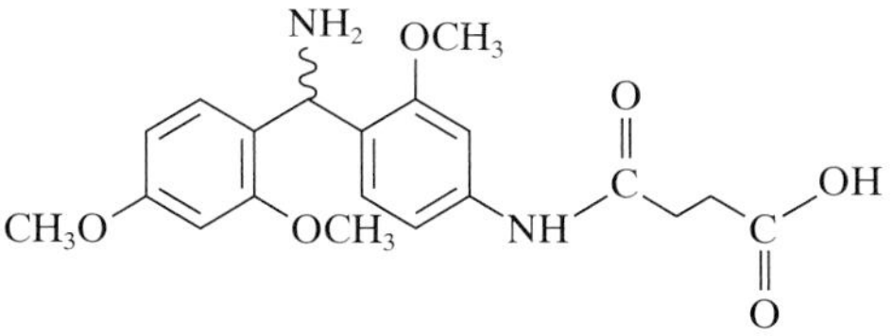	TFA	*n*
4-[4,4′-Bis(methylsulfinyl)-2-oxybenzhydrylamino]butanoic acid (SCAL) 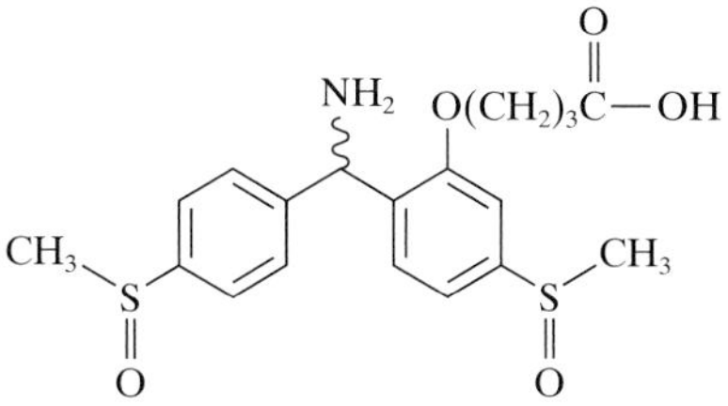	$(CH_3)_3SiBr/$ thioanisole/TFA	*o*
3-(α-Amino-2′-methoxybenzyl)-4-methoxyphenoxyacetic acid (DAL) 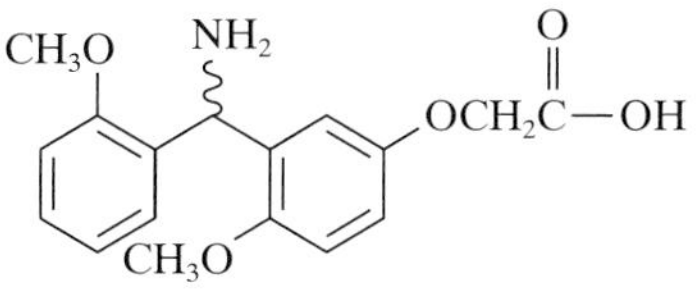	TFA	*p*

(*continued*)

TABLE III (continued)

[a] Handle and resin structures in this table are drawn with the point of attachment to the resin on the far right and the benzhydrylamino (free or protected, see cited papers for details) group on the top center. Use of these handles and resins follows the principles described in Scheme 4 (see also accompanying discussion in text).

[b] P. G. Pietta and G. R. Marshall, *Chem. Commun.,* 650 (1970); P. G. Pietta, P. F. Cavallo, K. Takahashi, and G. Marshall, *J. Org. Chem.* **39,** 44 (1974); and W. M. Bryan, *J. Org. Chem.* **51,** 3371 (1986).

[c] G. R. Matsueda and J. M. Stewart, *Peptides* **2,** 45 (1981); and M. Christensen, O. Schou, and V. S. Pedersen, *Acta Chem. Scand., Ser. B* **35,** 573 (1981).

[d] P. G. Pietta, P. F. Cavallo, G. Marshall, and M. Pace, *Gazz. Chim. Ital.* **103,** 483 (1973); and R. C. Orlowski, R. Walter, and D. Winkler, *J. Org. Chem.* **41,** 3701 (1976).

[e] B. Penke and J. Rivier, *J. Org. Chem.* **52,** 1197 (1987).

[f] R. Colombo, *J. Chem. Soc., Chem. Commun.,* 1012 (1981).

[g] A. Ajayaghosh and V. N. R. Pillai, *Tetrahedron Lett.* **36,** 777 (1995).

[h] S. A. Gaehde and G. R. Matsueda, *Int. J. Pept. Protein Res.* **18,** 451 (1981).

[i] W. Stüber, J. Knolle, and G. Breipohl, *Int. J. Pept. Protein Res.* **34,** 215 (1989).

[j] G. Breipohl, J. Knolle, and W. Stüber, *Tetrahedron Lett.* **28,** 5651 (1987); and W. Voelter, G. Breipohl, C. Tzougraki, and E. Jungfleisch-Turgut, *Collect. Czech. Chem. Commun.* **57,** 1707 (1992).

[k] S. Funakoshi, E. Murayama, L. Guo, N. Fujii, and H. Yajima, *J. Chem. Soc., Chem. Commun.,* 382 (1988); and S. Funakoshi, E. Murayama, L. Guo, N. Fujii, and H. Yajima, *Collect. Czech. Chem. Commun.* **53,** 2791 (1988).

[l] G. Breipohl, J. Knolle, and W. Stüber, *Int. J. Pept. Protein Res.* **34,** 262 (1989).

[m] H. Rink, *Tetrahedron Lett.* **28,** 3787 (1987).

[n] B. Penke, L. Nyerges, N. Klenk, K. Nagy, and A. Asztalos, in "Peptides: Chemistry, Biology, Interactions with Proteins" (B. Penke and A. Török, eds.), p. 121. de Gruyter, Berlin, 1988; and B. Penke and L. Nyerges, *Pept. Res.* **4,** 289 (1991).

[o] M. Pátek and M. Lebl, *Tetrahedron Lett.* **32,** 3891 (1991); and M. Patek and M. Lebl, in "Peptides: Chemistry, Structure and Biology" (R. S. Hodges and J. A. Smith, eds.), p. 146. Escom, Leiden, The Netherlands, 1994.

[p] M. Calmes, J. Daunis, D. David, and R. Jacquier, *Int. J. Pept. Protein Res.* **44,** 58 (1994).

peptide amide linker (PAL),[21,63] which gives rise to acid-labile tris-(alkoxy)benzylamides, 5-[9-(Fmoc)aminoxanthen-2 or 3-oxy]valeric acid (XAL),[35,64,65] which provide dilute acid-labile xanthenylamides, and derivatives of 3-nitro-4-aminomethylbenzoic acid (Nonb),[66] which give photolabile *o*-nitrobenzylamide anchors. 5-[4′-(Fmoc)aminomethyl-3′,5′-dimethoxyphenoxy]valeric acid (Fmoc-PAL-OH, **4**), was the first handle reagent compatible with Fmoc chemistry to be introduced, and it is used widely

[63] F. Albericio and G. Barany, *Int. J. Pept. Protein Res.* **30,** 206 (1987).

[64] R. J. Bontems, P. Hegyes, S. L. Bontems, F. Albericio, and G. Barany, in "Peptides: Chemistry and Biology" (J. A. Smith and J. E. Rivier, eds.), p. 601. Escom, Leiden, The Netherlands, 1992.

[65] Y. Han, S. L. Bontems, P. Hegyes, M. C. Munson, C. A. Minor, S. A. Kates, F. Albericio, and G. Barany, *J. Org. Chem.* **61,** 6326 (1996), and references cited therein.

[66] R. P. Hammer, F. Albericio, L. Gera, and G. Barany, *Int. J. Pept. Protein Res.* **36,** 31 (1990).

TABLE IV

Non-Benzhydrylamine-Based Handles and Resins for Synthesis of Peptide Amides[a]

Handle or resin	Cleavage conditions	Refs.
Dehydroalanine resin	HCl/acetic acid/H_2O	b
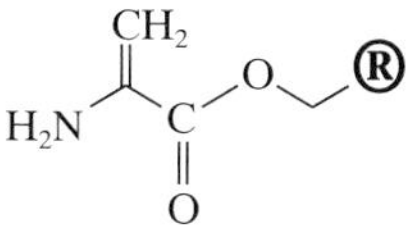		
Phenylethylamine resin (PEA)	HF	c
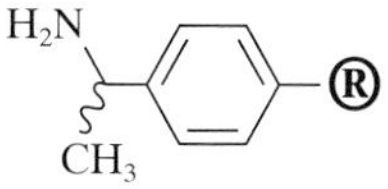		
α-Aminopropionyl polystyrene resin	$h\nu$ (350 nm)	d
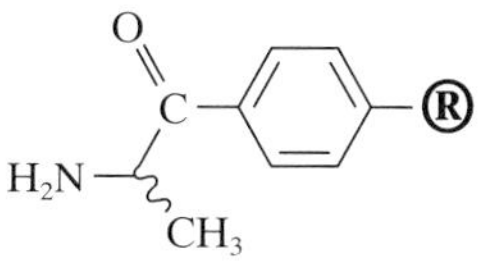		
5-[4-(1-Aminoethyl)-2-methoxy-5-nitrophenoxy]butyric acid	$h\nu$ (350 nm)	e
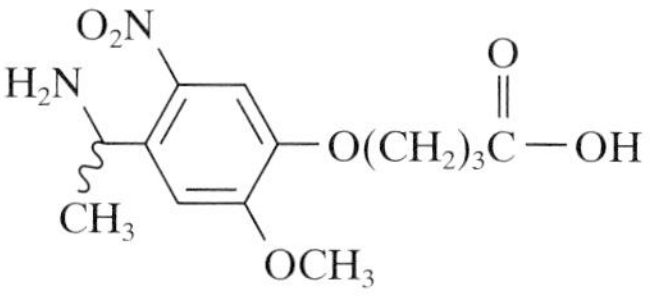		
N-MNP-4,4-diaminobutyramide resin	(i) $SnCl_2$ (ii) 100°, pH 7.0	f
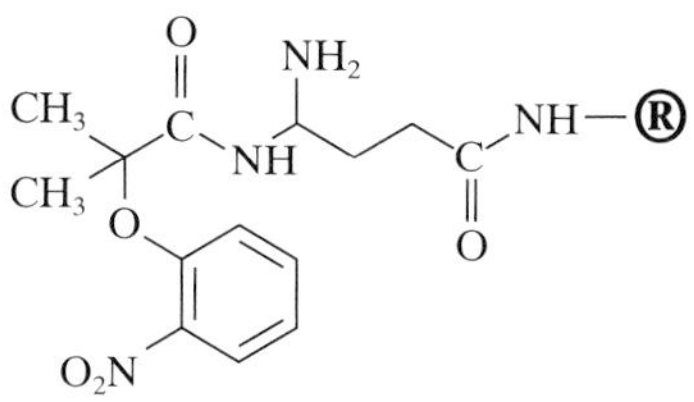		

(*continued*)

TABLE IV (*continued*)

Handle or resin	Cleavage conditions	Refs.
4-[1-Amino-2-(trimethylsilyl)ethyl]phenoxyacetic acid (SAL)	TFA	*g*
4-Amino-4-(2,4-dimethoxyphenyl)butyric acid	TFA	*h*
4-Benzyloxy-4′,4″-dimethoxytritylamine resin (BDMTA)	Dilute TFA	*i*
5-[(5-Amino-10,11-dihydrodibenzo[*a,d*]cyclohepten-2-yl)oxy]valeric acid (CHA)	Dilute TFA	*j*

owing to its reliability and superior characteristics for numerous applications.[21,67] 5-[9-(Fmoc)aminoxanthen-3-oxy]valeric acid (Fmoc-3-XAL₄-OH, **5**) is preferred for special cases, such as peptide amides with unusually sensitive tryptophan-containing sequences, and for retention of side-chain protection.[65] There follow experimental procedures for preparation of **4**

[67] M. S. Bernatowicz, S. B. Daniels, and H. Köster, *Tetrahedron Lett.* **30**, 4645 (1989).

TABLE IV (*continued*)

a Handle and resin structures in this table are drawn with the point of attachment to the resin on the far right and the amino (free or protected, see cited papers for details) group on either the left or top center. Use of these handles and resins follows the principles described in Scheme 4 (see also accompanying text discussion). The fifth, sixth, and seventh entries of Table I, namely, PAL, XAL, and Nonb, would also fit here and are not repeated.

b E. Gross, K. Noda, and B. Nisula, *Angew. Chem., Int. Ed. Engl.* **12**, 664 (1973); *Angew. Chem.* **85**, 672 (1973); and K. Noda, D. Gazis, and E. Gross, *Int. J. Pept. Protein Res.* **19**, 413 (1982).

c J. M. Stewart and G. R. Matsueda, U.S. Patent No. 3,954,709; *Chem. Abstr.* **85**, 160532 (1976).

d A. Ajayaghosh and V. N. R. Pillai, *Proc. Indian Acad. Sci., Chem. Sci.* **100**, 389 (1988).

e C. P. Holmes and D. G. Jones, *J. Org. Chem.* **60**, 2318 (1995).

f M. Pinori, G. Di Gregorio, and P. Mascagni, in "Innovation and Perspectives in Solid Phase Synthesis: Peptides, Proteins and Nucleic Acids: Biological and Biomedical Applications, 1994" (R. Epton, ed.), p. 635. Mayflower Worldwide, Birmingham, UK, 1994.

g H.-G. Chao, M. S. Bernatowicz, and G. R. Matsueda, *J. Org. Chem.* **58**, 2640 (1993); and H.-G. Chao, M. S. Bernatowicz, and G. R. Matsueda, in "Peptides: Chemistry, Structure and Biology" (R. S. Hodges and J. A. Smith, eds.), p. 138. Escom, Leiden, The Netherlands, 1994.

h M. Souček, in "Peptides 1990" (E. Giralt and D. Andreu, eds.), p. 129. Escom, Leiden, The Netherlands, 1991.

i J. Shao and W. Voelter, in "Peptides: Chemistry, Structure and Biology" (R. S. Hodges and J. A. Smith, eds.), p. 149. Escom, Leiden, The Netherlands, 1994.

j R. Ramage, S. L. Irving, and C. McInnes, *Tetrahedron Lett.* **34**, 6599 (1993); and M. Noda, M. Yamaguchi, E. Ando, K. Takeda, and K. Nokihara, *J. Org. Chem.* **59**, 7968 (1994).

(Scheme 5), starting from aldehyde **2** (a common intermediate with HAL, shown in Scheme 2), and for preparation of **5** (Scheme 6).

Preparation of 5-[4'-(9-fluorenylmethyloxycarbonyl)aminomethyl-3',5'-dimethoxyphenoxy]valeric Acid (Fmoc-PAL-OH, 4)

The amine precursor **7** (695 mg, 2.3 mmol) is suspended in a mixture of dioxane (9 ml) and 10% (w/v) aqueous Na_2CO_3 (9 ml), and Fmoc-azide (588 mg, 2.2 mmol) dissolved in dioxane (4.0 ml, plus 0.5 ml for rinsing) is added with good stirring. The pH is maintained between pH 9.5 and 9.8 by addition of further 10% (w/v) aqueous Na_2CO_3, and the progress of the reaction is followed by TLC ($CHCl_3$–methanol–acetic acid, 18:1:1, by volume). The starting amine **7** is a streak at the origin, Fmoc-azide has R_f 0.85, and product **4** has R_f 0.72. The always heterogeneous reaction mixture is stirred for 3 hr at 25°, poured into water (20 ml), extracted with ether (two times, 10 ml), and then carefully acidified to pH 3.0 by use of 6 *N* aqueous HCl (6.2 ml), with cooling. The aqueous solution is extracted with ethyl acetate (three times, 25 ml), and the combined organic phases are

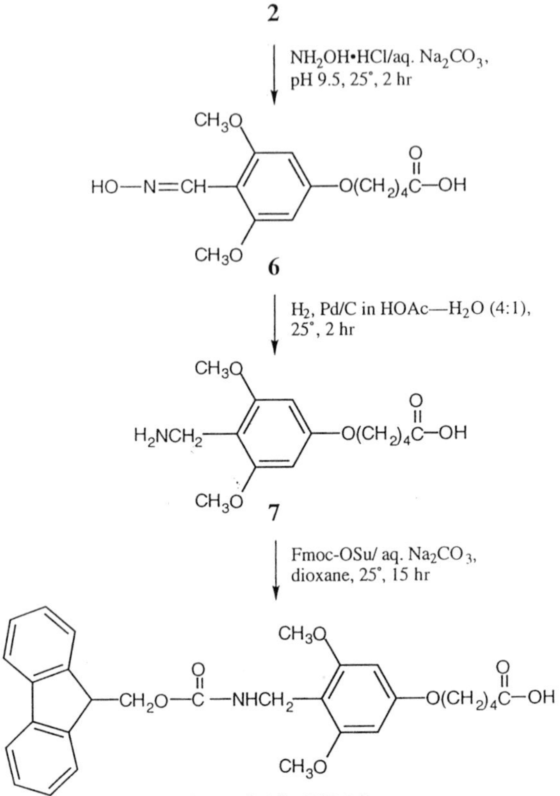

2

NH$_2$OH•HCl/aq. Na$_2$CO$_3$,
pH 9.5, 25°, 2 hr

HO—N=CH— [CH$_3$O / CH$_3$O ring] —O(CH$_2$)$_4$C—OH

6

H$_2$, Pd/C in HOAc—H$_2$O (4:1),
25°, 2 hr

H$_2$NCH$_2$— [CH$_3$O / CH$_3$O ring] —O(CH$_2$)$_4$C—OH

7

Fmoc-OSu/ aq. Na$_2$CO$_3$,
dioxane, 25°, 15 hr

—CH$_2$O—C—NHCH$_2$— [CH$_3$O / CH$_3$O ring] —O(CH$_2$)$_4$C—OH

Fmoc-PAL-OH (**4**)

SCHEME 5. An optimized synthesis of Fmoc-PAL-OH (**4**), following Albericio et al.[21] Compound **4** is then coupled onto amino-functionalized supports to create Fmoc-PAL-®. Structure **2** is in Scheme 1. See text for all corresponding experimental procedures.

washed with brine (three times, 15 ml), dried (MgSO$_4$), concentrated, and dried *in vacuo* over P$_2$O$_5$ to give an NMR and TLC pure white solid (730 mg, 65%).

Preparation of 5-[9-(9-Fluorenylmethyloxycarbonyl)aminoxanthen-3-oxy]valeric Acid (Fmoc-XAL-OH, 5)

A mixture of keto acid **10** (1.0 g, 3.2 mmol), NaOH (solid flakes) (0.62 g, 16 mmol), and *freshly* activated zinc (0.84 g, 13 mmol; see Ref. 65 for details) is refluxed in ethanol–water (19:1, v/v, 86 ml) for 6 hr to provide a light-pink solution. Zinc is removed by filtration; this is followed by washing with hot ethanol–water (19:1, v/v, three times, 5 ml), and the

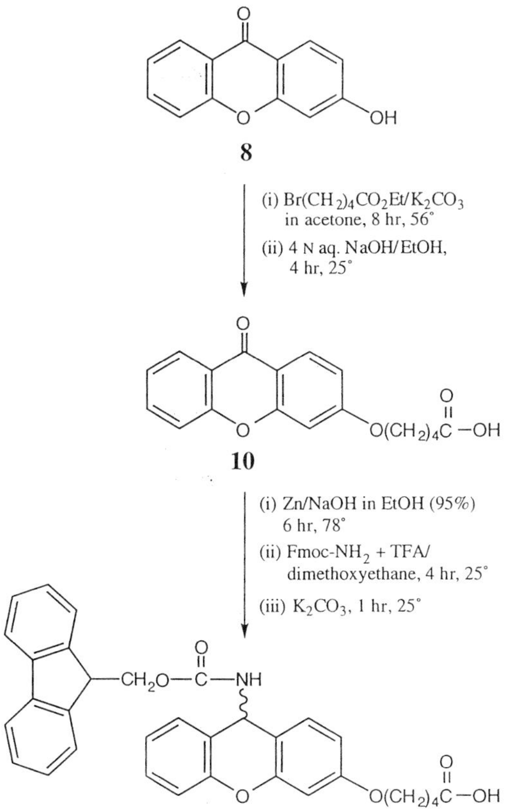

8

(i) Br(CH$_2$)$_4$CO$_2$Et/K$_2$CO$_3$
in acetone, 8 hr, 56°

(ii) 4 N aq. NaOH/EtOH,
4 hr, 25°

10

(i) Zn/NaOH in EtOH (95%)
6 hr, 78°

(ii) Fmoc-NH$_2$ + TFA/
dimethoxyethane, 4 hr, 25°

(iii) K$_2$CO$_3$, 1 hr, 25°

Fmoc-XAL-OH (**5**)

SCHEME 6. An optimized synthesis of Fmoc-3-XAL$_4$-OH (**5**), following Han *et al.*[65] Compound **5** is then coupled onto amino-functionalized supports to create Fmoc-XAL-®. See text for all corresponding experimental procedures.

combined filtrates are concentrated to give a light-pink solid that is dissolved in a mixture of TFA (2.0 ml, 26 mmol) and 1,2-dimethoxyethane (43 ml). Fmoc-NH$_2$ (0.92 g, 3.8 mmol) is added to this yellow mixture in one portion, and the resultant solution is stirred for 4 hr at 25°. Solid K$_2$CO$_3$ (0.88 g, 6.4 mmol) is then added, and within 1 hr at 25°, a white precipitate is formed. The resultant potassium salt is collected by filtration, washed with ethanol (three times, 10 ml) to remove excess Fmoc-NH$_2$, and then suspended in a mixture of water–ethanol (43 ml, 5:1, v/v) and acidified to about pH 5 by addition of 1 M aqueous KHSO$_4$ (1.4 ml). The resultant light-beige solid is collected by filtration, washed with water (three times, 10 ml), and dried *in vacuo* over P$_2$O$_5$: yield 1.33 g (77%).

Preparation of 5-[4'-Formyl-3',5'-dimethoxyphenoxy]valeric Acid Oxime (6)

Hydroxylamine hydrochloride (0.6 g, 8.6 mmol) is dissolved in water (24 ml), and solid Na_2CO_3 (1.14 g) is added until a pH of 9.5 is obtained. This basic hydroxylamine solution is added to 5-(4-formyl-3,5-dimethoxyphenoxy)valeric acid (2,[21] 1.2 g, 4.3 mmol), and the suspension is stirred for 2 hr with the pH maintained at pH 9.5 by addition of further Na_2CO_3. The resultant solution is acidified with 12 *N* aqueous HCl to pH 1 to provide the title product as a pale yellow precipitate, which is collected, washed with water (three times, 20 ml), and dried *in vacuo* over P_2O_5: yield 1.21 g (95%).

Preparation of 5-[4'-Aminomethyl-3',5'-dimethoxyphenoxy]valeric Acid (7)

The oxime precursor **6** (0.8 g, 2.7 mmol) is suspended in acetic acid–water (4:1, v/v, 40 ml) in a Parr hydrogenation vessel, and 10% Pd/C (344 mg) is added. Hydrogenation proceeds at 50 psi (3 atm) and 25° for 2 hr, at which time complete reaction is ascertained by TLC, R_f ($CHCl_3$–methanol–acetic acid, 18:1:1, by volume) 0.07 (origin). The solution is filtered and evaporated to dryness to give a dark gold oil (0.90 g, 96%), which solidifies on further standing.

Preparation of Ethyl 5-(9-Oxoxanthen-3-oxy)valerate (9)

Anhydrous K_2CO_3 (51.3 g, 372 mmol) is added to a solution of 3-hydroxyxanthone (**8**,[65] 20.5 g, 97 mmol) plus ethyl 5-bromovalerate (53.3 g, 316 mmol) in dry acetone (1.5 liters). The reaction mixture is heated to reflux for 8 hr, cooled, filtered to remove inorganic salts, washed with acetone (three times, 50 ml), and concentrated. The initial brownish crystals are washed with hexane (three times, 60 ml), providing white crystals which are dried *in vacuo* over P_2O_5: yield 28.5 g (86%).

Preparation of 5-(9-Oxoxanthen-3-oxy)valeric Acid (10)

Ethyl ester **9** (28.5 g, 84 mmol) is suspended in a mixture of ethanol–water (19:1, v/v, 210 ml) and 4 *N* aqueous NaOH (210 ml); a yellow homogeneous solution is obtained within 30 min and stirring is continued at 25° for a total of 4 hr. Partial evaporation to remove ethanol is followed by addition of water (200 ml) followed by 6 *N* aqueous HCl under cooling to bring the pH to 3.5. The resultant white precipitate is collected by filtration, washed with water (three times 50 ml), and dried *in vacuo* over P_2O_5: yield 26.0 g (99%).

Handles for Synthesis of C-Terminal Modified Peptides

Earlier sections of this article have rationalized why peptides for biological investigations and pharmaceutical development are required usually as their C-terminal acid or amide derivatives. Nevertheless, the myriad of naturally occurring peptides, and their biologically relevant synthetic analogs, extends to numerous cases where the carboxyl termini are modified as other functionalities, as elaborated further in the discussion that follows. Similarly, the importance of cyclic peptides (which have no termini), both of natural origin and as designed/constructed by peptide scientists, is well established.[67a,68–72]

C-Terminal modified peptides serve a variety of often overlapping purposes. For example, such peptides can be substrates used to study the active sites and catalytic mechanisms of target enzymes, or to facilitate sensitive assays for enzymatic processes, and as peptide inhibitors and analogs with specifically designed biological activities. In addition, they can serve as activated (and/or protected) segments for convergent syntheses.[2,23–27] For this last-mentioned case, the C-terminal modifications serve to activate the corresponding fragments for later coupling to form the larger target peptide. One strategy involves preparation of peptide hydrazides,[2,73,74] which can be converted to the corresponding azides.[75,76] Appropriately activated peptide thioacids and thioesters can also be used effectively in segment condensa-

[67a] C. Blackburn and S. A. Kates, *Methods Enzymol.* **289**, [9], 1997 (this volume).

[68] Y. A. Ovchinnikov and V. T. Ivanov, in "The Proteins" (H. Neurath and R. L. Hill, eds.), p. 307. Academic Press, New York, 1982.

[69] A. E. Tonelli, in "Cyclic Polymers" (J. A. Semlyen, ed.), p. 261. Elsevier, London, 1986.

[70] V. J. Hruby, F. Al-Obeidi, and W. Kazmierski, *Biochem. J.* **268**, 249 (1990).

[71] J. Rizo and L. M. Gierasch, *Annu. Rev. Biochem.* **61**, 387 (1992).

[72] S. A. Kates, N. A. Solé, F. Albericio, and G. Barany, in "Peptides: Design, Synthesis, and Biological Activity" (C. Basava and G. M. Anantharamaiah, eds.), p. 39. Birkhaeuser, Boston, 1994.

[73] R. Ramage, S. L. Irving, and C. McInnes, *Tetrahedron Lett.* **34**, 6599 (1993).

[74] R. Ramage, A. R. Brown, C. McInnes, S. L. Irving, S. G. Love, T. D. Pallin, T. W. Muir, G. Raphy, K. Shaw, F. O. Wahl, and J. Wilken, in "Innovation and Perspectives in Solid Phase Synthesis: Peptides, Proteins and Nucleic Acids: Biological and Biomedical Applications, 1994" (R. Epton, ed.), p. 1. Mayflower Worldwide, Birmingham, UK, 1994.

[75] J. Honzl and J. Rudinger, *Collect. Czech. Chem. Commun.* **26**, 2333 (1961).

[76] H. B. Milne and C. F. Most, *J. Org. Chem.* **33**, 169 (1968).

tion and ligation.[76a,77] A similar approach[78] involves linking two segments through a thioether bond formed by alkylating one segment, modified as a C-terminal thiol, with the other segment modified at its N terminus with a bromoacetyl group. Aldehyde end groups at the C terminus are sites for ligation with a variety of nucleophile-capped segments, resulting in the creation of nonnative peptidomimetic links.[76a]

For enzymological studies, peptidyl substrates may be modified to include functionalities capable of forming covalent bonds with the host enzyme.[79] The resultant enzyme–substrate complexes are then analyzed by a variety of techniques to identify the amino acid side chains involved in binding and catalysis, thus providing a detailed map of the active site. Functionalities most often employed for this purpose[80] include haloketones,[81-83] which are capable of alkylating the active site His in serine proteases, and aldehydes[84-87] and nitriles,[87-89] which are generally believed to mimic the geometry of the tetrahedral transition states and trigonal acyl-enzyme intermediates (respectively) formed during enzymatic catalysis.

Peptidyl substrates can be modified at the C terminus by addition of various chromophores, and then used for drug screening assays,[90] mapping enzyme active sites,[91,92] determining enzyme kinetics[93] and specifici-

[76a] S. B. H. Kent, *Methods Enzymol.* **289**, [13], 1997 (this volume); T. W. Muir, P. E. Dawson, M. C. Fitzgerald, and S. B. H. Kent, *Methods Enzymol.* **289**, [24], 1997 (this volume); J. P. Tam and J. C. Spetzler, *Methods Enzymol.* **289**, [28], 1997 (this volume).

[77] P. Lloyd-Williams, F. Albericio, and E. Giralt, "Chemical Approaches to the Synthesis of Peptides and Proteins." CRC, Boca Raton, Florida, 1997.

[78] D. R. Englebretsen, B. G. Garnham, D. A. Bergman, and P. F. Alewood, *Tetrahedron Lett.* **36**, 8871 (1995).

[79] E. Shaw, *Physiol. Rev.* **50**, 244 (1970).

[80] E. Shaw, *Adv. Enzymol.* **63**, 271 (1990).

[81] R. C. Thompson and E. R. Blout, *Biochemistry* **12**, 44 (1973).

[82] G. Schoellmann and E. Shaw, *Biochemistry* **2**, 252 (1963).

[83] E. Shaw, M. Mares-Guia, and W. Cohen, *Biochemistry* **4**, 2219 (1965).

[84] J. O. Westerick and R. Wolfenden, *J. Biol. Chem.* **247**, 8195 (1972).

[85] R. C. Thompson, *Biochemistry* **12**, 47 (1973).

[86] E. Sarubbi, P. F. Seneci, M. R. Angelastro, N. P. Peet, M. Denaro, and K. Islam, *FEBS Lett.* **319**, 253 (1993).

[87] É. Dufour, A. C. Storer, and R. Ménard, *Biochemistry* **34**, 9136 (1995).

[88] S. A. Thompson, P. R. Andrews, and R. P. Hanzlik, *J. Med. Chem.* **29**, 104 (1986).

[89] R. P. Hanzlik, J. Zygmunt, and J. B. Moon, *Biochim. Biophys. Acta* **1035**, 62 (1990).

[90] M. W. Pennington and P. Baur, *Lett. Pept. Sci.* **1**, 143 (1994).

[91] B. J. McRae, K. Kurachi, R. L. Heimark, K. Fujikawa, E. W. Davie, and J. C. Powers, *Biochemistry* **20**, 7196 (1981).

[92] B. J. McRae, T.-Y. Lin, and J. C. Powers, *J. Biol. Chem.* **256**, 12362 (1981).

[93] G. D. J. Green and E. Shaw, *Anal. Biochem.* **93**, 223 (1979).

ties,[94] and providing sensitive means for detecting specific enzymes.[95,96] Several chromophores are used, including naphthylamides,[95] p-nitroanilides,[94–98] coumarylamides,[95,96,99,100] and (indirectly by virtue of quantitation of released thiol) thioesters.[91–93,95,101] Internally quenched fluorogenic peptides, where both N and C termini are modified with halves of a fluorescent donor–acceptor pair, have been designed[90,102]; enzymatic cleavage of the scissile bond is accompanied by increased fluorescent emission.

Most research involving C-terminal modified peptides has focused on the development of peptide analogs with altered biological activity, either as effectors or as inhibitors. In fact, many of the peptide analogs already mentioned also fit under these categories. For example, peptide chloromethyl ketones were among the earliest C-terminal modifications studied and shown to be potent inhibitors of serine and cysteine proteinases.[79,80] Similarly, peptide aldehydes, which were first isolated from microbial culture filtrates,[103,104] possess inhibitory activity against a variety of proteolytic enzymes[80,105,106]; substitution of an aldehyde for the α-carboxyl remains an often employed strategy for peptide modification.[86,107–110] Alkylamides are

[94] A. J. Whitmore, R. M. Daniel, and H. H. Petach, *Tetrahedron Lett.* **36,** 475 (1995).

[95] M. J. Castillo, K. Nakajima, M. Zimmerman, and J. C. Powers, *Anal. Biochem.* **99,** 53 (1979).

[96] A. J. Barrett and H. Kirschke, *Methods Enzymol.* **80,** 535 (1981).

[97] L. A. Reiter, *Int. J. Pept. Protein Res.* **43,** 87 (1994).

[98] J. C. H. M. Wijkmans, A. Nagel, and W. Bloemhoff, *Lett. Pept. Sci.* **3,** 107 (1996).

[99] T. Morita, H. Kato, S. Iwanaga, K. Takada, T. Kimura, and S. Sakakibara, *J. Biochem.* (*Tokyo*) **82,** 1495 (1977).

[100] L. C. Alves, P. C. Almeida, L. Franzoni, L. Juliano, and M. A. Juliano, *Pept. Res.* **9,** 92 (1996).

[101] D. A. Farmer and J. H. Hageman, *J. Biol. Chem.* **250,** 7366 (1975).

[102] I. Y. Hirata, M. H. S. Cezari, C. R. Nakaie, P. Boschcov, A. S. Ito, M. A. Juliano, and L. Juliano, *Lett. Pept. Sci.* **1,** 299 (1994).

[103] H. Umezawa, "Enzyme Inhibitors of Microbial Origin." University Park Press, Tokyo, 1972.

[104] T. Aoyagi and H. Umezawa, *in* "Proteases and Biological Control" (E. Reich, D. B. Rifkin, and E. Shaw, eds.), p. 429. Cold Spring Harbor Laboratory Press, Cold Spring Harbor, New York, 1975.

[105] H. Umezawa, *Annu. Rev. Microbiol.* **36,** 75 (1982).

[106] J.-A. Fehrentz, M. Paris, A. Heitz, J. Velek, C.-F. Liu, F. Winternitz, and J. Martinez, *Tetrahedron Lett.* **36,** 7871 (1995).

[107] K. T. Chapman, *Bioorg. Med. Chem. Lett.* **2,** 613 (1992).

[108] R. T. Shuman, R. B. Rothenberger, C. S. Campbell, G. F. Smith, D. S. Gifford-Moore, and P. D. Gesellchen, *J. Med. Chem.* **36,** 314 (1993).

[109] J.-T. Woo, S. Sigeizumi, K. Yamaguchi, K. Sugimoto, T. Kobori, T. Tsuji, and K. Kondo, *Bioorg. Med. Chem. Lett.* **5,** 1501 (1995).

[110] S. W. Kaldor, M. Hammond, B. A. Dressman, J. M. Labus, F. W. Chadwell, A. D. Kline, and B. A. Heinz, *Bioorg. Med. Chem. Lett.* **5,** 2021 (1995).

tried frequently for the development of peptide analogs, although early studies with gastrin,[42] eledoisin,[111] and oxytocin[112] all reported significant decreases in the bioactivities of the C-terminal N-alkylamide derivatives relative to the native peptides. The first example of a highly active peptide alkylamide was reported by Fujino and co-workers,[113,114] who prepared various analogs of luteinizing hormone-releasing hormone (LHRH) and observed about five times more hormonal activity with the ethylamide derivative relative to native LHRH.

Despite their inherent biological potencies, the use of peptides in biological systems is often stymied for a variety of reasons.[115] Modifications at the C terminus, as well as the loss of termini via cyclization, represent avenues to more effective therapeutic agents, because they allow alteration of bioavailability of a peptide by protecting it from enzymatic degradation,[116–118] by improving its ability to cross various biological barriers,[119] such as the blood–brain barrier,[120] or by simply increasing its solubility.[121] Such modifications can also lead to increased enzymatic binding and substrate specificity.[117,122,123] Several hundred C-terminal modified analogs of LHRH have been prepared,[117,122] as well as analogs of such biologically important peptides as the enkephalins[116,119] and

[111] L. Bernardi, G. Bosisio, F. Chillemi, G. de Caro, R. de Castiglione, V. Erspamer, A. Glaesser, and O. Goffredo, *Experientia* **20,** 306 (1964).

[112] H. Takashima, W. Fraefel, and V. du Vigneaud, *J. Am. Chem. Soc.* **91,** 6182 (1969).

[113] M. Fujino, S. Kobayashi, M. Obayashi, S. Shinagawa, T. Fukuda, C. Kitada, R. Nakayama, I. Yamazaki, W. F. White, and R. H. Rippel, *Biochem. Biophys. Res. Commun.* **49,** 863 (1972).

[114] M. Fujino, S. Shinagawa, I. Yamazaki, S. Kobayashi, M. Obayashi, T. Fukuda, R. Nakayama, W. F. White, and R. H. Rippel, *Arch. Biochem. Biophys.* **154,** 488 (1973).

[115] S. A. St-Pierre, in "Neural and Endocrine Peptides and Receptors" (T. W. Moody, ed.), p. 653. Plenum, New York, 1986.

[116] K. B. Mathur, in "Advances in the Biosciences. Volume 38. Current Status of Centrally Acting Peptides" (B. N. Dhawan, ed.), p. 37. Pergamon, Oxford, 1982.

[117] M. J. Karten and J. E. Rivier, *Endocr. Rev.* **7,** 44 (1986).

[118] D. C. Heimbrook, W. S. Saari, N. L. Balishin, T. W. Fisher, A. Friedman, D. M. Kiefer, N. S. Rotberg, J. W. Wallen, and A. Oliff, *J. Med. Chem.* **34,** 2102 (1991).

[119] J. S. Morley, *Annu. Rev. Pharmacol. Toxicol.* **20,** 81 (1980).

[120] S. Bajusz, A. Patthy, Á. Kenessey, L. Gráf, J. I. Székely, and A. Z. Rónai, *Biochem. Biophys. Res. Commun.* **84,** 1045 (1978).

[121] S. A. Boyd, A. K. L. Fung, W. R. Baker, R. A. Mantei, Y.-L. Armiger, H. H. Stein, J. Cohen, D. A. Egan, J. L. Barlow, V. Klinghofer, K. M. Verburg, D. L. Martin, G. A. Young, J. S. Polakowski, D. J. Hoffman, K. W. Garren, T. J. Perun, and H. D. Kleinert, *J. Med. Chem.* **35,** 1735 (1992).

[122] D. H. Coy and A. V. Schally, *Ann. Clin. Res.* **10,** 139 (1978).

[123] B. J. Dhotre, S. Chaturvedi, and K. B. Mathur, *Indian J. Chem., Sect. B* **23B,** 828 (1984).

bombesin.[124–126] Many other C-terminal moieties, in addition to those already discussed, have been studied, including alcohols,[127–129] ethers,[118] esters,[130] thiols,[131] dialkylamides,[132,133] aminoalkylamides,[134,135] hydroxyalkylamides,[131,136,137] haloalkylamides,[137,138] hydrazides,[134,136,137,139] trifluoromethyl ketones,[140,141] α-ketoaldehydes,[142] semicarbazones[143–145] and related

[124] J. R. Best, P. Byrne, R. Cotton, A. S. Dutta, B. Fleming, A. Garner, J. J. Gormley, C. F. Hayward, P. F. McLachlan, and P. B. Scholes, *Drug Design Delivery* **5**, 267 (1990).

[125] L.-H. Wang, D. H. Coy, J. E. Taylor, N.-Y. Jiang, S. H. Kim, J.-P. Moreau, S. C. Huang, S. A. Mantey, H. Frucht, and R. T. Jensen, *Biochemistry* **29**, 616 (1990).

[126] R. T. Jensen and D. H. Coy, *Trends Pharmacol. Sci.* **12**, 13 (1991).

[127] Y. Kiso, M. Yamaguchi, T. Akita, H. Moritoki, M. Takei, and H. Nakamura, *Naturwissenschaften* **68**, 210 (1981).

[128] J. L. Krstenansky, M. H. Payne, T. J. Owen, M. T. Yates, and S. J. T. Mao, *Thromb. Res.* **54**, 319 (1989).

[129] H. Wenschuh, M. Beyermann, H. Haber, J. K. Seydel, E. Krause, M. Bienert, L. A. Carpino, A. El-Faham, and F. Albericio, *J. Org. Chem.* **60**, 405 (1995).

[130] D. C. Heimbrook, W. S. Saari, N. L. Balishin, A. Friedman, K. S. Moore, M. W. Riemen, D. M. Kiefer, N. S. Rotberg, J. W. Wallen, and A. Oliff, *J. Biol. Chem.* **264**, 11258 (1989).

[131] H. Kodama, H. Uchida, T. Yasunaga, M. Kondo, T. Costa, and Y. Shimohigashi, *J. Mol. Recognit.* **3**, 197 (1990).

[132] Y. Kiso, T. Miyazaki, T. Akita, H. Moritoki, M. Takei, and H. Nakamura, *FEBS Lett.* **136**, 101 (1981).

[133] F. C. Buonomo, J. S. Tou, and L. A. Kaempfe, *Life Sci.* **48**, 1953 (1991).

[134] Z. Grzonka, Z. Palacz, L. Baran, E. Przegalinski, and G. Kupryszewski, *Pol. J. Chem.* **55**, 1025 (1981).

[135] E. V. Grishin, T. M. Volkova, and A. S. Arseniev, *Toxicon* **27**, 541 (1989).

[136] K. Suzuki, H. Fujita, Y. Sasaki, M. Shiratori, S. Sakurada, and K. Kisara, *Chem. Pharm. Bull.* **36**, 4834 (1988).

[137] H. Fujita, Y. Sasaki, H. Kohno, Y. Ohkubo, A. Ambo, K. Suzuki, and M. Hino, *Chem. Pharm. Bull.* **38**, 2197 (1990).

[138] D. H. Coy, J. A. Vilchez-Martinez, E. J. Coy, N. Nishi, A. Arimura, and A. V. Schally, *Biochemistry* **14**, 1848 (1975).

[139] R. F. Geraghty, G. B. Irvine, C. H. Williams, and G. A. Cottrell, *Peptides* **15**, 73 (1994).

[140] R. A. Smith, L. J. Copp, S. L. Donnelly, R. W. Spencer, and A. Krantz, *Biochemistry* **27**, 6568 (1988).

[141] P. D. Edwards, *Tetrahedron Lett.* **33**, 4279 (1992).

[142] B. Walker, N. McCarthy, A. Healy, T. Ye, and M. A. McKervey, *Biochem. J.* **293**, 321 (1993).

[143] I. J. Galpin, A. H. Wilby, and R. J. Beynon, *in* "Peptides 1982" (K. Bláha and P. Maloň, eds.), p. 649. de Gruyter, Berlin, 1983.

[144] T. L. Graybill, R. E. Dolle, C. T. Helaszek, R. E. Miller, and M. A. Ator, *Int. J. Pept. Protein Res.* **44**, 173 (1994).

[145] A. Basak, F. Jean, N. G. Seidah, and C. Lazure, *Int. J. Pept. Protein Res.* **44**, 253 (1994).

groups,[146] thioesters,[91,92,147] and thioamides,[148–150] as well as boronic acids,[151] sulfonic acids,[152,153] and phosphonic acids.[154,155]

Because so much of the biological activity of a peptide can be controlled through changes at the C terminus, it is important to have efficient methods for the synthesis of such derivatives. Most preparative techniques to access peptides modified at the C terminus rely on solution-phase transformations carried out according to one of the following modes: (1) standard peptide synthesis (usually in the C → N direction) beginning with the incorporation of an amino acid building block appropriately modified at the C terminus; (2) extension of a previously synthesized peptide acid through its free C-terminal carboxyl by such a modified building block; or (3) conversion of a more readily available C-terminal functional group to a more sensitive one. The first two categories often entail relatively straightforward procedures, as is the case for many of the various substituted and unsubstituted alkylamides,[113,136,138,156–158] thioesters,[91,92,101,147] and chromophoric groups[90,95,102] listed earlier. As examples of the third category, aminolyses of peptide esters readily provide the corresponding alkylamides,[134,139,159] whereas aldehydes are often prepared[160] by reduction of peptide and amino

[146] K. Folkers, C. Y. Bowers, W. B. Lutz, K. Friebel, T. Kubiak, B. Schircks, and G. Rampold, *Z. Naturforsch.* **37B**, 1075 (1982).

[147] R. R. Cook, B. J. McRae, and J. C. Powers, *Arch. Biochem. Biophys.* **234**, 82 (1984).

[148] M. Thorsen, B. Yde, U. Pedersen, K. Clausen, and S.-O. Lawesson, *Tetrahedron* **39**, 3429 (1983).

[149] A. J. Douglas, B. Walker, D. T. Elmore, and R. F. Murphy, *Biochem. Soc. Trans.* **15**, 927 (1987).

[150] M. Kruszyński, G. Kupryszewski, K. Misterek, and S. Gumułka, *Pol. J. Pharmacol. Pharm.* **42**, 483 (1990).

[151] D. E. Zembower, C. L. Neudauer, M. J. Wick, and M. M. Ames, *Int. J. Pept. Protein Res.* **47**, 405 (1996).

[152] R. W. Roeske and T. P. Hrinyo-Pavlina, in "Peptides: Chemistry and Biology" (G. R. Marshall, ed.), p. 273. Escom, Leiden, The Netherlands, 1988.

[153] S. Bajusz, A. Z. Rónai, J. I. Székely, A. Turán, A. Juhász, A. Patthy, E. Miglécz, and I. Berzétei, *FEBS Lett.* **117**, 308 (1980).

[154] P. Mastalerz, L. Kupczyk-Subotkowska, Z. S. Herman, and G. Łaskawiec, *Naturwissenschaften* **69**, 46 (1982).

[155] A. Páldi, M. Móra, S. Bajusz, and L. Gráf, *Int. J. Pept. Protein Res.* **29**, 746 (1987).

[156] M. Fujino, S. Shinagawa, M. Obayashi, S. Kobayashi, T. Fukuda, I. Yamazaki, R. Nakayama, W. F. White, and R. H. Rippel, *J. Med. Chem.* **16**, 1144 (1973).

[157] E. Kasafírek, I. Šutiaková, M. Bartík, and A. Šturc, *Collect. Czech. Chem. Commun.* **53**, 2877 (1988).

[158] L. Zhang, W. Rapp, C. Goldammer, and E. Bayer, in "Innovation and Perspectives in Solid Phase Synthesis: Peptides, Proteins and Nucleic Acids: Biological and Biomedical Applications, 1994" (R. Epton, ed.), p. 717. Mayflower Worldwide, Birmingham, UK, 1994.

[159] S. Mammi and M. Goodman, *Int. J. Pept. Protein Res.* **28**, 29 (1986).

[160] J. Jurczak and A. Gołębiowski, *Chem. Rev.* **89**, 149 (1989).

esters[161–163] (peptide alcohols also result from similar procedures[161,164,165]), thioesters,[166,167] N-methoxy-N-methylamides,[110,140,168] or, in the case of Arg, the δ-lactam[108,169]; alternatively, aldehydes are obtained by oxidation of the respective alcohols.[107,109,170–172] As another example, thioamides can be prepared from amides by treatment with Lawesson's reagent [2,4-bis(4-methoxyphenyl)-1,3-dithia-2,4-diphosphetane 2,4-disulfide].[148–150] Nevertheless, preparation of several desirable C-terminal functionalities can be quite arduous, often requiring special coupling procedures such as those used to access p-nitroanilides,[94,97,173–177] or multistep procedures such as those reported to access mercaptoalkylamides[131] and boronic acids.[151]

Options for the preparation of C-terminal modified peptides on solid-phase resins are more limited. The most frequently used method generalizes the previously discussed ammonolytic cleavage of benzyl esters (Scheme 3) with other appropriate nucleophiles and/or more labile anchoring linkages. Cleavage of the completed resin-bound peptide by either amines or alcohols results in formation of either the corresponding peptide N-alkylamides[112,124,178–182] or esters,[124,181,183–185] respectively. Similarly, use of hydra-

[161] A. Ito, R. Takahashi, C. Miura, and Y. Baba, *Chem. Pharm. Bull.* **23**, 3106 (1975).

[162] P. Zlatoidsky, *Helv. Chim. Acta* **77**, 150 (1994).

[163] I. J. Galpin, A. H. Wilby, G. A. Place, and R. J. Beynon, *Int. J. Peptide Protein Res.* **23**, 477 (1984).

[164] K. Soai, H. Oyamada, and M. Takase, *Bull. Chem. Soc. Jpn.* **57**, 2327 (1984).

[165] G. Kokotos, *Synthesis* 299 (1990).

[166] P. T. Ho and K.-y. Ngu, *J. Org. Chem.* **58**, 2313 (1993).

[167] B. A. Malcolm, C. Lowe, S. Shechosky, R. T. McKay, C. C. Yang, V. J. Shah, R. J. Simon, J. C. Vederas, and D. V. Smith, *Biochemistry* **34**, 8172 (1995).

[168] J.-A. Fehrentz, A. Heitz, and B. Castro, *Int. J. Peptide Protein Res.* **26**, 236 (1985).

[169] S. Bajusz, E. Barabás, E. Széll, and D. Bagdy, *in* "Peptides: Chemistry, Structure, and Biology" (R. Walter and J. Meienhofer, eds.), p. 603. Ann Arbor Science, Ann Arbor, Michigan, 1976.

[170] R. C. Thompson, *Methods Enzymol.* **46**, 220 (1977).

[171] Y. Hamada and T. Shioiri, *Chem. Pharm. Bull.* **30**, 1921 (1982).

[172] A. Damodaran and R. B. Harris, *J. Protein Chem.* **14**, 431 (1995).

[173] K. Noda, M. Oda, M. Sato, and N. Yoshida, *Int. J. Pept. Protein Res.* **36**, 197 (1990).

[174] D. T. S. Rijkers, H. C. Hemker, G. H. I. Nefkens, and G. I. Tesser, *Recl. Trav. Chim. Pays-Bas* **110**, 347 (1991).

[175] H. Nedev, H. Naharisoa, and T. Haertlé, *Tetrahedron Lett.* **34**, 4201 (1993).

[176] V. F. Pozdnev, *Int. J. Pept. Protein Res.* **44**, 36 (1994).

[177] M. Schutkowski, C. Mrestani-Klaus, and K. Neubert, *Int. J. Pept. Protein Res.* **45**, 257 (1995).

[178] D. H. Coy, E. J. Coy, A. V. Schally, J. A. Vilchez-Martinez, L. Debeljuk, W. H. Carter, and A. Arimura, *Biochemistry* **13**, 323 (1974).

[179] T. J. Lobl and L. L. Maggiora, *J. Org. Chem.* **53**, 1979 (1988).

[180] M. Mergler and R. Nyfeler, *in* "Innovation and Perspectives in Solid Phase Synthesis: Peptides, Polypeptides, and Oligonucleotides, 1992" (R. Epton, ed.), p. 429. Intercept, Andover, UK, 1992.

zine[138,186-188] results in peptide hydrazides, and bifunctional amines, such as amino alcohols[180,189-191] and diamines,[180] readily provide hydroxyalkylamides and aminoalkylamides, respectively. Nucleophilic cleavage to release peptide derivatives works well with a variety of benzyl or phenacyl or related ester linkages; of these, the oxime resin developed by DeGrado and Kaiser[24,192,193] (shown in Table II) is the one used most widely. The oxime resin is stable to Boc chemistry chain assembly procedures and can be cleaved by nucleophiles as gentle as N^{α}-unprotected amino acids,[194,195] and their C-terminal modified derivatives,[196] hence allowing easy access to peptide alcohols[128,139] and esters,[192,196,197] among other derivatives. Peptides have also been cleaved effectively from oxime resins by use of sulfur nucleophiles to provide the corresponding thioacids,[198] and cyclic peptides can be obtained in some cases by head-to-tail nucleophilic attack by the peptide N terminus to cleave an (activated) anchoring linkage.[67a,72,199] An interesting synthesis of peptide alcohols occurs by reductive cleavage with borohydride

[181] E. Nicolás, J. Clemente, M. Perelló, F. Albericio, E. Pedroso, and E. Giralt, *Tetrahedron Lett.* **33**, 2183 (1992).

[182] N. Voyer, A. Lavoie, M. Pinette, and J. Bernier, *Tetrahedron Lett* **35**, 355 (1994).

[183] F. Baleux, J. Daunis, R. Jacquier, and B. Calas, *Tetrahedron Lett.* **25**, 5893 (1984).

[184] B. Calas, J. Mery, J. Parello, and A. Cave, *Tetrahedron* **41**, 5331 (1985).

[185] M. Mergler and R. Nyfeler, *in* "Innovation and Perspectives in Solid Phase Synthesis: Peptides, Proteins and Nucleic Acids: Biological and Biomedical Applications, 1994" (R. Epton, ed.), p. 599. Mayflower Worldwide, Birmingham, UK, 1994.

[186] T. Mizoguchi, K. Shigezane, and N. Takamura, *Chem. Pharm. Bull.* **18**, 1465 (1970).

[187] R. Epton, G. Marr, P. W. Small, and G. A. Willmore, *Int. J. Biol. Macromol.* **4**, 62 (1982).

[188] M. Mergler and R. Nyfeler, *in* "Peptides: Chemistry and Biology" (J. A. Smith and J. E. Rivier, eds.), p. 551. Escom, Leiden, The Netherlands, 1992.

[189] K. U. Prasad, T. L. Trapane, D. Busath, G. Szabo, and D. W. Urry, *Int. J. Pept. Protein Res.* **19**, 162 (1982).

[190] G. B. Fields, C. G. Fields, J. Petefish, H. E. Van Wart, and T. A. Cross, *Proc. Natl. Acad. Sci. U.S.A.* **85**, 1384 (1988).

[191] W. B. Edwards, C. G. Fields, C. J. Anderson, T. S. Pajeau, M. J. Welch, and G. B. Fields, *J. Med. Chem.* **37**, 3749 (1994).

[192] W. F. DeGrado and E. T. Kaiser, *J. Org. Chem.* **45**, 1295 (1980).

[193] S. H. Nakagawa and E. T. Kaiser, *J. Org. Chem.* **48**, 678 (1983).

[194] P. T. Lansbury, Jr., J. C. Hendrix, and A. I. Coffman, *Tetrahedron Lett.* **30**, 4915 (1989).

[195] J. C. Hendrix, J. T. Jarrett, S. T. Anisfeld, and P. T. Lansbury, Jr., *J. Org. Chem.* **57**, 3414 (1992).

[196] L. M. Siemens, F. W. Rottnek, and L. S. Trzupek, *J. Org. Chem.* **55**, 3507 (1990).

[197] W. F. DeGrado and E. T. Kaiser, *J. Org. Chem.* **47**, 3258 (1982).

[198] A. W. Schwabacher and T. L. Maynard, *Tetrahedron Lett.* **34**, 1269 (1993).

[199] G. Ösapay and J. W. Taylor, *J. Am. Chem. Soc.* **112**, 6046 (1990).

reagents of benzyl ester linkages.[49,200] All methods involving nucleophilic attack on an ester, including oximes, suffer from the risk of racemization, steric effects, and a number of residue-specific side reactions.[2,6,26,27,51] Furthermore, these methods may be incompatible with 9-fluorenylmethyloxycarbonyl (Fmoc) chemistry,[4–6] because of the susceptibility of some anchoring linkages to the base needed to remove Fmoc.

Alternative solid-phase approaches to C-terminal modified peptides follow principles described earlier in this article (Fig. 1) and involve the use of a handle or resin derivative designed to anchor the C terminus of the peptide through the functionality ultimately desired, in much the same way as described for peptide amide-forming handles (see Scheme 4 and accompanying discussion). In this way, when the completed peptide–resin is cleaved, the desired peptide is released without need for further solution-phase transformations (or only minor manipulations), and many of the problems associated with previously discussed techniques can be avoided. As examples, a number of suitable preformed thioester handles have been immobilized onto amino-functionalized resins with later cleavage giving thioacids,[201–205] and alcohols have been successfully anchored to, and cleaved from,[129] the Barlos chlorotrityl resin[37,38] (the structure of that resin is provided in Fig. 3). Further handles and resins have been developed along that theme to access various specific functional groups (Table V). Another effective technique requires appending a small bifunctional "spacer" molecule to the modified C terminus and then coupling the free end of this spacer to a standard linker or resin. The peptide thus anchored can be cleaved by standard means, and the spacer subsequently removed in solution. Functionalities prepared in this manner include peptide alcohols,[206–208] linked through succinic acid, and peptide sulfonic acids,[152] linked through 4-hydroxybenzoic acid.

An approach with considerable general applicability for the solid-phase synthesis of C-terminal modified peptides is the use of side-chain anchor-

[200] M. Mergler and R. Nyfeler, in "Peptides 1992" (C. H. Schneider and A. N. Eberle, eds.), p. 177. Escom, Leiden, The Netherlands, 1993.

[201] J. Blake, Int. J. Pept. Protein Res. 17, 273 (1981).

[202] D. Yamashiro and J. Blake, Int. J. Pept. Protein Res. 18, 383 (1981).

[203] J. Blake and C. H. Li, Proc. Natl. Acad. Sci. U.S.A. 78, 4055 (1981).

[204] D. Yamashiro and C. H. Li, Int. J. Pept. Protein Res. 31, 322 (1988).

[205] L. E. Canne, S. M. Walker, and S. B. H. Kent, Tetrahedron Lett. 36, 1217 (1995).

[206] W. Neugebauer and E. Escher, Helv. Chim. Acta 72, 1319 (1989).

[207] J. Swistok, J. W. Tilley, W. Danho, R. Wagner, and K. Mulkerins, Tetrahedron Lett. 30, 5045 (1989).

[208] W. Neugebauer, R. Brzezinski, and G. E. Willick, in "Peptides 1990" (E. Giralt and D. Andreu, eds.), p. 188. Escom, Leiden, The Netherlands, 1991.

TABLE V

HANDLES AND RESINS FOR SYNTHESIS OF C-TERMINAL MODIFIED PEPTIDES[a]

Handle	C Termini produced	Cleavage conditions	Refs.
p-Alkoxybenzyloxycarbonylhydrazide resin	Hydrazides	TFA	b
N-Alkylaminomethyl resin	N-Alkylamides (R = alkyl)	HF	c
3-Nitro-4-N-alkylaminomethylbenzamido resin	N-Alkylamides (R = alkyl)	$h\nu$ (350 nm)	d
(5-N-Alkylamino-10,11-dihydrodibenzo[a,d]cyclohepten-2-yl)oxymethyl resin	Hydrazides (R = NH$_2$) or N-alkylamides (R = alkyl)	Dilute TFA	e

Structure	Product	Cleavage	
9-N-Alkylaminoxanthenyl-3-oxymethyl resin	N-Alkylamides (R = alkyl)	Dilute TFA	f
4-[N-Alkylamino-(2',4'-dimethoxyphenyl)methyl]phenoxyacetamido resin	N-Alkylamides (R = alkyl)	TFA	f
4-(8-Aminooctylaminocarbonyloxymethyl)phenoxyacetic acid	N-Aminoalkylamides	TFA or H₂/Pd	g
N-Substituted tritylamine resin	Hydrazides (R = H₂N) or N-aminoalkyl-amides (R = aminoalkyl)	Dilute TFA	h

(continued)

TABLE V (continued)

Handle	C Termini produced	Cleavage conditions	Refs.
[4-(5-Aminopentylaminocarbonyloxy)-cis-2-butene-1-oxy]-acetamido resin $H_2N(CH_2)_5NH-C(=O)-O-\cdots-OCH_2C(=O)-NH-Ⓡ$	N-Aminoalkylamides	Pd(II)/Bu$_3$SnH	i
N-[bis-N-(3-Aminopropoxycarbonylmethyl)acetamidyl]-N-(tert-butyloxycarbonyl)aminoacetic acid $H_2N(CH_2)_3O-$ ... (Boc) ...$-C(=O)-OH$, $H_2N(CH_2)_3O-$	N-Hydroxyalkylamides	1. (i) TFA (ii) pH 8.5 2. $^-$OH (step 1 releases first equivalent of peptide, step 2 releases the second)	j
4-(2-Aminoethylmercapto)methylphenoxyacetic acid $H_2N(CH_2)_2SCH_2-$ ... $-OCH_2C(=O)-OH$	N-Mercaptoalkylamides	HF	k
N^a-(tert-Butyloxycarbonyl)lysylprolyl resin H_2N- ... (NH-Boc) ... $-C(=O)-O-Ⓡ$	N-Diketopiperazinyl-alkylamides	(i) TFA (ii) pH 7	l

p-Aminoanilidocarbonyloxymethyl resin (UPAA)	*p*-Nitroanilides	(i) HF (ii) NaBO₃/acetic acid	*m*
N-Terminally anchored, C-terminal modified amino acid substituted *p*-methylbenzhydrylamine resin	Various termini (R = appropriate C-terminally functionalized amino acid residue)	HF	*n*
N-Terminally anchored amino ester substituted 2-nitrobenzylamine resin	Esters (R = appropriate amino ester side chain)	$h\nu$ (350 nm)	*o*
N-(*tert*-Butyloxycarbonyl)amino aldehyde semicarbazidyl, *trans*-4-methylcyclohexanecarboxylic acid	Aldehydes (R = appropriate amino aldehyde side chain)	H_2CO/H^+	*p*

(continued)

TABLE V (continued)

Handle	C Termini produced	Cleavage conditions	Refs.
3-(N-Methoxyamino)propionic acid	Aldehydes	LiAlH₄	q
4-(α-Mercaptobenzyl)phenoxyacetic acid	Thioacids	HF	r

[a] Handle and resin structures in this table are drawn with the point of attachment to the resin on the far right and the point of attachment to the peptide on the far left or top center. Use of these handles and resins follows the general principles described in Scheme 4 (see also accompanying text discussion). For this table, in every case shown, cleavage of the temporary peptide–resin linkage results in the indicated functionality being transferred from the handle/resin to the completed peptide. The eleventh, thirteenth, fourteenth, and fifteenth entries include the C-terminal amino acid derivative as part of the handle/resin; thus, the first coupling cycle of the synthesis actually incorporates the second residue of the target peptide. In all other cases, the C-terminal residue of the target peptide is incorporated onto the handle/resin as the first step of the synthesis.

[b] S.-S. Wang. J. Am. Chem. Soc. 95, 1328 (1973).

[c] J. E. Rivier, L. H. Lazarus, M. H. Perrin, and M. R. Brown, J. Med. Chem. 20, 1409 (1977); and W. Kornreich, H. Anderson, J. Porter, W. Vale, and J. Rivier, Int. J. Pept. Protein Res. 25, 414 (1985).

[d] A. Ajayaghosh and V. N. R. Pillai, *Indian J. Chem., Sect. B* **27B**, 1004 (1988); and A. Ajayaghosh and V. N. R. Pillai, *J. Org. Chem.* **55**, 2826 (1990).

[e] R. Ramage, S. L. Irving, and C. McInnes, *Tetrahedron Lett.* **34**, 6599 (1993); and R. Ramage, A. R. Brown, C. McInnes, S. L. Irving, S. G. Love, T. D. Pallin, T. W. Muir, G. Raphy, K. Shaw, F. O. Wahl, and J. Wilken, *in* "Innovation and Perspectives in Solid Phase Synthesis: Peptides, Proteins and Nucleic Acids: Biological and Biomedical Applications, 1994" (R. Epton, ed.), p. 1. Mayflower Worldwide, Birmingham, UK, 1994.

[f] W. C. Chan and S. L. Mellor, *J. Chem. Soc., Chem. Commun.*, 1475 (1995).

[g] G. Breipohl, J. Knolle, and R. Geiger, *Tetrahedron Lett.* **28**, 5647 (1987).

[h] A. van Vliet, R. H. P. H. Smulders, B. H. Rietman, A.-M. E. Klink, D. T. S. Rijkers, I. F. Eggen, G. van de Werken, and G. I. Tesser, *in* "Peptides 1992" (C. H. Schneider and A. N. Eberle, eds.), p. 279. Escom, Leiden, The Netherlands, 1993; and A. van Vliet and G. I. Tesser, Presented at the 4th International Symposium on Innovations and Perspectives in Solid Phase Peptide Synthesis and Combinatorial Chemical Libraries, Edinburgh, UK, 1995, Poster P91.

[i] K. Kaljuste and A. Undén, *in* "Peptides: Chemistry, Structure and Biology" (R. S. Hodges and J. A. Smith, eds.), p. 143. Escom, Leiden, The Netherlands, 1994.

[j] P. Kočiš, V. Krchňák, and M. Lebl, *Tetrahedron Lett.* **34**, 7251 (1993).

[k] D. R. Englebretsen, B. G. Garnham, D. A. Bergman, and P. F. Alewood, *Tetrahedron Lett.* **36**, 8871 (1995).

[l] N. J. Maeji, A. M. Bray, and H. M. Geysen, *J. Immunol. Methods* **134**, 23 (1990); and A. M. Bray, N. J. Maeji, R. M. Valerio, R. A. Campbell, and H. M. Geysen, *J. Org. Chem.* **56**, 6659 (1991).

[m] D. J. Burdick, M. E. Struble, and J. P. Burnier, *Tetrahedron Lett.* **34**, 2589 (1993).

[n] J. Rivier, J. Porter, and C. Hoeger, *in* "Peptides: Chemistry, Biology, Interactions with Proteins" (B. Penke and A. Török, eds.), p. 75. de Gruyter, Berlin, 1988.

[o] M. Renil and V. N. R. Pillai, *Tetrahedron Lett.* **35**, 3809 (1994).

[p] A. M. Murphy, R. Dagnino, P. L. Vallar, A. J. Trippe, S. L. Sherman, R. H. Lumpkin, S. Y. Tamura, and T. R. Webb, *J. Am. Chem. Soc.* **114**, 3156 (1992).

[q] J.-A. Fehrentz, M. Paris, A. Heitz, J. Velek, C.-F. Liu, F. Winternitz, and J. Martinez, *Tetrahedron Lett.* **36**, 7871 (1995).

[r] L. E. Canne, S. M. Walker, and S. B. H. Kent, *Tetrahedron Lett.* **36**, 1217 (1995).

ing,[209] particularly for solid-phase preparation of cyclic peptides.[72,210,211] The peptide is anchored to the resin through a covalent bond between an appropriate side-chain functionality and an appropriate handle, leaving the carboxyl terminus available for modification and/or further manipulations. However, the method remains limited to a relatively few trifunctional amino acid residues. To circumvent this limitation, we have developed[212,213] a novel and general anchoring concept that involves attachment of the growing peptide to an appropriate handle through a backbone amide nitrogen (i.e., with the general structure **11**, shown as implemented with the PAL handle). This *backbone amide linker* (BAL) approach provides linkers that are cousins to handles developed by us[214,215] and others (see relevant entries in Table V) that yield N-alkylamide substituted peptides.

As a prelude to explaining BAL, we describe our studies[212,213] that refined and generalized our previously reported[216,217] reductive amination chemistries for PAL and allowed the development of handles suitable for the preparation of C-terminal modified peptide N-alkylamides. We reasoned that incorporation of various alkylamines in a reductive amination procedure (Scheme 7, R^1 = alkyl) would provide ready entry to 5-[4-(N-Fmoc-N-alkyl)aminomethyl-3,5-dimethoxyphenoxy]valeric acids [Fmoc-(R)PAL-OH, **12**], and that such handles when used in a corresponding way as PAL would transfer the alkylamine substituent to the peptide. Similarly, the BAL anchor is established by reductive amination of an amino acid residue (or an appropriately modified derivative; Scheme 7, R^1 = amino acid derivative). Following N-acylation by a second amino acid residue,

[209] F. Albericio, R. Van Abel, and G. Barany, *Int. J. Pept. Protein Res.* **35**, 284 (1990).

[210] S. A. Kates, N. A. Solé, C. R. Johnson, D. Hudson, G. Barany, and F. Albericio, *Tetrahedron Lett.* **34**, 1549 (1993), and references cited therein.

[211] J. Alsina, F. Rabanal, E. Giralt, and F. Albericio, *Tetrahedron Lett.* **35**, 9633 (1994).

[212] K. J. Jensen, M. F. Songster, J. Vágner, J. Alsina, F. Albericio, and G. Barany, *in* "Peptides: Chemistry, Structure and Biology" (T. P. Kaumaya and R. S. Hodges, eds.), p. 30. Mayflower Scientific, Kingswinford, UK, 1996.

[213] K. J. Jensen, M. F. Songster, J. Vágner, J. Alsina, F. Albericio, and G. Barany, *in* "Innovation and Perspectives in Solid Phase Synthesis and Combinatorial Chemical Libraries: Biomedical and Biological Applications, 1996" (R. Epton, ed.), p. 187. Mayflower Scientific, Kingswinford, UK, 1996.

[214] M. F. Songster, J. Vágner, and G. Barany, *Lett. Pept. Sci.* **2**, 265 (1996).

[215] M. F. Songster, J. Vágner, and G. Barany, *in* "Innovation and Perspectives in Solid Phase Synthesis and Combinatorial Chemical Libraries: Biomedical and Biological Applications, 1996" (R. Epton, ed.), p. 537. Mayflower Scientific, Kingswinford, UK, 1996.

[216] S. K. Sharma, M. F. Songster, T. L. Colpitts, P. Hegyes, G. Barany, and F. J. Castellino, *J. Org. Chem.* **58**, 4993 (1993).

[217] M. F. Songster and G. Barany, *in* "Innovation and Perspectives in Solid Phase Synthesis: Peptides, Proteins and Nucleic Acids: Biological and Biomedical Applications, 1994" (R. Epton, ed.), p. 685. Mayflower Worldwide, Birmingham, UK, 1994.

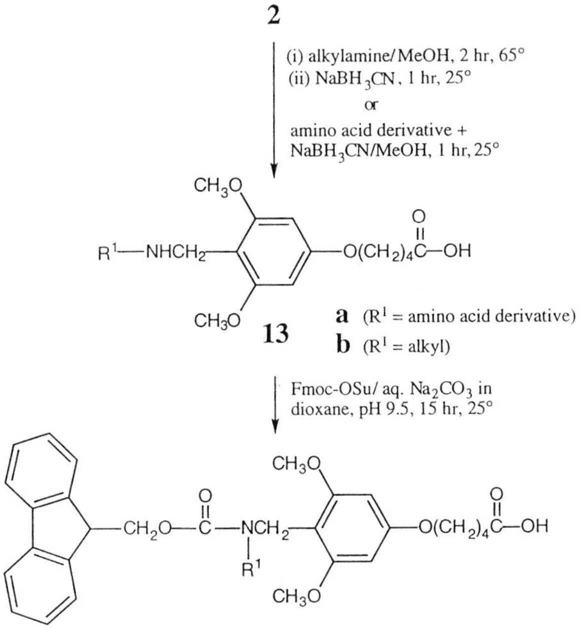
2
(i) alkylamine/MeOH, 2 hr, 65°
(ii) NaBH3CN, 1 hr, 25°
or
amino acid derivative +
NaBH3CN/MeOH, 1 hr, 25°
CH3O
R1—NHCH2
O(CH2)4C—OH
CH3O
13
a (R1 = amino acid derivative)
b (R1 = alkyl)
Fmoc-OSu/ aq. Na2CO3 in
dioxane, pH 9.5, 15 hr, 25°
CH3O
CH2O—C—NCH2
R1
O(CH2)4C—OH
CH3O

Fmoc-(BAL-OH)-AA-OR (**11**)
(R^1 = appropriately modified amino acid derivative)

Fmoc-(R)PAL-OH (**12**)
(R^1 = appropriate alkyl substituent)

SCHEME 7. Preparation of N^{α}-[4-(carboxylbutyloxy)-2,6-dimethoxybenzyl]-N^{α}-(9-fluorenylmethoxycarbonyl)amino acid ester [Fmoc-(BAL-OH)AA-OR, **11**], following Jensen *et al.*,[212,213] and 5-[4-(*N*-Fmoc-*N*-alkyl)aminomethyl-3,5-dimethoxyphenoxy]valeric acid [Fmoc-(R)PAL-OH, **12**], following Songster *et al.*[214,215] Compounds **11** and **12** are then coupled independently onto amino-functionalized supports to create Fmoc-(BAL-®)AA-OR and Fmoc-(R)PAL-®, respectively. Structure **2** is in Scheme 1. See text for all corresponding experimental procedures.

appropriately protected, a dipeptidyl unit is formed that is linked to the support through a backbone amide bond. Further chain growth then proceeds normally. Because BAL anchoring does not involve a C-terminal carboxyl as the point of attachment and initiation of solid-phase synthesis, it allows synthesis of C-terminal modified peptides, some of which could not be bound covalently by any other means. We have demonstrated the application of BAL preformed handles to prepare peptides with a variety of C-terminal functionalities, for example, not only acids, but also aldehydes, alcohols, and disubstituted amides, with others readily envisaged. Additionally, when used in conjunction with an orthogonally removable C-terminal carboxyl protecting group (e.g., allyl), it becomes possible to perform further modifications to the resin-bound peptide, for example, head-to-tail cyclizations (when a methyl ester is used, on-resin diketopiperazine formation can be achieved).

Preparation of N^α-[4-(Carboxylbutyloxy)-2,6-dimethoxybenzyl]-N^α-(9-fluorenylmethoxycarbonyl)amino Acid Esters [Fmoc-(BAL-OH) AA-OR, 11]

5-(4-Formyl-3,5-dimethoxyphenoxy)valeric acid (**2**,[21] 0.14 g, 0.50 mmol) and an appropriate amino acid derivative (0.6 mmol) are dissolved in methanol (5 ml) and stirred at 25° for 10 min. NaBH$_3$CN (47 mg, 0.75 mmol) is added and the resultant suspension is stirred at 25° for 1 hr. The heterogeneous reaction mixture is concentrated to dryness *in vacuo*, and the residual oil (secondary amine **13a**) is protected with Fmoc by a standard protocol similarly to that described earlier for compound **4**, to provide NMR pure products as oils.

Preparation of 5-[4-(N-Fmoc-N-alkyl)aminomethyl-3,5-dimethoxyphenoxy]valeric Acids [Fmoc-(R)PAL-OH, 12]

The alkylamine RNH$_2$ (27 mmol, demonstrated for R = CH$_3$, CH$_3$CH$_2$, C$_6$H$_5$ CH$_2$CH$_2$, etc.) is added to a solution of 5-(4-formyl-3,5-dimethoxyphenoxy)valeric acid (**2**,[21] 5.0 g, 18 mmol) in absolute methanol (150 ml) under N$_2$ at 0°. The reaction is allowed to warm to 25°, stirred at reflux for 2 hr, and then cooled to 25°. NaBH$_3$CN (1.7 g, 27 mmol) is added, and the reaction stirred for 1 hr at 25°. Solvent is removed under reduced pressure, and the residue (secondary amine **13b**) is protected with Fmoc by a standard protocol similarly to that described earlier for compound **4**, to yield title products as off-white solids or oils. The products are crystallized from ether–hexane.

Acknowledgments

We thank our co-workers and colleagues, in particular Drs. Fernando Albericio and Steve Kates, for sharing experiences with handles and other aspects of solid-phase peptide synthesis. Preparation of this review and supporting experimental work were supported by the National Institutes of Health (GM 42722).

[9] Solid-Phase Synthesis of Cyclic Homodetic Peptides

By Christopher Blackburn and Steven A. Kates

Introduction

There are two general classes of cyclic peptides: homodetic and heterodetic. Ring formation may occur to form the usual amide linkage between an amino and carboxylic acid function to give homodetic cyclic peptides, whereas any other linkages such as lactone, ether, thioether, and, most commonly, the disulfide bridge are referred to as heterodetic. The interest in cyclic homodetic peptides began in the late 1940s when the antibiotic gramicidin S was found to be a cyclic decapeptide.[1] Commercially available drugs such as the immunosuppressant cyclosporin indicate that research in this field continues. Many antibiotics and toxins have been shown to contain cyclic amino acid sequences that result in metabolic stability, increased potency, receptor selectivity, and bioavailability, and they are synthetically challenging. The constrained geometry of a cyclic peptide allows for conformational investigations such as determining the connection between primary and secondary structures of proteins and sequences related to protein folding and as selective metal ion chelating agents (ionophores). Initial work in this area focused on preparing peptides with a small ring size. Cyclic hexapeptides containing a proline to further restrict conformational possibilities were prepared and extensively studied. With the development of improved synthetic methods, larger ring sizes and more intriguing targets have been successfully attempted.

In the past, cyclic peptides have been prepared totally in solution[2] or by assembling the linear peptide in the solid phase,[3] cleaving the peptide from the resin, and cyclizing in solution. Unfortunately, there are still limitations to the cyclization of peptides via classic solution-phase synthesis despite efforts to develop practical and convenient methods.[4] For instance, a "doubling reaction" has been observed during ring closure.[3] As a result, literature methods rely on suitably protected linear precursors, which are selectively activated and cyclized in solution under highly dilute conditions.

---

[1] R. J. Consden, A. H. Gordon, A. J. P. Martin, and R. D. M. Synge, *Biochem. J.* **41,** 596 (1947).
[2] K. D. Kopple, *J. Pharm. Sci.* **61,** 1345 (1972).
[3] A. E. Tonelli, *in* "Cyclic Polymers" (J. A. Semlyen, ed.), p. 261. Elsevier Applied Science, London, 1986.
[4] D. S. Perlow, J. M. Erb, N. P. Gould, R. D. Tung, R. M. Freidinger, P. D. Williams, and D. F. Veber, *J. Org. Chem.* **57,** 4394 (1992).

METHODS IN ENZYMOLOGY, VOL. 289   0076-6879/97 $25.00

Cyclodimerizations and cyclooligomerizations may also occur even under high dilution usually as unwanted side reactions but occasionally to benefit for certain symmetrical targets, as temperature, solvent, and the concentration and structure of the linear peptide are other important factors.[2,4,5]

There are advantages to performing peptide cyclizations on solid phase rather than in solution. A pseudodilution phenomenon is achieved by having the linear peptide attached to a polymeric support. Intramolecular cyclization is favored over intermolecular because the solid support provides a large distance between chains.[3] Furthermore, reactions can be driven to completion by the use of excess soluble reagents that can be removed by simple filtration and washing, and in an automated process.

There are three classic modes of medium- and long-range lactam cyclization of peptides. A ring may be formed via cyclization of the amino terminus to the carboxyl terminus. This mode of ring closure is called head to tail or end to end. Alternatively, the functional groups of side chains, usually the amino group of lysine or ornithine, can provide a method for lactam formation with the carboxyclic acid of aspartic acid or glutamic acid. This type of cyclization is referred to as side chain-to-side chain or backbone-to-backbone cyclization. Lastly, cyclization can occur through a side chain such as the amino side-chain function of Lys/Orn to the carboxyl terminus. Alternatively, the carboxylic acid function of Asp/Glu can cyclize onto the amino terminus. Solid-phase syntheses of these classes of peptides have been attempted through different strategies outlined in Schemes 1, 2, and 3.

This article reviews the synthesis of cyclic homodetic peptides by solid-phase methods with an emphasis on solid supports, protecting group strategy, reagents, side reactions, and the emergence of cyclic peptide combinatorial libraries.[6]

Solid Supports

A wide range of polymeric supports has been successfully used to prepare cyclic peptides such as polystyrene (PS) resins and cellulose membranes.[7] With the development of milder orthogonal protection schemes

[5] M. Rothe, A. Sander, W. Fischer, W. Mastle, and B. Nelson, in "Peptides: Proceedings of the Fifth American Peptide Symposium" (M. Goodman and J. Meienhofer, eds.) p. 506. Wiley, New York, 1977.

[6] For a recent review on solid-phase synthesis of cyclic peptides, see S. A. Kates, N. A. Solé, F. Albericio, and G. Barany, in "Peptides: Design, Synthesis and Biological Activity" (C. Basava and G. M. Anantharamaiah, eds.), p. 39. Birkhaeuser, Boston, 1994.

[7] D. Winkler, A. Schuster, B. Hoffmann, and J. Schneider-Mergener, in "Peptides 1994: Proceedings of the Twenty-third European Peptide Symposium" (H. L. S. Maia, ed.), p. 485. Escom, Leiden, The Netherlands, 1995.

a. side chain to side chain
A-NH
O
C-X
handle
NH-B1
CO2-B2
X=O, NH
1. Removal of B1 and B2
2. Cyclization
3. Final deprotection and cleavage
b. N terminal to side chain
A-NH
O
C-X
handle
CO2B
X=O, NH
1. Removal of A and B
2. Cyclization
3. Final deprotection and cleavage
c. head to tail
A-NH
O
C-X
handle
X = O-N=R
or S
1. Removal of A
2. Cyclization and concomitant release of protected peptide from the resin
3. Removal of protecting groups in solution

SCHEME 1. C-terminal anchoring strategy for different classes of cyclic peptides.

and more efficient reagents for cyclization, polyethylene glycol–polystyrene graft supports (PEG–PS) have been shown to be excellent resins.[8] The PEG spacers allow complete solvation of the reactive sites, which increases coupling efficiency for both linear assembly and cyclization. Moreover, PEG–PS supports do exist that are compatible with both 9-fluorenylmethyloxycarbonyl (Fmoc)/*tert*-butyl (*t*Bu) and *tert*-butyloxycarbonyl (Boc)/ben-

[8] G. Barany, F. Albericio, N. A. Solé, G. W. Griffin, S. A. Kates, and D. Hudson, *in* "Peptides 1992: Proceedings of the Twenty-second European Peptide Symposium" (C. H. Schneider and A. N. Eberle, eds.), p. 267. Escom, Leiden, The Netherlands, 1993.

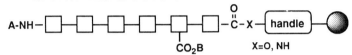

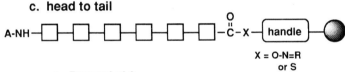

I. Through carboxylic acid

a. head to tail

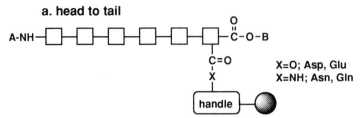

X=O; Asp, Glu
X=NH; Asn, Gln

1. Removal of A and B (usually two steps)
2. Cyclization
3. Final deprotection and cleavage

b. *N* terminal to side chain

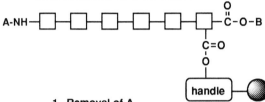

1. Removal of A
2. Cyclization and concomitant release of peptide from the resin
3. Removal of protecting groups in solution

II. Through other amino acid side chain

a. head to tail

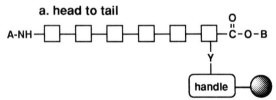

1. Removal of A and B (usually two steps)
2. Cyclization
3. Final deprotection and cleavage

SCHEME 2. Side-chain anchoring strategy for different classes of cyclic peptides.

a. head to tail

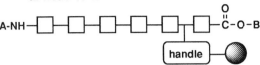

1. Removal of A and B (usually two steps)
2. Cyclization
3. Final deprotection and cleavage

b. side chain to C terminal

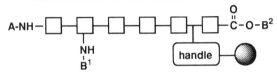

1. Removal of B¹ and B²
2. Cyclization
3. Final deprotection and cleavage

SCHEME 3. Backbone anchoring strategy for different classes of cyclic peptides.

zyl (Bzl) methods.[9] Finally, PepsynK (polyamide–kieselguhr), a resin developed for continuous-flow instrumentation, has also been successfully used.[10]

Resin Loading

The amount of dimerization and oligomerization is dependent on the substitution level of the polymeric support. Plaué has prepared 4-methylbenzhydrylamine (MBHA) resins with an initial loading of 0.1–0.6 mmol/g and examined the cyclization of a model compound.[11] Analysis of the high-performance liquid chromatography (HPLC) data clearly indicate that as the resin substitution increases from 0.1 to 0.6 mmol/g, the percentage of cyclodimer increases from 8 to 19%. PEG–PS supports with 0.40 and 0.20 mmol/g substitutions have been used in conjunction with an Fmoc/

[9] F. Albericio, J. Bacardit, G. Barany, J. M. Coull, M. Egholm, E. Giralt, G. W. Griffin, S. A. Kates, E. Nicolás, and N. A. Solé, in "Peptides 1994: Proceedings of the Twenty-third European Peptide Symposium" (H. L. S. Maia, ed.), p. 271. Escom, Leiden, The Netherlands, 1995.

[10] C. Mendre, R. Pascal, and B. Calas, Tetrahedron Lett. 35, 5429 (1994).

[11] S. Plaué, Int. J. Pept. Protein Res. 35, 510 (1990).

*t*Bu/allyl strategy for the preparation of a head-to-tail cyclic peptide.[12] The product purity from the on-resin cyclization is greater with the lower loaded PEG–PS support. In contrast, when the repeating unit of gramicidin S is constructed on an oxime resin with 0.54 and 0.17 mmol/g loadings, the resin of lower loading gives more dimer due to interchain reactions.[13] Cyclic monomeric peptide is formed at a faster rate and higher yield from the higher loaded resin.

Handles

Anchors that have been commonly employed for on-resin lactamization of cyclic peptides are shown in Fig. 1. The selection of a handle is governed by the protection strategy incorporated into a synthetic scheme. Initially, benzhydrylamine (BHA)– and MBHA–polystyrene resins were used because of their stability to strong acid (HF), which was commonly employed for final deprotection of side-chain protecting groups and release of the cyclic peptide from the resin.

The preparation of head-to-tail or a side chain-to-tail cyclic peptides requires "semipermanent" protection of the terminal carboxylic acid function (for definitions of the different protecting groups, see Protection Schemes). In standard linear assembly, the α-carboxylic acid is used for attachment to a solid support. Therefore, construction of these two classes of cyclic peptides demands side-chain anchoring of a trifunctional amino acid to a polymeric support. Side-chain anchoring on route to head-to-tail cyclic peptides has been accomplished by attachment of the ω-carboxyl groups of Asp or Glu to hydroxymethyl or benzhydrylamine resins.[14] The linear peptide was assembled on a phenylacetamido (PAM) resin starting with attachment of Boc-Asp(OH)-OFm (where Fm is 9-fluorenylmethyl) followed by peptide chain elongation with Boc/Bzl amino acids. Tromelin *et al.* reported the synthesis of a head-to-tail cyclic peptide on MBHA resin via side-chain anchoring of Boc-Asp(OH)-OFm through an amide link.[15] After acetylation of residual MBHA sites, the peptide chain was assembled using Boc/Bzl amino acids. Following removal of the N-terminal Boc group

[12] B. F. McGuinness, S. A. Kates, G. W. Grifin, L. W. Herman, N. A. Solé, J. Vágner, F. Albercio, and G. Barany, *in* "Peptides: Chemistry and Biology, Proceedings of the Fourteenth American Peptide Symposium" (P. T. P. Kaumaya and R. S. Hodges, eds.), p. 125. Mayflower Worldwide, Birmingham, UK, 1996.

[13] M. Bouvier and J. W. Taylor, *in* "Peptide: Chemistry and Biology, Proceedings of the Twelfth American Peptide Symposium" (J. A. Smith and J. E. Rivier, eds.), p. 535. Escom, Leiden, The Netherlands, 1992.

[14] P. Rovero, L. Quartara, and G. Fabbri, *Tetrahedron Lett.* **32**, 2639 (1991).

[15] A. Tromelin, M. H. Fulachier, G. Mourier, and A. Ménez, *Tetrahedron Lett.* **33**, 5197 (1992).

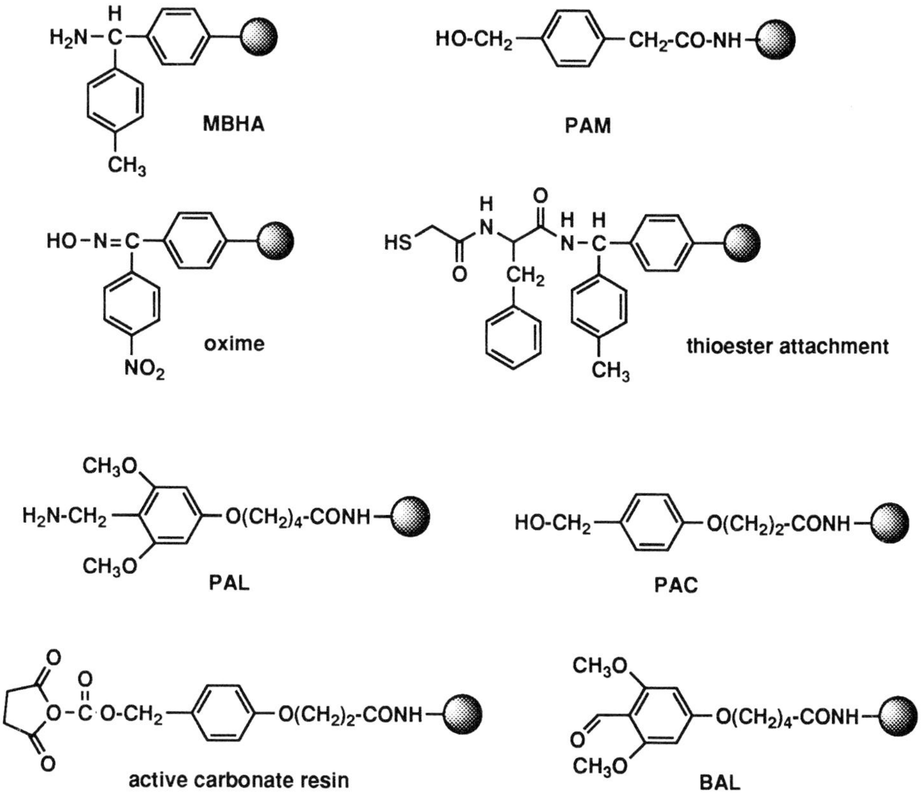

FIG. 1. Handles used for the preparation of cyclic peptides.

and the C-terminal Fmoc group and cyclization, HF cleavage afforded a cyclic peptide in which the Asn residue was derived from the Asp group initially attached to the resin.

The side-chain anchoring strategy for head-to-tail cyclic peptides in Fmoc/tBu-based chemistry was first conducted by McMurray[16] by attaching Fmoc-Glu(OH)-ODmb (2,4-dimethoxybenzyl) to a 4-hydroxymethylphen-oxyacetylaminomethyl–resin for the preparation of cyclo(Ala-Ala-Arg-DPhe-Pro-Glu-Asp-Asn-Tyr-Glu). Albericio and co-workers applied an Fmoc/tBu/allyl strategy to the preparation of the same sequence using four different starting resins: (a) Fmoc-Glu(OPAC–PEG–PS)-OAl (where PAC is p-alkoxybenzyl alcohol and Al is allyl) at positions Glu-6 and Glu-10, (b) Fmoc-Asp(OPAC–PEG–PS)-OAl at position Asp-7, and (c) Fmoc-

---

[16] J. S. McMurray, *Tetrahedron Lett.* **32**, 7679 (1991).

Asp(OPAL–PEG–PS)-OAl at position Asn-8. In strategy (c), Asn-8 was derived from an aspartic acid with its β-carboxylic acid bound to the resin via the 5-(4-aminomethyl-3,5-dimethoxyphenoxy)valeric acid (PAL) handle. All four strategies gave the desired peptide; the best yield and purity (71%) were obtained with the Asn-8 strategy, and the Glu-10 strategy gave the highest level of by-products.[17]

Side-chain anchoring an amino acid to a solid support has been accomplished most frequently through a carboxylic acid function, which restricts linear sequences to those containing Asx or Glx. As an extension to this approach, Alsina et al.[18] prepared an active carbonate resin to which the side chain of Lys was attached. The novel solid support was prepared in quantitative yield by treatment of PAC resin with N,N'-disuccinimidyl carbonate and 4-dimethylaminopyridine (DMAP) followed by coupling with Fmoc-Lys-OAl in the presence of N,N-diisopropylethylamine (DIEA).

Procedure. PAC–PS resin is suspended in N,N-dimethylformamide (DMF), and disuccinimidyl carbonate (10 equivalent) and DMAP (1 equivalent) are added under Ar for 2 hr. The resin is filtered and washed with DMF. Fmoc-Lys-OAl (10 equivalent) is dissolved in DMF, added to the resin in the presence of DIEA (20 equivalent), and allowed to react for 4 hr at 25°. The Fmoc group is removed with piperidine–DMF followed by peptide chain elaboration. Allyl removal is carried out using 1.9 M Pd(PPh$_3$)$_4$ in dimethyl sulfoxide (DMSO)–tetrahydrofuran (THF)–0.5 N aqueous HCl–morpholine (2:2:1:0.1, by volume), for 150 min at 25°. Following Fmoc group removal, the resin is treated with benzotriazol-1-yl-oxytris(dimethylamino)phosphonium hexafluorophosphate (BOP)/1-hydroxy-7-azabenzotriazole (HOAt)/DIEA (5:5:10 equivalents) in DMF for 2 hr at 25° to effect cyclization. The product is cleaved from the resin with concomitant removal of protecting group side chains with reagent R (TFA–thioanisole–1,2-ethanedithiol–anisole, 90:5:3:2, by volume) for 2 hr at 25°.

The above methods rely on the concept of side-chain anchoring an amino acid to a resin via a carboxylic acid or amine function, which restricts the sequences that can be cyclized on a solid support. The BAL handle (backbone amide linker) has been developed for Fmoc/tBu methods to attach a peptide chain via a backbone amide nitrogen.[19] An example of

---

[17] S. A. Kates, N. A. Solé, C. R. Johnson, D. Hudson, G. Barany, and F. Albericio, *Tetrahedron Lett.* **34**, 1549 (1993).

[18] J. Alsina, F. Rabanal, E. Giralt, and F. Albericio, *Tetrahedron Lett.* **35**, 9633 (1994).

[19] K. J. Jensen, M. F. Songster, J. Vágner, J. Alsina, F. Albericio, and G. Barany, *in* "Peptides: Chemistry and Biology, Proceedings of the Fourteenth American Peptide Symposium" (P. T. P. Kaumaya and R. S. Hodges, eds.), p. 30. Mayflower Worldwide, Birmingham, UK, 1996.

the use of this handle involves the attachment of Fmoc-(BAL-OH)Ala-OAl to MBHA resin. Using the BAL anchor, head-to-tail cyclic peptides devoid of a trifunctional amino acid residue can be readily constructed.

Anchoring the first amino acid to a PAM-type handle for the synthesis of head-to-tail cyclic peptides via a Boc/Bzl/OFm strategy has been investigated.[20] Coupling Boc-Asp(OH)-OFm to hydroxymethyl PAM resins with N,N'-diisopropylcarbodiimide (DIPCDI)/DMAP using various conditions gives low yields and high levels of racemization. Nucleophilic substitution of bromomethyl-Pam handles with the cesium and zinc salts of Boc-Asp(OH)-OFm substantially improves the anchoring yield and lowers the amount of racemization.

Protection Schemes

The success of solid-phase synthesis of cyclic peptides is based on the proper combination of temporary, semipermanent, and permanent protecting groups (Table I) and the corresponding methods for their removal (Fig. 2). Temporary groups are for N^α-protection; semipermanent groups are for Lys/Orn-Asp/Glu side-chain protection, and permanent groups are for side-chain protection and resin–peptide anchorage. Two or more classes of groups that are removed by differing chemical mechanisms in any order and in the presence of other classes is called an orthogonal protection scheme. This discussion is organized according to the following format: semipermanent/temporary/permanent/handle (anchoring strategy and class of cyclic peptide, based on Schemes 1, 2, and 3) accompanied with a brief experimental section.

Oxime/Boc/Bzl/oxime Handle (Scheme 1c)

A polystyrene-based resin frequently used is the p-nitrobenzophenone oxime–resin[21,22] which acts as both a semipermanent and permanent protecting group for the C terminus. This resin provides a method for preparation of head-to-tail cyclic peptides with concomitant cleavage from the resin via intramolecular nucleophilic displacement of the anchoring linkage by the amine terminus of a peptide.[23,24] This strategy requires N^α-Boc amino acids for linear chain elongation because oxime resins are not compatible with secondary amines used for N^α-Fmoc removal. To avoid premature cleavage of the peptide from the resin, coupling of amino acids is accom-

[20] M.-L. Valero, E. Giralt, and D. Andreu, *Tetrahedron Lett.* **37**, 4229 (1996).
[21] W. F. DeGrado and E. T. Kaiser, *J. Org. Chem.* **45**, 1295 (1980).
[22] W. F. DeGrado and E. T. Kaiser, *J. Org. Chem.* **47**, 3258 (1982).
[23] G. Ösapay, M. Bouvier, and J. W. Taylor, *Tetrahedron Lett.* **31**, 6121 (1990).
[24] G. Ösapay and J. W. Taylor, *J. Am. Chem. Soc.* **112**, 6046 (1990).

TABLE I
PROTECTING GROUPS FOR SOLID-PHASE CYCLIC PEPTIDES

Semipermanent	Structure	Removal	Temporary	Permanent	Final cleavage
tBu, Boc		TFA	Fmoc	Bzl	HF
Fm-Fmoc		Piperidine	Boc	Bzl	HF
Allyl-Aloc		Pd(0)	Fmoc/Boc	tBu/Bzl	TFA/HF
Dmb		1% TFA	Fmoc	tBu	TFA

	Structure	Deprotection			
Dde	(2-acetyl-5,5-dimethylcyclohexane-1,3-dione; CH$_3$, CH$_3$, H$_3$C, CH$_3$, O, O)	Hydrazine	Fmoc	tBu	TFA
Dmab-N-ivDde	(CH$_3$, CH$_3$, H$_3$C, CH$_3$, O, O, N, H; —CH$_2$— aryl)	Hydrazine	Fmoc	tBu	TFA
CAM	—CH$_2$—C(=O)—NH$_2$	NaOH	Fmoc	tBu	TFA
TMSE	H$_3$C—Si(CH$_3$)(CH$_3$)—CH$_2$—CH$_2$—	Tetrabutylammon- ium fluoride (TBAF)	Fmoc	tBu	TFA
Adpoc-OPp	O=C—O—C(CH$_3$)$_2$-adamantyl ; CH$_3$—C(CH$_3$)—O— phenyl	1% TFA	Fmoc	tBu	TFA

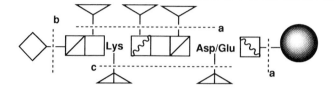

a. Side-chain protecting groups and resin–peptide anchorange ("permanent")
b. N$^{\alpha}$-Protecting group ("temporary")
c. Lys-Asp/Glu side-chain protecting groups ("semipermanent")

FIG. 2. Protection scheme for synthesis of cyclic peptides.

plished with BOP in conjunction with *in situ* neutralization using DIEA. Although final products have been obtained in high purity, a limitation is that the lability of the oxime residue can result in peptide lost at each step, resulting in low yields.

Procedure. The linear peptide chain is assembled on Boc-Leu-oxime resin (0.11 mmol/g). The Boc protecting group is removed by treatment of the resin with trifluoroacetic acid (TFA)–CH$_2$Cl$_2$ (3 : 3, v/v) for 30 min. After washing, the appropriate Boc-amino acid (5 equivalent), BOP (5 equivalent), and DIEA (5 equivalent) in DMF are added to the resin and allowed to couple for 2 hr. Couplings of Boc-Asn-OH and Boc-Gln-OH are repeated using 2.5 equivalents of reactants. After the final coupling, the Boc group is removed from the N terminus as described above and the resin treated with DIEA (1.5 equivalent) in CH$_2$Cl$_2$ for 24 hr. The protected cyclic peptide is isolated from the solution phase and purified by chromatography.

The selection of the amino acid to be anchored to an oxime resin influences the yield and purity of the desired target. Cyclo[Arg(Tos)-Gly-Asp(OcHex)-Phg] (where Tos is 4-toluenesulfonyl, cHex is cyclohexyl, and Phg is phenylglycine) has been synthesized and the monomer to dimer ratio studied as a function of the order of attachment of the amino acids to the resin.[25] The highest yield of monomer is achieved when Phg is the C terminus residue, whereas Arg(Tos) gives the lowest yield.

Oxime resins in conjunction with phenacyl (Pac) and allyl (Al) protecting groups have been used for the preparation of protected side chain-to-side chain cyclic peptides.[26] The linear peptide is anchored to the oxime

[25] N. Nishino, M. Xu, H. Mihara, T. Fujimoto, Y. Ueno, and H. Kumagai, *Tetrahedron Lett.* **33,** 1479 (1992).
[26] A. Kapimiolu and J. W. Taylor, *Tetrahedron Lett.* **34,** 7031 (1993).

resin via the β-carboxyl group of Boc-Asp(OH)-OX, where X is phenacyl or allyl ester. Fmoc-Lys(Boc)-OH is incorporated as the final amino acid following peptide elongation. Deprotection of the side chain with TFA–CH_2Cl_2 (1:3, v/v) followed by neutralization with DIEA–CH_2Cl_2 (1:19, v/v), releases the peptide from the resin and gives the cyclized, fully protected peptide.

As an alternative approach, Richter *et al.* have used the reactivity of thioesters toward nucleophilic attack for the preparation of head-to-tail cyclic peptides without Asx or Glx residues.[27] *S*-Tritylthioglycolic acid has been attached to Phe-MBHA resin and deprotected to give a handle terminating in a thiol group. Coupling of the first amino acid to the free thiol is achieved using DIPCDI and DIEA in CH_2Cl_2 with minimal epimerization. A series of linear penta-, hexa-, and heptapeptides is assembled from the thiol function using Boc chemistry and cyclized in the presence of DIEA (3 equivalent) and DMAP (0.1 equivalent) in *N,N*-dimethylacetamide (DMA) for 2–7 days.

Fmoc-Fm/Boc/Bzl/BHA Handle (Scheme 1b)

Felix *et al.* have described the Boc/Bzl/Fmoc approach for the construction of side chain-to-side chain cyclic peptides.[28] Linear peptides are prepared on BHA–resin by Boc/Bzl chemistry using Boc-Asp(OFm)-OH and Boc-Lys(Fmoc)-OH as precursors to the lactam residue. The Fm-based side-chain protecting groups have been shown to be stable to DIEA–CH_2Cl_2 used in neutralization steps. Selective removal of these groups using piperidine–DMF (1:4, v/v), cyclization, and HF cleavage gives the desired product.

Procedure. Boc-[Asp^8(OFm),Lys^{12}[Asp^8,Ala^{15}]GRF(1–29)]-BHA–resin (0.4 mmol) is deprotected at the Asp side chain using piperidine–DMF (1:4, v/v) (once for 1 min and once for 20 min) and then washed successively with DMF, methanol, CH_2Cl_2, and methanol (1 min each). The peptide is cyclized by addition of BOP (2.5 equivalent) in DMF containing triethylamine (2.8 equivalent) three times for 4 hr until the resin gives a negative ninhydrin test. The cyclic peptide is cleaved from the resin by treatment with HF–1,2-ethanedithiol at 0° for 2 hr. The product is isolated by evaporation of HF, washing with ethyl acetate, extraction with TFA, evaporation, and trituration with ether.

[27] L. S. Richter, J. Y. K. Tom, and J. P. Burnier, *Tetrahedron Lett.* **35,** 5547 (1994).
[28] A. M. Felix, E. P. Heimer, C. T. Wang, T. J. Lambros, A. Fournier, T. F. Mowles, S. Maines, R. M. Campbell, B. B. Wegrzynski, V. Toome, D. Fry, and V. S. Madison, *Int. J. Pept. Protein Res.* **32,** 441 (1988).

Boc-tBu/Fmoc/Bzl/BHA Handle (Scheme 1c)

Schiller *et al.* have developed the first method for the preparation of side chain-to-side chain cyclic lactam peptides on BHA–resin incorporating N^α-Fmoc-amino acids with Bzl side-chain protecting groups for linear assembly and with Boc and *t*Bu protection for amino- and carboxylate-containing side chains for cyclization.[29] Removal of the Boc and *t*Bu side-chain protecting groups with TFA–CH_2Cl_2, neutralization using DIEA–CH_2Cl_2, cyclization, and subsequent HF treatment remove benzyl side-chain protecting groups and cleave the peptide from the solid support.

Later, Schiller *et al.* reported a procedure enabling the preparation of demanding and larger cyclic peptides.[30] The linear sequences are synthesized on MBHA–resin using N^α-Boc-amino acids and changing to N^α-Fmoc-amino acids with Boc and *t*Bu side-chain protection for Orn and Asp, respectively, for the lactam region. Removal of the side-chain protecting groups with TFA is followed by neutralization and cyclization. Removal of the N^α-Fmoc group allows for the completion of the chain elongation using Boc/Bzl chemistry. The cyclic peptide is obtained from the resin on HF treatment.

Procedure. The linear peptide is constructed on MBHA–resin with the terminal N^α-Fmoc of Orn retained. Treatment with TFA–CH_2Cl_2 (1:1, v/v) to remove the side chains of Orn and Asp residues is followed by neutralization using DIEA–CH_2Cl_2 (1:9, v/v). After washing with DMF, cyclization is conducted by addition of *N,N'*-dicyclohexylcarbodiimide (DCC) (5 equivalent) and 1-hydroxybenzotriazole (HOBt) (5 equivalent) in DMF over a period of several days adding fresh DCC and HOBt every 48 hr. The N-terminal Fmoc group is removed with piperidine–CH_2Cl_2 (1:1, v/v) and the final residue coupled in the usual manner. The peptide is released from the solid support with removal of the Bzl side-chain protecting groups with HF.

Dmb/Fmoc/tBu/PAC Handle (Scheme 2a)

The utility of 2,4-dimethoxybenzyl (Dmb) ester protection for the α-carboxylic acid of Glu in conjunction with side-chain anchoring has been reported.[16] Following linear peptide chain elaboration, removal of the Dmb group is accomplished with TFA–CH_2Cl_2 (1:99, v/v), conditions which are compatible with *t*Bu and 2,2,5,7,8-pentamethylchroman-6-sulfonyl (Pmc) side-chain protecting groups but not with Trt (trityl) protection for Cys

---

[29] P. W. Schiller, T. M. Nguyen, and J. Miller, *Int. J. Pept. Protein Res.* **25**, 171 (1985).
[30] P. W. Schiller, T. M.-D. Nguyen, L. A. Maziak, B. C. Wilkes, and C. Lemieux, *J. Med. Chem.* **30**, 2094 (1987).

and His residues. N^{α}-Fmoc removal, cyclization, release of the peptide from the resin and removal of tBu side chains with prolonged treatment with TFA–phenol, and gel filtration to remove oligomers give a monomeric cyclic peptide.

Procedure. Fmoc-Glu(OH)-ODmb is coupled to PAC–PS resin (0.45 mmol/g) using a DMAP-catalyzed symmetric anhydride procedure. After assembly of the desired linear sequence, the Dmb group is removed by treatment with TFA–CH_2Cl_2 (1 : 99, v/v, six times for 5 min). The Fmoc group is removed (piperidine–DMF, 1 : 4, v/v) and the peptide cyclized by addition of BOP/HOBt (3 equivalent) and N-methylmorpholine (NMM) (6 equivalent) for 5 hr. After DMF washing, the resin is cleaved by treatment with TFA–phenol (95 : 5, v/v), to give the cyclic peptide.

Alloc-Al/Fmoc/tBu/PAC-PAL Handle (Scheme 2a)

The synthesis of cyclic head-to-tail peptides by a three-dimensional orthogonal protecting strategy was accomplished independently by Trzeciak and Bannwarth[31] and by Albericio and co-workers[17] with an Fmoc/tBu/allyl scheme. Allyl-based protecting groups for peptide synthesis have been introduced independently by Lyttle and Hudson[32] and by Loffet and Zhang[33] and are removed smoothly and quantitatively with mild palladium-catalyzed transfer without the undesired back-alkylation side reactions that result when carbocations are generated. Furthermore, owing to their lower steric hindrance and less hydrophobic character, allyl derivatives couple more efficiently than do tBu-based derivatives.

The Fmoc/tBu/allyl strategy has become an attractive approach and is incorporated in a variety of sequences. The antibiotic bacitracin has been synthesized by initially attaching the side chain of L-Fmoc-Asp(OH)-OAl to PAL–PS.[34] Linear assembly, allyl removal with Pd(PPh$_3$)$_4$, acetic acid and triethylamine, cyclization, addition of the N-terminal thiazoline dipeptide, and cleavage afford the cyclic peptide.

The preparation of cyclic peptides on solid phase with Fmoc/tBu/allyl-based methods has been automated.[35] Standard allyl removal uses a sus-

---

[31] A. Trzeciak and W. Bannwarth, *Tetrahedron Lett.* **33**, 4557 (1992).

[32] M. H. Lyttle and D. Hudson, *in* "Peptides: Chemistry and Biology, Proceedings of the Twelfth American Peptide Symposium" (J. A. Smith and J. E. Rivier, eds.), p. 583. Escom, Leiden, The Netherlands, 1992.

[33] A. Loffet and H. X. Zhang, *in* "Innovation and Perspectives in Solid Phase Synthesis: Peptides, Polypeptides and Oligonucleotides" (R. Epton, ed.), p. 77. Intercept, Andover, UK, 1992.

[34] J. Lee, J. H. Griffin, and T. Nicas, *J. Org. Chem.* **61**, 3983 (1996).

[35] S. A. Kates, S. B. Daniels, and F. Albericio, *Anal. Biochem.* **212**, 303 (1993).

pended palladium catalyst, which is not feasible on a batch and continuous-flow peptide synthesizer. Solvent conditions have been examined and optimized to solubilize the catalyst, prevent undesired Fmoc deblocking, and be compatible with sensitive amino acids (Trp and Met) and with glyco- and sulfopeptides. $CHCl_3$–acetic acid–NMM (37:2:1, by volume), has proved to be most effective for the use in automated instruments, and the results give comparable purity to that obtained by performing the cyclization manually.

Procedure. Automated side-chain deblocking of the allyl group is performed by placing a vial containing $Pd(PPh_3)_4$ (125 mg) under Ar in the amino acid module following the last amino acid in the sequence. The catalyst is dissolved to a final concentration of 0.14 M in $CHCl_3$–acetic acid–NMM (37:2:1, by volume), delivered to a 0.3 mmol/g substituted peptidyl–resin (500 mg), and recycled through the column for 2 hr. The column is washed with a solution of 0.5% DIEA and 0.5% sodium diethyldithiocarbamate in DMF (10 min, 6 ml/min) and DMF (10 min, 6 ml/min). The N^α-Fmoc group of the final amino acid is removed, and benzotriazol-1-yl-N-oxytrispyrrolidinophosphonium hexafluorophosphate (PyBOP)/HOBt (4 equivalent) is dissolved to a final concentration of 0.3 M in a solution of 0.6 M DIEA in DMF and delivered to the solid support. The cyclization is allowed to occur until the resin gives a negative ninhydrin test, and the resin is washed with DMF (15 sec, 3 ml/min) and CH_2Cl_2 (6 min, 3 ml/min). The cyclic peptide is cleaved from the resin using TFA–anisole–2-mercaptoethanol (95:3:2, by volume), for 2 hr and the filtrate collected. The resin is washed with TFA and the combined filtrates treated with ether and cooled to $-70°$. After removal of the supernatant, the precipitated cyclic peptide is washed with ether, dissolved in acetic acid, and lyophilized.

After assembly of a linear peptide on MBHA–resin (0.4 mmol/g) the N^α-allyloxycarbonyl (Alloc) protecting group is removed by addition of $PdCl_2(PPh_3)_2$ (0.04 equivalent) and acetic acid (3.5 equivalent) in CH_2Cl_2 (3 ml/g of resin).[36] Tributyltin hydride (3 equivalent) is then added and allowed to react for 5 min before washing with CH_2Cl_2, triethylamine–CH_2Cl_2 (1:9, v/v), CH_2Cl_2, methanol, and CH_2Cl_2.

Dde/allyl/Fmoc/tBu/PAL Handle (Scheme 1a)

A branched, cyclic peptide has been prepared by Bloomberg *et al.*[37] with an Fmoc/tBu/allyl/1-(4,4-dimethyl-2,6-dioxocyclohex-1-ylidine)ethyl (Dde)

---

[36] O. Dangles, F. Guibé, G. Balavoine, S. Lavielle, and A. Marquet, *J. Org. Chem.* **52**, 4984 (1987).

[37] G. B. Bloomberg, D. Askin, A. R. Gargaro, and M. J. A. Tanner, *Tetrahedron Lett.* **34**, 4709 (1993).

protection scheme. The first branch is assembled by sequential coupling of Fmoc-Lys(Dde)-OH, Fmoc-Ala-OH, Fmoc-Glu(OAl)-OH, and Fmoc-Cys(Trt)-OH to PAL–PEG–PS followed by acetylation. The Dde function is removed with hydrazine, and the second branch is added. Allyl removal of the carboxyl side chain of Glu, N^α-Fmoc removal, cyclization, and cleavage from the resin give the crude peptide, which is purified by HPLC.

Procedure. Following assembly of the appropriate linear peptide the Dde group is removed by flowing a 1.5% solution of hydrazine in DMF through the resin for 9 min. Elongation from the side chain is carried out with Fmoc-amino acids. Allyl removal is achieved by suspending the resin in DMSO–THF–0.5 M HCl–NMM (20:10:10:1, by volume) (5 ml/100 mg of resin) and Pd(PPh$_3$)$_4$ (0.15 to 1.0 equivalent based on resin substitution) for 3 hr in an inert atmosphere. The Fmoc group is removed and cyclization is carried out by treatment of the resin with 1.5 equivalents of N-[(1H-benzotriazol-1-yl)(dimethylamino)methylene]-N-methylmethanaminium hexafluorophosphate N-oxide (HBTU)/HOBt/NMM (1:1:2, by volume) in DMF (2 ml/100 mg resin) for 2 hr. The cyclic peptide is cleaved from the resin using TFA–1,2-ethanedithiol–phenol–water–triisobutylsilane (92:3:2:2:1, by volume) and the product isolated by precipitation with ether.

N-ivDde-Dmab/Fmoc/tBu/PAC–PAL Handle (Scheme 1a,b)

To complement the Dde protecting group, 4-{N-[1-(4,4-dimethyl-2,6-dioxocyclohexylidine)-3-methylbutyl]amino}benzyl (ODmab) ester protection for the carboxylic acid function of Asp and Glu residues and the more sterically hindered 1-(4,4-dimethyl-2,6-dioxocyclohexylidine)-3-methylbutyl (N-ivDde) group for amine protection have been reported.[38] Both protecting groups are orthogonal to Fmoc/tBu methods, and cleavage is achieved within minutes on treatment with hydrazine–DMF (1:49, v/v).

Procedure. Fmoc-Asp(ODmab)-OH is coupled to 2-chlorotrityl resin and the linear sequence assembled using Fmoc/tBu method. The Dmab group is removed by reaction with hydrazine–DMF (1:49, v/v) for 9 min. Cyclization is carried out by addition of DIPCDI/HOAt in DMF and the product cleaved from the resin using TFA–CH$_2$Cl$_2$ (1:99, v/v).

CAM/Fmoc/tBu/PAC Handle (Scheme 1b)

Starting with Fmoc-Trp–PepsynK resin, Mendre *et al.* have assembled a head-to-side chain cyclic peptide.[10] To avoid aspartimide formation, dimer

---

[38] D. J. Evans, B. W. Bycroft, W. C. Chan, and P. D. White, *in* "Peptides 1996: Proceedings of the Twenty-fourth European Peptide Symposium" (R. Ramage, ed.), in press. Mayflower Worldwide, Birmingham, UK, 1996.

Fmoc-Asp([Z]-Dpr-CAM)-OH (CAM, carboxyamidomethyl) is prepared in solution and incorporated into the linear sequence. Hydrolysis of the CAM group is achieved with NaOH; cyclization and cleavage from the resin with concomitant removal of *t*Bu-based side-chain protecting groups give the desired target.

Procedure. The CAM side-chain protecting group is removed by treatment with NaOH (3 equivalent) in 2-propanol–water (7:3, v/v) for 1 hr at 25° followed by neutralization with 1.75 N acetic acid (5 equivalent) and washed with acetic acid in 2-propanol–water (7:3, v/v), 2-propanol–water (7:3, v/v), 2-propanol and diethyl ether. The linear peptide is cyclized by three successive treatments with *N*-[(1*H*-benzotriazol-1-yl)(dimethylamino)methylene]-*N*-methylmethanaminium   tetrafluoroborate-*N*-oxide (TBTU)/HOBt (3 equivalent) and DIEA (6 equivalent) for 2 hr. After washing, side-chain protecting groups are removed and the product cleaved from the resin by treatment with reagent K (TFA–water–phenol–thioanisole-1,2-ethanedithiol, 34:2:2:1:1, by volume) for 1.5 hr at 25°.

TMSE/Fmoc/tBu/PAL Handle (Scheme 1b)

The trimethylsilylethyl ester (TMSE) group has been used for side-chain carboxyl protection in a three-dimensional orthogonal scheme for preparing cyclic peptides on Rink amine resin.[39] Fmoc-Asp(OTMSE)-OH is first coupled to the resin following standard Fmoc/*t*Bu elongation to elaborate the linear sequence. Treatment of the resin-bound peptide with 1 M tetrabutylammonium fluoride (TBAF) in DMF removes both the TMSE side chain and the N^α-Fmoc group; cyclization and cleavage give the desired target.

Procedure. Fmoc-Asp(OTMSE)-OH is coupled to Rink amide resin using DCC/HOBt, and residual amino groups are capped by treatment with acetic anhydride. After construction of the linear sequence, treatment with 1 M TBAF in DMF for 20 min removes both the TMSE and N^α-terminal Fmoc protecting groups. Cyclization is carried out by treatment with BOP (10 equivalent) and DIEA (5 equivalent) in DMF for 4 hr. Treatment with reagent K removes the side-chain remaining protecting groups and releases the cyclic peptide from the resin.

Adpoc-OPp/Fmoc/tBu/PAC-PAL Handle (Scheme 1a,b)

2-Phenyl-2-propyl (Opp) ester and 2-(1′-adamantyl)-2-propoxycarbonyl (Adpoc) carbamate protection for carboxylic acids and amines, respec-

---

[39] C. K. Marlowe, *Bioorg. Med. Chem. Lett.* **3**, 437 (1993).

tively, have been introduced as a third level of orthogonality in Fmoc/*t*Bu methods.[40] Linear sequences are assembled on *p*-alkoxybenzyl alcohol and 5-amino-10,11-dihydro-5*H*-dibenzo[*a,d*]cycloheptenyl-2-oxyacetyl-Nle-methylbenzhydrylamine resins for C-terminal acids and amides, respectively, and the Opp and Adpoc protecting groups are removed with TFA–CH_2Cl_2 (1 : 99, v/v). Neutralization, on-resin cyclization, and final cleavage yield the desired targets in high purity.

Procedure. Adpoc and Opp deprotections are conducted by mixing the resin with TFA–CH_2Cl_2 (1 : 99, v/v) six times for 10 min. The resin is washed with CH_2Cl_2 (twice, 1 min), DIEA–CH_2Cl_2 (1 : 49, v : v, twice, 1 min), and DMF (three times, 1 min). The peptide is cyclized with TBTU (1.2 equivalent) and DIEA (1.8 equivalent) in DMF three times for 1.5 hr. Treatment of the resin with TFA–water–*p*-cresol–1,2-ethanedithiol–2-methylindole (75 : 5 : 5 : 12.5 : 2.5, by volume) gives the crude peptide.

Reagents for Cyclization

Once a protection scheme has been chosen and successfully incorporated into a synthetic strategy, the penultimate step for on-resin cyclization requires an efficient coupling reagent (Fig. 3). Initially, carbodiimides were used for ring closure. In the first report for the assembly of side chain-to-side chain cyclic peptides, representative opioid analogs were treated with DCC (5 equivalent) and HOBt (5 equivalent) in DMF over a period of several days, adding fresh DCC and HOBt every 48 hr, to give 6–22% yields.[29]

A comparison of different activators for lactamization was studied by Felix *et al.*[41] Cyclization of a linear peptidyl–resin with 6 equivalents of DCC/HOBt in DMF required 6 days for a negative ninhydrin test, but the cleaved product was contaminated with side products. Alternatively, treatment with 6 equivalent of BOP in DMF containing excess DIEA gave a purer product in 4 hr. Bis(2-oxo-3-oxazolidinyl)phosphinic chloride (BOP-Cl)- and diphenylphosphoryl azide (DPPA)-mediated cyclizations gave even less efficient cyclization in comparison to DCC as measured by quantitative ninhydrin. In a similar study, Hruby *et al.* investigated the cyclization

[40] F. Dick, U. Fritschi, G. Haas, O. Hässler, R. Nyfeler, and E. Rapp, *in* "Peptides 1996: Proceedings of the Twenty-fourth European Peptide Symposium" (R. Ramage, ed.), in press. Mayflower Worldwide, Birmingham, UK, 1996.

[41] A. M. Felix, C. T. Wang, E. P. Heimer, and A. Fournier, *Int. J. Pept. Protein Res.* **31,** 231 (1988).

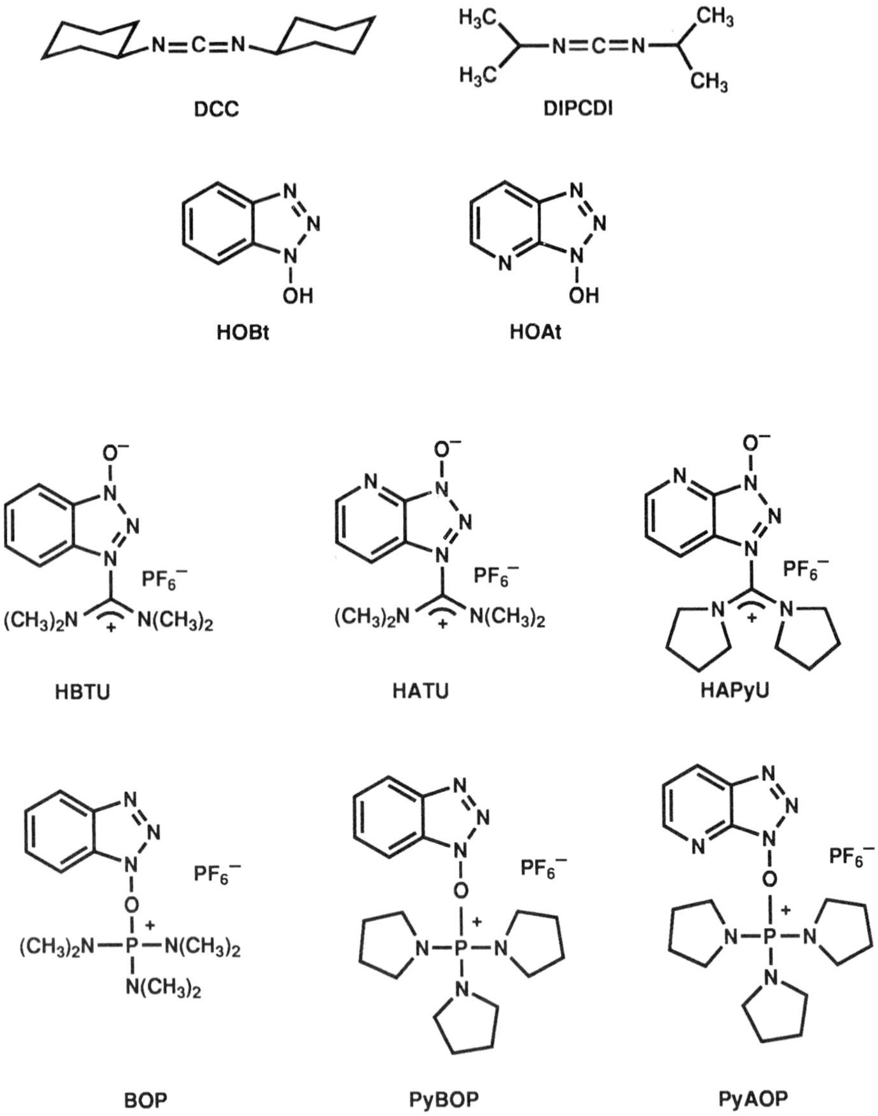

Fig. 3. Coupling reagents used for on-resin cyclization.

of model peptides.[42] To facilitate ring closure and minimize cyclodimerizations, D-amino acids were incorporated in the design, and linear peptides were assembled on chloromethyl and MBHA resins at loadings of 0.2 to 0.45 mmol/g. For on-resin cyclizations, BOP was found to be superior to DIPCDI/HOBt and DPPA, and to solution methods in terms of product yield, purity, and reaction rate.

Plaué found that BOP/HOBt/DIEA-mediated cyclization of a model peptide occurred in 2 hr, whereas with DIPCDI/HOBt lactamization required 3 days for completion.[11] The same ratio of cyclic monomer to dimer was obtained, but the amount of polymeric material formed was greater when using the BOP-containing mixture. For the formation of large loop peptides, cyclodimerization occurs at high resin loading if the cyclization rate is slow (DIPCDI). Polymerization occurs at high loading when cyclization is fast (BOP).

McMurray et al. studied the side reactions of head-to-tail cyclizations of resin-bound peptides.[43] BOP/HOBt/DIEA-mediated cyclizations gave more racemization and oligomerization than DIPCDI/HOBt. In addition, oligomerization and racemization were resin dependent but solvent independent with DIPCIDI (BOP in CH_2Cl_2 gave greater amounts of oligomer). Story and Aldrich found that HBTU-mediated on-resin cyclizations gave poor results due to formation of a tetramethylguanidinium (Tmg) Schiff base derivative of the ω-amino group of α,β-diaminopropionic acid (Dap).[44] BOP- and DIPCDI/HOBt-activated cyclizations gave higher yields, but the reaction proceeded more slowly.

Applications of 1-hydroxy-7-azabenzotriazole (HOAt) and its uronium {N-[(dimethylamino)-1H-1,2,3-triazolo[4,5-b]pyridin-1-ylmethylene]-N-methylmethanaminium hexafluorophosphate N-oxide (HATU)} and phosphonium [7-azabenzotriazol-1-yloxytris(pyrrolidino)phosphonium hexafluorophosphate (PyAOP)] derivatives have been thoroughly examined for both solution and solid-phase peptide synthesis.[45] These reagents enhance coupling rates and yields, reduce racemization, and are suitable for the synthesis of difficult peptides such as those which incorporate hindered amino acids and hydrophobic peptides or involve the coupling of protected segments. To study the utility of azabenzotriazole derivatives, the linear

[42] V. J. Hruby, F. Al-Obeidi, D. G. Sanderson, and D. D. Smith, in "Innovation and Perspectives in Solid Phase Synthesis: Peptides, Polypeptides and Oligonucleotides" (R. Epton, ed.), p. 197. SPCC, Birmingham, U.K., 1990.

[43] J. S. McMurray, C. A. Lewis, and N. U. Obeyesekere, Pept. Res. 7, 195 (1994).

[44] S. C. Story and J. V. Aldrich, Int. J. Pept. Protein Res. 43, 292 (1994).

[45] S. A. Kates, L. A. Carpino, and F. Albericio, in "Peptides: Chemistry and Biology, Proceedings of the Fourteenth American Peptide Symposium" (P. T. P. Kaumaya and R. S. Hodges, eds.), p. 893. Mayflower Worldwide, Birmingham, UK, 1996.

TABLE II
COMPARISON OF COUPLING REAGENTS
FOR THE LACTAMIZATION STEP OF
CYCLO(Asp-Tyr-DTrp-Val-DTrp-DTrp-Arg)

Coupling method	Linear (%)	Cyclic (%)	Purity (%)
PyAOP/HOAt	5	74	61
PyBOP/HOBt	5	69	51
HATU/HOAt	5	70	50
HBTU/HOBt	5	63	44

heptapeptide cyclo(Asp-Tyr-DTrp-Val-DTrp-DTrp-Arg) was cyclized. For this model, azabenzotriazole-based reagents performed slightly better than the benzotriazole analogs, with PyAOP producing the peptide in highest yield (74%) and purity (61%) (Table II).[46]

Aldrich and co-workers surveyed several activating reagents for their efficiency of resin-bound lactam formation in the cyclic dynorphin A model.[47] Uronium-based reagents HBTU, HATU, and 1-(1-pyrrolidinyl-1H-1,2,3-triazolo[4,5-b]pyridin-1-ylmethylene) pyrrolidinium hexafluoro-phosphate N-oxide (HAPyU) gave low yields of desired product and a high formation of the linear alkyl guanidinium by-products even when a low concentration of reagent was used. HATU and HAPyU gave better results than HBTU, but the dipyrrolidinylguanidinium (Dpg) by-product from HAPyU was more difficult to remove during purification. On the contrary, the phosphonium-based reagents BOP, PyBOP, and PyAOP gave satisfactory yields of cyclic peptide, but the reactions progressed more slowly (3 days). PyXOP cyclizations were faster than those with BOP, but all phosphonium reagents gave similar yields of cyclic peptide, undesired linear starting peptide, linear peptide imide, and cyclodimer.

Zhang and Taylor examined solvent conditions for solid-phase cyclizations.[48] Solvent mixtures were chosen that combine high polarity with excellent resin swelling properties. For BOP/DIEA lactamization on a p-chloro-methylbenzyl–polystyrene resin, optimal monomer–dimer ratios were obtained with DMSO–N-methylpyrrolidone (NMP) (1:4, v/v) > THF–NMP (35:65, v/v), DMF, and DMF–CH$_2$Cl$_2$ (1:1, v/v) > CH$_2$Cl$_2$–DMF (96:4, v/v).

[46] S. A. Kates, C. A. Minor, H. Shroff, R. C. Haaseth, S. Triolo, A. El-Faham, L. A. Carpino, and F. Albericio, in "Peptides 1994: Proceedings of the Twenty-third European Peptide Symposium" (H. L. S. Maia, ed.), p. 248. Escon, Leiden, The Netherlands, 1995.
[47] S. Arttamangkul, B. Arbogast, D. Barofsky, and J. V. Aldrich, Lett. Pept. Sci. 3, 357 (1996).
[48] W. Zhang and J. W. Taylor, Tetrahedron Lett. 37, 2173 (1996).

Cyclic Peptide Libraries

Combinatorial libraries serve as a source of diverse potential therapeutic candidates and have become an important tool in the drug discovery field. Pioneering work involving the creation and screening of linear peptide libraries has expanded to the preparation of cyclic peptide libraries.[49] Spatola et al. coupled the side chain of Boc-Asp-OFm to a benzyl alcohol resin (Merrifield) using Mitsunobu conditions and then constructed combinatorial libraries of cyclic pentapeptides.[50] The linear chain was assembled using Boc-amino acids with the exception of the last amino acid, which was added as the Fmoc derivative. To avoid partial racemization of the C-terminal Asp residue during chain elongation, 1 equivalent of DIEA in the couplings was used. Removal of Fmoc and Fm groups was accomplished using piperidine–DMF (1:4, v/v). Initially cyclizations were accomplished using BOP/HOBt, but the final lactam bond was formed faster using BOP/HOAt activation. Cleavage of peptide mixtures from the resin was carried out using HF.

Mihara et al. prepared cyclic peptide mixtures using oxime–resin. The cyclic peptides contained ε-aminocaproic acid (Aca) as a flexible residue to enhance the cyclization step.[51] The synthesis of cell adhesion inhibitors cyclo[Arg(Tos)-Gly-Asp(OcHex)-Ser(Bzl)-Aca] was accomplished by BOP/HOBt-induced coupling of Boc-amino acids to Aca-oxime resin. After removal of the N-terminal Boc group, cyclization was effected by treatment with acetic acid–DIEA. A combinatorial library based on this structure in which Ser was replaced by Lys, Gln, Ala, Glu, Pro, Val, Tyr, Phe, and Trp was constructed via both the "mix and split" and mixture methods. For the mix and split technique, the cyclic peptides were formed in almost equal amount throughout the coupling and cyclization reactions. The mixed protocol (0.3 equivalent of each amino acid) gave similar results except lower yields for Pro and Val.

Side Reactions

Although fluoride-labile protecting groups and linkers have been developed as an extra level of orthogonality, a side reaction was observed when peptidyl–resins were treated with tetrabutylammonium fluoride. For Glu(OtBu)- and Asp(OtBu)-containing peptides, fluoride ions were found

[49] K.-H. Wiesmüller, S. Feiertag, B. Fleckenstein, S. Kienle, D. Stoll, M. Herrmann, and G. Jung, in "Peptides and Nonpeptide Libraries—A Handbook" (G. Jung, ed.), p. 203. VCH, Germany, 1996.
[50] A. F. Spatola, K. Darlak, and P. Romanovskis, Tetrahedron Lett. 37, 591 (1996).
[51] H. Mihara, S. Yamabe, T. Niidome, and H. Aoyagi, Tetrahedron Lett. 36, 4837 (1995).

to convert α-peptides to their corresponding γ- or β-rearranged sequence.[52] Although HPLC, matrix-assisted laser desorption/time-of-flight (MALDI-TOF) mass spectrometry, and amino acid analysis are powerful techniques for peptide identification, the use of capillary zone electrophoresis (CZE) was required to detect this phenomenon.

In the synthesis of small loop peptides using Fmoc/Bzl amino acids by Plaué,[11] HF cleavage afforded the desired cyclic peptide and a second product that was identified to be the result of imide formation between Asp-146 and Phe-147. To avoid the succinimide side reaction, cyclohexyl side-chain protection of Asp-146 was incorporated into the linear sequence.

With Fmoc/Fm protecting groups, following standard treatment with piperidine–DMF (1:4, v/v), a wash with 0.4% concentrated HCl in DMF,[17] HOBt in DMF,[43] or a tertiary amine[50] was recommended prior to cyclization. This step ensured complete removal of piperidine and avoided the formation of C-terminal piperidylamides as a by-product in the cyclization step. If DBU–piperidine–DMF (1:1:48, by volume) was used, the acid wash was not required.

Epimerization during linear assembly was identified to be problematic for strategies that include side-chain anchoring of Asp residues. Coupling with preactivated pentafluorophenyl esters or reducing the amount of base to 1 equivalent for BOP/HOBt significantly reduced the level of racemization.[50] The degree of racemization during on-resin cyclization is affected by the amino acid sequence.[43]

[52] S. A. Kates and F. Albericio, *Lett. Pept. Sci.* **1**, 213 (1994).

[10] Disulfide Bond Formation in Peptides

By IOANA ANNIS, BALAZS HARGITTAI, and GEORGE BARANY

Introduction

Disulfide bridges represent important evolutionarily conserved structural motifs in many biologically important peptides and proteins, including hormones, enzymes, growth factors, toxins, and immunoglobulins.[1-5] Intra-

[1] J. M. Thornton, *J. Mol. Biol.* **151**, 261 (1981).
[2] T. E. Creighton, *BioEssays* **8**, 57 (1988).
[3] N. Srinivasan, R. Sowdhamini, C. Ramakrishnan, and P. Balaram, *Int. J. Pept. Protein Res.* **36**, 147 (1990).

molecular disulfides serve to covalently cross-link portions of the polypeptide chain that are apart in the linear sequence but come together in three dimensions. Formation of these bridges does not appear to be the driving force per se for folding, but the bridges are believed to play a major role in stabilizing the bioactive conformations as well as promoting entropic destabilization of the denatured state.[6] Artificial introduction of disulfide bridges into natural or designed peptides and small proteins has been carried out with the goal to improve biological activities/specificities,[5,7,8] and stabilities.[2,4,5,9–12] The capability to link two (or more) separate chains via intermolecular disulfides has numerous implications for biological research, including conjugation of peptides to carriers for immunological studies, preparation of standards corresponding to proteolytic fragments isolated during structural elucidation work on large proteins, development of discontinuous epitopes, and generation of active site models.[5,13–15]

Reproducing (or engineering) the disulfide arrangement(s) found in (or suggested by) nature represents a significant challenge to peptide and protein chemists.[5,16–19] The extensive literature in the field describes the total synthesis of materials of >100 residues (however, usually <60 residues), composed of a single or two polypeptide chain(s), and including two to ten

[4] S. F. Betz, *Protein Sci.* **2,** 1551 (1991).

[5] D. Andreu, F. Albericio, N. A. Solé, M. C. Munson, M. Ferrer, and G. Barany, *in* "Methods in Molecular Biology, Volume 35: Peptide Synthesis Protocols" (M. W. Pennington and B. M. Dunn, eds.), p. 91. Humana Press, Totowa, New Jersey, 1994.

[6] E. Barbar, G. Barany, and C. Woodward, *Folding Design* **1,** 65 (1996).

[7] V. J. Hruby, *Life Sci.* **31,** 189 (1981).

[8] V. J. Hruby, F. Al-Obeidi, and W. Kazmierski, *Biochem J.* **268,** 249 (1990).

[9] R. Wetzel, *Trends Biochem. Sci.* **12,** 478 (1987).

[10] J. Clarke and A. R. Fersht, *Biochemistry* **32,** 4322 (1993).

[11] N. M. Young, C. R. MacKenzie, S. A. Narang, R. P. Oomen, and J. E. Baenziger, *FEBS Lett.* **377,** 135 (1995), and references cited therein.

[12] J. H. Ko, W. H. Jang, E. K. Kim, H. B. Lee, K. D. Park, J. H. Chung, and O. J. Yoo, *Biochem. Biophys. Res. Commun.* **221,** 631 (1996), and references cited therein.

[13] E. Wünsch, L. Moroder, S. Göhring-Romani, H.-J. Musiol, W. Göhring, and G. Bovermann, *Int. J. Pept. Protein Res.* **32,** 368 (1988).

[14] B. Ponsati, E. Giralt, and D. Andreu, *Anal. Biochem.* **181,** 389 (1989).

[15] H. Kataoka, J. P. Li, A. S. T. Lui, S. J. Kramer, and D. A. Schooley, *Int. J. Pept. Protein Res.* **39,** 29 (1992).

[16] G. Barany and R. B. Merrifield, *in* "The Peptides–Analysis, Synthesis, Biology" (E. Gross and J. Meienhofer, eds.), Vol. 2, p. 1. Academic Press, New York, 1979.

[17] E. E. Büllesbach, *Kontakte (Darmstadt)* **1,** 21 (1992).

[18] Y. Kiso and H. Yajima, *in* "Peptides—Synthesis, Structures, and Applications" (B. Gutte, ed.), p. 39. Academic Press, San Diego, 1995.

[19] L. Moroder, D. Besse, H. J. Musiol, S. Rudolph-Böhner, and F. Siedler, *Biopolymers* **40,** 207 (1996).

half-cystines. The disulfide bridges participate in eight-membered or larger intramolecular rings, or form interchain cross-links. Sulfur–sulfur bonds are created most often by oxidation (air or stronger reagents) of precursors with free or protected sulfhydryls (Scheme 1, left and center pathways, respectively); chemistries for unsymmetrical "directed" formation of sulfur–sulfur bonds can be applied as well (Scheme 1, right pathway). Regioselective pairing of disulfides can occur spontaneously or by controlled chemistries involving orthogonally removable Cys* protecting groups. The relevant transformations must be carried out in ways that minimize the breaking or *scrambling* of disulfide bridges (Scheme 2). Methods directed toward intramolecular disulfide bridges (Scheme 1) suffer from intermolecular dimerization and/or oligomerization (Scheme 3; mitigated to some extent by carrying out oxidations under high *dilution*), whereas methods directed toward intermolecular disulfide bridges are limited by *disproportionation* of the desired unsymmetric heterodimers to symmetric homodimer species (Scheme 4).

In general, the more complex peptide targets have been obtained in relatively low yields, only after extensive optimization of experimental protocols for synthesis, purification, and oxidation/folding.[5,16–19] Applica-

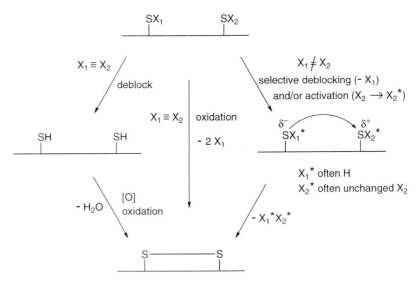

SCHEME 1. Synthetic approaches to intramolecular disulfide bridges. X_1 and X_2 designate sulfhydryl protecting groups that are stable to chain assembly.

* Unless stated otherwise, amino acid symbols denote the L-configuration, and all solvent ratios and percentages are volume/volume.

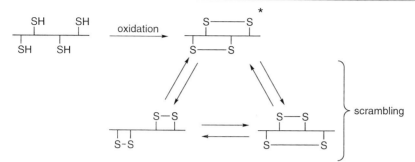

* desired regioisomer, for purpose of illustration

SCHEME 2. Scrambling of disulfide bridges, as side reaction during oxidation.

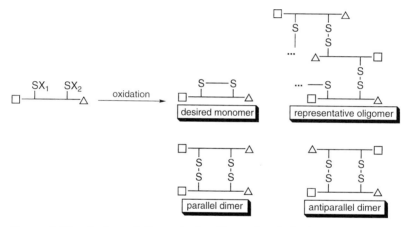

SCHEME 3. Dimerization and oligomerization side reactions during intramolecular disulfide formation. X_1 and X_2 are sulfhydryl protecting groups as defined in Scheme 1.

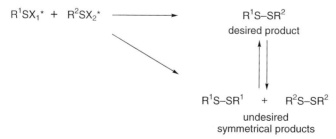

$$R^1SX_1^* + R^2SX_2^* \longrightarrow R^1S\text{–}SR^2$$
desired product

$$R^1S\text{–}SR^1 + R^2S\text{–}SR^2$$
undesired
symmetrical products

SCHEME 4. Disproportionation of desired unsymmetric heterodimer to undesired symmetric homodimers. X_1^* and X_2^* as defined in Scheme 1.

tions of disulfide bridging as the basis for intramolecular cross-links in molecules prepared by combinatorial library methods require particularly general, reliable, high-yield chemistries.[20-22] Furthermore, when peptide-derived compounds are drug candidates, the exact protocols used for disulfide formation, and for subsequent formulation to allow administration to animal or human subjects, may introduce unwelcome surprises even for relatively simple structures.

This chapter aims to provide a highly selective albeit state-of-the-art menu of experimental options that can be tried to make intra- or intermolecular disulfide-containing targets of varying complexities. Reactions can be carried out with peptides in solution or while a peptide remains anchored to a polymeric support, in the latter case taking advantage of the *pseudodilution* phenomenon[16,23] which favors intramolecular cyclization. We take it as given that the linear substrates (free or *S*-protected), which are to be converted to the disulfide form, can be prepared (purified as required) in a reasonable state of chemical homogeneity. Furthermore, these substrates need to be sufficiently flexible to access thermodynamically favored conformations in which to-be-paired half-cystine residues come into reasonable spatial proximity. The environment (steric; neighboring charges; buried versus surface) about the Cys residues affects their pK_a values, and consequently the ease with which they oxidize. The rates, yields, and specificities of oxidations can be influenced as well by the reagents used, solvents, presence of denaturants, and/or the parent *S*-protecting groups. For regioselective schemes, some of the indicated disulfide-forming chemistries can be used in series, provided again that the desired regiomers have the thermodynamic and kinetic characteristics to resist scrambling under the reaction conditions. There exists a substantial empirical base of knowledge on a number of these issues, but as yet few clear correlations with predictive power have been drawn. Previous reviews[5,16-19] are worth consulting for additional perspectives.

Cysteine Protection

Many ways have been proposed to protect the nucleophilic β-thiol of Cys during solid-phase peptide synthesis (Table I). Deprotection of Cys

---

[20] K. T. O'Neil, R. H. Hoess, S. A. Jackson, N. S. Ramachandran, S. A. Mousa, and W. F. DeGrado, *Proteins: Struct. Funct. Genet.* **14**, 509 (1992).

[21] S. E. Salmon, K. S. Lam, M. Lebl, A. Kandola, P. S. Khattri, S. Wade, M. Pátek, P. Kocis, V. Krchnák, D. Thorpe, and S. Felder, *Proc. Natl. Acad. Sci. U.S.A.* **90**, 11708 (1993).

[22] S. Cheng, W. S. Craig, D. Mullen, J. F. Tschopp, D. Dixon, and M. D. Pierschbacher, *J. Med. Chem.* **37**, 1 (1994).

[23] S. Mazur and P. Jayalekshmy, *J. Am. Chem. Soc.* **101**, 677 (1978).

TABLE I
Cysteine Protecting Groups for Peptide Synthesis[a]

Lability	Protecting group	Structure	Compatibility	Stability	Removal conditions
Dilute acid					
	S-Xan[b]		Fmoc	Base	Dilute TFA/scavengers;[c] I$_2$,[d] Tl(III)[d]
	S-Tmob[e]		Fmoc	Base, nucleophiles	Dilute TFA/scavengers;[c] I$_2$,[d] Tl(III)[d]
Moderate acid					
	S-Mmt[f]		Fmoc	Base, nucleophiles	TFA/scavenger;[c] Hg(II), Ag(I), I$_2$,[d] Tl(III),[d] RSCl

(continued)

TABLE I (*continued*)

Lability	Protecting group	Structure	Compatibility	Stability	Removal conditions
	S-Trt[a]		Fmoc	Base, nucleophiles	TFA/scavenger,[c] Hg(II), Ag(I), I$_2$,[d] Tl(III),[d] RSCl
Concentrated acid	S-Mob[a]	—S—CH$_2$— OCH$_3$	Boc, Fmoc	TFA, base, RSCl, I$_2$	HF(0°), Tl(III),[d] Hg(II), Ag(I)
	S-Meb[a]	—S—CH$_2$— CH$_3$	Boc, Fmoc	TFA, Ag(I), base, RSCl	HF(0°), Tl(III)[d]
	S-tBu[a]	—S—C(CH$_3$)$_3$	Boc, Fmoc	HF(0°), base, I$_2$, Ag(I)	HF(20°), Hg(II), NpsCl, HF/scavengers
Metal ions[g]	S-Acm[a,h]	—S—CH$_2$—NH—C(O)—CH$_3$	Boc, Fmoc	HF(0°), TFA, base	Hg(II), Ag(I), I$_2$,[d] Tl(III),[d] RSCl
	S-Phacm[i]	—S—CH$_2$—NH—C(O)—CH$_2$—	Boc, Fmoc	HF, base	Hg(II), Tl(III),[d] I$_2$,[d] penicillin amidohydrolase

204

Base				
S-Fm[a]		Boc	HF, I_2	NH_3 in MeOH, piperidine, dilute DBU
S-Dnpe[j]		Boc	HF	Piperidine, dilute DBU
Reducing agents[k]				
S-StBu[a]		Boc, Fmoc	TFA, HF (partial), base, RSCl	RSH, Bu_3P, other reducing agents
S-Snm[l,m,n]		Boc	HF, TFA	RSH, other reducing agents
S-Scm[a]		None[o]	HF, TFA	RSH, other reducing agents
S-Npys[p,q,r]		Boc	HF, HOBt	RSH, Bu_3P

(*continued*)

TABLE I (continued)

[a] The protecting group structure is drawn to include the sulfur (shown in **bold**) of the cysteine that is being protected. Unless indicated otherwise, conditions or reagents outlined under Removal conditions are intended for quantitative removal and/or oxidative cleavage; the products from these reactions may be free thiols, mercaptides, or disulfides. For metal-mediated removals, the counterion and solvent are sometimes critical; for acidolytic cleavages with HF or similar strong acids, the nature and amounts of scavengers added may affect the stability and/or lability of the protecting group significantly. References provided are highly selective, with only the most relevant papers being cited. For more extensive information about any particular protecting group, the reader is advised to consult the general reviews cited in the text, especially G. Barany and R. B. Merrifield, *in* "The Peptides—Analysis, Synthesis, Biology" (E. Gross and J. Meienhofer, eds.), Vol. 2, p. 1. Academic Press, New York, 1979; and D. Andreu, F. Albericio, N. A. Solé, M. C. Munson, M. Ferrer, and G. Barany, *in* "Methods in Molecular Biology, Volume 35: Peptide Synthesis Protocols" (M. W. Pennington and B. M. Dunn, eds.), p. 91. Humana Press, Totowa, New Jersey, 1994.

[b] Y. Han and G. Barany, *J. Org. Chem.* **62**, 3841 (1997).

[c] These protecting groups give relatively stable carbocations, which can realkylate cysteine and/or irreversibly modify tryptophan; to prevent this, scavengers are added to the acid deprotection cocktail. Trialkylsilanes have been found to be particularly effective in this regard. For details, see D. A. Pearson, M. Blanchette, M. L. Baker, and C. A. Guindon, *Tetrahedron Lett.* **30**, 2739 (1989); M. C. Munson, C. García-Echeverría, F. Albericio, and G. Barany, *J. Org. Chem.* **57**, 3013 (1992); and M. C. Munson and G. Barany, *J. Am. Chem. Soc.* **115**, 10203 (1993).

[d] Reaction with iodine or Tl(III) generates the disulfide directly.

[e] M. C. Munson, C. García-Echeverría, F. Albericio, and G. Barany, *J. Org. Chem.* **57**, 3013 (1992).

[f] K. Barlos, D. Gatos, O. Hatzi, N. Koch, and S. Koutsogiasnni, *Int. J. Pept. Protein Res.* **47**, 148 (1996).

[g] When metal ions are used as deprotecting agents, treatment with excess 2-mercaptoethanol or hydrogen sulfide is necessary to generate the free thiol. Sometimes the metal ion binds very tightly and is difficult to remove completely from the peptide.

[h] Under acidic conditions or under conditions of Acm deprotection, $S \rightarrow N$ or $S \rightarrow O$ migration of the Acm group can occur. Glycerol and glutamine have been proposed as useful scavengers. For more details, see H. Lamthanh, H. Virelizier, and D. Frayssinhes, *Pept. Res.* **8**, 316 (1995); M. Royo, J. Alsina, E. Giralt, U. Słomczyńska, and F. Albericio, *J. Chem. Soc., Perkin Trans.* **1**, 1095 (1995); and H. Lamthanh, H. Virelizier, and D. Frayssinhes, *Pept. Res.* **8**, 316 (1995).

[i] M. Royo, J. Alsina, E. Giralt, U. Słomczyńska, and F. Albericio, *J. Chem. Soc., Perkin Trans.* **1**, 1095 (1995).

[j] M. Royo, C. García-Echeverría, E. Giralt, R. Eritja, and F. Albericio, *Tetrahedron Lett.* **33**, 2391 (1992).

[k] Reactions with other thiols (including peptide thiols) give mixed disulfide intermediates, as covered in more detail later.

[l] A. L. Schroll and G. Barany, *J. Org. Chem.* **54**, 244 (1989).

[m] L. Chen, H. Bauerová, J. Slaninová, and G. Barany, *Pept. Res.* **9**, 114 (1996).

[n] L. Chen, I. Zoulíková, J. Slaninová, and G. Barany, *J. Med. Chem.* **40**, 864 (1997).

[o] Must be introduced indirectly.

[p] R. Matsueda, T. Kimura, E. T. Kaiser, and G. R. Matsueda, *Chem. Lett.* **6**, 737 (1981).

[q] M. S. Bernatowicz, R. Matsueda, and G. R. Matsueda, *Int. J. Pept. Protein Res.* **28**, 107 (1986).

[r] F. Albericio, D. Andreu, E. Giralt, C. Navalpotro, E. Pedroso, B. Ponsati, and M. Ruiz-Gayo, *Int. J. Pept. Protein Res.* **34**, 124 (1989).

can be carried out while the peptide is still anchored to the polymeric support, or concurrent to cleavage from the support, or after it is in solution; and it can be conducted either to generate the free thiol (preferably under an inert atmosphere, to minimize inadvertent or premature oxidation), or to provide directly a disulfide bond. In the former case, the orthogonal deblocking modes are acid [relative rates: 9H-xanthen-9-yl (Xan) > 2,4,6-trimethoxybenzyl (Tmob) > p-methoxytrityl (Mmt) > triphenylmethyl (Trt) > 4-methoxybenzyl (Mob) > 4-methylbenzyl (Meb) >> $tert$-butyl (tBu)]; base [for 9-fluorenylmethyl (Fm) and 2-(2,4-dinitrophenyl)ethyl (Dnpe), with the latter somewhat preferred on the basis of solubility considerations; unless a reducing agent is present, disulfide forms directly]; thiols, trialkylphosphines, or other reducing agents [for 3-nitro-2-pyridinesulfenyl (Npys), S-methyloxycarbonylsulfenyl (Scm), S-[(N-methyl-N-phenylcarbamoyl)sulfenyl] (Snm), as well as the sluggishly cleavable S-$tert$-butylmercapto (S-tBu)]; electrophilic metals such as Ag(I) or Hg(II) salts [for acetamidomethyl (Acm) and its relatives, and many of the acidolyzable groups]; or enzymes [for phenylacetamidomethyl (Phacm) with penicillin G acylase; a reducing agent must be present or else a symmetrical homodimer forms directly]. Alternatively, oxidations with iodine (or other sources of I$^+$) in acidic aqueous or alcoholic media; thallium(III) trifluoroacetate [Tl(tfa)$_3$] in trifluoroacetic acid (TFA) or N,N-dimethylformamide (DMF); or mixtures of dialkyl or diaryl sulfoxides with TFA or chlorosilanes, among other methods, serve to intentionally convert the S-protected species (Xan, Tmob, Trt, Acm, Mob, tBu; listed in approximate order of reactivity from most reactive to least) *directly* to the corresponding disulfide.

Formation of Disulfides from Free Thiol Precursors

The conceptually simplest approach to disulfide formation involves complete deprotection of the precursor linear peptide to form the free (poly)-thiol derivative, which is oxidized in the hope that the appropriately folded product with the desired pairings will predominate. With careful attention to experimental conditions, for example, pH, ionic strength, temperature, time, and concentration, this ideal is often realized, but mispairing/misfolding can occur despite the best efforts at optimization (Scheme 2). The strategy under discussion intersects with the experimental logic of protein renaturation studies,[5,24-29] and it has the advantage of requiring only one

[24] R. R. Hantgan, G. G. Hammes, and H. A. Scheraga, *Biochemistry* **13,** 3421 (1974).
[25] A. Karim Ahmed, S. W. Schaffer, and D. B. Wetlaufer, *J. Biol. Chem.* **250,** 8477 (1975).
[26] R. Jaenicke and R. Rudolph, *in* "Protein Structure: A Practical Approach" (T. E. Creighton, ed.), Chap. 9, p. 191. IRL Press, Oxford, 1989, and references cited therein.
[27] D. M. Rothwarf and H. A. Scheraga, *Biochemistry* **32,** 2680 (1993).

type of protecting group for all Cys residues. It is generally helpful to pretreat the crude material with appropriate reducing agents, followed by purification (under acidic conditions to minimize premature oxidation and/or disulfide exchange); this is to ensure that only the linear monomeric species is subjected to the oxidation/folding. High dilution is recommended to avoid aggregation and dimers, oligomers, or intractable polymers; however, the formation of such intermolecular by-products cannot always be avoided, and "recycling' of nonmonomeric species through alternations of reduction/reoxidation steps is also quite difficult to achieve in practice.[26]

Air Oxidation

The easiest oxidations are carried out in the presence of atmospheric oxygen, at high dilution, and generally under slightly alkaline conditions. This widely used approach may be subject to one or more of the following limitations: (1) dimerization or worse, despite precautions to carry out reactions at low concentrations of substrate; (2) inadequate solubility for basic or hydrophobic peptides; (3) very long times (up to 5 days) sometimes required for complete reaction[30]; (4) difficulty in controlling oxidations, because the rate depends on trace amounts of metal ions[24]; and (5) accumulation of side products due to oxidation of Met residues.[31]

Procedure. In general,[5,16,32,33] the peptide (free thiol form) is dissolved in an appropriate buffer at a concentration of 0.01–0.10 mM (concentrations as low as 1 μM[34] and as high as 1 mM[35] have been reported), and the solution is stirred in open atmosphere, or with oxygen bubbling through it. Buffers used for air oxidations typically have pH values of 6.5–8.5, with the rate of oxidation higher with increasing pH (assuming that the substrate is soluble at the pH value chosen). The most widely used aqueous buffers are 0.1–0.2 M Tris-HCl or Tris–acetate, pH 7.7–8.7; 0.01 M phosphate buffers, pH 7–8; 0.2 M ammonium acetate, pH 6–7; and 0.01 M ammonium

---

[28] M. Ruoppolo and R. B. Freedman, *Biochemistry* **34,** 9380 (1995).
[29] R. Kuhelj, M. Dolinar, J. Pungercar, and V. Turk, *Eur. J. Biochem.* **229,** 533 (1995).
[30] K. Akaji, T. Tatsumi, M. Yoshida, T. Kimura, Y. Fujiwara, and Y. Kiso, *J. Am. Chem. Soc.* **114,** 4137 (1992).
[31] K. Akaji, Y. Nakagawa, Y. Fujiwara, K. Fujino, and Y. Kiso, *Chem. Pharm. Bull.* **41,** 1244 (1993).
[32] C. Kellenberger, H. Hietter, and B. Luu, *Pept. Res.* **8,** 321 (1995).
[33] Unpublished work from our laboratory (1995–1997).
[34] H. Tamamura, T. Murakami, S. Horiuchi, K. Sugihara, A. Otaka, W. Takada, T. Ibuka, M. Waki, N. Yamamoto, and N. Fujii, *Chem. Pharm. Bull.* **43,** 853 (1995).
[35] J. G. Adamson and G. A. Lajoie, *in* "Peptides—Chemistry, Structure and Biology: Proceedings of the Thirteenth American Peptide Symposium" (R. S. Hodges and J. A. Smith, eds.), p. 44. Escom, Leiden, The Netherlands, 1993.

bicarbonate, pH 8. For the oxidation of certain peptides, the use of organic cosolvents (methanol, acetonitrile, dioxane) and the addition of tertiary amines (N-methylmorpholine, triethylamine, and N,N-diisopropylethylamine) is recommended.[32,33] Guanidine hydrochloride (Gdm-HCl, 2–8 M) is sometimes added to aid in solubility and to increase the conformational flexibility of peptides; this is reported to result in overall improvements of the rates and yields of air oxidations.[35] Addition of $CuCl_2$ (0.1–1 μM) has also been reported to improve certain oxidations.[26] Reaction times are found to vary from a few hours to a few days.

On-resin air oxidation has also been described.[32,33,36–38] Cys residues that are protected by S-Tmob,[38,39] S-Xan,[40] S-Mmt,[32] S-Fm,[37] or S-mercaptobenzyl[36] are selectively deblocked by mild acid, base, or thiolysis, in a way that retains the peptide free thiol on the polymeric support. The peptide–resin is then incubated at 25° with triethylamine* (0.02–0.175 M, 2–10 equivalent) in N-methylpyrrolidone (NMP) for 5–36 hr,[32,33,38] or dioxane–methanol (1:1)[36] for 4 hr, while air or oxygen is gently bubbled through the suspension [in the case where S-Fm protection is used, the Fm removal conditions (piperidine–DMF, 1:1, 3 hr, 25°) provide the desired disulfide directly]. Both intra- and intermolecular disulfides have been made in these ways.

Oxidation in Presence of Redox Buffers

When poly(thiol) precursors are oxidized in the presence of mixtures of low molecular weight disulfides and thiols, overall rates and yields are often better than in the case of straightforward air oxidation. This is because the mechanism changes from direct oxidation (free radical intermediates) to thiol–disulfide exchange (thiolate intermediate), which facilitates the reshuffling of incorrect disulfides to the natural ones. Mixtures of oxidized and reduced glutathione, cysteine, cysteamine, or 2-mercaptoethanol are commonly used. As before, high dilution of the peptide or protein substrate

---

[36] R. Buchta, E. Bondi, and M. Fridkin, *Int. J. Pept. Protein Res.* **28,** 289 (1986).
[37] F. Albericio, R. P. Hammer, C. García-Echeverría, M. A. Molins, J. L. Chang, M. C. Munson, M. Pons, E. Giralt, and G. Barany, *Int. J. Pept. Protein Res.* **37,** 402 (1991), and references cited therein.
[38] M. C. Munson, M. Lebl, J. Slaninová, and G. Barany, *Pept. Res.* **6,** 155 (1993).
[39] L. Chen, I. Zoulíková, J. Slaninová, and G. Barany, *J. Med. Chem.* **40,** 864 (1997).
[40] Y. Han and G. Barany, *J. Org. Chem.* **62,** 3841 (1997).
* In earlier publications from our laboratory which were inspired by a precedent from the organosulfur chemistry literature [E. Wenschuh, M. Heydenreich, R. Runge, and S. Fischer, *Sulfur Lett.* **8,** 251 (1989)], CCl_4 was used equimolar to the tertiary amine. Our more recent studies have shown that CCl_4 is not necessary.

is necessary to maximize yields of the desired intramolecular disulfide-bridged species, and avoid formation of oligomers.[5,26]

Procedure. Using a buffer of 0.1–0.2 M Tris-HCl, pH 7.7–8.7, plus 1 mM EDTA, both reduced (1–10 mM) and oxidized (0.1–1 mM) glutathione (alternatively, cysteine/cystine, cysteamine/cystamine, 2-mercaptoethanol/2-hydroxyethyl disulfide, dithiothreitol reduced/oxidized) are dissolved.[5,24–29] Typically, the optimal molar ratio of the reduced to oxidized compound is 10:1, but ratios of up to 1:1 have been reported. The poly-(thiol) peptide, which has been previously reduced and purified, is dissolved in the aforementioned redox buffer at a concentration of 0.05–0.1 mM. In cases where formation of aggregates competes with the oxidation process, addition of a nondenaturing chaotropic agent is recommended (1–2 M Gdm-HCl or urea). Oxidations proceed at 25–35° and can be monitored by high-performance liquid chromatography (HPLC); 16 hr to 2 days are common reaction times. The oxidized peptide is concentrated by lyophilization and purified by gel filtration on a Sephadex G-10 or G-25 column, developed with aqueous buffers at acidic pH. If intractable precipitates form during this procedure, an alternative folding/oxidation is recommended in which the peptide is oxidized against a series of redox buffers with a slow pH gradient from 2.2 to 8.[29,41]

Dimethyl Sulfoxide-Mediated Oxidation

In contrast to air oxidation (see previous section), oxidation of thiols to disulfides promoted by dimethyl sulfoxide (DMSO) can be carried out with an extended pH range (3–8).[42,43] This is advantageous, because the substrates undergoing oxidation often have improved solubility characteristics under those conditions. DMSO is miscible with water, so a relatively high concentration can be used. A higher DMSO concentration leads to faster reaction, but also to reduced selectivity. Problems in removing DMSO from the final products have been observed in some cases.[44] Within the range of pH 3–8, side reactions involving oxidation of nucleophilic side chains, for example, Met, Trp, and Tyr, have not been observed.[43] DMSO is even effective as an oxiziding agent in the presence of 1 N aqueous HCl, as demonstrated by the conversion of Cys(Ag) residues [obtained on deprotection of Cys(Acm) with silver triflate (AgOTf)] to disulfides.[45]

[41] J.-M. Sabatier, H. Darbon, P. Fourquet, H. Rochat, and J. Van Rietschoten, *Int. J. Pept. Protein Res.* **30**, 125 (1987).
[42] T. J. Wallace, *J. Am. Chem. Soc.* **86**, 2018 (1964).
[43] J. P. Tam, C.-R. Wu, W. Liu, and J.-W. Zhang, *J. Am. Chem. Soc.* **113**, 6657 (1991).
[44] M. C. Munson and G. Barany, *J. Am. Chem. Soc.* **115**, 10203 (1993).
[45] H. Tamamura, A. Otaka, J. Nakamura, K. Okubo, T. Koide, K. Ikeda, and N. Fujii, *Tetrahedron Lett.* **34**, 4931 (1993).

Procedure (Slightly Acidic pH). The crude peptide, as obtained directly from the cleavage/deblocking steps, is dissolved in acetic acid and water (as required).[43] The solution is diluted to a final peptide concentration of 0.5–1.6 mM, and a final concentration of 5% acetic acid in water. The solution is adjusted to pH 6 with $(NH_4)_2CO_3$. DMSO (10–20% by volume) is added, and oxidation is allowed to proceed for 1–4 hr at 25° (monitor by HPLC). The final reaction mixture is diluted two-fold (v/v) with buffer A (5% CH_3CN, 0.05% TFA in H_2O) and loaded onto a preparative reversed-phase HPLC column. The desired product is purified by HPLC, eluted with a linear gradient of buffer A and buffer B (60% CH_3CN, 0.04% TFA in H_2O).

Procedure (Slightly Basic pH). The crude peptide is dissolved to a final concentration of approximately 1 mM in 0.01 M phosphate buffer, pH 7.5, and DMSO (1% by volume) is added.[44] The reaction at 25° is monitored by HPLC, and, after completion, usually 3–7 hr, it is quenched by lyophilization.

Potassium Ferricyanide-Mediated Oxidation

Potassium ferricyanide is a relatively mild inorganic oxidizing reagent that is used widely for the conversion of bis(thiols) to disulfides, for example, in the oxytocin and somatostatin families. The reagent has also been used for the formation of the first disulfide bridges in orthogonal schemes. Because $K_3Fe(CN)_6$ is slightly light sensitive, reactions are best conducted in the dark. Oxidation side products are possible when Met or Trp residues are present in the substrate.[46,47]

Procedure. The peptide (free thiol form) is dissolved, to a concentration of 0.1–1 mg/ml (= 0.1–1 mM), in a suitable buffer (acidic or basic, depending on the solubility characteristics of the peptide).[48,49] This peptide solution is added slowly to an aqueous $K_3Fe(CN)_6$ solution (0.01 M), under nitrogen, at 25°. The amount of oxidant used should be in 20% excess over theory, and the pH of the reaction solution should be kept constant at pH 6.8–7.0 by controlled addition of 10% aqueous NH_4OH. Addition times vary between 6 and 24 hr, with purer products noted on slower addition. For the formation of intermolecular homodimers, the peptide thiol solution must be more concentrated (>1 mg/ml), and addition is carried out in-

[46] P. Sieber, K. Eisler, B. Kamber, B. Riniker, W. Rittel, F. Märki, and M. De Gasparo, *Hoppe-Seyler's Z. Physiol. Chem.* **359**, 113 (1978).

[47] A. Misicka and V. J. Hruby, *Pol. J. Chem.* **68**, 893 (1994).

[48] J. Rivier, R. Kaiser, and R. Galyean, *Biopolymers* **17**, 1927 (1978), and references cited therein.

[49] W. R. Gray, F. A. Luque, R. Galyean, E. Atherton, R. C. Sheppard, B. L. Stone, A. Reyes, J. Alford, M. McIntosh, B. M. Olivera, L. J. Cruz, and J. Rivier, *Biochemistry* **23**, 2796 (1984).

versely, that is, oxidizing solution added to peptide solution. On completion of reactions, the solution is adjusted to pH 5 with 50% aqueous acetic acid, and the solution is filtered first through celite, and then (under mild suction) through a weakly basic anion exchange column (AG-3), to remove ferro- and ferricyanide ions. The column is washed with water, and the filtrate, along with the washes, are applied to a weakly acidic cation-exchange column, or a Sephadex column equilibrated with 50% acetic acid. Yields of 20–50% purified product are typically obtained.

For peptides containing Trp and/or Met, which have side chains suscepti- ble to oxidation in the presence of excess $K_3Fe(CN)_6$, Misicka and Hruby[47] proposed a modification of the above procedure: The peptide solution and the oxidant solution (about 10 mM each), are added simultaneously, very slowly, and at the same rate, to a reaction mixture. This modification allows for peptide and ferricyanide to be at the proper ratios and the highly dilute concentrations believed to allow for optimal intramolecular cyclization without side reactions.

$K_3Fe(CN)_6$ has also been used for the polymer-supported formation of disulfide bonds.[33,50] On swelling the resin in DMF, a 0.1–0.5 M $K_3Fe(CN)_6$ solution in H_2O–DMF (1 : 1 to 1 : 10) is added, and the suspension is agitated overnight at 25°. The resin is washed several times with H_2O, DMF, and CH_2Cl_2.

Formation of Disulfides from S-Protected Precursors

A number of oxidizing reagents take S-protected Cys derivatives directly to the corresponding disulfides (Scheme 1, middle pathway). This approach is used for intra- or intermolecular pairing of two Cys residues that originally have the same protecting group, and it can be generalized in experiments aimed at the construction of multiple disulfides.

Iodine Oxidation

Conversion of S-Trt or S-Acm with iodine is a widely applied example of the general approach. Rates depend on the solvent used, and as shown by Kamber et al.,[51] remarkable selectivities can be achieved (Table II). The main caution with iodine is to avoid overoxidation of the thiol functionality to the corresponding sulfonic acid, as well as to minimize modification of

[50] R. Eritja, J. P. Ziehler-Martin, P. A. Walker, T. D. Lee, K. Legesse, F. Albericio, and B. E. Kaplan, *Tetrahedron* **43**, 2675 (1987).
[51] B. Kamber, A. Hartmann, K. Eisler, B. Riniker, H. Rink, P. Sieber, and W. Rittel, *Helv. Chim. Acta* **63**, 899 (1980).

TABLE II
HALF-TIMES (t_h) FOR IODINE OXIDATION OF MODEL PEPTIDES IN VARIOUS SOLVENTS[a]

Group	Solvent	Halftimes (t_h) For S-Trt	For S-Acm
Group I $t_h(S\text{-Trt}) < t_h(S\text{-Acm})$	Methanol	3–5 sec	1 min
	Methanol–H_2O (4:1)	<1 sec	4–6 sec
	Acetic acid	70–80 sec	40–45 min
	Acetic acid–H_2O (4:1)	1–3 sec	50–60 sec
	Dioxane	1 min	1.5–2 hr
	Dioxane–H_2O (4:1)	5–10 sec	5–10 min
	Methanol–$CHCl_3$ (1:1)	2–4 sec	15 min
Group II $t_h(S\text{-Trt}) \ll t_h(S\text{-Acm})$	$CHCl_3$, CH_2Cl_2	1–2 sec	1.5–2 hr
	$(CF_3)_2CHOH$–$CHCl_3$ (1:1)	1–2 sec	>2 hr
	HFIP–$CHCl_3$ (3:1)	<1 sec	>2 hr
	CF_3CH_2OH–$CHCl_3$ (1:1)	5–6 sec	>2 hr
	TFE–$CHCl_3$ (3:1)	4–5 sec	>2 hr
Group III $t_h(S\text{-Trt}) > t_h(S\text{-Acm})$	DMF	25–35 sec	2–3 sec
	DMF–H_2O (4:1)	30–40 sec	3–5 sec

[a] Modified from B. Kamber, A. Hartmann, K. Eisler, B. Riniker, H. Rink, P. Sieber, and W. Rittel, *Helv. Chim. Acta* **63**, 899 (1980). Reactions were carried out at 20–25°, with 5 mM peptide [Boc-Cys(Trt/Acm)-Gly-Glu(OtBu)-OtBu] and 15 mM iodine. Group II solvents allow selective and quantitative oxidation of S-Trt in the presence of S-Acm. With group I solvents, cooxidation of two different linear peptides, one protected with S-Trt and the other with S-Acm, showed a surprising preference for open-chain asymmetrical heterodimer formation. The corresponding experiments with group III solvents gave the expected random statistical mixtures.

other sensitive amino acid side chains (Tyr, Met, Trp).[5,51,52] As usual, the level of side reactions can be kept low by use of appropriate solvents and scavengers, and by careful control of pH and reaction time. Other reagents, such as N-iodosuccinimide[53] or cyanogen iodide,[54] have been proposed as alternative sources of iodonium ion (I^+).

Procedure. The peptide is dissolved in acetic acid–water (4:1), to a final concentration of about 2 mM.[33,51] Iodine (10 equivalent) is added to this solution in one portion. The reaction mixture is stirred for 10 min to 1 hr at 25°, and, after complete oxidation is revealed by HPLC monitoring, the reaction is quenched by diluting to twice the volume with water and extracting the iodine with CCl_4 (5–6 times, equal volume each time). The

[52] K. Akaji, T. Tatsumi, M. Yoshida, T. Kimura, Y. Fujiwara, and Y. Kiso, *J. Am. Chem. Soc.* **114**, 4137 (1992), and references cited herein.
[53] H. Shih, *J. Org. Chem.* **58**, 3003 (1993).
[54] P. Bishop and J. Chmielewski, *Tetrahedron Lett.* **33**, 6263 (1992).

aqueous phase is lyophilized, and the product is purified by preparative HPLC.

Iodine oxidation is also used for on-resin intramolecular disulfide bond formation.[33,39,40,55] The peptide–resin (protected with S-Acm, S-Xan, S-Tmob, or S-Trt) is swelled in DMF. Iodine (10–20 equivalent) is added, and the mixture is stirred gently at 25° for 1–4 hr. On completion of the oxidation, the peptide–resin is washed and drained. Yields by this procedure are in the 60–80% range.

Thallium(III) Trifluoroacetate

Thallium(III) trifluoroacetate is a mild oxidant that can be used as an alternative to iodine, and it sometimes gives better yields and purities of disulfides.[56] The major limitation of this reagent is its high toxicity. TFA is the solvent of choice for reasons both of solubilization and chemistry, but DMF is an acceptable solvent for on-resin oxidations[37] in conjunction with TFA-labile anchoring linkages. His and Tyr survive exposure to Tl(tfa)$_3$, but Met and Trp need to be protected.[56–58] Anisole should be added as a scavenger for the alkyl cations generated during the deprotection/oxidation. When starting with bis(Acm) sequences, only a slight excess of Tl(tfa)$_3$ should be used. Bis(Trt) sequences cannot be oxidized at all with this reagent, whereas bis(Tmob) sequences tolerate a range of reagent excess.[59] As with other metals, thallium can be difficult to remove entirely from sulfur-containing peptides.

Procedure. The peptide is dissolved in TFA–anisole (19 : 1), to a final concentration of about 1.1 mM.[44] The solution is chilled to 0°, and Tl(tfa)$_3$ (1.2 equivalent) is added. After stirring for 5–18 hr at 4°, the reaction mixture is concentrated, and the crude product is precipitated with diethyl ether (~3.5 ml for each micromole peptide). The product is triturated with diethyl ether for 2 min, followed by centrifugation and removal of the ether by decantation. The trituration/centrifugation cycle is repeated three times to ensure the removal of thallium salts. Yields of 35–50% crude cyclic peptide have been reported.

[55] L. Chen, H. Bauerová, J. Slaninová, and G. Barany, *Pept. Res.* **9**, 114 (1996).

[56] N. Fujii, A. Otaka, S. Funakoshi, K. Bessho, T. Watanabe, K. Akaji, and H. Yajima, *Chem. Pharm. Bull.* **35**, 2339 (1987), and references cited therein.

[57] H. Yajima, N. Fujii, S. Funakoshi, T. Watanabe, E. Murayama, and A. Otaka, *Tetrahedron* **44**, 805 (1988).

[58] W. B. Edwards, C. G. Fields, C. J. Anderson, T. S. Pajeau, M. J. Welch, and G. B. Fields, *J. Med. Chem.* **37**, 3749 (1994).

[59] M. C. Munson, C. García-Echeverría, F. Albericio, and G. Barany, *J. Org. Chem.* **57**, 3013 (1992), and references cited therein.

This reagent can also be used to form disulfide bonds on solid support.[37,44,58,60] The bis(S-Acm) peptide–resin [if Trp is present, use N^{in}-*tert*-butyloxycarbonyl (Boc) derivative] is swelled in DMF and treated with Tl(tfa)$_3$ (1.5–2 equivalent) in DMF–anisole (19 : 1, ~0.35 ml/25 mg resin). After 1–18 hr at 0°, the resin is washed with DMF and CH$_2$Cl$_2$ to remove the excess Tl reagent. Up to 80–95% yields have been reported.

Chlorosilane–Sulfoxide Oxidation

Kiso and co-workers and Fujii and co-workers, independently, proposed a novel oxidizing milieu, a mixture of chlorosilanes and sulfoxides.[61–63] These reagents can cleave several S-protecting groups (Acm, tBu, Mob, Meb) and form disulfides directly. Depending on the protecting groups present in the substrate, certain chlorosilane–sulfoxide combinations are more effective than others (see Table III). The reactions are fast, and no disulfide exchange was reported when the method was applied in an orthogonal scheme. The method is compatible with most amino acid side chains, although Trp must be used as its N^{in}-formyl derivative (later removed by rapid base treatment without affecting the disulfides), because the side-chain indole can become fully chlorinated under the oxidizing conditions.[30,31,52] (An interesting variation involves the use of DMSO/acid combinations, which rely on DMSO acting as an oxidant to deblocked Cys in the presence of the acid, for example, DMSO/TFA for S-Mob[64] and S-Trt.[65])

Procedure. Depending on the Cys protecting groups present in the substrate peptide, the optimal oxidizing milieu is chosen after inspection of Table III.[52,61–63,66,67] The peptide (S-Acm, S-tBu, S-Mob, S-Meb protected) is dissolved in TFA to a concentration of 1–10 mM. Sulfoxide (10 equivalent), chlorosilane (100–250 equivalent), and anisole (100 equivalent) are added. The reaction generally proceeds for 10–30 min at 25°, and is quenched by addition of solid NH$_4$F (300 equivalent). The crude product is precipitated with a large excess of dry diethyl ether, and the precipitate

[60] R. H. Angeletti, L. Bibbs, L. F. Bonewald, G. B. Fields, J. S. McMurray, W. T. Moore, and J. T. Stults, in "Techniques in Protein Chemistry VII" (D. Marshak, ed.), p. 261. Academic Press, San Diego, 1995.
[61] K. Akaji, T. Tatsumi, M. Yoshida, T. Kimura, Y. Fujiwara, and Y. Kiso, *J. Chem. Soc., Chem. Commun.,* 167 (1991).
[62] T. Koide, A. Otaka, H. Suzuki, and N. Fujii, *Syn. Lett.,* 345 (1991).
[63] K. Akaji, H. Nishiuchi, and Y. Kiso, *Tetrahedron Lett.* **36,** 1875 (1995).
[64] T. Koide, A. Otaka, and N. Fujii, *Chem. Pharm. Bull.* **41,** 1030 (1993).
[65] A. Otaka, T. Koide, A. Shide, and N. Fujii, *Tetrahedron Lett.* **32,** 1223 (1991).
[66] K. Akaji, K. Fujino, T. Tatsumi, and Y. Kiso, *Tetrahedron Lett.* **33,** 1073 (1992).
[67] A. Adeva, J. A. Camarero, E. Giralt, and D. Andreu, *Tetrahedron Lett.* **36,** 3885 (1995).

TABLE III
YIELDS OF CYSTINE FROM Cys(R) WITH VARIOUS SULFOXIDE–SILYL COMPOUND/TFA SYSTEMS[a]

Sulfoxide	R	Yields (%)			
		TMSCl	CH_3SiCl_3	$SiCl_4$	TMSOTf
DMSO	Acm	**98**	75^b	68^b	<5
	Mob	**90**	39^b	41^b	$<5^b$
	Meb	24^b	17^b	31^b	<5
	tBu	77^b	16^b	26^b	<5
$CH_3S(O)Ph$	Acm	**94**	34^b	38^b	8
	Mob	63^b	31^b	34^b	23^b
	Meb	77^b	25^b	34^b	<5
	tBu	**93**	31^b	37^b	14^b
Ph_2SO	Acm	10	**87**	**90**	<5
	Mob	11	**87**	**91**	$<5^b$
	Meb	9	**86**	**87**	<5
	tBu	14	**90**	**96**	<5

[a] Abbreviated from T. Koide, A. Otaka, H. Suzuki, and N. Fujii, *Syn. Lett.*, 345 (1991). Reactions were run at 4° for 1–4 hr. Entries in bold indicate that the combination is recommended for disulfide formation.
[b] Unknown by-products were detected on amino acid analysis.

is collected by centrifugation. The product is purified on a Sephadex G-15 column eluted with 4 N aqueous acetic acid. The collected fractions are lyophilized, and the product can be purified further by preparative HPLC. Yields of 20–40% have been reported.

This method has been applied for disulfide formation on a solid support.[68] A bis(Acm)–peptide–MBHA–resin, assembled by Boc chemistry, is reacted with PhS(O)Ph, CH_3SiCl_3 (10 and 250 equivalent, respectively, added every 10 min), and anisole (100 equivalent) in TFA–CH_2Cl_2 (1:1) for 30 min at 25°. Upon completion of the oxidation, the resin is washed and the oxidized peptide is cleaved with hydrogen fluoride (HF)–p-cresol (9:1) for 1 hr at 0°. The final yield was 3%, which is considered by the authors to be relatively high compared to other reported yields for similar cases.

Directed Methods for Unsymmetrical Formation of Disulfides

The formation of an asymmetric disulfide bridge by cooxidation of two different peptide chains is expected to give a statistical mixture of the

[68] J. A. Camarero, E. Giralt, and D. Andreu, *Tetrahedron Lett.* **36**, 1137 (1995).

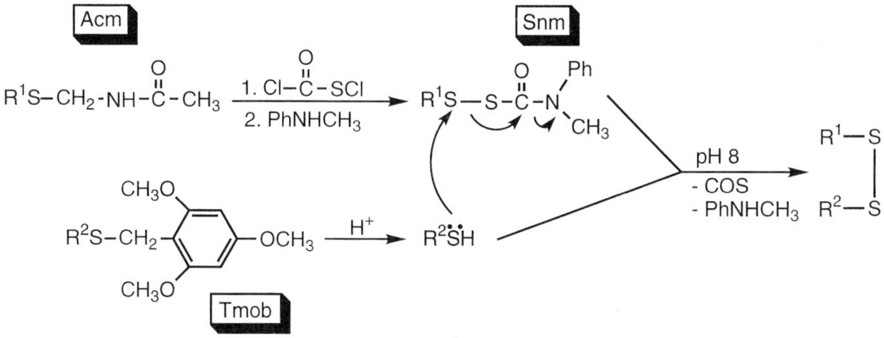

SCHEME 5. Directed formation of intermolecular disulfide bridge using S-Acm and S-Tmob protecting groups, and S-Snm activating group.

desired heterodimer as well as the undesired homodimer. To control regio-selectivity, directed disulfide formation methods are used.[5,16,17,69,70] These approaches involve activation of the thiol function from one peptide, followed by addition of the second peptide in the free thiol form.

The reasonably stable, isolable S-methyloxycarbonylsulfenyl (Scm) derivatives are made readily by reaction of S-Acm, S-Trt, S-Xan, or S-Tmob protected peptides with methyloxycarbonylsulfenyl chloride in chloroform and/or methanol.[71,72] The resultant activated peptides undergo facile thiol-mediated heterolytic fragmentation, an irreversible process, to yield unsymmetrical disulfides, along with carbonyl sulfide and methanol.[69,71,72] Problems with S-Scm stem from its lability to base [including 9-fluorenyl-methyloxycarbonyl (Fmoc) removal conditions] and the risk of $S \rightarrow N$ migration. A better behaved protecting/activating group is the S-[(N-methyl-N-phenylcarbamoyl)sulfenyl] (Snm) function[39,55,72] (Scheme 5) which consistently gives higher yields and purities in disulfide-forming chemistry by comparison to Scm controls. S-Snm is derived from S-Acm in two steps, and it is stable to the orthogonal deprotection of S-Tmob under acidic conditions. The resultant free thiol then attacks S-Snm in a directed disulfide formation process, which can be conducted in solution or on-resin.

[69] S. J. Brois, J. F. Pilot, and H. W. Barnum, *J. Am. Chem. Soc.* **92**, 7629 (1970).
[70] R. G. Hiskey, *in* "The Peptides—Analysis, Synthesis, Biology" (E. Gross and J. Meienhofer, eds.), Vol. 3, p. 137. Academic Press, New York, 1981.
[71] B. Kamber, *Helv. Chim. Acta* **56**, 1370 (1973).
[72] A. L. Schroll and G. Barany, *J. Org. Chem.* **54**, 244 (1989), and references cited therein.

Procedure

ON-RESIN. A peptide–resin with N-terminal Snm-protected 2-mercapto-propionic acid (Mpa) and internal Cys(Tmob) is treated with TFA–CH_2Cl_2–Triethylsilane-phenol–H_2O (7:90:1:1:1) two times for 15 min each time at 25° to remove the S-Tmob group and expose the free thiol.[39,55] The directed disulfide formation is conducted in NMM–DMF (1:99) with gentle stirring for 2 hr at 25°. The resin is then washed (DMF and CH_2Cl_2), and final deprotection/cleavage [TFA–CH_2Cl_2–Triethylsilane-phenol–H_2O, 92:5:1:1:1, 25°, 1 hr] followed by HPLC purification gives the desired intramolecular cyclized disulfide in isolated yield of 20–33%.

IN SOLUTION. The same peptide–resin described above[39,55] is cleaved with TFA–CH_2Cl_2–Triethylsilane-phenol–H_2O (92:5:1:1:1) for 1 hr at 25° to provide soluble peptide (free internal thiol and terminal S-Snm protected thiol). The peptide is dissolved to a final concentration of 1–2 mM using a combination of 0.01 M phosphate buffer, pH 8, and acetonitrile (final ratio ~1:1) at 25°, and gently stirred. The cyclization time is about 10 min, as judged by HPLC, with an isolated yield of about 56%.

As another example of a dual purpose S-protecting/activating group, the S-(3-nitro-2-pyridyl) (Npys) group has been used with considerable success.[73–77] Npys is compatible with stepwise incorporation in Boc chemistry, but its considerable lability to piperidine precludes use with Fmoc protocols (Npys group is lost and disulfide homodimers are noted). Directed disulfide formation follows by inter- or intramolecular attack from a thiol, and is driven by the low pK_a of the aromatic thiol. The reaction takes place in aqueous systems at moderately acidic (1 N aqueous acetic acid) to neutral conditions.[74,75]

Regioselective Formation of Disulfides

The methods reported so far can be carried out in series to form multiple disulfide bridges. The general approach involves graduated deprotection (often orthogonal) and/or cooxidation of pairwise half-cystine residues, as specified by the original protection scheme. Much elegant work has been

[73] R. J. Ridge, G. R. Matsueda, E. Haber, and R. Matsueda, *Int. J. Pept. Protein Res.* **19**, 490 (1982).

[74] M. S. Bernatowicz, R. Matsueda, and G. R. Matsueda, *Int. J. Pept. Protein Res.* **28**, 107 (1986).

[75] F. Albericio, D. Andreu, E. Giralt, C. Navalpotro, E. Pedroso, B. Ponsati, and M. Ruiz-Gayo, *Int. J. Pept. Protein Res.* **34**, 124 (1989).

[76] B. Ponsati, M. Ruiz-Gayo, E. Giralt, F. Albericio, and D. Andreu, *J. Am. Chem. Soc.* **112**, 5345 (1990).

[77] J. W. Drijfhout and W. Bloemhoff, *Int. J. Pept. Protein Res.* **37**, 27 (1991).

reported in primary publications and reviews,[5,16–19] and a full treatment is beyond the scope of this article. A representative procedure, adapted from our work on α-conotoxin SI, is given here (Scheme 6).[33,44]

Procedure

IN SOLUTION (SCHEME 6, LEFT SIDE). Following standard Fmoc solid-phase assembly of the linear protected precursor peptide on a tris(alkoxy)-benzylamide (PAL) support, TFA–phenol–H_2O–thioanisole–1,2-ethanedithiol (82.5:5:5:5:2.5) is used for 2 hr at 25° to deprotect the acid-labile protecting groups and cleave the peptide from the resin. The bis(thiol), bis(Acm) intermediate is cyclized overnight at 25° using DMSO in 0.01 M phosphate buffer, pH 7.5.[33,43,44] The second disulfide bond is formed with thallium(III)trifluoroacetate (1.2 equivalent) overnight at 4°.[33,44]

ON-RESIN (SCHEME 6, RIGHT SIDE). Starting with the same protected peptide–resin described above, TFA–CH_2Cl_2–triethylsilane (1:98.5:0.5) is used for 2 hr at 25° to remove selectively the *S*-Xan protecting group; negligible cleavage of the PAL anchoring linkage occurs under those conditions. The resin-bound bis(thiol), bis(Acm) intermediate is cyclized on the resin with triethylamine (2.0 equivalent) in NMP for 4 hr at 25°.[44] The resulting peptide–resin is cyclized further with thallium(III)trifluoroacetate (1.2 equivalent) overnight at 4°.[33,44] The final product is cleaved from the resin using TFA–CH_2Cl_2–triethylsilane–H_2O–anisole (94.5:4:0.5:0.5:0.5).

Conclusions and Summary

The goal of this review has been to present different chemical approaches for the formation of disulfide bonds in synthetic peptides and small proteins. Three general types of approaches have been described: (1) oxidation starting from the unprotected thiols; (2) oxidation starting from protected thiols; and (3) directed methods for formation of unsymmetrical disulfides. Individual or sequential disulfide-forming reactions can be carried out in solution or on a polymeric support. Overall yields and purities of products depends on protecting group combinations chosen, precise reaction conditions, and the targeted structure. Although no procedure can be guaranteed to give outstanding results for all cases, there are sufficient options available to support an optimistic view that one or more approaches can be optimized.

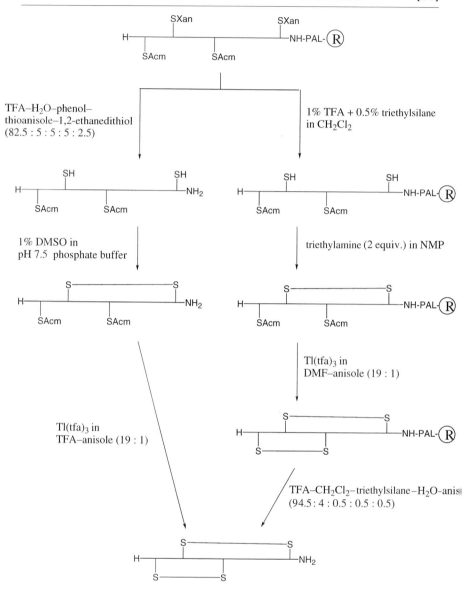

SCHEME 6. Orthogonal solution and solid-phase synthesis of α-conotoxin SI, H-Ile-Cys-Cys-Asn-Pro-Ala-Cys-Gly-Pro-Lys-Tyr-Ser-Cys-NH₂.

Acknowledgments

We thank co-workers and colleagues, in particular Dr. Lin Chen, for sharing experiences regarding peptide disulfide chemistry. Preparation of this review and underlying experimental work were supported by the National Institutes of Health (GM 43552). I. A. gratefully acknowledges support via a National Science Foundation Graduate Fellowship.

[11] Direct Synthesis of Glycosylated Amino Acids from Carbohydrate Peracetates and Fmoc Amino Acids: Solid-Phase Synthesis of Biomedicinally Interesting Glycopeptides

By JAN KIHLBERG, MIKAEL ELOFSSON, and LOURDES A. SALVADOR

Introduction

Most proteins found in nature carry carbohydrate residues that are covalently attached to amino acid side chains through O- or N-glycosidic linkages. The carbohydrate units affect the properties of the parent protein in many and diverse ways as reviewed elsewhere in greater detail.[1–3] For example, glycosylation of a protein can confer protection against proteolysis and influence uptake, distribution, and excretion, and it can also determine the biological function of the protein. These effects can in some cases be indirect consequences of the glycosylation because conformational changes may be induced in the protein by the carbohydrate moieties. Investigations have also revealed that attachment of carbohydrates to peptides, which are not glycosylated in nature, can influence the pharmacological and pharmacokinetic properties of peptides functioning as enzyme inhibitors,[4] neuropeptides,[5,6] and hormones.[7]

---

[1] H. Lis and N. Sharon, *Eur. J. Biochem.* **218,** 1 (1993).
[2] A. Varki, *Glycobiology* **3,** 97 (1993).
[3] R. A. Dwek, *Chem. Rev.* **96,** 683 (1996).
[4] A. W. Harrison, J. F. Fisher, D. M. Guido, S. J. Couch, J. A. Lawson, D. M. Sutter, M. V. Williams, G. L. DeGraaf, J. E. Rogers, D. T. Pals, and D. W. DuCharme, *Bioorg. Med. Chem.* **2,** 1339 (1994).
[5] R. E. Rodriguez, F. D. Rodriguez, M. P. Sacristán, J. L. Torres, G. Valencia, and J. M. G. Antón, *Neurosci. Lett.* **101,** 89 (1989).
[6] R. Polt, F. Porreca, L. Z. Szabò, E. J. Bilsky, P. Davis, T. J. Abbruscato, T. P. Davis, R. Horvath, H. I. Yamamura, and V. J. Hruby, *Proc. Natl. Acad. Sci. U.S.A.* **91,** 7114 (1994).
[7] J. Kihlberg, J. Åhman, B. Walse, T. Drakenberg, A. Nilsson, C. Söderberg-Ahlm, B. Bengtsson, and H. Olsson, *J. Med. Chem.* **38,** 161 (1995).

In view of the important biological functions of glycoproteins and glyco-peptides large efforts have been focused on the development of methodology for synthesis of glycopeptides. Several reviews have covered the progress made in this field.[8-14] Convergent synthesis of glycopeptides by attachment of an oligosaccharide to a peptide has found very limited use for preparation of O-linked glycopeptides[15,16] but can be accomplished for N-linked glycopeptides.[17-19] However, the alternative approach according to which glycosylated amino acids are used as building blocks in the stepwise assembly of glycopeptides has been shown to be more reliable and efficient for both categories of glycopeptides.[10] It is also well suited for synthesis on solid phase. Development of efficient methods for synthesis of glycosylated amino acids is therefore of central importance to achieve success in glyco-peptide synthesis. Protective groups for the glycosylated amino acid building blocks have to be chosen considering both the lability of glycosidic bonds toward acids and the tendency of peptides glycosylated on serine and threonine to undergo β-elimination when treated with strong base. Preferably, the fluoren-9-ylmethoxycarbonyl (Fmoc) group[20] is used for protection of the α-amino group, whereas the hydroxyl groups of the carbohydrate may be protected with acyl groups or with acid-labile silyl and isopropylidene groups, or even left unprotected.

This article summarizes the use of peracetylated carbohydrates (Fig. 1) in Lewis acid-catalyzed glycosylations of 3-mercaptopropionic acid and different Fmoc amino acids. The glycosylated building blocks are prepared without protection of the carboxyl groups of the glycosyl acceptors to give

[8] H. Kunz, *Angew. Chem. Int. Ed. Engl.* **26**, 294 (1987).
[9] H. G. Garg, K. von dem Bruch, and H. Kunz, *Adv. Carbohydr. Chem. Biochem.* **50**, 277 (1994).
[10] M. Meldal, *in* "Neoglycoconjugates: Preparation and Applications" (Y. C. Lee and R. T. Lee, eds.), p. 145. Academic Press, San Diego, 1994.
[11] M. Meldal, *Curr. Opin. Struct. Biol.* **4**, 710 (1994).
[12] T. Norberg, B. Lüning, and J. Tejbrant, *Methods Enzymol.* **247**, 87 (1994).
[13] Y. Nakahara, H. Iijima, and T. Ogawa, *in* "Synthetic Oligosaccharides" (P. Kovác, ed.), p. 249. American Chemical Society, Washington, DC, 1994.
[14] H. Paulsen, *Angew. Chem. Int. Ed. Engl.* **29**, 823 (1990).
[15] M. Hollósi, E. Kollát, I. Laczkó, K. F. Medzihradszky, J. Thurin, and L. Otvos, Jr., *Tetrahedron Lett.* **32**, 1531 (1991).
[16] D. M. Andrews and P. W. Seale, *Int. J. Pept. Protein Res.* **42**, 165 (1993).
[17] S. T. Cohen-Anisfeld and P. T. Lansbury, Jr., *J. Am. Chem. Soc.* **115**, 10531 (1993).
[18] D. Vetter, D. Tumelty, S. K. Singh, and M. A. Gallop, *Angew. Chem. Int. Ed. Engl.* **34**, 60 (1995).
[19] J. Offer, M. Quibell, and T. Johnson, *J. Chem. Soc., Perkin Trans. 1*, 175 (1996).
[20] L. A. Carpino and G. Y. Han, *J. Org. Chem.* **37**, 3404 (1972).

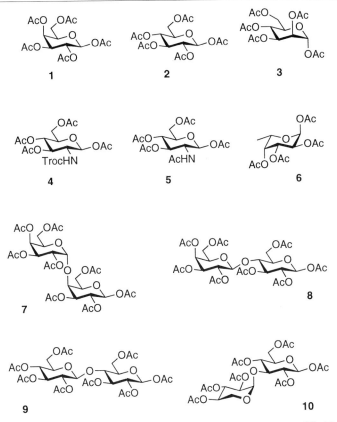

FIG. 1. Peracetylated carbohydrates are useful glycosyl donors as exemplified by the use of **1–10** in glycosylations of 3-mercaptopropionic acid and Fmoc amino acids.

aliphatic and phenolic *O*- and *S*-glycosides in one step. The procedure uses commercial or readily available starting materials and does not require extensive experience in synthetic carbohydrate chemistry. Because the glycosylated building blocks carry *O*-acetyl groups and an N^α-Fmoc group they are suitable for direct use in synthesis of glycopeptides. The latter part of this article therefore covers Fmoc solid-phase glycopeptide synthesis, including a discussion of methods for Fmoc cleavage and the choice of protective groups for the carbohydrate moiety of glycosylated amino acids. Finally, some applications of glycopeptides in pharmacology and immunology are outlined.

Glycosylation of 3-Mercaptopropionic Acid and Fmoc Amino Acids
Using Carbohydrate Peracetates as Glycosyl Donors

General

1,2-trans-Peracetates of simple mono- and oligosaccharides are often
commercially available, whereas those that are not may be prepared from
unprotected saccharides or glycosyl bromides,[21] as well as from 2-(trimethylsilyl)ethyl glycosides.[22] In an early report it was shown that 1,2-*trans*-perace-
tylated carbohydrates were useful glycosyl donors in boron trifluoride
etherate-promoted glycosylations of primary alcohols such as 2,2,2-trichlor-
oethanol.[23] 1,2-*trans*-Peracetates of xylose, glucose, and galactose have been
used in Lewis acid-promoted glycosylations of derivatives of serine and
threonine protected both at the α-amino and the α-carboxyl group.[24–26]
The glycosylated amino acids prepared by this approach, however, required
deprotection of the α-carboxyl group,[24,25] or deprotection of both the α-
amino and α-carboxyl groups followed by reprotection of the α-amino
group,[26] to be useful for synthesis of glycopeptides. Direct glycosylation
of Fmoc amino acids and the spacer 3-mercaptopropionic acid, without
protection of their carboxyl groups, avoids such additional protective group
manipulations.

Glycosylated Derivatives of 3-Mercaptopropionic Acid

Glycosylation of 3-mercaptopropionic acid (4 equivalent) with β-D-
galactose pentaacetate (**1**) was performed in dry dichloromethane with
boron trifluoride etherate (1.5 equiv) as promoter both in the presence and
absence of molecular sieves.[27,28] In the presence of molecular sieves both
the mercaptopropionate **11** and the desired glycoside **12** were obtained
when the reaction had reached equilibrium (Fig. 2). However, thin-layer
chromatography (TLC) revealed that if the sieves were omitted **11** was
formed initially, and then rearranged within 1 hr so that **12** could be obtained

[21] M. L. Wolfrom and A. Thompson, *Methods Carbohydr. Chem.* **2**, 211 (1963).
[22] K. Jansson, S. Ahlfors, T. Frejd, J. Kihlberg, G. Magnusson, J. Dahmén, G. Noori, and K. Stenvall, *J. Org. Chem.* **53**, 5629 (1988).
[23] G. Magnusson, G. Noori, J. Dahmén, T. Frejd, and T. Lave, *Acta Chem. Scand.* **B35**, 213 (1981).
[24] H. G. Garg, T. Hasenkamp, and H. Paulsen, *Carbohydr. Res.* **151**, 225 (1986).
[25] B. G. de la Torre, J. L. Torres, E. Bardaji, P. Clapés, N. Xaus, X. Jorba, S. Calvet, F. Albericio, and G. Valencia, *J. Chem. Soc., Chem. Commun.*, 965 (1990).
[26] F. Filira, L. Biondi, F. Cavaggion, B. Scolaro, and R. Rocchi, *Int. J. Pept. Protein Res.* **36**, 86 (1990).
[27] M. Elofsson, B. Walse, and J. Kihlberg, *Tetrahedron Lett.* **32**, 7613 (1991).
[28] M. Elofsson, S. Roy, B. Walse, and J. Kihlberg, *Carbohydr. Res.* **246**, 89 (1993).

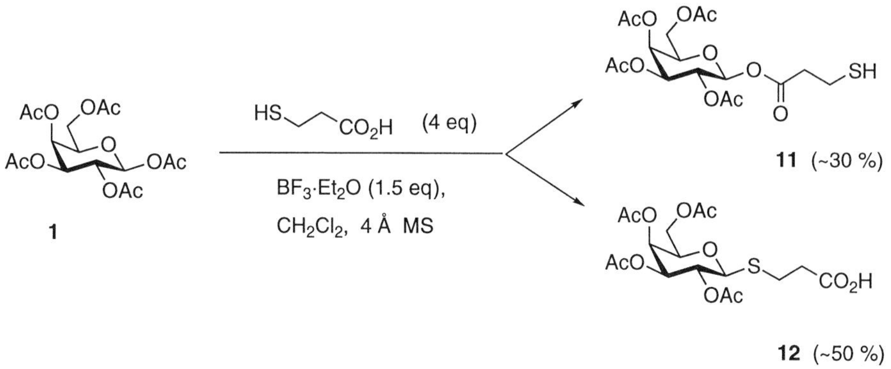

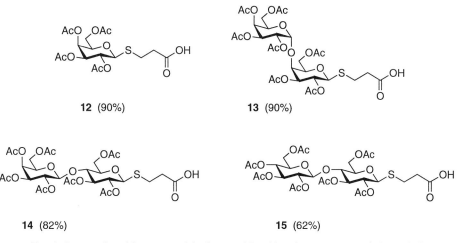

FIG. 2. Boron trifluoride etherate-promoted glycosylation of 3-mercaptopropionic acid with **1** in the presence of molecular sieves (MS) gave a mixture of **11** and **12**[27] [L. A. Salvador, M. Elofsson, and J. Kihlberg, *Tetrahedron* **51**, 5643 (1995)]. In the absence of molecular sieves only **12** was obtained when the reaction had reached equilibrium.

in 90% yield after purification by flash column chromatography. Most likely this rearrangement is prevented by adsorption of boron trifluoride etherate by the molecular sieves. Using boron trifluoride etherate as promoter in the absence of sieves the mercaptopropionic acid glycosides of the disaccharides galabiose, lactose, and cellobiose (**13–15,** Fig. 3) were obtained from the corresponding disaccharide β-D-octaacetates **7–9** in 62–90% yields. On attempted use of the 1,2-cis-linked, α-peracetate of galabiose a larger excess

FIG. 3. Spacer glycosides prepared by boron trifluoride etherate-promoted glycosylation of 3-mercaptopropionic acid using carbohydrate 1,2-*trans*-peracetates as glycosyl donors.

of boron trifluoride etherate had to be employed as promoter, and a 1:1 anomeric mixture of mercaptopropionic acid glycosides was obtained.[27] 1,2-Cis-linked peracetates of saccharides composed of common monosaccharides such as Glc, Gal, Man, GlcNAc, and GalNAc are therefore less useful as glycosyl donors, in agreement with previous observations.[29]

The 3-mercaptopropionic acid glycosides **12–15** have been used in solid-phase synthesis of helper T-cell-stimulating neoglycopeptides.[28,30] In addition compounds **12, 14,** and a tetrasaccharidic mercaptopropionic acid glycoside have been deacetylated and subsequently coupled to proteins or aminated microtiter plate wells.[31–33]

O-Linked Glycosylated Fmoc Amino Acids
with 1,2-Trans Anomeric Configuration

The glycosylated amino acids **16–30** (Fig. 4) have all been prepared by boron trifluoride etherate-promoted glycosylation of the corresponding Fmoc amino acids using the carbohydrate peracetates **1–10** (Fig. 1) as glycosyl donors. Compounds **16** and **17** were prepared as model compounds,[27,34] whereas the remaining glycosylated amino acids are found in glycopeptides and glycoproteins of biological importance. For instance, incorporation of *trans*-4-hydroxyproline carrying mono- or disaccharide residues (cf. **18, 21,** and **28**) in the opioid agonist morphiceptin substantially increased the analgetic activity.[5] Cell surface layer (S-layer) glycoproteins of anaerobic eubacteria carry oligosaccharides in which β-D-galactose and β-D-glucose residues are linked to the phenolic hydroxyl group of tyrosine (cf. **19** and **22**).[35,36] In bovine blood clotting factor a novel connection of a trisaccharide to the polypeptide chain by a β-D-glucosyl-*O*-Ser linkage (cf. **20** and **30**) has been identified.[37] Furthermore, incorporation of a β-D-glucosylated Ser residue (cf. **20**) in an enkephalin analog allowed the analgetic neoglycopeptide to cross the blood–brain barrier.[6] The glycosyl-

[29] H. Paulsen and M. Paal, *Carbohydr. Res.* **135**, 53 (1984).
[30] C. V. Harding, J. Kihlberg, M. Elofsson, G. Magnusson, and E. R. Unanue, *J. Immunol.* **151**, 2419 (1993).
[31] L. Moroder, *Biol. Chem. Hoppe-Seyler* **369**, 381 (1988).
[32] M. Elofsson, J. Broddefalk, T. Ekberg, and J. Kihlberg, *Carbohydr. Res.* **258**, 123 (1994).
[33] U. Nilsson, R. T. Striker, S. J. Hultgren, and G. Magnusson, *Bioorg. Med. Chem.* **4**, 1809 (1996).
[34] L. A. Salvador, M. Elofsson, and J. Kihlberg, *Tetrahedron* **51**, 5643 (1995).
[35] P. Messner, R. Christian, J. Kolbe, G. Schultz, and U. B. Sleytr, *J. Bacteriol.* **174**, 2236 (1992).
[36] K. Bock, J. Schuster-Kolbe, E. Altman, G. Allmaier, B. Stahl, R. Christian, U. B. Sleytr, and P. Messner, *J. Biol. Chem.* **269**, 7137 (1994).
[37] S. Hase, H. Nishimura, S. Kawabata, S. Iwanaga, and T. Ikenaka, *J. Biol. Chem.* **265**, 1858 (1990).

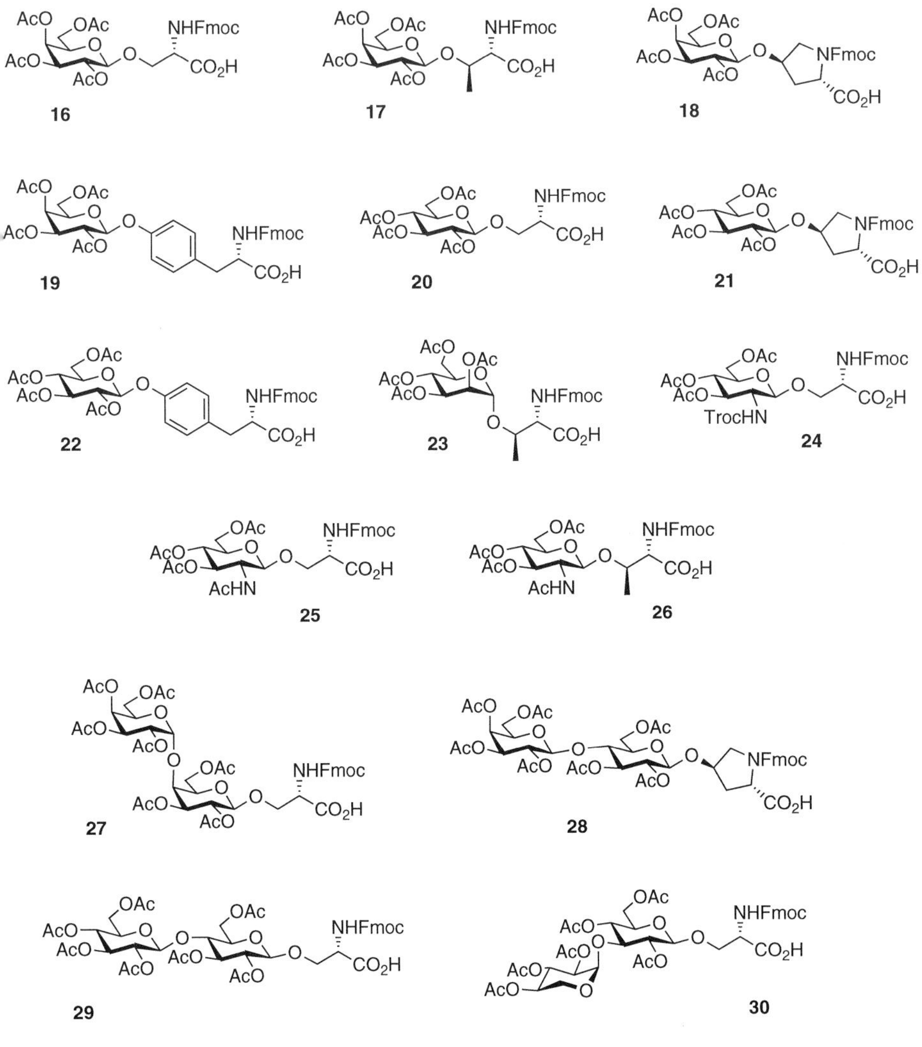

(Troc = 2,2,2-trichloroethoxycarbonyl)

FIG. 4. O-Linked glycosylated amino acids prepared in one step under boron trifluoride etherate promotion from the corresponding Fmoc amino acid and carbohydrate 1,2-*trans*-per-acetate.

ated analog of human insulin-like growth factor-I, expressed from yeast together with the native form, was found to contain an α-D-mannosyl-O-Thr linkage (cf. **23**).[38] Multiple 2-acetamido-2-deoxy-β-D-glucopyranosyl-O-Ser and threonine residues (cf. **24–26**) are common in glycoproteins found in the cytoplasm and the nucleus.[39] Finally, T-cell immunogenic glycopeptides containing the disaccharides galabiose and cellobiose linked to Ser (cf. **27** and **29**) have been shown to give a carbohydrate-specific helper T-cell response in mice.[40]

The glycosylated amino acids shown in Fig. 4 were obtained under standardized conditions by coupling of a carbohydrate 1,2-*trans*-peracetate to an Fmoc amino acid (~1.2 equiv) using boron trifluoride etherate (3 equiv) as promoter (Table I[28,34,41–44]). The reactions were performed at room temperature, with dichloromethane or acetonitrile as solvent, and the products were in general isolated by preparative reversed-phase high-performance liquid chromatography (HPLC) after an aqueous workup. As many carbohydrate peracetates and Fmoc amino acids are commercially available, the simplicity of the procedure allows glycosylated amino acid building blocks to be prepared in research groups that lack experience in the different variables (protective groups, types of glycosyl donors, promoters, etc.) that usually need to be fine-tuned by the synthetic carbohydrate chemist. As shown in Table I the yields in the glycosylations range from 34 to 70%. The reproducibility of the procedure appears to be good as no or only a slight variation was observed in the two cases when a glycosylated amino acid was prepared independently in different laboratories. Thus compound **18**[34,41] was obtained in 51 and 67% yields, whereas **20**[34,42] was obtained in 35 and 37% yields (cf. Table I). Interestingly, disaccharide peracetates often gave higher yields than when monosaccharides were used as glycosyl donors. For example, substantially higher yields were obtained for **29**[43] and **30**[44] (52 and 62%, respectively), in which glycosylglucose residues are linked to Ser, than when glucose pentaacetate (**2**) was coupled to Ser to give **20** (37%). A somewhat higher yield was obtained in the synthesis of **27**[28] (64%) than in the synthesis of **16**[34] (53%), whereas a slightly lower yield was obtained for **28**[41] (45%) as compared to **21**[41] (51%).

[38] P. Gellerfors, K. Axelsson, A. Helander, S. Johansson, L. Kenne, S. Lindqvist, B. Pavlu, A. Skottner, and L. Fryklund, *J. Biol. Chem.* **264**, 11444 (1989).

[39] G. W. Hart, R. S. Haltiwanger, G. D. Holt, and W. G. Kelly, *Annu. Rev. Biochem.* **58**, 841 (1989).

[40] B. Deck, M. Elofsson, J. Kihlberg, and E. R. Unanue, *J. Immunol.* **155**, 1074 (1995).

[41] G. Arsequell, N. Sàrries, and G. Valencia, *Tetrahedron Lett.* **36**, 7323 (1995).

[42] W. Steffan, M. Schutkowski, and G. Fischer, *J. Chem. Soc., Chem. Commun.*, 313 (1996).

[43] M. Elofsson, B. Walse, and J. Kihlberg, *Int. J. Pept. Protein Res.* **47**, 340 (1996).

[44] J. Tejbrant, *Chem. Commun., Stockholm Univ.* **No. 7** (1992).

TABLE I
GLYCOSYLATION OF Fmoc AMINO ACIDS USING BORON
TRIFLUORIDE ETHERATE AS PROMOTER[a]

Carbohydrate peracetate	Fmoc amino acid	Solvent	Reaction time (hr)	Product	Yield[b] (%)	Ref.
1(Galβ)	Ser	CH_3CN	1	**16**	53	34
1(Galβ)	Thr	CH_3CN	1	**17**	50	34
1(Galβ)	Hyp	CH_3CN	1.5	**18**	51	34
1(Galβ)	Hyp	CH_3CN	o.n.[c]	**18**	67	41
1(Galβ)	Tyr	CH_2Cl_2	4	**19**	34	34
2(Glcβ)	Ser	CH_2Cl_2	18.5	**20**	37	34
2(Glcβ)	Ser	CH_3CN	2	**20**	35	42
2(Glcβ)	Hyp	CH_3CN	o.n.[c]	**21**	51	41
2(Glcβ)	Tyr	CH_2Cl_2	8	**22**	41	34
3(Manα)	Thr	CH_2Cl_2	20	**23**[d]	42	34
4(GlcNTrocβ)	Ser	CH_2Cl_2	5	**24**	70	34
5(GlcNAcβ)	Ser	CH_2Cl_2	48–150	**25**	55[e]	f
5(GlcNAcβ)	Thr	CH_2Cl_2	48–150	**26**	53[e]	f
7[Galα(1–4)Galβ]	Ser	CH_3CN	1.25	**27**	64	28
8[Galβ(1–4)Glcβ]	Hyp	CH_3CN	o.n.[c]	**28**	45	41
9[Glcβ(1–4)Glcβ]	Ser	CH_2Cl_2	4	**29**	52	43
10[Xylα(1–3)Glcβ]	Ser	CH_2Cl_2	—[g]	**30**	62	44

[a] Unless otherwise stated the carbohydrate peracetates were reacted with 1.0–1.2 molar equivalents of Fmoc amino acid and 3 equivalents of $BF_3 \cdot (C_2H_5)_2O$. Detailed reaction conditions are given in the text and in the cited literature reference.

[b] Yields are based on the carbohydrate peracetate after chromatography of the product as described in the cited literature reference.

[c] Reaction overnight.

[d] 9 Equivalents of $BF_3 \cdot (C_2H_5)_2O$ were used.

[e] It is not clear whether 1 or 2 equivalents of **5** were used, as compared to the amino acid, and the actual yield may therefore be only half of the given value.

[f] G. Arsequell, L. Krippner, R. A. Dwek, and S. Y. C. Wong, *J. Chem. Soc., Chem. Commun.*, 2383 (1994).

[g] Not stated.

Dichloromethane appears to be preferable to acetonitrile as solvent because use of acetonitrile resulted in substantial O-acetylation of the Fmoc amino acid, for instance, in the preparation of **18** and **23**.[34] In addition, use of acetonitrile as solvent in glycosylations of Fmoc-tyrosine resulted in loss of stereoselectivity. However, acetonitrile had to be used in cases when the Fmoc amino acid was sparingly soluble in dichloromethane, and, as revealed by the examples in Table I, it could often be used without major losses in the overall yield.

Because of the anomeric effect[45] α-D-mannose pentaacetate (**3**) is less reactive than the β-peracetates of glucose and galactose. Consequently a larger excess of boron trifluoride etherate (9 equiv) and a prolonged reaction time (20 hr) were required to obtain the desired α-mannoside **23** from pentaacetate **3** and Fmoc-Thr.[34] The glycoester 2,3,4,6-tetra-O-acetyl-D-mannopyranosyl-N^{α}-Fmoc-threoninoate was formed as an intermediate in this glycosylation and then rearranged slowly to **23**. The unprotected carboxyl group of the Fmoc amino acid thus competes with the hydroxyl group for attack at C-1 of the glycosyl donor just as in glycosylations of 3-mercaptopropionic acid.

The lower nucleophilicity of a phenolic as compared to an aliphatic hydroxyl group makes glycosylations of Tyr more difficult than of Ser and Thr. Consequently, glycosylation of N^{α}-Fmoc-Tyr with β-D-galactose pentaacetate (**1**) required a longer time to reach completion, and it gave **19** in a yield (34%) that was lower than in the galactosylations of Fmoc-Ser and -Thr, which gave **16** and **17** in 53 and 50% yields, respectively.[34] In contrast to the glycosylations of aliphatic hydroxyl groups in amino acids, the anomeric ratio of **19** was significantly affected by the solvent. In dichloromethane, almost complete stereoselectivity (β/α, 45:2) was obtained, whereas the α-glycoside corresponding to **19** was the predominant product in acetonitrile (β/α, 1:2). An appreciable amount of O-acetylated Fmoc-Tyr was also formed in acetonitrile. A similar loss of stereoselectivity has previously been observed on replacement of dichloromethane with acetonitrile as solvent in the silver triflate-promoted reactions of allyl and pentafluorophenyl esters of Fmoc-Tyr with perbenzoylated or peracetylated glucosyl bromides.[46]

Carbohydrate 1,2-*trans*-peracetates have also been used for glycosylation of Fmoc amino acid pentafluorophenyl (Pfp) esters under boron trifluoride etherate promotion, and compounds **31–33** (Fig. 5) were prepared in this way.[7,43] With the exception of the preparation of **31**, somewhat higher yields (~10%) were obtained when Fmoc-serine was replaced by Fmoc-Ser pentafluorophenyl ester as glycosyl acceptor. Glycosylated Fmoc amino acid pentafluorophenyl esters are readily purified by flash column chromatography on silica gel which may facilitate scaling up of their synthesis, as compared to glycosylated Fmoc amino acids having an unprotected carboxyl group. When the carboxyl group is unprotected purification has predominantly been achieved using reversed-phase HPLC. However, a method for gram-scale purification of compounds **18** and **21** on reversed-phase HPLC has been developed and may be generally applica-

[45] E. Juaristi and G. Cuevas, *Tetrahedron* **48**, 5019 (1992).
[46] K. Jensen, M. Meldal, and K. Bock, *J. Chem. Soc., Perkin Trans. 1*, 2119 (1993).

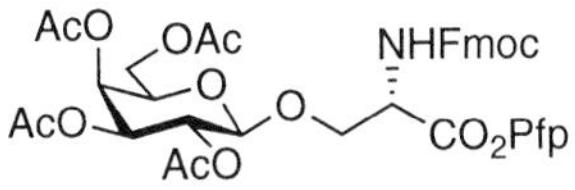

31 (42%)

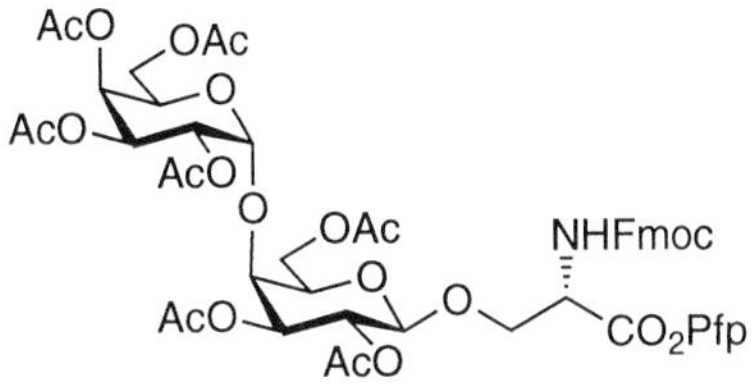

32 (75%)

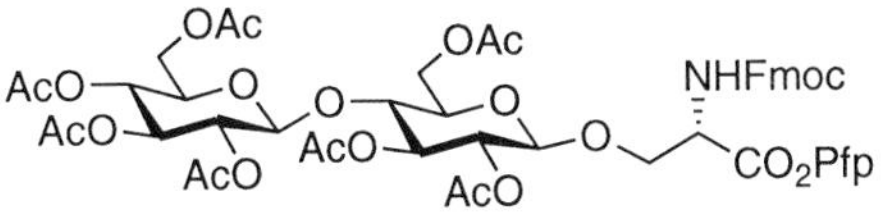

33 (61%)

FIG. 5. Glycosylated Fmoc amino acid pentafluorophenyl esters prepared under boron trifluoride etherate promotion using carbohydrate 1,2-*trans*-peracetates as glycosyl donors.

ble.[47] Furthermore, compounds **18, 21, 25, 26,** and **28** have been purified by chromatography on silica gel.[41,48]

Glycosyl halides, trichloroacetimidates, and thioglycosides are glycosyl donors that have found wide application in synthetic carbohydrate chemistry.[49] However, substantially lower yields of the glycosylated derivatives of Fmoc-Ser pentafluorophenyl ester (**32** and **33**) were obtained when acetylated glycosyl trichloroacetimidates, thioglycosides, and glycosyl bromides were employed as glycosyl donors instead of carbohydrate 1,2-*trans*-peracetates.[43] A large decrease in yield was also observed on attempted synthesis of **16** by glycosylation of Fmoc-Ser with 2,3,4,6-tetra-*O*-acetyl-α-D-galactopyranosyl bromide, as compared to when the peracetate **1** was used as glycosyl donor (28 and 53% yields, respectively).[27] These and other[24] results therefore imply that carbohydrate peracetates may be the glycosyl donors of choice for preparation of 1,2-*trans*-*O*-glycosylated Fmoc amino acid building blocks.

O-Fucosylated Fmoc Amino Acids with 1,2-Cis Anomeric Configuration

Glycosylation of Fmoc-Ser and -Thr with L-fucose tetraacetate (**6**) under boron trifluoride etherate promotion initially gave the expected 1,2-*trans*-glycosides as the major products.[50] However, if the reaction time was prolonged to 2 days, or if 6 instead of 3 equivalents of boron trifluoride

[47] J. L. Torres, E. Pagans, and P. Clapes, *Lett. Pept. Sci.* **3**, 61 (1996).
[48] G. Arsequell, L. Krippner, R. A. Dwek, and S. Y. C. Wong, *J. Chem. Soc., Chem. Commun.,* 2383 (1994).
[49] F. Barresi and O. Hindsgaul, *J. Carbohydr. Chem.* **14**, 1043 (1995).
[50] M. Elofsson, S. Roy, L. A. Salvador, and J. Kihlberg, *Tetrahedron Lett.* **37**, 7645 (1996).

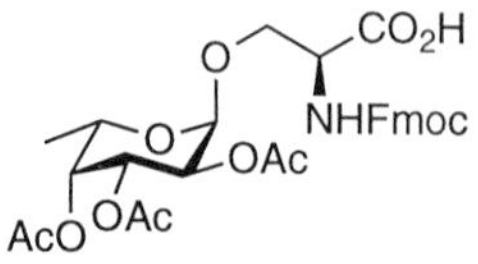

34 (44% in CH_3CN) 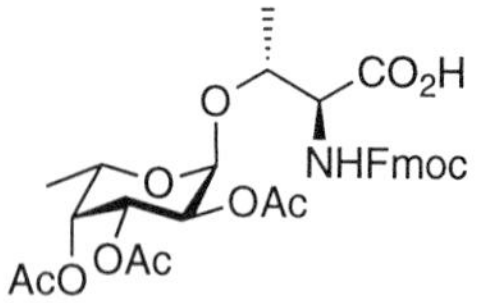   **35** (35% in CH_2Cl_2)

Fig. 6. 1,2-Cis-linked glycosylated amino acids prepared from fucose tetraacetate under boron trifluoride etherate promotion.[50]

etherate were used as promoter in the reaction, rearrangement to the thermodynamically more stable 1,2-cis-linked α-fucosides **34** and **35** occurred (Fig. 6). As expected, the rearrangement proceeded at a higher rate in acetonitrile than in the less polar dichloromethane, but in the synthesis of **35** use of acetonitrile as solvent led to substantial formation of O-acetylated Fmoc-Thr, which decreased the overall yield.

S-Linked Glycosylated Fmoc Amino Acids
with 1,2-Trans Anomeric Configuration

Fmoc amino acids that contain mercapto groups may also be glycosylated using carbohydrate 1,2-*trans*-peracetates as glycosyl donors.[27,34] For the S-linked galactosides of Cys and homocysteine (**36** and **37**, Fig. 7), use of tin(IV) chloride as promoter in dichloromethane was found to give significantly better yields than when tin(IV) chloride in acetonitrile or boron trifluoride etherate in dichloromethane were employed (20–25% improved yields with $SnCl_4$ in CH_2Cl_2). Tin(IV) chloride was also used as promoter in model glycosylations of hydroxylated Fmoc amino acids and then gave yields that differed little from those obtained with boron trifluoride etherate.[27,34] However, the high reactivity of tin(IV) chloride toward moisture constitutes a drawback that motivates use only to obtain improved yields in the glycosylation of mercapto groups.

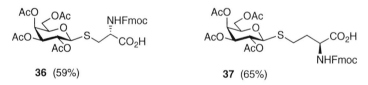

36 (59%)   **37** (65%)

Fig. 7. S-Linked glycosylated amino acids obtained from galactose pentaacetate and the corresponding Fmoc amino acid under tin(IV) chloride promotion.[27,34]

Mechanistic Considerations

The synthesis of 1,2-*trans*-glycosides from glycosyl halides that have participating acyl protective groups at O-2 is considered to involve orthoesters as intermediates.[51,52] Rearrangement of the orthoester can occur via a cyclic 1,2-*cis*-dioxocarbenium ion, which directs formation of the 1,2-*trans*-glycoside, or via a pathway leading to acylation of the glycosyl acceptor. The observation of O-acetylation of the acceptor in the preparations of **18, 19,** and **23,**[34] as well as the predominant formation of 1,2-*trans*-glycosides, suggests that orthoesters are intermediates also in the boron trifluoride etherate-mediated glycosylations of Fmoc amino acids. It is noteworthy that, when formed, 1,2-*trans*-glycosides of amino acids do not undergo Lewis acid-mediated anomerization to the thermodynamically more stable α-glycosides. However, glycosides of the 6-deoxy sugar L-fucose are more labile under acidic conditions, thereby allowing rearrangement to the more stable α-glycosides **34** and **35,** as discussed above. A high stability has also been observed[53] for O-acetylated dibromoisobutyl and bromoethyl glycosides in contrast to ordinary alkyl glycosides that undergo more facile anomerization. When an acetamido group is present at C-2 of a glycosyl donor, such as **5,** an orthoester cannot be formed, but the glycosylation instead proceeds with an analogous oxazoline as a discrete intermediate.[54] This pathway was used in the preparation of compounds **25** and **26.**[48]

Use of Glycosylated Fmoc Amino Acids
in Solid-Phase Glycopeptide Synthesis

General

The use of glycosylated amino acids as building blocks in Fmoc solid-phase glycopeptide synthesis has been documented in a large number of cases (reviewed in Refs. 10–12). The general picture that emerges from these studies is that glycosylated amino acids may be coupled just as efficiently to the N-terminus of a resin-bound peptide as an ordinary amino acid, and that further extension of the peptide chain then proceeds without problems. This conclusion appears to be independent of both the amino acid and the carbohydrate moiety as well as of the mode of protection used

[51] J. Banoub and D. R. Bundle, *Can. J. Chem.* **57,** 2091 (1979).
[52] P. J. Garegg, P. Konradsson, I. Kvarnström, T. Norberg, S. C. T. Svensson, and B. Wigilius, *Acta Chem. Scand.* **B39,** 569 (1985).
[53] G. Magnusson, S. Ahlfors, J. Dahmén, K. Jansson, U. Nilsson, G. Noori, K. Stenvall, and A. Tjörnebo, *J. Org. Chem.* **55,** 3932 (1990).
[54] W. P. Stöckl and H. Weidmann, *J. Carbohydr. Chem.* **8,** 169 (1989).

for the carbohydrate unit. Even amino acids that carry oligosaccharides as large as heptasaccharides couple well, and the carbohydrate residue does therefore not impose a steric hindrance that substantially affects the rate of the coupling.[55,56] Two methods for activation and coupling of glycosylated amino acids have predominantly been used for synthesis of glycopeptides, the N,N'-diisopropylcarbodiimide/1-hydroxybenzotriazole method and the pentafluorophenyl (Pfp) ester method.[10,12] In view of the effort invested in preparation of glycosylated amino acids it is desirable to use the smallest possible amount in each coupling. Examples reveal that use of a 1- to 1.5-fold excess of the activated glycoamino acid, as compared to the capacity of the peptide–resin, is sufficient to obtain complete and fast coupling.[43,55,57] Moreover, even an equivalent amount of the glycosylated amino acid has been used without a serious decrease in the overall yield.[58,59] When a minimal amount of glycosylated amino acid is employed, monitoring of the progress of the coupling becomes important. This has been performed spectrophotometrically using either 3,4-dihydro-3-hydroxy-4-oxo-1,2,3-benzotriazine[60] (Dhbt-OH) or bromphenol blue[61] as an indicator of unreacted amino groups on the solid phase.

The additional complexity of a glycopeptide, as compared to an ordinary peptide, imposes some limitations on the synthetic transformations that may be used for preparation of glycopeptides. All O-glycosidic bonds are labile toward acids, especially if the carbohydrate is deoxygenated, and furthermore carbohydrates that are O-glycosidically linked to Ser and Thr may undergo β-elimination on treatment with base. In Fmoc solid-phase peptide synthesis, cleavage from the resin with simultaneous deprotection of the amino acid side chains is usually performed under acidic conditions with trifluoroacetic acid. As discussed in greater detail below, protection of carbohydrate residues in glycopeptides with acyl groups confers stability toward treatment with trifluoroacetic acid, however, unprotected saccharides may be sufficiently stable during cleavage and deprotection. This fact allows standard linkers to the solid phase to be used in Fmoc solid-phase synthesis of glycopeptides just as for peptides. Thus the

[55] I. Christiansen-Brams, A. M. Jansson, M. Meldal, K. Breddam, and K. Bock, *Bioorg. Med. Chem.* **2**, 1153 (1994).
[56] L. Urge, D. C. Jackson, L. Gorbics, K. Wroblewski, G. Graczyk, and L. Otvos, Jr., *Tetrahedron* **50**, 2373 (1994).
[57] S. Peters, T. Bielfeldt, M. Meldal, K. Bock, and H. Paulsen, *Tetrahedron Lett.* **33**, 6445 (1992).
[58] J. Broddefalk, K.-E. Bergquist, and J. Kihlberg, *Tetrahedron Lett.* **37**, 3011 (1996).
[59] M. Elofsson, L. A. Salvador, and J. Kihlberg, *Tetrahedron* **53**, 369 (1997).
[60] L. R. Cameron, J. L. Holder, M. Meldal, and R. C. Sheppard, *J. Chem. Soc., Perkin Trans. 1*, 2895 (1988).
[61] M. Flegel and R. C. Sheppard, *J. Chem. Soc., Chem. Commun.*, 536 (1990).

p-hydroxymethylphenoxyacetic acid linker[62] may be used for synthesis of glycopeptides having a C-terminal carboxyl group, and the Rink[63,64] [*p*-(α-amino-2,4-dimethoxybenzyl)phenoxyacetic acid] or the PAL[65] [5-(4-aminomethyl-3,5-dimethoxyphenoxy)valeric acid] linker may be used in synthesis of glycopeptides having a C-terminal amide. Deprotection of the N^α-Fmoc group, and removal of *O*-acyl protective groups from the carbohydrate moieties of glycopeptides, is performed under basic conditions. This has caused concern that base-catalyzed side reactions, such as β-elimination and epimerization of peptide stereocenters, might be important in glycopeptide synthesis.[8,66] Suitable conditions for Fmoc cleavage and the choice of protective groups for the carbohydrate residues of glycopeptides are therefore discussed below.

Conditions for Cleavage of Fmoc Group

In solid-phase peptide synthesis the Fmoc group is usually removed with piperidine in *N,N*-dimethylformamide (DMF), but other bases such as morpholine and 1,8-diazabicyclo[5.4.0]undec-7-ene (DBU) have also been used. To remain below the sensitivity limit for β-elimination of carbohydrate moieties linked to Ser and Thr, the weaker base morpholine (pK_a 8.3) was advocated for use in synthesis of glycopeptides instead of piperidine (pK_a 11.1).[8] The choice of base for Fmoc deprotection of glycopeptides has now been investigated in greater detail.[67–69] These studies clearly show that neither piperidine nor DBU cause any β-elimination when used in synthesis of glycopeptides, and the fear of β-elimination has thus been exaggerated. Furthermore, when morpholine and piperidine were compared in the synthesis of glycopeptide **38** (Fig. 8a) Fmoc cleavage with morpholine was less efficient, which resulted in a crude product contaminated by several by-products (Fig. 8b).[67,69] In a study of the use of trimethylsilyl protection of the carbohydrate hydroxyl groups of glycopeptides, treatment with DBU led to the decomposition, and it might therefore be desir-

[62] E. Atherton, C. J. Logan, and R. C. Sheppard, *J. Chem. Soc., Perkin Trans. 1*, 538 (1981).

[63] H. Rink, *Tetrahedron Lett.* **28**, 3787 (1987).

[64] M. S. Bernatowicz, S. B. Daniels, and H. Köster, *Tetrahedron Lett.* **30**, 4645 (1989).

[65] F. Albericio, N. Kneib-Cordonier, S. Biancalana, L. Gera, R. I. Masada, D. Hudson, and G. Barany, *J. Org. Chem.* **55**, 3730 (1990).

[66] H. Kunz and W. K.-D. Brill, *Trends Glycosci. Glycotechnol.* **4**, 71 (1992).

[67] J. Kihlberg and T. Vuljanic, *Tetrahedron Lett.* **34**, 6135 (1993).

[68] M. Meldal, T. Bielfeldt, S. Peters, K. J. Jensen, H. Paulsen, and K. Bock, *Int. J. Pept. Protein Res.* **43**, 529 (1994).

[69] T. Vuljanic, K.-E. Bergquist, H. Clausen, S. Roy, and J. Kihlberg, *Tetrahedron* **52**, 7983 (1996).

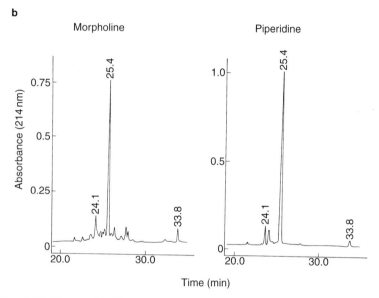

FIG. 8. (a) Glycopeptide **38** corresponds to amino acids 312–327 of the V3 loop from the HIV-3B isolate of human immunodeficiency virus.[69] A cysteine has been added at the C-terminus to allow conjugation to carrier proteins. (b) Analytical reversed-phase HPLC chromatograms of crude **38** obtained from syntheses in which the Fmoc groups of the six N-terminal residues were cleaved either with morpholine or with piperidine.[67,69]

able to avoid the use of DBU in combination with silyl protection of glycopeptides.[70]

Choice of Protective Groups for Carbohydrate Moieties of Glycopeptides

Several studies show that O-glycosidic linkages of common monosaccharides (e.g., GlcNAc, GalNAc, Glc, Gal, Man) having unprotected hydroxyl groups are sufficiently stable to allow treatment with trifluoroacetic acid

[70] I. Christiansen-Brams, M. Meldal, and K. Bock, *Tetrahedron Lett.* **34**, 3315 (1993).

for a limited period of time (≤2 hr), such as encountered in cleavage of glycopeptides from a solid phase.[19,56,71,72] Some of these studies[56,71] also reveal that degradation of the oligosaccharide moiety occurs if the treatment with trifluoroacetic acid is prolonged, or if nucleophiles such as water are added, which is usually the case in cleavage and deprotection of peptides and glycopeptides. Unprotected O-linked oligosaccharides composed of the common monosaccharides therefore appear to be on the verge of decomposition in trifluoroacetic acid medium. This may explain why some workers have found that a particular saccharide moiety is degraded during treatment of a glycopeptide with trifluoroacetic acid, whereas others have found it to be stable.[19,73] In addition, it cannot be ruled out that the structure of the peptide part of a glycopeptide influences the stability of the saccharide moiety. Only a slight increase in the acid lability of an O-linked saccharide necessitates protection of the hydroxyl groups with electron-withdrawing acetyl or benzoyl groups that stabilize the O-glycosidic bonds during cleavage from the solid phase. For instance, acyl protection has been found to be required for glycosides of the 6-deoxy sugar L-fucose,[74,75] which undergoes acid-catalyzed hydrolysis only 5–6 times faster than glycosides of the corresponding nondeoxygenated monosaccharide, galactose.[76] Because glycosides of sialic acid are known to be more labile toward acid than the ordinary monosaccharides, protection of sialic acid residues with acetyl groups has been suggested to be a suitable precaution in solid-phase synthesis of sialoglycopeptides.[59] In conclusion, even though acyl protection of the carbohydrate moieties of glycopeptides is not always necessary, it should be considered as a suitable precaution for most glycopeptides in order to prevent decomposition during acid catalyzed cleavage and deprotection. Furthermore, many routes for preparation of glycosylated amino acids, such as the one described in this article, provide acyl protective groups "free of charge."

Carbohydrates that are linked to serine and threonine residues in glycoproteins undergo facile β-elimination on treatment with aqueous alkali (pH ~10, 45°, 0.5–6 days), and this reaction is used analytically in structural

[71] L. Urge, L. Otvos, Jr., E. Lang, K. Wroblewski, I. Laczko, and M. Hollosi, *Carbohydr. Res.* **235**, 83 (1992).

[72] K. B. Reimer, M. Meldal, S. Kusumoto, K. Fukase, and K. Bock, *J. Chem. Soc., Perkin Trans. 1*, 925 (1993).

[73] H. Kunz, H. Waldman, and J. März, *Liebigs Ann. Chem.*, 45 (1989).

[74] C. Unverzagt and H. Kunz, *Bioorg. Med. Chem.* **2**, 1189 (1994).

[75] S. Peters, T. L. Lowary, O. Hindsgaul, M. Meldal, and K. Bock, *J. Chem. Soc., Perkin Trans. 1*, 3017 (1995).

[76] B. Capon, *Chem. Rev.* **69**, 407 (1969).

determinations of glycoproteins.[77] On the basis of this and other observations it was anticipated that β-elimination and epimerization of peptide stereocenters might constitute serious problems on removal of O-acyl protective groups from glycopeptides.[8] Recently, a systematic investigation of conditions used for de-O-acylation of the carbohydrate moieties of glycopeptides was performed.[78] This investigation revealed that neither β-elimination nor epimerization of peptide stereocenters was encountered on treatment of the model glycopeptide **39** with hydrazine hydrate in methanol,[79] saturated methanolic ammonia,[80] or dilute methanolic sodium methoxide, that is, under conditions in common use for deacetylation of glycopeptides (Fig. 9). However, it should be borne in mind that removal of acetyl groups has occasionally been reported to be accompanied by various side reactions, including β-elimination and cysteine-induced degradation of the peptide backbone.[59,69,75,81] When glycopeptide **39** was treated with a more concentrated sodium methoxide solution during a prolonged period of time, that is, when **39** was exposed to conditions resembling those used for removal of O-benzoyl groups, both β-elimination and epimerization were encountered to give compounds **41–43**.[78] This agrees well with previous observations which suggest that the more drastic conditions required for removal of benzoyl groups are more likely to be accompanied by base-induced side reactions than conditions used for deacetylation.[72,80,82]

The fact that side reactions may occur on removal of O-acyl protective groups from glycopeptides has stimulated the use of alternative protective groups for the carbohydrate moieties of glycopeptides. A further reason for such a development is that peptide chain termination by transfer of an acetyl group from a glycosylated amino acid to the N-terminus of the resin-bound peptide during couplings has been reported in a few cases.[59,83] Trimethylsilyl (TMS) protective groups have been used for 1-aminoalditols and mono- to heptasaccharidic glycosylamines which were N-glycosidically linked to Fmoc Asn pentafluorophenyl ester (e.g., **44**, Fig. 10).[55,70] The more stable $tert$-butyldimethylsilyl (TBDMS), $tert$-butyldiphenylsilyl (TBDPS), and isopropylidene groups have been used for the glycosylated

[77] J. Montreuil, S. Bouquelet, H. Debray, B. Fournet, G. Spik, and G. Strecker, *in* "Carbohydrate Analysis: A Practical Approach" (M. F. Chaplin and J. F. Kennedy, eds.), p. 143. IRL, Oxford, 1986.
[78] P. Sjölin, M. Elofsson, and J. Kihlberg, *J. Org. Chem.* **61**, 560 (1996).
[79] P. Schultheiss-Reimann and H. Kunz, *Angew. Chem. Suppl.,* 39 (1983).
[80] H. Paulsen, M. Schultz, J. Klamann, B. Waller, and M. Paal, *Liebigs Ann. Chem.,* 2028 (1985).
[81] H. Paulsen, W. Rauwald, and U. Weichert, *Liebigs Ann. Chem.,* 75 (1988).
[82] B. Erbing, B. Lindberg, and T. Norberg, *Acta Chem. Scand.* **B32**, 308 (1978).
[83] L. Otvos, Jr., K. Wroblewski, E. Kollat, A. Perczel, M. Hollosi, G. D. Fasman, H. C. J. Ertl, and J. Thurin, *Pept. Res.* **2**, 362 (1989).

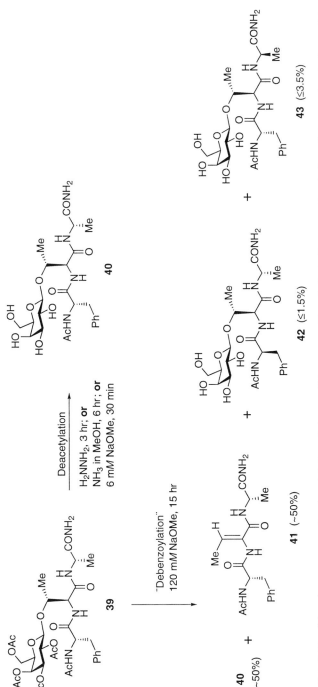

Fig. 9. Conditions in common use for deacetylation of glycopeptides cause neither β-elimination nor epimerization of stereocenters in the model glycopeptide **39**.[78] However, under the more basic conditions required for removal of benzoyl groups, both side reactions were encountered, and **41–43** were formed.

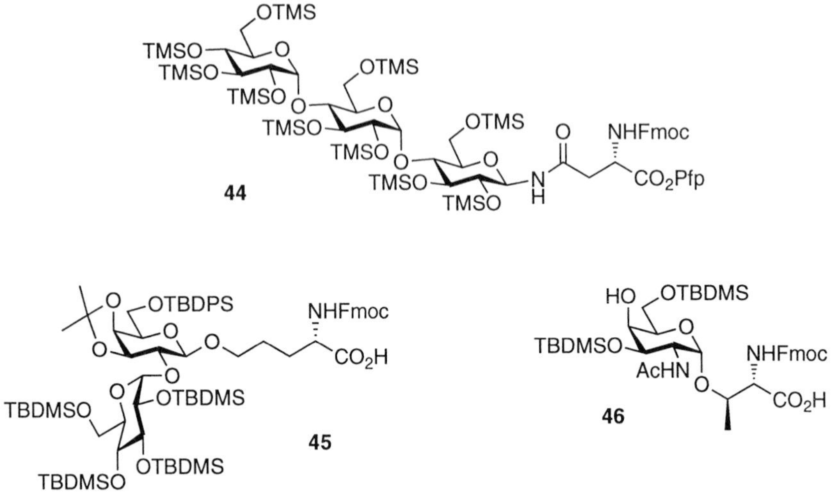

FIG. 10. Glycosylated amino acids, such as **44–46**, which carry acid-labile protective groups, have been prepared and used in solid-phase glycopeptide synthesis.

amino acid building blocks **45** and **46** (Fig. 10).[58,59] These building blocks were then employed in preparation of a glycopeptide analog of collagen[58] and a glycopeptide from human immunodeficiency virus (HIV) gp120 containing the Tn [α-D-GalNAc(1→O)-Thr] epitope.[59] As desired, the silyl and isopropylidene protective groups were rapidly and completely removed during glycopeptide cleavage and deprotection with trifluorocetic acid, thus eliminating the need for a separate step for deprotection of the carbohydrate moiety. It should, however, be pointed out that such acid-labile protective groups cannot be used for labile deoxygenated saccharides.

Benzyl ethers are often used as protective groups in carbohydrate synthesis; however, they are generally removed by hydrogenolysis, and their use is therefore restricted[84] to glycopeptides that lack Cys and Met. This has limited the use of benzyl ethers in glycopeptide synthesis, and only a few glycopeptides, which all contained O-benzylated sialic acid residues, have been prepared and then deprotected by hydrogenolysis.[85–87]

---

[84] M. Bodanszky and J. Martinez, *Synthesis*, 333 (1981).

[85] Y. Nakahara, H. Iijima, S. Sibayama, and T. Ogawa, *Tetrahedron Lett.* **31,** 6897 (1990).

[86] Y. Nakahara, H. Iijima, S. Shibayama, and T. Ogawa, *Carbohydr. Res.* **216,** 211 (1991).

[87] Y. Nakahara, H. Iijima, and T. Ogawa, *Tetrahedron Lett.* **35,** 3321 (1994).

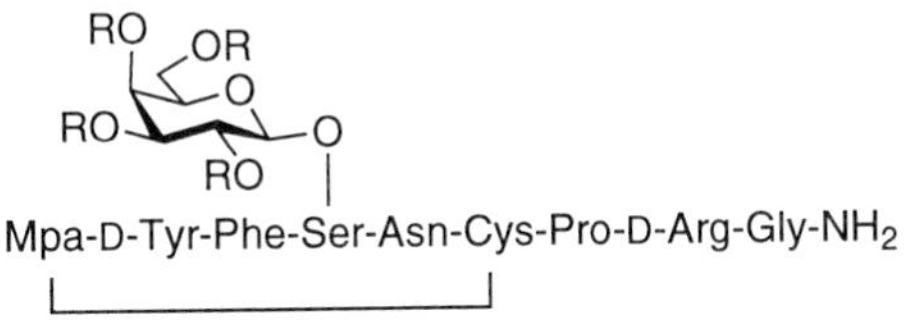

Mpa-D-Tyr-Phe-Ser-Asn-Cys-Pro-D-Arg-Gly-NH₂   Tyr-D-Met-Gly-Phe-Hyp-NH₂

49

47 R = Ac
48 R = H

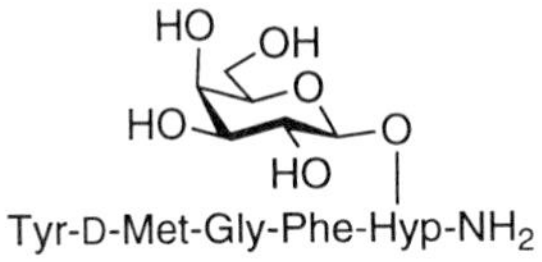

Asp-Tyr-Gly-Ile-Ser-Gln-Ile-Asn-Ser-Arg-NH₂   Phe-Ala-Pro-Ser-Asn-Tyr-Pro-Ala-Leu

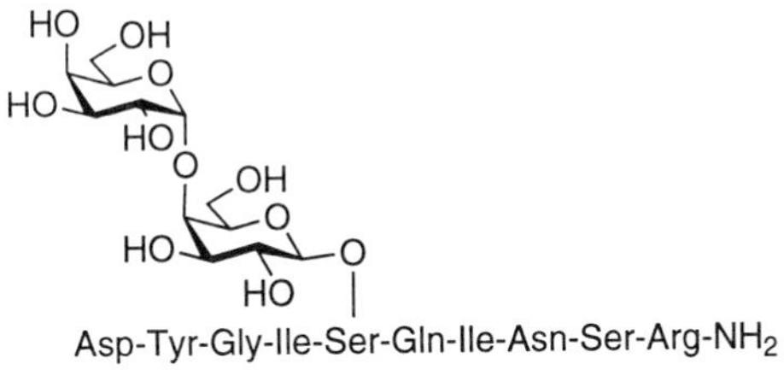

50 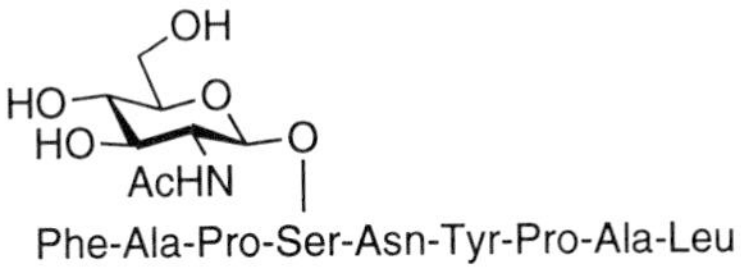   51

FIG. 11. Glycopeptides **47–49** are glycosylated analogs of peptide hormones that were prepared from building blocks **16** and **18**. Glycosylation was found to lead to improved pharmacokinetic and pharmacological properties.[5,7] The glycosylated amino acids **27** and **24** were used to prepare glycopeptides **50** and **51**, which elicited specific T cells in mice[40] [J. S. Haurum, G. Arsequell, A. C. Lellouch, S. Y. C. Wong, R. A. Dwek, A. J. McMichael, and T. Elliot, *J. Exp. Med.* **180**, 739 (1994)]. Mpa, 3-Mercaptopropionic acid.

Examples of Use of Glycopeptides in Pharmacology and Immunology

Glycosylated Peptide Hormones

Several studies indicate that glycosylation can be used in efforts to overcome the intrinsic shortcomings[88] of small peptides that are used as drugs, or which potentially could be used as drugs. Glycosylation has thus been shown to overcome proteolytic degradation,[7,89] rapid excretion,[4] and poor transport across membranes.[6,7] The galactosylated serine **31** (Fig. 5) has been used as a building block in the synthesis of glycopeptides **47** and **48** (Fig. 11), which are analogs of the antidiuretic drug [1-deamino-8-D-arginine]vasopressin (DDAVP).[7] The glycosylated analogs had significantly higher bioavailabilities than DDAVP on intraintestinal administration in

[88] J. J. Plattner and D. W. Norbeck, *in* "Drug Discovery Technologies" (C. R. Clark and W. R. Moos, eds.), p. 92. Ellis Hardwood-Halstead Press, Chichester, UK, 1989.
[89] M. F. Powell, T. Stewart, L. Otvos, Jr., L. Urge, F. C. A. Gaeta, A. Sette, T. Arrhenius, D. Thomson, K. Soda, and S. M. Colon, *Pharm. Res.* **10**, 1268 (1993).

rats, owing to both increased absorption from the small intestine and increased stability toward enzymatic degradation. In another investigation the galactosylated hydroxyproline **18** (Fig. 4) was incorporated in the enkephalinamide analog **49,** thereby increasing the analgetic activity 1000- to 10,000-fold as compared to the nonglycosylaed enkephalinamide.[5] Strikingly, attachment of glucose to another enkephalin analog enabled the neuropeptide to cross the blood–brain barrier, presumably via an active glucose transport mechanism.[6] The roles of the carbohydrate residues in these few selected examples are thus analogous to some of the properties that carbohydrate residues confer on proteins.

Studies of T-Cell Response to Glycopeptides

Peptides have a central role in activation of the immune system toward foreign protein antigens.[90] Proteins are processed into peptides in antigen presenting cells and then displayed to T cells as bimolecular complexes with major histocompatibility (MHC) molecules. Depending on the type of cell responsible for the processing and presentation, a complex between a MHC molecule and a peptide will stimulate helper T cells or cytotoxic T cells, both of which have critical functions in directing the immune response toward the foreign protein antigen. Even though most proteins are actually glycoproteins it was, until relatively recently, not known if glycoproteins could be processed into glycopeptides by antigen presenting cells and if glycopeptides could be bound by MHC molecules and presented to T cells. From an applied perspective it would be most interesting if glycopeptides would stimulate a T-cell response. If this could be achieved the immune system could then be directed toward carbohydrates that are important antigens on tumor cells, infectious virus, and bacteria but that usually elicit a weak, T-cell-independent immune response.

The galabiosylated Ser **27** (Fig. 4) has been used to prepare glycosylated derivatives of the helper T-cell stimulating peptide HEL(52–61).[28,43] It was found that the neoglycopeptide **50** (Fig. 11), in which the galabiose residue is located in the center of HEL(52–61), was bound by MHC molecules and elicited helper T cells with specificity for the galabiose residue, when used for immunizations of mice.[40] When the galabiose moiety instead was located in the N-terminal part of HEL(52–61), no T cells with specificity for the carbohydrate moiety could be obtained.[30,40] In a similar investigation the building block **25** (Fig. 4) was incorporated[91] in the center of a peptide from Sendai virus nucleoprotein to give the neoglycopeptide **51** which

[90] V. H. Engelhard, *Sci. Am.* **Aug.,** 44 (1994).
[91] G. Arsequell, J. S. Haurum, T. Elliott, R. Dwek, and A. C. Lellouch, *J. Chem. Soc., Perkin Trans. 1,* 1739 (1995).

elicited carbohydrate-specific cytotoxic T cells in mice.[92] In these two examples the recognition by the T-cell receptor was shown to involve the MHC molecule, in addition to the carbohydrate. Most likely the recognition also involved amino acid residues in the glycopeptide. However, a more recent report describes that immunization with neoglycopeptides elicited a cytotoxic T-cell response that was directed only toward the carbohydrate moiety.[93] Importantly, this suggests that neoglycopeptides may be used to direct the immune system toward carbohydrate epitopes of importance in medicine such as different tumor-associated antigens.

Experimental

General

Thin-layer chromatography is performed on silica gel 60 F_{254} (Merck, Darmstadt, Germany) with detection by UV light and then charring with sulfuric acid. Immediately before being used as solvents in the glycosylations, CH_2Cl_2 is dried by distillation from calcium hydride, and CH_3CN is passed through a column of neutral aluminum oxide (activity 1). Flash column chromatography is performed on silica gel (Matrex, 60 Å, 35–70 μm, Grace Amicon, Beverly, MA) with distilled solvents. Analytical HPLC separations are performed on a Kromasil C_8 column (100 Å, 5 μm, 4.6 × 250 mm, Hichrom, Reading, UK) with a flow rate of 1.5 ml/min and detection at 214 nm using a Beckman System Gold HPLC. Preparative reversed-phase HPLC separations are performed on a Kromasil C_8 column (100 Å, 5 μm, 20 × 250 mm, Hichrom) with a flow rate of 10–16 ml/min and detection at 214 nm. Isocratic mixtures or linear gradients of solvent systems (with solvent A being 0.1% (v/v) aqueous trifluoroacetic acid and B being 0.1% trifluoroacetic acid in CH_3CN) are used for both analytical and preparative reversed-phase HPLC separations. Preparative normal-phase HPLC separations are performed on a Kromasil silica column (100 Å, 5 μm, 20 × 250 mm, Hichrom) with a flow rate of 20 ml/min and detection at 254 nm.

N^α-Fmoc-L-Ser, -L-Thr, and -L-Tyr may be purchased from Bachem Feinchemikalien (Bubendorf, Switzerland). The preparations of N^α-Fmoc-*trans*-4-hydroxy-L-proline,[94] N^α-Fmoc-L-Cys,[34] and N^α-Fmoc-L-homocysteine[34] have been described previously. The peracetates of β-D-galactose

[92] J. S. Haurum, G. Arsequell, A. C. Lellouch, S. Y. C. Wong, R. A. Dwek, A. J. McMichael, and T. Elliot, *J. Exp. Med.* **180**, 739 (1994).
[93] U. M. Abdel-Motal, L. Berg, A. Rosén, M. Bengtsson, C. J. Thorpe, J. Kihlberg, J. Dahmén, G. Magnusson, K.-A. Karlsson, and M. Jondal, *Eur. J. Immunol.* **26**, 544 (1996).
[94] L. Lapatsanis, G. Milias, K. Froussios, and M. Kolovos, *Synthesis,* 671 (1983).

(1), β-D-glucose (2), α-D-mannose (3), β-D-glucosamine (5), and β-D-lactose (8) may be purchased from Sigma (St. Louis, MO). The preparations of 1,3,4,6-tetra-O-acetyl-2-deoxy-2-(2′,2′,2′-trichloroethoxycarbonylamino)-β-D-glucopyranose[95] (4), 1,2,3,4-tetra-O-acetyl-L-fucopyranose[21,50] (6), 1,2, 3,6-tetra-O-acetyl-4-O-(2,3,4,6-tetra-O-acetyl-α-D-galactopyranosyl)-β-D-galactopyranose[22] (7), 1,2,3,6-tetra-O-acetyl-4-O-(2,3,4,6-tetra-O-acetyl-β-D-glucopyranosyl)-β-D-glucopyranose[21,43] (9), and 1,2,4,6-tetra-O-acetyl-3-O-(2,3,4-tri-O-acetyl-α-D-xylopyranosyl)-β-D-glucopyranose[96] (10) have been described previously.

Glycosylation of 3-Mercaptopropionic Acid Using Carbohydrate Peracetate as Glycosyl Donor

Boron trifluoride etherate (139 μl, 1.11 mmol) is added to a solution of the carbohydrate 1,2-*trans*-peracetate (0.737 mmol) and 3-mercaptopropionic acid (257 μl, 2.95 mmol) in dry dichloromethane (10 ml), at room temperature. After 3 hr the solution is diluted with dichloromethane (30 ml) and washed with 1 *M* aqueous hydrogen chloride (40 ml). The aqueous phase is extracted with dichloromethane (two times, 20 ml each time), and the combined organic phases are dried with Na_2SO_4, filtered, and concentrated. The residue is then purified by flash column chromatography on silica gel. This procedure is used for preparation of the 3-(glycosylthio) propionic acids **12–15**[28] (cf. Fig. 3) which are purified using the following solvent systems: heptane–ethyl acetate–acetic acid (10 : 10 : 1, by volume) for **12, 13,** and **15;** toluene–methanol–acetic acid (100 : 5 : 2, by volume) for **14.**

1,2-Trans-O-Glycosylation of $N^α$-Fmoc Amino Acids Using Carbohydrate Peracetate as Glycosyl Donor

Boron trifluoride etherate [49 μl, 390 μmol, except for **23** where 154 μl $BF_3 \cdot (C_2H_5)_2O$ is used] is added to the carbohydrate 1,2-*trans*-peracetate (130 μmol) and the $N^α$-Fmoc amino acid (156 μmol) in dry CH_2Cl_2 (2–2.5 ml) or CH_3CN (1.5 ml) under a nitrogen atmosphere at room temperature (see Table I). The reaction is monitored by TLC and analytical HPLC. When no further reaction progress is observed, the mixture is diluted with CH_2Cl_2 (8 ml), washed with 1 *M* aqueous HCl (1 ml) and water (1 ml), dried, and concentrated. The residue is purified in two portions by preparative HPLC. This procedure is used for preparation of the glycosylated amino acids **16–20, 22–24, 27,** and **29** (see Fig. 4).[34]

[95] U. Ellervik and G. Magnusson, *Carbohydr. Res.* **280,** 251 (1995).
[96] B. Lüning, T. Norberg, and J. Tejbrant, *J. Carbohydr. Chem.* **11,** 933 (1992).

1,2-Cis-O-Glycosylation of N^α-Fmoc Amino Acids Using Fucose Tetraacetate (6) as Glycosyl Donor

Boron trifluoride etherate (110 μl, 0.90 μmol) is added to a solution of **6** (50 mg, 150 μmol) and N^α-Fmoc-Ser-OH (59 mg, 180 μmol) in dry CH_3CN (3 ml). After 4 hr the solution is diluted with CH_2Cl_2 (10 ml) and washed with water (10 ml). The aqueous phase is extracted with CH_2Cl_2 (twice, 10 ml) and the combined organic phases are dried with Na_2SO_4, filtered, and concentrated. The crude product is purified in 30- to 40-mg portions by preparative reversed-phase HPLC (isocratic conditions using solvents A and B at 1:1, v/v) to give **34**[75] (Fig. 6).[50] Compound **35**[97] is prepared under similar conditions[50] II and is purified by normal-phase HPLC using a linear gradient of ethanol in hexane [0 to 60% over 250 min, with both eluants containing 2% (v/v) acetic acid].

Use of Glycosylated Amino Acids in Solid-Phase Glycopeptide Synthesis

The glycosylated amino acids are used in solid-phase synthesis of glycopeptides according to standard protocols[98] for Fmoc solid-phase synthesis of peptides. Detailed procedures for synthesis in a simple, manual setup or in a fully automated continuous flow peptide synthesizer have been published.[28,43]

[97] H. Hietter, M. Schultz, and H. Kunz, *Synlett,* 1219 (1995).
[98] E. Atherton and R. C. Sheppard, "Solid Phase Peptide Synthesis: A Practical Approach." IRL at Oxford Univ. Press, Oxford, 1989.

[12] Synthesis of Phosphopeptides Using Modern Chemical Approaches

By JOHN WILLIAM PERICH

Introduction

Studies into the chemical phosphorylation of amino acid derivatives can be documented back to the 1950s when several groups investigated Ser, Thr, and Tyr phosphorylation on the premise that these phosphorylated residues were the possible source of protein-bound phosphate in phosphoproteins.[1] Although early chemical studies employed harsh phosphorylation

[1] A. W. Frank, *Crit. Rev. Biochem.* **16**, 51 (1984).

0076-6879/97 $25.00

reagents such as phosphoryl trichloride/pyridine and orthophosphoric acid/ phosphorus pentoxide for the phosphorylation of amino acids and their simple peptide derivatives,[1,2] the subsequent introduction of protected monofunctional dialkyl phosphorochloridate and phosphoramidite reagents provided new avenues for the mild preparation of phosphorylated biomolecules with improved chemical efficiency. This article details the use of these modern phosphorylation methods for the synthesis of phosphorylated peptides and their peptide mimetics.

Synthesis of Phosphotyrosine-Peptides Using Protected Fmoc-Tyr(PO$_3$R$_2$)-OH Derivatives

The impetus for the application of synthetic phosphotyrosine [Tyr(P)]-containing peptides in biochemical and physiological studies arose during the 1970s with the recognition that protein tyrosyl phosphorylation was involved in cell regulation through complex signal transduction processes. Although early solid-phase Tyr(P)-peptide syntheses were cumbersome and required chemical expertise,[3] the routine synthesis of Tyr(P)-containing peptides has now become straightforward through the commercial availability of protected tert-butyloxycarbonyl (Boc)-Tyr(PO$_3$R$_2$)-OH and fluorenylmethyloxycarbonyl (Fmoc)-Tyr(PO$_3$R$_2$)-OH derivatives and the development of improved solid-phase synthesis and deprotection methods.

With respect to Boc solid-phase peptide synthesis, only Boc-Tyr-(PO$_3$Me$_2$)-OH[4] (where Me is methyl) has proved suitable for use in peptide assembly due to the stability of the methyl phosphate groups to routine Boc-cleavage treatments [trifluoroacetic acid (TFA)]. Unfortunately, the incompatibility of the acid-sensitive benzyl (Bzl) phosphate groups of Boc-Tyr(PO$_3$Bzl$_2$)-OH (Chem-Impex Int., Peninsula Laboratories) with the Boc cleavage step has limited the use of this derivative for the generation of the Tyr(P) residue as the N-terminal residue in Fmoc solid-phase synthesis.[5] Although Boc-Tyr(PO$_3$Me$_2$)-OH is now commercially available (Chem-Impex Int., Peninsula Laboratories), this derivative can be readily prepared in good overall yield by dimethyl phosphorochloridate/pyridine phosphorylation of Boc-Tyr-ONBzl followed by palladium-mediated hydrogenolytic conversion of Boc-Tyr(PO$_3$Me$_2$)-ONBzl to Boc-Tyr(PO$_3$Me$_2$)-OH.[4] This derivative can be readily incorporated into peptide assembly using Merri-

[2] G. Folsch, Sven. Kem. Tidskr. 79, 38 (1967).
[3] R. M. Valerio, P. F. Alewood, R. B. Johns, and B. E. Kemp, Int. J. Pept. Protein Res. 33, 428 (1989).
[4] J. W. Perich, P. F. Alewood, and R. B. Johns, Aust. J. Chem. 44, 233 (1991).
[5] K. Ramalingam, S. R. Eaton, W. L. Cody, J. A. Loo, and A. M. Doherty, Lett. Pept. Sci. 1, 73 (1994).

field, phenylacetamido (PAM), or 4-methylbenzhydrylamine (MBHA) polystyrene resins, with the incorporation of Boc-Tyr(PO$_3$Me$_2$)-OH being compatible with the majority of commonly used coupling reagents {N,N'-dicyclohexyl (DCC), (benzotriazol-1-yloxy)tris(pyrrolidino)phosphonium hexafluorophosphate (PyBOP), N-[(1H-benzotriazol-1-yl)(dimethylamino)methylene]-N-methylmethanaminium hexafluorophosphate N-oxide (HBTU), etc}. Because partial dephosphorylation of Tyr(PO$_3$R$_2$) residues occurs on HF deprotection.[6–8] peptide deprotection is best performed by initial acidolytic cleavage of the Tyr(PO$_3$Me$_2$)-peptide from its resin support followed by cleavage of the methyl phosphate groups using either trimethylsilyl bromide (TMSBr)/TFA/thioanisole, trifluoromethanesulfonic acid (TFMSA)/TFA/dimethyl sulfide (DMS), or trifluoromethanesulfonic acid trimethylsilyl ester (TMSOTf)/TFA/DMS based procedures.[8–10] A list of several Tyr(P)-peptides prepared by the use of Boc-Tyr(PO$_3$Me$_2$)-OH in Boc solid-phase synthesis is provided in Table I.

With respect to Fmoc solid-phase peptide synthesis, the commercial availability of Fmoc-Tyr(PO$_3$Me$_2$)-OH (Bachem, Novabiochem), Fmoc-Tyr(PO$_3$Bzl$_2$)-OH,* and Fmoc-Tyr(PO$_3$$tBu_2$)-OH (Chem-Impex Int.; where tBu is $tert$-butyl) has now provided a simple procedure for the synthesis of large, complex Tyr(P)-peptides. The synthesis of Fmoc-Tyr(PO$_3$Me$_2$)-OH can be readily accomplished by a three-step procedure that involves conversion of Fmoc-Tyr-OtBu to Fmoc-Tyr(PO$_3$Me$_2$)-OtBu using (MeO)$_2$PNiPr$_2$ (Novabiochem; where iPr is 2-propyl) followed by acidolytic cleavage (TFA) of the $tert$-butyl ester protecting group (see following experimental section). This route is preferred over a previously reported route[12] in which the efficiency of dithionite reduction of the Maq group from Fmoc-Tyr(PO$_3$Me$_2$)-OMaq is a function of the quality of commercial sodium dithionite, which has been found to deteriorate rapidly on exposure to oxygen.[13] In the case of acid-labile phosphate groups, both Fmoc-Tyr-

[6] B. W. Gibson, A. M. Falick, A. L. Burlingame, L. Nadasdi, A. C. Nguyen, and G. L. Kenyon, *J. Am. Chem. Soc.* **109**, 5343 (1987).

[7] E. A. Kitas, J. W. Perich, R. B. Johns, and G. W. Tregear, *Tetrahedron Lett.* **29**, 3591 (1988).

[8] E.-S. Lee and M. Cushman, *J. Org. Chem.* **59**, 2086 (1994).

[9] R. M. Valerio, J. W. Perich, E. A. Kitas, P. F. Alewood, and R. B. Johns, *Aust. J. Chem.* **42**, 1519 (1989).

[10] A. Otaka, K. Miyoshi, M. Kaneko, H. Tamamura, N. Fujii, M. Nomizu, T. R. Burke, Jr., and P. P. Roller, *J. Org. Chem.* **60**, 3967 (1995).

* Obtained by treatment of Boc-Tyr(PO$_3$Bzl$_2$)-OH (Chem-Impex Int.) with HCOOH and introduction of Fmoc group using Fmoc-Su.[11]

[11] D. M. Andrews, J. Kitchin, and P. W. Seale, *Int. J. Pept. Protein Res.* **38**, 469 (1991).

[12] E. A. Kitas, J. W. Perich, J. D. Wade, R. B. Johns, and G. W. Tregear, *Tetrahedron Lett.* **30**, 6229 (1989).

[13] R. C. Cambie, J. B. J. Milbank, and P. S. Rutledge, *Synth. Commun.* **26**, 715 (1996).

TABLE I

Phosphotyrosine-Peptides Prepared Using Boc-Tyr(PO₃Me₂)-OH

$$\text{Phosphotyrosine-Peptides Prepared Using Boc-Tyr(PO}_3\text{Me}_2\text{)-OH}$$

Peptide	Protein	Resin	Resin cleavage	Methyl phosphate cleavage	Ref.
LRRAY(P)VLG		Merrifield	HBr/TFA/anisole	45% HBr/acetic acid	7
FTSTEPQT(P)QPGENL	pp60src	Merrifield	HF/MS/anisole	TFMSA/TFA/DMS/m-cresol	8
TEPQY(P)QPGE	Src protein	Merrifield	TMSOTf/thioanisole/TFA/EDT/m-cresol + DMS/TMSOTf		10
FLT(P)EY(P)VATRWTRAPEIMLN-NH₂	MAP kinase	MBHA	TMSOTf/thioanisole/TFA/EDT/m-cresol + DMS/TMSOTf		10

(PO_3Bzl_2)-OH and Fmoc-Tyr(PO_3tBu_2)-OH can be prepared up to the 20 mmol scale by a three-step procedure that involves conversion of Fmoc-Tyr-OPac (where Pac is phenacyl) to Fmoc-Tyr(PO_3R_2)-OPac (R = Bzl or tBu) using either $(BzlO)_2PNR'_2$ and $(tBuO)_2PNR'_2$ [R' = C_2H_5 (Et) or iPr] (Aldrich, Milwaukee, WI; Novabiochem) followed by zinc/acetic acid reduction of the phenacyl group (see below).[14] However, it is absolutely crucial that all traces of acid are thoroughly washed from the ethereal solution prior to solvent evaporation so as to prevent acid-catalyzed decomposition of the acid-labile *tert*-butyl and benzyl phosphate groups during long-term storage. The product is obtained as a crisp white foam on rapid vacuum drying of the oil, and it is recommended that large quantities of both derivatives should be stored in small portions (0.5 g) at $-78°$ under nitrogen so as to minimize acid-catalyzed autodecomposition.

The synthesis of both Fmoc-Tyr(PO_3tBu_2)-OH and Fmoc-Tyr-(PO_3Bzl_2)-OH can also be accomplished at the 3 to 5 mmol scale by a "one-pot" procedure (see Fig. 1) that utilizes either $(BzlO)_2PNR'_2$ or $(tBuO)_2PNR'_2$ for the key phosphorylation step.[15,16] This procedure involves initial protection of the carboxyl group by treatment of Fmoc-Tyr-OH with *tert*-butyldimethylsilyl chloride followed by phosphitylation of the tyrosyl hydroxyl group using $1H$-tetrazole/$(RO)_2PNR'_2$. On *tert*-butyl hydroperoxide (tBuOOH) oxidation of the dialkyl phosphite triester intermediate, final acidolytic cleavage of the *tert*-butyldimethylsilyl carboxyl-protecting group is effected during the sodium metabisulfite treatment (pH 3.5).

Synthesis of Fmoc-Tyr(PO_3Me_2)-OH

Dimethyl N,N-diisopropylphosphoramidite (Novabiochem; 0.232 g, 1.2 mmol) in dry tetrahydrofuran (THF, 1 ml) is added to a solution of Fmoc-Tyr-OtBu (0.459 g, 1.0 mmol) and $1H$-tetrazole (0.154 g, 2.2 mmol) in dry THF (4 ml) and the solution stirred for 60 min at 20°. The reaction solution is cooled to $-15°$ (ice–salt bath), and m-chloroperoxybenzoic acid (0.24 g, 1.2 mmol) in dichloromethane (DCM, 2 ml) is added. After 30 min, 10% $Na_2S_2O_5$ (30 ml) and ethyl acetate (30 ml) are added, the solution transferred to a separating funnel, and the aqueous layer discarded. The organic solution is washed with 10% $Na_2S_2O_5$ (two times, 20 ml each time), 5% $NaHCO_3$ (two times, 20 ml), and 1 M HCl (20 ml). The solution is dried (Na_2SO_4), filtered, and the solvent evaporated under reduced pressure to

[14] R. M. Valerio, A. M. Bray, N. J. Maeji, P. O. Morgan, and J. W. Perich, *Lett. Pept. Sci.* **2**, 33 (1995).
[15] J. W. Perich and E. C. Reynolds, *Synlett*, 577 (1991).
[16] J. W. Perich, M. Ruzzene, L. A. Pinna, and E. C. Reynolds, *Int. J. Pept. Protein Res.* **43**, 39 (1993).

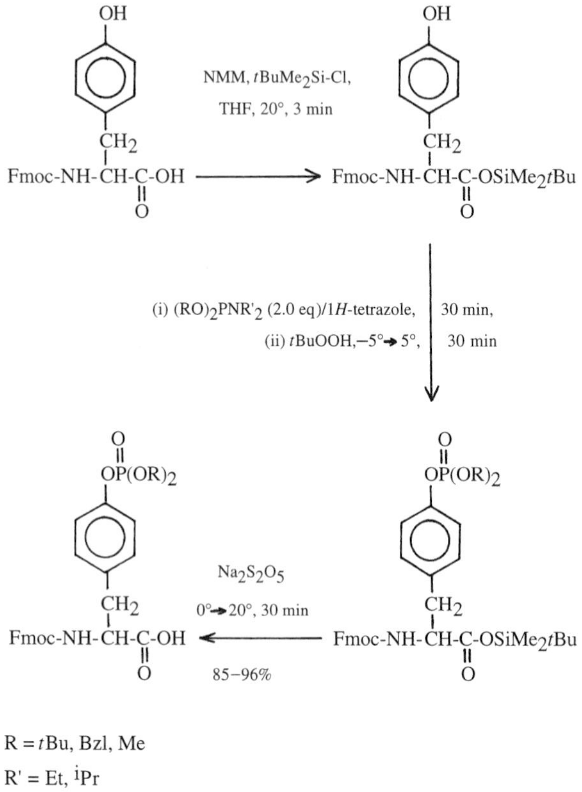

OH
NMM, tBuMe$_2$Si-Cl,
THF, 20°, 3 min

CH$_2$
Fmoc-NH-CH-C-OH
||
O

→

OH

CH$_2$
Fmoc-NH-CH-C-OSiMe$_2t$Bu
||
O

(i) (RO)$_2$PNR'$_2$ (2.0 eq)/1H-tetrazole,   30 min,
(ii) tBuOOH, −5° → 5°,   30 min

O
||
OP(OR)$_2$

CH$_2$
Fmoc-NH-CH-C-OH
||
O

Na$_2$S$_2$O$_5$
0° → 20°, 30 min
←
85–96%

O
||
OP(OR)$_2$

CH$_2$
Fmoc-NH-CH-C-OSiMe$_2t$Bu
||
O

R = tBu, Bzl, Me

R' = Et, ^{i}Pr

Fig. 1. One-pot synthesis of Fmoc-Tyr(PO$_3$R$_2$)-OH derivatives (R = tBu, Bzl, or Me).

give a light yellow oil. The oil is dissolved in TFA (3 ml), and the mixture is allowed to stand for 1 hr. The solvent is evaporated under reduced pressure and the residual oil dissolved in diethyl ether (10 ml). The solution is washed with 1 M HCl (two times, 10 ml) and extracted with 5% NaHCO$_3$ (three times, 8 ml). The combined base extracts are washed with diethyl ether (20 ml) and acidified with 3 M HCl. The aqueous solution is extracted with DCM (three times, 20 ml), dried (Na$_2$SO$_4$), and filtered. The solvent is evaporated under reduced pressure and the residue dried under high vacuum to give Fmoc-Tyr(PO$_3$Me$_2$)-OH as a crisp, white foam.

Large-Scale Synthesis of Fmoc-Tyr(PO$_3t$Bu$_2$)-OH

Di-*tert*-butyl *N,N*-diethylphosphoramidite (7.47 g, 30.0 mmol) in dry THF (10 ml) is added to a stirred solution of Fmoc-Tyr-OPac (9.90 g, 19.0

mmol) and 1H-tetrazole (4.97 g, 71.0 mmol) in dry THF (85 ml) and the solution stirred for 30 min at 20°. The reaction solution is cooled to −15° (ice–salt bath), and 70% *tert*-butyl hydroperoxide (8.4 ml) in water (30 ml) is added dropwise such that the solution temperature is maintained below 4°. After vigorous stirring at 4° for 30 min, diethyl ether (250 ml) and water (500 ml) are added, the solution transferred to a separating funnel, and the aqueous layer discarded. The organic solution is washed with 10% $Na_2S_2O_5$ (twice, 50 ml), 5% $NaHCO_3$ (three times, 50 ml), 10% citric acid (twice, 50 ml), and saturated NaCl (once, 50 ml). The solution is dried (Na_2SO_4) and filtered, and the solvent is evaporated under reduced pressure to give crude Fmoc-Tyr(PO_3tBu_2)-OPac as a thick oil.

The oil is dissolved in ethyl acetate (60 ml), acetic acid (150 ml), and water (30 ml), and fresh acid-washed zinc dust (8.54 g) is added. After the solution is vigorously stirred for 2 hr at 20°, unreacted zinc is removed by suction filtration, and ethyl acetate (200 ml) and water (1000 ml) are added to the filtrate. The organic solution is washed with water (three times, 400 ml) and the solvent evaporated under reduced pressure. The residual oil is dissolved in diethyl ether (100 ml), extracted with 5% $NaHCO_3$ (six times, 40 ml), and the combined base extracts then acidified to pH 3.5 by the careful addition of solid citric acid. The aqueous solution is extracted with diethyl ether (twice, 100 ml) and then washed with water (50 ml) and saturated NaCl (50 ml). The solution is dried (Na_2SO_4) and the solvent evaporated udner reduced pressure to give Fmoc-Tyr(PO_3tBu_2)-OH (8.94 g, 79%) as a white foam. Thin-layer chromatography (TLC), R_f 0.30, single spot in $CHCl_3$/methanol/acetic acid (90:8:2, by volume). Ion spray mass spectrometry (MS): found, 596.0, [M + H]; expected, 596.2 for $[C_{32}H_{38}NO_8P]$ + H.

One-Pot Synthesis of Fmoc-Tyr(PO₃R₂)-OH Derivatives

N-Methylmorpholine (NMM; 0.152 g, 1.5 mmol) in dry THF (1 ml) and *tert*-butyldimethylsilyl chloride (0.214 g, 1.43 mmol) in THF (1 ml) are successively added to a solution of Fmoc-Tyr-OH (0.605 g, 1.5 mmol) in THF (3 ml) at 20°. After 3 min, 1H-tetrazole (0.49 g, 7.0 mmol) is added in one portion followed by the addition of a solution of the dialkyl N,N-diethylphosphoramidite (3.0 mmol) in THF (1 ml) at 20°. Afer 30 min, the reaction mixture is cooled to −5° and 14% aqueous *tert*-butyl hydroperoxide (2.6 ml, 4.0 mmol) added. After 30 min at 5°, an aqueous solution of 10% $Na_2S_2O_5$ (10 ml) is added at 0° and the reaction mixture rapidly stirred for 30 min at 20°. The solution is transferred to a separating funnel using diethyl ether (40 ml) and the aqueous phase discarded. The organic phase is washed with 10% $Na_2S_2O_5$ (two times, 15 ml) and extracted with 5% $NaHCO_3$

solution (three times, 12 ml). (*Note:* The sodium salt of the benzyl derivative partially separates as a thick brown oil from the aqueous solution.) The combined aqueous extract is washed with diethyl ether (30 ml), acidified to pH 3.5 (pH electrode) with 30% citric acid, and then extracted with CH_2Cl_2 (three times, 30 ml). The solvent is evaporated under reduced pressure, and the residue is dissolved in diethyl ether (30 ml), washed with water (five times, 30 ml), dried (Na_2SO_4), and filtered. The solvent is evaporated under reduced pressure and dried under high vacuum to give both Fmoc-Tyr(PO_3tBu_2)-OH and Fmoc-Tyr(PO_3Bzl_2)-OH as crisp, white foams.

Peptide Synthesis and Deprotection

The three protected Fmoc-Tyr(PO_3R_2)-OH derivatives (R = Me, Bzl, *t*Bu) can be readily incorporated into Fmoc solid-phase peptide synthesis using phosphonium- and uronium-coupling reagents[12,14,16–20] with benzo-triazol-1-yl-oxytris-(dimethylamino)phosphonium hexafluorophosphate (BOP)/PyBOP and HBTU being used most frequently employed. A feature in the use of *tert*-butyl phosphate protection is that these groups are inert to piperidine-mediated dealkylation during Fmoc cleavage steps. Although both dimethyl and dibenzyl phosphorotriester groups are susceptible to piperidine-mediated delkylation, the continual generation of the alkyl phosphorodiester during repeated Fmoc cleavage steps has not been observed to be severely detrimental to overall peptide assembly.[12,18] However, the use of DCC or diisopropylcarbodiimide (DIPCDI) is not recommended because these coupling reagents react with the phosphorodiester functionality and cause undesired activation at this site.

On completion of peptide assembly, the deprotection of the Tyr-(PO_3Me_2)-peptide–resin can be effected by initial release of the peptide from its resin support by routine TFA/scavenger treatments followed by acidolytic or silylitic cleavage of the methyl groups from the Tyr(PO_3Me_2)-peptide (see earlier section). With the use of Wang resin as the polymer support, complete peptide deprotection and resin cleavage can be accomplished in a single step using 1 *M* THSBr/thioanisole in TFA.[12] This derivative has found wide application in peptide synthesis with some successful syntheses including [Tyr(P)][4]-angiotensin II[19] and several Tyr(P)-peptide sequences corresponding to viral proteins pp60[v-src] and p90[gag-yes],[12] p56[lck],[20]

[17] J. W. Perich and E. C. Reynolds, *Int. J. Pept. Protein Res.* **37,** 572 (1991).
[18] E. A. Kitas, J. D. Wade, R. B. Johns, J. W. Perich, and G. W. Tregear, *J. Chem. Soc., Chem. Commun.,* 338 (1991).
[19] E. A. Kitas, R. B. Johns, C. N. May, G. W. Tregear, and J. D. Wade, *Pept. Res.* **6,** 205 (1993).
[20] E. A. Kitas, R. Knorr, A. Trezeciak, and W. Bannwarth, *Helv. Chim. Acta* **74,** 1314 (1991).

insulin receptor kinase,[20] epidermal growth factor (EGF) kinase,[20] Ig-α ARH1, and Ig-β ARH1.[21] Although these Tyr(P)-peptides range from 6- to 26-mers and contain a single Tyr(P)-residue, one form of the Ig-α ARH1 peptide has been prepared containing two Tyr(P)-residues.

A feature in the use of the acid-sensitive benzyl and *tert*-butyl phosphate groups is that simultaneous cleavage of these protecting groups from the Tyr(PO$_3$R$_2$)-peptide–resin (R = *t*Bu, Bzl) is effected during the acidolytic (TFA) resin cleavage step and thereby does not require any further specialized chemical treatments. Thus, the ready incorporation of these two derivatives in Fmoc solid-phase synthesis has provided a simple and straightforward synthesis of Ser-Ser-Ser-Tyr(P)-Tyr(P),[16] [Tyr(P)]^4-neurotensin 8–13,[16] [Tyr(P)]^4-angiotensin II,[16] Ala-Glu-Tyr(P)-Thr-Ala,[17] Glu-Ile-Tyr(P)-Glu-Glu-Glu-Glu-Glu-Ser-Ala,[22] and an α-helix 17-mer peptide.[23] In addition, these derivatives have also provided a simple methodology for the synthesis of complex Tyr(P)-peptides that include peptide sequences corresponding to viral protein p85$^{gag-fes}$,[14,18] EGF receptor kinase,[18] cdc2,[24] Fcγ-receptor,[24] and the triply phosphorylated Ile-Tyr(P)-Glu-Thr-Asp-Thr(P)-Tyr(P)-Arg-Lys[25] peptide.*

Synthesis of Phosphotyrosine-Peptides Using Fmoc-Tyr(PO$_3$H$_2$)-OH

A major development during the early 1990s is the improvement in the efficiency of incorporating side-chain free Boc-Tyr(PO$_3$H$_2$)-OH or Fmoc-Tyr(PO$_3$H$_2$)-OH into solid-phase synthesis, which is facilitated by a change

---

[21] M. R. Clark, S. A. Johnson, and J. C. Cambier, *EMBO J.* **13**, 1911 (1994).

[22] F. Meggio, J. W. Perich, O. Marin, and L. A. Pinna, *Biochem. Biophys. Res. Commun.* **182**, 1460 (1992).

[23] J. D. Wade, J. W. Perich, M. J. McLeish, L. Otvos, Jr., and G. W. Tregear, *Lett. Pept. Sci.* **2**, 71 (1995).

[24] J. W. Perich, T. Johnson, and H.-C. Cheng, unpublished data.

[25] O. Marin, F. Meggio, J. W. Perich, and L. A. Pinna, *Int. J. Biochem. Cell Biol.* in press.

* In addition to the above derivatives, Fmoc-Tyr(PO$_3$All$_2$)-OH (where All is allyl) has been used for the synthesis of the p60^{src} peptide TEPQY(P)QPGE, which employed Pd(PPh$_3$)$_4$ cleavage of the allyl phosphate groups.[20] The acid-labile Dmpse derivative Fmoc-Tyr(PO$_3$ Dmpse$_2$)-OH [where Dmpse is 2-(methyldiphenylsilyl)ethyl] has also been used for the synthesis of three MB-1 Tyr(P)-octapeptide amides, with the Dmpse groups cleaved during the final TFA deprotection step.[26] In a departure from alkyl protection of the phosphate functionality, the two phosphoramidate derivatives Fmoc-Tyr[P(O)(NHR)$_2$]-OH (R = Pr or *i*Pr) have been used for the synthesis of Gly-Val-Tyr(P)-Ala-Ala-Ser-Gly, with the phosphoramidate functionality being converted to the phosphate group during the 95% TFA/ water (4 hr) cleavage step.[27]

[26] H.-G. Chao, M. S. Bernatowicz, P. D. Reiss, and G. R. Matsueda, *J. Org. Chem.* **59**, 6687 (1994).

[27] M. Ueki, J. Tachibana, Y. Ishii, J. Okumura, and M. Goto, *Tetrahedron Lett.* **37**, 4953 (1996).

from DIPCDI coupling[28] to BOP/1-hydroxybenzotriazole (HOBt)/diiso-propylethylamine (DIEA) coupling.[29] The generation of aberrant phospho-peptides obtained using the DCC- or DIPCDI-mediated coupling procedure is most likely attributable to activation of the phosphate group by the carbodiimide reagent and its subsequent participation in undesirable side reactions. The Fmoc derivative Fmoc-Tyr(PO$_3$H$_2$)-OH can be readily uti-lized in manual or automated peptide synthesizers using either phos-phonium- or uronium-based coupling reagents, namely, BOP, PyBOP, HBTU, N-[(1H-benzotriazol-1-yl)(dimethylamino)methylene]-N-methyl-methanaminium tetrafluoroborate N-oxide (TBTU), and TPTU, and it requires the use of additional base (NMM or DIEA) to ensure neutraliza-tion of the phosphate group. Although BOP/HOBt/NMM is found to be more suitable than either TPTU or N-[(dimethylamino)-1H-1,2,3-tria-zolo[4,5-b]pyridin-1-ylmethylene]-N-methylmethanaminium   hexafluoro-phosphate N-oxide (HATU) for the coupling of this derivative to a test peptide–resin,[30] the slow coupling of this anionic amino acid derivative necessitates the use of longer reaction times and double couplings.

Despite the simplicity of employing Fmoc-Tyr(PO$_3$H$_2$)-OH in Fmoc solid-phase peptide synthesis, several studies have shown that the coupling of Fmoc-Tyr(PO$_3$H$_2$)-OH is not as straightforward as initially demon-strated[14,29] and that some particular Tyr(P)-peptide syntheses are even prone to total failure. With the use of PyBOP, the observation of aberrant high molecular weight Tyr(P)-peptides and diphosphate-bridged Tyr(P)-peptide dimers [2M − 16]$^+$ indicates that this reagent may cause undesired activation of the phosphate group during the coupling step.[31,31a] Activation of the phosphate group by PyBOP is consistent with a past study in which the formation of a reactive phosphate–HOBt intermediate was proposed on the basis that a C-alkylphosphoric acid underwent rapid PyBOP-mediated mono- and diesterification in the presence of excess alcohol.[32]

Phosphate activation by HBTU does not seem to be as problematic, judging from the fact that the synthesis of the three insulin receptor pep-tides Glu-Ile-Tyr-Glu-Thr-Asp-Tyr-Tyr(P)-Ala, Glu-Ile-Tyr-Glu-Thr-Asp-Tyr(P)-Tyr-Ala, and Glu-Ile-Tyr(P)-Glu-Thr-Asp-Tyr-Tyr-Ala all pro-ceeded in high yield and with minimal by-product formation.[25] However,

[28] G. Zardeneta, D. Chen, S. T. Weintraub, and R. J. Klebe, *Anal. Biochem.* **190**, 347 (1990).
[29] E. A. Ottinger, L. L. Shekels, D. A. Bernlohr, and G. Barany, *Biochemistry* **32**, 4354 (1993).
[30] C. García-Echeverría, *Lett. Pept. Sci.* **2**, 369 (1995).
[31] J. W. Perich, T. Johnson, and H.-C. Cheng, submitted for publication.
[31a] C. García-Echeverría, *Lett. Pept. Sci.* **2**, 93 (1995).
[32] J. M. Campagne, J. Coste, L. Guillou, A. Heitz, and P. Jouin, *Tetrahedron Lett.* **34**, 4181 (1993).

the HBTU-mediated coupling of contiguous $Tyr(PO_3H_2)$ residues is not straightforward, as the synthesis of the triply phosphorylated insulin receptor sequence Glu-Ile-Tyr(P)-Glu-Thr-Asp-Tyr(P)-Tyr(P)-Ala is contaminated with 10% Glu-Ile-Tyr(P)-Glu-Thr-Asp-Tyr-Tyr(P)-Ala. A further limitation in the use of $Fmoc-Tyr(PO_3H_2)$-OH in solid-phase peptide synthesis is that the preparation of particular Tyr(P)-peptides have inexplicably failed irrespective of the coupling reagent used. For example, the syntheses of the lyn C-terminal peptide Asp-Asp-Phe-Phe-Thr-Ala-Thr-Glu-Gly-Gln-Tyr(P)-Gln-Gln-Gln-Pro (PyBOP coupling),[24] the insulin receptor Glu-Ile-Tyr(P)-Glu-Thr-Asp-Tyr(P)-Tyr(P)-Ala (PyBOP coupling),[33] and the RAF-1 peptide sequence Lys-Asn-Lys-Ile-Arg-Pro-Arg-Gly-Gln-Arg-Asp-Ser-Ser-Tyr(P)-Tyr(P)-Trp-Glu-Ile-Glu-Ala-Ser-Glu-Val (HBTU coupling)[33] all give rise to a multitude of aberrant peptides with no product corresponding to the target Tyr(P)-peptide.

Summary

Although $Fmoc-Tyr(PO_3H_2)$-OH was initially favored for the synthesis of Tyr(P)-peptides because of its rapid commercialization, the more recent commercial availability of both $Fmoc-Tyr(PO_3tBu_2)$-OH and Fmoc-Tyr-(PO_3Dmpse_2)-OH has made these latter derivatives the reagents of choice for use in peptide synthesis. In contrast to $Fmoc-Tyr(PO_3H_2)$-OH, these derivatives are more readily incorporated into peptide assembly without steric complications or side reactions, and they afford improved reliability and peptide efficiency. In addition, these derivatives are amenable to contiguous couplings, and multiple derivatives can be incorporated as desired for the generation of complex multiphosphorylated peptide sequences.

Synthesis of Phosphotyrosine-Peptide Mimetics Using Protected Fmoc-Phe($CH_2PO_3R_2$)-OH Derivatives

In view of the increased interest in peptide mimetics, attention has been directed to the chemical synthesis of the $CH_2PO_3H_2$, $CHFPO_3H_2$, and $CF_2PO_3H_2$ analogs of the *O*-phosphotyrosyl residue (see Fig. 2). A key feature of these peptide mimetics is that the carbon–phosphorus linkage is inert to phosphatase action and thereby provides a range of nonhydrolyzable substrates for the study of signal transduction processes. Since 1990, the synthesis of several Phe($CH_2PO_3H_2$)- and ($CHFPO_3H_2$)-peptides has

---

[33] O. Marin and J. W. Perich, unpublished data.

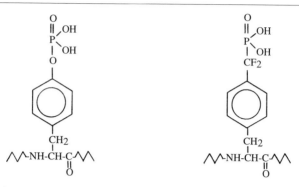

FIG. 2. Structures of the Tyr(PO$_3$H$_2$) residue and its Phe(CF$_2$PO$_3$H$_2$) mimetic.

been reported by the use of protected Fmoc-Phe(CH$_2$PO$_3$R$_2$)-OH (R = *t*Bu, Me, Et)[34-36] and Fmoc-DL-Phe(CHFPO$_3$*t*Bu$_2$)-OH[37] in Fmoc solid-phase peptide synthesis using routine peptide assembly and deprotection methods (see Table II).[35-41] However, syntheses of Phe(CF$_2$PO$_3$H$_2$)-peptide analogs are now favored because the pK_a of the CF$_2$PO$_3$H$_2$ group (~5.71) is closer to that of the *O*-phosphotyrosyl residue (~6.22)[42] and because the fluorine atoms have been found to promote increased hydrogen bonding. The better performance of the Phe(CF$_2$PO$_3$H$_2$) residue has been borne out in three biochemical studies in which the (CF$_2$PO$_3$H$_2$)-peptides.[40,41,43] clearly outperformed their corresponding Phe(CH$_2$PO$_3$H$_2$)- and (CHFPO$_3$H$_2$)-peptides.

[34] T. R. Burke, Jr., P. Russ, and B. Lim, *Synthesis*, 1019 (1991).

[35] M. Cushman and E.-S. Lee, *Tetrahedron Lett.* **33**, 1193 (1992).

[36] O. M. Green, *Tetrahedron Lett.* **35**, 8081 (1994).

[37] T. R. Burke, Jr., M. S. Smyth, M. Nomizu, A. Otaka, and P. P. Roller, *J. Org. Chem.* **58**, 1336 (1993).

[38] S. E. Shoelson, S. Chatterjee, M. Chaudhuri, and T. R. Burke, Jr., *Tetrahedron Lett.* **32**, 6061 (1991).

[39] M. Nomizu, A. Otaka, T. R. Burke, Jr., and P. P. Roller, *Tetrahedron* **50**, 2691 (1994).

[40] T. R. Burke, Jr., M. S. Smyth, A. Otaka, M. Nomizu, and P. P. Roller, *Biochemistry* **33**, 6490 (1994).

[41] T. Gilmer, M. Rodriguez, S. Jordan, R. Crosby, K. Alligood, M. Green, M. Klimery, C. Wagner, D. Kinder, P. Charifson, A. M. Hassell, D. Willard, M. Luther, D. Rusnak, D. D. Sternbach, M. Mehrotra, M. Peel, L. Shampine, R. Davis, J. Robbins, I. R. Patel, D. Kassel, W. Burkhart, M. Moyer, T. Bradshaw, and J. Berman, *J. Biol. Chem.* **269**, 31711 (1994).

[42] M. S. Smyth, H. Ford, Jr., and T. R. Burke, Jr., *Tetrahedron Lett.* **33**, 4137 (1992).

[43] T. R. Burke, Jr., H. K. Kole, and P. P. Roller, *Biochem. Biophys. Res. Commun.* **204**, 129 (1994).

TABLE II

PHOSPHOTYROSINE-PEPTIDE MIMETICS PREPARED USING N-PROTECTED DERIVATIVES[a]

Peptide	Ref.	Amino acid	Ref.	Acidolytic conditions for phosphate group cleavage
D- and L-RDIXETDYYRK	38	Fmoc-DL-Phe(CH$_2$PO$_3$tBu$_2$)-OH	34	TFA
D- and L-RENEXMPMAPEIH	38	Fmoc-DL-Phe(CH$_2$PO$_3$tBu$_2$)-OH	34	TFA
D- and L-GXVPML	37	Fmoc-DL-Phe(CH$_2$PO$_3$tBu$_2$)-OH	34	TFA
D- and L-cyclo(GPXVPML	39	Fmoc-DL-Phe(CH$_2$PO$_3$tBu$_2$)-OH	34	TFA
DL-AcDADEXL-NH$_2$	40	Fmoc-DL-Phe(CH$_2$PO$_3$tBu$_2$)-OH	34	TFA
DRVXIHPFHL	35	Boc-L-Phe(CH$_2$PO$_3$Me$_2$)-OH	35	TFMSA/TFA/DMS/m-cresol
Ac-DGVXTGLSTRQGETXETLK-NH$_2$	36	Fmoc-L-Phe(CH$_2$PO$_3$Et$_2$)-OH	36	TMSI/MeCN
AcXEEIE	41	Unknown		Unknown
D- and L-GYVPML	37	Fmoc-DL-Phe(CHFPO$_3$tBu$_2$)-OH	37	TFA
AcYEEIE	41	Unknown		Unknown
D- and L-GZVPML	44,49	Boc-DL-Phe(CF$_2$PO$_3$Et$_2$)-OH	44	TMSOTf/DMS/TFA/EDT/m-cresol
D- and L-Ac-DZIIPL-NH$_2$	49	Boc-DL-Phe(CF$_2$PO$_3$Et$_2$)-OH	44	TMSOTf/DMS/TFA/EDT/m-cresol
D- and L-GZVPML	44,49	Fmoc-DL-Phe(CF$_2$PO$_3$Et$_2$)-OH	44	TMSOTf/DMS/TFA/EDT/m-cresol
D- and L-Ac-NZIDLD-NH$_2$	49	Fmoc-DL-Phe(CF$_2$PO$_3$Et$_2$)-OH	44	TMSOTf/DMS/TFA/EDT/m-cresol
Ac-NZVNIE-OH	40,49	Fmoc-DL-Phe(CF$_2$PO$_3$Et$_2$)-OH	44	TMSOTf/DMS/TFA/EDT/m-cresol
Ac-DADEZL-NH$_2$	43	Fmoc-L-Phe(CF$_2$PO$_3$Et$_2$)-OH	46–48	TMSOTf/DMS/TFA/EDT/m-cresol
Ac-NZVNIE-NH$_2$	40	Fmoc-L-Phe(CF$_2$PO$_3$Et$_2$)-OH	46–48	TMSOTf/DMS/TFA/EDT/m-cresol
Ac-QZEEIP-NH$_2$	40	Fmoc-L-Phe(CF$_2$PO$_3$Et$_2$)-OH	46–48	TMSOTf/DMS/TFA/EDT/m-cresol
Ac-DZVPML-NH$_2$	40	Fmoc-L-Phe(CF$_2$PO$_3$Et$_2$)-OH	46–48	TMSOTf/DMS/TFA/EDT/m-cresol
Ac-ZINQ-NH$_2$	46	Fmoc-L-Phe(CF$_2$PO$_3$H$_2$)-OH	46,47	Not required
Ac-EZINQ-NH$_2$	46	Fmoc-L-Phe(CF$_2$PO$_3$H$_2$)-OH	46,47	Not required
AcZEEIE	41	Unknown		Unknown

[a] X, Phe(CH$_2$PO$_3$H$_2$); Y, Phe(CHFPO$_3$H$_2$); Z, Phe(CF$_2$PO$_3$H$_2$).

The synthesis of several $Phe(CF_2PO_3Et_2)$-peptides has, to date, been accomplished by the use of either $Fmoc-Phe(CF_2PO_3Et_2)-OH^{44-48}$ or $Fmoc-Phe(CF_2PO_3H_2)-OH^{46,47}$ in Fmoc solid-phase peptide synthesis. In their studies, Burke and colleagues prepared DL-Gly-$Phe(CF_2PO_3H_2)$-Val-Pro-Met-Leu by the use of $Fmoc-DL-Phe(CF_2PO_3Et_2)-OH$ in Fmoc solid-phase synthesis (DIPCDI/HOBt couplings) followed by cleavage of the ethyl phosphonate groups from the $Phe(CF_2PO_3Et_2)$-peptide by TMSOTf/DMS/ TFA/1,2-ethanedithiol (EDT)/m-cresol treatment.[44,49] However, as Gordeev et al.[46] encountered difficulties in the final cleavage of ethyl phosphonate groups, this group converted $Fmoc-Phe(CF_2PO_3Et_2)-OH$ to $Fmoc-Phe(CF_2PO_3H_2)-OH$ by BSTFA/TMSI treatment and demonstrated the use of this derivative for the high yielding Fmoc solid-phase synthesis (HBTU/HOBt/DIEA couplings) of $Ac-Glu-Phe(CF_2PO_3H_2)$-Ile-Asn-Gln-NH_2.

In a comparative study using $Fmoc-Phe(CF_2PO_3Et_2)-OH$ and $Fmoc-Phe(CF_2PO_3H_2)-OH$ for the Fmoc solid-phase synthesis [$Phe(CF_2PO_3H_2)$]^4-angiotensin II, Perich[50] found that the latter derivative provided the more efficient synthesis of the target $Phe(CF_2PO_3H_2)$-peptide. The use of $Fmoc-Phe(CF_2PO_3H_2)-OH$ in peptide synthesis is particularly straightforward because this derivative is now commercially available (AminoTech) and does not require the use of a complex deprotection step. Although peptide synthesis using $Fmoc-Phe(CF_2PO_3Et_2)-OH$ is complicated due to the slow acidolytic cleavage of ethyl phosphate groups, it is expected that the future availability of either $Fmoc-Phe(CF_2PO_3Me_2)-OH$ or $Fmoc-Phe-(CF_2PO_3Bzl_2)-OH$ should provide improved synthetic efficiency for the preparation of this interesting class of peptide mimetics and also permit the ready synthesis of multiphosphorylated peptides.

Solid-Phase Synthesis of Phosphoserine- and Phosphothreonine-Peptides

The synthesis of phosphoserine [Ser(P)]- and phosphothreonine [Thr(P)]-containing peptides using phosphorylated amino acid derivatives

---

[44] T. R. Burke, Jr., M. S. Smyth, A. Otaka, and P. P. Roller, *Tetrahedron Lett.* **34,** 4125 (1993).
[45] J. Wrobel and A. Dietrich, *Tetrahedron Lett.* **34,** 3543 (1993).
[46] M. F. Gordeev, D. V. Patel, P. L. Barker, and E. M. Gordon, *Tetrahedron Lett.* **35,** 7585 (1994).
[47] D. Solas, R. L. Hale, and D. V. Patel, *J. Org. Chem.* **61,** 1537 (1996).
[48] M. S. Smyth and T. R. Burke, Jr., *Tetrahedron Lett.* **35,** 551 (1995).
[49] A. Otaka, T. R. Burke, Jr., M. S. Smyth, M. Nomizu, and P. P. Roller, *Tetrahedron Lett.* **34,** 7039 (1993).
[50] J. W. Perich, unpublished data.

has been accomplished by the use of (a) Boc-Ser(PO_3R_2)-OH derivatives (R = aryl, Me) in Boc solid-phase synthesis or (b) Fmoc-Ser(PO_3Bzl,H)-OH derivatives in Fmoc solid-phase synthesis.

Use of Boc-Ser(PO_3Ph_2)-OH

The synthesis of Ser(P)- and Thr(P)-peptides can be accomplished by the use of Boc-Ser(PO_3Ph_2)-OH and Boc-Thr(PO_3Ph_2)-OH (Chem-Impex Int.) in Boc solid-phase peptide synthesis using either Merrifield or PAM polystyrene resins as the polymer support.[51–55] The Boc-protected derivatives can be readily incorporated into peptide assembly using routine coupling reagents (DCC, PyBOP, or HBTU), and no major steric problems have been encountered with the coupling of the more sterically hindered Boc-Thr(PO_3Ph_2)-OH derivative. However, the HF deprotection of Ser-(PO_3Ph_2)-peptide and Thr(PO_3Ph_2)-peptide resins is complicated because both the Ser(PO_3Ph_2) and Thr(PO_3Ph_2) residues are subject to extensive HF-mediated dephosphorylation.[55] Nevertheless, HF deprotection can be employed on the proviso that low product yields can be tolerated and that the peptide sequence of the target peptide is relatively simple and contains a single phosphorylated residue.[53,55] In practice, HF treatments should be limited to 30 to 45 min, and a greater chance of a successful deprotection can be expected in the case of Thr(PO_3Ph_2) residues, as this residue undergoes dephosphorylation at a slower rate than the Ser(PO_3Ph_2) residue. After isolation of the Ser(PO_3Ph_2)- or Thr(PO_3Ph_2)-peptide by semipreparative high-performance liquid chromatography (HPLC), the phenyl groups are cleaved by hydrogenolysis (1 atm) using 1.1 equivalent of platinum oxide per phenyl group in TFA/acetic acid solution.[56] Because of the hydrogenation step, this synthetic approach is not suitable for the synthesis of phosphopeptides containing Phe, Tyr, Trp, His, Met, or Cys residues.

In addition to HF deprotection, high pressure hydrogenolysis (40 psi, 50°) in DMF solution has also been used as an alternative procedure for the cleavage of Ser(PO_3Ph_2)-peptides from their polymer supports.[54] However, this deprotection step has not been widely adopted by peptide chemists, owing to the need for specialized high pressure hydrogenation equipment

[51] J. W. Perich, R. M. Valerio, and R. B. Johns, *Tetrahedron Lett.* **27,** 1377 (1986).

[52] A. Arendt, K. Palczewski, W. T. Moore, R. M. Caprioli, J. H. McDowell, and P. A. Hargrave, *Int. J. Pept. Protein Res.* **33,** 468 (1989).

[53] D. W. Litchfield, A. Arendt, F. J. Lozeman, E. G. Krebs, P. A. Hargrave, and K. Palczewski, *FEBS Lett.* **261,** 117 (1990).

[54] J. W. Perich, R. M. Valerio, P. F. Alewood, and R. B. Johns, *Aust. J. Chem.* **44,** 771 (1991).

[55] J. W. Perich, E. Terzi, E. Carnazzi, R. Seyer, and E. Trifilieff, *Int. J. Pept. Protein Res.* **44,** 305 (1994).

[56] J. W. Perich, P. F. Alewood, and R. B. Johns, *Aust. J. Chem.* **44,** 233 (1991).

and chemical expertise. This problem has been somewhat simplified with the finding that hydrogenolytic peptide–resin cleavages can be accomplished using 1 atm of hydrogen under modified hydrogenation conditions. For example, the synthesis of Glu-Ser(P)-Leu-Ser-Ser-Ser-Glu-Glu is accomplished in moderate yield by the one-step hydrogenolytic cleavage (1 atm, 50°) of the Ser(PO$_3$Ph$_2$)-peptide–resin (Merrifield) using 10 equivalents of palladium acetate and 2 equivalents of platinum oxide/phenyl group in a minimum volume of TFA.[50] The only requirements of this deprotection procedure are a miniature glass hydrogenolysis vessel and a simple hydrogenation apparatus.

Use of Boc-Ser(PO$_3$Me$_2$)-OH

The Boc solid-phase synthesis of Ser(P)- and Thr(P)-peptides is also possible using alkyl phosphate protection. Although early work demonstrated the use of Boc-Ser(PO$_3$Et$_2$)-OH for the Boc solid-phase synthesis (DCC/HOBt coupling) of Glu-Ser(PO$_3$Et$_2$)-Leu and employed HBr/TFA for cleavage of the peptide from its polystyrene resin support,[53] the removal of ethyl phosphate groups using lengthy 45% HBr/acetic acid or TMSBr deprotection procedures caused extensive peptide degradation. Better results were obtained by Otaka *et al.*[10] by the use of Boc-Ser(PO$_3$Me$_2$)-OH[55] (Chem-Impex Int., Peninsula Laboratories) for the Boc solid-phase peptide syntheses (DIC/HOBt coupling method) of both Arg-Arg-Val-Ser(P)-Val-Ala-Ala-Glu and Lys-Arg-Thr(P)-Leu-Arg-Arg-Leu-Leu using a TFMSA-based deprotection procedure. This procedure involved initial liberation of the Ser(PO$_3$Me$_2$)-peptide from the peptide–resin by TFMSA/DMS/TFA treatment (1.5 hr), with the methyl phosphate groups being cleaved *in situ* by the further addition of TFMSA and DMS and continuing deprotection for a further 3 hr.[10]

Fmoc-Ser(PO$_3$Bzl,H)-OH

Although the use of protected Fmoc-Ser(PO$_3$R$_2$)-OH derivatives in Fmoc solid-phase synthesis had been precluded due to piperidine-mediated β-elimination, this situation changed in 1994 with the demonstration that Fmoc-Ser(PO$_3$Bzl,H)-OH (Novabiochem) can be incorporated into peptide synthesis and that base-mediated β-elimination of the Ser(PO$_3$Bzl,H) residue is suppressed by ionization of the phosphorodiester functionality during subsequent piperidine treatments.[57] In a coupling study, HBTU has been found to be the more efficient of the phosphonium and uronium coupling reagents for the incorporation of this derivative and subsequent peptide

---

[57] T. Wakamiya, K. Saruta, J. Yasuoka, and S. Kusumoto, *Chem. Lett.,* 1099 (1994).

assembly.[58] However, some care has to be exercised with the routine use of this derivative because the incorporation of the Fmoc-Ser(PO$_3$Bzl,H)-OH can often be incomplete; Johnson et al.[59] found that coupling of this phosphorodiester derivative was only 85% complete after two BOP/HOBt/DIEA coupling steps.

Synthesis of Phosphoserine- and Phosphothreonine-Peptide Mimetics

Attention has been directed to the synthesis of peptide mimetics of the Ser(P) and Thr(P) residue. The CH$_2$isostere of Fmoc-Ser(PO$_3$Me$_2$)-OH, Fmoc-Abu(PO$_3$Me$_2$)-OH, is readily prepared from Boc-Asp-OtBu by a seven-step procedure[60] and can be incorporated into Fmoc solid-phase synthesis using PyBOP or HBTU coupling reagents. The Abu(PO$_3$Me$_2$)-peptide is readily cleaved from its resin support using 95% TFA, and final removal of the methyl phosphate groups is effected by TMSBr/thioanisole/TFA treatment. This procedure is used for the synthesis of the kemptide analog Leu-Arg-Arg-Ala-Abu(P)-Leu-Gly and the αs_1-related peptide Ile-Val-Pro-Asn-Abu(P)-Val-Glu-Glu.[61] Alternatively, Shapiro et al. also prepared Fmoc-Abu(PO$_3$All$_2$)-OH[62] from Schöllkopf's bislactam ether and used this derivative for the Fmoc solid-phase synthesis (DIPCDI/HOBt) of 16 neuromodulin-related Abu(P)-peptides (hexa- to nonapeptides)[63] and a calcineurin-related nonadecapeptide; the allyl phosphate groups are cleaved by Pd(PPh$_3$)$_4$ treatment on the peptide–resin prior to TFA cleavage of Abu(PO$_3$H$_2$)-peptide from the resin support. In addition, synthetic procedures have also been devised for the synthesis of both the CF$_2$-amino acid derivatives Boc-Ala(CF$_2$PO$_3$Et$_2$)-OH[64,65] and Boc-Gly[CH(CH$_3$)CF$_2$-PO$_3$Et$_2$]-OH.[65] The Ser(P) residue mimetic Boc-Ala(CF$_2$PO$_3$Et$_2$)-OH has been used for the solid-phase synthesis (Merrifield resin, DIPCDI/HOBt couplings) of Arg-Arg-Val-Ala(Cf$_2$PO$_3$H$_2$)-Val-Ala-Ala-Glu[64] with peptide deprotection being effected by a two-step method which involves initial treatment with TMSOTf/thioanisole/TFA/EDT/m-cresol (2 hr) followed

[58] S. N. Eagle, N. J. Ede, W. R. Sampson, B. Gubbins, B. M. Krywult, C. A. Pham, F. Erciyas, K. M. Stewart, K. H. Ang, I. A. James, and A. M. Bray, unpublished data.
[59] T. Johnson, L. C. Packman, C. B. Hyde, D. Owen, and M. Quibell, J. Chem. Soc., Perkin Trans. 1, 719 (1996).
[60] J. W. Perich, Synlett, 595 (1992).
[61] J. W. Perich, Int. J. Pept. Protein Res. 44, 288 (1994).
[62] G. Shapiro, D. Buechler, V. Ojea, E. Pombo-Villar, M. Ruiz, and H.-P. Weber, Tetrahedron Lett. 34, 6255 (1993).
[63] G. Shapiro, D. Buechler, A. Enz, and H.-P. Weber, Tetrahedron Lett. 35, 1173 (1994).
[64] A. Otaka, K. Miyoshi, T. R. Burke, Jr., P. P. Roller, H. Kubota, H. Tamamura, and N. Fujii, Tetrahedron Lett. 36, 927 (1995).
[65] D. B. Berkowitz, M. Eggen, Q. Shen, and R. K. Shoemaker, J. Org. Chem. 61, 4666 (1966).

by *in situ* addition of DMS (2 hr) to facilitate cleavage of the ethyl phosphate groups. Although these derivatives should provide an avenue to the preparation of some interesting Ser(P)- and Thr(P)-peptide mimetics, it is clear that peptide synthesis efficiency will be increased by the use of more labile phosphate protecting groups requiring milder acidolytic cleavage conditions.

Synthesis of Phosphopeptides Using Global Phosphorylation Approach

The development of the global or postassembly phosphorylation approach for the synthesis of phosphopeptides arose from the need of having a simple chemical procedure for the rapid preparation of Ser(P), Thr(P)-, and Tyr(P)-containing peptides. This method involves the initial Fmoc solid-phase synthesis of a Xxx-containing peptide–resin (Xxx = Ser, Thr, Tyr) by the incorporation of Fmoc-Xxx-OH using phosphonium (BOP, PyBOP) or uronium reagents (HBTU, TBTU) followed by its postassembly phosphite triester phosphorylation using various dialkyl *N,N*-diethylphosphoramidites.[66] Two key features of this approach are that, first, the coupling of side-chain Fmoc-Xxx-OH using the above reagents can be accomplished with little or no side-chain acylation and, second, $(RO)_2PNR'_2$ reagents are highly reactive reagents for effecting phosphitylation of resin-bound Ser, Thr, or Tyr residues.

This phosphorylation approach has been used for the synthesis of a variety of phosphopeptides[11,20,66–76] and has found particular application for the synthesis of Tyr(P)-peptides owing to the readier phosphitylation of the tyrosyl hydroxyl group. The commercial availability of both $(tBuO)_2PNR'_2$ and $(BzIO)_2PNR'_2$ (R' = Et, *i*Pr) has overcome the difficult chemical aspects of this approach and has made this phosphorylation method within the scope of most peptide chemists. The Tmse reagent

---

[66] J. W. Perich and R. B. Johns, *Tetrahedron Lett.* **29,** 2369 (1988).
[67] H. B. A. de Bont, J. H. van Boom, and R. M. J. Liskamp, *Tetrahedron Lett.* **31,** 2497 (1990).
[68] J. W. Perich, D. Le Nguyen, and E. C. Reynolds, *Tetrahedron Lett.* **32,** 4033 (1991).
[69] G. Stærkær, M. H. Jakobsen, C. E. Olsen, and A. Holm, *Tetrahedron Lett.* **32,** 5389 (1991).
[70] J. W. Perich, *Int. J. Pept. Protein Res.* **40,** 134 (1992).
[71] E. Larsson and B. Luning, *Tetrahedron Lett.* **35,** 2737 (1994).
[72] G. Shapiro, R. Swoboda, and U. Stauss, *Tetrahedron Lett.* **35,** 869 (1994).
[73] M. Rodriguez, R. Crosby, K. Alligood, T. Gilmer, and J. Berman, *Lett. Pept. Sci.* **2,** 1 (1995).
[74] R. Hoffman, W. O. Wachs, R. G. Berger, H. R. Kalbitzer, D. Waidelich, E. Bayer, W. Wagner-Redeker, and M. Zeppezauer, *Int. J. Pept. Protein Res.* **45,** 26 (1995).
[75] L. Poteur and E. Trifilieff, *Lett. Pept. Sci.* **2,** 271 (1995).
[76] J. W. Perich, F. Meggio, and L. A. Pinna, *Bioorg. Med. Chem. Lett.* **4,** 143 (1996).

(TmseO)$_2$PNR$'_2$ can be readily prepared by the treatment of Cl$_2$PNR$'_2$ (Aldrich) with excess 2,2,2-trimethylsilylethanol in the presence of triethyla-mine.[55,67] However, as distillative isolation of (TmseO)$_2$PNR$'_2$ is compli-cated owing to extensive disproportionation of this reagent under moderate temperature,[56] the reagent is best purified by silica gel chromatography,[67] with its purity established by ^{31}P nuclear magnetic resonance (NMR) spec-trometry prior to use.

In practice, the phosphorylation procedure requires thorough vacuum drying of the peptide–resin so as to minimize unwanted consumption of the phosphoramidite reagent by any resin-absorbed water. In addition, 1H-tetrazole should be vacuum dried prior to each use, and the phosphoramid-ite reagent should be regularly checked by ^{31}P NMR spectroscopy to ascer-tain the level of oxidation on prolonged storage. Commercially available anhydrous DMF (Aldrich) has proved to be a satisfactory solvent for direct use in phosphitylation reactions and obviates the tedious distillation and drying procedures required for the purification of lower quality grades of this solvent.

In the use of the global phosphorylation method, the phosphitylation of simple, noncomplex Tyr- and Ser-peptide–resins can generally be accom-plished using 10 equivalents of (tBuO)$_2$PNR$'_2$, (BzlO)$_2$PNR$'_2$, or (TmseO)$_2$PNR$'_2$ under dilute reaction conditions [approximately 0.2 M (RO)$_2$PNR$'_2$ in DMF] followed by MCPBA, tBuOOH, or aqueous iodine/ THF oxidation of the phosphite triester intermediate. However, in the case of peptide sequences containing Cys or Met residues, the use of tBuOOH is preferred so as to minimize oxidation of these residues. Because the rate of phosphitylation of peptide–resins is markedly affected by steric factors, the reaction time can vary from one to several hours and is best ascertained by trial phosphorylation reactions using 10–50 equivalents of phosphora-midite reagent per hydroxyl group.

In the phosphorylation of larger and more complex Xxx-containing peptide–resins in which there is increased steric hindrance at the target site (particularly when Xxx is Thr), improved phosphitylative conversions can be effected using 20–50 equivalents of (RO)$_2$PNR$'_2$ in a minimal volume of reaction solvent [generally dimethylformamide (DMF)]. However, a complication in the use of both (tBuO)$_2$PNR$'_2$ and (TmseO)$_2$PNR$'_2$ under these modified reaction conditions is that the resultant high molar concen-trations of 1H-tetrazole causes acid-mediated conversion of both the acid sensitive di-*tert*-butyl and bis-2,2,2-trimethylsilylethyl phosphite triester in-termediates to the alkyl H-phosphonate derivative. In these cases, aqueous iodine/THF must be used in the oxidation step as this treatment, unlike tBuOOH, also effects oxidative conversion of the undesired alkyl H-phos-phonate to the alkyl phosphorodiester.[56]

Phosphitylations using $(BzlO)_2PNR'_2$ are far more straightforward because the dibenzyl phosphite triester intermediate is not prone to conversion to the H-phosphonate in the presence of high solution concentrations of $1H$-tetrazole (150 mg/ml). Although aqueous iodine/THF, MCPBA, and tBuOOH can all be used for the oxidation step in these phosphorylation reactions, MCPBA is unsuitable for the oxidation of peptide sequences containing Cys and/or Met residues. The final acidolytic deprotection of $Xxx(PO_3R_2)$-containing peptide–resins (R = tBu, Bzl, Tmse) is readily accomplished by treatment with TFA in the presence of appropriate scavengers for 1 to 2 hr (see Novabiochem peptide synthesis handbook for details, 1995/1996 catalog). However, for the deprotection of $Ser(PO_3Bzl_2)$- and $Thr(PO_3Bzl_2)$-peptide–resins, the slower cleavage of the benzyl phosphate requires the use of longer deprotection times (up to 4 hr).

In the event where global phosphorylation of complex peptide–resins is hampered by peptide aggregation or excessive steric hindrance in lengthy sequences, two novel approaches have been developed to improve phosphorylation of target sites. In the first situation, Johnson et al.[59] employed 2-hydroxy-4-methoxybenzy (Hmb)-backbone protection at Leu-182 of MAP kinase ERK2(178–188) to disrupt peptide aggregation effects and facilitate the sequential phosphorylation of Thr-183 and Tyr-185 using $(BzlO)_2PNiPr_2$ and aqueous iodine/THF. To overcome excessive steric problems encountered in long peptide sequences, Mostafavi et al.[77] prepared the 32-mer peptide sequence from phospho-urodilatin by the use of the global phosphorylation approach for the generation of the protected 14-mer Ser-(PO_3tBu_2)-containing segment $[(tBuO)_2PNEt_2$ and tBuOOH] and its subsequent TBTU segment coupling (solution) to a protected 18-mer segment. Although the global phosphorylation approach was initially considered to have limited use, the two latter examples clearly illustrate that the future application of modern synthetic strategies to this approach will provide novel and efficient routes for the synthesis of larger and more complex phosphorylated peptides.

Synthetic Procedure for Global Phosphorylation

Dry $1H$-tetrazole (0.316 g, 4.5 mmol) is added to a suspension of vacuum-dried peptide–resin (0.075 mmol) and dibenzyl N,N-diethylphosphoramidite (0.476 g, 1.50 mmol) in anhydrous DMF (1.0 ml) and the vessel rotated for 3 to 24 hr. The solution is removed by vacuum suction, the resin is washed with DCM (three times, 3 ml), and a solution of (a) 14% *tert-*

[77] H. Mostafavi, S. Austermann, W.-G. Forssmann, and K. Adermann, *Int. J. Pept. Protein Res.* **48**, 200 (1996).

butyl hydroperoxide (in either THF, CH_3CN, or DMF) or (b) 85% m-chloroperoxybenzoic acid (0.129 g, 0.75 mmol) in DCM (3 ml) is added. The reaction vessel is rotated for 1 hr and the solvent removed by vacuum suction. The peptide–resin is washed with DCM (five times, 3 ml), dried under high vacuum, and then deprotected with the appropriate TFA/scavenger cocktail solution according to the amino acyl constituents of the peptide sequence. The solvent is then evaporated under reduced pressure and the phosphorylated peptide precipitated by the addition of diethyl ether.

Synthesis of Phosphotyrosine-Peptides via "On-line" Phosphorylation

The "on-line" phosphorylation approach arose from the realization that improved Tyr-phosphorylation would be effected if phosphite triester phosphorylation is performed after the immediate incorporation of Fmoc-Tyr-OH to the peptide chain rather than at the completion of peptide assembly. A feature of this approach is that the Tyr-residue is more accessible for phosphitylation by the phosphoramidite reagent $[(RO)_2PNEt_2,$ R = tBu, Bzl, Tmse] because it is not hindered by the steric constraints of a potentially bulky N-terminal peptide sequence. However, as with the global phosphorylation approach, the phosphorylation of the Fmoc-Tyr-peptide–resin with $(tBuO)_2PNEt_2$ or $(TmseO)_2PNEt_2$ must be performed under nonconcentrated conditions so as to minimize $1H$-tetrazole-mediated conversion of the generated dialkyl phosphite triester to its alkyl H-phosphonate.

The utility of the on-line phosphorylation approach has been demonstrated[78,79] by the use of $(tBuO)_2PNEt_2$ for the high yielding preparation of both the Fcγ-peptide Glu-Ala-Glu-Asn-Thr-Ile-Thr-Tyr(P)-Ser-Leu-Leu-Lys-His-Pro-Glu-Ala-Leu and the Stat 91 peptide Gly-Pro-Lys-Gly-Thr-Gly-Tyr(P)-Ile-Lys-Thr-Glu-Leu-Ile-Ser. In these two syntheses, $(tBuO)_2$ $PNEt_2$ is selected as the phosphoramidite reagent on the basis that it provides the *in situ* preparation of the Fmoc-Tyr(PO_3tBu_2) residue and that the di-*tert*-butyl phosphorotriester, unlike the dibenzyl or bis-2,2,2-trimethylsilylethyl phosphorotriester, is inert to piperidine-mediated dealkylation during repetitive cleavage of the Fmoc group. In general peptide synthesis, a particular feature of this approach is that it is suitable for the synthesis of Tyr(P)-peptides in which the Trp and the oxidation-prone Met and Cys residues are located N terminal to the Tyr(P)-residue and thereby avoid any contact with oxidation reagents.

[78] J. W. Perich, *Lett. Pept. Sci.* **3**, 127 (1996).
[79] W. D. F. Meutermans and P. F. Alewood, *Tetrahedron Lett.* **37**, 4765 (1996).

In the event of sterically hindered sequences, phosphitylations are best performed by the use of $(BzlO)_2PNR'_2$ under concentrated reaction conditions, with the level of phosphorylation being established from trial deprotections prior to continuance of peptide assemblage. Although the benzyl phosphate group is subject to piperidine-mediated cleavage,[18] this side reaction can be minimized by changing to 1 M 1,8-diazabicyclo[5.4.0]undec-7-ene (DBU)/DMF for Fmoc cleavage. With respect to peptide efficiency, preliminary studies have indicated that the continual emergence of the phosphorodiester during peptide assemblage is not particularly detrimental to the final peptide purity.[18]

Conclusion

Since the early 1980s, several synthetic approaches have been developed for the efficient preparation of phosphorylated peptides and may provide a confusing array of possibilities for biochemical researchers. In the case of Tyr(P)-peptides, these peptides are best prepared by the use of Fmoc-Tyr(PO$_3$tBu$_2$)-OH and Fmoc-Tyr(PO$_3$Bzl$_2$)-OH in Fmoc solid-phase peptide synthesis, whereas the synthesis of the corresponding Phe(CF$_2$PO$_3$H$_2$)-peptide is best accomplished using Fmoc-Phe(CF$_2$PO$_3$H$_2$)-OH. The solid-phase synthesis of Ser(P)- and Thr(P)-peptides has not been so straightforward because of the susceptibility of the phosphorylated residues to piperidine treatment. In these cases, preparation of the phosphorylated peptides is best done by the use of the global phosphorylation approach or by the incorporation of Fmoc-Ser(PO$_3$Bzl,H)-OH and Fmoc-Thr(PO$_3$-Bzl,H)-OH during peptide assembly. Nevertheless, it is expected that future work will resolve many current problems in this area and thereby provide efficient procedures for the synthesis of complex Ser(P)- and Thr(P)-peptides.

[13] Protein Synthesis by Chemical Ligation of Unprotected Peptides in Aqueous Solution

By Tom W. Muir, Philip E. Dawson, and Stephen B. H. Kent

Introduction

The field of synthetic protein chemistry has been reenergized by the emergence in 1992 of a novel strategy involving the direct chemical reaction in aqueous solution of large, unprotected peptide building

0076-6879/97 $25.00

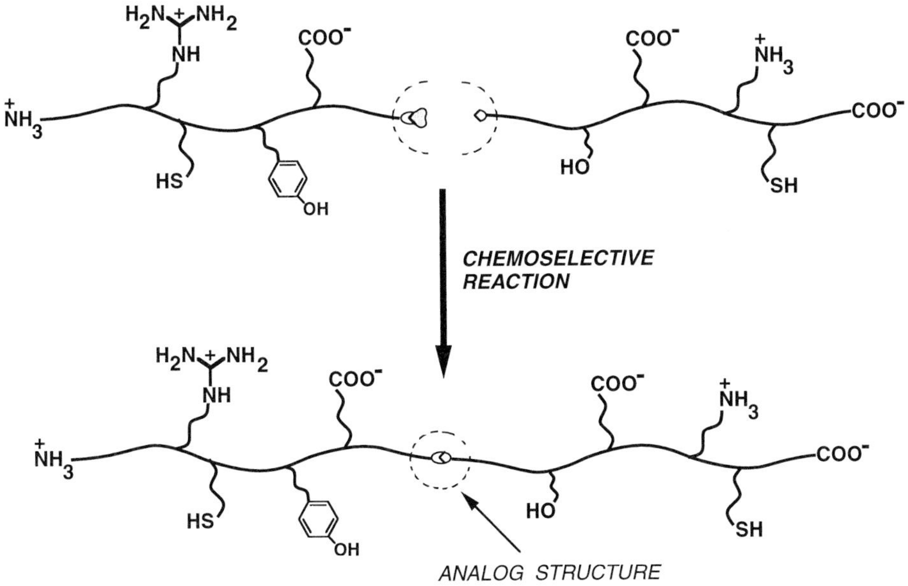

FIG. 1. Chemical ligation of unprotected peptide segments in aqueous solution. [See M. Schnölzer and S. B. H. Kent, *Science* **256,** 221 (1992).]

blocks.[1] This "chemical ligation" approach was initially based on the premise that an unnatural moiety can be used to covalently join two *completely unprotected peptides* to give a functional protein molecule. The approach relies on "chemoselective reaction" between synthetic peptide segments bearing unique, mutually reactive groups.[1] A number of different chemistries have been developed for this purpose,[1–5] all of which give rise to an unnatural covalent structure at the ligation site. The selectivity of these ligation reactions means that unprotected peptide segments (i.e., containing no protecting groups for any of the normally reactive functional groups found in peptides) can be used in chemical synthesis. Furthermore, the ligation reactions can be carried out in aqueous solution, and they are rapid and quantitative. Obviously, chemical ligation of unprotected peptide segments at once enables a dramatic increase in the size of synthetically

[1] M. Schnölzer and S. B. H. Kent, *Science* **256,** 221 (1992).
[2] A. Nefzi, X. Sun, and M. Mutter, *Tetrahedron Lett.* **36,** 229 (1995).
[3] K. Rose, *J. Am. Chem. Soc.* **116,** 30 (1994).
[4] C.-F. Liu and J. P. Tam, *Proc. Natl. Acad. Sci. U.S.A.* **91,** 6584 (1994).
[5] J. Shao and J. P. Tam, *J. Am. Chem. Soc.* **117,** 3893 (1995).

accessible molecules and permits the ready preparation of the large polypeptides found in proteins.

The principles underlying protein synthesis by chemical ligation of unprotected peptides in aqueous solution are shown in Fig. 1. The use of unprotected peptide segments has a number of advantages. Unprotected peptides are readily prepared in good yield and high purity using, for example, optimized stepwise solid-phase peptide synthesis (SPPS).[6] Because they are unprotected, the peptides are readily characterized by high resolution techniques such as analytical high-performance liquid chromatography (HPLC) and electrospray mass spectrometry (ESMS) and can be handled with ease using conventional means. Furthermore, the use of unprotected peptide segments overcomes the poor solubility, inadequate purification techniques, limited characterization methods, and other problems associated with the fully protected segments used in the classic convergent approach to the synthesis of large polypeptides/proteins in solution.[7,8] Because of this, the chemical ligation approach has transformed the practice of synthetic protein chemistry, making a wide range of systems synthetically accessible in a straightforward fashion.[9-15]

The first total protein synthesis by the chemical ligation of unprotected peptides in aqueous solution was of the human immunodeficiency virus type-1 protease (HIV-1 PR) (Fig. 2).[1] The 99-residue monomer polypeptide chain was obtained by thioester-forming chemical ligation of two approximately 50-residue unprotected peptide segments. The folded, homodimeric synthetic $[(COS)^{51-52,151-152}]$HIV-1 PR had enzymatic properties identical

[6] M. Schnölzer, P. Alewood, A. Jones, D. Alewood, and S. B. H. Kent, *Intl. J. Pept. Protein Res.* **40**, 180 (1992).

[7] H. Yajima and N. Fujii, *J. Am. Chem. Soc.* **103**, 5867 (1981).

[8] T. Kimura, M. Takai, T. Morikawa, and S. Sakakibara, *in* "Peptide Chemistry 1981" (T. Shioiri, ed.), p. 131. Protein Research Foundation, Osaka, Japan, 1981.

[9] H. F. Gaertner, K. Rose, R. Cotton, D. Timms, R. Camble, and R. E. Offord, *Bioconj. Chem.* **3**, 262 (1992).

[10] M. Baca, T. W. Muir, M. Schnölzer, and S. B. H. Kent, *J. Am. Chem. Soc.* **117**, 1881 (1995).

[11] L. E. Canne, A. R. Ferre-D'Amare, S. K. Burley, and S. B. H. Kent, *J. Am. Chem. Soc.* **117**, 2998 (1995).

[12] M. J. Williams, T. W. Muir, M. H. Ginsberg, and S. B. H. Kent, *J. Am. Chem. Soc.* **116**, 10797 (1994).

[13] M. Mutter, G. G. Tuchscherer, C. Miller, K.-H. Altmann, R. I. Carey, D. F. Wyss, A. M. Labardt, and J. E. Rivier, *J. Am. Chem. Soc.* **114**, 1463 (1992).

[14] P. E. Dawson and S. B. H. Kent, *J. Am. Chem. Soc.* **115**, 7263 (1993).

[15] T. W. Muir, M. J. Williams, M. H. Ginsberg, and S. B. H. Kent, *Biochemistry* **33**, 7701 (1994).

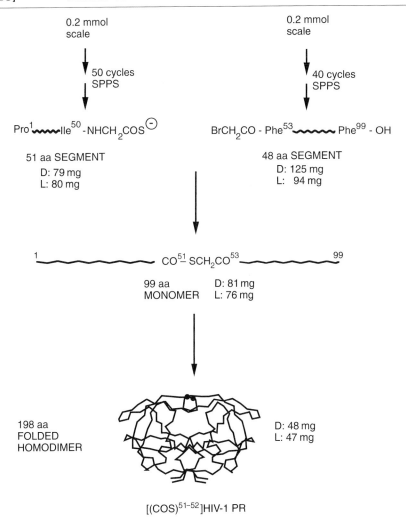

0.2 mmol
scale

0.2 mmol
scale

50 cycles
SPPS

40 cycles
SPPS

Pro^1〜〜Ile^{50}-NHCH$_2$COS$^{\ominus}$

BrCH$_2$CO - Phe^{53}〜〜 Phe^{99} - OH

51 aa SEGMENT
D: 79 mg
L: 80 mg

48 aa SEGMENT
D: 125 mg
L: 94 mg

1 〜〜〜 CO^{51}–SCH$_2$CO^{53} 〜〜〜 99

99 aa   D: 81 mg
MONOMER   L: 76 mg

198 aa
FOLDED
HOMODIMER

D: 48 mg
L: 47 mg

[(COS)$^{51-52}$]HIV-1 PR

FIG. 2. Total synthesis of HIV-1 protease by thioester-forming chemical ligation. The folded, homodimeric synthetic [(COS)$^{51-52,151-152}$]HIV-1 PR had enzymatic properties identical to native backbone HIV-1 PR. [See M. Schnölzer and S. B. H. Kent, *Science* **256**, 221 (1992).]

L-HIV-1 PROTEASE, NATIVE SEQUENCE
EXPRESSED IN *E. coli* [P. FITZGERALD, 1990]

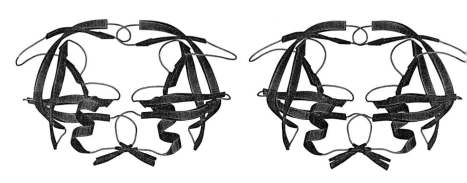

D-HIV-1 PROTEASE [SHOWN AS MIRROR IMAGE]
Aba^{67,95}, (COS)$^{51-52}$ SEQUENCE, CHEMICAL SYNTHESIS:
D-AA's, THIOESTER LIGATION [M. MILLER, 1993]

FIG. 3. Crystal structure of the mirror-image enzyme molecule D-[(COS)$^{51-52,151-152}$]HIV-1 PR, prepared by total synthesis using thioester-forming chemical ligation. The backbone fold of the synthetic enzyme is shown as a Molscript representation, emphasizing the secondary structural features (helix, β sheet). The D-PR enzyme molecule is shown after a mirror image transformation, for comparison with the L-PR enzyme molecule prepared by recombinant DNA-based bacterial expression. The two proteins have identical three-dimensional structures, within experimental uncertainty. [From M. Miller, M. Baca, J. K. M. Rao, and S. B. H. Kent, *J. Mol. Struct.* in press (1996).]

to native backbone HIV-1 PR.[16-21] Subsequently, synthetic access by chemical ligation was used to carry out unique so-called backbone engineering studies of HIV-1 PR that showed that backbone hydrogen bonds were

[16] The enzyme HIV-1 protease (HIV-1 PR) had previously been prepared by total chemical synthesis using stepwise solid-phase methods.[17] The HIV-1 PR prepared by chemical synthesis had the Cys-67 and Cys-95 residues in each monomer 99-residue polypeptide chain replaced by L-$^\alpha$NH$_2$-*n*-butyric acid (Aba) to facilitate handling, and it was used to determine the original crystal structures of the HIV-1 PR complexed with canonical classes of inhibitors.[18-20] These data formed the basis of the successful structure-based drug design programs that led to the highly effective protease inhibitor class of therapeutics for treatment of acquired immunodeficiency syndrome (AIDS).[21]

[17] J. Schneider and S. B. H. Kent, *Cell* (*Cambridge, Mass.*) **54,** 363 (1988).

[18] M. Miller, B. K. Sathyanarayana, M. V. Toth, G. R. Marshall, L. Clawson, L. Selk, J. Schneider, S. B. H. Kent, and A. Wlodawer, *Science* **246,** 1149 (1989).

[19] A. L. Swain, M. Miller, J. Green, D. H. Rich, J. Schneider, S. B. H. Kent, and A. Wlodawer, *Proc. Natl. Acad. Sci. U.S.A.* **87,** 8805 (1990).

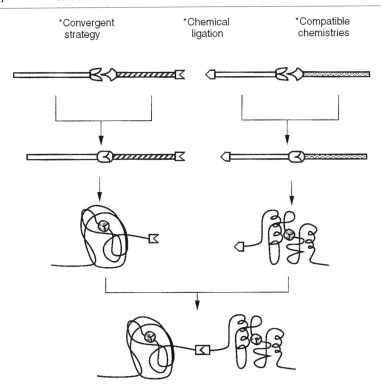

FIG. 4. Convergent strategy for chemical ligation of multiple unprotected peptide segments, to yield large synthetic polypeptide chains.

critical to the enzymatic activity of the protein.[22] Large amounts (~50 mg each) of the mirror image proteins L- and D-[(COS)$^{51-52,151-152}$]HIV-1 PR were prepared for crystallization and X-ray diffraction studies.[23] A high-resolution crystal structure of the chemically synthesized mirror image enzyme D-[(COS)$^{51-52,151-152}$]HIV-1 PR resulting from this work is shown in Fig. 3.

Chemical ligation of multiple unprotected peptide segments has also been employed in a convergent synthetic strategy, to prepare synthetic

[20] M. Jaskolski, A. G. Tomasselli, T. K. Sawyer, D. G. Staples, R. L. Heinrikson, J. Schneider, S. B. H. Kent, and A. Wlodawer, *Biochemistry* **30**, 1600 (1991).
[21] J. Cohen, *Science* **271**, 755 (1996).
[22] M. Baca and S. B. H. Kent, *Proc. Natl. Acad. Sci. U.S.A.* **90**, 11638 (1993).
[23] R. deLisle Milton, S. Milton, and S. B. H. Kent, *in* "Techniques in Protein Chemistry IV" (R. H. Angeletti, ed.), p. 257. Academic Press, New York, 1993.

a

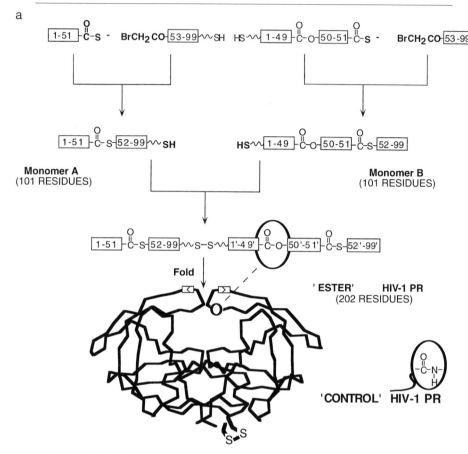

FIG. 5. (a) Total synthesis of a "covalent dimer" form of the HIV-1 PR by chemical ligation. Thioester-forming ligation was used to prepare each 101-residue monomer; the two polypeptides were joined by means of direct disulfide formation using noncoded thiol structures at the end of each chain, as shown. (b) The resulting synthetic protein (~22 kDa) had full enzymatic activity. [See M. Baca, T. W. Muir, M. Schnölzer, and S. B. H. Kent, *J. Am. Chem. Soc.* **117**, 1881 (1995).]

proteins with molecular masses exceeding 20 kDa.[10,11] The convergent chemical ligation strategy is schematically illustrated in Fig. 4. This approach was used to prepare a covalent dimer form of HIV-1 PR; this synthetic protein of molecular mass of 21,786 Da contained multiple noncoded covalent structures and had full enzymatic activity (Fig. 5).[10] The same synthetic strategy was used to prepare a covalent dimer form of the HIV-1 PR "backbone engineered" to delete a critical H-bonding element in only

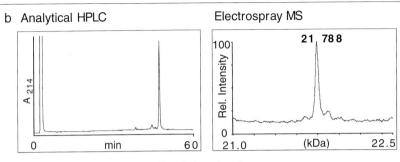

b Analytical HPLC   Electrospray MS

FIG. 5. (*continued*)

one of the two flap structures of the enzyme; the resulting asymmetrically engineered protein molecule had full intrinsic catalytic activity. This suggested that the HIV-1 PR uses only one flap in catalysis, in a manner analogous to the single flap cell-encoded aspartyl proteinases.[24] Convergent chemical ligation of multiple unprotected peptide segments was also used in the preparation of covalent heterodimers of basic helix–loop–helix zipper (b/HLH/Z) transcription factors (Fig. 6). It had been proposed that noncovalent heterodimers of the transcription factors c-Myc and Max played a key role in the regulation of gene expression. A covalent Myc–Max heterodimer (20,610 Da) prepared by total synthesis using a convergent chemical ligation strategy was found to be preorganized (i.e., more folded in free solution than the noncovalent Max homodimer) and showed potent, specific DNA sequence recognition.[11]

In 1994 we introduced a second-generation ligation chemistry, known as native chemical ligation, that allows the straightforward preparation of proteins with native backbone structures from fully unprotected peptide building blocks.[25] This important extension of the chemical ligation method of protein synthesis makes use of the remarkable intramolecular acylating power of the thioester functionality.[26–28] The principles underlying native chemical ligation[25] are shown in Fig. 7. The first step is the chemoselective reaction of an unprotected synthetic peptide–α-thioester with the thiol side chain of another unprotected peptide segment containing an amino-terminal Cys residue; this gives a thioester-linked intermediate as the initial

[24] M. Baca, Ph.D. Thesis, The Scripps Research Institute, 1994.
[25] P. E. Dawson, T. W. Muir, I. Clark-Lewis, and S. B. H. Kent, *Science* **266,** 776 (1994).
[26] T. Wieland, *in* "The Roots of Modern Biochemistry" (V. D. Kleinkauf Jaeniche, ed.), p. 213. de Gruyter, Berlin and New York, 1988.
[27] T. Wieland, E. Bokelmann, L. Bauer, H. U. Lang, and H. Lau, *Liebigs Ann. Chem.* **583,** 129 (1953).
[28] T. Wieland and G. Schneider, *Liebigs Ann. Chem.* **580,** 159 (1953).

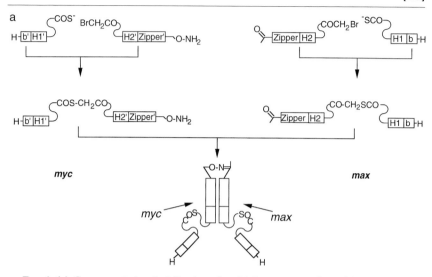

FIG. 6. (a) Convergent chemical ligation of multiple unprotected peptide segments was used to prepare (b) covalent Myc–Max heterodimer (20,610 Da) by total synthesis. [See L. E. Canne, A. R. Ferre-D'Amare, S. K. Burley, and S. B. H. Kent, *J. Am. Chem. Soc.* **117**, 2998 (1995).]

covalent product (cf. Ref. 1). Without change in the reaction conditions, this intermediate undergoes spontaneous, rapid intramolecular reaction to form a native peptide bond at the ligation site (Fig. 7).[25] The target full-length polypeptide is obtained in the desired final form without further manipulation. Native chemical ligation of unprotected peptide segments can be performed in aqueous solution in the presence of all the functionalities commonly found in proteins. Even in the presence of Cys residue free sulfhydryl groups in one or both segments, the ligation reaction is completely regioselective. This was originally illustrated by the total chemical synthesis of the protein interleukin-8, a 72-amino acid chemokine containing four Cys residues that form two functionally critical disulfide bonds.[25] This thioester-mediated amide bond-forming ligation chemistry has more recently been repeated and confirmed by the synthesis of model peptides.[29]

Protein synthesis by native chemical ligation of unprotected peptides in aqueous solution has been applied to a range of systems. The practicality of the method was illustrated in the total synthesis of the small protein

[29] J. P. Tam, Y.-A. Lu, C.-F. Liu, and J. Shao, *Proc. Natl. Acad. Sci. U.S.A.* **92**, 12485 (1995).

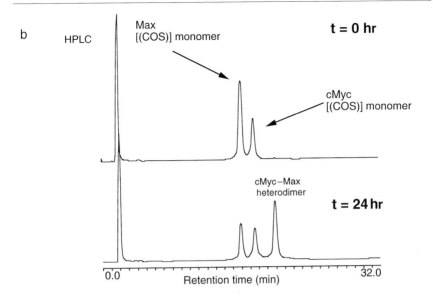

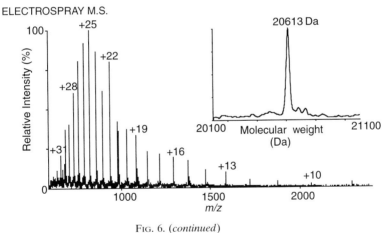

FIG. 6. (*continued*)

OMTKY3 (50 amino acid residues, 3 disulfides), a serine proteinase inhibitor.[30] Native chemical ligation has been used to prepare multiple analogs of the microbial ribonuclease barnase (110 amino acids), an intensively studied model system for protein folding.[31] The regioselectivity of the native

[30] W. Lu, M. A. Qasim, and S. B. H. Kent, *J. Am. Chem. Soc.* **118**, 8518 (1996).
[31] P. E. Dawson, M. J. Churchill, M. R. Ghadiri, and S. B. H. Kent, *J. Am. Chem. Soc.* **119**, 4325 (1997).

Peptide-1   Peptide-2

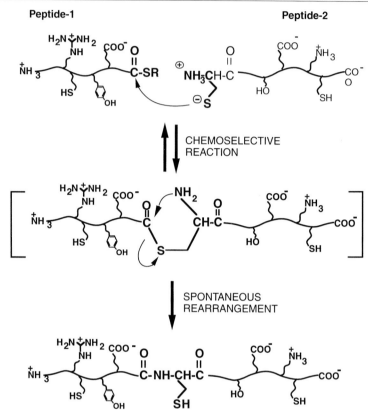

FIG. 7. Mechanism of native chemical ligation. The first step involves thiol exchange to generate a thioester-linked intermediate. This undergoes spontaneous intramolecular rearrangement to generate a native peptide bond at the site of ligation and to regenerate the Cys side chain. [From P. E. Dawson, T. W. Muir, I. Clark-Lewis, and S. B. H. Kent, *Science* **266**, 776 (1994).]

chemical ligation strategy was vividly demonstrated in the total synthesis of the enzyme human secretory phospholipase A$_2$ (sPLA$_2$). This protein has a polypeptide chain of 124 amino acids, containing 14 Cys residues. Native chemical ligation of two approximately 60-residue unprotected peptide segments yielded high-purity polypeptide that folded in good yield to give homogeneous sPLA$_2$ protein of mass 13,784 Da, containing 7 disulfide bonds, and with full enzymatic activity (Fig. 8).[32] In another example of

[32] T. M. Hackeng, C. M. Mounier, C. Bon, P. E. Dawson, J. H. Griffin, and S. B. H. Kent, *Proc. Natl. Acad. Sci. U.S.A.* **94**, in press (1997).

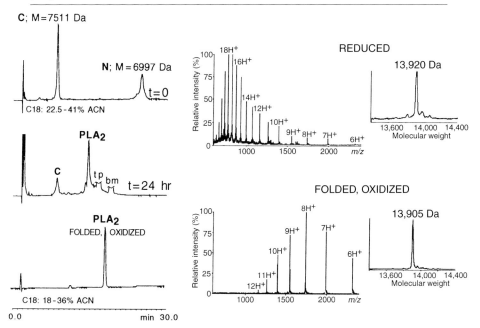

FIG. 8. Total synthesis of human secretory phospholipase A_2 by native chemical ligation. Precise, regiospecific ligation was observed even though 14 Cys residues were present in the unprotected peptide segments. The folded synthetic protein showed full enzymatic activity. [See T. M. Hackeng, C. M. Mounier, C. Bon, P. E. Dawson, J. H. Griffin, and S. B. H. Kent, *Proc. Natl. Acad. Sci. U.S.A.* **94,** in press (1997).]

the utility of protein synthesis by native chemical ligation of unprotected peptides in aqueous solution, the contribution of *intermolecular* backbone H-bonding to a protein–protein interaction was investigated by total synthesis of $[(COO)^{17-18}]OMTKY3$. Measurement of the binding of this backbone H-bond deleted protein to a panel of Ser proteinases showed that deletion of a single intermolecular H-bond lowered protein–protein binding by 1.5 ± 0.3 kcal/mol.[33] Native chemical ligation of unprotected peptides in aqueous solution has proved to be a practical and robust route to the total synthesis of a wide range of proteins.[34]

In this article we report additional evidence for the mechanism of the native chemical ligation reaction, we describe improved reaction conditions based on these insights, and we illustrate novel strategies that allow native

[33] W. Lu, M. A. Qasim, M. Laskowski, and S. B. H. Kent, *Biochemistry* **36,** 673 (1997).
[34] P. Dawson, T. Muir, M. Fitzgerald, and S. B. H. Kent, *in* "Peptides 1996: Proceedings of the Twenty-fourth European Peptide Symposium, Edinburgh, September 1996" (R. Ramage, ed.), in press, 1997.

chemical ligation to be used in the chemical synthesis of large polypeptides and proteins. Simple procedures are described that allow three or more peptide segments to be ligated, either by sequential native chemical ligation or by a convergent strategy employing a combination of native and oxime forming chemical ligations. This greatly extends the scope of the native chemical ligation reaction and will allow a wider application of this method for the total chemical synthesis of proteins.

Technical Aspects

Mechanism of Peptide–α-Thioester Formation

Synthetic peptides containing the α-thiocarboxyl functionality ($^\alpha$COSH) can be generated directly from thioester resins.[35–37] The $^\alpha$COSH group within these peptides can act as a nucleophile at pH 3–6, thus providing a unique window of reactivity relative to all nucleophiles normally found within peptides and proteins. A peptide–$^\alpha$COSH can be cleanly converted to a peptide–α-thioester by one of two methods. Nucleophilic reaction of the thioacid with an alkyl halide such as benzyl bromide in acidic aqueous solvents gives the corresponding peptide alkyl thioester in quantitative yield.[25,31] Alternatively, the peptide–$^\alpha$COSH is reacted with a symmetrical disulfide such as 5,5'-dithiobis(2-nitrobenzoic acid) (DTNB), again to give the corresponding peptide thioester.[25]

The second of these synthetic approaches is of particular interest from a mechanistic viewpoint. In their pioneering work in this area, Yamashiro and co-workers suggested that a 1-acyl-2-aryl(alkyl) disulfide species (-CO-S-S-R) (see Fig. 9a, **A**) was the final product of the reaction between a peptide–α-thioacid and a symmetrical disulfide.[36] A similar claim was made using native chemical ligation in model peptide ligations, although this assertion was not supported by experimental data.[29] In our hands, reaction of a peptide–$^\alpha$COSH with a molar excess of a symmetrical disulfide such as Ellman's reagent always gives the thioester derivative as the final isolated product (Fig. 9a, **A**).[25]

To further investigate the mechanism of this reaction, we reacted the model peptide Leu-Tyr-Arg-Ala-Gly-$^\alpha$COSH (**1**) with 1.1 equivalents of Ellman's reagent at pH 6.5. This was an artificially low amount of Ellman's reagent (normally 10 equivalents are used; see Appendix) to allow us to see any intermediates formed in the reaction. After 1 min, the earliest time

---

[35] J. Blake and C. H. Li, *Proc. Natl. Acad. Sci. U.S.A.* **78,** 4055 (1981).
[36] D. Yamashiro and C. H. Li, *Intl. J. Pept. Protein Res.* **31,** 322 (1988).
[37] L. E. Canne, S. M. Walker, and S. B. H. Kent, *Tetrahedron Lett.* **36,** 1217 (1995).

point taken, the reaction mixture was injected onto an analytical reversed-phase HPLC C_{18} column and the eluted peaks collected and immediately analyzed by ESMS (Fig. 9b). The reaction was extremely fast, as no peptide thioacid starting material **1** was observed after this short period of time. Two peptidic reaction products were generated. The first was identified by the measured mass relative to the starting peptide thioacid starting material **1** as the previously observed[25] 2-nitrobenzoic acid thioester derivative **2**; significantly, the second product **3** had a mass consistent with the 1-acyl-2-aryl disulfide derivative.[36] The ratio of products **2** and **3** in the 1-min time point was approximately 2 : 1. After 60 min of reaction the thioester derivative **2** was by far the major product, with only a small amount (<10%) of the 1-acyl-2-aryl disulfide derivative **3** present in the mixture (Fig. 9b).

The transient generation of the 1-acyl-2-aryl disulfide derivative **3** detected under these nonforcing conditions supports a mechanism in which the thioacid group of **1** initially attacks the Ellman's reagent, forming a mixed disulfide with an acyl group on one side and an aryl group on the other (Fig. 9a, Step 1). This was previously reported[29,36] as the final product of the reaction. However, the combination of HPLC analysis of the reaction mixture and electrospray mass spectrometric analysis[38] of all components showed that this was only a transient intermediate, even under these artificially nonforcing conditions. We hypothesize that the highly reactive 1-acyl-2-aryl disulfide species **3** reacts with the single equivalent of 2-nitro-4-mercaptobenzoic acid produced as a by-product during the first step to give the S-(2-nitrobenzoic acid) peptide thioester **2** as the final product (Fig. 9a, Step 2). Under reaction conditions typically used (see Appendix),[25] the peptide–α-thioester was the only product detected; it is stable to handling and is readily purified by reversed-phase HPLC as originally reported.[25] Reported work supports this mechanism.[39]

Modulation of Thioester Reactivity

We have previously shown that the rate of the native chemical ligation reaction is related to the chemical nature of the α-thioester group,[25] with faster reactions being obtained using electron-withdrawing aryl thioesters (see also Ref. 36). Under suitable conditions, peptide–α-thioesters undergo transthioesterification reactions when exposed to thiol-containing compounds.[27] In principle, such transthioesterification reactions could be used to modulate the chemical reactivities of peptide–α-thioesters during native chemical ligation (Fig. 10). Consequently, the *in situ* interconversion of

---

[38] M. Schnölzer, A. Jones, P. F. Alewood, and S. B. H. Kent, *Anal. Biochem.* **204**, 335 (1992).
[39] L. Zhang and J. P. Tam, *Tetrahedron Lett.* **38**, 3 (1997).

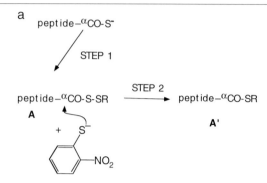

FIG. 9. Reaction of a peptide–αCOSH with Ellman's reagent (DTNB). (a) Two products of this reaction have been proposed: the disulfide species -αCOSSNB (**A**) and the -αCOSNB thiolester (**A′**). To investigate this question, the peptide Leu-Tyr-Arg-Ala-Gly-αCOSH (**1**) was reacted with 5,5′-dithiobis(2-nitrobenzoic acid) (Ellman's reagent) at an artificially low mole ratio of 1:1.1. (b) The upper HPLC trace is of purified **1** and can be considered a zero point. The middle HPLC was obtained after 1 min of reaction with 1.1 equivalents of Ellman's reagent at pH 6.5. Note the disappearance of the starting material **1** and the appearance of the 2-nitrobenzoic acid thioester derivative **2** and 1-acyl-2-aryl disulfide derivative **3** (2:1 ratio). The peak labeled nonpeptide was presumed to be the 2-nitro-4-mercaptobenzoic acid by-product of the reaction. The lower HPLC trace was obtained after 1 hr of reaction. Note that the ratio of **2** to **3** has changed significantly in favor of the former, and that the nonpeptide peak is now almost gone. These observations are consistent with a mechanism in which the transient species **3** reacts with 2-nitro-4-mercaptobenzoic acid to give final product **2**.

peptide–α-thioesters under the conditions of native chemical ligation was investigated.[31]

Peptides containing the highly reactive 2-nitrobenzoic acid α-thioester group can be converted to the less active benzyl α-thioester through the addition of an excess of benzyl mercaptan (2% by volume) in aqueous buffer at pH 6.5. Accordingly, transthioesterification of Leu-Tyr-Arg-Ala-Gly-αCOSNB to yield Leu-Tyr-Arg-Ala-Gly-αCOS-benzyl proceeded to an extent of approximately 50% in less than 30 min, without significant hydrolysis.[31] Similarly, peptides containing the benzyl α-thioester group can be converted to a more reactive thioester by addition of thiophenol. For example, as shown in Fig. 10, the peptide Ala-Glu-Ile-Ala-Ala-αCOS-benzyl (**4**) was converted to the corresponding phenyl thioester **5** by reaction with thiophenol (2% by volume), under aqueous conditions at pH 6.5.

The straightforward modulation of the thioester reactivity made possible by the transthioesterification reaction allows great flexibility in the synthesis and use of peptide–α-thioesters. Peptides can be isolated as relatively stable, easy to handle, low-activation thioesters, such as the previously reported

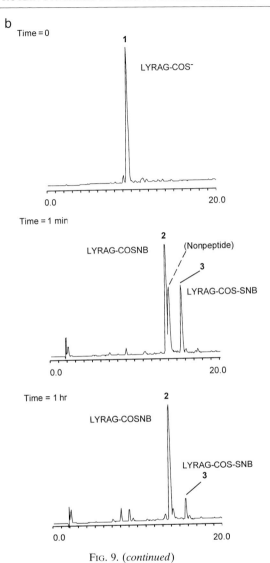

FIG. 9. (*continued*)

alkylthioesters.[40] Then, thiols such as benzyl mercaptan or thiophenol can be added to a ligation mixture to regulate the reactivity of the peptide–α-thioester component and improve reaction yield.[31] Thiol-modulated

[40] H. Hojo, C. Maegawa, S. Yoshimura, and S. Aimoto, *in* "Peptide Chemistry 1989" (N. Yanaihara, ed.), p. 297. Protein Research Foundation, Osaka, Japan, 1989.

a

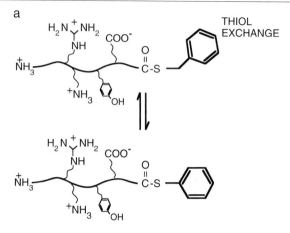

b

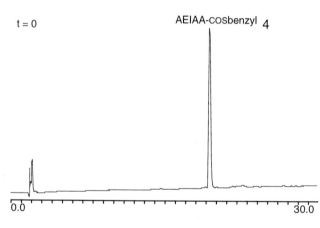

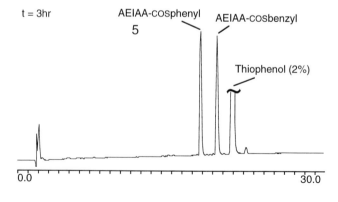

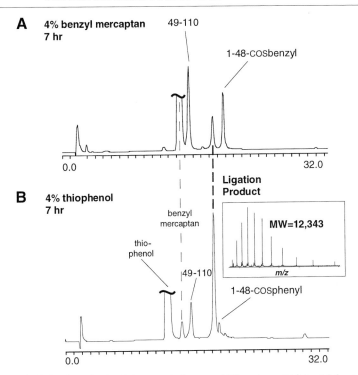

FIG. 11. (A) Total synthesis of the protein barnase (110 amino acids) by thiol-modulated native chemical ligation. (B) Use of thiophenol accelerated the reaction and provided a high yield of the desired ligation product without side reactions. [See P. E. Dawson, M. J. Churchill, M. R. Ghadiri, and S. B. H. Kent, *J. Am. Chem. Soc.* **119**, 4325 (1997).]

chemical ligation gives rapid and clean reaction to form long polypeptide chains in near-quantitative yields, as shown in Fig. 11.

Sequential Native Chemical Ligation

Highly optimized SPPS allows polypeptides of up to approximately 60 amino acids in length to be routinely constructed in good yield and high

FIG. 10. Thiol modulation of native chemical ligation of unprotected peptides in aqueous solution. (a) Peptides containing the benzyl α-thioester group can be converted to a more reactive thioester by addition of thiophenol. For example (b), the peptide Ala-Glu-Ile-Ala-Ala-αCOS-benzyl (**4**) was converted to the corresponding phenyl thioester **5** by reaction with thiophenol under aqueous conditions at pH 6.5. [See P. E. Dawson, M. J. Churchill, M. R. Ghadiri, and S. B. H. Kent, *J. Am. Chem. Soc.* **119**, 4325 (1997).]

purity.[6] This means that protein domains or modules of around 100–120 amino acids in length can be reliably generated from two peptide building blocks in a single chemical ligation step.[12,25] Commonly occurring protein modules in this size range include immunoglobulin domains, fibronectin modules (types 1–3), and SH2/3 signal transduction adaptor domains. The construction of larger protein targets (≥100 residues) via the simple ligation of just two unprotected synthetic peptide segments becomes more and more problematic with size due to increasing difficulties associated with the direct stepwise solid-phase synthesis of peptide segments longer than 60 residues.[41] In principle, this difficulty can be avoided by a multiple-ligation strategy using three or more peptide building blocks. In such an approach, readily accessible unprotected synthetic peptide segments (~50 residues) would be stitched together either in a linear fashion using the same ligation chemistry multiple times (cf. Ref. 36) or in a convergent strategy using two or more compatible ligation chemistries.[2,10,11]

Given the convenience and efficiency of native chemical ligation,[25,30–34] it was decided to investigate the feasibility of a multiple native chemical ligation approach to the synthesis of larger protein targets. To explore this possibility, an analog of the membrane proximal component of the extracellular domain from the β subunit (hβc) of granulocyte–macrocyte colony-stimulating factor receptor was used as a model system. This 95-amino acid residue protein module is homologous to the widespread fibronectin type 3 (F3) β-sandwich domain,[42] a structural motif readily accessible to total synthesis via chemical ligation.[12] Because of our familiarity with this type of system, we used the hβc extracellular domain of this receptor as a model system to investigate the feasibility of sequential native chemical ligation.

An explicit limitation of the native chemical ligation technique is that the new peptide bond must be formed at an Xxx-Cys ligation site in the target polypeptide.[25] Thus, for the purpose of this model study, two pairs of residues were changed in the wild-type hβc F3 sequence, namely, Ser-45 and Ser-76, which were both changed to Gly to improve the kinetics of the two native chemical ligation steps, and Lys-46 and Arg-77, which were each replaced with Cys to accommodate native chemical ligation. The effect of these substitutions on folding or biological activity of the domain is unknown and is not relevant to the present study of feasibility of the synthetic strategy. Other work is under way to expand the range of se-

[41] S. B. H. Kent, *Annu. Rev. Biochem.* **57,** 957 (1988).
[42] J. F. Bazan, *Proc. Natl. Acad. Sci. U.S.A.* **87,** 6934 (1990).

quences suitable for use as native chemical ligation sites, and the first results have been reported.[43]

Construction of $[Gly^{45,76}Cys^{46,77}](1-95)h\beta c$ (9) was achieved using sequential native chemical ligation involving the assembly of three synthetic peptides, $h\beta c[Gly^{45}](1-45)^\alpha COSNB$ (6), N^α-Msc$[Cys^{46}(SNB),Gly^{76}](46-76)$ $^\alpha COSNB$ (7), and $[Cys^{77}](77-95)h\beta c$ (8) (Fig. 12). A key feature of sequential native chemical ligation strategy is the reversible N^α-protection of the central peptide segment 7 with the base-labile 2-(methylsulfonyl) ethyloxycarbonyl (Msc) group.[25,44] In the absence of such N^α-protection, peptide 7 would cyclize due to the presence of both an N-terminal Cys residue and a C-terminal thioester functionality within the same molecule.[45] This problem can be simply avoided by reversibly protecting either the Cys sulfhydryl group or the N^α-amino group. The acid-stable Msc group has previously been used in directed fragment condensation approaches involving conventional coupling and the use of minimally protected peptide fragments.[36]

As illustrated in Fig. 12, the target sequence 9 was constructed by sequential native chemical ligation in a C-terminal to N-terminal direction. Reversed-phase HPLC was used to follow the progress of the ligation reactions; the covalent structure of each component in a ligation mixture was deduced by direct analysis of the isolated HPLC peak by ESMS (Fig. 13).[38] In the first step, peptides 7 and 8 were reacted; N^α-protection of 7 with the Msc group[25] ensured that only the desired ligation reaction took place, that is, between the C terminus of 7 and the N terminus of 8. Analysis of the ligation mixture after 15 hr indicated that the ligation reaction had essentially gone to completion, although the majority of the ligation product had oxidized to form an intramolecular disulfide between Cys-46 and Cys-77. Formation of this disulfide bond presumably involved highly favored thiolysis of the S-(5-thio-2-nitrobenzoic acid) derivative of Cys-46 by the sulfhydryl of Cys-77. Addition of the reducing agent tris(2-carboxyethyl) phosphine (TCEP) subsequently gave the desired product, in fully reduced form (Fig. 13, middle). The N^α-Msc-protecting group was then removed by brief exposure (~1 min) to pH 12–13, giving the fully unprotected peptide $[Cys^{46,77},Gly^{76}](46-95)h\beta c$ (Fig. 13, bottom). Note that the TCEP reduction and base deprotection steps were performed directly on the crude ligation solution without any intervening purification steps.

In the final native chemical ligation step, the α-thioester-containing

[43] L. E. Canne, S. J. Bark, and S. B. H. Kent, *J. Am. Chem. Soc.* **118**, 5891 (1996).
[44] Tesser and I. C. Balvert-Geers, *Int. J. Pept. Protein Res.* **7**, 295 (1975).
[45] L. Zhang and J. P. Tam, *J. Am. Chem. Soc.* **119**, 2363 (1997).

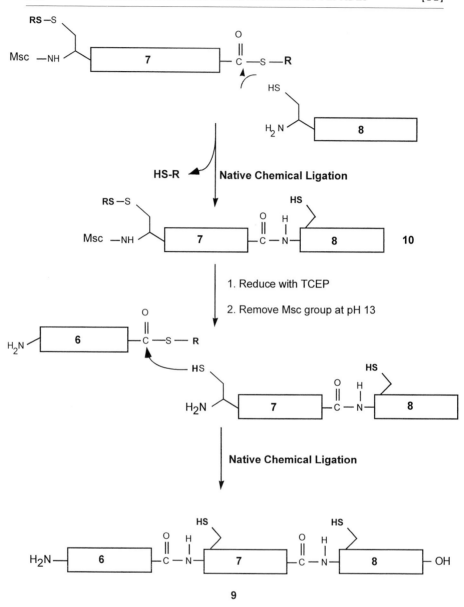

FIG. 12. Synthesis of [Gly⁴⁵,⁷⁶,Cys⁴⁶,⁷⁷](1–95)hβc (9) by sequential native chemical ligation. In the first native chemical ligation step, peptides 7 and 8 are reacted to give a ligation product that contains an amide bond at the site of ligation. The Nᵅ-Msc group is then removed by brief exposure to high pH, generating fully unprotected peptide. A second native chemical ligation reaction is performed by reacting with the peptide–α-thioester 6 to give the peptide product 9.

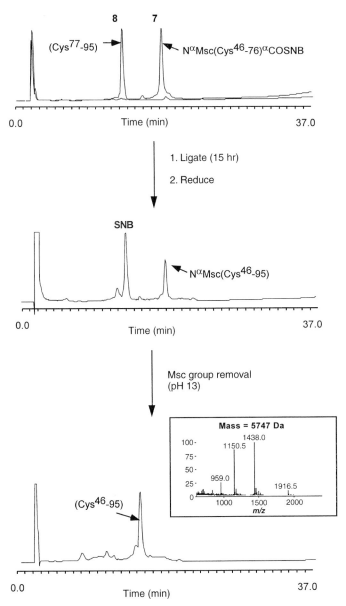

FIG. 13. Synthesis of [Cys^{46,77},Gly^{76}](46–95)hβc by native chemical ligation. (*Top*) HPLC traces of unprotected peptides fragments **7** {N$^\alpha$Msc[Cys^{46}(SNB),Gly^{76}](46–76)$^\alpha$COSNB} (later eluting product) and **8** {[Cys^{77}](77–95)hβc} (earlier eluting product). (*Middle*) HPLC trace showing formation of ligation product after 15 hr of reaction at pH 6.5 and subsequent reduction with TCEP. The other major peak corresponds to 5-thio(2-nitrobenzoic acid). (*Bottom*) Formation of deprotected peptide following removal of the N$^\alpha$-Msc group by brief exposure to pH 13. The inset is the electrospray mass spectrum of the deprotected product [calculated mass of 5749.4 Da (average isotope composition)].

peptide **6** was mixed with peptide [Cys^{46,77},Gly^{76}](46–95)hβc. Benzyl mercaptan was included in the ligation buffer (5%, v/v) to reverse the unproductive reaction of the sulfhydryl of Cys-77 in peptide [Cys^{46,77},Gly^{76}](46–95) hβc with the thioester group of peptide **6**. Note that inclusion of this thiol also resulted in conversion of 2-nitrobenzoic acid thioester of peptide **6** to the corresponding benzyl thioester derivative.[31] Under these reducing conditions the 95-amino acid product **9** was obtained in excellent yield (Fig. 14).

Our experimental results with model protein (1–95)hβc suggest that sequential native chemical ligation would be a practical route to proteins well in excess of 100 residues in length. While the present example involves only two ligation steps, the approach can in principle be extended to three or more consecutive ligations, provided the N^α-amino protection strategy outlined above is followed.

Convergent Native Chemical Ligation

The chemical ligation of unprotected peptides is uniquely suited to the generation of covalently linked protein assemblies of unusual topology such as those containing multiple N or C termini.[3,5,11,13–15] Native chemical ligation can be adapted to the generation of proteins of unusual topology in the following way. A second, compatible ligation chemistry is first used to link together the N or C termini of the peptide building blocks. This allows subsequent native chemical ligation steps to be performed on each arm of the resulting template. The viability of such a strategy was investigated using a combination of native chemical ligation[25] with the oxime-forming ligation chemistry.[3] It was anticipated that the aminooxy and ketone reactive groups underlying the oxime chemistry would be compatible with the thioester and Cys thiol groups used in native chemical ligation, because of our results in other systems using similar chemistries.[10,11]

The synthetic strategy employed in the synthesis of the model peptide **14**, which has unnatural topology, is illustrated in Fig. 15. A key feature of this strategy was the reversible protection of the N-terminal Cys residues in the bifunctional model peptides **10** and **11** by different, mutually compatible means. Specifically, the sulfhydryl group of the N-terminal Cys in peptide **10** was protected as the reductively removable *S*-(5-thio-2-nitrobenzoic acid) (SNB) mixed disulfide, whereas the N^α-amino group of peptide **11** was protected with the base-labile Msc group. This orthogonal mode of Cys protection ensured that subsequent native chemical ligation steps could be selectively performed on each of the two N termini of construct **12**. In addition, previous studies suggested that under acidic aqueous conditions an unprotected N-terminal Cys residue could react with a ketone or alde-

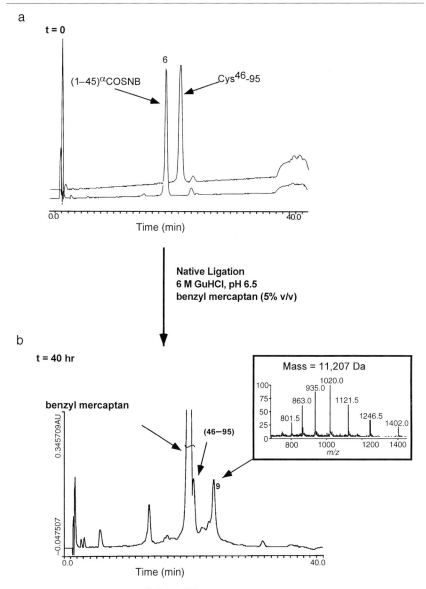

FIG. 14. Synthesis of **9** {[Gly[45,76],Cys[46,77]](1–95)hβc} by native chemical ligation. (a) HPLC traces of the purified unprotected peptides **6** {hβc[Gly[45]](1–45)[α]COSNB} (earlier eluting product) and [Cys[46,77],Gly[76]](46–95)hβc (later eluting product). (b) HPLC trace showing the formation of the ligation product **9** after 40 hr of reaction at pH 6.5. The inset shows the electrospray mass spectrum of **9** [calculated mass of 11,208.6 Da (average isotope composition)].

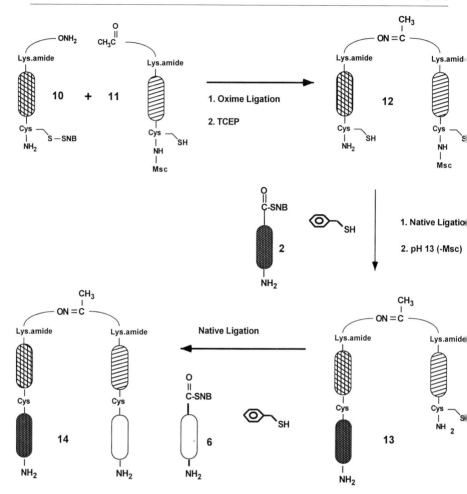

FIG. 15. Synthesis of peptide topological isomer **14** by parallel native chemical ligation. In the first step the aminooxy-containing peptide **10** is reacted with the ketone-containing peptide **11**. When the ligation reaction is complete, the SNB-protected group is removed *in situ* by treatment with TCEP, generating the oxime-linked product **12.** The single unprotected N-terminal Cys residue within **12** is then free to participate in a native chemical ligation reaction with peptide **2.** Note that the N^{α}-Msc group is then removed *in situ* to afford the fully unprotected ligation product **13.** In the second native chemical ligation step peptide **13** is reacted with peptide **6** to give the final ligation product **14.**

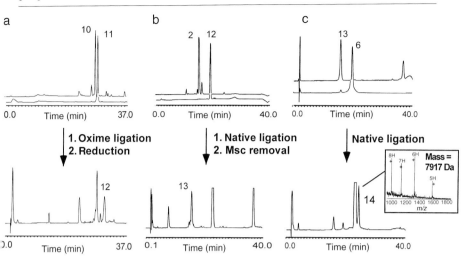

FIG. 16. Synthesis of peptide topological isomer **14** by parallel native chemical ligation. (a) Oxime-forming ligation. The upper HPLC traces are of purified peptides **10** and **11**. Lower HPLC trace shows the formation of the ligation product **12** after 24 hr of reaction at pH 4.6, followed by reduction with TCEP. (b) Native chemical ligation. Upper HPLC traces are of purified peptides **2** and **12**. Lower HPLC trace shows formation of ligation product **13** after 16 hr of reaction at pH 6.5 and subsequent *in situ* removal of the N^{α}-Msc group from the initial ligation product. (c) Native chemical ligation. Upper HPLC traces are of purified peptide starting materials **13** and **6**. Lower HPLC trace shows formation of ligation product **14** after 15 min of reaction at pH 6.5. The inset in the lower trace shows the electrospray mass spectrum of **14** [expected mass of 7920.0 Da (average isotope composition)].

hyde to produce a thiazolidine ring system.[46] Model peptide **11** contains the ketone levulinic acid which might be expected to react with N-terminal Cys residue in either peptide **10** or **11**. Protection of either the sulfhydryl or N^{α}-amino groups of the N-terminal Cys in these peptides has the added benefit of preventing this undesired side reaction from occurring.

As depicted in Fig. 15, the target peptide topological isomer **14** was assembled by chemical ligation in three steps from four synthetic peptides. In the first oxime-forming ligation step, peptides **10** and **11** were mixed together in 6 *M* guanidine, 0.1 *M* sodium acetate at pH 4.6. Under these conditions the ligation reaction was observed to be over 90% complete after 24 hr, after which *in situ* deprotection of the *S*-(5-thio-2-nitrobenzoic acid) group in the initial ligation product was achieved by direct addition of TCEP to the ligation solution. The resulting fully reduced oxime-containing product **12** was purified by preparative HPLC (Fig. 16a).

[46] C.-F. Liu and J. P. Tam, *J. Am. Chem. Soc.* **116**, 4149 (1994).

The second ligation step involved mixing purified **12** with peptide **2** in 6 M guanidine, 0.1 M phosphate at pH 6.5 containing benzyl mercaptan (5% by volume). Benzyl mercaptan was included in the ligation buffer to reverse the unproductive reaction of the sulfhydryl of the N^α-Msc-protected Cys in peptide **12** with the thioester group of **2**. Addition of benzyl mercaptan to the ligation mixture also resulted in conversion of peptide **2** into the corresponding benzyl thioester **4**.[31] After 16 hr of reaction, the N^α-Msc protecting group was removed from the ligation product by brief exposure (~1 min) of the crude ligation solution to pH 12–13, and the fully unprotected product **13** (Fig. 16b) was purified by semipreparative HPLC. The oxime linkage within **13** was found to be stable to the deprotection conditions used.

In the final step, a second native chemical ligation reaction was performed by reacting purified peptide **13** with peptide thioester **6** in 6 M guanidine, 0.1 M phosphate at pH 6.5, again containing benzyl mercaptan (5% by volume). The benzyl mercaptan was included on this occasion to reverse unproductive reaction of **6** with the internal Cys residue of **13**, and as before it also resulted in conversion of **6** to the corresponding benzyl thioester.[31] As described, the ligation reaction was found to be extremely clean and rapid as indicated by the large amount of product **14** present after only 15 min of reaction (Fig. 16c). The single product produced from the reaction was analyzed by ESMS, which indicated a covalent structure consistent with that of the expected peptide topological isomer **14**. Thus, four unprotected peptide segments have been chemically ligated to give the desired model polypeptide product with a molecular mass of 7917 Da.

We have shown that by combining the native chemical ligation[25] and the oxime-forming[3] ligation approaches it was possible to construct unsymmetrical peptide assemblies with multiple N termini. In the present example native chemical ligation was used to attach a different peptide, one on each arm, of the oxime-containing peptide **12**. The oxime linkage was found to be completely stable during the native chemical ligation steps. The required selectivity was achieved by using differential protection on the two N-terminal Cys residues in **12**, thereby providing the necessary control over which N termini is available for ligation.

Summary

In these studies we have shown that reaction of a peptide–α-thioacid with an aryl disulfide gives a product with mass consistent with a 1-acyl-2-aryl disulfide species as the initial product, but that this is rapidly converted *in situ* to give a peptide–α-thioester, which can be isolated for use in native chemical ligation. In addition, we have demonstrated that peptide–

α-thioesters can be readily interconverted using transthioesterification reactions, permitting the reactivity of peptide thioesters to be modulated with thiol-reducing agents during native chemical ligation. Finally, novel synthetic strategies have been developed that allow sequential and convergent native chemical ligations to be performed, allowing the assembly of protein analogs from multiple unprotected peptide building blocks. A wide range of protein systems has been accessed through total synthesis by native chemical ligation of unprotected peptides in aqueous solution.[34]

Appendix: Experimentation

Materials

tert-Butyloxycarbonyl (Boc)-amino acids and 2-[1*H*-benzotriazolyl]-1,1,3,3-tetramethyluronium hexafluorophosphate (HBTU) are obtained from Novabiochem (San Diego, CA). Preloaded Boc-aminoacyl-OCH$_2$-Pam-copoly(styrene–divinylbenzene)resins and *N,N*-diisopropylethlamine (DIEA) are purchased from Applied Biosystems (Foster City, CA). *N,N*-Dimethylformamide (DMF) is obtained from Mallinckrodt Chemical Co. HPLC-grade acetonitrile is purchased from EM Science (Gibbstown, NJ). Trifluoroacetic acid (TFA) is purchased from Halocarbon (River Edge, NJ). Hydrogen fluoride (HF) is purchased from Matheson Gas. TCEP is obtained from Strem Chemicals.

Reversed-Phase High-Performance Liquid Chromatography

Analytical and semipreparative gradient HPLC are performed on a Rainin dual-pump high-pressure mixing system with 214 nm detection. Semipreparative HPLC is run on a Vydac C$_{18}$ column (10 μm, 10 × 250 mm) at a flow rate of 3 ml/min. Analytical HPLC is performed on a Vydac C$_{18}$ column (5 μm, 4.6 × 150 mm) at a flow rate of 1 ml/min. Preparative HPLC is performed on a Waters Prep 4000 (Waters, Milford, MA) system fitted with a Waters 486 tunable absorbance detector using a Vydac C$_{18}$ column (15–20 μm, 50 × 250 mm) at a flow rate of 30 ml/min. All runs used linear gradients of 0.1% aqueous TFA versus [90% acetonitrile plus (0.1% TFA in H$_2$O)].

Mass Spectrometry

Electrospray mass spectrometric analysis is routinely applied to all synthetic peptides and components of reaction mixtures. ESMS is performed on a Sciex API-III triple quadrupole electrospray mass spectrometer as previously described.[38] Calculated masses are obtained using the program MacProMass (Sunil Vemuri and Terry Lee, City of Hope, Duarte, CA).

Solid-Phase Peptide Synthesis

All peptides are synthesized according to the *in situ* neutralization/ HBTU activation protocol for Boc solid-phase peptide synthesis as previously described.[6] Short model peptides are synthesized by manual methods, whereas assembly of the longer hβc fragments (**6–8**) is achieved by established machine-assisted synthesis on a custom-modified Applied Biosystems 430A peptide synthesizer.[6] Coupling yields are monitored by the quantitative ninhydrin determination of residual free amine.[47] Peptides with C-terminal Gly-$^\alpha$COSH or Ala-$^\alpha$COSH groups are constructed on either Gly-thioester or Ala-thioester supports, respectively.[37] Peptide α-carboxamides are constructed on a 4-methylbenzhydrylamine–resin, and all other peptides are synthesized on appropriate Boc-aminoacyl-OCH$_2$-Pam-copoly(styrene–divinylbenzene) resins. Side-chain protection is as previously described,[6] except for peptides assembled on thioacid-generating resins where it is necessary to use Boc-His(Bom)-OH and unprotected Boc-Trp-OH. In all cases, side-chain protecting groups are removed and the peptides cleaved from the resin by treatment with liquid HF containing 4% *p*-cresol, for 1 hr at 0°. Crude peptide products are precipitated and washed with diethyl ether before being dissolved in degassed aqueous acetic acid (10–50%) and lyophilized.

Leu-Tyr-Arg-Ala-Gly-$^\alpha$COSNB (2). Following chain assembly on Boc-Gly-SCH-(phenyl)phenyl-OCH$_2$CONHCH$_2$–resin, the peptide is cleaved from the solid support using HF, dissolved in dilute aqueous acetic acid, and lyophilized. The crude peptide–$^\alpha$COSH **1** is then dissolved in 6 *M* guanidine hydrochloride (GuHCl), 0.1 *M* sodium phosphate at pH 6.5 (10 mg/ml) and 10 molar equivalents of a solution of 10 m*M* DTNB in 0.1 *M* sodium phosphate added. After stirring for 45 min the product **2** is purified by preparative HPLC.

Ala-Glu-Ile-Ala-Ala-$^\alpha$COS-benzyl (4). The crude lyophilized peptide-$^\alpha$COSH obtained from HF cleavage is taken up in 6 *M* GuHCl, 0.1 *M* sodium phosphate at pH 6.5 at a concentration of 10 mg/ml. Solid benzyl bromide (10 equivalents) is added and the solution stirred at room temperature for 30 min. The product **4** is then purified by preparative HPLC and lyophilized.

hβc[Gly^{45}](1–45)$^\alpha$COSNB (6). The crude dry peptide–$^\alpha$COSH obtained after HF cleavage and lyophilization is dissolved in 6 *M* GuHCl, 0.1 *M*

---

[47] V. K. Sarin, S. B. H. Kent, J. P. Tam, and R. B. Merrifield, *Anal. Biochem.* **117,** 147 (1981).

sodium phosphate at pH 6.5 (10 mg/ml) and 10 molar equivalents of a solution of 10 mM DTNB in 0.1 M sodium phosphate added. The solution is stirred at room temperature for 45 min and the desired product **6** purified by preparative HPLC.

N^α-*Msc[Cys^{46}(SNB),Gly^{76}](46–76)$^\alpha$COSNB (7)*. After completion of stepwise chain assembly on Boc-Gly-SCH-(phenyl)phenyl-OCH$_2$ CONHCH$_2$–resin, the amino-terminal Boc group is removed with neat TFA (two times, 1 min each) and the deprotected peptide is neutralized with 10% DIEA in DMF (two times, 1 min). The 2-(methylsulfonyl) ethyloxycarbonyl (Msc) group is then introduced at the free N^α terminus by coupling as the preformed 4-nitrophenyl carbonate (2 mmol) for 3 hr. Following cleavage with HF and lyophilization, the crude dry peptide is dissolved in 6 M GuHCl, 0.1 M sodium phosphate at pH 6.5 (10 mg/ml) and 10 molar equivalents of a solution of DTNB (10 mM) in 0.1 M sodium phosphate added. This solution is stirred at room temperature for 45 min and the desired product **7** purified by preparative HPLC.

[Cys^{77}](77–95)hβc (8). Synthesis of the peptide is carried out on pre-formed Boc-Thr(Bzl)-OCH$_2$-Pam–resin. Following chain assembly the N-terminal Boc group is removed by treatment with neat TFA and the resulting TFA salt neutralized by treatment with 10% DIEA in DMF. The formyl groups are then removed from the Trp residues by treatment with ethanolamine as previously described.[6] Following cleavage with anhydrous HF, the crude peptide **8** is purified by preparative HPLC.

Cys(SNB)-Arg-Leu-Ser-Val-Leu-Gly-Lys[N$^\varepsilon$-(COCH$_2$ONH$_2$)]-CONH$_2$ (10). Synthesis of the peptide is carried out on Boc-Lys[fluorenyl-methyloxycarbonyl (Fmoc)]-4MeBHA-resin. Following chain assembly the N^ε-Fmoc group on the C-terminal Lys residue is removed by treatment with 20% piperidine in DMF (two times, 2 min). The free N^ε-amino group of the Lys is then (aminooxy)acetylated as described previously.[11] Briefly, [(N-(2-Cl-Z)-amino)oxy]acetic acid (2 mmol) is dissolved in 5 ml of DMF and activated as the succinimide ester by addition of N-hydroxysuccinimide (2 mmol) and diisopropylcarbodiimide (DIPCDI, 2 mmol). After an activa-tion period of 15 min the solution is added to the peptide–resin and coupled for 1.5 hr. Following removal of the N-terminal Boc group and neutraliza-tion of the TFA salt, peptide is cleaved with HF and the crude peptide lyophilized. The crude peptide is then dissolved in 6 M GuHCl, 0.1 M sodium phosphate at pH 6.5 (10 mg/ml) and 10 molar equivalents of a solution of 10 mM DTNB (in 0.1 M sodium phosphate) added. The solution is stirred at room temperature for 45 min and the desired product **10** purified by preparative HPLC.

N$^\alpha$-(Msc)Cys-Arg-Leu-Ser-Val-Leu-Gly-Lys[N$^\varepsilon$-(COCH$_2$CH$_2$ COCH$_3$)]-CONH$_2$ (11). The peptide is synthesized on Boc-Lys(Fmoc)-

4MeBHA–resin using standard Boc protocols. On completion of the stepwise assembly, the Fmoc group is removed by treatment with 20% piperidine in DMF (two times, 2 min). The N^ε-amino group of the C-terminal Lys is then 4-oxopentolyated. 4-Oxopentanoic acid (levulinic acid) (2 mmol) is activated as the symmetrical anhydride by dissolving in DMF (5 ml), to which is added DIPCDI (1.0 mmol). After activating for 15 min, the symmetrical anhydride is added to the peptide–resin and coupled for 1.5 hr. Following removal of the N-terminal Boc group and subsequent neutralization, the Msc group is introduced at the free N^α terminus by coupling as the preformed 4-nitrophenyl carbonate (2 mmol) for 3 hr. The target peptide **11** is then cleaved from the solid suport and purified by preparative HPLC.

Purification and Characterization of Peptide Segments

Crude peptides are dissolved in aqueous acetonitrile containing 0.1% TFA and purified by either semipreparative or preparative HPLC followed by lyophilization. The purity and covalent structure of all peptides is characterized by ESMS.[38]

Model Studies on Thioacid and Thioester Groups

Purified model peptides are reacted under the conditions described, and in each case the reaction is followed by analytical reversed-phase HPLC. Peaks are collected on the basis of UV absorbance (214 nm) and analyzed by ESMS.

Lys-Tyr-Arg-Ala-Gly-$^\alpha$COS-SNB (3). Lys-Tyr-Arg-Ala-Gly-$^\alpha$COSH (**1**) (0.5 mg, 1.0 equivalents) is dissolved in 0.1 M sodium phosphate at pH 6.5 and DTNB (0.4 mg, 1.2 equivalents) added. After reaction for 1 min the mixture is subjected to analysis by HPLC and the two major components characterized by ESMS (see Fig. 9b). On the basis of mass analysis, the early eluting peak is found to be the 2-nitrobenzoic acid thioester peptide **2** and the late eluting peak the 1-acyl-2-aryl disulfide derivative **3** [found 791.5 Da $\pm$ 0.5; calculated 791.3 Da (monoisotopic), 791.7 Da (average isotope composition)].

Ala-Glu-Ile-Ala-Ala-$^\alpha$COS-phenyl (5). Peptide Ala-Glu-Ile-Ala-Ala-$^\alpha$COSH (**4**) (0.2 mg) is dissolved in 0.1 ml of 6 M GuHCl, 0.1 M sodium phosphate at pH 6.5 (10 mg/ml) containing thiophenol (2 μl). After stirring for 45 min the product **5** is purified by HPLC [found 579 Da $\pm$ 0.0; calculated 579.2 Da (monoisotopic), 579.5 Da (average isotope composition)].

Sequential and Convergent Native Chemical Ligation Studies

All ligation reactions are monitored by reversed-phase analytical HPLC and by ESMS.

[Cys^{46,77},Gly^{76}](46–95)hβc. Reaction is initiated by combining **7** (10.80 mg, 2.6 μmol) and **8** (6.60 mg, 2.9 μmol) in 3.00 ml of 6 *M* GuHCl, 0.1 *M* sodium phosphate at pH 6.5. The reaction is stirred at room temperature for 15 hr after which the reducing agent TCEP (10 mg) is added and the solution stirred for a further 30 min to give fully reduced ligation product. The N^α-Msc protecting group is then removed by raising the crude ligation solution to pH 13 with 1 *N* NaOH. After 1 min the pH is lowered to pH 5.0 by addition of 1 *N* HCl. The deprotected ligation product is purified by preparative HPLC (25–50% B over 60 min) to give 5.0 mg (0.8 μmol, 31%) of white solid following lyophilization [found 5747.5 Da ± 0.7; calculated 5745.7 Da (monoisotopic), 5749.4 Da (average isotope composition)].

[Gly^{45,76},Cys^{46,77}](1–95)hβc **(9)**. The reaction is initiated by combining **6** (1.1 mg, 0.19 μmol) and [Cys^{46,77},Gly^{76}](46–95)hβc (1.0 mg, 0.17 μmol) in 200 μl of 6 *M* GuHCl, 0.1 *M* sodium phosphate at pH 6.5 containing benzyl mercaptan (5%, v/v). The reaction is stirred at room temperature for 40 hr after which desired ligation product **9** is purified by semipreparative HPLC (25–50% B over 60 min) to give 1.4 mg (0.12 μmol, 73%) of white solid following lyophilization [found 11205.1 Da ± 1.5; calculated 11201.3 Da (monoisotopic), 11208.6 Da (average isotope composition)].

Ligation Product **(12)**. The ligation reaction is carried out by combining **10** (12.2 mg, 10.6 μmol) and **11** (12 mg, 10.6 μmol) in 1.2 ml of 6 *M* GuHCl, 0.1 *M* sodium acetate at pH 4.6. The reaction is stirred at room temperature for 24 hr after which the reducing agent TCEP (5 mg) is added and the mixture stirred for a further 30 min. The desired ligation product **12** is subsequently purified by preparative HPLC (25–50% B over 60 min) to give 6.0 mg (3 μmol, 28.3%) of white solid [found 2050.5 Da ± 0.5; calculated 2050.0 Da (monoisotopic), 2051.1 Da (average isotope composition)].

Ligation Product **(13)**. The ligation reaction is initiated by combining peptides **2** (0.96 mg, 1.2 μmol) and **12** (2.60 mg, 1.2 μmol) in 0.3 ml of 6 *M* GuHCl, 0.1 *M* sodium phosphate at pH 6.5 containing benzyl mercaptan (5%, v/v). The reaction is stirred for 16 hr after which the N^α-Msc protecting group is removed from ligation product by raising the crude ligation solution to pH 13 with 1 *N* NaOH. After 1 min the pH is lowered to pH 5.0 by addition of 1 *N* HCl. The fully deprotected ligation product **13** is then purified by semipreparative HPLC (20–40% B over 30 min) to give 0.8 mg (0.3 μmol, 25%) of white solid [found 2461.1 Da ± 1.1; calculated 2460.3 Da (monoisotopic), 2461.8 Da (average isotope composition)].

Ligation Product **(14)**. The reaction is performed on an analytical scale and is initiated by combining **6** (0.23 mg, 41 nmol) and **13** (0.1 mg, 41 nmol) in 50 μl of 6 *M* GuHCl, 0.1 *M* sodium phosphate at pH 6.5 containing benzyl mercaptan (5%, v/v). The reaction is stirred at room temperature and the ligation product **14** identified using a combination of HPLC and

ESMS [found 7917.6 Da ± 1.5; calculated 7914.9 Da (monoisotopic), 7920.0 Da (average isotope composition)].

Acknowledgments

We gratefully thank Lynne Canne, Sharon Walker, and Michael Williams for helpful advice and discussions during the course of this work. This work was supported by funds from the National Institutes of Health, Grants GM48870, HL31950, and GM48897 (S. B. H. K.).

[14] Synthesis of Proteins by Subtiligase

By ANDREW C. BRAISTED, J. KEVIN JUDICE, and JAMES A. WELLS

Introduction

Driven by a desire to extend the power of site-directed mutagenesis, several groups have developed methodologies to enable incorporation of unnatural amino acids into proteins.[1-3] It has been hoped that the broad repertoire of synthetic chemistry could be incorporated into much larger proteins. A biosynthetic methodology for the site-specific incorporation of unnatural amino acids has proved to be useful in producing single-site variants of many proteins.[2] At present this approach is limited by poor protein expression levels *in vitro,* and by an inability to make multiple substitutions. Solid-phase peptide synthesis (SPPS) techniques have been extended to allow the production of small proteins (less than 100 amino acids); however, the synthesis of proteins this size or larger remains a significant challenge.[4]

The ability to specifically ligate peptides represents a critical link between *in vivo* expression of proteins and total chemical synthesis. Several larger proteins have been synthesized using segment condensation approaches; this approach holds promise as an improvement over total linear synthesis.[1,5-7] Uncatalyzed segment condensation reactions have been car-

[1] D. Y. Jackson, J. Burnier, C. Quan, M. Stanley, J. Tom, and J. A. Wells, *Science* **265,** 243 (1994).
[2] V. W. Cornish, D. Mendel, and P. G. Schultz, *Angew Chem. Int. Ed. Engl.* **34,** 621 (1995).
[3] M. Baca, P. F. Alewood, and S. B. H. Kent, *Protein Sci.* **2,** 1085 (1993).
[4] R. C. Milton, S. C. F. Milton, and S. B. H. Kent, *Science* **256,** 1445 (1992).
[5] P. E. Dawson, T. W. Muir, I. Clark-Lewis, and S. B. H. Kent, *Science* **266,** 776 (1994).
[6] H. F. Gaertner, R. E. Offord, R. Cotton, D. Timms, R. Camble, and K. Rose, *J. Biol. Chem.* **269,** 7224 (1994).
[7] J. P. Tam, Y.-A. Lu, C.-F. Liu, and J. Shao, *Proc. Natl. Acad. Sci. U.S.A.* **92,** 12485 (1995).

ried out on fully side-chain protected peptides, followed by deprotection to yield the native sequence.[8] More recently methods for uncatalyzed segment condensations of unprotected fragments have been reported, incorporating either unique chemical linkages or native peptide bonds.[1,5–7] The benefits of using unprotected fragments include improved solubility in aqueous conditions and no final deprotection steps. Segment condensation approaches offer the important advantage that individual fragments can be highly purified prior to coupling, and any impurities can be removed after each coupling. Furthermore, the potential of synthetic chemistry can be readily applied, as nonnatural structures can be incorporated at multiple sites, semisynthetic proteins can be constructed, and significant quantities of final product can be produced. A powerful strategy using enzyme-catalyzed segment condensation has been developed.[1,9] This methodology, which allows for efficient coupling of unprotected peptides, has been applied to total protein synthesis,[1] semisynthesis,[10] and peptide cyclizations.[11] A practical version of this approach is described here.

Catalysis of Peptide Ligation Reactions

The principle of microscopic reversibility dictates that proteases can catalyze both proteolysis and ligation. Under physiological conditions, though, the equilibrium lies strongly in favor of proteolysis. Van't Hoff first proposed in 1898 that by shifting this equilibrium one might convert a hydrolase to a ligase, thus providing a catalytic approach to segment condensation reactions.[12] This equilibrium point can be shifted by altering reaction conditions such as solvent polarity, temperature, and pH; varieties of these experiments have been carried out and have in fact induced proteases to work backward and function as peptide ligases. In practice, however, this approach has significant limitations. Although the product ratio can be shifted in favor of aminolysis by using organic solvents, the proteases tend to be insoluble and are often less stable. Furthermore, control of the reaction equilibrium typically must be optimized for each step.[12]

Engineering of the protease to favor ligation relative to hydrolysis offers many advantages. Studies originally reported by Chu and Mautner found

[8] R. G. Denkewalter, D. F. Veber, F. W. Holly, and R. Hirschmann, *J. Am. Chem. Soc.* **91,** 502 (1969).
[9] L. Abrahmsén, J. Tom, J. Burnier, K. A. Butcher, A. Kossiakoff, and J. A. Wells, *Biochemistry* **30,** 4151 (1991).
[10] T. K. Chang, D. Y. Jackson, J. P. Burnier, and J. A. Wells, *Proc. Natl. Acad. Sci. U.S.A.* **91,** 12544 (1994).
[11] D. Y. Jackson, J. P. Burnier, and J. A. Wells, *J. Am. Chem. Soc.* **117,** 819 (1995).
[12] W. Kullmann, "Enzymatic Peptide Synthesis." CRC Press, Boca Raton, Florida, 1987.

that the ratio of aminolysis to hydrolysis is favorable for thioesters and even more so for selenol esters.[13] Kaiser and co-workers demonstrated the practicality of this work by preparing a subtilisin variant, thiolsubtilisin, where the active site Ser was chemically converted to Cys (S221C).[14] Using activated esters to acylate the active site Cys in the presence of amine nucleophiles, it was possible to efficiently synthesize amide bonds. The ratio of aminolysis to hydrolysis is 600-fold greater for thiolsubtilisin relative to subtilisin; the variant selenolsubtilisin was later prepared by Hilvert and co-workers and shown to be 14,000-fold more effective for aminolysis than subtilisin.[15]

These subtilisin variants are, however, limited by low k_{cat} values and the slow hydrolysis of the amide product. The rates of aminolysis for either thiol- or selenolsubtilisin are about 10^2- to 10^4-fold below the esterase activity of subtilisin. The low catalytic efficiencies are believed to be related to steric congestion within the active site caused by the significantly larger sulfur or selenium atom. In addition, to acylate these enzymes highly activated esters must be used; these substrates are both difficult to prepare and undergo rapid uncatalyzed hydrolysis in aqueous solution.

Subtiligase

Engineering of Subtilisin to Subtiligase

The potential to transform thiolsubtilisin into a more efficient peptide ligase led to the consideration of other mutations to relieve the steric crowding from the sulfur atom within the active site. A previously characterized Pro to Ala mutation (P225A) had been shown to move the hydroxyl group of the active site Ser away from the oxyanion hole and from the catalytic His by about 0.5–1.0 Å. Therefore, the double mutant S221C/P225A (subtiligase), designed to capture the favorable ratio of aminolysis yet without steric crowding, was prepared. The success of the design was demonstrated both by measuring peptide ligase activity and by determining the X-ray crystal structure.[9]

Structure of Subtilisin S221C/P225A (Subtiligase)

Analysis of the difference map between wild-type subtilisin and the double mutant (subtiligase) indicated that the α helix, which includes the

[13] S. H. Chu and H. G. Mautner, *J. Org. Chem.* **31**, 308 (1966).
[14] T. Nakatsuka, T. Sasaki, and E. T. Kaiser, *J. Am. Chem. Soc.* **109**, 3808 (1987).
[15] Z.-P. Wu and D. Hilvert, *J. Am. Chem. Soc.* **111**, 4513 (1989).

P225A mutation, had in fact moved in the double mutant. Comparison of the displacement of the residues preceding P225 showed a concerted shift of the N-terminal portion of the helix. This end of the helix, which supports C221, has moved 0.3 Å away from the active site. This movement appears to be sufficient to better accommodate the longer C–S bond (1.8 Å compared to 1.4 Å for the wild-type C–O bond) now in the active site. Superimposing the structure of subtiligase on the structure of wild-type subtilisin complexed with a boronic acid inhibitor also provides information about the active site.[9] From this perspective, the $S\delta$ is 1.5 Å from the boron atom, which mimics the carbonyl carbon of the substrate. Although this is still 0.3 Å too close to allow for optimal positioning of the substrate, it is significantly more favorable than without the P225A mutation. Overall, the perturbation of the active site appears to be subtle, creating space for the appropriate orientation of the Cys side chain.

Kinetic Analysis of S221C/P225A (Subtiligase)

Individually, the S221C and P225A mutations lead to significant reductions in both amidase and esterase activity. Together, in the double mutant S221C/P225A, the amidase activity is reduced to below the level of detection, whereas the esterase activity is enhanced relative to the single mutant S221C. The ratio of aminolysis to hydrolysis in the double mutant is 500-fold improved over the wild-type enzyme, enabling ligation yields in the range of 95% without background proteolysis.

The rate and success of the ligation reaction are dependent on the nature of the substrate used to acylate the enzyme. Design of appropriate substrates requires a reactive ester, but not so reactive that background hydrolysis of the substrate becomes limiting. In addition, consideration must be given to the nature of the ester substrate, as efficient binding to the enzyme is essential for acylation. It is known that peptide substrates bind to the native subtilisin in an extended β-sheet conformation, so two simple mimics were designed and tested. A glycolate ester is isosteric with Gly at P_1', whereas a lactate ester is isosteric with Ala. Both of these substrates efficiently acylate the enzyme as the simple amides (Fig. 1A). By incorporating either Phe or Lys (Fig. 1B) at the P_2' position the value of k_{cat}/K_m is increased, predominately by lowering the K_m. These results suggest that binding of the ester substrate to both of the enzyme sites is in fact important for acylation of the enzyme. As the glycolate substrates generally showed about 5-fold greater activity than the lactate esters,[9] the glycolate esters have been used in further studies.

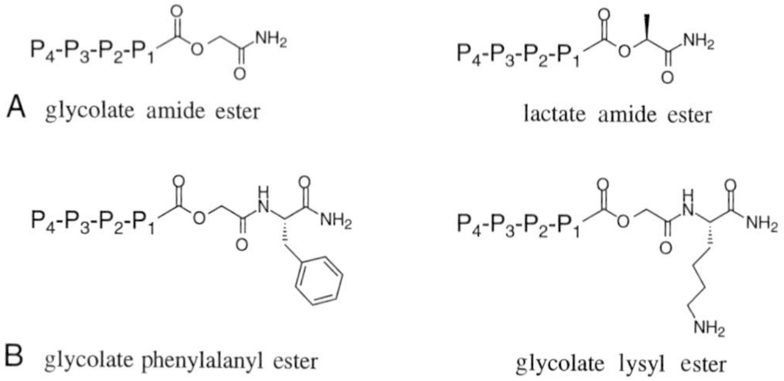

A glycolate amide ester

lactate amide ester

B glycolate phenylalanyl ester

glycolate lysyl ester

FIG. 1. (A) Comparison of the glycolate amide ester with the lactate amide ester. Both substrates acylate subtiligase; the glycolate esters are more reactive and are generally preferred. (B) Structures of the glycolate phenylalanyl and lysyl esters. Both the phenylalanyl and lysyl glycolate esters acylate subtiligase efficiently; the latter class is more soluble and thus more generally used.

Sequence Specificity of Subtiligase

The substrate specificity of wild-type subtilisin BPN' has been extensively studied.[16,17] Substrates are bound over a seven-residue sequence, from P_4 to P_3'. The preferred residues at each of these subsites in subtilisin remain consistent in the double mutant subtiligase (Fig. 2). For acylating the enzyme, hydrophobic side chains are preferred at P_4, whereas positions P_3 and P_2 are less discriminating. Hydrophobic residues are best at P_1, whereas Pro and β-branched residues are poor. On the nucleophile side, P_1' will accept most residues, with the exception of β-branched and acidic residues (Table I).

Using Table I and the rules outlined above it should be possible to target general areas of the primary sequence as possible ligation junctures. In practice, we have found that prediction of good ligation sequences can be challenging, as ligation efficiency depends on the entire substrate sequence in ways that do not always reflect the sum of the subsite specificities. For this reason, it is usually best to perform test ligations of a given site by preparing the corresponding tetramer peptides and thus determine the ligation efficiency directly. In general, the ligation of the tetramer sequences is representative of how the full-length peptides will perform. If the tetramer sequences ligate well but the full-length peptides do not, the most likely causes are either insolubility, aggregation, or secondary structure of the peptides.

[16] D. A. Estell, T. P. Graycar, J. V. Miller, D. B. Powers, J. P. Burnier, P. G. Ng, and J. A. Wells, Science 233, 659 (1986).
[17] H. Grøn, M. Meldal, and K. Breddam, Biochemistry 31, 6011 (1992).

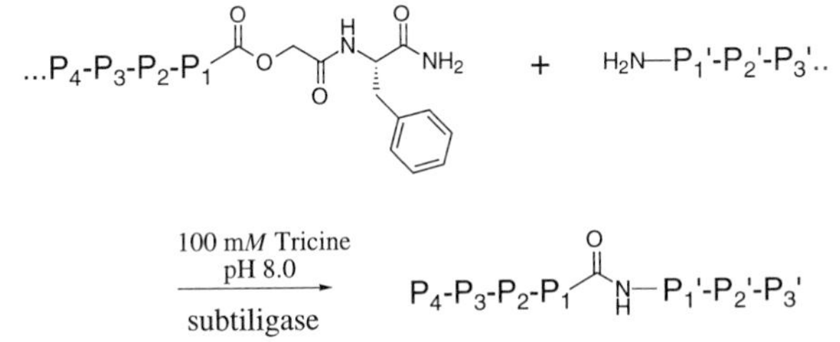

FIG. 2. Subtiligase binds to the acceptor and donor substrates in much the same way that wild-type subtilisin binds to its substrate. The substrates must be in an extended conformation, and specificity is determined by the individual residues at each subsite, P_4 through P_3'.

Examples

Total Synthesis of Proteins: RNase A

The usefulness of subtiligase for total protein synthesis has been demonstrated with the synthesis of RNase A.[1] Using a sequential segment condensation approach, both wild-type RNase A and a series of mutants containing unnatural amino acids were synthesized in milligram quantities. Peptide fragments from RNase A were synthesized as the glycolate phenylalanyl esters using standard Boc (*tert*-butyloxycarbonyl) chemistry on a solid support as described in the Methods section. The N terminus of peptides derived from internal sections of the sequence were protected with the isonicotinyl carbamate group (iNoc), which is stable to HF cleavage but can be selectively removed from the free peptide using zinc metal in acetic acid. Thus, two peptides can be ligated together and then the new N terminus can be deprotected for the next ligation step.

The protein RNase A was divided into six fragments from 11 to 31 residues in length, and the peptides were ligated sequentially in the C to N direction (Fig. 3). The average yield for the ligation step was 66%, and the deprotection of the iNoc group averaged 86%.

In addition to wild-type RNase A, a series of mutants were prepared. The two critical histidines, His-12 and His-119, which act as a general base and a general acid, respectively, in the cleavage of RNA substrates, were mutated to the His analog 4-fluorohistidine (fHis) either individually or together. The pK_a of 4-fHis is about 3.3 units lower than His, and thus the pH rate profile for RNase A should be indicative of the catalytic function of these histidines. As anticipated, the pH rate maximum shifts dramatically in the His12fHis, His119fHis double mutant to a maximum rate at approximately pH 4.0, as opposed to 6.5 for the wild-type enzyme. More surprisingly, the maximal rate of the double mutant is only down by a factor

TABLE I

Amino Acid Preferences at Each of the Subsites of Subtiligase

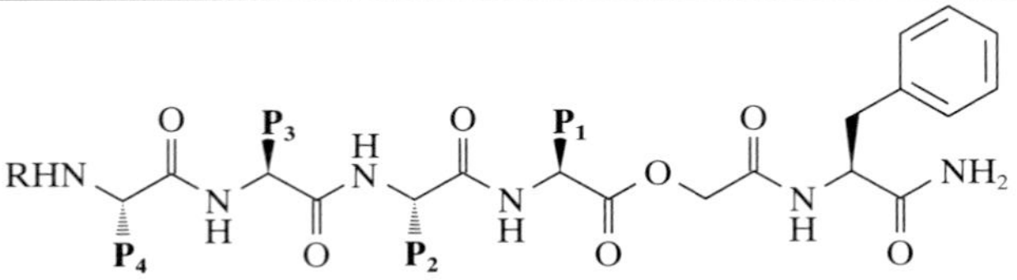

Glycolate ester

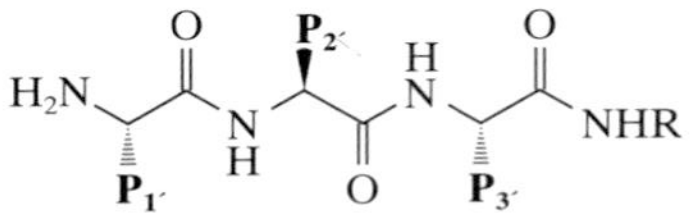

Nucleophile

Preference	P_4	P_3	P_2	P_1	P_1'	P_2'	P_3'
Preferred	Hydrophobic	Variable	Variable	Hydrophobic, positive	Hydrophobic, polar	Hydrophobic	Variable
Best	W, Y, F, M, L, I, A	—	—	M, Y, F, A, K, R, L	—	—	—
Avoid	—	—	—	I, V, T, P, G	β-Branched, acidic, Pro	Acidic, Pro	—

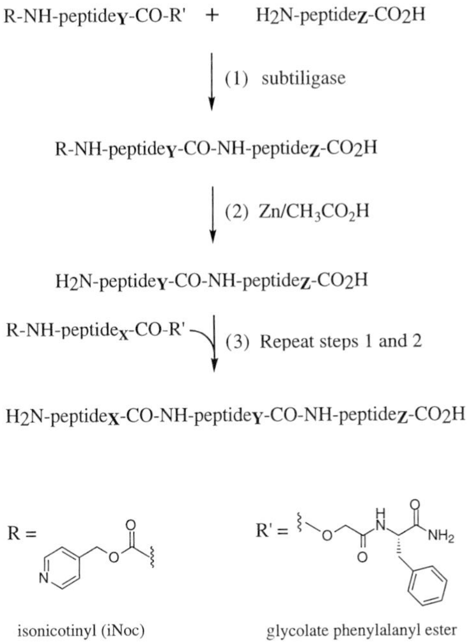

R-NH-peptide$_Y$-CO-R'   +   H$_2$N-peptide$_Z$-CO$_2$H

$\downarrow$ (1) subtiligase

R-NH-peptide$_Y$-CO-NH-peptide$_Z$-CO$_2$H

$\downarrow$ (2) Zn/CH$_3$CO$_2$H

H$_2$N-peptide$_Y$-CO-NH-peptide$_Z$-CO$_2$H

R-NH-peptide$_X$-CO-R' $\searrow$ $\downarrow$ (3) Repeat steps 1 and 2

H$_2$N-peptide$_X$-CO-NH-peptide$_Y$-CO-NH-peptide$_Z$-CO$_2$H

R =   R' =

isonicotinyl (iNoc)   glycolate phenylalanyl ester

FIG. 3. Strategy for the blockwise synthesis of proteins or large peptides using subtiligase. The N-terminal donor peptide (glycolate ester) is protected on the N terminus with the iNoc group while the C-terminal acceptor peptide is fully deprotected. After subtiligase-catalyzed ligation (step 1), the ligation product can be purified and the iNoc protecting group removed (step 2). The next N-terminal glycolate ester fragment can then be ligated and the cycle repeated as necessary.

of 3, suggesting that proton transfer is more important for the catalytic mechanism than the actual pK_a of the residue.

The synthesis of RNase A in 10% overall yield on a milligram scale is an important advance in the total synthesis of proteins. The sequential assembly of peptide fragments allows for the facile incorporation of fragments containing unnatural residues or alternative backbones, an important ability for deciphering reaction mechanisms and engineering novel proteins. Synthetically, high purity products are accessible, as any side products can be removed during the synthesis by purification of each individual peptide and of the ligation products. This is in stark contrast to the synthesis of proteins by SPPS where deletion or other errors are carried on throughout the synthesis, leading to a mixture of chemically similar products that can be difficult to separate.

Semisynthesis of Proteins Using Subtiligase

Another important application of subtiligase is in the semisynthesis of proteins. The high specificity of the enzyme for the N-terminal amine of proteins allows for site-specific derivatization of proteins. For example, both biotinylated human growth hormone (hGH) and a mercurated Cys derivative were prepared by ligating the corresponding peptide onto the N terminus of hGH.[10] Other possibilities include incorporation of affinity handles, isotopic labels, or polyethylene glycol (PEG) polymers. The N-terminal specific point of attachment offers many advantages over traditional nonspecific derivatization methods.

The success of N-terminal ligations is determined by two factors, the suitability of the N-terminal sequence as a nucleophile and whether the N terminus of the protein is accessible to subtiligase in an extended conformation. Ligation of the biotinylated peptide onto the N terminus of Met-hGH (sequence Met-Phe-Pro-Thr-Ile-Pro . . .) proceeds with 95% efficiency, whereas ligation of the same biotinylated peptide onto hGH (sequence Phe-Pro-Thr-Ile-Pro . . .) only yields 2% of the ligation product. This experiment emphasizes the importance of having an appropriate N-terminal sequence and the dramatic influence of Pro at P_2'.

Improving Stability of Subtiligase: Stabiligase

Extensive mutagenesis work on the native enzyme subtilisin has generated a number of interesting mutants that have enhanced stability under denaturing conditions.[18,19] Five of these mutations were incorporated into subtiligase to create a heptamutant termed stabiligase.[10] In contrast to subtiligase or subtilisin, which suffer greater than a 50% loss of activity in 1 M guanidine hydrochloride, stabiligase retains greater than 50% activity even in 4 M guanidine hydrochloride. A version of hGH with the N-terminal eight residues missing has a favorable sequence for subtiligase (Leu-Phe-Asp-Asn . . .), yet efforts to ligate the biotinylated peptide with subtiligase yielded only 5% of the ligation product. The low efficiency is attributed to the likelihood that the N terminus is in an α-helical conformation; in wild-type hGH this sequence is within helix 1. By performing the same ligation reaction under mildly denaturing conditions [0.1% sodium dodecyl sulfate (SDS), 55°] the yield was improved to 45% using stabiligase. The use of protein engineering to tailor the properties of subtiligase offers great potential, particularly with respect to stability and specificity mutants. Engi-

[18] C.-H. Wong, S.-T. Chen, W. J. Hennen, J. A. Bibbs, Y.-F. Wang, J. L.-C. Liu, M. W. Pantoliano, M. Whitlow, and P. N. Bryan, *J. Am. Chem. Soc.* **112**, 945 (1990).
[19] K. Chen and F. H. Arnold, *Proc. Natl. Acad. Sci. U.S.A.* **90**, 5618 (1993).

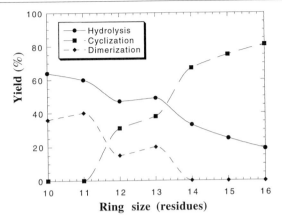

FIG. 4. Subtiligase-catalyzed peptide cyclizations as a function of peptide length. Hydrolysis of the ester is the predominate reaction until the peptide is at least 13 residues in length.

neering of the subtilisin binding site has led to the development of proteases with unique specificities.[20,21] Application of this strategy to subtiligase has generated variants with enhanced specificity for specific P_1 residues,[9] but this area is still largely unexplored.

Cyclization of Linear Peptide Esters

Cyclic peptides provide an important link in the transition from the discovery of peptide ligands, which bind to proteins, to the development of small molecule drug candidates. Through cyclization, conformational restriction often leads to higher affinity ligands and more precise structure–activity relationships. Synthetic peptides can be cyclized using methodologies developed for solid-phase synthesis, or through solution-phase chemistry. Difficulties are often encountered owing to the large entropic cost of cyclizing peptides greater than 10 residues in length, and in the case of solution-phase chemistry intermolecular oligomerization can occur.

Subtiligase has been shown to be effective for cyclization of peptides (see Fig. 4) greater than 12 residues in length.[11] The same subsite specificity requirements apply, although the chain length requirement appears to indicate the minimal size of the loop that can offer both the N and C terminus in an extended conformation. A comparison of ring size versus the yield of uncyclized, dimerized, and cyclized product emphasizes the efficiency

---

[20] P. Carter, B. Nilsson, J. P. Burnier, D. Burdick, and J. A. Wells, *Proteins* **6**, 240 (1989).
[21] M. D. Ballinger, J. Tom, and J. A. Wells, *Biochemistry* **35**, 13579 (1996).

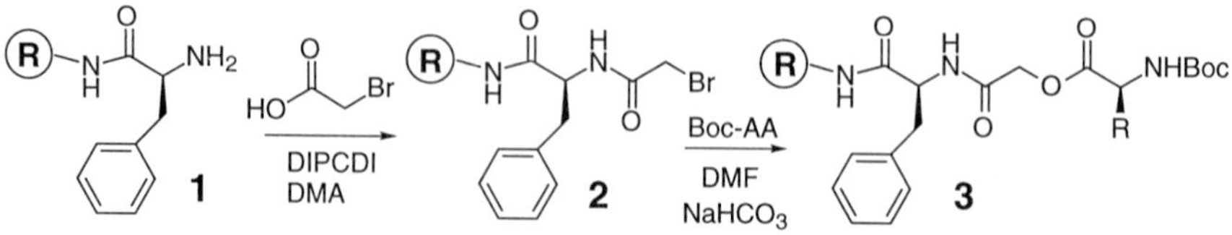

SCHEME 1. Preparation of resin for the synthesis of glycolate esters through Boc chemistry.

with which subtiligase catalyzes the formation of cyclic peptides greater than 12 residues in length.

Methods: Preparation of Substrates

Methodologies for the solid-phase synthesis of glycolate esters by either Boc[1] or Fmoc (fluorenylmethyloxycarbonyl)[22,23] chemistry have been developed. Both approaches are facile and efficient; the resulting peptide esters are stable and require no special handling.

tert-Butyloxycarbonyl Chemistry

Phenylalanyl-*p*-methylbenzhydrylamine (MBHA) resin (1) (10 g, 6.3 mmol) (Advanced ChemTech) is mixed with bromoacetic acid (4.38 g, 31.5 mmol) and 1,3-diisopropylcarbodiimide (DIPCDI, 31.5 mmol, 3.97 g) for 1 hr at 25° in dimethylacetamide (DMA, 25 ml) to generate the bromoacetyl derivative (2) (Scheme 1). The resin is washed extensively with DMA, and then a solution of the appropriate C-terminal butyloxycarbonyl (Boc) protected amino acid (3 equivalents) in dimethylformamide (DMF, 25 ml) is added. Stirring with sodium bicarbonate (37.8 mmol, 3.18 g) at 50° for 24 hr generates the corresponding Boc-amino acid-glycolate-Phe ester resin (3). The resin is washed extensively with DMA and dichloromethane (DCM) and then dried under vacuum. The resin can be stored at room temperature for at least 2 months. Peptide synthesis using standard butyloxycarbonyl SPPS protocols can be carried out either manually or with an automated peptide synthesizer. Elongation of the peptide chain proceeds without noticeable loss of peptide through the potentially labile ester linkage.

If the peptide glycolate ester is to be used for testing a ligation site, the N terminus can be capped with either a succinyl group (Suc) or an acetyl group. For peptides that will be sequentially ligated, the N terminus of the

[22] J. K. Judice, A. K. Namenuk, and J. P. Burnier, *Bioorg. Med. Chem. Lett.* **6**, 1961 (1996).
[23] D. J. Suich, M. D. Ballinger, J. A. Wells, and W. F. DeGrado, *Tetrahedron Lett.* **37**, 6653 (1996).

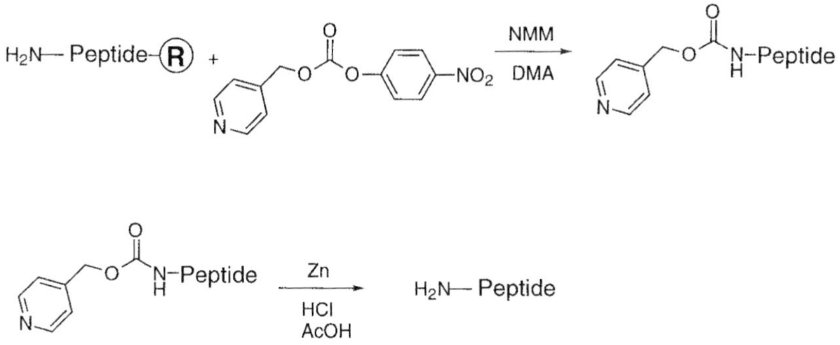

SCHEME 2. Incorporation and removal of the iNoc protecting group.

peptide is protected using the isonicotinyl carbamate (iNoc) group to prevent self-coupling (Scheme 2).[24] Following removal of the final N-terminal
Boc group, the iNoc group is incorporated by stirring the resin with iNoc
nitrophenyl carbonate (3 equivalents) and N-methylmorpholine (NMM, 6
equivalents) in DMA at 25° for 24 hr to give the fully protected peptide.
The iNoc group is stable to HF cleavage; removal of the iNoc group from
the otherwise deprotected peptide is accomplished by treatment with HCl-
activated zinc dust (10 equivalents) in acetic acid for 2 hr. The reaction is
monitored by analytical high-performance liquid chromatography (HPLC),
and on completion the zinc dust is removed by filtration, the acetic acid
removed by evaporation, and the product purified by preparative reversed-
phase HPLC.

Cleavage of the peptide from the resin is accomplished under standard
conditions, in anhydrous hydrofluoric acid with 5% anisole and 5% ethyl
methyl sulfide. The use of free thiols as scavengers should be avoided, as
thiolysis of the glycolate ester linkage may occur.

The glycolate ester peptides are purified by reversed-phase HPLC using
an acetonitrile gradient in water with 0.1% trifluoroacetic acid (TFA). After
lyophilization, the glycolate ester peptides can be stored in a desiccator for
extended periods without degradation.

As an alternative to the glycolate Phe amide ester, we have found that
the glycolate Lys amide ester works equally well as an acyl donor for
subtiligase yet offers superior solubility for more hydrophobic sequences.
Preparation of the glycolate Lys amide esters follows an identical procedure
with the substitution of Boc-Lys(Cl-Z) for Boc-Phe.

[24] D. F. Veber, W. J. Paleveda, Y. C. Lee, and R. Hirschmann, *J. Org. Chem.* **42**, 3286 (1977).

Fluorenylmethyloxycarbonyl Chemistry

For rapid synthesis of glycolate ester peptides Fmoc chemistry offers the advantage of facile cleavage with TFA. The following procedure has been developed for the preparation of a glycolate derivatized resin suitable for Fmoc chemistry[22]; an alternative approach where the P_1 residue is first converted to the glycolate ester before attachment to the resin has also been reported.[23]

A 10 g (0.5 mEq/g, 5 mmol) portion of Rink resin (Advanced Chem-Tech) is swollen with 100-ml washes of DCM and DMA. The resin is then treated with 20% piperidine in DMA for 15 min, then washed 5 times with DMA, once with DCM, and resuspended in 10 ml of DCM. Fmoc-Phe (5.8 g, 15 mmol) is activated in a separate vessel with 6.6 g of benzotriazol-1-yloxytris(dimethylamino)phosphonium hexafluorophosphate (BOP; 15 mmol) and 3.3 ml (30 mmol) of *N*-methylmorpholine (NMM) in 10 ml DMA; after 10 min, this solution is added to the resin and the resulting suspension agitated for 1 hr to generate the Fmoc-Phe derivatized resin (**4**, Scheme 3). After washing with DMA and DCM as before, the resin is treated with 20% piperidine, washed as before, and then resuspended in 20 ml DCM. To this is added a solution of 3.5 g (30 mmol) of acetoxy acetic acid in 10 ml of DCM, followed by 30 ml of a 1 *M* solution of 1,3-diisopropylcarbodiimide (DIPCDI) in DCM to form the acetoxyacetamide (**5**). After 30 min the resin is washed as before and resuspended in 40 ml DMA. To this is added 5 ml (100 mmol) of hydrazine monohydrate. This suspension is agitated for 6–12 hr to deprotect the hydroxyl group, providing **6**; the resin is then washed as before and resuspended in DCM. To this is added a solution of the amino acid corresponding to the P_1 residue (20

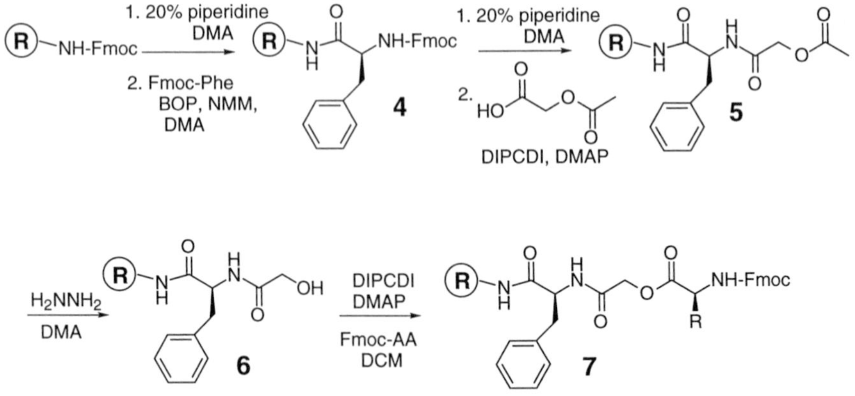

SCHEME 3. Preparation of resin for the synthesis of glycolate esters through Fmoc chemistry.

mmol, 4 equivalents) in DCM, followed by 20 ml of 1 M DIPCDI in DCM and 0.1 mol% DMAP for 1 hr to give **7**. Resin substitution can be checked at this point.[25] Peptide synthesis is then continued under normal Fmoc chemistry conditions.

This resin can be prepared up to the alcohol (**6**) and stored for at least 6 months. The corresponding glycolate Lys amide resin can be prepared if solubility of the peptide–glycolate ester is a potential problem. Peptides are again capped at the N terminus or protected using the iNoc group if the peptide will be extended further. Cleavage of the glycolate ester peptide is accomplished with 5% triisopropylsilane in anhydrous TFA for 1 to 3 hr. Purification of the peptides is again carried out using standard reversed-phase HPLC techniques with a gradient of acetonitrile in water containing 0.1% TFA. The glycolate ester peptides are stable for extended periods when stored in a desiccator.

Peptide Ligations

Standard Protocol

Ligations of the activated esters with the appropriate nucleophiles take place under very mild conditions. Typically the only side reaction is the hydrolysis of the glycolate ester. When possible, the glycolate ester is used in sufficient excess to drive the reaction to completion. The standard ligation buffer is 100 mM Tricine at pH 8.0; the buffer should be freshly prepared and degassed to minimize oxidation of the active site Cys. Nucleophilic buffers such as Tris should be avoided to prevent reaction with the activated ester or acyl enzyme intermediate. A stock solution of the glycolate ester can be prepared in either dimethyl sulfoxide (DMSO) or DMA; DMF must not be used as it will inhibit the enzyme. The stock solution should be of high enough concentration that the reaction will not contain greater than 5% organic solvent. Alternatively, the glycolate ester peptide can be added as a lyophilized powder directly to the solution containing the enzyme and the nucleophile peptide.

A typical ligation reaction begins with 100 μl of freshly prepared and degassed 100 mM Tricine (pH 8.0) at 25° containing 1 mM nucleophilic peptide and 0.5 mol% subtiligase. A 5-fold excess of the glycolate ester peptide is added, either as a lyophilized powder or as a concentrated stock in dimethyl sulfoxide. The reaction is incubated at 25° for 1 to 2 hr; reaction progress can be monitored by analytical reversed-phase HPLC or by SDS–

---

[25] J. Meienhofer, M. Waki, E. P. Heimer, T. J. Lambros, R. C. Makofske, and C.-D. Chang, *Int. J. Pept. Protein Res.* **13**, 35 (1979).

polyacylamide gel electrophoresis (SDS–PAGE). When the ligation is complete, the product is purified either by preparative HPLC or by standard protein chromatography methods (see Fig. 5).

Troubleshooting

Ligation junctures should be evaluated in the context of tetrameric model peptides prior to undertaking syntheses of longer fragments. If inefficient ligation is observed there are several common causes, each with available solutions and/or diagnostic protocols. With some glycolate ester/nucleophile pairs, hydrolysis of the ester can proceed more rapidly than ligation such that the ester is consumed before the nucleophile. Addition of several more equivalents of the ester is frequently adequate to drive the reaction to completion.

Another potential complication is that the enzyme may be inhibited by the ligation product. This can be overcome by adding more subtiligase. Because the enzyme has no detectable proteolysis activity, this does not adversely affect product yield or purity.

If neither of the two approaches above result in efficient ligation, it can be informative to determine whether the source of difficulty comes from the ester or the nucleophile. First, the ability of the glycolate ester to acylate subtiligase can be probed by comparing the rates of hydrolysis (i.e., the rate of appearance of the peak analogous to D in Fig. 5) in the presence and absence of subtiligase. If hydrolysis is not markedly faster in the presence of the enzyme, the glycolate ester in question is not a good substrate. In this

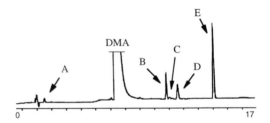

FIG. 5. HPLC trace of a test ligation between Suc-Ala-Ala-Pro-Phe-glycolate lysyl ester and nucleophile H_2N-Met-Phe-Ala. An equimolar mixture of both the donor and the acceptor peptides (10 mM each) was treated with subtiligase (14 μM) in 100 mM Tricine (pH 8.0) for 30 min at 25°. Peak A, Glycolate lysyl amide fragment; B, unreacted H_2N-Met-Phe-Ala; C, unreacted Suc-Ala-Ala-Pro-Phe-glycolate lysyl ester; D, hydrolyzed ester Suc-Ala-Ala-Pro-Phe-CO_2H; E, ligation product Suc-Ala-Ala-Pro-Phe-Met-Phe-Ala-CO_2H. The large DMA peak is from the stock solution of the Suc-Ala-Ala-Pro-Phe-glycolate lysyl ester. High-performance liquid chromatography was conducted on a C_{18} reversed-phase column with a gradient of 0 to 50% acetonitrile (0.1% TFA) in water (0.1% TFA) over 17 min and detection at 214 nm. All peaks were identified by electrospray mass spectroscopy.

case the ligation propensity of the nucleophile can be assessed by pairing it with an ester component known to acylate the enzyme efficiently; if ligation is observed, the problem at the site in question lies solely with the original glycolate ester.

If, however, the glycolate ester is hydrolyzed efficiently by subtiligase, then the nucleophile may be unreactive. This can be confirmed by testing the glycolate in question with a nucleophile known to perform well in other sites. If ligation is still not observed the problem is more obscure and may be difficult to assess. In either case—poor nucleophile or poor glycolate ester—an alternative ligation juncture should be evaluated.

Summary

Application of protein engineering strategies to the redesign of the active site of subtilisin has successfully generated an efficient peptide ligase, subtiligase. The novel enzyme subtiligase has been shown to have many uses, from the total synthesis of RNase A to the semisynthesis of a variety of other proteins. Although the enzyme is in an early stage of development, it shows great promise. Subtiligase will certainly be a useful and important addition to the available strategies for the synthesis of proteins via segment condensation.

[15] Convergent Solid-Phase Peptide Synthesis

By FERNANDO ALBERICIO, PAUL LLOYD-WILLIAMS, and ERNEST GIRALT

Introduction

There are two main strategies for the chemical synthesis of peptides.[1] Chain elongation in linear synthesis is carried out by repetitive N^α-amino group deprotection and protected amino acid coupling steps.[2–4] Convergent synthesis, however, involves the synthesis and coupling of protected peptide

---

[1] P. Lloyd-Williams, F. Albericio, and E. Giralt, "Chemical Approaches to the Synthesis of Peptides and Proteins." CRC Press, Boca Raton, Florida, 1997.

[2] R. B. Merrifield, *Angew. Chem. Int. Ed. Engl.* **24,** 799 (1985).

[3] G. B. Fields, Z. Tian, and G. Barany, *in* "Synthetic Peptides: A User's Guide" (G. A. Grant, ed.), p. 77. Freeman, New York, 1992.

[4] R. B. Merrifield, *in* "Peptides: Synthesis, Structures and Applications" (B. Gutte, ed.), p. 93. Academic Press, New York, 1995.

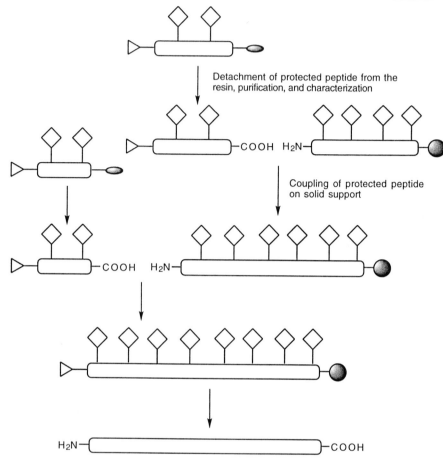

FIG. 1. Convergent solid-phase peptide synthesis strategy.

segments (see Fig. 1).[5,6] Both strategies can be carried out in solution or on a solid support, although solution and solid-phase methodologies can coexist in convergent synthesis. Protected peptide segments can be elaborated on solid supports and coupled in solution or vice versa. In this chapter, however, we discuss only the solid-phase synthesis and coupling of protected peptides. The reader should also consult the bibliographies found in the references cited.

[5] P. Lloyd-Wiliams, F. Albericio, and E. Giralt, *Tetrahedron* **49**, 11065 (1993).
[6] H. Benz, *Synthesis*, 337 (1994).

Convergent solid-phase peptide synthesis (CSPPS) involves (i) solid-phase synthesis of protected peptides (these must retain the protecting groups of the N^α-amino and the side-chain functions after cleavage from the resin), (ii) purification and characterization of the protected peptides, and (iii) their solid-phase coupling. The final steps of detachment of the free peptide from the solid support and formation of disulfide bridges, when necessary, are common to those carried out in a linear strategy and are discussed elsewhere in this volume.

Planning a Synthesis

Before starting a synthesis several factors must be considered, including the protecting group strategy [tert-butoxycarbonyl (Boc)/benzyl (Bzl) or 9-fluorenylmethoxycarbonyl (Fmoc)/tert-butyl (tBu)], the size of the different protected peptides, the residues at the C- and N-terminal positions, and the support to be used for solid-phase segment couplings.

Both the Boc/Bzl and Fmoc/tBu approaches have been used equally successfully for the preparation of protected peptides, although the milder conditions used in the latter have led to its being progressively more favored. The principal advantage of Fmoc/tBu is that both protecting groups are compatible with the use of linkers (C-terminal carboxylic protecting groups) labile to extremely dilute acid. The manipulations associated with the cleavage of protected peptides from these linkers are synthetically straightforward when compared to the application of ultraviolet (UV) light or the use of transition metal complexes that are sometimes necessary in the Boc/Bzl strategy. From the point of view of solubility, no significant differences have been reported between both kinds of protected peptides, although no systematic study has been done.

Regarding the protection of the amino acid side chains, although there are some (Thr, Ser, Tyr, Gln, Asn, His, Trp) whose protection is not mandatory, the tendency is to use maximal protection. This ensures that the risk of side reactions such as acylation of Ser, Thr, and Tyr and the alkylation of Trp is substantially decreased. The protection of His is important, because the imidazole can be acylated either by a protected peptide or by any acetylation reagent used after coupling (see below). During base treatment a transacylation reaction can then take place, leading either to insertion of an extra peptide or to capping of the growing chain with the acetyl moiety.[7] Furthermore, unprotected Asn and Gln can lead to aggregation of the peptide chains via inter- or intrachain hydrogen bonds,

---

[7] C. Celma, F. Albericio, E. Pedroso, and E. Giralt, *Pept. Res.* **5**, 62 (1992).

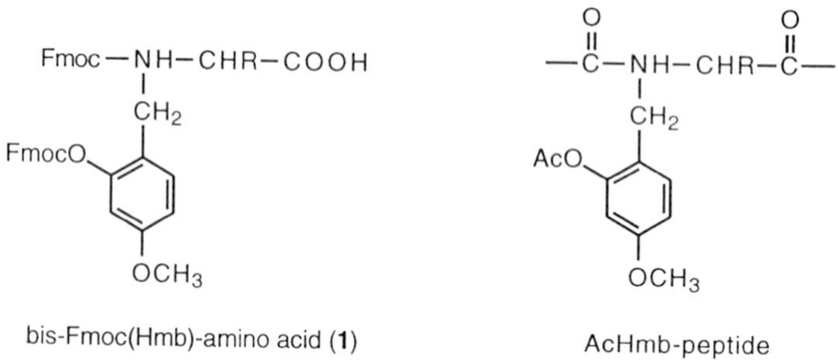

bis-Fmoc(Hmb)-amino acid (1)   AcHmb-peptide

FIG. 2. Structure of Hmb derivatives.

resulting in poor solubility of the protected peptide and/or collapse of the peptide–resin. Both effects will adversely affect the coupling yield.

In the Fmoc/*t*Bu approach, another way to increase the solubility of the protected peptides is through backbone protection using the 2-acetoxy-4-methoxybenzyl (AcHmb) group.[8] The advantage of AcHmb protection in a convergent strategy is 2-fold.[9] The presence of the AcHmb group on the protected peptide segment will improve its solubility, facilitating its purification and coupling. However, AcHmb groups on the fragment bound to the resin reduce interchain association, believed to be the main cause for incomplete deprotection and coupling. This group is incorporated into a protected peptide using the corresponding *N,O*-bis-Fmoc-*N*-(2-hydroxy-4-methoxybenzyl)amino acids (1, Fig. 2) and the phenol of the backbone protecting group is acetylated before the cleavage of the peptide from the resin.

The size of the protected peptides and the residue at the C-terminal position are intrinsically linked. Furthermore, in many cases the necessity of having a determined residue at the C-terminal position dictates the size of the protected peptide. As is considered below, the coupling of a protected peptide segment is much more demanding than for a single protected amino acid. First of all, the risk of epimerization at the C-terminal residue increases notably. This is mainly due to the fact that urethane protection (Boc and Fmoc) of single amino acids slows down the formation of the corresponding oxazolone, which is one of the possible intermediates in epimerization processes. Even if the oxazolone is formed, the urethane group destabilizes the carbanion that would be generated by proton abstraction. In the case of the coupling of a protected peptide the substituent at the N^α-amino

---

[8] T. Johnson, M. Quibell, and R. C. Sheppard, *J. Pept. Sci.* **1**, 11 (1995).

[9] M. Quibell and T. Johnson, *in* "Peptides 1994: Proceedings of the Twenty-third European Peptides Symposium" (H. L. S. Maia, ed.), p. 173. Escom, Leiden, The Netherlands, 1995.

group of the C-terminal residue is an acyl group, which promotes oxazolone formation. For this reason, Gly, where epimerization cannot occur, and Pro, where for steric reasons oxazolonium salt formation is strongly disfavored, are the residues of choice to be selected as the C-terminal amino acid. However, Pro and Gly have a significant drawback, in that they can lead to substantial amounts of diketopiperazine (DKP) formation during the deprotection of the second amino acid of the sequence.[10,11] This intramolecular cyclization will be favored by the presence of good leaving groups at the peptide–resin anchorage. This is the case for benzyl, allyl, and oxime type resins. In the case of benzyl resins, the presence of an additional electron withdrawing group (as in photolabile resins) on the aromatic ring will also exacerbate the problem. However, DKP formation can be suppressed when the C terminus has a poor leaving group (as in fluorenylmethyl-based resins) or a large steric requirement (as in the trityl-based resins).

In other cases, for Boc chemistry, the formation of DKP can be diminished using coupling methods in which neutralization of the dipeptide–resin is done *in situ* during coupling of the third amino acid, rather than in a separate wash step with tertiary amine. A convenient protocol consists of removing the Boc group with trifluoroacetic acid (TFA)–dichloromethane (DCM) and carrying out the coupling with the third protected amino acid with (benzotriazol-1-yloxy)tris(pyrrolidino)phosphonium hexafluorophosphate (PyBOP) or (7-azabenzotriazol-1-yloxy)tris(pyrrolidino)phosphonium hexafluorophosphate (PyAOP) in the presence of N,N-diisopropylethylamine (DIEA).[12] For Fmoc/tBu strategy, the second amino acid has to be introduced with the N^{α}-amino function protected with the Trt group, as this protecting group can be selectively removed with very dilute acid solution (0.2–1% TFA in DCM) in the presence of tBu type protecting groups.[13] Although the formation of DKP can also be catalyzed by acids, the extent of this side reaction is more severe in the presence of bases. Therefore, more care must be taken in the Fmoc/tBu strategy, which involves treatment with a secondary amine.

If it is not possible to have Gly or Pro as C-terminal residues, there is no clear rule, although it appears that residues lacking β-substitution should be the most resistant to epimerization.[14] However, the process is sequence

[10] B. F. Gisin and R. B. Merrifield, *J. Am. Chem. Soc.* **94**, 3102 (1972).
[11] E. Giralt, R. Eritja, and E. Pedroso, *Tetrahedron Lett.* **22**, 3779 (1981).
[12] M. Gairí, P. Lloyd-Williams, F. Albericio, and E. Giralt, *Tetrahedron Lett.* **31**, 7363 (1990).
[13] J. Alsina, E. Giralt, and F. Albericio, *Tetrahedron Lett.* **37**, 4195 (1996).
[14] N. L. Benoiton, Y. C. Lee, and F. M. F. Chen, *in* "Peptides, Chemistry and Biology: Proceedings of the Twelfth American Peptide Symposium" (J. A. Smith and J. E. Rivier, eds.), p. 496. Escom, Leiden, The Netherlands, 1992.

dependent and is favored by long reaction times. Besides epimerization, other side reactions can also intervene during coupling. Thus, unprotected Asn or Gln and Thr(Bzl) can lead to cyanoderivatives and α-aminocrotonic acid, respectively.[15] Each individual coupling should therefore be treated on a case-by-case basis. Similar considerations apply to selecting the most suitable residue for the N-terminal position. The unsuccessful coupling of a protected peptide to a peptide resin with N-terminal Pro has been reported,[16] but in our own laboratory we have obtained excellent yields in a similar coupling.[17] When Pro is at the N terminus, monitoring of the coupling is more difficult, because the ninhydrin test cannot be used.[18] When Gln and Glu(OBzl) are at the N terminus, pyroglutamyl formation can occur, terminating the peptide chain.[15,17] Again, the selection of amino acids lacking β-branching should facilitate segment coupling and disfavor possible side reactions.

The prediction of the optimum size for protected peptides is difficult. The two most desirable characteristics of protected peptides for CSPPS are purity and solubility, and, in principle, both depend on size. For some large peptides it is difficult to obtain pure material and sometimes even to assess purity. Although the solubility of protected peptides tends to be more dependent on sequence than on length, peptides of more than 18 residues are often very poorly soluble. However, the purification and successful coupling of a 21-amino acid protected peptide having three AcHmb backbone protecting groups has been reported.[19] Peptides having between 8 and 15 residues appear to be optimal for the CSPPS strategy.[17]

In CSSPS, the nature of the support and the degree of functionalization are important. Polystyrene (PS),[20] polyacrylamide,[21] and polyethylene glycol (PEG)-grafted polystyrene[22] have given best results. Levels of substitution (typically in the range 0.04–0.3 mmol/g), lower than in the stepwise solid-phase peptide synthesis, are normally used.[19,23] This reduction of substitution level is normally done during assembly or incorporation of the first segment. Thus, the first amino acid or the first protected segment (see

---

[15] S. Sakakibara, *Biopolymers* (*Pept. Sci.*) **37**, 17 (1995).

[16] S. Nakagawa, H. S. H. Lau, F. J. Kézdy, and E. T. Kaiser, *J. Am. Chem. Soc.* **107**, 7087 (1985).

[17] A. Grandas, F. Albericio, J. Josa, E. Giralt, E. Pedroso, J. M. Sabatier, and J. van Rietschoten, *Tetrahedron* **45**, 4637 (1989).

[18] E. Kaiser, R. L. Colescott, C. D. Bossinger, and P. I. Cook, *Anal. Biochem.* **34**, 595 (1970).

[19] M. Quibell, L. C. Packman, and T. Johnson, *J. Am. Chem. Soc.* **117**, 11656 (1995).

[20] G. Barany and R. B. Merrifield, in "The Peptides: Analysis, Synthesis, Biology" (E. Gross and J. Meienhofer, eds.), Vol. 2, p. 1. Academic Press, New York, 1979.

[21] F. Albericio, M. Pons, E. Pedroso, and E. Giralt, *J. Org. Chem.* **54**, 360 (1989).

[22] N. Kneib-Cordonier, F. Albericio, and G. Barany, *Int. J. Pept. Protein Res.* **35**, 527 (1990).

[23] K. Barlos, D. Gatos, and W. Schäfer, *Angew. Chem. Int. Ed. Engl.* **30**, 590 (1991).

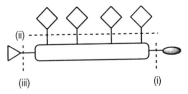

(i) Resin peptide-linker ("pseudo-permanent")
(ii) Side-chain protecting groups ("permanent")
(iii) N^α-Protecting group ("temporary")

FIG. 3. Protection scheme for CSPPS.

below) are incorporated using less than the amount required to react with the most active sites on the resin, and the resin is acetylated to block the remainder.

Solid-Phase Synthesis of Protected Peptides

For CSPPS, three types of protection are required (see Fig. 3). The resin–peptide linker must be stable to the repetitive treatments necessary for the removal of the N^α-amino protecting group of the growing peptide chain, and it must allow the detachment of the protected peptide from the resin without provoking the premature deprotection of any of the protecting groups. This requirement differentiates CSPPS from linear SPPS, because in the latter the peptide is detached from the resin with concomitant removal of the side-chain protecting groups. The peptide–resin anchorage used for the generation of protected peptide segments may be considered to be pseudopermanent protection of the C terminus. The functional groups of the amino acid side chains must be protected with groups that are stable to both the removal of the N^α-amino protecting group and the cleavage of the protected peptide from the resin. Such side-chain protecting groups are called permanent. Finally, the protecting group of the N^α-amino function, referred normally as temporary, must be stable to the detachment of the peptide from the resin.

Although a strictly three-dimensional orthogonal scheme[24,25] is not required, it is necessary that at least the temporary and pseudopermanent protection be orthogonal, because either may then be removed in the presence of the other. Thus, the Boc group as N^α-amino protection precludes the use of an acid-labile peptide–resin anchorage, as the conditions used to remove the Boc group would lead to detachment of peptide from

[24] G. Barany and R. B. Merrifield, *J. Am. Chem. Soc.* **99**, 7363 (1977).
[25] G. Barany and F. Albericio, *J. Am. Chem. Soc.* **107**, 4936 (1985).

the resin. Conversely the conditions required to detach peptides from more acid-stable peptide–resin linkers will remove the Boc group. For such N^α-amino protection, linkers that can be cleaved with light, metal complexes, or nucleophiles are required. However, the use of the Fmoc group as temporary protection is compatible with acid-labile resin–peptide linkers, as well as those cleavable by light and metal complexes.

Acid-Labile Resins

In the Fmoc/*t*Bu strategy, the most convenient peptide–resin linkers are ones that are highly acid-labile. From now on we discuss only those most commonly used. For an exhaustive listing of linkers, see elsewhere.[26] The detachment of protected peptides from these resins can be achieved with acid solutions as dilute as 1% (v/v) TFA in DCM, or sometimes even less. These conditions do not affect *t*Bu-based protecting groups. The most widely used handles and functionalized resins (for a definition of handles, as well as the advantages of their use compared to functionalized resins, see Ref. 26) of this kind are the 4-hydroxymethylphenoxybutyric acid (HMPB, Novabiochem, Läufelfingen, Switzerland) handle (**2**),[27] the 2-methoxyl-4-alkoxybenzyl alcohol (SASRIN, Bachem Feinchemikalien, Bubendorf, Switzerland) resin (**3**),[28,29] and the chlorotrityl (ClTR-Cl, Novabiochem, Läufelfingen, Switzerland) (**4**)[30,31] resin (Fig. 4).

The first two are formed by introducing electron-donating substituents into the benzyl groups that serve as the peptide–resin anchor. The introduction of the first Fmoc-amino acid has to be done by activation of the carboxylic acid with carbodiimides in the presence of 4-dimethylaminopyridine (DMAP),[32] 2,6-dichlorobenzoyl chloride,[33] or alternatively as acid fluorides,[34] with the concomitant risk of racemization that this entails. To

[26] M. Songster and G. Barany, *Methods Enzymol.* **289**, [8], 1997 (this volume).

[27] B. Riniker, A. Flörsheimer, H. Fretz, and B. Kamber, *in* "Peptides 1992: Proceedings of the Twenty-second European Peptides Symposium" (C. H. Schneider and A. N. Eberle, eds.) p. 34. Escom, Leiden, The Netherlands, 1993.

[28] M. Mergler, R. Nyfeler, R. Tanner, J. Gosteli, and P. Grogg, *Tetrahedron Lett.* **29**, 4005 (1988).

[29] M. Mergler, R. Nyfeler, R. Tanner, J. Gosteli, and P. Grogg, *Tetrahedron Lett.* **29**, 4009 (1988).

[30] K. Barlos, D. Gatos, S. Kapolos, G. Papaphotiu, W. Schäfer, and Y. Wenqing, *Tetrahedron Lett.* **30**, 3943 (1989).

[31] K. Barlos, D. Gatos, J. Kallitsis, G. Papaphotiu, P. Sotiriu, Y. Wenqing, and W. Schäfer, *Tetrahedron Lett.* **30**, 3947 (1989).

[32] E. Atherton, N. L. Benoiton, E. Brown, R. C. Sheppard, and B. J. Williams, *J. Chem. Soc., Chem. Commun.*, 336 (1981).

[33] P. Sieber, *Tetrahedron Lett.* **28**, 6147 (1987).

[34] D. Granitza, M. Beyermann, H. Wenschuh, H. Haber, L. A. Carpino, G. A. Truran, and M. Bienert, *J. Chem. Soc., Chem. Commu.*, 2223 (1995).

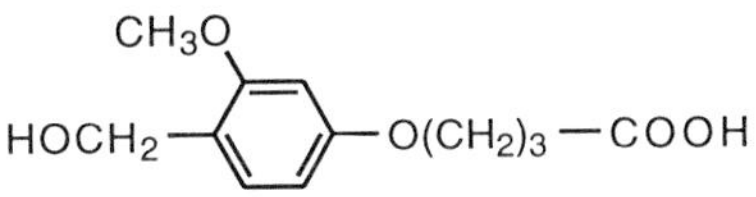

CH₃O

HOCH₂—⟨ ⟩—O(CH₂)₃—COOH

HMPB handle (2)

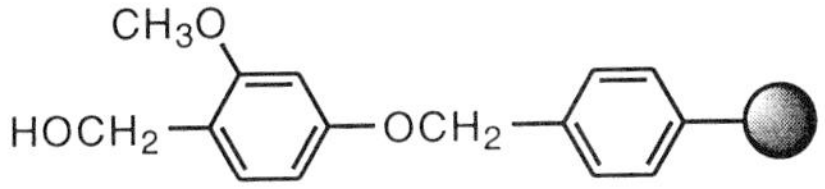

CH₃O

HOCH₂—⟨ ⟩—OCH₂—⟨ ⟩—●

SASRIN resin (3)

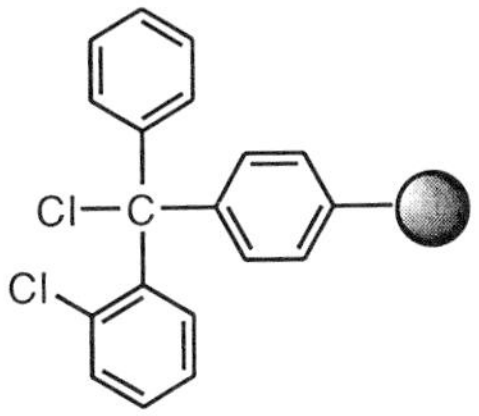

Cl—C—⟨ ⟩—●
Cl

CITR-Cl resin (4)

FIG. 4. Structures of some highly acid-labile handles and resins.

avoid the formation of DKPs, a modified cycle using the trityl (Trt) group as N^α-amino protection for the second residue may be used, as described above. In this case, the Trt group must be removed with 0.2% TFA in DCM to avoid premature cleavage of the peptide from the resin.[13] Cleavage from CITR-Cl, a chlorotrityl-based resin, can be accomplished with less than 1% in TFA in DCM, and even with acetic acid–2,2,2-trifluoroethanol (TFE)–DCM (1:1:8, by volume), or 1,1,1,3,3,3-hexafluoro-2-propanol (HFIP)–DCM (1:4, v/v).[35] This last method has the advantage of avoiding contamination of the protected peptide segment with a carboxylic acid, which can lead to capping of the free N^α-amino group of protected peptide segments in coupling reactions.[36] The incorporation of the first residue to the resin is performed by the displacement of the chloride by the diisopropylethylammonium carboxylate. This reaction takes place with very low levels of racemization.[37] Because of the large steric impediment of this resin, there is no DKP formation.[38]

Representative Experimental Procedure for Cleavage of Protected Peptides from Highly Acid-Labile Resins. The following procedure is adapted

[35] R. Bollhagen, M. Schmiedberger, K. Barlos, and E. Grell, *J. Chem. Soc., Chem. Commun.,* 2559 (1994).
[36] M. Gairí, P. Lloyd-Williams, F. Albericio, and E. Giralt, *Int. J. Pept. Protein Res.* **46,** 119 (1995).
[37] K. Barlos, O. Chatzi, D. Gatos, and G. Stavropoulos, *Int. J. Pept. Protein Res.* **37,** 513 (1991).
[38] P. Rovero, S. Viganò, S. Pegoraro, and L. Quartara, *Lett. Pept. Sci.* **2,** 319 (1995).

from Refs. 35, 39, and 40. Although the optimal protocol for the cleavage of protected peptides from ClTR-Cl-, SASRIN-, and HMPB-based solid supports would be a continuous-flow one, this can be effectively applied only when the resin is PEG–PS. The uneven shrinkage of polystyrene resins in TFA–DCM under moderate pressure makes these resins incompatible with the continuous-flow approach. A general alternative for all classes of solid supports is a batchwise procedure of several short cleavage steps with neutralization of TFA in the effluent.

The peptide–resin (0.5 g) is preswollen, in a 10-ml polypropylene syringe fitted with a polyethylene disk and a stopcock, with washes of DCM (5 ml, three times, 10 min). Cleavage is carried out at 25°, by alternating washes of 1% (v/v) TFA in DCM and DCM (2.5 ml of each solution, ten times, 20 sec) and filtering into methanol–pyridine (9 : 1, v/v, 20 ml). This operation is repeated three times. For peptides containing Trp and Cys, 5% 1,2-ethanedithiol (EDT) is added to the TFA solution. The combined cleavage portions are concentrated to one-third of the original volume, and the peptide is precipitated with water (usually about one-half of the organic volume). The precipitate is filtered and dried in vacuum over P_2O_5. Using these conditions, all common protecting groups including the Trt of His are stable.

Detachment of protected peptides from ClTR-Cl resin using HFIP–DCM (1 : 4, v/v) can also be performed in a polypropylene syringe fitted with a polyethylene disk and a stopcock. Quantitative cleavage is obtained after 3 min at 25°. Neutralization of the effluents with methanol–pyridine (9 : 1, v/v) is advisable when His(Trt) is present. The combined filtrates are evaporated to dryness.

Photolabile Resins

Photolysis[41] of the peptide–resin bond is a mild method that is, in principle, compatible with both the Fmoc/*t*Bu and the Boc/Bzl strategies. The photolabile handles most widely used are those that contain the *o*-nitrobenzyl (Nb, **5, 6**)[22,25,42–44] and the phenacyl (**7**) moieties (Fig. 5).[45–47]

[39] B. Riniker, A. Flörsheimer, H. Fretz, P. Sieber, and B. Kamber, *Tetrahedron* **49**, 9307 (1993).
[40] I. Dalcol, F. Rabanal, M.-D. Ludevid, F. Albericio, and E. Giralt, *J. Org. Chem.* **60**, 7575 (1995).
[41] V. N. R. Pillai, *Synthesis*, 1 (1980).
[42] D. H. Rich and S. K. Gurwara, *J. Chem. Soc., Chem. Commun.*, 610 (1973).
[43] E. Giralt, F. Albericio, E. Pedroso, C. Granier, and J. van Rietschoten, *Tetrahedron* **38**, 1193 (1982).
[44] C. P. Holmes, D. G. Jones, B. T. Frederick, and L.-C. Dong, *in* "Peptides: Chemistry, Structure and Biology" (P. T. P. Kaumaya and R. S. Hodges, eds.), p. 44. Mayflower Scientific, Birmingham, U.K., 1996.
[45] S. S. Wang, *J. Org. Chem.* **41**, 3258 (1976).
[46] F. S. Tjoeng, J. P. Tam, and R. B. Merrifield, *Int. J. Pept. Protein Res.* **14**, 262 (1979).

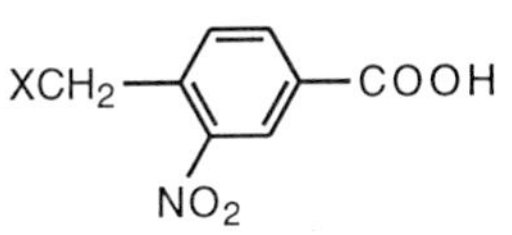

XCH₂—⟨benzene⟩—COOH

X = Br, OH
Nb handle (5)

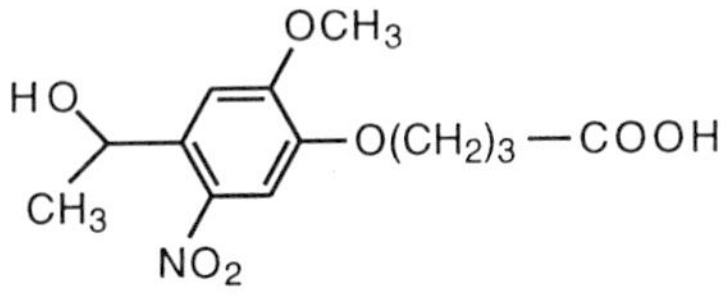

α-methyl-6-nitroveratryl handle (6)

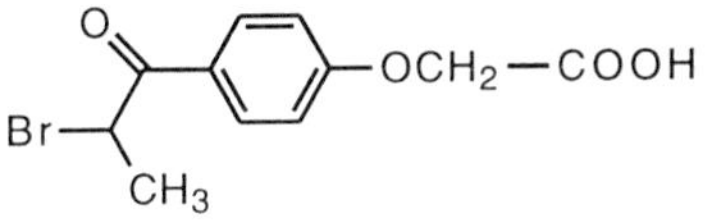

phenacyl handle (7)

FIG. 5. Structures of photolabile handles.

This technique is compatible with all encoded amino acids except Met. We have found that under the conditions of photolytic cleavage, oxidation of Met to the corresponding sulfoxide occurred in significant quantities.[22,48] An alternative is to use protected Met sulfoxide instead of unprotected Met. Boc-Met sulfoxide is completely stable under prolonged photolysis conditions, with no evidence of further oxidation to the corresponding sulfone being obtained. An additional advantage of protected peptides containing Met sulfoxide is that they are often more soluble in common solvents, facilitating their manipulation.[49] Once the protein/peptide has been fully assembled, the final reduction of the sulfoxide to the free Met can easily be done.[50] A drawback of the use of Met sulfoxide is that it leads to the formation of diastereomeric peptides, which in our experience give usually, but not always, a complex high-performance liquid chromatography (HPLC) profile.[36]

The main limitation of the use of the Nb handle (5) is that only relatively small amounts (up to 700 mg) of peptide–resin can be photolyzed at one time. This is believed to be a consequence of the formation of azo and azoxy compounds from the o-nitrosobenzaldehyde produced in the photolysis

[47] N. A. Abraham, G. Fazal, J. M. Ferland, S. Rakhit, and J. Gauthier, *Tetrahedron Lett.* **32,** 577 (1991).
[48] P. Lloyd-Williams, F. Albericio, and E. Giralt, *Int. J. Pept. Protein Res.* **37,** 58 (1991).
[49] P. Lloyd-Williams, M. Gairí, F. Albericio, and E. Giralt, *Tetrahedron* **49,** 10069 (1993).
[50] E. Nicolás, M. Vilaseca, and E. Giralt, *Tetrahedron* **51,** 5701 (1995).

reaction.[51] These compounds are a deep red color and act as an internal light filter. This effect is more pronounced with increasing amounts of resin. We have found that optimum results are achieved by suspending up to 700 mg of peptide–resin in a mixture of TFE in DCM or toluene (1 : 4, v/v), degassing, and irradiating at 360 nm, with stirring, for 7–15 hr. In our hands,[52] it gives reproducible cleavage yields of up to 85%. Traditionally, the incorporation of the first amino acid has been performed using the cesium salt method,[53] but in some cases yields can be low and poorly reproducible. Alternatively, the bromomethyl nitrobenzyl handle can be converted to the corresponding hydroxy derivative and the first amino acid attached by esterification. This can be done with the handle in solution to give the preformed handle or, alternatively, after previously attaching the handle to the resin.[22,52] The formation of DKPs is a problem with this type of solid support, and the program described above, involving *in situ* neutralization, must be used to avoid it. The peptide–resin bond is not completely stable to treatment with piperidine. The release of the peptide from the resin depends on the steric hindrance of the first amino acid. Thus, for Gly it is approximately 3% per 20-min treatment with piperidine–*N,N*-dimethylformamide (DMF) (1 : 4, v/v) cycle, but for Val or Phe it appears to be negligible.[54] The synthesis of long peptides with nonhindered first amino acids is not advisable.

The preparation of the α-methyl-6-nitroveratryl handle (**6**) has been described.[44] The incorporation of two additional alkoxy groups onto the benzene ring and a methyl on the benzylic carbon allows the photolytic release of peptides in good yield in short times (1 hr). The preparation of methionine containing peptides with low (<4%) amounts of methionine sulfoxide peptides has been described using this handle.

The use of phenacyl handles (**7**) is compatible only with the Boc/Bzl strategy because the peptide–resin anchor is not stable to the nucleophilic conditions used in the Fmoc/*t*Bu strategy. Besides the formation of DKPs, another side reaction can take place on incorporation of the second residue. Attack of the free amino group of the first amino acid onto the carbonyl group of the phenacyl resin leads to a cyclic Schiff base.[46] Both types of side reactions can be minimized using the *in situ* neutralization protocol described above. Photolysis is carried out as for nitrobenzyl resins.

Representative Experimental Procedure for Cleavage of Protected Peptides from Photolabile Resins. This procedure is adapted from Ref. 49. A

[51] A. Patchornik, B. Amit, and R. B. Woodward, *J. Am. Chem. Soc.* **92**, 6333 (1970).
[52] P. Lloyd-Williams, M. Gairí, F. Albericio, and E. Giralt, *Tetrahedron* **47**, 9867 (1991).
[53] B. F. Gisin, *Helv. Chim. Acta* **56**, 1476 (1973).
[54] E. Nicolás, J. Clemente, T. Ferrer, F. Albericio, and E. Giralt, *Tetrahedron* **53**, 3179 (1997).

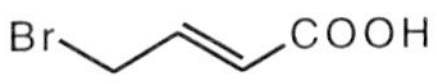

$$Br \diagdown \diagup COOH \qquad HO \diagdown \diagup O-(CH_2)_n-COOH$$

$$n = 1,5$$

(8) 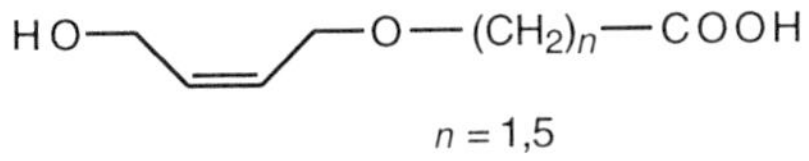   (9)

$$Br \diagdown \diagup O \left(\diagdown \diagup O \right)_3 \diagdown COOH$$

(10)

Fig. 6. Structures of allyl-based handles.

two-neck cylindrical reaction vessel is silylated before photolysis to prevent resin adhering to its walls. This is done by rinsing it 3 or 4 times with trimethylsilyl chloride $[(CH_3)_3SiCl]$–toluene (1:9, v/v), followed by washing with absolute ethanol and drying. Peptide–resin (0.5–0.7 g) is suspended in TFE–DCM or toluene (1:4, v/v) (100 ml) in the reaction vessel. Prior to photolysis the peptide–resin suspension is degassed by evacuation at water-pump pressure and purging with Ar three times in succession. The resin is photolyzed at 360 nm in a Rayonnet RPR-100 apparatus for 1 to 15 hr, maintaining vigorous magnetic stirring, and keeping the temperature below 40°, using the incorporated fan. The crude reaction product is filtered and the resin washed with TFE–DCM (1:4, v/v), DCM, DMF, and methanol (three times, 5 ml in each case). The combined filtrates are then evaporated to dryness.

Allyl-Based Resins

Allyl-based solid supports (**8, 9, 10,** Fig. 6) are potentially very useful,[55–58] as the cleavage of the protected peptide from the resin can be achieved under practically neutral conditions by Pd^0-catalyzed allyl transfer to weakly basic nucleophiles. These conditions are in principle compatible with both of the major CSSPS strategies.

Reproducible, high yielding cleavage of protected peptides from allyl-based resins requires strict control of the reaction conditions. Two general protocols have been developed.[59] The first involves treating a suspension

[55] H. Kunz and B. Dombo, *Angew. Chem. Int. Ed. Engl.* **27**, 711 (1988).
[56] B. Blankemeyer-Menge and R. Frank, *Tetrahedron Lett.* **29**, 5871 (1988).
[57] F. Guibé, O. Dangles, G. Balavoine, and A. Loffet, *Tetrahedron Lett.* **30**, 2641 (1989).
[58] O. Seitz and H. Kunz, *Angew. Chem. Int. Ed. Engl.* **34**, 803 (1995).
[59] P. Lloyd-Williams, A. Merzouk, F. Guibé, F. Albericio, and E. Giralt, *Tetrahedron Lett.* **35**, 4437 (1994).

of the peptide–resin in dimethyl sulfoxide (DMSO)–tetrahydrofuran (THF)–0.5 M HCl (2 : 2 : 1, by volume) with morpholine as nucleophile, in the presence of [(Ph$_3$P)$_4$Pd]. However, the use of a nucleophilic secondary amine as the allyl acceptor in the cleavage of Fmoc-protected peptides leads to deprotection of the Fmoc group. Furthermore, the presence of DMSO in the medium causes the oxidation of methionine to the sulfoxide. In these cases, morpholine can be substituted by N-methylaniline (NMA), and chloroform (TCM)–acetic acid (9 : 1, v/v) can be used as solvent.[60] Cleavage yields appear to depend on the quality of the catalyst and can sometimes be improved by increasing the amount used. Alternatively, an hydrostannolytic allyl transfer method based on the use of PdCl$_2$(Ph$_3$P)$_4$ in the presence of tributyltin hydride, using DMF–DCM (1 : 1, v/v) as solvent, can be applied. Because no base is required, this cleavage protocol is compatible with the Fmoc group. Palladium-catalyzed allyl transfer cleavages are best carried out under Ar atmosphere.

Incorporation of the first amino acid onto bromomethyl-type resins derived from handles **8** and **10** can be done by the cesium salt method,[53] although in some cases yields can be low and poorly reproducible. For handle **10,** the incorporation of the first amino acid onto the handle in solution and anchoring of this preformed handle onto the solid support are recommended.[58] However, incorporation of the first amino acid can be done using hydroxymethyl resins derived from handles **9** through carbodiimide–DMAP- or 2,4,6-mesitylene-sulfonyl-3-nitro-1,2,4-triazolide (MSNT')–pyridine-mediated couplings. Alternatively, protected amino acids activated as the acid chloride, in the case of unprotected side chain Fmoc-amino acids,[56,59,61] can be used. For this type of handle, DKP formation, mainly in the Fmoc/tBu strategy, is a risk and must be avoided as described earlier.

Representative Experimental Procedures for Cleavage of Protected Peptides from Allyl-Based Resins. The procedures are adapted from Ref. 59. For method A, peptide–resin (100 mg, ~0.075 mmol) and crystalline, yellow Pd(Ph$_3$)$_4$ (30 mg, 0.026 mmol) are suspended in a previously degassed mixture of DMSO–THF–0.5 M HCl (2 : 2 : 1, by volume, 5 ml) in a silylated two-neck round-bottom flask and stirred vigorously under Ar. The catalyst dissolves, giving the suspension a yellow color. NMA (385 μl, 3.5 mmol) is added and the mixture stirred under Ar for 12 hr, at 25°. Filtration followed by washing the resin with DMF and TCM (three times, 1 ml in each case) and removing the solvent under high vacuum gives the crude

[60] S. A. Kates, S. B. Daniels, and F. Albericio, *Anal. Biochem.* **212,** 303 (1993).
[61] P. Lloyd-Williams, G. Jou, F. Albericio, and E. Giralt, *Tetrahedron Lett.* **32,** 4207 (1991).

peptide. For Met-containing peptides, TCM–acetic acid (9:1, v/v) can be used as a solvent.

For method B, peptide–resin (100 mg, ~0.075 mmol) and $PdCl_2(Ph_3P)_4$ (2 mg, 2.5 μmol) are suspended in a previously degassed mixture of DMF–DCM (1:1, v/v, 5 ml) in a silylated two-neck round-bottom flask and stirred vigorously under Ar. The catalyst dissolves giving the suspension a yellow color. Tributyl hydride (75 μl, 0.25 μmol) in DCM (1 ml) is added over 30 min and the mixture stirred for a further 10 min, at 25°. After filtration, the resin is washed with DMF–DCM (1:1, v/v, three times, 1 ml), and the filtrate is extracted repeatedly (four times, 5 ml) with pentane to eliminate tin by-products. HCl (1 M, 5 ml) is added to convert the tin carboxylate to the peptide free acid, followed by water (5 ml) to precipitate the crude peptide, which is isolated by centrifugation and filtration.

In this kind of cleavage gel-filtration chromatography through Sephadex LH-60, using DMF as eluent, is advisable to completely remove any tin by-products.

Fluorenylmethyl-Based Resins

Because the Boc and Fmoc groups are orthogonal, one approach for preparing Boc/Bzl-protected peptides is to use fluorenylmethyl-based handles. These allow cleavage of protected peptides from the solid support by a β-elimination mechanism on treatment with secondary amines such as piperidine or morpholine. Several such handles (**11, 12**) have been prepared and used successfully (Fig. 7).[62–64] We have reported[65] the use of **13**, which circumvents some of the problems associated with **11** and **12a**, such as the premature release of protected peptides on treatment with the DIEA used to neutralize the resin after Boc group removal with TFA. The problem is due to the electron-withdrawing effect provided by substituents such as -CONH (in **11**) or -CH$_2$CONH (in **12a**), which increases the acidity of the hydrogen at position 9 of the fluorene ring, facilitating the elimination process. In contrast, the base lability of resins derivatized with handle **13** has been fine tuned by means of an electron-donating N-acylamido group.

Although excellent yields (>95%) have been obtained after treatment of peptide–resins with both piperidine–DMF and morpholine–DMF (1:4, v/v), the morpholine-based cleavages give products with cleaner HPLC

[62] M. Mutter and D. Bellof, *Helv. Chim. Acta* **67**, 2009 (1984).
[63] Y.-Z. Liu, S.-H. Ding, J.-Y. Chu, and A. M. Felix, *Int. J. Pept. Protein Res.* **35**, 95 (1990).
[64] W. Lin, L. Chen, Y. Z. Liu, and C. I. Niu, *in* "Peptides, Biology and Chemistry: Proceedings of the 1992 Chinese Peptide Symposium" (Y. C. Du, J. P. Tam, and Y. S. Zhang, eds.), p. 299. Escom, Leiden, The Netherlands, 1993.
[65] F. Rabanal, E. Giralt, and F. Albericio, *Tetrahedron* **51**, 1449 (1995).

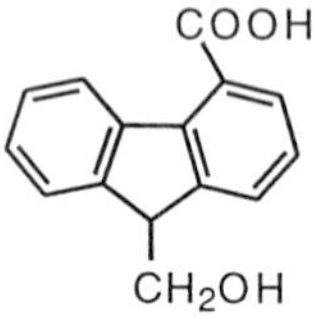

FIG. 7. Structures of fluorenylmethyl-based handles.

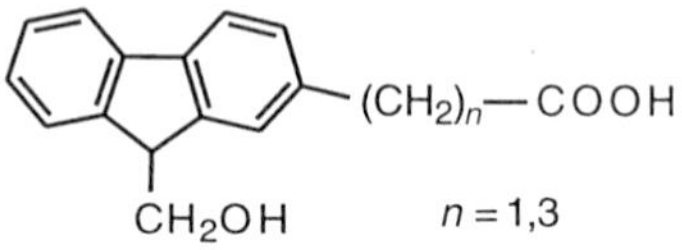

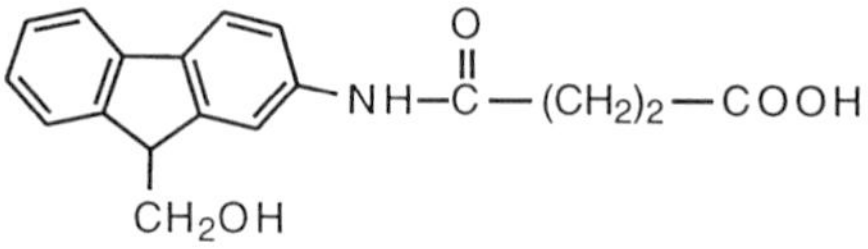

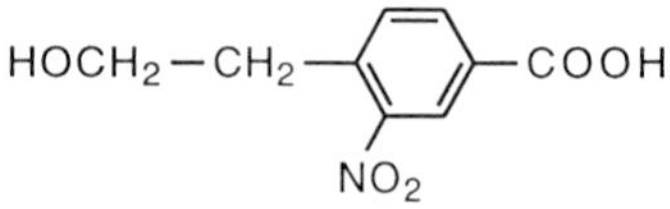

profiles than those afforded by piperidine treatments. Using these conditions, aspartimide formation has not been detected during the cleavage of Asp-containing peptides, although an exhaustive study has not been done. To minimize this risk, morpholine is preferred to piperidine for cleavage and cyclohexyl ester is preferred to benzyl ester protection for the β-carboxylic group of aspartic acid. DKP formation has not been detected during the preparation of several peptides containing Pro and Gly at the C-terminal position. The handle **14**, with similar characteristics to **11–13**, has been also used for the synthesis of protected peptides.[66]

Representative Experimental Procedure for Cleavage of Protected Peptides from Fluorenylmethyl Resin Based on Handle 13. This procedure is adapted from Ref. 65. Peptide–resin (1 g) is preswollen, in a 20-ml polypropylene syringe fitted with a polyethylene disk and a stopcock, with washes of DMF (10 ml, three times, 10 min). The cleavage is carried out by treatment of peptide–resins with a freshly prepared solution of morpholine–DMF (1 : 4, v/v, 10 mol) for 2 hr, at 25°, with occasional agitation. After filtration the resin is washed with DMF (three times, 5 ml), and the combined cleavage portions are evaporated to dryness under high vacuum.

Oxime-Based Resins

The *p*-nitrobenzophenone oxime resin (**15**, Fig. 8) allows the release of peptides by several methods, including ammonolysis, hydrazinolysis, and

[66] F. Albericio, E. Giralt, and R. Eritja, *Tetrahedron Lett.* **32**, 1515 (1991).

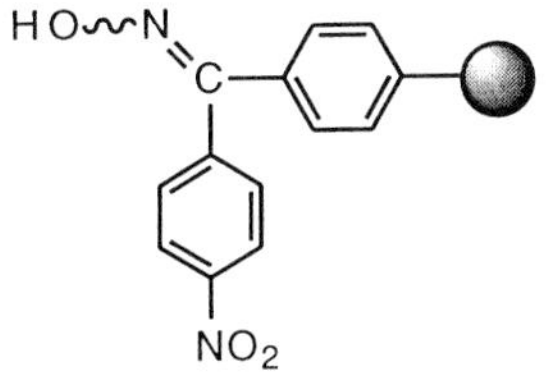

Oxime resin (**15**)

FIG. 8. Structure of oxime resin.

aminolysis using a suitable amino acid ester.[67-69] An alternative and more convenient procedure for the cleavage reaction is by transesterification of the peptide–resin with hydroxypiperidine (HOPip). This gives rise initially to the hydroxypiperidine ester of the protected peptide. Treatment of this ester with Zn in acetic acid then gives rise to the corresponding protected peptide with the free carboxylic acid at the C terminus. This resin is compatible only with the Boc/Bzl strategy, as it is labile to nucleophiles. Furthermore, a premature loss of peptide from the resin on neutralization with base after acidolytic removal of the N^α-Boc group can also occur. This loss is much more important after the incorporation of the second residue, owing to DKP formation. For all couplings, an *in situ* neutralization protocol is recommended.

Representative Experimental Procedure for Cleavage of Protected Peptides from Oxime-Based Resins. This procedure is adapted from Ref. 70. Protected peptide–oxime resin (1 g) is preswollen, in a silylated screw-cap tube reaction vessel with a Teflon-lined cap, a sintered glass frit, and a stopcock, with washes of DCM (10 ml, three times, 10 min). The cleavage is carried out by treatment of peptide–resins with HOPip (3 equivalents, freshly recrystallized from hexane) for 3–16 hr, at 25°, with mechanical shaking. The resin is filtered and washed with DCM, DMF, and methanol (three times, 5 ml, in each case). The combined filtrates are evaporated under vacuum. The residue is triturated with ether or hexane to give the crude peptide 1-piperidyl ester. This is then dissolved in acetic acid–water (9 : 1, v/v, 15 ml), and Zn dust (30 equivalents) is added. After vigorous stirring for 15–30 min, at 25°, Zn is removed by filtration and washed with acetic acid–H$_2$O (9 : 1, v/v, three times, 5 ml). The combinated filtrates are

[67] W. F. DeGrado and E. T. Kaiser, *J. Org. Chem.* **47**, 3258 (1982).
[68] E. T. Kaiser, H. Mihara, G. A. Laforet, J. W. Kelly, L. Walters, M. A. Findeis, and T. Sasaki, *Science* **241**, 187 (1989).
[69] J. C. Hendrix, J. T. Jarrett, S. T. Anisfield, and P. T. Lansbury, *J. Org. Chem.* **57**, 3414 (1992).
[70] S. H. Nakagawa and E. T. Kaiser, *J. Org. Chem.* **48**, 678 (1983).

evaporated under vacuum. Gel-filtration chromatography using Sephadex LH-60 in DMF is advisable for the removal of residual Zn and of by-products originating from HOPip.

Purification of Protected Peptides

One of the key steps in a CSPPS strategy is the purification of protected peptides to ensure that the species to be coupled are free from deletion or modified peptides and other impurities. Use of the techniques that separate peptides on the basis of the charge they carry are normally not useful because protected peptides have the majority of their reactive functional groups blocked. The application of other methods such as countercurrent distribution and normal-phase or reversed-phase liquid chromatography (RP-LC) is complicated by the unpredictable, but generally poor, solubility of the protected peptide segments in the most commonly used solvents. Thus, the principal requisite for any purification procedure is solubility of the protected peptide in question in a solvent compatible with the purification procedure.

Although protected peptides are usually not soluble in aqueous media and only poorly so in most of the commonly used organic solvents, many are at least moderately soluble in polar aprotic solvents such as DMF, DMSO, and N-methylpyrrolidone (NMP). Furthermore, some are soluble in mixtures of fluorinated (HFIP and TFE) and chlorinated (TCM and DCM) solvents.[15,71] The use of HFIP and TFE for purification may be prohibited by their cost, and for the coupling step TFE is more convenient than HFIP because it reacts only slowly to form TFE esters.[15,71] The solubility of protected peptides in less polar solvents such as THF can be improved by the addition of chaotropic salts,[72,73] although their use has been reported to have a detrimental effect on the coupling yield and to favor epimerization in some cases.[27] Finally, as was pointed out earlier, the use of the AcHmb backbone-protecting group significantly increases the solubility of protected peptides in DMF (3- to 12-fold) and makes them soluble in DCM.[9,74]

Although countercurrent distribution can be useful for the purification of protected peptides,[27,75] liquid chromatography (MPLC or HPLC, both in normal- or reversed-phase modes) is the most widely available option

[71] H. Kuroda, Y.-N. Chen, T. Kimura, and S. Sakakibara, *Int. J. Pept. Protein Res.* **40,** 294 (1992).
[72] J. M. Stewart and W. A. Klis, *in* "Innovation and Perspectives in Solid-Phase Synthesis and Related Technologies. Peptides, Polypeptides and Oligonucleotides. Macroorganic Reagents and Catalysts" (R. Epton, ed.), p. 1. SPCC, Birmingham, U.K., 1990.
[73] A. Thaler, D. Seebach, and F. Cardinaux, *Helv. Chim. Acta* **74,** 617 (1991).
[74] M. Quibell, L. C. Packman, and T. Johnson, *J. Chem. Soc., Perkin Trans. 1,* 1219 (1996).
[75] R. J. Hill, *Methods Enzymol.* **11,** 173 (1967).

in many laboratories. If the protected peptides are soluble in mixtures of acetonitrile and water, then standard analytical HPLC conditions (aqueous acetonitrile eluent containing an ion-pairing modifier reagent) can be used. For protected peptides that are not sufficiently soluble under these conditions, we have found that RP-LC purification can often be done by the addition of DMF to the mobile phase. Best results are achieved by adding the same amount to each of the eluents.[36,76] Some increase in viscosity and column pressure is observed, and UV detection is also complicated owing to the strong absorbance of DMF between 220 and 270 nm. If a suitable chromophore such as the Fmoc group or the Tyr residue is present then detection at 300 or 280 nm may be possible; if not the monitoring can be carried out by analytical HPLC. Some hydrophobic segments that are soluble in TCM or TCM–TFE mixtures can be purified by RP-LC, loading a TCM solution onto the column and eluting successively with mixtures of methanol–water, methanol, and methanol–TCM. For protected peptides having only the C-terminal carboxylic acid free, the use of an ion-pairing reagent can often be dispensed with, and no deleterious effect with regard to recovery of chromatographic efficiency is observed. In those cases where separation efficiency is improved by using such a reagent, if propionic or acetic acid are used care must be taken to remove them completely from the peptide prior to coupling; otherwise, capping of amino groups may be observed. Purification of His(Trt) peptides should be done in the presence of 0.1% pyridine to avoid loss of the Trt group. Such loss can occur even when acetonitrile–water mixtures without carboxylic acid modifier are used.[40]

Normal-phase LC is a less potent alternative and has been recommended only when impurities in the crude product were derived from the loss of a side-chain protecting group. If the crude product exhibits significant amount of deletion peptides, then purification by RP-LC techniques is required.[74] In this case C_8 or even C_4 columns are the most suitable for peptides of around 10 residues. For longer peptides, the less polar diphenyl-based phase is recommended.

Characterization of Protected Peptides

Characterization of the protected peptide segments is an important aspect of CSPPS. Amino acid analysis gives no information on whether protecting groups are still present. Nuclear magnetic resonance (NMR)

---

[76] P. Lloyd-Williams, F. Albericio, M. Gairí, and E. Giralt, in "Innovation and Perspectives in Solid-Phase Synthesis" (R. Epton, ed.), p. 195. Mayflower Worldwide, Kingswinford, UK, 1997.

spectroscopy is one of the most powerful structure elucidation techniques in organic chemistry, but for large peptides the interpretation of NMR spectra, in either one- or two-dimensional experiments, can be complicated. Mass spectrometry (MS), especially in the fast atom bombardment (FAB)[77,78] and matrix assisted laser desorption/time-of-flight (MALDI-TOF)[79] modes, is a useful technique for the structure elucidation of this type of compound and can provide accurate information, in a short time, on the structure of purified protected peptides. Thus, those purified protected peptides that show single peaks in at least two different HPLC elution systems, and correct amino acid and mass spectrometric analyses, may be considered to be pure and suitable for use in coupling reactions.

Coupling of Protected Peptides on Solid Support

The case of the first protected peptide segment (the extreme C terminal) must be differentiated from the rest of the segments. The first segment can be incorporated either by carrying out stepwise solid-phase synthesis of the desired sequence or by incorporating a previously synthesized, purified, and characterized segment. If the first approach is chosen, conventional SPPS methods are used. After incorporation of the last amino acid and before the coupling of the second segment, cleavage and analysis by HPLC, amino acid analysis, and MS of the resulting free peptide are mandatory before continuing. The incorporation of a protected peptide should, in principle, give rise to a purer final product, but unfortunately the coupling of protected peptide segments onto polymeric resins does not always proceed in high yield.[19,80] Incorporation of a protected peptide by amide bond formation can be performed relatively easily, but incorporation by esterification, however, is more difficult. In the latter case, the first protected peptide can be prepared with a handle already incorporated at the C terminus, allowing attachment to the solid support by amide bond formation.[19] This is illustrated in Fig. 9.

Usually, the coupling of a protected peptide segment is much more demanding than a single protected amino acid. First, as pointed out earlier, the risk of epimerization at the C-terminal residue increases notably. Furthermore, large excesses of the segment cannot normally be used to drive the coupling to completion, for economic reasons. Finally, the poor solubil-

[77] A. Grandas, E. Pedroso, A. Figueras, J. Rivera, and E. Giralt, *Biomed. Environ. Mass Spectrom.* **15,** 681 (1988).

[78] C. Celma and E. Giralt, *Biomed. Environ. Mass Spectrom.* **19,** 235 (1990).

[79] M. Schmidt, E. Krause, M. Beyermann, and M. Bienert, *Pept. Res.* **8,** 238 (1995).

[80] E. Pedroso, A. Grandas, M. A. Saralegui, E. Giralt, C. Granier, and J. van Rietschoten, *Tetrahedron* **38,** 1183 (1982).

IRAA = internal reference amino acid

F_{IG}. 9. Incorporation of the first protected peptide using a handle.

ity of peptide segments often means that dilute solution must be used. All these factors, together with the intrinsic difficulty of making a peptide bond between two large molecules, mean that the coupling reactions may require many hours or even days, and in many cases the yield is not quantitative.

Coupling methods for protected peptide segments have evolved in parallel to those used for the coupling of single amino acids. Thus, early syntheses included the use of azide[43,81] and oxidation–reduction[82,83] methods, which were replaced by carbodiimides [N,N'-dicyclohexylcarbodiimide (DCC), N,N'-diisopropylcarbodiimide (DIPCDI), and 1-ethyl-3-(3'-dimethylaminopropyl)carbodiimide hydrochloride (EDC)] in the presence of hydroxylamines such as N-hydroxysuccinimide (HOSu), and in particular 1-oxo-2-hydroxydihydrobenzotriazine (HODhbt) and 1-hydroxybenzotriazole

[81] J. Meienhofer, in "The Peptides: Analysis, Synthesis, Biology" (E. Gross and J. Meienhofer, eds.), Vol. 1, p. 197. Academic Press, New York, 1979.
[82] R. Matsueda, H. Maruyama, E. Kitazawa, H. Takahagi, and T. Mukaiyama, Bull. Chem. Soc. Jpn. 46, 3240 (1973).
[83] T. Mukaiyama, R. Matsueda, and M. Ueki, in "The Peptides: Analysis, Synthesis, Biology" (E. Gross and J. Meienhofer, eds.), Vol. 2, p. 383. Academic Press, New York, 1979.

(HOBt) to reduce epimerization.[17,84–86] These reagents react *in situ* with the *O*-acylisourea or symmetrical anhydride to give the corresponding active esters, which are the main active species in the coupling reaction. More recent developments include the phosphonium (BOP, PyBOP)[22,87,88] and then aminium {*N*-[(1*H*-benzotriazol-1-yl)(dimethylamino)methylene]-*N*-methylmethanaminium hexafluorophosphate and tetrafluoroborate *N*-oxide (HBTU and TBTU)}[89,90] salts based mainly on HOBt. These have also been used for the coupling of protected peptide segments. Since 7-aza-1-hydroxybenzotriazole (HOAt) has been proposed as an additive in carbodiimide couplings instead of HOBt itself, phosphonium (PyAOP) and aminium {*N*-[(dimethylamino)-1*H*-1,2,3-triazolo[4,5-*b*]pyridin-1-ylmethylene]-*N*-methylmethanaminium hexafluorophosphate *N*-oxide (HATU) and the pyrrolidino derivative (HAPyU)} salts derived from HOAt have been also used for this kind of couplings.[40,91,92]

Aminium salts should be used with caution, because they can react with the amino component leading to a guanidino species.[93,94] This side reaction is not important during the coupling of single protected amino acids, because the activation is a fast step and the aminium salt is rapidly consumed. However, the coupling of protected peptide segments can be a rather time-consuming process, and the aminium salt can react with the amino component.

In those couplings that are base assisted, it is important to pay attention to the nature of the base. Hitherto, only DIEA and NMM have been used for the coupling of protected peptides, but more hindered bases such as 2,4,6-trimethylpyridine or collidine (TMP), 2,3,5,6-tetramethylpyridine (TEMP), and 2,6-di-*tert*-butyl-4-(dimethylamino)pyridine (DBDMAP) should be explored in CSPPS, as preliminary results obtained in solution

[84] J. C. Sheehan and G. P. Hess, *J. Am. Chem. Soc.* **77**, 1067 (1955).
[85] W. König and R. Geiger, *Chem. Ber.* **103**, 788 (1970).
[86] W. König and R. Geiger, *Chem. Ber.* **103**, 2034 (1970).
[87] B. Castro, J. R. Dormoy, G. Evin, and C. Selve, *Tetrahedron Lett.*, 1219 (1975).
[88] J. Coste, D. Le-Nguyen, and B. Castro, *Tetrahedron Lett.* **31**, 205 (1990).
[89] A. Surovoy, J. Metzger, and G. Jung, in "Innovation and Perspectives in Solid-Phase Synthesis: Peptides, Polypeptides and Oligonucleotides" (R. Epton, ed.), p. 467. Intercept Andover, UK, 1992.
[90] R. Knorr, A. Trzeciak, W. Bannwarth, and D. Gillessen, *Tetrahedron Lett.* **30**, 1927 (1989).
[91] L. A. Carpino, *J. Am. Chem. Soc.* **115**, 4397 (1993).
[92] L. A. Carpino, A. El-Fahan, C. A. Minor, and F. Albericio, *J. Chem. Soc., Chem. Commun.*, 201 (1994).
[93] H. Gausepohl, U. Pieles, and R. W. Frank, in "Peptides: Chemistry and Biology, Proceedings of the Twelfth American Peptides Symposium" (J. A. Smith and J. E. Rivier, eds.), p. 523. Escom, Leiden, The Netherlands, 1992.
[94] S. C. Story and J. V. Aldrich, *Int. J. Pept. Protein Res.* **43**, 292 (1994).

are promising. Their use affords good coupling yields and maintains the configurational integrity of the C-terminal residue.[95,96] Of major importance in the case of phosphonium and aminium salt coupling is the avoidance of any preactivation time, as during this time epimerization of the residue at the C-terminal can occur.[97] For the same reason, the first hour of the coupling is best carried out at 4°. The choice of solvent is often dictated by the solubility of the protected peptide, but the use of nonpolar solvents when possible leads to lower epimerization levels.[71,97] Systematic acetylation after the coupling of each protected segment is optional. Some researchers advocate this, but others hold that this extra step can cause other side reactions.[7] One of the advantages of the CSPPS strategy is that the difference in size and properties of the possible deletion peptides (differing in at least one peptide segment) is sufficinet to facilitate the final purification.

Representative Experimental Procedure for Coupling of Protected Peptides. This procedure is adapted from Refs. 17 and 40. For those couplings running for up to 4 hr, a polypropylene syringe fitted with a polyethylene disk and a stopcock can be used. Occasional manual stirring may be sufficient. However, for longer couplings, a silylated screw-cap tube reaction vessel with a Teflon-lined cap, a sintered glass frit, and a stopcock and continuous mechanical shaking is preferable. Coupling should first be tested on a small amount of peptide–resin (2–3 μmol) in a 1-ml syringe before carrying out the large-scale experiment. The process of dissolving the protected peptide segment may need to be started several hours before initiating the coupling reaction and may require special agitation such as ultrasound. The most concentrated protected peptide solution possible should be prepared. The excess of peptide segment required depends on the coupling, but between 2 and 5 equivalents are most widely used. Protected peptide–resins are preswollen in DMF (1 ml/0.1 g of resin, three times for 10 min plus once for 1 hr). After removal of the N^{α}-amino protecting group with the appropriate reagent [TFA–DCM (4:6, v/v) for Boc chemistry or piperidine–DMF (1:4, v/v) for Fmoc chemistry, with an extended time of reaction, ~1 hr] and washings, a mixture of the protected peptide, coupling reagent (carbodiimide, aminium or phoshonium salt, 1 equivalent with respect to the peptide), HOBt/HOAt (1 equivalent), and base (1 equivalent), if aminium or phosphonium salts are used, is added to the resin at 0°. If carbodiimide-based couplings are carried out, some researchers prefer to perform a preactivation step of the protected peptide with the carbodi-

[95] L. A. Carpino and A. El-Faham, *J. Org. Chem.* **59**, 685 (1994).
[96] L. A. Carpino, D. Ionescu, and A. El-Faham, *J. Org. Chem.* **61**, 2460 (1996).
[97] L. Carpino, A. El-Faham, and F. Albericio, *Tetrahedron Lett.* **35**, 2279 (1994).

imide and HOBt/HOAt for 10–60 min at 0°. Acetylation, when required, is done with acetic anhydride–DIEA (1 : 1, v/v, 30 equivalents) in the minimum amount of DMF. The ninhydrin test can be done on a few beads of the peptide–resin and analyzed with the help of a microscope.

Monitoring of Coupling Reaction

Monitoring of the coupling of protected peptides on solid supports is rather difficult. The ninhydrin test and amino acid analysis become less useful as the length of the peptide chain increases. The first is less sensitive, and the information that the second can provide is sometimes limited, as it can be difficult to judge the extent of incorporation of a new segment if it contains residues that are already present in the peptide sequence. We have used solid-phase automatic sequencing to calculate yields, by analyzing aliquots of the different peptide resins after each of the segment couplings.[17,22] Results from "one segment previews" will give the coupling yield, and those from "one amino acid previews" will assess the homogeneity of the protected peptide. Unfortunately, the availability of this technique in many laboratories is still limited. Finally, an indirect method for monitoring the coupling of protected peptides consists of removal of an aliquot of the resin after each coupling, cleavage of the peptide, and analysis of the free peptide by HPLC and MS.

Acknowledgment

Work in the authors' laboratories is supported by funds from CICYT (PB95-1131) and Generalitat de Catalunya (Centre de Referència en Biotecnología).

[16] Synthetic Peptide Libraries

By MICHAL LEBL and VIKTOR KRCHNÁK

Introduction

When two independent presentations at the 13th American Peptide Symposium in 1991 described the techniques for the synthesis and screening of synthetic peptide libraries,[1-4] only a few scientists were convinced that

[1] K. S. Lam, S. E. Salmon, E. M. Hersh, V. J. Hruby, W. M. Kazmierski, and R. J. Knapp, *Nature* (*London*) **354**, 82 (1991).

[2] R. A. Houghten, C. Pinilla, S. E. Blondelle, J. R. Appel, C. T. Dooley, and J. H. Cuervo, *Nature* (*London*) **354**, 84 (1991).

this technology may eventually change the approach to the discovery of new drugs in major pharmaceutical companies, and, moreover, that even small biotechnology companies and university laboratories will join in the quest for new drug leads. These techniques were considered an oddity, which, being limited to peptides, would never play a significant role in drug discovery.* However, several companies based on the combinatorial library concept were founded (Selectide, Houghten Pharmaceuticals, Pharmacopeia, Sphinx, and others), but when Ellman and colleagues[10] and a Parke-Davis group[11] showed that the same library principles can also be applied to small organic molecules, the real "gold rush" started. The plethora of scientific data accumulated in a relatively short time has already answered the question as to whether combinatorial chemistry will work. The scientific community has responded (Professor Arno Spatola organized the first combinatorial techniques course at The University of Louisville in spring of 1996), and it can be expected that new chemists will not be surprised when asked to synthesize half a million compounds for a screening project "next week."

The very origin of all multiple and combinatorial syntheses resides with the frustration of Merrifield with the tediousness of peptide synthesis in

---

[3] R. A. Houghten, J. H. Cuervo, C. Pinilla, J. R. Appel, C. T. Dooley, and S. E. Blondelle, in "Peptides: Chemistry and Biology" (J. A. Smith and J. E. Rivier, eds.), p. 560. Escom, Leiden, The Netherlands, 1992.

[4] K. S. Lam, S. E. Salmon, E. M. Hersh, V. J. Hruby, F. Al-Obeidi, W. M. Kazmierski, and R. J. Knapp, in "Peptides: Chemistry and Biology" (J. A. Smith, and J. E. Rivier, eds.), p. 492. Escom, Leiden, The Netherlands, 1992.

* This attitude has changed slightly after papers describing synthetic library techniques were published in Nature.[1,2] It is appropriate to mention here that the scientific community was ready to accept the library techniques in 1991. A study of multiple solid-phase peptide synthesis on glass chips using photolithography was already published,[5] as well as the techniques using genetic engineering for the display of modified proteins on surfaces of phage.[6-8] However, in 1986, when Mario Geysen submitted a paper describing synthesis and screening of peptide mixtures bound to plastic pins, several prestigious scientific journals refused to publish his paper (it was later published in Molecular Immunology[9]).

[5] S. P. A. Fodor, R. J. Leighton, M. C. Pirrung, L. Stryer, A. T. Lu, and D. Solas, Science 251, 767 (1991).

[6] S. E. Cwirla, E. A. Peters, R. W. Barrett, and W. J. Dower, Proc. Natl. Acad. Sci. U.S.A. 87, 6378 (1990).

[7] J. J. Devlin, L. C. Panganiban, and P. E. Devlin, Science 249, 404 (1990).

[8] J. K. Scott and G. P. Smith, Science 249, 386 (1990).

[9] H. M. Geysen, S. J. Rodda, and T. J. Mason, Mol. Immunol. 23, 709 (1986).

[10] B. A. Bunin and J. A. Ellman, J. Am. Chem. Soc. 114, 10997 (1992).

[11] S. H. DeWitt, J. K. Kiely, C. J. Stankovic, M. C. Schroeder, D. M. R. Cody, and M. R. Pavia, Proc. Natl. Acad. Sci. U.S.A. 90, 6909 (1993).

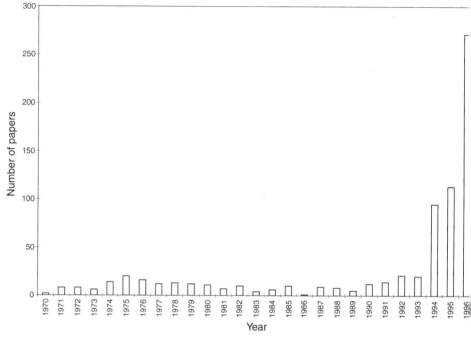

FIG. 1. Number of articles describing nonpeptide solid-phase chemistry.

solution.[12,13] The concept of synthesizing peptides on the solid phase seems obvious: a large excess of reagents is used to drive reactions to near completion, any soluble material is readily removed by filtration, and, owing to the fact that individual steps include either addition of liquid to the solid phase or separation of solid and liquid phases, the whole process can easily be automated. The advantages of the solid-phase concept are even more useful for multiple syntheses and the generation of combinatorial libraries. However, it took years to fine-tune all aspects of this fundamental synthesis concept. Merrifield's solid-phase synthesis was readily accepted by peptide chemists. Organic chemists, however, adopted it less enthusiastically. Only more recently has solid-phase synthesis become promoted among organic and medicinal chemists. Figure 1 illustrates the number of articles describing solid-phase synthesis of organic compounds (other than peptides) published between 1970 and 1996.

In this article we describe the currently available techniques for multiple and library synthesis of peptides. We focus on synthetic techniques that

---

[12] R. B. Merrifield, *J. Am. Chem. Soc.* **85**, 2149 (1963).
[13] S. Ostergaard and A. Holm, *Mol. Diversity* in press (1997).

can be immediately applied in most laboratories, that is, techniques that do not require major capital investment and instrumentation. All that is needed is the appropriate mind-set for the use of library techniques. We do not cover preparation of biological libraries, issues connected with library screening against particular targets, or results obtained in peptide library applications, and we also do not discuss various theoretical methods used for design of libraries.[14] We do not exhaustively review related literature, as this has been done by numerous authors.[15-26] Furthermore, a dynamic database of literature in the field of molecular diversity is available on the Internet.[27] Several reviews are dedicated to peptide libraries specifically (including peptide libraries generated by biological methods).[28-36] Practical

[14] W. A. Warr, *J. Chem. Inform. Comput. Sci.* **37**, 134 (1997).

[15] M. Rinnova and M. Lebl, *Collect. Czech. Chem. Commun.* **61**, 171 (1996).

[16] L. A. Thompson and J. A. Ellman, *Chem. Rev.* **96**, 555 (1996).

[17] J. S. Fruchtel and G. Jung, *Angew. Chem. Int. Ed.* **35**, 17 (1996).

[18] S. H. DeWitt and A. W. Czarnik, *Acc. Chem. Res.* **29**, 114 (1996).

[19] R. W. Armstrong, A. P. Combs, P. A. Tempest, S. D. Brown, and T. A. Keating, *Acc. Chem. Res.* **29**, 123 (1996).

[20] J. A. Ellman, *Acc. Chem. Res.* **29**, 132 (1996).

[21] W. C. Still, *Acc. Chem. Res.* **29**, 155 (1996).

[22] X. Williard, I. Pop, L. Bourel, D. Horvath, R. Baudelle, P. Melnyk, B. Deprez, and A. Tartar, *Eur. J. Med. Chem.* **31**, 87 (1996).

[23] V. Krchnák and M. Lebl, *Mol. Diversity* **1**, 193 (1996).

[24] C. Pinilla, J. Appel, C. T. Dooley, S. E. Blondelle, J. Eichler, B. Dorner, J. Ostresh, and R. A. Houghten, *in* "Peptide and Non-Peptide Libraries: A Handbook for the Search of Lead Structures" (G. Jung, ed.), p. 139. VCH, Weinheim, Germany, 1996.

[25] B. Dorner, S. E. Blondelle, C. Pinilla, J. Appel, C. T. Dooley, J. Eichler, J. M. Ostresh, E. Perez-Paya, and R. A. Houghten, *in* "Combinatorial Libraries: Synthesis, Screening and Application Potential" (R. Cortese, ed.), p. 1. de Gruyter, Berlin, 1996.

[26] J. M. Ostresh, S. E. Blondelle, B. Dorner, and R. A. Houghten, *Methods Enzymol.* 220 (1996).

[27] M. Lebl and Z. Leblova, Internet World Wide Web address: http://vesta.pd.com (1997).

[28] C. Pinilla, J. R. Appel, and R. A. Houghten, *in* "Immunological Recognition of Peptides in Medicine and Biology" (N. D. Zegers, W. J. A. Boersma, and E. Claassen, eds.), p. 1. CRC Press, Boca Raton, Florida, 1995.

[29] M. A. Gallop, R. W. Barrett, W. J. Dower, S. P. A. Fodor, and E. M. Gordon, *J. Med. Chem.* **37**, 1233 (1994).

[30] R. A. Houghten, *Trends Genet.* **9**, 235 (1993).

[31] J. K. Scott and L. Craig, *Curr. Opin. Biotechnol.* **5**, 40 (1994).

[32] C. Pinilla, J. R. Appel, S. E. Blondelle, C. T. Dooley, J. Eichler, J. M. Ostresh, and R. A. Houghten, *Drug Dev. Res.* **33**, 133 (1994).

[33] J. Eichler, J. R. Appel, S. E. Blondelle, C. T. Dooley, B. Dorner, J. M. Ostresh, E. Perez-Paya, C. Pinilla, and R. A. Houghten, *Med. Res. Rev.* **5**, 481 (1995).

[34] J. E. Fox, *Mol. Biotechnol.* **3**, 249 (1995).

[35] P. M. Dean, *Exp. Opin. Ther. Patents* **5**, 887 (1995).

[36] V. J. Hruby, *in* "The Practice of Medicinal Chemistry," p. 135. Academic Press, San Diego, 1996.

aspects of peptide library syntheses have also been covered in earlier review articles.[37–39]

Solid Support

Selection of the solid support depends on the type of library to be synthesized, as well as on the way it will be screened. The standard resin beads used for one-bead–one-compound libraries[1] are large enough to carry a sufficient amount of compound for bioassay; however, they should be homogeneously substituted. Furthermore, the resin beads should possess good mechanical properties; excessive fragility and tendency to form clusters (which may substantially lower the number of structures created) are the most common undesirable features. When bead-binding assays are used for the screening, the resin should swell both in organic solvents and aqueous media.

Polydimethylacrylamide or polyoxyethylene grafted polystyrene (TentaGel[40,41]) beads fulfill the above criteria. TentaGel, owing to its uniformity in size as well as its nonstickiness, is now the resin of choice for solid-phase library (both peptide and organic) synthesis based on the one-bead–one-compound principle. Polyethylene glycol–polystyrene (PEG–PS) resin[42] (Perseptive Biosystems, Bedford, MA) has a composition and properties similar to TentaGel. The major difference between the two resins is the location of the chemically reactive groups (anchor groups) in relation to the polystyrene matrix. Whereas TentaGel has the anchor groups at the end of long polyoxyethylene chains, far away from the polystyrene core, which promotes presentation of the library compounds for on-resin binding assays, PEG–PS may have the anchor groups in close proximity to the polystyrene core. In this case, the polyoxyethylene chains in PEG–PS do not serve as a spacer between the library compound and the polymer, but rather as a "modifier" of the polymer properties. ArgoGel (Argonaut

[37] M. Lebl, V. Krchňák, S. E. Salmon, and K. S. Lam, Methods (San Diego) 6, 381 (1994).
[38] K. S. Lam and M. Lebl, Methods (San Diego) 6, 372 (1994).
[39] R. A. Hougthen, Methods (San Diego) 6, 354 (1994).
[40] W. Rapp, L. Zhang, R. Habich, and E. Bayer, in "Peptides 1988" (G. Jung and E. Bayer, eds.), p. 199. de Gruyter, Berlin, 1989.
[41] W. Rapp, in "Peptide and Non-Peptide Libraries: A Handbook for the Search of Lead Structures" (G. Jung, ed.), p. 425. VCH, Weinheim, Germany, 1996.
[42] G. Barany, F. Albericio, S. Biancalana, S. L. Bontems, J. L. Chang, R. Eritja, M. Ferrer, C. G. Fields, G. B. Fields, M. H. Lyttle, H. A. Sole, Z. Tian, R. J. Van Abel, P. B. Wright, S. Zalipsky, and D. Hudson, in "Peptides: Chemistry and Biology" (J. A. Smith and J. E. Rivier, eds.), p. 603. Escom, Leiden, The Netherlands, 1992.

Technologies, San Carlos, CA)[43] differs from TentaGel by having branched polyoxyethylene chains at the attachment point to the polystyrene core, resulting in a higher substitution compared to TentaGel (0.4 versus 0.2–0.3 mmol/g). Meldal's group[44-46] developed a polymer based on the copolymerization of acrylamide and polyoxyethylene that allows penetration of macromolecular targets (e.g., enzymes) into the interior of the beads (penetration of TentaGel by proteins is limited[47]). The more recently described highly cross-linked hydrophilic polymer CLEAR[48] may find its application in library synthesis soon. The compatibility of the resin with aqueous media and size uniformity are not critical for libraries that are cleaved from the resin and screened in solution, permitting the use of conventional resins, such as chloromethyl and benzhydrylamine polystyrene resins.

A variety of carriers in addition to resin beads have been used for library synthesis. The first peptide library was synthesized on polyacrylamide-grafted polypropylene "pins."[9] This support was found to be very useful for multiple peptide synthesis.[49] Larger scale synthesis was achieved by attaching higher substituted "crowns" to the pins.[50] Alternative solid support materials for peptide library synthesis include segmental planar carriers such as paper sheets[51-56] and cotton disks or squares.[57-64] Soluble

[43] O. Gooding, P. D. J. Hoeprich, J. W. Labadie, J. A. J. Porco, P. van Eikeren, and P. Wright, *in* "Molecular Diversity and Combinatorial Chemistry Libraries and Drug Discovery" (I. M. Chaiken and K. D. Janda, eds.), p. 199. American Chemical Society, Washington, D.C., 1996.

[44] M. Meldal, *Tetrahedron Lett.* **33**, 3077 (1992).

[45] M. Meldal, F. I. Auzanneau, O. Hindsgaul, and M. M. Palcic, *J. Chem. Soc., Chem. Commun.*, 1849 (1994).

[46] M. Meldal, F. I. Auzanneau, and K. Bock, *in* "Innovation and Perspectives in Solid Phase Synthesis" (R. Epton, ed.), p. 259. Mayflower Worldwide, Birmingham, UK, 1994.

[47] J. Vagner, G. Barany, K. S. Lam, V. Krchnák, N. F. Sepetov, J. A. Ostrem, P. Strop, and M. Lebl, *Proc. Natl. Acad. Sci. U.S.A.* **93**, 8194 (1996).

[48] M. Kempe and G. Barany, *J. Am. Chem. Soc.* **118**, 7083 (1996).

[49] A. M. Bray, R. M. Valerio, A. J. Dipasquale, J. Greig, and N. J. Maeji, *J. Pept. Sci.* **1**, 80 (1995).

[50] N. J. Maeji, A. M. Bray, R. M. Valerio, and W. Wang, *Pept. Res.* **8**, 33 (1995).

[51] R. Frank and R. Doring, *Tetrahedron* **44**, 6031 (1988).

[52] R. Frank, *Bioorg. Med. Chem. Lett.* **3**, 425 (1993).

[53] R. Frank, *Tetrahedron* **48**, 9217 (1992).

[54] W. Tegge, R. Frank, F. Hofmann, and W. R. G. Dostmann, *Biochemistry* **34**, 10569 (1995).

[55] R. Frank, S. Hoffmann, M. Kiess, H. Lahmann, W. Tegge, C. Behn, and H. Gausepohl, *in* "Peptide and Non-Peptide Libraries: A Handbook for the Search of Lead Structures" (G. Jung, ed.), p. 363. VCH, Weinheim, Germany, 1996.

[56] R. Frank, *J. Biotechnol.* **41**, 259 (1995).

[57] J. Eichler and R. A. Houghten, *Biochemistry* **32**, 11035 (1993).

[58] M. Lebl and J. Eichler, *Pept. Res.* **2**, 297 (1989).

[59] J. Eichler, M. Bienert, A. Stierandova, and M. Lebl, *Pept. Res.* **4**, 296 (1991).

PEG carriers[65-67] have been proposed for library synthesis.[68,69] These supports are soluble in a variety of organic and aqueous solvents, enabling synthesis as well as screening of libraries in solution, but can be readily precipitated for removal of excess reagents by filtration.

Chemistry

This section focuses on the specifics of peptide library synthesis compared to the synthesis of individual peptides. In general, library synthesis requires greater emphasis on simplicity and reproducibility of the synthesis process.

Coupling

Typically, standard coupling reagents, such as active esters and carbodiimides, are used for library synthesis.[70-72] Difficult couplings can be driven to completion by more reactive coupling reagents such as TFFA[73] and N-[(dimethylamino)-1H-1,2,3-triazolo[4,5-b]pyridin-1-ylmethylene]-

[60] V. Pokorny, P. Mudra, J. Jehnicka, K. Zenísek, M. Pavlik, Z. Voburka, M. Rinnova, A. Stierandova, A. W. Lucka, J. Eichler, R. A. Houghten, and M. Lebl, *in* "Innovation and Perspectives in Solid Phase-Synthesis" (R. Epton, ed.), p. 643. Mayflower Worldwide, Birmingham, UK, 1994.

[61] M. Lebl, A. Stierandova, J. Eichler, M. Patek, V. Pokorny, J. Jehnicka, P. Mudra, K. Zenísek, and J. Kalousek, *in* "Innovation and Perspectives in Solid Phase Peptide Synthesis" (R. Epton, ed.), p. 251. Intercept, Andover, UK, 1992.

[62] M. Schmidt, J. Eichler, J. Odarjuk, E. Krause, M. Beyermann, and M. Bienert, *Bioorg. Med. Chem. Lett.* **3**, 441 (1993).

[63] J. Eichler, M. Bienert, N. F. Sepetov, P. Stolba, V. Krchnák, O. Smekal, V. Gut, and M. Lebl, *in* "Innovations and Perspectives in Solid Phase Synthesis" (R. Epton, ed.), p. 337. SPCC, Birmingham, UK, 1990.

[64] J. Eichler, C. Pinilla, S. Chendra, J. R. Appel, and R. A. Houghten, *in* "Innovation and Perspectives in Solid Phase Synthesis" (R. Epton, ed.), p. 227. Mayflower Worldwide, Birmingham, UK, 1994.

[65] V. N. R. Pillai, M. Mutter, E. Bayer, and I. Gatfield, *J. Org. Chem.* **45**, 5364 (1980).

[66] E. Bayer and M. Mutter, *Nature (London)* **237**, 512 (1972).

[67] M. Mutter and E. Bayer, *Peptides* **2**, 285 (1979).

[68] H. Han, M. M. Wolfe, S. Brenner, and K. D. Janda, *Proc. Natl. Acad. Sci. U.S.A.* **92**, 6419 (1995).

[69] A. M. van der Steen, H. Han, and K. D. Janda, *Mol. Diversity* **2**, 89 (1996).

[70] G. A. Grant, *in* "Synthetic Peptides: A User's Guide" p. 185. Freeman, New York, 1992.

[71] J. M. Stewart and J. D. Young, "Solid Phase Peptide Synthesis." Pierce, Rockford, Illinois, 1984.

[72] R. B. Merrifield, *in* "Peptides: Synthesis, Structures, and Applications" (B. Gutte, ed.), p. 94. Academic Press, San Diego, 1995.

[73] L. A. Carpino and A. El-Faham, *J. Am. Chem. Soc.* **117**, 5401 (1995).

N-methylmethanaminium hexafluorophosphate N-oxide (HATU).[74] Couplings in peptide library synthesis are performed in the usual way (i.e., in a battery of bubblers, in closed plastic vials, in polypropylene syringes,[75] in tea bags,[76] or in microtiter plates[77]).

Once the length of a peptide exceeds about four amino acids, the tendency to acylate the amino groups is more influenced by the character of the acylated peptide than the acylating agent.[78-81] Because libraries are synthesized in most cases from 20 proteinogenic amino acids, the library will also contain the most difficult sequences. At the same time, the library will also contain sequences prone to side reactions, for example, alkylation of Trp nucleus during deprotection of 2,2,5,7,8-pentamethylchroman-6-sulfonyl (Pmc)-protected Arg-X-Trp peptides.[82] Such complications are unavoidable at the present time. Because we cannot know *a priori* whether a certain biological response is caused by desired peptide or a side product, the reproducibility of the chemical process is of ultimate importance.

Monitoring

Monitoring of acylation requires modified techniques compared to the synthesis of individual peptides. In the "split synthesis" method[1,2,83] different peptides are synthesized on each resin bead, and therefore coupling kinetics may be different for each bead. Whether the coupling reaction is complete in a resin sample is not relevant, as it is necessary to know whether all beads in the sample were coupled. The classic ninhydrin test[84] does not provide this information. However, the reaction can be followed at the level of individual beads using nondestructive methods, such as bromphenol blue monitoring.[85] Using this method, incompletely coupled (blue) individual beads can be identified among a vast majority of completely coupled (colorless) beads.

[74] L. A. Carpino, A. Elfaham, and F. Albericio, *Tetrahedron Lett.* **35**, 2279 (1994).
[75] V. Krchňák and J. Vagner, *Pept. Res.* **3**, 182 (1990).
[76] R. A. Houghten, *Proc. Natl. Acad. Sci. U.S.A.* **82**, 5131 (1985).
[77] V. Krchňák, A. S. Weichsel, M. Lebl, and S. Felder, *Bioorg. Med. Chem. Lett.* submitted (1997).
[78] V. Krchňák, Z. Flegelova, and J. Vagner, *Int. J. Pept. Protein Res.* **42**, 450 (1993).
[79] R. C. de L. Milton, S. C. F. Milton, and P. A. Adams, *J. Am. Chem. Soc.* **112**, 6039 (1990).
[80] W. J. van Woerkom and J. W. van Nispen, *Int. J. Pept. Protein Res.* **38**, 103 (1991).
[81] S. B. H. Kent, *Annu. Rev. Biochem.* **57**, 957 (1988).
[82] A. Stierandova, N. F. Sepetov, G. V. Nikiforovich, and M. Lebl, *Int. J. Pept. Protein Res.* **43**, 31 (1994).
[83] A. Furka, F. Sebestyen, M. Asgedom, and G. Dibo, *Int. J. Pept. Protein Res.* **37**, 487 (1991).
[84] E. Kaiser, R. L. Colescott, C. D. Bossinger, and P. I. Cook, *Anal. Biochem.* **34**, 595 (1969).
[85] V. Krchňák, J. Vagner, P. Safár, and M. Lebl, *Collect. Czech. Chem. Commun.* **53**, 2542 (1988).

Equally valuable is information about the total amount of 9-fluorenyl-methyloxycarbonyl (Fmoc) group incorporated (and subsequently cleaved) relative to the entire mass of the resin used for the synthesis. We strongly recommend spectrophotometric quantitation of the cleaved Fmoc group after each step. The deprotection is usually performed after the resin is recombined, and one measurement can detect the major synthetic problems (premature cleavage of the peptide from the resin, use of improperly protected amino acids, etc.).

Protection, Deprotection, and Cleavage

Library synthesis involves handling of large numbers of reaction vessels at the same time. Because of the ease of manipulation and the simplicity of deprotection in multiple vessels in parallel, the Fmoc/*tert*-butyl (*t*Bu) protection strategy is mostly favored over the *tert*-butyloxycarbonyl/benzyl (Boc/Bzl) protection scheme. When a typical linker for either strategy is used to anchor the C-terminal amino acid, trifluoroacetic acid (TFA) or HF is applied to remove the side-chain protecting groups and at the same time cleave the peptide from the resin.

To facilitate the TFA cleavage process, reaction vessels, such as the wells of a microtiter plate, can be placed in an evacuated centrifuge, and the TFA can be evaporated simultaneously from all wells. Different multiple parallel processing methods must be applied when evaporation would harm the product. In this case, the TFA solution can be applied in small increments to reaction vessels, equipped at the bottom with a frit, and placed on top the grid of receiving vessels that contain a precipitating solvent (e.g., ether). This arrangement minimizes exposure of the products to TFA and cationic species generated from protecting groups and may, in some cases, eliminate the need to use scavengers.

Cleavage with liquid hydrogen fluoride (HF) is practical when an apparatus equipped with multiple reaction vessels is available.[86] Handling of HF is not very popular in many laboratories, but its low boiling point, combined with the well-understood chemistry of cleavage,[87] makes it a very flexible reagent. Furthermore, the very stable bond between the peptides and the polymeric carrier enables the use of a wide range of reagents for peptide and nonpeptide synthesis, as well as peptide modifications (see below).

In general, it is beneficial for library syntheses to remove the side-chain protecting groups prior to the peptide cleavage, thus avoiding contamina-

[86] R. A. Houghten, M. K. Bray, S. T. DeGraw, and C. J. Kirby, *Int. J. Pept. Protein Res.* **27,** 673 (1986).

[87] J. P. Tam, W. F. Heath, and R. B. Merrifield, *J. Am. Chem. Soc.* **105,** 6442 (1983).

tion of the library with cleaved protecting groups and scavengers. One way to separate side-chain deprotection from final cleavage of the peptides from the resin is the use of TFA–cleavable side-chain protecting groups in combination with a TFA-stable, HF-cleavable peptide–resin linkage.[75] Alternatively, the high/low HF procedure[87] can be used.

Application of gaseous reagents is an elegant method of multiple sample processing. Ester bonds between the peptides and the solid support can be cleaved by gaseous ammonia[88,89] or ammonia under high pressure.[90] Hydrogen chloride, hydrogen fluoride, and trifluoroacetic acid[91,92] can be applied in the form of gas as well. The use of gaseous reagents, as well as light, for peptide cleavage has one significant advantage; the cleaved peptides (or other compounds) remain physically located inside the resin beads, and individual beads can be handled either in the dry state or as a suspension in a nonextracting solvent (e.g., petroleum ether, ether) without cross-contamination of the cleaved compounds. The peptides can later be physically released from the beads for the bioassay (e.g., into the wells of the assay plate or onto the surface of an agar layer containing a cell culture) on addition of an extracting solvent, such as an aqueous buffer, methanol, or N,N-dimethylformamide (DMF), or by slow diffusion of the extracting solvent into the beads (e.g., on the surface of agar).[91,93]

Linkers for Library Synthesis

The decision whether to use a cleavable linker for the attachment of the first amino acid to the solid support depends on the intended use of the library. For on-bead binding assays, the use of linkers is not only unnecessary, but also contraindicated, as it may interfere with the binding of the macromolecular target to the peptide on the bead. Spacers between the peptide and the solid support, however, are beneficial in that they may improve the recognition of the peptide by the target molecule. Some resins

[88] A. M. Bray, N. J. Maeji, A. G. Jhingran, and R. M. Valerio, *Tetrahedron Lett.* **32,** 6163 (1991).
[89] A. M. Bray, R. M. Valerio, and N. J. Maeji, *Tetrahedron Lett.* **34,** 4411 (1993).
[90] M. Flegel, L. Lepsa, Z. Panek, I. Blaha, and M. Rinnova, *in* "Peptides: Chemistry, Structure and Biology" (P. T. P. Kaumaya and R. S. Hodges, eds.), p. 119. Mayflower Scientific, Kingswinford, UK, 1995.
[91] C. K. Jayawickreme, G. F. Graminski, J. M. Quillan, and M. R. Lerner, *Proc. Natl. Acad. Sci. U.S.A.* **91,** 1614 (1994).
[92] V. Krchnák, A. S. Weichsel, D. Cabel, and M. Lebl, *in* "Molecular Diversity and Combinatorial Chemistry. Libraries and Drug Discovery" (I. M. Chaiken and K. D. Janda, eds.), p. 99. American Chemical Society, Washington, DC, 1996.
[93] S. E. Salmon, R. H. Liu-Stevens, Y. Zhao, M. Lebl, V. Krchnák, K. Wertman, N. Sepetov, and K. S. Lam, *Mol. Diversity* **2,** 57 (1996).

are already equipped with long inert spacers (e.g., polyoxyethylene in TentaGel).

When the library compounds are intended to be screened in solution, the peptides have to be cleaved from the solid support. An ideal linker for combinatorial synthesis of peptide libraries should meet several criteria: (i) it must be stable to all chemical reactions during the library synthesis; (ii) it should enable quantitative cleavage of the peptides from the resin; (iii) the cleavage reagents should not alter the library compounds; and (iv) the cleavage conditions should not require complicated equipment and, preferably, should render the released compounds in a state ready for screening.

The linkers primarily used for standard peptide synthesis were designed in a way that the cleavage cocktail not only cleaves the peptides from the resin, but at the same time removes all side-chain protecting groups. For peptide libraries, however, where the intent is not to isolate the final products, but rather to test the library for various biological activities, the concurrent removal of protecting groups and cleavage of peptides from the resin is not necessarily beneficial. Therefore, several linkers were devised that enable deprotection of side chains while leaving the peptides on the resin. This step, which usually requires the use of scavengers, can eliminate any contamination of peptides by products of protecting groups, as the resin can be extensively washed prior to peptide cleavage. The peptides can then be cleaved under relatively mild conditions, preferably in aqueous media, providing solutions ready for the bioassay.

The library compounds can be cleaved from resin beads in one single step, or the release of compounds can be gradual, in several steps. Multiply cleavable linkers are particularly useful for one-bead–one-compound libraries, in that they enable the repeated release of library compounds for different levels of library screening, thus eliminating deconvolution steps.

Singly Cleavable Linkers

Linkers used most frequently for the synthesis of libraries were those developed for individual peptide synthesis. The most popular linkers for peptide acids and amides are acid-labile, and their core structures are based on benzyl structures (Fig. 2). Whereas benzyl esters yield carboxylates on cleavage from the resin, benzyl amide linkers were designed to generate C-terminal carboxamides. Benzyl (and other) esters can also be advantageously cleaved by various nucleophiles.

The original Merrifield resin[12] is still widely used for the generation of peptide acids using the Boc/Bzl protection scheme. Its acid lability was

FIG. 2. Benzyl-based linkers cleavable by acids.

increased for Fmoc/*t*Bu strategy by substituting the benzyl ring with electron donating groups (e.g., Wang[94] and Sheppard[95] linkers). Additional
methoxy groups further increase the acid lability, as documented by
the SASRIN[96] and 5-(4-aminomethyl-3,5-dimethoxyphenoxy)valeric acid
(PAL)[97] resins.

Because of the extremely high stability of benzyl amides to acids, benzhydrylamine type linkers[98] were developed for the synthesis of peptide
amides. The *p*-methylbenzhydrylamine linker,[99] which requires liquid HF
for peptide cleavage, is widely used for mixture libraries. Owing to its
stability to a vast range of reagents, it is compatible with a variety of
organic reactions used to modify resin-bound peptides. The acid stability
of benzhydrylamine linkers was reduced by methoxy substituents,[100] which
makes this linker cleavable by TFA. The most acid-labile linker is the
trityl linker,[101,102] to which acids, alcohols, thiols, as well as amines can be
attached. In addition to liquid cleavage cocktails, it can be cleaved by HCl
gas or TFA vapors.[92]

Safety-catch linkers contain electron-withdrawing groups that increase
their acid stability. Prior to the release of the peptides, these groups are
chemically transformed into electron-donating groups, which makes the
linker more acid labile, thus enabling the use of Boc-protecting groups in
combination with TFA cleavage of the peptides from the resin. In the
SCAL linker,[103,104] the electron-withdrawing methyl sulfoxide is reduced

[94] S. S. Wang, *J. Am. Chem. Soc.* **95**, 1328 (1973).
[95] R. C. Sheppard and B. J. Williams, *Int. J. Pept. Protein Res.* **20**, 451 (1982).
[96] M. Mergler, R. Tanner, J. Gosteli, and P. Grogg, *Tetrahedron Lett.* **29**, 4005 (1988).
[97] F. Albericio, N. Kneib-Cordonier, S. Biancalana, L. Gera, R. I. Masada, D. Hudson, and
G. Barany, *J. Org. Chem.* **55**, 3730 (1990).
[98] J. P. Tam, R. D. DiMarchi, and R. B. Merrifield, *Tetrahedron Lett.* **22**, 2851 (1981).
[99] G. R. Matsueda and J. M. Stewart, *Peptides* **2**, 45 (1981).
[100] H. Rink, *Tetrahedron Lett.* **28**, 3787 (1987).
[101] K. Barlos, D. Gatos, I. Kallitsis, D. Papaioannou, and P. Sitiriou, *Liebigs Ann. Chem.*,
1079 (1988).
[102] K. Barlos, O. Chatzi, D. Gatos, and G. Stavropoulos, *Int. J. Pept. Protein Res.* **37**, 513 (1991).
[103] M. Patek and M. Lebl, *Tetrahedron Lett.* **31**, 5209 (1990).
[104] M. Patek and M. Lebl, *Tetrahedron Lett.* **32**, 3891 (1991).

to a methylthio group after the peptide has been assembled, prior to its release from the resin with TFA. A similar principle was applied to a benzyl ester linker by Kiso et al.[105]

Nucleophilic cleavage of ester bonds, for example, those formed from hydroxymethylbenzoic acid as a linker, can be used to release peptides directly into the screening solution. The formation of diketopiperazines (DKPs) from dipeptide esters, a well-documented side reaction in peptide synthesis, was used for the design of linkers that are cleavable under very mild conditions (e.g., pH 7).[106] Peptides cleaved in this manner, however, still have the DKP moiety attached to them. To overcome this limitation, Hoffmann and Frank developed a new type of safety-catch linker based on the intramolecular catalytic hydrolysis of an ester bond. This linker enables the release of free peptide acids into a neutral aqueous buffer.[107]

Multiply Cleavable Linkers

Peptide libraries often contain millions of individual peptides, which cannot be screened one by one. Although mixture libraries are screened in large pools of up to millions of peptides, a two-step procedure for the screening in solution of one-bead–one-peptide libraries was designed that involves the release of peptides from the beads in two separate steps.[37,108]

Three different types of multiply releasable linkers have been proposed. The first one, used at Selectide, is based on two different types of cleavage of an ester bond.[37,108,109] One ester bond is cleaved through DKP formation (see above), and the second by alkali or ammonia. These two reactions, however, do not yield exactly the same compounds, because the DKP moiety is still attached to the peptide cleaved in the first step.

To overcome this limitation, a second generation of multiply cleavable linkers was developed, in which the DKP remained on the resin. Peptides are attached to the linker via an ester bond of Fmoc-Gly-NH-$(CH_2)_3$-OH (Fmoc-Gly-HOPA), and after release to an aqueous solution, both copies of peptides have an identical C terminus, that is, the hydroxypropylamide of Gly (Gly-HOPA). The first linker of this generation was based on the Glu-Pro dipeptide. Glu provided a side-chain function for the attachment to the resin, and Pro enhanced the tendency to form the DKP. In another

[105] Y. Kiso, T. Fukui, S. Tanaka, T. Kimura, and K. Akaji, *Tetrahedron Lett.* **35**, 3571 (1994).
[106] A. M. Bray, N. J. Maeji, R. M. Valerio, R. A. Campbell, and H. M. Geysen, *J. Org. Chem.* **56**, 6659 (1991).
[107] S. Hoffmann and R. Frank, *Tetrahedron Lett.* **35**, 7763 (1994).
[108] S. E. Salmon, K. S. Lam, M. Lebl, A. Kandola, P. S. Khattri, S. Wade, M. Patek, P. Kocis, V. Krchnák, D. Thorpe, and S. Felder, *Proc. Natl. Acad. Sci. U.S.A.* **90**, 11708 (1993).
[109] M. Lebl, M. Patek, P. Kocis, V. Krchnák, V. J. Hruby, S. E. Salmon, and K. S. Lam, *Int. J. Pept. Protein Res.* **41**, 201 (1993).

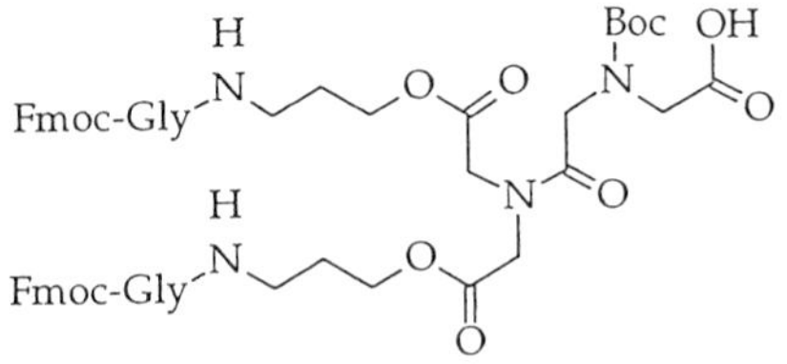

FIG. 3. Two doubly cleavable linkers (1, 2) based on the Ida–Ida motif.

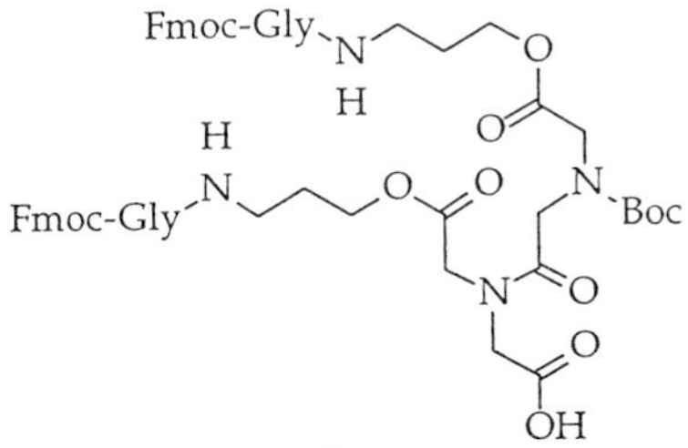

linker, the core dipeptide was formed from two iminodiacetic acid residues (Ida).[110] The advantages of Ida include the presence of two chemically equivalent carboxyl groups, a high tendency to cyclize via DKP (a characteristic feature of N-substituted amino acids), and its low price. Two Fmoc-Gly-HOPAs are either coupled to both carboxyl groups of the C-terminal Ida (Fig. 3, linker 1), or each Ida is linked to one Fmoc-Gly-HOPA (linker 2). The remaining carboxyl group was used to attach the linker to the resin. The chemistry of both releases is shown in Fig. 4. A third, noncleavable copy of the peptide is attached to enable structure determination by sequencing from the resin beads. To avoid the presence of the hydroxypropylamide of Gly at the C terminus of released peptides, a modified linker (Fig. 5, linker 3) was designed. It includes an additional ester linkage,[111] which is hydrolyzed after the release of the peptides from the resin beads.

Pharmacopeia scientists, who followed the Lam one-bead–one-compound library approach, used a photolabile linker of the o-nitrobenzyl type[112] (Fig. 6, linker 4). The original linker, which yields a reactive resin-

---

[110] P. Kocis, V. Krchnák, and M. Lebl, *Tetrahedron Lett.* **34**, 7251 (1993).

[111] V. Krchnák, N. F. Sepetov, P. Kocis, M. Patek, K. S. Lam, and M. Lebl, *in* "Combinatorial Libraries: Synthesis, Screening and Application Potential" (R. Cortese, ed.), p. 27. de Gruyter, Berlin, 1996.

[112] J. J. Baldwin, J. J. Burbaum, I. Henderson, and M. H. J. Ohlmeyer, *J. Am. Chem. Soc.* **117**, 5588 (1995).

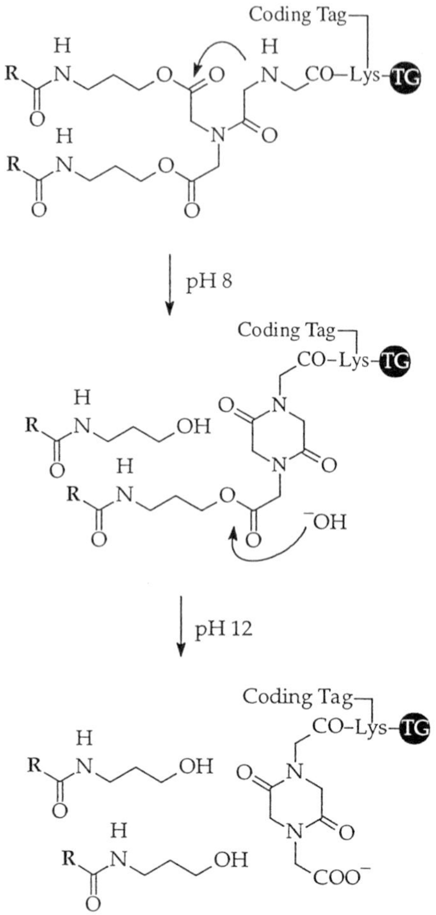

FIG. 4. Chemistry of both releases from a doubly cleavable linker. TG, TentaGel.

bound aldehyde, has been modified to produce less reactive ketone derivatives on cleavage (Fig. 6, linkers 5 and 6).[113,114]

The third type of linkers for gradual release are based on the varying stability of different benzyl structures toward acids. Partial release of peptides from a benzhydrylamine linker by TFA vapors was used by Jayawickreme et al.[91] Cardno and Bradley[115] used a combination of benzyl linkers

[113] C. P. Holmes and D. G. Jones, J. Org. Chem. 60, 2318 (1995).
[114] B. B. Brown, D. S. Wagner, and H. M. Geysen, Mol. Diversity 1, 4 (1995).
[115] M. Cardno and M. Bradley, Tetrahedron Lett. 37, 135 (1996).

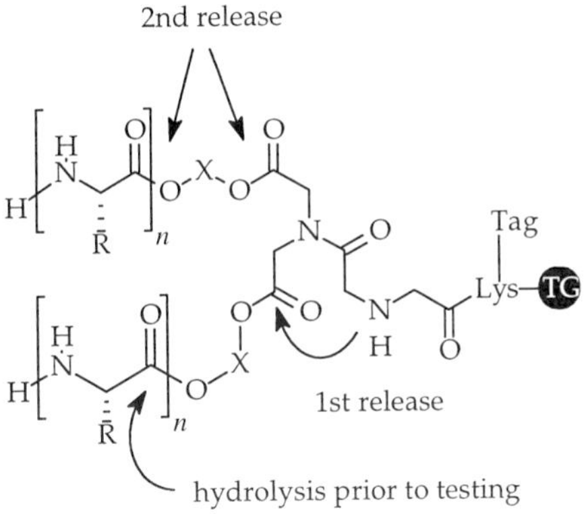

FIG. 5. A doubly cleavable linker (3) that yields peptide acids.

(Fig. 7). The most acid-labile linker can be cleaved with 1% TFA, whereas 95% TFA is necessary to cleave the second linker. The third copy of the same compound remains attached to the resin for structure analysis or on-bead binding assays. It can, however, be cleaved by HF, if needed. Alternative designs of linkers allowing gradual release of library compounds has been reviewed.[116]

Multiple Synthesis Techniques

Solid-phase synthesis is amenable to parallel processing.[117] Individual cycles of peptide synthesis are in most cases composed of identical steps, the only difference being the kind of amino acid used for acylation of the growing peptide chain. Therefore, it was not surprising that most library synthesis methods were adapted from existing techniques for multiple parallel synthesis.

The surface of plastic (polypropylene) pins grafted with an easily functionalizable polymeric matrix (polystyrene or polyacrylamide) was used as the solid support for the first method of multiple peptide synthesis.[118] The pins are attached to a plate and arranged in the 96-well microtiter plate

[116] D. Madden, V. Krchnák, and M. Lebl, *Perspect. Drug Disc. Design* **2,** 269 (1995).
[117] G. Jung and A. G. Beck-Sickinger, *Angew. Chem., Int. Ed. Engl.* **31,** 367 (1992).
[118] H. M. Geysen, R. H. Meloen, and S. J. Barteling, *Proc. Natl. Acad. Sci. U.S.A.* **81,** 3998 (1984).

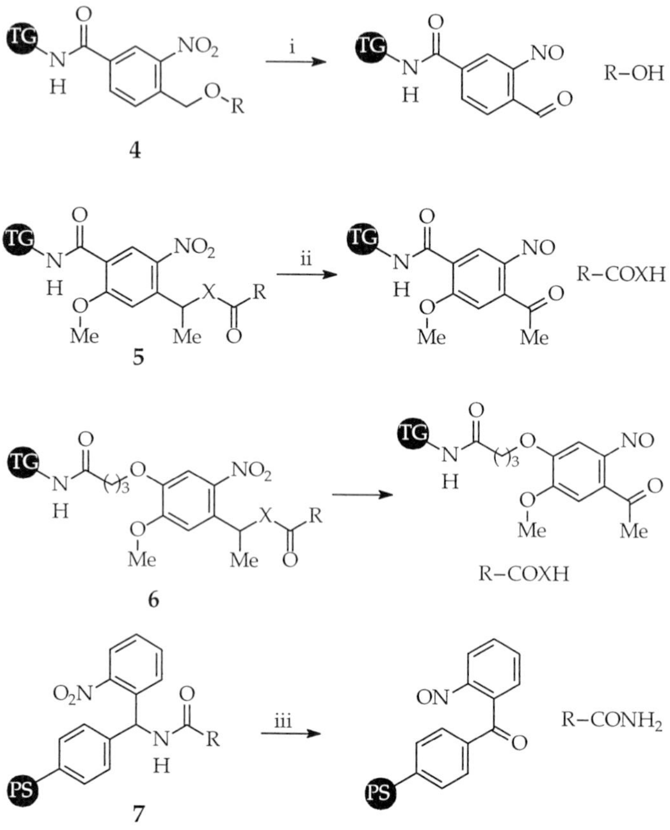

(i) hv (365 nm), MeOH, 16 hr, 25°   X = O or NH
(ii) hv (365 nm), pH 7.4, 3 hr, 25°
(iii) hv (350 nm), MeOH/DMF

FIG. 6. Photocleavable linkers (4–7).

format, so that the wells of microtiter plates can serve as the reaction vessels for all synthesis steps. Originally, the peptides remained on the pins and were tested immobilized for binding to antibodies. Later, the substitution of the pins was increased to micromole amounts, and linkers were introduced, enabling cleavage and isolation of the peptides from the pins.[9,118–124]

[119] N. J. Maeji, A. M. Bray, R. M. Valerio, M. A. Seldon, J. X. Wang, and H. M. Geysen, Pept. Res. 4, 142 (1991).
[120] R. M. Valerio, A. M. Bray, R. A. Campbell, A. J. Dipasquale, C. Margellis, S. J. Rodda, H. M. Geysen, and N. J. Maeji, Int. J. Pept. Protein Res. 42, 1 (1993).

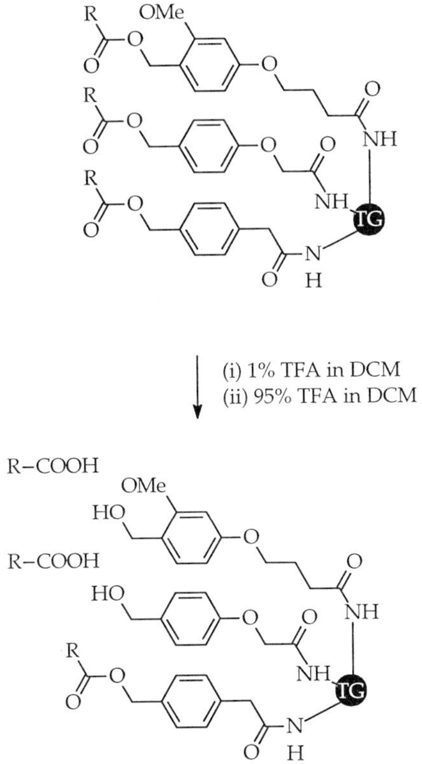

Fig. 7. Acid-sensitive multiply cleavable linker assembly.[115]

Another pioneering multiple peptide synthesis approach, termed the "tea bag" method, was introduced by Houghten.[76] The idea of this technique is very simple: If the solid support can be compartmentalized in separate packets (polypropylene mesh bags, Fig. 8), all common processes of peptide synthesis, such as wash and deprotection steps, can be performed simultaneously on many resin packets without cross-contamination. For the synthesis of peptides composed of the 20 proteinogenic amino acids, only 20 containers with activated amino acids, and two common containers

[121] J. X. Wang, A. M. Bray, A. J. Dipasquale, N. J. Maeji, and H. M. Geysen, *Int. J. Pept. Protein Res.* **42**, 384 (1993).

[122] H. M. Geysen and T. J. Mason, *Bioorg. Med. Chem. Lett.* **3**, 397 (1993).

[123] A. M. Bray, N. J. Maeji, and H. M. Geysen, *Tetrahedron Lett.* **31**, 5811 (1990).

[124] N. J. Maeji, R. M. Valerio, A. M. Bray, R. A. Campbell, and H. M. Geysen, *React. Polym.* **22**, 203 (1994).

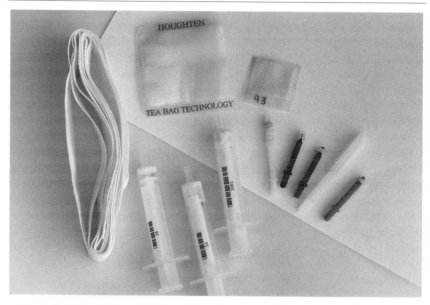

FIG. 8. Tools for simple multiple peptide synthesis: T-bag,[76] plastic syringe,[75] paper sheet,[53] cotton strip,[59] and plastic pin.[118]

for deprotection and wash steps, respectively, are needed. Up to several hundred labeled resin packets can be processed simultaneously. For common steps, they are placed in one big container, followed by sorting for individual couplings. Sorting of the bags can be facilitated by adding a radiofrequency tag to each bag.[125,126] This procedure is repeated several times, until the peptide sequences in all bags are completed. The resin used for this method is typically 4-methylbenzhydrylamine (MBHA) resin, and the peptides are deprotected and cleaved from the resin using the low/high HF method[87] in combination with multiple cleavage apparatus, which enables the simultaneous cleavage of up to 120 peptides.[86]

Polypropylene syringes equipped with a sintered polypropylene disks[75] are probably the simplest and least expensive disposable reaction vessel for solid-phase synthesis, enabling multiple peptide synthesis to be carried out in any laboratory. One 10-ml plastic syringe can hold up to 500 mg of resin. We have found the use of syringes as convenient reaction vessels for

[125] R. W. Armstrong, P. A. Tempest, and J. F. Cargill, *Chimia* **50**, 258 (1996).
[126] E. J. Moran, S. Sarshar, J. F. Cargill, M. M. Shahbaz, A. Lio, A. M. M. Mjalli, and R. W. Armstrong, *J. Am. Chem. Soc.* **117**, 10787 (1995).

the synthesis of peptide and especially nonpeptide libraries.[127] The synthesis protocol follows the common procedure for batchwise peptide synthesis using either Boc/Bzl or Fmoc/tBu protection strategy. The syringes containing the resin are manually filled with the solvent or reagent of the particular synthesis step, shaken, and emptied. During library synthesis using the split/mix protocol, deprotection and all wash steps are done in one common reaction vessel (big syringe, classic glass bubbler, etc.), followed by distributing the resin slurry into as many syringes as amino acids are used in the next coupling. The syringes are typically kept on a tumbler during the couplings, which can be accelerated in a hot (70°) DMF bath[128,129] or by sonication.[130]

Manual parallel handling of a large number of syringes, although feasible, is not very convenient. Therefore, we constructed a block holding 42 syringes, termed MultiBlock (http://www.5z.com/csps.htm), to facilitate and accelerate multiple synthesis in syringes. The MultiBlock (Fig. 9) consists of five parts: a Teflon block that holds 42 polypropylene syringes equipped with a plastic frit; a vacuum adapter that connects each reactor to a vacuum line, enabling rapid washing of the resin under continuous flow; two Teflon plates with flexibly attached 42 stoppers to seal the syringes during reactions; and a glass cover to allow mixing of resin for split/mix library synthesis. The MultiBlock is made from Teflon, polypropylene, glass, and stainless steel. A number of manufacturers have introduced their own versions (http://www.charybtech.com),[131-133] most of them designed for general organic synthesis on the solid phase with the option to be coupled to an automatic liquid delivery station.

A method for the concurrent synthesis of 96 peptides in the wells of commercially available polypropylene deep-well microtiter plates was reported by Schnorrenberg and Gerhardt.[134] The wells of microtiter plates have no frits or filters at the bottom. Solvent and solutions are introduced by a pipetting device. For the removal of liquid the tip of their washing device is protected by a narrow stainless steel net. One well is handled at

[127] V. Krchnák, A. S. Weichsel, D. Cabel, Z. Flegelova, and M. Lebl, *Mol. Diversity* **1**, 149 (1996).

[128] K. Barlos, *Liebigs Ann. Chem.* **11**, 1950 (1986).

[129] J. P. Tam and Y. A. Lu, *J. Am. Chem. Soc.* **117**, 12058 (1995).

[130] J. Vagner, P. Kocna, and V. Krchnák, *Pept. Res.* **4**, 1 (1991).

[131] J. F. Cargill and R. R. Maiefski, *Lab. Robotics Automation* **8**, 139 (1996).

[132] J. R. Harness, *in* "Molecular Diversity and Combinatorial Chemistry: Libraries and Drug Discovery" (I. M. Chaiken and K. D. Janda, eds.), p. 188. American Chemical Society, Washington, DC, 1996.

[133] H. V. Meyers, G. J. Dilley, T. L. Durgin, T. S. Powers, N. A. Winssinger, H. Zhu, and M. R. Pavia, *Mol. Diversity* **1**, 13 (1995).

[134] G. Schnorrenberg and H. Gerhardt, *Tetrahedron* **45**, 7759 (1989).

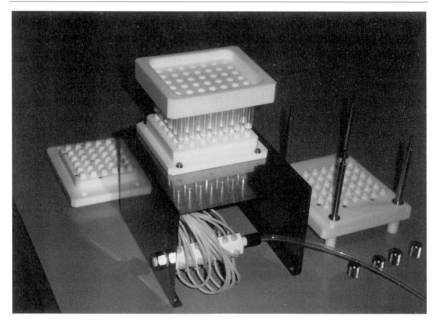

FIG. 9. MultiBlock (with permission of CSPS, San Diego, CA).

a time. An alternative approach has been described.[77] The solvent is aspirated by suction from the surface of the liquid after the resin beads have settled. Vacuum is applied to stainless steel needles that are slowly immersed into the wells. Solid-phase synthesis in modified microtiter plates where each well is equipped with a frit has also been reported.[133]

Planar segmental supports, such as cellulose membranes[51–55] and cotton,[58–64] have been shown to be versatile carriers for multiple synthesis. The substitution of these carriers must be sufficiently low, so that the amount of reagent solutions soaked into the carrier is high enough to ensure complete couplings. This principle of "inclusion volume coupling," which was later also tested on conventional resins in combination with applying centrifugation for liquid removal,[135] enables (i) the use of higher concentrations of activated amino acids, resulting in increased coupling rates, (ii) drastically decreased consumption of solvents, and (iii) the construction of multiple peptide synthesizers with virtually no reaction vessels, as the solid carriers themselves serve as the reaction vessels.[60]

[135] J. Eichler, R. A. Houghten, and M. Lebl, *J. Pept. Sci.* **2**, 240 (1996).

A synthesis method based on the attachment of the growing peptide chain to a glass surface, and using photolithographic techniques, was developed at Affymax.[5,136-139] A photolabile amino protecting group is cleaved on defined locations on the glass surface, and the whole surface is subjected to coupling with an activated amino acid. After completion of the coupling, other surface locations are deprotected, and the second amino acid attached. This process is repeated until all amino acids are attached to their particular locations, thus creating a surface covered with different peptides. That technology requires highly sophisticated instrumentation, which makes its widespread application unlikely. Various automated multiple peptide synthesizers have been constructed, some of which are commercially available.[34,134,140-150]

One-Bead–One-Compound Libraries

Having millions of peptides mixed together in one vessel, yet at the same time having each resin bead carry only one peptide, is achievable using the so-called split/mix, portioning/mixing, or divide–couple–recombine technique. This method was first reported in 1988 by Furka *et*

[136] C. Y. Cho, E. J. Moran, S. R. Cherry, J. C. Stephans, S. P. A. Fodor, C. L. Adams, A. Sundaram, J. W. Jacobs, and P. G. Schultz, *Science* **261**, 1303 (1993).

[137] L. F. Rozsynai, D. R. Benson, S. P. A. Fodor, and P. G. Schultz, *Angew. Chem., Int. Ed. Engl.* **31**, 759 (1992).

[138] J. W. Jacobs and S. P. A. Fodor, *Trends Biotechnol.* **12**, 19 (1994).

[139] C. P. Holmes, C. L. Adams, S. P. A. Fodor, and P. Yu-Yang, in "Perspectives in Medicinal Chemistry" (B. Testa, E. Kyburz, W. Fuher, and R. Giger, eds.), p. 489. VHCA, Basel, 1992.

[140] H. Gausepohl and R. W. Frank, in "Peptides 1992" (C. H. Schneider and A. N. Eberle, eds.), p. 310. Escom, Leiden, The Netherlands, 1993.

[141] R. N. Zuckermann, M. A. Siani, and S. C. Banville, *Lab. Robotics Automation* **4**, 183 (1992).

[142] H. Gausepohl, C. Boulin, M. Kraft, and R. W. Frank, *Pept. Res.* **5**, 315 (1992).

[143] G. Schnorrenberg, *Chim. Oggi.* **10**, 33 (1992).

[144] K. Nokihara and R. Yamamoto, in "Peptides: Chemistry and Biology" (J. A. Smith and J. E. Rivier, eds.), p. 507. Escom, Leiden, The Netherlands, 1992.

[145] J. E. Fox, *Biochem. Soc. Trans.* **20**, 851 (1992).

[146] V. Krchnák, D. Cabel, and M. Lebl, *Pept. Res.* **9**, 45 (1996).

[147] T. Luu, S. Pham, and S. Deshpande, *Int. J. Pept. Protein Res.* **47**, 91 (1996).

[148] J. A. Boutin, P. Hennig, P. H. Lambert, S. Bertin, L. Petit, J. P. Mahieu, B. Serkiz, J. P. Volland, and J. L. Fauchere, *Anal. Biochem.* **234**, 126 (1996).

[149] T. Geiser, H. Beilan, B. J. Bergot, and K. M. Otteson, in "Macromolecular Sequencing and Synthesis: Selected Methods and Applications" (D. H. Schlesinger, ed.), p. 199. Alan R. Liss, New York, 1988.

[150] H. H. Saneii and J. D. Shannon, in "Innovation and Perspectives in Solid Phase Synthesis" (R. Epton, ed.), p. 335. Intercept, Andover, UK, 1994.

al.[83,151,152] and independently used for the generation of peptide libraries by Houghten *et al.*[2] and Lam *et al.*[1] It is the method of choice for one-bead–one-compound libraries, which are based on the fact that each resin bead has only one peptide sequence attached to it.[1]

The first chemical reaction is performed in as many reaction vessels as building blocks (amino acids) are to be incorporated at the first position of the library. On completion of the coupling, all resin portions are combined, mixed thoroughly, and divided into as many reaction vessels as building blocks (amino acids) are to be incorporated at the second position of the library. This process is repeated until the entire library sequence is assembled. Because at each coupling only one amino acid is coupled to each bead, only one peptide is generated on each bead. One can drive any particular (condensation) reaction almost to completion without being concerned about the different coupling rate of amino acids because the different reactions are physically separated. This feature ensures the equimolarity of synthesized peptides.

Automation

Peptide libraries can be synthesized manually or by using an automated synthesizer.[148,153–155] The distribution of the resin is achieved either by volume distribution of a homogeneous (nonsedimenting) suspension of beads in an isopycnic solution,[154] by continuous stirring of the suspension during the distribution,[153] or by a combination of mechanical and gas stirring for creation of a homogeneous suspension.[148] Another design uses gas/mechanical stirring followed by sedimentation of the suspension in a symmetrical distribution vessel.[155]

When using amino acid mixtures for the coupling, standard multiple peptide synthesizers can be used for the synthesis of peptide libraries (see above). The only difference between the synthesis of individual peptides and libraries is the fact that amino acid mixtures, rather than individual

[151] A. Furka, F. Sebestyen, M. Asgedom, and G. Dibo, *in* "Highlights of Modern Biochemistry, Proceedings of the 14th International Congress of Biochemistry" p. 47. VSP, Utrecht, The Netherlands, 1988.
[152] A. Furka, F. Sebestyen, M. Asgedom, and G. Dibo, Poster presented at Tenth International Symposium on Medicinal Chemistry, Budapest, 1988.
[153] H. H. Saneii, J. D. Shannon, R. M. Miceli, H. D. Fischer, and C. W. Smith, *Pept. Chem.* **31**, 117 (1993).
[154] R. N. Zuckermann, J. M. Kerr, M. A. Siani, and S. C. Banville, *Int. J. Pept. Protein Res.* **40**, 497 (1992).
[155] Z. Bartak, J. Bolf, J. Kalousek, P. Mudra, M. Pavlik, V. Pokorny, M. Rinnova, Z. Voburka, K. Zenísek, V. Krchnák, M. Lebl, S. E. Salmon, and K. S. Lam, *Methods (San Diego)* **6**, 432 (1994).

amino acids, are incorporated in one, several, or all positions of the sequence.

Equimolarity

The most reliable way to ensure equimolarity of peptides in a library is using split synthesis.[1,2,83] Because the couplings of different amino acids are physically separated, differences in coupling rates between amino acids are not a problem. Owing to statistical distribution, however, in order for the library to contain all members with 99% confidence, the number of beads used for the synthesis should be at least five times higher than the number of peptides in the library. Some of the members will be underrepresented and some will be overrepresented, with five being the average occurrence of each member.

Although being optimal for the generation of completely randomized libraries, the split synthesis method may be less appropriate for the synthesis of peptide libraries composed of mixtures containing defined and mixture positions, such as iterative or positional scanning libraries (see below). Unless the defined positions are located exclusively at the N terminus of the sequence, the high number of reaction vessels needed for the synthesis makes the use of split synthesis impractical. (For example, the split synthesis of a library with one position defined with one of 20 amino acids, and 20 amino acids used for mixture positions, requires the use of 400 reaction vessels.)

Alternatives to split synthesis are coupling of mixtures of amino acids in a predetermined molar ratio, which compensates for the different coupling rates of the amino acids,[9,57,156–158] and repeated couplings of subequimolar amounts of equimolar mixtures of amino acids.[159–161] The goal of those techniques is the equimolar incorporation of all building blocks, which may be complicated by the fact that the coupling depends not only on the character of the incoming activated amino acid, but also on the sequence

[156] W. J. Rutter and D. V. Santi, U.S. Patent 5,010,175 (1991).

[157] C. Pinilla, J. R. Appel, P. Blanc, and R. A. Houghten, *BioTechniques* **13**, 901 (1992).

[158] J. M. Ostresh, J. M. Winkle, V. T. Hamashin, and R. A. Houghten, *Biopolymers* **34**, 1681 (1994).

[159] R. Frank, *in* "Innovation and Perspectives in Solid Phase Synthesis" (R. Epton, ed.), p. 509. Mayflower Worldwide, Birmingham, UK, 1994.

[160] A. Kramer, R. Volkmer-Engert, R. Malin, U. Reineke, and J. Schneider-Mergener, *Pept. Res.* **6**, 314 (1993).

[161] P. C. Andrews, J. Boyd, R. R. Ogorzalek-Loo, R. Zhao, C. Q. Zhu, K. Grant, and S. Williams, *in* "Techniques in Protein Chemistry V" (J. W. Crabb, ed.), p. 485. Academic Press, San Diego, 1994.

of peptide attached to the solid support. One library technique is based on the incorporation of mixtures that are depleted of or enriched in certain amino acids.[162]

Directed Libraries

Split synthesis results in a statistical distribution of library members, with most library members being represented more than once, which may bias the evaluation of the screening of one-bead–one-compound libraries. Progressively dividable materials, such as membrane-type carriers, are well-suited solid supports for the synthesis of libraries with nonstatistical distribution of library members.[163] The synthesis of this kind of library starts with *n* pieces of the carrier, to which *n* different building blocks are coupled (first randomization). Each of the *n* pieces is then divided into *m* parts, and these smaller parts are distributed into *m* reaction vessels in which *m* reactions are performed (second randomization). This process can be repeated as often as is practical regarding the size and number of the carrier segments. The result of this process is a library of as many support-bound compounds as pieces of the carrier have been produced during the synthesis, in which each library member is represented once and only once. Each part of the membrane can be prelabeled, and determination of structure can then be avoided. This approach has been applied by Pfizer scientists,[164] who used a classic beaded resin sealed between layers of porous polypropylene sheets as the solid support.

One-Bead–One-Mixture Libraries

Varying Complexity of Library Positions

Because the synthesis of complete one-bead–one-peptide libraries of longer sequences (e.g., >8 amino acids) is unrealistic owing to the large amount of resin required, coupling of amino acid mixtures rather than individual amino acids at selected positions of the library has been proposed as an alternative.[165] The least complex mixtures or individual amino acids were used for the N-terminal positions, and the C-terminal positions were the most complex ones, yielding a library with up to millions of peptides

[162] J. Blake and L. Litzi-Davis, *Bioconjugate Chem.* **3,** 510 (1992).
[163] M. Stankova, S. Wade, K. S. Lam, and M. Lebl, *Pept. Res.* **7,** 292 (1994).
[164] J. Steele, Second Annual Solid-Phase Synthesis Conference, February 6–7, 1997, Coronado, CA, 1997.
[165] V. Hornik and E. Hadas, *React. Polym.* **22,** 213 (1994).

per bead. Beads that reacted positively in a binding assay were sequenced, but only the N-terminal residues could be identified, and secondary libraries were synthesized in order to deconvolute the structure of the active peptides. The density of peptides on the bead surface can be an important factor in the binding assay. Wallace *et al.*[166] could detect binding only after the peptide library was displayed as an octamer complex.

Library of Libraries

The one-bead–one-motif library or "library of libraries" approach[167] represents a combination of the one-bead–one-compound and positional scanning (see below) library concepts. The idea is based on the assumption that a measurable biological signal can be obtained if a limited number of critical amino acids (usually three) are present at critical positions in the peptide sequence. The rest of the peptide sequence can be filled with an "average" amino acid, represented by a mixture of amino acids. The first of this type of library was a hexapeptide library prepared from the 20 proteinogenic amino acids with three defined and three mixture positions in all possible combinations, which yields a total of 160,000 mixtures. Because the defined positions were introduced through split synthesis, each peptide on a given bead had the same positions defined, and also the same amino acids at the defined positions, thus creating a one-bead–one-motif library. Consequently, each bead carried a mixture of $20^3 = 8000$ peptides. The principal advantage of this library format is the fact that it enables the identification of all possible combinations of (in this case three) key residues within a given sequence length, rather than single key residues in the positional scanning library format, which may not always be sufficient to be detected in the bioassay. The one-bead–one-"entity" library approach was the method of choice for the generation of this library, as the separate synthesis of 160,000 peptide mixtures would not be practical, unless ultra-high-throughput synthesis instrumentation is available.

An alternative is the library of libraries with variable length. At the beginning of synthesis and after each acylation, one-quarter of the resin is removed, the protecting group is cleaved off, and the mixture of amino acids is coupled to the remaining part. After this coupling, one-third of the resin is separated, and the remainder undergoes coupling with the mixture of amino acids. The next coupling is performed with half of the resin from the previous coupling. All removed portions of the resin are then combined

[166] A. Wallace, S. Altamura, C. Toniatti, A. Vitelli, E. Bianchi, P. Delmastro, G. Ciliberto, and A. Pessi, *Pept. Res.* **7**, 27 (1994).
[167] N. F. Sepetov, V. Krchnák, M. Stankova, S. Wade, K. S. Lam, and M. Lebl, *Proc. Natl. Acad. Sci. U.S.A.* **92**, 5426 (1995).

with the main part, and a randomization is performed. Synthesis of a library of libraries with a three-amino acid motif, by this method, consists of three randomization steps and four stages of multiple couplings of amino acid mixtures. As a result, each solid-phase particle of the library goes through three mandatory randomization steps (mixing the resin, separation into n parts, and coupling of individual amino acid) and as many as 12 acylations with the mixture of amino acids. This library, containing peptides of lengths from 3 to 5 residues, consists of 256 positional motif sublibraries. Among sublibraries of up to hexapeptides, all positional motifs are presented. The synthetic scheme of this example did not allow more than three successive acylations with the amino acid mixture, and therefore motifs in which "pharmacophore" positions are separated by more than three adjacent "structural unit" positions could not be represented. However, the scheme utilizing partial removal and separate processing of the resin from synthetic process can create unlimited possibilities of "library of libraries."

Libraries of Organized Mixtures

The characteristic feature of organized peptide mixtures is the presence of a single amino acid at certain position(s) in a sequence and a mixture of amino acids at remaining positions. This systematic array eliminates the need for any kind of structure determination at the library stage (i.e., sequencing of peptides or coding for nonpeptide compounds), as identification of the structures of the active compounds is inherent to the library deconvolution process (see below).

Iterative Libraries

The first iterative library was composed of 400 separate hexapeptide mixtures, represented as O_1O_2XXXX, in which the first two positions (O_1 and O_2) were individually defined and represented all possible combinations of two of the 20 proteinogenic amino acids (AA, AC, ..., through YW, YY).[2] The remaining four positions (X) were mixtures of 19 amino acids (Cys was excluded), so that each mixture was composed of $19^4 = 130,321$ individual peptides. This library was synthesized using the tea bag method in combination with the divide–couple–recombine synthesis.

In the initial screening of this type of library in a given bioassay, the most active peptide mixtures are identified, followed by an iterative process of synthesis and screening, during which all positions of the active mixtures are successively defined. If, for example, the mixture HKXXXX was the most active mixture in a given bioassay, His and Lys at positions one and two, respectively, would be kept unchanged, and the next set of 20 peptide mixtures with the third position defined (HKOXXX) would be synthesized and tested in order to identify the most effective amino acids at the third position. After repeating this process twice more, in the last step of the iterative process, the sixth position of the active peptide mixtures would be defined by synthesizing and testing individual peptides.

Positional Scanning Libraries

The positional scanning (PS) library format[32,157,168,169] enables the identification of active compounds directly from the initial library screening data, thus avoiding the iterative synthesis and screening process associated with the above library format. A typical PS library is composed of n (n = number of diversity positions) sublibraries. Accordingly, a PS hexapeptide library is composed of six independent sublibraries (O_1XXXXX, XO_2XXXX, XXO_3XXX, $XXXO_4XX$, $XXXXO_5X$, $XXXXXO_6$), in which one position (O) is individually defined, and the other five positions are mixtures of amino acids. Thus, each of the sublibraries, while addressing a specific position, represents the same collection of peptides.

Screening of all sublibraries in a given bioassay provides information about the most effective amino acids at each position for the biological effect of interest, as well as about the relative specificity of each position (i.e., the fewer amino acids found to be active at a position, the more specific that position). The synthesis of all possible combinations of the most active amino acids at each position yields a range of individual peptides, which are then tested in order to determine their individual activities. If, for example, two amino acids were found to be highly effective at each position of the above hexapeptide PS library, 2^6 = 64 individual peptides would be synthesized and tested based on that screening data.

Alternatively, each of the sublibraries can serve as the starting point for the iterative synthesis and screening process described above. It has to be kept in mind that amino acids identified in different positions need not be a part of the same motif, or they may represent only the most significant amino acid(s) of the same motif, which can be placed anywhere in the peptide sequence.[170–172] For example, if Arg is the key amino acid in the potential enzyme ligand (substrate, inhibitor), and the minimal length of

[168] C. Pinilla, J. R. Appel, and R. A. Houghten, *Methods Mol. Biol.* **66,** 171 (1996).
[169] J. R. Appel, S. Muller, N. Benkirane, R. A. Houghten, and C. Pinilla, *Pept. Res.* **9,** 174 (1996).
[170] S. M. Freier, D. A. M. Konings, J. R. Wyatt, and D. J. Ecker, *J. Med. Chem.* **38,** 344 (1995).
[171] D. A. M. Konings, J. R. Wyatt, D. J. Ecker, and S. M. Freier, *J. Med. Chem.* **39,** 2710 (1996).
[172] A. Wallace, K. S. Koblan, K. Hamilton, D. J. Marquis-Omer, P. J. Miller, S. D. Mosser, C. A. Omer, M. D. Schaber, R. Cortese, A. Oliff, J. B. Gibbs, and A. Pessi, *J. Biol. Chem.* **271,** 31306 (1996).

the ligand is a dipeptide, Arg can be identified as key residue in almost any position of hexapeptide library.[57]

Orthogonal Libraries

The characteristic feature of orthogonal libraries is the fact that each library member is present in two (or three) different mixtures, and any two (or three) mixtures of the library have one, and only one, peptide in common. This approach was first described by Deprez et al.[173] with the synthesis of a two-dimensional (2D, n = 2) orthogonal library consisting of two libraries representing the exact same set of peptides in two different arrangements. These libraries are prepared by coupling defined groups of amino acids as mixtures to each position of the sequence. The amino acids in each group are different for the two libraries and are arranged so that each group of amino acids used for the synthesis of library A contained one, and only one, amino acid of each group used for the synthesis of library B. Screening of both libraries in a bioassay of interest enables the identification of the most effective amino acid groups at each position of the peptide for that particular interaction. If, for example, the group containing Ala, Cys, Asp, and Glu would be the most effective at a given position in library A, these four amino acids would be present in a different group in library B, and screening of library B would reveal which of them is actually the best at that particular position.

Alternatively, by coupling groups of amino acids, orthogonal libraries can also be prepared through the synthesis of individual peptides, which are mixed as intermediates at certain stages of the synthesis. This is preferably done using an automated synthesizer in the 96-well microtiter plate format by alternatingly mixing the resins of rows or columns of the plate,[174] which greatly facilitates the generation of higher dimensional orthogonal libraries.

Libraries with Defined Structural Features

Cyclic Peptide Libraries

The conformational flexibility of peptides can be decreased by cyclization, which may increase the affinity of peptides to their biological receptors. Ligands for the gpIIbIIIa receptor were found using a cyclic disulfide peptide library, which owing to low concentration of library components, would

---

[173] B. Deprez, X. Williard, L. Bourel, H. Coste, F. Hyafil, and A. Tartar, J. Am. Chem. Soc. **117**, 5405 (1995).

[174] S. Felder and R. Kris, Mol. Diversity **2**, in press (1997).

not have been found in linear libraries.[108] Cyclization changing the preference of the biological receptor toward a particular target was shown in the study of streptavidin binding of cyclic libraries of various sizes.[175] Peptide libraries built around a cyclic peptide template, to which amino and other carboxylic acids were attached,[176] yielded an inhibitor of α-glucosidase.[177] Eichler *et al.*[177] studied an array of cyclic disulfide and lactam libraries and compared their activities with the linear analogs of the libraries. Only one lactam library was found to be active in an α-glucosidase inhibition assay.

Cyclic lactam peptide libraries have been intensively studied,[178-188] and an optimized strategy for their syntheses was developed. A cyclic pentapeptide library based on a known endothelin antagonist was synthesized, and the sequence of the known ligand was correctly identified.[187] The cyclic β-turn mimetic library by Virgilio and Ellman,[189] in which the ε-amino group of the C-terminal Lys was acylated with acrylic acid, the N terminus was acylated with iodobenzoic acid, and palladium-mediated cyclization provided clean products in high yield,[190] is closer to a nonpeptide than a peptide library.

[175] K. S. Lam, M. Lebl, S. Wade, S. Stierandova, P. S. Khattri, N. Collins, and V. J. Hruby, *in* "Peptides, 1994" (R. S. Hodges and J. A. Smith, eds.), p. 1005. Escom, Leiden, The Netherlands, 1994.

[176] J. Eichler, A. W. Lucka, and R. A. Houghten, *Pept. Res.* **7**, 300 (1994).

[177] J. Eichler, A. W. Lucka, C. Pinilla, and R. A. Houghten, *Mol. Diversity* **1**, 233 (1996).

[178] K. Darlak, P. Romanovskis, and A. F. Spatola, *in* "Peptides: Chemistry, Structure, and Biology" (R. S. Hodges and J. A. Smith, eds.), p. 981. Escom, Leiden, The Netherlands, 1994.

[179] D. Tumelty, D. Vetter, and V. V. Antonenko, *J. Chem. Soc., Chem. Commun.*, 1067 (1994).

[180] D. Winkler, A. Schuster, B. Hoffmann, and J. Schneider-Mergener, *in* "Peptides 94" (H. L. S. Maia, ed.), p. 485. Escom, Leiden, The Netherlands, 1995.

[181] C. G. Bradshaw, E. Magnenat, and A. Chollet, *in* "Peptides 94" (H. L. S. Maia, ed.), p. 485. Escom, Leiden, The Netherlands, 1995.

[182] D. Tumelty, M. C. Needels, V. V. Antonenko, and P. R. Bovy, *in* "Peptides: Chemistry, Structure and Biology" (P. T. P. Kaumaya and R. S. Hodges, eds.), p. 121. Mayflower Scientific, Kingswinford, UK, 1996.

[183] A. F. Spatola, Y. Crozet, P. Romanovskis, and E. Valente, *in* "Peptides: Chemistry, Structure and Biology" (P. T. P. Kaumaya and R. S. Hodges, eds.), p. 281. Mayflower Scientific, Kingswinford, UK, 1996.

[184] D. Winkler, R. D. Stigler, J. Hellwig, B. Hoffmann, and J. Schneider-Mergener, *in* "Peptides: Chemistry, Structure and Biology" (P. T. P. Kaumaya and R. S. Hodges, eds.), p. 315. Mayflower Scientific, Kingswinford, UK, 1996.

[185] A. F. Spatola and P. Romanovskis, *in* "Peptide and Non-Peptide Libraries: A Handbook for the Search of Lead Structures" (G. Jung, ed.), p. 327. VCH, Weinheim, Germany, 1996.

[186] J. J. Chen, L. M. Teesch, and A. F. Spatola, *Lett. Pept. Sci.* **3**, 17 (1996).

[187] A. F. Spatola and Y. Crozet, *J. Med. Chem.* **39**, 3842 (1996).

[188] A. F. Spatola, K. Darlak, and P. Romanovskis, *Tetrahedron Lett.* **37**, 591 (1996).

[189] A. A. Virgilio and J. A. Ellman, *J. Am. Chem. Soc.* **116**, 11580 (1994).

[190] M. Hiroshige, J. R. Hauske, and P. Zhou, *J. Am. Chem. Soc.* **117**, 11590 (1995).

Conformationally Defined Peptide Libraries

A conformationally defined peptide library was designed and synthesized by replacing five positions of an amphipathic α-helical 18-mer peptide (Tyr-Lys-Leu-Leu-Lys-Lys-Leu-Leu-Lys-Lys-Leu-Lys-Lys-Leu-Leu-Lys-Lys-Leu), either on the hydrophilic or the hydrophobic face of the helix, with one defined and four mixture positions. This library was tested for its catalytic activity in the decarboxylation of oxaloacetate. The catalytic activities of individual peptides identified were found to correlate well with their ability to fold into an α-helical conformation.[191–193] A similar approach was taken by Bianchi *et al.* by randomizing five positions in the α-helical portion of a 26-residue zinc finger motif.[194] Consensus sequences for the binding to an antilipopolysaccharide antibody were found through the screening of this library. Stabilization of the two-stranded α-helical coiled coil was achieved by incorporation of a lactam bridge.[195]

"Reversed" Peptide Libraries

Peptides are typically synthesized starting from the C terminus, because the synthesis of peptides in the N $\rightarrow$ C direction results in significant racemization at each coupling step. Biological receptors may recognize the C or N terminal or both ends of their peptide ligands. Although this is not an issue with respect to peptide libraries in solution (both termini are accessible), it becomes problematic when using support-bound libraries for the screening against receptors that recognize their ligands at the C terminus. Therefore, methods to "reverse" resin-bound peptide libraries have been developed.[196–198] This is done by on-resin cyclization of the peptides, followed by opening the ring in a manner that exposes the C terminus (Fig. 10).

[191] E. Perez-Paya, R. A. Houghten, and S. E. Blondelle, *J. Biol. Chem.* **270,** 1048 (1995).
[192] S. E. Blondelle, E. Takahashi, R. A. Houghten, and E. Perez-Paya, *Biochem. J.* **313,** 141 (1996).
[193] E. Perez-Paya, R. A. Houghten, and S. E. Blondelle, *J. Biol. Chem.* **271,** 4120 (1996).
[194] E. Bianchi, A. Folgori, A. Wallace, M. Nocotra, S. Acali, A. Phalipon, G. Barbato, R. Bazzo, R. Cortese, F. Felici, and A. Pessi, *J. Mol. Biol.* **247,** 154 (1995).
[195] M. E. J. Houston, A. Wallace, E. Bianchi, A. Pessi, and R. S. Hodges, *J. Mol. Biol.* **262,** 270 (1996).
[196] M. Lebl, V. Krchnák, N. F. Sepetov, V. Nikolaev, A. Stierandova, P. Safár, B. Seligmann, P. Strop, D. Thorpe, S. Felder, D. F. Lake, K. S. Lam, and S. E. Salmon, *in* "Innovation and Perspectives in Solid Phase Synthesis" (R. Epton, ed.), p. 233. Mayflower Worldwide, Birmingham, UK, 1994.
[197] C. P. Holmes and C. M. Rybak, *in* "Peptides: Chemistry, Structure, and Biology" (R. S. Hodges and J. A. Smith, eds.), p. 992. Escom, Leiden, The Netherlands, 1994.
[198] R. S. Kania, R. N. Zuckermann, and C. K. Marlowe, *J. Am. Chem. Soc.* **116,** 8835 (1994).

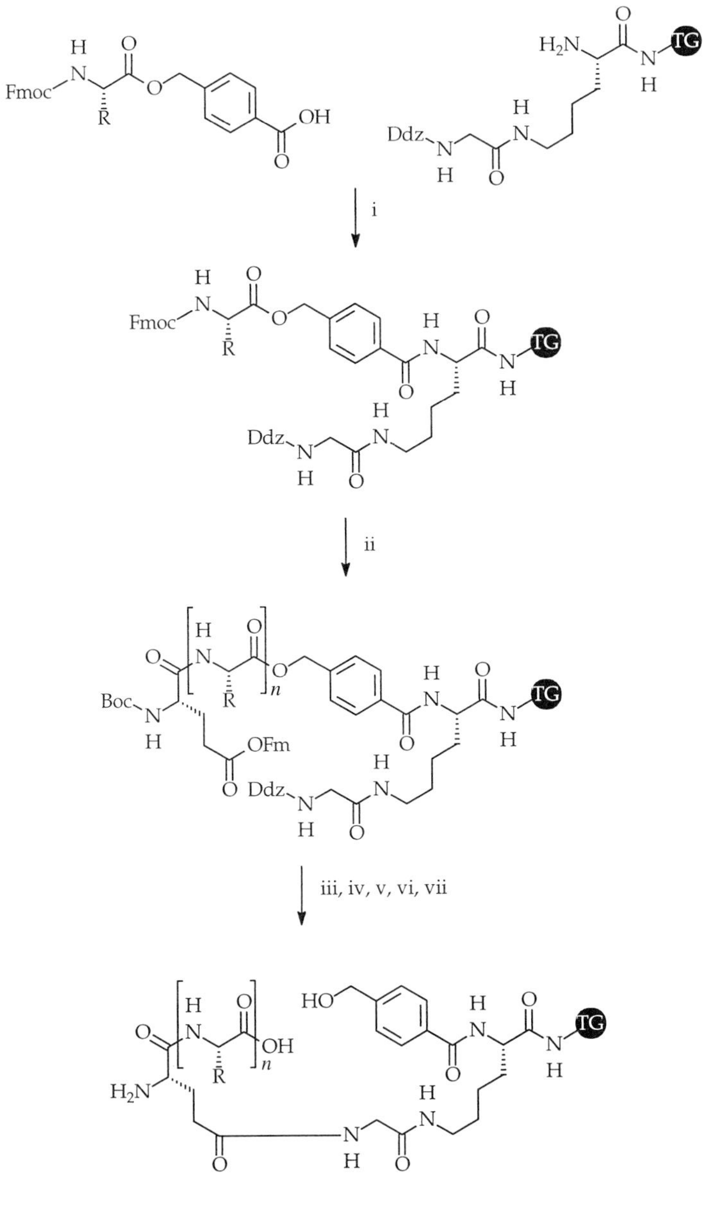

(i)   DCC, HOBt, DMF, overnight
(ii) peptide synthesis, standard Fmoc protocol
(iii) 50% piperidine in DMF
(iv) 3% TFA in DCM
(v)   DIPCDI, HOBt
(vi) 50% TFA in DCM
(vii) 0.5% NaOH, 30 min

FIG. 10. Reversing the peptide chain on the resin.

Increasing Diversity of Peptide Libraries

Libraries Containing Nonpeptidic Components

The diversity of peptide libraries is determined by the amino acid chains, while the peptide backbone is constant. The introduction of other than α-amino acids, however, enables the diversification also of the peptide backgone.[116] The peptide backbone can serve as a scaffold to which a variety of building blocks can be attached via coupling to various trifunctional amino acids, such as aminoglycine, diaminopropionic acid, diaminobutyric acid, ornithine, Lys, iminodiacetic acid, Asp, Glu, Ser, Thr, Hyp, and Cys.[196,199–201] Iminodiacetic acid was shown to be a suitable backbone unit for the construction of peptidelike libraries.[202] Iminodiacetic acid anhydride and similar symmetrical anhydrides were also used as templates for library synthesis[203] or for structural coding.[204] Attachment of carboxylic acids to the α- or ω-amino groups of various diamino acids was the basis for the construction of a so-called α,β,γ library (the scheme of the synthesis of this library is given in Fig. 11), in which both the peptide backbone and the amino acid side chains were diversified.[205] Figure 12 illustrates two extreme structures from this library.

Rivier and co-workers[206] prepared libraries of oligoamides, termed betides, in which one amino group of aminoglycine residues was used for the backbone, and the second amino group was derivatized (acylated, alkylated). In such "betidamino" acids, each N'-acyl/alkyl group can mimic naturally occurring amino acid side chains or introduce other functionalities. A potent gonadotropin-releasing hormone antagonist was discovered using

[199] M. Lebl, V. Krchňák, P. Safár, A. Stierandova, N. F. Sepetov, P. Kocis, and K. S. Lam, *in* "Techniques in Protein Chemistry V" (J. W. Crabb, ed.), p. 541. Academic Press, San Diego, 1994.

[200] M. Lebl, V. Krchňák, A. Stierandova, P. Safár, P. Kocis, V. Nikolaev, N. F. Sepetov, R. Ferguson, B. Seligmann, K. S. Lam, and S. E. Salmon, *in* "Peptides: Chemistry, Structure, and Biology" (R. S. Hodges and J. A. Smith, eds.), p. 1007. Escom, Leiden, The Netherlands, 1994.

[201] R. M. Valerio, A. M. Bray, and K. M. Stewart, *Int. J. Pept. Protein Res.* **47**, 414 (1996).

[202] P. Safár, A. Stierandova, and M. Lebl, *in* "Peptides 94" (H. L. S. Maia, ed.), p. 471. Escom, Leiden, The Netherlands, 1995.

[203] D. L. Boger, C. M. Tarby, P. L. Myers, and L. H. Caporale, *J. Am. Chem. Soc.* **118**, 2109 (1996).

[204] Z. J. Ni, D. Maclean, C. P. Holmes, M. M. Murphy, B. Ruhland, J. W. Jacobs, E. M. Gordon, and M. A. Gallop, *J. Med. Chem.* **39**, 1601 (1996).

[205] V. Krchňák, A. S. Weichsel, D. Cabel, and M. Lebl, *Pept. Res.* **8**, 198 (1995).

[206] J. E. Rivier, G. C. Jiang, S. C. Koerber, J. Porter, L. Simon, A. G. Craig, and C. A. Hoeger, *Proc. Natl. Acad. Sci. U.S.A.* **93**, 2031 (1996).

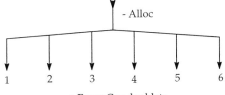

Boc-βAla-Gly-βAla-Gly-Lys(Alloc)-TG

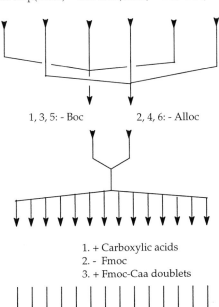

- Alloc

1   2   3   4   5   6

+ Fmoc-Caa doublets
- Boc

Boc-Dap(Alloc)   Boc-Dab(Alloc)   Boc-Orn(Alloc)

1, 3, 5: - Boc   2, 4, 6: - Alloc

1. + Carboxylic acids
2. - Fmoc
3. + Fmoc-Caa doublets

1. - Alloc
2. - Boc

FIG. 11. Scheme of the synthesis of an α,β,γ library.[205]

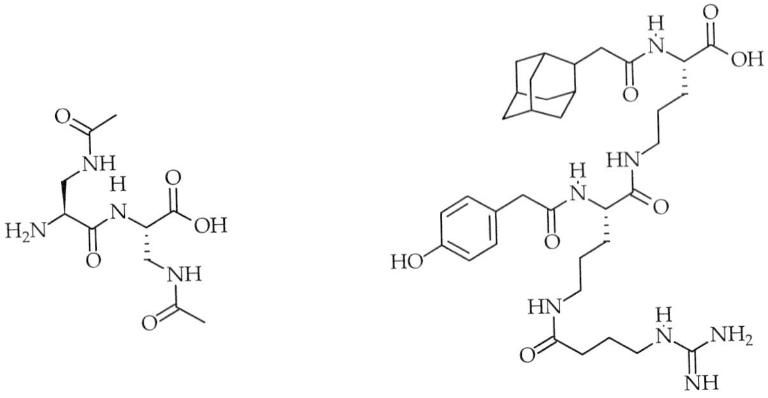

FIG. 12. Examples of "extreme" structures from an α,β,γ library.

a betide library. Interestingly, few differences in activities were found between the D- and L-nonalkylated betidamino acids.

Potent and specific zinc endopeptidase inhibitors were identified using a library of peptides modified at the N terminus with Z-Phe(PO$_2$CH$_2$).[207] In the Pfizer study of the endothelin antagonist developed by Fujisawa, Terrett et al.[208] kept N-terminal substitution intact and randomized all amino acids by an array of natural and unnatural α- and non-α-amino acids. The peptide chain was used as the "biasing element," targeting the binding pocket of Src SH3 domain in the library constructed by acylation of the peptide chain by the array of three building blocks containing amino and carboxyl functionalities and capped by a set of carboxylic acids.[209,210]

Peptide mimetics composed of α-aza-amino acids (termed azatides) have been applied to solid-phase synthesis.[211] A set of Boc protected alkyl hydrazines has been prepared, either by reduction of protected hydrazones or by alkylation of hydrazine with an alkyl halide followed by Boc protection. These hydrazine derivatives were converted to activated species by bis(pentafluorophenyl) carbonate and used in a stepwise manner to build

[207] J. Jiracek, A. Yiotakis, B. Vincent, A. Lecoq, A. Nicolaou, F. Checler, and V. Dive, J. Biol. Chem. **270,** 21701 (1995).

[208] N. K. Terrett, D. Bojanic, D. Brown, P. J. Bungay, M. Gardner, D. W. Gordon, C. J. Mayers, and J. Steele, Bioorg. Med. Chem. Lett. **5,** 917 (1995).

[209] A. P. Combs, T. M. Kapoor, S. Feng, J. K. Chen, L. F. Daude-Snow, and S. L. Schreiber, J. Am. Chem. Soc. **118,** 287 (1996).

[210] S. B. Feng, T. M. Kapoor, F. Shirai, A. P. Combs, and S. L. Schreiber, Chem. Biol. **3,** 661 (1996).

[211] H. Han and K. D. Janda, J. Am. Chem. Soc. **118,** 2539 (1996).

azatides. Vinylogous sulfonyl peptide libraries[212] were used for studies of synthetic receptors.[213] Libraries of synthetic receptors were generated by attaching randomized dipeptides to macrocyclic tetramine cyclen[214] or steroid scaffolds.[215]

Peptoid Libraries

Peptoids differ from peptides in the location of the side chains.[216–220] Whereas in peptides the side chains are connected to the α-carbon of the amino acid residues, in peptoids they are linked to the amide nitrogen of the backbone. Peptoids are resistant toward proteolytic degradation, and the diversity of peptoid libraries can be increased by introducing many amines as compared to the limited number of amino acids used for peptide libraries. Peptoids are easily synthesized from bromoacetic acid and amines,[221] or from preformed N-alkylamino acids.[216] Specific ligands for the α_1-adrenergic receptor and μ-opioid receptors were found using peptoid libraries.[219]

The combination of amino acids and N-alkylated Gly in the same molecule was successfully applied in several model studies.[164,222,223] Ostergaard

[212] C. Gennari, H. P. Nestler, B. Salom, and W. C. Still, *Angew. Chem., Int. Ed. Engl.* **34,** 1763 (1995).

[213] C. Gennari, H. P. Nestler, B. Salom, and W. C. Still, *Angew. Chem., Int. Ed. Engl.* **34,** 1765 (1995).

[214] M. T. Burger and W. C. Still, *J. Org. Chem.* **60,** 7382 (1995).

[215] R. Boyce, G. Li, H. P. Nestler, T. Suenaga, and W. C. Still, *J. Am. Chem. Soc.* **116,** 7955 (1994).

[216] R. J. Simon, R. S. Kaina, R. N. Zuckermann, V. D. Huebner, D. A. Jewell, S. Banville, S. Ng, L. Wang, S. Rosenberg, C. K. Marlowe, D. C. Spellmeyer, R. Tan, A. D. Frankel, D. V. Santi, F. E. Cohen, and P. A. Bartlett, *Proc. Natl. Acad. Sci. U.S.A.* **89,** 9367 (1992).

[217] S. M. Miller, R. J. Simon, S. Ng, R. N. Zuckermann, J. M. Kerr, and W. H. Moos, *Bioorg. Med. Chem. Lett.* **4,** 2657 (1994).

[218] R. J. Simon, E. J. Martin, S. M. Miller, R. N. Zuckermann, J. M. Blaney, and W. H. Moos, *in* "Techniques in Protein Chemistry V" (J. W. Crabb, ed.), p. 533. Academic Press, San Diego, 1994.

[219] R. N. Zuckermann, E. J. Martin, D. C. Spellmeyer, G. B. Stauber, K. R. Shoemaker, J. M. Kerr, G. M. Figliozzi, D. A. Goff, M. A. Siani, R. J. Simon, S. C. Banville, E. G. Brown, L. Wang, L. S. Richter, and W. H. Moos, *J. Med. Chem.* **37,** 2678 (1994).

[220] S. M. Miller, R. J. Simon, S. Ng, R. N. Zuckermann, J. M. Kerr, and W. H. Moos, *Drug Dev. Res.* **35,** 20 (1995).

[221] R. N. Zuckermann, J. M. Kerr, S. B. H. Kent, and W. H. Moos, *J. Am. Chem. Soc.* **114,** 10646 (1992).

[222] V. Nikolaiev, A. Stierandova, V. Krchnák, B. Seligmann, K. S. Lam, S. E. Salmon, and M. Lebl, *Pept. Res.* **6,** 161 (1993).

[223] M. Lebl, V. Krchnák, N. F. Sepetov, B. Seligmann, P. Strop, S. Felder, and K. S. Lam, *Biopolymers* (*Pept. Sci.*) **37,** 177 (1995).

and Holm called the molecules composed partly from peptides and partly from peptoids "peptomers."[164]

Libraries from Libraries

Chemically modified peptide libraries have been generated through alkylation and/or reduction of the peptide bonds of existing peptide libraries,[224] thus dramatically changing the physicochemical properties of the peptides and greatly extending the range and repertoire of chemical diversity. The components of such transformed libraries, which have been termed "libraries from libraries," are stable toward proteolytic degradation, as they lack the characteristic peptide bond-CO-NH-. The chemical modification of peptide libraries is reflected by the library screening results, as illustrated by the following example.[225] A tetrapeptide library was divided into four aliquots. One aliquot remained untreated, one aliquot was reduced, the third aliquot N-benzylated, and the fourth aliquot N-benzylated and reduced, generating four different libraries from the same parent peptide library. These four libraries were screened in antimicrobial and receptor binding assays, yielding very different results depending on the character of the library backbone (see Fig. 13).

Library Analysis

Analysis of the synthesized library should make sure that all components are present and that they are present in the expected amounts. Important issues in the analysis of one-bead–one-compound libraries include examination of "statistically significant" sample sizes (how many beads should be analyzed), sensitivity of the analytical method (only picomolar quantities of compounds are available on the bead), and throughput of the method. Microsequencing and amino acid analysis are appropriate methods for peptide library characterization. Sequencing can detect incomplete couplings and incomplete side-chain deprotection. The sensitivity of modern automatic microsequencers (in the high femtomolar range) enables the detection of impurities in the range of 1 to 2%.

Drawbacks of an alternative technique, mass spectrometry, are (i) poor quantitation (signal intensity depends on ionizability of each component of the mixture, which can be dramatically different) and (ii) the fact that the compound has to be detached from the bead prior to analysis. The most

[224] J. M. Ostresh, G. M. Husar, S. E. Blondelle, B. Dorner, P. A. Weber, and R. A. Houghten, Proc. Natl. Acad. Sci. U.S.A. 91, 11138 (1994).
[225] R. A. Houghten, S. E. Blondelle, C. T. Dooley, B. Dorner, J. Eichler, and J. M. Ostresh, Mol. Diversity 2, 41 (1996).

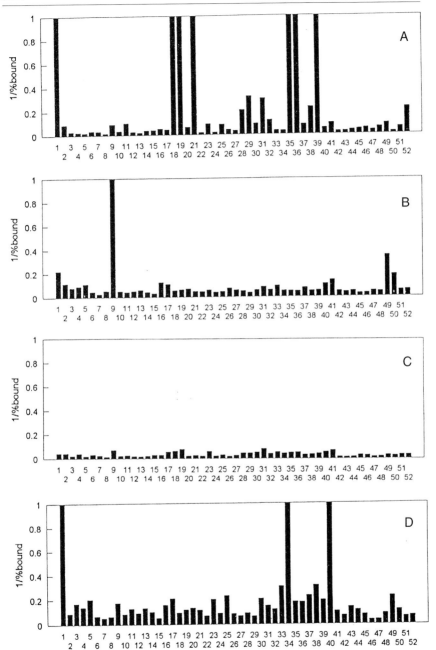

FIG. 13. Primary binding assay screening results from different "libraries from libraries."
(A) Tetrapeptide library (mother library); (B) reduced library (pentamine library); (C) peptide
perbenzylated library; (D) perbenzylated and reduced library.[225]

convenient methods for compound detachment are exposure to gaseous reagents (ammonia, hydrogen fluoride, cyanogen bromide, trifluoroacetic acid) or photolytic cleavage. In both cases, the beads do not have to be treated in separate vessels, because the detached compounds remain physically inside the bead until they are extracted directly prior to, or during, analysis.[226–230] Mass spectrometry can identify all types of impurities and, therefore, can be used to optimize the library synthesis.[231] Mass spectrometric analysis of hundreds of samples can be performed automatically.[232–234]

Drawbacks of an alternative technique, mass spectrometry, are (i) poor quantitation (signal intensity depends on ionizability of each component of the mixture, which can be dramatically different) and (ii) the fact that the compound has to be detached from the bead prior to analysis. The most convenient methods for compound detachment are exposure to gaseous reagents (ammonia, hydrogen fluoride, cyanogen bromide, trifluoroacetic acid) or photolytic cleavage. In both cases, the beads do not have to be treated in separate vessels, because the detached compounds remain physically inside the bead until they are extracted directly prior to, or during, analysis.[226–230] Mass spectrometry can identify all types of impurities and, therefore, can be used to optimize the library synthesis.[231] Mass spectrometric analysis of hundreds of samples can be performed automatically.[232–234]

Infrared spectroscopy is a valuable technique for the evaluation of the solid-phase transformation of functional groups, in particular techniques that work with individual beads.[235–237] Its application to the analysis of

[226] C. L. Brummel, I. N. W. Lee, Y. Zhou, S. J. Benkovic, and N. Winograd, *Science* **264,** 399 (1994).

[227] R. A. Zambias, D. A. Boulton, and P. R. Griffin, *Tetrahedron Lett.* **35,** 4283 (1994).

[228] R. S. Youngquist, G. R. Fuentes, M. P. Lacey, and T. Keough, *Rapid Commun. Mass Spectrom.* **8,** 77 (1994).

[229] B. J. Egner, G. J. Langley, and M. Bradley, *J. Org. Chem.* **60,** 2652 (1995).

[230] B. J. Egner, M. Cardno, and M. Bradley, *J. Chem. Soc., Chem. Commun.,* 2163 (1996).

[231] R. S. Youngquist, G. R. Fuentes, M. P. Lacey, and T. Keough, *J. Am. Chem. Soc.* **117,** 3900 (1995).

[232] S. W. Fink, W. L. Thompson, and J. R. B. Slayback, *Spectroscopy* **11,** 26 (1996).

[233] S. S. Smart, T. J. Mason, P. S. Bennell, N. J. Maeji, and H. M. Geysen, *Int. J. Pept. Protein Res.* **7,** 47 (1996).

[234] G. Jung, A. G. Beck-Sickinger, N. Zimmermann, J. Metzger, R. Spohn, S. Stevanovic, K. Deres, and K. H. Wiesmuller, *in* "Innovation and Perspectives in Solid Phase Peptide Synthesis" (R. Epton, ed.), p. 227. Intercept, Andover, UK, 1992.

[235] B. Yan, G. Kumaravel, H. Anjaria, A. Wu, R. C. Petter, C. F. Jewell, and J. R. Wareing, *J. Org. Chem.* **60,** 5736 (1995).

[236] B. Yan and G. Kumaravel, *Tetrahedron* **52,** 843 (1996).

[237] B. Yan, J. B. Fell, and G. Kumaravel, *J. Org. Chem.* **61,** 7467 (1996).

peptide libraries is rather limited,[238] as is the application of nuclear magnetic resonance (NMR) spectroscopy, which was shown to provide information about incorporation of individual amino acids into the mixtures.[148]

Amino acids analysis is also a useful method for the assessment of the "statistical purity" of peptide libraries, provided library samples with predictable properties are used. For example, five beads from a tetrapeptide library are essentially as likely to contain any amino acid only once as they are likely to not contain any particular amino acid at all. However, 5000 beads from the same library (about 5 mg of resin, assuming an average bead size of 130 μm) must represent equimolar amounts of all amino acids for the library to be "statistically pure." Using this method, major synthetic problems, such as omission of particular amino acids, or loss of the product from the carrier during synthesis or deprotection, can be easily detected. Amino acid analysis is quantitative and applicable even to peptides containing unnatural amino acids, or for the analysis of nonpeptide libraries containing amino acids as scaffolds or building blocks.

Similarly useful are multiple sequencing methods.[239] Sequencing by Edman degradation, however, is limited to peptides that have a free a N terminus and contain only α-amino acids.

The comparison of computer-generated mass distribution profiles[240] with experimentally obtained mass spectra of libraries[161,239] can be applied to samples of both one-bead–one compound, and mixture libraries. The theoretical profile may need to be corrected for the different proton affinities of amino acids. Peptide library evaluation by fast atom bombardment (FAB) mass spectrometry should take into consideration the high proton affinity of Arg.[148] Evaluation of mass distribution reveals incomplete couplings (shift toward lower molecular weights), incomplete deprotection, and unwanted library modification, such as oxidation, acylation, and alkylation (shift toward higher molecular weights) (see, e.g., Fig. 14). Information about classes of ions fragmenting into a common daughter ion, and about compound classes losing the same neutral mass, may be obtained using tandem mass spectrometry. This technique is also helpful in determining the completeness of the removal of specific protecting groups and for the detection of side products generated during the synthesis.[241]

[238] K. Russell, D. C. Cole, F. M. McLaren, and D. E. Pivonka, *J. Am. Chem. Soc.* **118,** 7941 (1996).
[239] J. W. Metzger, S. Stevanovic, J. Brunjes, K. H. Wiesmuller, and G. Jung, *Methods (San Diego)* **6,** 425 (1994).
[240] M. Lebl, V. Krchnák, and G. Lebl, Peptide Companion. Software, CSPS, P.O. Box 22567, San Diego, CA 92192-2567, 1995.
[241] J. W. Metzger, C. Kempter, K. H. Wiesmuller, and G. Jung, *Anal. Biochem.* **219,** 261 (1994).

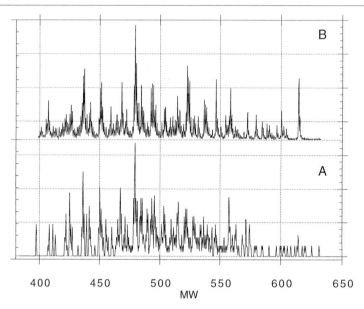

400   450   500   550   600   650
                            MW

FIG. 14. Comparison of (A) calculated and (B) experimental mass spectra of a library of 480 members. (Data courtesy of Michael Griffith.)

Electrospray mass spectroscopy was used to characterize peptide libraries synthesized by various methods.[117,234,239,241–246] Boutin et al.[148] compared FAB mass spectrometry, capillary electrophoresis, and NMR for library characterization, and they concluded that existing analytical techniques can provide valuable information on the quality of synthetic libraries.

Structure Determination of Hits from Libraries

One-Bead–One-Compound Libraries

Direct Sequencing. The picomolar amounts of peptides contained in individual beads are a sufficient quantity for Edman degradation. Only

[242] J. W. Metzger, K. H. Wiesmuller, V. Gnau, J. Brunjes, and G. Jung, *Angew. Chem., Int. Ed. Engl.* **32,** 894 (1993).

[243] S. Stevanovic, K. H. Wiesmuller, J. Metzger, A. G. Beck-Sickinger, and G. Jung, *Bioorg. Med. Chem. Lett.* **3,** 431 (1993).

[244] T. Carell, E. A. Wintner, A. J. Sutherland, J. J. Rebek, Y. M. Dunayevskiy, and P. Vouros, *Chem. Biol.* **2,** 171 (1995).

[245] Y. M. Dunayevskiy, P. Vouros, T. Carell, E. A. Wintner, and J. Rebek, *Anal. Chem.* **67,** 2906 (1995).

[246] Y. M. Dunayevskiy, P. Vouros, E. A. Wintner, G. W. Shipps, T. Carell, and J. J. Rebek, *Proc. Natl. Acad. Sci. U.S.A.* **93,** 6152 (1996).

minimal modification of the standard sequencing protocols of automated sequencers is required for using most solid support beads. Difficulties in sequencing peptides from Sepharose beads[247] may be overcome by mechanically disintegrating the beads and/or by applying the isothiocyanate solution in several cycles. A potential problem in bead sequencing is the uniqueness of the individual bead with respect to the always possible malfunctioning of the sequencer, or power failure; when the sequence information on one bead is lost, it cannot be recovered. A solution to this problem is cutting the beads in half and subjecting only one part to sequencing, while retaining the second half as a backup. The techniques of single bead manipulation are very simple. A good dissecting microscope, a petri dish, and an injection needle are needed. The needle is sharp enough to cut the beads, and its tip can be used lift the bead and transfer it to the sequencing support. "Packing" the bead into the fibers of a support filter prevents it from "jumping out." The originally used micromanipulator[1] was soon abandoned in favor of micropipette and needle techniques.

Peptides containing other than α-amino acids, or other nonsequenceable building blocks, can be sequenced until the first nonsequenceable building block is reached. Sequencing beyond that point can be enabled by coupling a subequimolar amount of the nonsequenceable building block, thus creating a sequenceable omission peptide on the same bead.[248,249] Alternatively, a sequenceable amino acid (typically Gly) can be cocoupled together with the nonsequenceable building block, thus creating a mixture of two peptides on one bead. The cocoupling of the second amino acid has also been used for the "tagging" of D-amino acids in the sequence, even though chiral sequencing is a possible alternative.[250]

For the sequencing of peptides containing more than the 20 proteinogenic amino acids, the chromatographic gradient for the separation of the phenylthiohydantoin (PTH)-amino acids can be modified, enabling the separation of unnatural and side-chain modified trifunctional amino acids,[223] as well as N-alkylated Gly.[13] Sequencing of cyclic peptides with a free α-amino group is feasible, because the amino acid at the N-terminal part of the cycle is not detected until the amino acid at the C-terminal side of the cycle is reached, which is detected as a modified amino acid (i.e., a cystine for disulfides and side-chain dipeptides for lactams). Sequencing of "reversed" peptides (see above) is more difficult.[196–198] Instead of applying

[247] V. Krchnák, A. Weichsel, S. Felder, and M. Lebl, unpublished results (1992).

[248] K. S. Lamand and S. E. Salmon, U.S. Patent 5,510,240 (1995).

[249] K. S. Lam, S. E. Salmon, V. J. Hruby, E. M. Hersh, and F. Al-Obeidi, WO Patent 546,845 (1992).

[250] M. Pavlok, Z. Voburka, T. Vanek, M. Rinnova, I. Bleha, L. Doleckova, and I. Kluh, *in* "Peptides 94" (H. L. S. Maia, ed.), p. 418. Escom, Leiden, The Netherlands, 1995.

C-terminal sequencing, the peptide can be attached (or reattached) to the resin via an amino acid side chain, thus exposing the α-amino group for sequencing (see Fig. 10).

The screening of one-bead–one-compound libraries often results in the identification of many positive beads. When a structural consensus can be expected (e.g., in small libraries, or libraries with fixed structural elements), it may be more efficient to simultaneously sequence groups of positive beads, rather than individual beads. Besides yielding information on specific positions of the sequence (only one or few amino acids are detected in that particular cycle of sequencing), as well as nonspecific positions (many or all amino acids are detected), this method provides structure–activity relationship information.[223,251] In a similar manner, sequencing can be used to detect mixture and defined positions in "libraries of libraries" (see above).

"Tagging" by addition of an unnatural amino acid has been used to identify mixtures composed of L- or D-amino acids by sequencing,[167] and nonsequenceable amino acids were "coded" by sequenceable amino acids.[165]

Coding and Decoding. Direct sequencing is not possible if other than α-amino acids and/or other nonsequenceable building blocks are used to synthesize a library. Therefore, several strategies for the coding of library compounds on single beads have been devised. The first, proposed by Brenner and Lerner,[252] was based on the coding of the structures of library components by oligonucleotides, which can be readily sequenced. Although nucleic acid coding has been successively used,[253–255] its application is limited owing to incompatibility of nucleic acid chemistry with many reaction conditions in organic synthesis. Two laboratories developed an alternative coding strategy for nonsequenceable one-bead–one-compound libraries[222,256] based on peptides as coding structures, which can be easily sequenced by Edman degradation. Coding subunits other than the building blocks used for library synthesis are attached to the solid support in a separate reaction. Other coding methods utilize different analytical techniques for the decoding of the coding tag, such as gas chromatography for

[251] M. Lebl, K. S. Lam, P. Kocis, V. Krchnák, M. Patek, S. E. Salmon, and V. J. Hruby, *in* "Peptides 1992" (C. H. Schneider and A. N. Eberle, eds.), p. 67. Escom, Leiden, The Netherlands, 1993.

[252] S. Brenner and R. A. Lerner, *Proc. Natl. Acad. Sci. U.S.A.* **89,** 5381 (1992).

[253] J. Nielsen and K. D. Janda, *Methods (San Diego)* **6,** 361 (1994).

[254] J. Nielsen, S. Brenner, and K. D. Janda, *J. Am. Chem. Soc.* **115,** 9812 (1993).

[255] M. C. Needels, D. G. Jones, E. H. Tate, G. L. Heinkel, L. M. Kochersperger, W. J. Dower, R. W. Barrett, and M. A. Gallop, *Proc. Natl. Acad. Sci. U.S.A.* **90,** 10700 (1993).

[256] J. M. Kerr, S. C. Banville, and R. N. Zuckermann, *J. Am. Chem. Soc.* **115,** 2529 (1993).

electrophoric tags (silylated halocarbons).[257,258] Ni *et al.*[204] used the very simple chemistry of polyamide formation in combination with protected iminodiacetic acid and a set of secondary amines.[203] Coding can be extended by applying the so-called digital coding principle, which is based on the use of more than one coding subunit for the coding of individual building blocks.[223,257,259]

A controlled ratio of stable isotopes in the tagging molecules (^{13}C and ^{15}N in Gly and Ala) was used as a very elegant method of encoding a combinatorial library.[259] Owing to the fact that the code can be constructed not only from different components, but that also their ratio can be changed and determined with high confidence, thousands of codes can be constructed from very limited numbers of coding blocks. The only limitation may be the price of isotopically labeled molecules.

Undesirable interaction between the coding structure and the biological target can be prevented by physically separating the coding from the library structures in the resin beads. This is done by isolating the coding structure in the bead interior, which is not accessible for most macromolecular targets (e.g., enzymes, antibodies), and assigning the library structure exclusively to the bead surface, where it can be recognized by the target molecule.[47,260] The method used to selectively address bead surface and interior volume is referred to as "bead shaving," because it involves the selective removal of protected amino acids from the bead surface using an enzyme, thus exposing a free amino group on the surface, while the bead interior remains amino-protected. Potential interference of the coding structures with the target molecule can be avoided completely by tagging the resin beads with a radio frequency transmitting chip.[126,261]

Mass Spectrometry. Sequencing of peptides by mass spectrometry was developed in several laboratories.[262,263] It can complement peptide sequencing[264] or be used to determine nonsequenceable components of peptides containing unnatural amino acids. The interpretation of sequencing data

[257] M. H. J. Ohlmeyer, R. N. Swanson, L. W. Dillard, J. C. Reader, G. Asouline, R. Kobayashi, M. Wigler, and W. C. Still, *Proc. Natl. Acad. Sci. U.S.A.* **90**, 10922 (1993).

[258] H. P. Nestler, P. A. Bartlett, and W. C. Still, *J. Org. Chem.* **59**, 4723 (1994).

[259] H. M. Geysen, C. D. Wagner, W. M. Bodnar, C. J. Markworth, G. J. Parke, F. J. Schoenen, D. S. Wagner, and D. S. Kinder, *Chem. Biol.* **3**, 679 (1996).

[260] J. Vagner, V. Krchnák, N. F. Sepetov, P. Strop, K. S. Lam, G. Barany, and M. Lebl, *in* "Innovation and Perspectives in Solid Phase Synthesis" (R. Epton, ed.), p. 347. Mayflower Worldwide, Birmingham, UK, 1994.

[261] K. C. Nicolaou, X. Y. Xiao, Z. Parandoosh, A. Senyei, and M. P. Nova, *Angew. Chem., Int. Ed. Engl.* **34**, 2289 (1995).

[262] K. Biemann and S. A. Martin, *Mass Spectrom. Rev.* **6**, 1 (1987).

[263] K. Biemann, *Methods Enzymol.* **193**, 455 (1990).

[264] J. W. Metzger, *Angew. Chem., Int. Ed. Engl.* **33**, 723 (1994).

can be substantially simplified using deuterium exchange experiments,[265] as the determination of exchangeable protons can decrease the number of possible sequences by an order of magnitude.[240] Partial (~10%) capping of the growing peptide chain after each coupling is the basis of another approach to peptide sequence determination.[231] Thus, the beads in this library contained all N-terminal truncation analogs in addition to the complete library peptides, and the synthetic history could easily be assessed based on the mass differences between the truncated analogs. Residues with the same molecular weight can be differentiated by using a mixture of different capping reagents. If the capping reagent contains an isotope with a typical isotopic "signature," such as bromine, the individual shorter sequences can be easily identified even at a level close to the noise of an experiment.[266]

Libraries Composed of Separate Compound Mixtures

The identification of active individual compounds from libraries of this type (i.e., iterative, positional scanning, and orthogonal libraries) does not require any additional analytical techniques, because owing to the systematic arrangement of the mixtures making up these libraries (i.e., the implementation of defined and mixture positions), the identification of individual compounds that are responsible for an observed biological activity of a library is inherent to the library screening process (see above). This is one of the principal advantages of this class of libraries.

Experimental Procedures

Syntheses of One-Bead–One-Compound Peptide Libraries

Synthesis of Noncleavable Library. The library is synthesized[38] on 130-μm TentaGel (Rapp Polymere, Tubingen, Germany) resin beads. Alternatively, ArgoGel (Argonaut Technologies) polydimethylacrylamide beads, of Pepsyn Gel Resin (Cambridge Research Biochemicals, Northwitch, UK) can be used. In general, any resin that is compatible with organic solvents, as well as aqueous media, is adequate. Spacers, such as aminocaproic acid, aminobutyric acid, and/or β-Ala, may be attached to the resin prior to assembling the library. The resin beads are divided into 19 aliquots contained in 19 polypropylene vials or plastic syringes equipped with a plastic

---

[265] N. F. Sepetov, O. L. Issakova, M. Lebl, K. Swiderek, D. C. Stahl, and T. D. Lee, *Rapid Commun. Mass Spectrom.* **7,** 58 (1993).
[266] N. Sepetov, O. Issakova, V. Krchnák, and M. Lebl, U.S. Patent 5,470,753 (1995).

frit at the bottom (CSPS, San Diego, CA). (The number of amino acids we use in randomization, equal to the number of reaction vessels, is limited only by the patience of the chemist; the highest number that we have used is 384 syringes.)

Nineteen Fmoc-amino acids (all proteinogenic amino acids except Cys) are added separately into each of the resin aliquots using a minimal amount of DMF. The amino acids are added in 3-fold excess, and coupling is initiated by adding a 3-fold excess of benzotriazol-1-yl-oxytris-(dimethylamino)phosphonium hexafluorophosphate (BOP) and N,N-diisopropylethylamine (DIEA) [or N,N'-diisopropylcarbodiimide (DIPCDI) and 1-hydroxybenzotriazole (HOBt). A trace amount of bromphenol blue is added to the reaction mixture.[85,267] The vials are tightly sealed (syringes are capped) and rocked gently for approximately 30 min at room temperature, or until all beads turn from blue to colorless. Completion of the coupling is confirmed by the ninhydrin test.[84] When the coupling is incomplete, the beads are allowed to settle, and the supernatant is gently removed; alternatively, in the case of synthesis in a syringe, it is expelled from the syringe. Fresh activated Fmoc-amino acid is added, and the reaction proceeds for 1 hr.

The resin pools are mixed in a siliconized cylindrical glass vessel fitted at the bottom with a frit. Dried nitrogen is bubbled through to mix the resin. After washing eight times with DMF, piperidine–DMF (1:4, v/v) is added. After 10 min of bubbling with nitrogen, the piperidine is removed and the resin washed 10 times with DMF. The amount of released dibenzofulvene–piperidine adduct is determined by measuring the absorbance at $\lambda = 302$ nm. A stable level of substitution determined in this manner throughout the library synthesis serves as one of the quality control criteria.

The resin is again divided into 19 aliquots for the coupling of the next 19 amino acids. After the coupling steps are completed, the Fmoc group is removed with piperidine–DMF (1:4, v/v), and the resin is washed with DMF and DCM. The side-chain protecting groups are removed by treatment with reagent K[268] (trifluoroacetic acid–phenol–water–phenol–ethanedithiol, 82.5:5:5:5:2.5; v/w/v/w/v) for 5 + 120 min. This treatment is performed in the common container (glass bubbler) or, in cases that separate pools of resin are required, in individual syringes. The resin is washed thoroughly with TFA, DCM, DMF, DMF–water (1:1, v/v), and 0.01% HCl in water, and stored in DMF at 4°. Again, small individual pools can be conveniently stored in the plastic syringes in which the whole

[267] V. Krchnák, J. Vagner, and M. Lebl, *Int. J. Pept. Protein Res.* **32**, 415 (1988).
[268] D. S. King, C. G. Fields, and G. B. Fields, *Int. J. Pept. Protein Res.* **36**, 255 (1990).

synthesis is performed. Larger library batches (up to 80 g) are stored with protected side chains in 0.2% HOBt/DMF at 4°.

To verify the quality of the library, several randomly chosen beads are sequenced, and the average amount of peptide per bead is determined. This value is confirmed by quantitative amino acid analysis of a random sample from the library (~1 mg). Whereas amino acid analysis is used to determine the overall amino acid composition of the library, sequence analysis confirms random distribution of amino acids at each position.

If the library is synthesized in MultiBlock (http:/www.z5.com/csps), the randomization of the resin is achieved by closing the block with glass plate, inverting it so that all resin is placed in the common area, shaking it mechanically, and inverting it again. Resin is uniformly distributed into the individual reaction compartments (syringes), and the next step of the synthesis can be performed. Any number of reaction chambers can be stoppered during the mixing, and therefore a number of different mixing schemes can be achieved in the MultiBlock.

Synthesis of Libraries for Two-State Solution Testing

TentaGel (5 g, 0.23 mmol/g, 130 μm average particle size) is swollen in DMF (swollen resin volume 25 ml), and Fmoc-Lys(Boc)/DIPCDI/HOBt (3 equivalents each) in DMF is coupled.[37] After 2 hr the resin is washed five times with DMF and once with DCM, and the Boc group is removed with TFA–DCM (1:1, v/v, 1 + 20 min). After washing with DCM (five times) and DMF (four times), the resin is neutralized with DIEA–DMF (1:49, v/v), washed with DMF (three times), and the linker **1** (Fig. 3, 3 equivalents) is activated by DIPCDI and HOBt (3 equivalents each) in DMF and coupled overnight. The resin is washed with DMF (five times), and the Fmoc group is removed with piperidine–DMF (1:4, v:v, 20 min). After washing with DMF (three times) and distribution of the resin into *m* reaction vessels (plastic vials or fritted syringes), individual Fmoc protected amino acids are coupled to each part of the resin using DIPCDI and HOBt (3 equivalents each). The reaction is monitored using bromphenol blue.[85,267] When complete coupling is observed in all reaction vessels (all resin particles are decolorized), completeness of the coupling is verified using the ninhydrin test.[84] All resin portions are combined, washed with DMF (five times), and the Fmoc group is removed as described above. This procedure (separate couplings and deprotection after combining the resin) is repeated $n - 1$ times (n = number of library positions). The side-chain protecting groups are cleaved by reagent K[268] for 2 hr, and washed with TFA (three times), DCM (five times), DMF containing 0.1% HCl (four times), and 0.1% HCl in water (five times). The library has to be stored in an acidic solution in order to prevent premature loss of peptides.

Quality Control of Doubly Releasable Library. Dried resin (5–10 mg) is shaken overnight in 2–5 ml of 0.1 M HEPES buffer (pH 8.5) in a polypropylene syringe equipped at the bottom with a polypropylene or Teflon frit and a polypropylene plunger. The absorbance of the solution (diluted, if necessary) at $\lambda = 280$ nm is measured, and the amount of released peptide is calculated according to the following formula:

$$\text{Release (mmol/g)} = \frac{\text{absorbance} \times \text{volume} \times \text{dilution}}{(1197n/x \pm 5559m/y)\text{mass}}$$

where mass is the quantity of library beads in grams, x is the number of amino acids in positions where Tyr is used, y is the number of amino acids in positions where Trp is used, n is the number of positions in the library where Tyr is used, and m is the number of positions in the library where Trp is used. If other amino acids with absorbance at $\lambda = 280$ nm are used in library construction, the above formula must be modified.

A solution of 0.2% NaOH is drawn into the syringe containing the library sample, and the syringe is shaken for 4 hr. The solution is expelled from the syringe and the absorbance measured at $\lambda = 280$ nm. The same calculation is performed using the formula shown above using coefficients 1507 and 5377 instead of 1197 and 5559, respectively. The amount of released peptide in each step should not differ by more then 10% from the theoretical value, which is calculated according to

$$\text{Theoretical release (mmol/g)} = \frac{\text{Subst.}}{1 + \text{Subst.}\,(3\text{MW} + 686)/1000}$$

where Subst. is the original substitution of the resin (in mmol/g), MW is the average molecular weight of the library peptides, and 686 is the molecular weight of the Ida linker (without Fmoc groups), plus one Lys residue, minus one molecule of water. The average molecular weight of a natural amino acid is 119.7 (19 amino acids, Cys excluded). Therefore, the average molecular weight of a pentapeptide library made from these 19 amino acids is 598.5. Starting with a resin substitution of 0.2 mmol/g, 0.134 mmol of pentapeptide should be released at each step using 1 g of dried library resin.

Two-Stage Release Assay in 96-Well Microassay Plates. Library beads are transferred into pH 4.5 buffer containing 1.0% carboxymethylcellulose (to retard sedimentation), shaken, and rapidly pipetted into the upper chambers of a vacuum-control 96-well filtration manifold (Model 09601, Millipore, South San Francisco, CA). Approximately 500 beads are placed in each filtration well, so that each plate contains approximately 48,500 unique peptides. The filtration plates serve as "master" plates for retaining subsets of peptides in unique locations. The transfer buffer is removed by vacuum filtration, and the first stage release of peptides is accomplished by dispensing the appropriate buffer or tissue culture medium (neutral pH) to each well and incubating overnight. The released peptides are vacuum filtered into 96-well microassay test plates where the biological activity is determined. In some experiments the released peptides are distributed into several replicate plates for multiple simultaneous assay against different molecular targets. Wells identified as "positive" are marked, and the beads of origin are recovered from the corresponding well(s) of the filtration master plate with the aid of a low power stereomicroscope. The recovered beads are transferred one by one (one bead per well) into individual microwells of 96-well filtration plates. Cleavage of the ester (second) linker is then accomplished by addition of 0.2% NaOH and overnight incubation followed by pH adjustment. Alternatively, the second stage release may be achieved by overnight incubation in ammonia vapors in a desiccator or dedicated pressurized chamber. After drying, the appropriate buffer is added, and the plates are gently shaken for several hours. Thereafter, the peptide-containing buffer is filtered into the test plates for bioassay. The individual peptide beads corresponding to each positive well in the second stage assay are recovered and submitted for microsequencing.

Synthesis of Directed Libraries

Various procedures for synthesizing directed libraries have been developed.[163]

Synthesis of Directed Library on Cotton String. A cotton ribbon (5 m long, 3 cm wide) is treated for 1 hr in TFA–DCM (1:3, v/v), washed with DCM (three times), neutralized with DIEA–DCM (1:19, v/v, 5 min), and washed with DCM and DMF (three times). Fmoc-β-Ala (2 mmol) is coupled overnight by DIPCDI (2 mmol) and HOBt (2 mmol) activation with addition of N-methylimidazole (3.5 mmol). The cotton is washed with DMF, and the substitution level is determined by spectrophotometric determination of the cleaved Fmoc group [typical value is 0.4 mmol/g (68 nmol/cm)]. Five pieces of the cotton string (25 cm each) are placed in five polypropylene syringes, and Fmoc protected amino acids [Phe, Tyr(tBu), Ala, Leu, Gly] are coupled by the DIPCDI/HOBt protocol. After the coupling is complete (monitored by bromphenol blue method[85]), the strings are washed with DMF, placed in one syringe (no frit needed), and the Fmoc groups cleaved. After washing with DMF and a 2% solution of HOBt in DMF, the mixture of 19 Fmoc-amino acids (Cys is excluded) with the molar ratios (determined in pilot experiments) adjusted for different reactivities[9,57,156–158] is coupled to all cotton pieces. After completion of coupling the cotton is washed, the

Fmoc groups cleaved, and the washed cotton string divided into five syringes in the following way. From each string 5 cm of the cotton is cut and placed in a different syringe. In this way all syringes have only one 5-cm piece of cotton cut from each 25-cm string and none have more than one. Coupling of five amino acid derivatives (same as above) is performed, the cotton washed, Fmoc groups removed, and cotton subdivided again. In this case 1-cm pieces are cut from all 5-cm pieces and placed in five syringes. Coupling of the same five amino acids is performed, the Fmoc groups are cleaved, and side-chain protecting groups are cleaved by TFA–DCM–anisole (50:45:5, v/v/v) for 2 hr. Cotton pieces are washed by DCM and methanol and dried. Quantitative amino acid analysis of a sample of one string has revealed a substitution level of 400 nmol/cm.

Synthesis of Directed Library on Functionalized Cross-Linked Teflon Membrane. Hydrophilic aminopropyl functionalized membrane (UV cross-linked aminopropylmethacrylamide, N,N-dimethylacrylamide, and methyl-enebisacrylamide on Teflon membrane, 16 × 16 cm, Perseptive Biosystems) with approximate 35 nmol/cm^2 substitution is placed into a 50-ml Falcon tube and acylated by Fmoc-β-Ala using the DIPCDI/HOBt procedure in DMF. Fmoc-Gly, Fmoc-β-Ala, and Fmoc-Gly are coupled consecutively. After deprotection, the membrane is divided into two parts, and Fmoc-Phe and Fmoc-Leu are coupled to them, respectively. After coupling completion (bromphenol blue monitoring[85]) and deprotection, the pieces are divided again into two halves and recombined for the coupling of Fmoc-Gln and Fmoc-Phe. For the next coupling the membrane is divided again into two pieces and recombined for coupling of Fmoc-Proc and Fmoc-Gly. The pieces resulting from these couplings (8 × 4 cm) are now divided into 19 strips (8 × 0.21 cm) and the strips are placed into 19 small polypropylene tubes. Nineteen natural amino acids (excluding Cys) are used for coupling in this stage. After coupling completion and Fmoc deprotection, the strips are cut into 19 pieces (4 × 2.1 mm) and divided into 19 vessels again. The same set of 19 Fmoc-amino acids is used for the last coupling. All pieces are combined, Fmoc groups are removed, and side-chain protecting groups are cleaved by reagent K.[268] Membrane pieces are washed with TFA (two times), DCM (five times), methanol (three times), and water (five times). The substitution level based on measurement of the absorbance of the last Fmoc release is 43.3 nmol/cm^2.

Synthesis of Dual Defined Iterative Hexapeptide Library

Nineteen (or any number corresponding to the number of used building blocks, depending on the capability of the chemist) individually labeled porous polypropylene mesh packets are charged with 20 g of MBHA resin each.[269] Each of 19 of the 20 genetically coded Boc-amino acids (Cys excluded) is activated by DIPCDI and coupled to one of the 19 resin packets. The coupling reaction is monitored for completion by using bromphenol blue[85] or the ninhydrin test.[84] The resin packets are washed with DCM and dried; the resins of all packets are recombined and mixed thoroughly. This one-position resin is referred to as X-resin.

The X-resin is divided into 19 equal portions and placed into new polypropylene mesh packets. The Boc group is removed with TFA–DCM (11:9, v/v), and the resin is washed with DCM and 2-propanol, neutralized with DIEA–DCM (1:19, v/v), and washed with DCM. The 19 amino acids are activated by DIPCDI and coupled to the resin packets to generate 361 (19^2) dipeptides. These dipeptides are termed OX-resins. Mixing of all OX-resins affords XX-resin.

The coupling steps are repeated twice more to generate a 130,321 tetrapeptide mixture resin (XXXX-resin). The XXXX-resin is divided into 400 equal aliquots and placed in labeled polypropylene mesh packets. The Boc group is removed and the resin neutralized. Two amino acids are coupled to each of the packets using standard coupling procedures. The result is a hexapeptide mixture resin (O_1O_2XXX-resin) with two defined (O) and four mixture (X) positions.

The 400 separate peptide mixtures are deprotected and cleaved using the high/low HF method[87] in a multivessel apparatus.[86] Peptides are extracted from the resins with water or a mixture of acetic acid and water, the solution is lyophilized twice, and peptide mixtures are dissolved in water at 1 to 5 mg/ml. The peptide library is stored 1 to 2 weeks at 4°, or frozen for prolonged storage. Sonication facilitates the solubilization of peptide mixtures with hydrophobic amino acids at the defined positions.

Synthesis of Positional Scanning Hexapeptide Library

The positional scanning hexapeptide library[269] is composed of six sublibraries (O_1XXXXX-NH$_2$, XO_2XXXX-NH$_2$, XXO_3XXX-NH$_2$, XXXO_4XX-NH$_2$, XXXXO_5X-NH$_2$, and XXXXXO_6-NH$_2$), where one position (O) is individually defined with one of 19 amino acids and the remaining five positions (X) are mixtures of 19 amino acids. Thus, the entire library is made up of 114 (19 × 6) distinct peptide mixtures. Amino acids are mixed for coupling in a molar ratio that ensures equimolar incorporation

[269] C. Pinilla, J. R. Appel, and R. A. Houghten, *in* "Current Protocols in Immunology" (J. E. Coligan, ed.), Wiley, New York, 1994.

TABLE I
COMPOSITION OF AMINO ACIDS MIXTURES
FOR COUPLING[a]

Amino acid	Molar ratio
Boc-Ala	1.18
Boc-Arg(Tos)	2.26
Boc-Asn	1.86
Boc-Asp(OBzl)	1.22
Boc-Gln	1.85
Boc-Glu(OBzl)	1.26
Boc-Gly	1.00
Boc-His(Dnp)	1.24
Boc-Ile	6.02
Boc-Lys(2-Cl-Z)	2.16
Boc-Leu	1.72
Boc-Met(O)	0.80
Boc-Phe	0.88
Boc-Pro	1.50
Boc-Ser(Bzl)	0.97
Boc-Thr(Bzl)	1.66
Boc-Trp(For)	1.32
Boc-Tyr(2-Br-Z)	1.44
Boc-Val	3.91

[a] Amino acids with different side-chain protection groups require an adjusted molar ratio.

of amino acids into peptides (Table I). One hundred fourteen polypropylene
mesh bags are labeled and loaded with 400 mg MBHA resin each. Nineteen
Boc protected amino acids are activated by DIPCDI and coupled to bags
96 to 114, whereas the mixture of 19 amino acids is coupled to bags 1 to
95. Resins bags 96 to 114 have a defined position at position 6. The other
bags have mixture positions there. After Boc removal, 19 individual amino
acids are coupled to bags 77 to 95, and the amino acid mixtures to the
remaining bags. Resin bags 77 to 95 have a defined position at position 5.
This procedure is repeated through the sixth coupling. The peptides are
cleaved, extracted, and lyophilized as described in the previous protocol.
Peptide mixtures are dissolved in water at 10 to 20 mg/ml and stored 1 to
2 weeks at 4° or are frozen for prolonged storage.

The higher final concentration of peptide mixtures in this library compared to the dual defined peptide library compensates for the presence of
19 times more peptides when compared to the latter peptide mixtures (five
versus four mixture positions). The relative concentration of individual
peptides within the mixtures decrease 19 times.

Synthesis of Library on Cellulose Paper Sheet

A sheet of chromatographic paper Whatman Chr1 (Maidstone, UK) is marked with a pencil (spot positions), and the sheet is dried under vacuum overnight.[53] A solution of 0.2 M Fmoc-amino acid (Pro or β-Ala), 0.24 M DIPCDI, and N-methylimidazole in DMF is soaked into the paper sheet, and the reaction is run for 3 hr in a closed container. The paper sheet is washed three times with DMF and treated with piperidine–DMF (1:4, v/v) for 20 min. After three washes with DMF and two washes with ethanol, the sheet is dried in a desiccator. Solutions of Fmoc-amino acid HOBt esters (0.5 ml, 0.3 M) are then spotted onto pencil marked spots, and reactions proceed for 20 min on a plastic tray covered with glass plate. The paper sheet is then washed twice with acetic anhydride–DMF (1:49, v/v) and treated with the same solution in the presence of 1% DIEA for 30 min. After washing with DMF (four times), deprotection by piperidine–DMF (1:4, v/v, 5 min), and washing by DMF (four times), the sheet is washed in bromphenol blue solution (0.01%) to reveal blue spots with available amino groups. The sheet is washed with ethanol (two times) and dried by cold air from a hair dryer between two layers of Whatman 3MM paper. Blue spots are used as a target for spotting activated solutions of amino acids (0.3 M, 0.5–1 ml). The paper sheet is optionally respotted after 15 min (if disappearance of blue color is slow). Final deprotection is performed by immersing the dry sheet into a solution of TFA–DCM–diisobutylsilane–water (50:453:2, v/v/v/v) for 2 hr. After washes with DCM (four times), DMF (three times), and ethanol (two times), the sheet is dried and ready for binding assay or storage ($-20°$ in a sealed plastic bag). In the case of peptide synthesis on a linker cleavable by intramolecular DKP formation, the sheet has to be washed exclusively by acidic solutions so as to not lose the synthesized peptides prematurely. The spots can be cut or punched out of the dried sheet into polypropylene tubes or into wells of a microtiter plate, and after addition of neutral buffer the peptides are released into solution.

Synthesis of Peptide Library on Soluble Support

The library is synthesized on a poly(ethylene glycol) methylether (MeO-PEG) (molecular weight 5000) support.[69] The support is divided into four aliquots, and the first Boc-protected amino acid (Leu, Phe, Tyr, and Gly) is attached to the support by the N,N'-dicyclohexylcarbodiimide/4-dimeth-ylaminopyridine (DCC/DMAP) procedure. Excess reagents are removed by precipitation with diethyl ether or ice-cold ethanol, or by ultrafiltration. Aliquots are taken from each portion and stored for later deconvolution, all portions are mixed, and the Boc protecting group is removed by a

TFA–DCM (1 : 1, v/v) mixture. The product is separated into four portions, and Boc-protected amino acids are coupled by HBTU in the presence of DIEA in a micture of DMF and DCM. Completeness of the coupling is followed by the ninhydrin reaction.[84] Uncoupled amino groups are acetylated by acetic anhydride. Recombination, separation into aliquots, and amino acid couplings are repeated another three times. After the last coupling, the aliquots are kept separate, and the final removal of Boc and BrZ groups is performed by iodotrimethylsilane.[270] Aqueous solution of aliquots are evaluated directly by biological assays.

Synthesis of Orthogonal Library

The synthesis of an orthogonal library is documented by an example of a tripeptide library composed of 12 amino acids in each step. Orthogonal mixtures are formed in the dipeptide stage, and each mixture contains 12 peptides. The synthesis is performed in a 96-well plate[77] according to the standard Fmoc/tBu protocol.

RAM TentaGel resin is distributed into 144 wells of 2 plates. The Fmoc group is removed by piperidine–DMF, the resin is washed, and 12 Fmoc-protected amino acids (activated by DIPCDI/HOBt) are distributed by "rows," where each row contains one amino acid. The plate is left overnight with occasional shaking on an orbital shaker. The resin is washed with DMF, Fmoc groups are cleaved, and the resin is washed with DMF. To prepare all combinatorial dipeptides, the second amino acid (activated by DIPCDI/HOBt) is distributed by "columns," that is, each column receives one amino acid. After the condensation is complete, the resin beads are washed with DMF, Fmoc groups are cleaved, and the resin is washed with DMF.

The orthogonal mixtures are made the following way. One-half of the resin beads are withdrawn from all wells in the first row, and the mixture (of 12 dipeptides) is redistributed into 12 wells of the first row of a new 96-well plate. This operation is repeated for all 12 rows, resulting in 12 rows of wells filled with a mixture of dipeptides. The second half of the resin beads from all wells in the first column is mixed and redistributed into 12 wells of a new plate. Again, this operation is repeated for all columns. The result of this mixing is 2 times 12 rows containing mixtures of dipeptides, with the first 12 rows containing the "horizontal" mixtures and the next 12 rows "vertical" mixtures.

For the last condensation the first of 12 activated amino acids is distributed into the first row of both "horizontal" and "vertical" mixtures. After

---

[270] R. S. Lott, V. Chauhan, and C. H. Stammer, *J. Chem. Soc., Chem. Commun.*, 495 (1979).

acylation the resin beads are washed, Fmoc groups cleaved, and the resin beads washed with DMF and methanol and dried. Cleavage of peptides is accomplished by the standard TFA procedure.

Synthesis of α,β,γ Library

The library is synthesized on TentaGel resin using the following procedures.[205] The Fmoc group is removed with piperidine–DMF (1:4, v/v) 5 + 20 min, then the resin is washed with DMF six times. All washes are collected, the absorbance at $\lambda = 302$ nm is measured, and the Fmoc release is quantified. For coupling, 3-fold molar excesses of protected amino acid and HOBt is dissolved in DMF, DIPCDI is added, and the solution is drawn into a syringe with the resin. Completeness of each condensation reaction is checked by the ninhydrin test. The chloranil test is used in cases of coupling to secondary amino groups. Coupling time varies between 1.5 and 40 hr. For Z removal, the resin is washed three times with DMF, three times with DCM, and two times with thioanisole–TFA (1:9, v/v) for 1 min. The resin is left in thioanisole–TFA (1:9, v/v) overnight. The resin is then washed five times with DCM, neutralized with DIEA–DCM (1:19, v/v), and washed three times with DCM and DMF. For Boc deprotection, the resin is washed with DCM, treated with TFA–DCM–anisole (9:9:2, v/v/v) for 5 + 20 min, and washed with DCM six times. For Alloc deprotection, the resin is washed five times with DMF, a mixture of DMF–acetic acid–*N*-methylmorpholine (10:2:1, v/v/v) is added, argon is bubbled for 15 min, tetrakis(triphenylphosphine)palladium(0) is added, and the reaction is allowed to proceed for 3 hr. The resin is washed five times each with DMF, DCM, and DMF.

The library synthesis consists of 50 synthetic steps:

1–9. Synthesis of Boc-β-Ala-Gly-β-Ala-Gly-Lys(Alloc)-TentaGel.
10. Remove Alloc.
11. Divide resin into six aliquots.
12. Couple six coding pairs of Fmoc-amino acids.
13. Remove Boc.
14. Couple Boc-Dap(Alloc)-OH to portions 1 and 2. Couple Boc-Dab(Alloc)-OH to portions 3 and 4. Couple Boc-Orn(Alloc)-OH to portions 5 and 6.
15. Combine portions 1, 3, and 5.
16. Remove Boc.
17. Combine portions 2, 4, and 6.
18. Remove Alloc.
19. Combine all resin.
21. Divide into 46 portions.

22. Couple 46 acids.
23. Remove Fmoc.
24. Couple coding doublets of Fmoc amino acids.
25. Mix the resin.
26. Remove Alloc.
27. Remove Boc.
28. Divide resin into six portions.
29. Couple Boc-Dap(Alloc)-OH to portions 1 and 2. Couple Boc-Dab(Alloc)-OH to portions 3 and 4. Couple Boc-Orn(Alloc)-OH to portions 5 and 6.
30. Remove Fmoc.
31. Couple six coding pairs of Fmoc-amino acids.
32. Combine portions 1, 3, and 5.
33. Remove Boc.
34. Combine portions 2, 4, and 6.
35. Remove Alloc.
36. Combine all resin.
37. Divide into 46 portions.
38. Couple 46 acids.
39. Remove Fmoc.
40. Couple coding doublets of Fmoc-amino acids.
41. Mix resin.
42. Remove Alloc.
43. Remove Boc.
44. Divide resin into 50 portions.
45. Couple 50 acids.
46. Remove Fmoc.
47. Couple coding doublets of Fmoc-amino acids.
48. Remove Fmoc.
49. Mix resin.
50. Remove Z side-chain protecting group.

Acylation Monitoring by Bromphenol Blue

Couplings performed in neutral solution (DIPCDI/HOBt, preformed anhydrides, active esters) can be monitored by addition of trace amounts of bromphenol blue (BB).[85] The sensitivity of the method can be significantly diminished by application of large amount of bromphenol blue, and therefore no more indicator than the amount equal to 1% of available amino groups should be applied. Usually several drops of 0.1% solution of BB in DMF or *N*-methylpyrrolidone (NMP) (if the dissolution provides blue solution, it can be decolorized by the addition of HOBt) are added into the last wash before coupling, or directly to the solution of activated acid. Blue-colored beads turn green, greenish yellow, and eventually yellow. The speed of some couplings can be puzzling. Most are complete within 2 to 5 min; in library synthesis, however, care should be taken about the slowest couplings, which may require much longer exposure. BB monitoring allows the evaluation of coupling at the level of individual beads. In this case a sample of the reaction slurry is placed on the petri dish and inspected under a microscope. It is easy to detect one incompletely coupled bead in the middle of tens of thousands of beads.

Successful application of BB monitoring requires the absence of quaternary ammonium salts on the resin, as resin containing these residual functionalities never becomes BB negative. Before using a new batch of solid support, the resin should be peracetylated (if it is not fully protected) and treated with BB solution. If blue coloration is observable, BB cannot be used for monitoring. In the presence of sulfonium salts (modified side chains of Met), the resin would be greenish even without the presence of free amino group.

Acknowledgments

We thank Jutta Eichler and Richard Houghten for help with the manuscript. Dr. Eichler's understanding of "Czenglish" and advice on mixture-based techniques made the article more readable. We thank Dr. Houghten for letting us use original data.

Section II
Analytical Techniques

[17] Edman Sequencing as Tool for Characterization of Synthetic Peptides

By GREGORY A. GRANT, MARK W. CRANKSHAW, and JOHN GORKA

Introduction

Sequence analysis by Edman chemistry is a useful tool in the characterization of synthetic peptides. As the name implies, its purpose is to determine the sequence of amino acid residues in a peptide. Thus, it is particularly useful for the determination of residue deletions in the sequence of a synthetic peptide. It is also useful for the determination of the location of stable modified amino acids in a sequence and to determine the identity of amino acid residues with equal or similar mass. As with any technique, it has its limitations, and thus a familiarity with these is necessary to avoid misleading or incorrect conclusions. Because peptides are most commonly synthesized on solid supports, most analytical methods can be applied only after the peptide is cleaved from the resin. Edman sequence analysis, however, can be performed either before or after cleavage. It can, in fact, be used to monitor the progress of the synthesis as it is occurring by analyzing a very small amount of resin removed from the reaction vessel of the synthesizer while the synthesis itself continues on.

Overview of Edman Sequencing

Edman Chemistry

Repetitive degradation of proteins and peptides from their amino terminus by reaction of the free α-amino group with phenyl isothiocyanate (PITC) (Fig. 1) was first reported by Edman in 1950[1] and automated by Edman and Begg in 1967.[2] Coupling of PITC to the peptide amino groups takes place in the presence of base at pH 9–10. A primary or secondary amino group is required for coupling to take place, so most modifications of the amino terminus, such as by acylation, will render the peptide refractory to the Edman chemistry. Subsequent to coupling, cleavage of the first amino acid from the peptide is produced by treatment with an anhydrous acid, usually trifluoroacetic acid, which promotes the attack of the PITC sulfur on the carbonyl carbon of the amino acid. This reaction is particularly

---

[1] P. Edman, *Acta Chem. Scand.* **4**, 283 (1950).
[2] P. Edman and G. Begg, *Eur. J. Biochem.* **1**, 80 (1967).

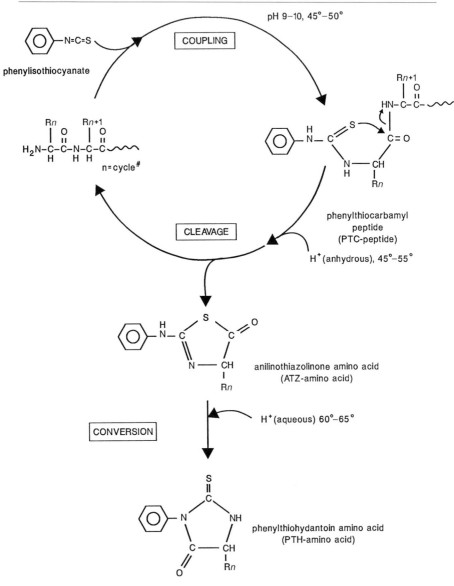

FIG. 1. Edman chemistry for the sequential degradation of proteins and peptides. The reaction is carried out in successive cycles with the product of each cycle being a PTH-amino acid. The PTH-amino acid identified at each cycle corresponds to the position from the amino terminus that the amino acid occupied in the peptide.

dependent on the appropriate chain length leading to a five-membered ring intermediate and thus proceeds efficiently only for α-amino acids. The attack on the carbonyl carbon and formation of the anilinothiazolinone derivative lead to peptide bond cleavage and the generation of a new amino terminus on the peptide that can subsequently react with PITC to repeat the process. Because excess PITC is removed before treatment with acid, only one amino acid is cleaved at a time. This provides the basis for sequence determination through repetitive reaction with PITC. Conversion of the cleaved anilinothiazolinone derivative to the more stable phenylthiohydantoin derivative takes place in aqueous acid, usually 25% (v/v) trifluoroacetic acid. During this conversion, the anilinothiazolinone ring opens to produce a phenylthiocarbamyl (PTC) derivative and then closes to a rearranged ring structure called the phenylthiohydantoin (PTH)-amino acid. If the conversion conditions are not optimized, some PTC-amino acid may persist and will subsequently be seen on high-performance liquid chromatography (HPLC). It is the generation of one PTH-amino acid per peptide chain per reaction cycle that, when subsequently identified, yields the sequence of the peptide.

Both acid treatments of the Edman chemistry occur under conditions that are more extreme than those used to remove many of the protecting groups used in 9-fluorenylmethyloxycarbonyl (Fmoc) synthesis. Thus, these groups will be removed to varying degrees during sequencing if they are present (see Table III and the section on the Analysis of Resin-Bound Peptides).

Identification of Phenylthiohydantoin-Amino Acids

Identification of the released amino acids from repetitive Edman degradation has been a major undertaking since development of the chemistry. The introduction of HPLC in the late 1970s revolutionized this aspect by providing a technique that was relatively rapid and sensitive and that could identify all of the commonly encountered amino acids in a single analysis.[3] As of this writing, all commercially available protein/peptide sequencers use reversed-phase chromatographic methods for the identification of the PTH-amino acids released during sequencing. Figure 2 shows a typical chromatogram of the common PTH-amino acids from one such instrument.

Analysis by HPLC, however, does not unequivocally identify the cleaved amino acid residue. Rather, it relies on the comparison of the elution position of the PTH-amino acid generated during sequencing from a reversed-phase HPLC column to the elution position of standard PTH-amino

[3] M. W. Hunkapiller and L. E. Hood, *Methods Enzymol.* **91**, 486 (1983).

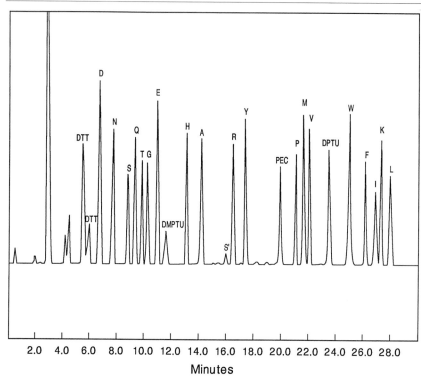

2.0 4.0 6.0 8.0 10.0 12.0 14.0 16.0 18.0 20.0 22.0 24.0 26.0 28.0

Minutes

FIG. 2. A reversed-phase HPLC chromatogram of a standard mixture of PTH-amino acids from a Perkin-Elmer Applied Biosystems Division Model 477 protein sequencer. The abbreviations used are the one-letter code for the amino acids. Also DTT, dithiothreitol; DMPTU, dimethylphenylthiourea; PEC, pyridylethyl cysteine; and DPTU, diphenylthiourea. The DTT doublet is reduced and oxidized DTT, and the prime (S') indicates a DTT derivative of a Ser degradation product. Note that the elution times vary somewhat from those presented in Table III. This is normal, and Table III should be used only as a guide of relative position. The large elution differences for His and Arg seen here and as listed in Table III represent normal variability for these residues arising from differences in elution buffer composition from one sequencer to another.

acids run on the same column under the same conditions. Thus, although it is critical that appropriate standards be available, this does not usually pose a problem for synthetic peptides because their putative sequence is known and standards of stable side-chain protected PTH-amino acids can be easily formed using the Edman chemistry and the appropriate α-amino deprotected amino acid derivative. These standards can be produced manually[4,5] or by simply subjecting the amino acid derivative to a single cycle of Edman chemistry on the sequencer.[6] Because modern sequencers can sequence sample loads in the low picomole to femtomole range, quantity is seldom a problem for synthetic peptides. However, it is important not to overload the sequencer, particularly for samples containing side-chain protected amino acids, because high levels of PTH-amino acids that may have low solubility in the sequencer solvents can precipitate and occlude the lines of the sequencer or the in-line HPLC. Given the fact that a typical bead from solid-phase peptide synthesis may produce an initial yield in the sequencer of 50–150 pmol of PTH-amino acid,[6] it is very easy to overload the sequencer.

Peptide Washout and Quantitation

Unless a peptide is covalently attached to a support during Edman sequencing, it may wash out of the reaction vessel before the end is reached. This is mainly due to the repeated organic extractions that the peptide is subjected to as the sequencing proceeds. In general, the level of washout tends to increase with the hydrophobicity of the peptide and decrease with increasing peptide length. In addition, as the peptide is shortened during sequencing, the level of washout may increase. Nonetheless, except for long peptides (>15–20 residues), it is usually possible to identify the penultimate residue and often the C-terminal residue. However, this is not guaranteed, and even with peptides that are still attached to the synthetic resin it may be difficult to identify the C-terminal residue. Thus, if a modification or deletion has occurred at the C-terminal residue for relatively short peptides, or close to the C terminus in the case of longer peptides, Edman degradation in itself may not produce usable data in the C-terminal region. Furthermore, Edman degradation does not produce highly quantitative results because not all amino acids are equally stable to the chemistry or extracted from the reaction vessel with the same efficiency.[7] In addition, the initial yield of the sequencing is unpredictable, and the repetitive yield of even the same amino acid at different points in the sequence can vary to some degree.

The common amino acid residues that are usually recovered without appreciable degradation or loss as a consequence of the Edman chemistry

[4] D. H. Schlesinger, *Methods Enzymol.* **91**, 494 (1983).
[5] D. M. Steiman, R. J. Ridge, and G. R. Matsueda, *Anal. Biochem.* **145**, 91 (1985).
[6] J. Pohl, *in* "Methods in Molecular Biology, Peptide Analysis Protocols" (B. M. Dunn and M. W. Pennington, eds.), p. 107. Humana Press, Totowa, New Jersey, 1994.
[7] M. W. Hunkapiller, *in* "Proteins, Structure and Function" (J. J. L'Italien, ed.), p. 363. Plenum, New York and London, 1987.

are Asp, Glu, Gly, Ala, Tyr, Pro, Val, Ile, Leu, Phe, and Lys. Thr, Ser, and Cys are subject to degradation by β-elimination. Although low levels of undegraded Thr and Ser can usually be seen, Cys usually is not seen at all unless it is derivatized to a stable adduct prior to sequencing. Some of the β-elimination products can be identified as relatively stable adducts of dithiothreitol (DTT) by including it in the appropriate sequencer solvents. Note that because the elimination product of Ser and Cys are the same, they cannot be distinguished on this basis. Met and Trp are particularly sensitive to oxidation. Inclusion of DTT in the sequencer solvents will convert any Met sulfoxide formed back to Met so that its yield is usually quite high. Trp can be converted to a number of oxidation products that are variably stable and ill-defined such that its identification can be difficult. Asn and Gln usually undergo a small degree of deamidation (see Table I, cycle 9) which does not interfere with their identification if this is recognized. His and Arg, whose PTH derivatives still retain charge character, are more difficult to extract from the reaction vessel because of this. Because there can be a significant variation in residue yield, Edman sequencing can be very poor at determining the presence of modifications or deletions that occur at low levels and particularly if they are heterogeneously dispersed throughout the peptide. This problem is usually compounded by the fact that the Edman chromatograms almost always show low to moderate levels of background and carryover from previous cycles.

Background and Carryover

Background amino acids are introduced into Edman sequencing either through amino acid contamination from the laboratory environment or more commonly as a consequence of the sequencing chemistry itself. Amino acid contamination from the environment manifests itself as a high level of background at the beginning of a sequence run that gradually diminishes or washes out. In addition, the internal peptide bonds of a peptide will undergo some low level of cleavage at each cycle producing free amino termini that can couple with PITC, resulting in a gradually increasing level of background signal. Depending on the sequence of the peptide, this has been estimated to normally be as much as 0.1% at each cycle.[8] Asp–Pro bonds, and to a lesser extent Asp–X bonds, are more acid labile than most other amino acid amide linkages and may produce higher levels of premature bond cleavage during sequencing, leading to higher background levels.

[8] W. F. Brandt, A. Henschen, and C. von Holt, *in* "Methods in Protein Sequence Analysis" (M. Elzinga, ed.), p. 101. Humana Press, Clifton, New Jersey, 1982.

TABLE I
SEQUENCE ANALYSIS OF PEPTIDE MIXTURE CONTAINING DELETION

| Cycle | Sequence | Yield (pmol)[c] | | | | | | | | Calculated % preview | |
		Met	Asp	Arg	Asn	Glu	Ile	Pro	Gln	Eq. (1)	Eq. (2)
1	Met	**2338**	117	4	24	36	8	25	14	5.7	—[a]
2	Asp	303	**1948**	13	34	47	50	32	11	—	—
3	Asp	16	**1941**	38	62	44	69	60	10	3.0	2.1
4	Arg	4	417	**1240**	**270**	60	61	65	32	22.6	18.7
5	Asn		168	530	**926**	**191**	61	58	37	25.1	21.8
6	Glu		61	81	246	**569**	**240**	58	32	29.9	25.2
7	Ile		26	11	29	118	**564**	**127**	29	27.9	22.4
8	Pro		25	8	10	36	147	**329**	**92**	25.3	20.4
9	Gln		**63**	4	6	51	32	111	**271**	25.5	—[b]
10	Asp		**184**	4	5	19	12	43	87		

[a] Cycle j + 1 (Asp-3) is not included because Asp is also the expected residue.
[b] No carryover value; cycle 11 not analyzed.
[c] Numbers that are bold and underlined are yields of amino acids expected at that cycle. Numbers that are bold and italic are the preview yields at that cycle.

Carryover or lag results from the fact that some PITC-coupled peptide will be left behind at each cycle because the chemistry of removal of the amino-terminal residue is never 100% complete.[9] When cleaved in subsequent cycles, it produces a signal that is out of phase or lags behind the major signal. This is usually most pronounced in the cycle immediately following the occurrence of the incomplete cleavage, but it can persist at some level throughout the rest of the sequence. Carryover is often seen most dramatically following proline residues because cleavage of the imino acid is relatively inefficient. Examples of carryover can be seen for most residues in the sequence presented in Table I.

Edman Analysis of Synthetic Peptides

Edman sequencing can be used to analyze both cleaved peptides[10-12] and peptides that have not yet been cleaved from the synthesis resin.[4-6,12-16] Although analysis of both are similar, on-resin sequencing presents some additional features that are discussed separately. In both cases, the major utility of the procedure is the detection of deletion sequences that have resulted from incomplete couplings or incomplete deprotection of the α-amino group. For unpurified cleavage products and for on-resin sequencing, the process has come to be known as preview analysis[13,14] because, when a mixture is present, residues expected at one cycle will show up in the previous cycle owing to the presence of some percentage of the deletion sequence in the mixture (see Table I).

Analysis of Cleaved Peptides

Sequence analysis of cleaved peptides is done in exactly the same manner as sequence analysis of any protein or peptide regardless of origin. For

[9] M. W. Hunkapiller, in "Methods in Protein Sequence Analysis" (K. A. Walsh, ed.), p. 367. Humana Press, Clifton, New Jersey, 1982.

[10] M. J. Geisow and A. Aitken, in "Protein Sequencing, A Practical Approach" (J. B. C. Findlay and M. J. Geisow, eds.), p. 85. IRL Press, Oxford, 1989.

[11] G. A. Grant, in "Synthetic Peptides: A User's Guide" (G. A. Grant, ed.), p. 185. Freeman, New York, 1992.

[12] G. A. Grant and M. W. Crankshaw, in "Methods in Molecular Biology, Protein Sequencing Protocols" (B. J. Smith, ed.), pp. 197–215. Humana Press, Totowa, New Jersey, 1996.

[13] G. R. Matsueda, E. Haber, and M. N. Margolies, Biochemistry 20, 2571 (1981).

[14] S. B. H. Kent, M. Riemen, M. LeDoux, and R. B. Merrifield, in "Methods in Protein Sequence Analysis" (M. Elzinga, ed.), p. 205. Humana Press, Clifton, New Jersey, 1982.

[15] D. J. McCormick, B. J. Madden, and R. J. Ryan, in "Proteins, Structure and Function" (J. J. L'Italien, ed.), p. 403. Plenum, New York and London, 1987.

[16] C. G. Fields, V. L. VanDrisse, and G. B. Fields, Pept. Res. 6, 39 (1993).

that reason, the specific methods for doing so are not repeated here because they are provided by the instrument manufacturer as standard procedure and are routinely available in all resource laboratories offering such services. Because cleaved peptides are not expected to retain side-chain protecting groups under normal circumstances, analysis for these are not usually a concern at this stage. However, for the occasional reticent protecting group, the presence of which is usually first suggested by mass spectrometric (MS) analysis, the same procedure as for on-resin sequencing, discussed below, can be used.

In practice, mainly due to time and expense considerations, purification of the cleaved synthetic peptide mixture is often done first. If, as a result of purification, a major homogeneous product is obtained that has the correct expected mass, there is often nothing to be gained by sequencing the peptide. Even if more than one peptide is obtained in good yield, the correct peptide can usually be identified by mass analysis. However, unless MS/MS capabilities are available and appropriate daughter ions can be formed, the location of the deleted residue in the sequence of the peptide may not be clear if the same residue appears more than once in the sequence or if it involves the presence of equal mass ions such as Ile and Leu or Gln and Lys. If the location of the deletion is in question, perhaps because the synthesis proves to be particularly problematic and attempts need to be made to increase the synthetic efficiency at the appropriate cycle, sequence analysis can provide useful information.

If a homogeneous deletion peptide is sequenced, the deletion will simply show up as an absence of the expected residue. If the crude cleavage product is sequenced before any purification is attempted, the results will more fully reflect what has occurred in the synthesis, and all areas involving deleted residues will show up as a preview of the next residue. Thus, multiple occurrences can be monitored in a single sequence analysis.

Analysis of Resin-Bound Peptides

Sequencing of resin-bound peptides can be approached in much the same manner as cleaved peptides. The major difference encountered with resin-bound peptides is the presence of protecting groups still on the amino acid side chains. Because these protected amino acids are usually more hydrophobic in nature than their deprotected counterparts, they will tend to elute later on reversed-phase HPLC. Hence, the gradient used to analyze the PTH-amino acids must be extended. Similarly, appropriate cycles for the reaction cartridge and conversion flask must be used that provide for appropriate cleavage and extraction parameters for resin-bound peptides.[17]

[17] Applied Biosystems User Bulletin No. 13, pp. 1–18. (1985).

The specific changes to be made will differ depending on the type of sequencer used but generally follow the same pattern of extending cleavage and extraction time and lengthening the HPLC gradient. An example of this for a Perkin-Elmer Applied Biosystems (Foster City, CA) Model 477 sequencer is summarized in Table II. Although the HPLC gradient programs presented in Table II do not produce exactly the same gradient at the beginning of the run where the native PTH-amino acids will elute, the elution profiles are very similar for both programs.

The side-chain protecting groups usually encountered in *tert*-butyloxycarbonyl (Boc) synthesis are designed to be stable to trifluoroacetic acid (TFA) treatment whereas those used for Fmoc synthesis are designed to be labile to TFA. Thus, the stability of the protecting group to the conditions of the sequencing chemistry will affect their analysis. Table III lists the elution positions of many of the common side-chain protected amino acids used in peptide synthesis and provides notes on their relative stability. They usually fall into three classes: those that are completely stable to the chemistry, those that are only partially stable, and those that are completely or nearly completely unstable. In most instances, the unstable derivatives revert back to the native amino acid and will often show signals for both native amino acid and derivative. For example, PTH-His(Bum), which elutes at 27.8 min, converts to PTH-His relatively quickly with successive Edman cycles. So, early in a run it may be identified predominately as PTH-His(Bum) but later in the run, predominately as PTH-His. Because the elution conditions used on one sequencer are never exactly the same as another, the elution positions of the PTH-amino acids will vary to some extent. Thus, Table III should be used only as a guide to relative elution position.

In addition, the linkage of the peptide to the resin is usually much less stable when employing Fmoc chemistry than for Boc chemistry. This may result in better retention of the peptide in the sequencer (less washout) when acid-stable linkages are used, but otherwise does not essentially change the analysis. Polybrene may be included with labile resin linkages to try to increase retention of the peptide, but in practice its effect is variable.

One may approach the loading of resin-bound peptides either qualitatively or quantitatively. That is, resin can either be loaded by applying a fixed volume of suspended beads, or individual beads can be counted under magnification. In either case, the resin beads are suspended at approximately 1 mg/ml in methanol, acetonitrile, or 40% (v/v) methanol in dichloromethane. Immediately following vigorous vortexing, 20 μl of suspension is loaded onto a TFA-treated glass fiber filter that is placed in the top glass cartridge block of a Perkin-Elmer Applied Biosystems sequencer. A cartridge seal is placed over this, followed by the bottom glass cartridge block.

TABLE II
SEQUENCING AND CHROMATOGRAPHY PROGRAMS FOR CLEAVED AND RESIN-BOUND PEPTIDES[a]

Parameter	Cleaved peptides			Resin-bound peptides		
	Reaction cartridge	Conversion flask	HPLC gradient	Reaction cartridge	Conversion flask	HPLC gradient
Cycle 1	BEGIN-1	BEGIN-1	NORMAL-1	BGN REZ-1	BGN REZ-1	REZ-1
Cycle 2 to n	NORMAL-1	NORMAL-1	NORMAL-1	REZ-1	REZ-1	REZ-1
Temperature	48°	64°	55°[b]	53°	64°	55°[b]
HPLC gradient[c]		Time	% B		Time	% B
		0	12		0	12
		0.4	12		0.4	12
		18.0	38		38.0	63
		25.0	38		39.0	90
		25.1	90		41.0	90
		28.1	90			
Data collection time		29 min			42 min	

[a] For use with a Perkin-Elmer Applied Biosystems Division Model 477 sequencer as supplied by the manufacturer.
[b] Column temperature.
[c] Flow rate 0.21 ml/min. See Ref. 12 for solvent compositions and additional details.

TABLE III

RELATIVE ELUTION POSITIONS OF PTH-AMINO ACIDS ON HPLC[a]

Retention time (min)	PTH-amino acid	Notes
5.8	Tyr(P)	Stable
6.1	PTC-Gly	Stable, partial conversion product of ATZ-Gly
6.2	Asp	Stable, slight conversion to Asp derivative
7.1	Asn	~5–10% recovered as PTH-Asp
8.1	PTC-Ala	Stable, partial conversion product of ATZ-Ala
8.2	Ser	Typically ~0–20% recovered, see Ser derivative
8.8	Gln	~5–20% recovered as PTH-Glu
9.3	Thr	Typically ~5–25% recovered, see Thr derivative
9.7	Gly	Stable
10.5	Glu	Stable, slight conversion to Glu derivative
11.0	Dimethylphenylthiourea (DMPTU)	Edman chemistry by-product
12.5	Cys-acetamidomethyl (Acm)	Stable
13.0	Hyp(cis)	cis- or trans- Hydroxyl, Edman chemistry produces both
13.8	Ala	Stable
14.0	Hyp(trans)	cis- or trans- Hydroxyl, Edman chemistry produces both
14.6	His[b]	Stable but may not be completely extracted from sequencer
15.1	PTC-Met	Stable, partial conversion product of ATZ-Met
15.2	Ser derivative	Breakdown product of serine due to Edman chemistry
16.8	Cys-3-nitro-2-pyridinesulfenyl (Npys)	Sensitive to DTT, converts to Cys with successive cycles
17.4	Asp-O-allyl (OAl)	Partially stable, converts to Asp
17.7	Tyr	Stable

18.8	Ser derivative	Breakdown product of Ser due to Edman chemistry
19.0	PTC-Val	Stable, partial conversion product of ATZ-Val
19.1	Ser-allyloxycarbonyl (Aloc)	Partially stable, converts to Ser derivatives
19.8	Arg[b]	Stable but may not be completely extracted from sequencer
19.9	PTC-Lys	Stable, partial conversion product of ATZ-Lys
20.7	PTC-Leu	Stable, partial conversion product of ATZ-Leu
20.7	PTC-Ile	Stable, partial conversion product of ATZ-Ile
20.8	Tyr-dimethoxyphosphoryl [OP(OCH$_3$)$_2$]	Rapidly converted to Tyr(P) in first few cycles
21.1	Pro	Stable
21.9	Met	Usually stable but susceptible to oxidation
22.3	Val	Stable
22.6	Arg-diallyloxycarbonyl (Aloc)$_2$ derivative	Product of conversion of Arg(Aloc)$_2$, may be Arg(Aloc)
23.3	Thr-allyloxycarbonyl (Aloc)	Stable
23.7	Asp derivative	Associated with conversion to PTH-Asp
23.9	Diphenylthiourea (DPTU)	Edman chemistry by-product
23.9	Arg-4-toluenesulfonyl (Tos)	Stable
23.3	Glu derivative	Associated with conversion to PTH-Glu
24.1	Lys-allyloxycarbonyl (Aloc)	Stable
24.7	His-2,4-dinitrophenyl (Dnp)	Stable
25.1	Glu-O-allyl (OAl)	Stable, slight conversion to Glu derivative
25.3	Trp	Typically only partially recovered at inconsistent levels
25.8	Asp-O-tert-butyl (OtBu)	Very unstable, converts to Asp
26.1	Phe	Stable
26.1	Trp-N^{in}-formyl (CHO)	20–40% Recovered as PTH-Trp
26.2	Phe-p-nitrophenyl	Stable
26.2	Cys-allyl (Al)	Stable
26.7	Ile	Stable

(continued)

TABLE III (*continued*)

Retention time (min)	PTH-amino acid	Notes
27.1	Lys(ptc)	Stable, both amino groups are derivatized with PITC
27.4	Leu	Stable
27.5	Norleucine	Stable
27.8	His-*tert*-butoxymethyl (Bum)	Converted to PTH-His with successive cycles
28.2	Ser-benzyl (Bzl)	Partially recovered, remainder converted to Ser derivatives
28.5	Cys-allyloxycarbonyl (Aloc)	Stable
28.9	Arg-4-methoxy-2,3,6-trimethylbenzenesulfonyl (Mtr)	Partially recovered, converted to PTH-Arg with successive cycles
29.2	Arg-mesitylene-2-sulfonyl (Mts)	Typically 1–5% recovered as PTH-Arg
29.5	Cys-*tert*-butyl (*t*Bu)	Very unstable, converts to Cys
30.0	Asp-O-benzyl (OBzl)	~5–20% Recovered as PTH-Asp
30.1	Arg-diallyloxycarbonyl(Aloc)$_2$	Partially stable
30.5	Trp derivative	One of multiple Trp degradation products
30.6	His-3-benzyloxymethyl (3-Bom)	Stable
30.6	Thr-benzyl (Bzl)	Partially stable, products include PTH-Thr and Thr derivatives
31.2	Tyr-allyl (Al)	Stable
31.3	Glu-O-benzyl (OBzl)	~10–40% Recovered as PTH-Glu
31.7	Tyr-*Tert*-butyl (*t*Bu)	Very unstable, converts to Tyr
32.1	Cys-4-methoxybenzyl (Mob)	Partially stable, products include Ser derivatives
32.5	Thr-*tert*-butyl (*t*Bu)	Very unstable, converts to Thr
33.4	Asp-O-cyclohexyl (OcHex)	~5–20% Recovered as PTH-Asp
33.6	Lys-chlorobenzyloxycarbonyl (ClZ)	Partially stable, some recovered as PTH-Lys(ptc)
33.6	Lys-2-chlorobenzyloxycarbonyl (2ClZ)	Stable, perhaps 1–2% recovered as PTH-Lys(ptc)
33.8	Thr-benzyl (Bzl) derivative	Produced from Thr-benzyl (Bzl)

33.9	Lys-2,4-dinitrophenyl (Dnp)	Stable
34.0	Trp derivative	One of multiple Trp degradation products
34.3	Hydroxyproline-4-benzyl (4-Bzl)	Unstable, converts to Hypro
34.3	Glu-O-cyclohexyl (OcHex)	~10–40% Recovered as PTH-Glu
34.6	Trp derivative	One of multiple Trp degradation products
34.9	Tyr-2,6-dichlorobenzyl (2,6-diClBzl)	Stable, perhaps 1–5% recovered as PTH-Tyr
35.1	Cys-4-methylbenzyl (Meb)	Partially stable, products include Ser derivatives
35.4	Hydroxyproline-4-benzyl (4-Bzl)	Unstable, converts to Hypro
37.8	Lys-9-fluorenylmethyloxycarbonyl (Fmoc)	Converted to PTH-Lys(ptc) with successive cycles
38.6	Tyr-2-bromobenzyloxycarbonyl (2-BrZ)	Stable, perhaps 1–5% recovered as PTH-Tyr
38.7	Glu-O-9-fluorenylmethyl (OFm)	~10–40% Recovered as PTH-Glu
38.9	Trp derivative	One of multiple Trp degradation products

Side-Chain-Protected Amino Acids Completely or Nearly Completely Unstable to Edman Chemistry

His-allyloxycarbonyl (Aloc)	Converts to His
His-tosyl (Tos)	Converts to His
His-benzyloxycarbonyl (Z)	Converts to His
Asp/Glu-O-tert-butyl (OtBu)	Converts to Asp/Glu
Ser/Thr/Tyr/Hypro-tert-butyl (tBu)	Converts to Ser/Thr/Tyr/Hypro
Lys/His-tert-butyloxycarbonyl (Boc)	Converts to Lys/His
His/Cys/Asn/Gln-triphenylmethyl (Trt)	Converts to His/Cys/Asn/Gln
Arg-2,2,5,7,8-pentamethylchroman-6-sulfonyl (Pmc)	Converts to Arg
Arg-2,2,4,6,7-pentamethyldihydrobenzofuran-5-sulfonyl (Pbf)	Converts to Arg
Asn/Gln/Cys-2,4,6-trimethoxybenzyl (Tmob)	Converts to Asn/Gln/Cys

[a] Elution positions are relative and may vary somewhat depending on system and column. Some of these derivatives were originally done with different conditions and are now represented on a typical resin-bound sequencing cycle (Applied Biosystems "REZ" cycle). Table composed from material from Refs. 14, 15, 16, 17, and C. G. Fields, A. Loffet, S. A. Kates, and G. B. Fields, *Anal. Biochem.* **203**, 245 (1992).

[b] Position is variable depending on ionic strength and elution buffer composition.

The cartridge blocks are held together, inverted, and placed in the sequencer as usual. This method routinely gives initial yields that range from 300 to 1000 pmol for Fmoc synthesis where the resin substitution is 0.4 to 0.6 mmol/g. If desired, the number of beads can be decreased by removing individual beads using the point of a 23-gauge needle, a wetted micropipette tip, or microforceps under 5× magnification with extreme care to not puncture the glass fiber filter. If either the filter or cartridge seal are punctured, beads can enter the sequencer lines and cause them to clog. When using p-hydroxymethylphenoxymethyl polystyrene (HMP) resins, it may be useful to stain the resin beads using a 0.1% (w/v) bromophenol blue in methanol solution[6] because HMP resin remains relatively colorless after synthesis. However, 2-chlorotrityl and Rink amide resins have sufficient color after synthesis, typically dark yellow orange, as to not require staining.

Deletion or Preview Analysis

When a peptide of length n also contains some amount of a peptide that has had a residue deleted from its sequence (length $n - 1$) but is otherwise identical, sequence analysis will show the presence of residue j of the intended sequence at cycle $j - 1$ as depicted in the following mixture of two peptides which contains a population of peptide with Arg-4 deleted:

```
        1           5           10

              j-1 j j+1         n

        M D D R N E I P Q D

        M D D N E I P Q D
```

When the level of deletion is significant relative to the amount of main sequence, it can be easily seen by simple inspection of the chromatograms or of a tabulation of residue yields such as that shown in Table I. Qualitatively, it is not difficult to see that a significant deletion of Arg-4 has occurred due to the relatively high levels of Asn in cycle 4. In general, the most commonly encountered examples of deletions are those where relatively large levels of deletion occur at only a single cycle so that it is not usually necessary to determine the degree of preview more precisely. This can be expressed quantitatively as follows:

$$(\%P_x)_j = (100)X_{(j-1)}/X_{(j-1)} + X_j \tag{1}$$

where $(\%P_x)_j$ is the percentage of preview of residue x which is expected to be in the jth cycle from the N terminus, $X_{(j-1)}$ is the amount of residue x in cycle $j - 1$, and X_j is the amount of residue x in cycle j. Often this simple calculation provides enough information to proceed. If the amount of carryover is significant, an additional term, $X_{(j+1)}$, can be added to account for the amount of residue x that has been carried over to cycle $j + 1$:

$$(\%P_x)_j = (100)X_{(j-1)}/X_{(j-1)} + X_j + X_{(j+1)}. \qquad (2)$$

Equations (1) and (2) do not consider the contributions that background levels of amino acids may make to the calculated value of the preview. In most cases, especially with relatively short synthetic peptides, these background values are small and do not significantly contribute to the calculation or the practical outcome and subsequent decision on how to proceed. If carryover is relatively high throughout the sequence, it may not be possible to obtain values of preview for all such residues in a sequence since at some point preview and carryover will become indistinguishable.

In the sequence shown in Table I, Asp appears three times. Preview values for Asp-2 and Asp-10 can be calculated, but it is not possible to calculate the preview of Asp-3 in cycle 2 because Asp is also the main signal in cycle 2.

Table I compares the level of preview for a sequence run using both Eqs. (1) and (2). Equation (2) yields slightly lower preview values because the contribution of carryover in the subsequent cycle is also considered. Regardless of which method is used, the main conclusion from Table I is that a major deletion of approximately 20% occurs at residue 4. At first inspection, it also appears that there may be another few percent of deletion at cycle 5 and cycle 6. However, they do not occur to the extent seen in cycle 4 and may not be real because, although the calculated amount of preview appears to increase at these points, it is not sustained in later cycles. As the effect of each deletion should be sustained in subsequent cycles, this rise is probably due to another cause such as an anomalous yield of one residue compared to another. It must also be kept in mind that as sequencing proceeds further into a peptide that is not covalently bound to a solid support, the effect of washout can have an increasingly significant effect on the quantitation of preview. For example, if proportionately more washout occurs in a subsequent cycle compared to the cycle just before it, the calculated preview value will appear higher than it would if no washout occurred.

Special Considerations

Amino Terminus

Because the Edman chemistry requires a primary or secondary amino group for PITC coupling, chemical modification at the nitrogen can render the peptide impossible to analyze by Edman sequencing. Therefore, the Boc or Fmoc groups must first be removed, and a capping step, such as acetylation with acetic anhydride, should not be used. Capping at each cycle except the last will allow sequencing of the full-length peptide, but any termination peptides generated during synthesis will not be detected by Edman sequencing. Amino-terminal Gln residues (and sometimes Glu residues) can easily cyclize under acidic conditions (see below), so these residues should be avoided at the amino terminus if sequence analysis is contemplated.

Need for α-Amino Acids and Main-Chain Amide Linkage

Edman degradation was originally designed to sequence naturally occurring proteins that contain α-amino acids linked together by amide bonds. Both the α-configuration of the amino acid and the amide linkage at the carboxyl terminus of the residue being cleaved are required for successful PITC-mediated cyclization. When peptides contain non-α-amino acids or residue linkages other than amides, such as alkyl linkages, Edman sequence analysis will stop when they are encountered.

Cysteine-Containing Peptides

It is virtually impossible to detect Cys residues by Edman sequence analysis unless they have first been modified. There are a variety of reagents that are available for this purpose,[18] and their use is often a matter of personal preference. Gambee *et al.*[19] compare the efficacy of several such reagents for this purpose using a cleaved synthetic peptide and suggest that derivitization with either acrylamide or bromopropylamine may give superior results.

On-resin analysis of Cys-containing peptides proves to be somewhat problematic in those cases where the side-chain protecting group used in synthesis is unstable or only partially stable to the Edman chemistry. This includes most of those used with Fmoc chemistry such as the Cys(Trt) group and some of those used with Boc chemistry such as Cys(Mob) or Cys(Meb).

If Cys residues have been oxidized to form disulfide bonds, Edman sequence analysis will proceed through the Cys positions. However, the first Cys of the disulfide encountered will not be extracted from the sequencer

[18] M. W. Crankshaw and G. A. Grant, *in* "Current Protocols in Protein Science" (J. E. Coligan, B. M. Dunn, H. L. Pleogh, D. W. Speicher, and P. T. Wingfield, eds.), p. 15.1.1. Wiley, New York, 1996.

[19] J. Gambee, P. C. Andrews, K. DeJongh, G. Grant, B. Merrill, S. Mische, and J. Rush, *in* "Techniques in Protein Chemistry VI" (J. W. Crabb, ed.), p. 209. Academic Press, San Diego, 1995.

because it is still covalently coupled to the remaining Cys. This will result in a blank cycle at this position. When the second Cys is finally encountered, the product is PTH-bis(cystine) that elutes very close to the position for PTH-Tyr. In addition, PTH-bis(cystine) is labile to the sequencing chemistry and only partially recovered. Recovery appears to be best when the Cys residues appear early in the sequence.[20] Reports of 20–30% yield within the first 5–10 cycles are common.

Formation of Cyclic Imides

Both Asn and Gln residues, and to a much lesser extent Asp and Glu residues, can form cyclic imides that have significant effects on sequence analysis. The mechanism and consequences of each are very different and are discussed separately.

The length of the Asn (Asp) side chain makes it particularly suited for interaction with the main chain amide group of the following residue to form a cyclic imide (succinimidyl) intermediate under basic conditions.[21,22] When this occurs there are several consequences that can be observed as depicted in Fig. 3. The formation of the cyclic intermediate is generally rate limiting. At this point, racemization about the Asn/Asp α-carbon can also occur via resonance stabilized epimerization of the succinimidyl ring. The succinimidyl ring is a relatively short-lived species that can hydrolyze to generate either an Asp-containing peptide with normal α peptide linkage or an Asp-containing peptide with β peptide linkage (iso-Asp). An example of the relative distribution of the resultant species from selected synthetic peptides is shown in Table IV. Note that formation of the iso-Asp peptide is favored by about 3 to 1 and that a significant degree of racemization can occur. In general, the Asn-containing peptides exhibit faster rates of degradation than do the Asp-containing peptides. The most susceptible peptides are those where Gly follows the Asp/Asn residue, and the half-life of degradation generally increases as the side chain gets larger and less polar. Because the amide nitrogen of Asn peptides is lost during imide formation, the linear products are always the Asp-containing peptide. Thus, conversion of the iso-Asp peptide back to the imide is very slow.

The practical consequence of the formation of iso-Asp peptides is that Edman degradation will stop when they are encountered because the β-Asp peptide linkage does not favor the ring cyclization step necessary for removal of the amino-terminal residue in Edman sequencing (See Fig. 1). Table V illustrates what happens to the sequencing signal when iso-Asp

---

[20] M. Haniu, C. Acklin, W. C. Kenney, and M. F. Rohde, *Int. J. Pept. Protein Res.* **43**, 81 (1994).
[21] T. Geiger and S. Clarke, *J. Biol. Chem.* **262**, 785 (1987).
[22] R. C. Stephenson and S. Clarke, *J. Biol. Chem.* **264**, 6164 (1989).

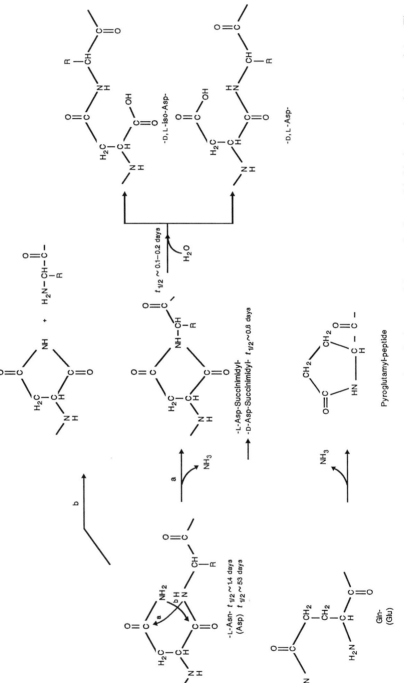

FIG. 3. Formation of cyclic imides in peptides. The top two reactions depict potential routes for internal imide formation of Asn and Asp. The bottom reaction depicts the potential cyclization of Gln, and occasionally Glu, when they occur at the amino terminus of a peptide.

TABLE IV
DEGRADATION OF SYNTHETIC PEPTIDES BY CYCLIC IMIDE FORMATION

Peptide sequence[a]	Rate of degradation via cyclic imide,[b] $t_{1/2}$ (days)	% Relative product distribution		
		Iso-Asp	Normal-Asp	Peptide cleavage[c]
-Asn-Gly-	1.4	73 (L-, 53) (D-, 20)	27 (L-, 20) (D-, 7)	0
-Asn-Ser-	8.0			
-Asn-Ala-	20.2			
-Asp-Gly-	53.0	79	21	0
-Asn-Leu-	70.0	64	22	14
-Asn-Pro-	106	0	0	100
-Asp-Ser-	168.0			
-Asp-Ala-	266			
-Asp(Me)-Gly-	0.0026			
-Asp(Me)-Ser-	0.0096			
-Asp(Me)-Ala-	0.027			
-Asp(Me)-Phe-	0.086			

[a] Residues X and Y, respectively, within the sequence of the synthetic peptide Val-Tyr-Pro-X-Y-Ala.
[b] pH 7.4, 37°.
[c] The cleavage product was identified as Val-Tyr-Pro-Asx.

residues are encountered both as part of a mixture and when it is the only peptide present. Sample 1 contains an approximately equimolar mixture of an iso-Asp linkage and a normal Asp linkage, whereas sample 2 is completely in the iso-Asp form. In both cases, sequencing stops when the iso-Asp is encountered but continues through the normal Asp linkage.

A second reaction that can be encountered in Asn-containing peptides is also illustrated in Fig. 3. If the side-chain amide nitrogen attacks the main chain carbonyl carbon rather than the alternate pathway, chain cleavage can result. Because this is generally a much slower process than iso-Asp formation, it is not often seen with freshly synthesized peptides. However, as shown in Table IV, it can be a significant consequence over time in peptides where iso-Asp formation is not favored as in the case of Asn-Pro. Because the base-catalyzed reaction requires extraction of the NH proton, it does not occur with the imino acid Pro in the second position. With time, the much slower alternative reaction is favored.

Of particular concern in peptide synthesis is the observation that peptides containing esters of Asp undergo cyclic imide-mediated degradation at a much faster rate than those with the free acid or amide (Asn) side chain (see Table IV). This has potential implications in the use and selection

TABLE V
SEQUENCE ANALYSIS OF ISOASPARTIC
ACID-CONTAINING PEPTIDES[a]

Cycle	Amino acid residue	Sample 1 (pmol)	Sample 2 (pmol)
1	Tyr	15.87	27.18
2	Glu	11.64	16.80
3	Val	28.80	29.00
4	Asp	18.80	21.22
5	Val	28.82	25.98
6	Ser	20.49	6.24
7	Ala	16.21	20.00
8	Gly	11.30	12.08
9	Asn	6.63	ND[b]
10	Gly	7.38	ND
11	Ala	7.21	ND
12	Gly	5.81	ND
13	Ser	1.21	ND
14	Val	6.23	ND
15	Gln	2.87	ND

[a] Reproduced from P. L. Derby, K. H. Aoki, V. Katta, and M. F. Rohde, in "Techniques in Protein Chemistry VII" (D. R. Marshak, ed.), p. 109. Academic Press, San Diego, 1996.
[b] ND, not detected.

of side-chain protecting groups in peptide synthesis.[23–26] In Fmoc synthesis it has also been shown that piperidine can react with the succinimidyl ring to form both α- and β-piperidides.[26,27] This results in an adduct that is 67 mass units larger than expected and can be easily identified by MS. Like the iso-Asp product, the α-piperidide will block further sequence analysis when it is reached. However, Edman analysis will proceed through the β-piperidide, yielding the PTH-Asp(β-piperidide) product.

Although there have been isolated reports, Gln- or Glu-containing peptides are not usually observed to degrade in a manner similar to Asn/Asp by way of cyclic imide formation. Under acidic conditions, however, Gln,

[23] G. B. Fields and R. L. Noble, Int. J. Pept. Protein Res. **35**, 161 (1990).
[24] J. P. Tam, M. W. Rieman, and R. B. Merrifield, Pept. Res. **1**, 6 (1988).
[25] J. Van Rietschoten, P. Fourquet, J. M. Sabatier, and C. Grainier, in "Peptides" (D. Theodoropoulous, ed.), p. 179. de Gruyter, Berlin, 1986.
[26] J. L. Lauer, C. G. Fields, and G. B. Fields, Lett. Pept. Sci. **1**, 197 (1994).
[27] M. Quibell, D. Owen, L. C. Packman, and T. Johnson, J. Chem. Soc., Chem. Commun., 2343 (1994).

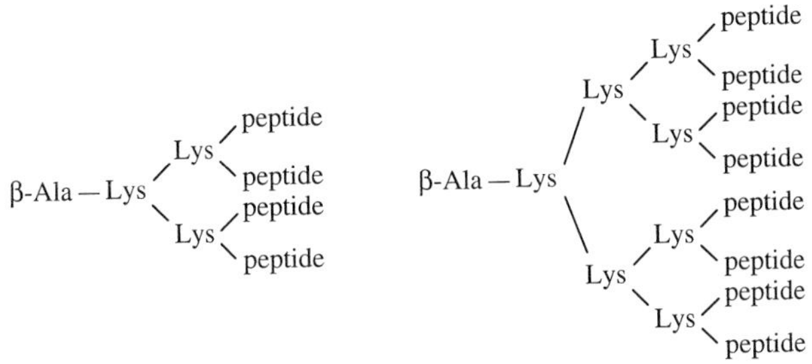

Fig. 4. Schematic representation of the general structure of four- and eight-branched multiple antigenic peptides (MAPs). During synthesis, the peptide is attached to the insoluble support through the carboxyl terminus of a spacer residue such as β-Ala, and synthesis proceeds as usual in the C- to N-terminal direction.

and to a much lesser extent Glu, at the amino terminus of a peptide can cyclize to a pyroglutamate structure (Fig. 3), which will render the peptide refractory to sequence analysis.[22] In addition, base-catalyzed pyroglutamate formation of amino-terminal Fmoc-Glu(Bzl) has been reported to occur during piperidine deprotection.[25]

Multiple Antigenic Peptides

Multiple antigenic peptides (MAPs) were first described by Tam and co-workers[28] as a method for producing antigenic peptides for the production of specific antibodies. These peptides are assembled on a branched Lys core attached to the solid-phase support. The most common forms of MAP peptides used are either tetramerically or octamerically branched, containing either three or seven Lys residues, respectively, allowing for either four or eight identical peptides to be assembled simultaneously on the α- and ε-amines of the terminal Lys of the branched Lys core. The general structure of MAPs are shown in Fig. 4. Peptides synthesized on this scaffolding have sufficient molecular weight and localized concentration of epitopes by themselves to produce an immune response without the need to be conjugated to a carrier protein to enhance their immunogenicity.

The successful use of MAPs to produce antipeptide antibodies has been previously described.[29–31] With the addition of orthogonal protection

---

[28] D. N. Posnett, H. McGrath, and J. Tam, *J. Biol. Chem.* **263,** 1719 (1988).

[29] J. Tam, *Proc. Natl. Acad. Sci. U.S.A.* **84,** 5409 (1988).

[30] C. Auriault, I. Woloczuk, H. Gras-Masse, M. Maguerite, M. D. Boulanger, A. Capron, and A. Tartar, *Pept. Res.* **4,** 6 (1991).

[31] C. Y. Wang, D. J. Looney, M. L. Li, A. M. Walfield, J. Ye, B. Hosein, J. P. Tam, and F. Wong-Stall, *Science* **254,** 285 (1991).

chemistries during synthesis, MAPs have also been used to produce dual-epitope antibodies.[32]

Characterization of MAPs has proved to be difficult. Reversed-phase HPLC under standard conditions (0.1% TFA with a linear 1%/min acetonitrile gradient) generally yields a broad (2–5 min) peak. Electrospray mass spectrometry (ESMS) of MAPs also tends to yield broad, ambiguous spectra that do not allow for a precise assignment of the molecular weight. It is generally believed that this is due to a higher than normal incidence of deletion sequences and incomplete side-chain deprotection as a result of the highly localized concentration of growing peptide chain on the polylysine core. In this respect, four-branched MAPs tend to give better spectra than eight-branched MAPs. Mass spectrometric analysis is thus compromised by the presence of four to eight peptide chains per molecule, each of which may have only a small percentage of modification at any particular residue but in the aggregate contribute to broad spectra. This feature of MAPs usually does not tend to compromise their ability to form antigens to the proper peptide, however, because the correct sequence is usually present in high enough concentration that a significant amount of the specific antibody is produced among the polyclonal population.

In this respect, Edman sequencing of a MAP may provide a more useful analysis. However, Edman analysis of a MAP will not necessarily show as clearly the location of individual synthesis problems as it would for a single free peptide. Rather, it will tend to give the average of the multiple chains in the MAPs. Overall, it will identify whether the peptide was basically assembled correctly, and it will also identify any major areas in the peptide where incomplete couplings have occurred. In fact, the sequence shown in Table I is that of an eight-branched MAP. We have found in practice that although there may be some minor problems on individual chains, if the sequence analysis indicates a reasonable sequence, one may conclude that the MAP is sufficient for the production of the desired antibody.

Summary

Sequence analysis of synthetic peptides using Edman chemistry can be very useful for the elucidation of certain types of synthetic problems, such as residue deletions and the presence of common stable derivatives, and for following the progress of the synthesis itself. However, it can also be a relatively poor technique for assessing quantitative aspects and the type and degree of adduct formation that arise from the synthetic chemistry. For these latter considerations, techniques such as mass spectrometry can

---

[32] J. Tam and Y. Lu, *Biochemistry* **86**, 9084 (1989).

often give more precise and informative data about the integrity of a synthetic peptide. Thus, sequence analysis is best applied judiciously and then used in combination with other methods. Furthermore, proper interpretation of the results of sequence analysis of synthetic peptides relies on a thorough knowledge of the sequencing process.

[18] Amino Acid Analysis

By ALAN J. SMITH

Introduction

Amino acid analysis is one of several methods that are available for characterizing synthetic peptides. It is certainly the oldest and best documented of the methods that are in common use. Other characterization methods include high-performance liquid chromatography (HPLC), capillary electrophoresis (CE), automated Edman protein sequencing, and mass spectrometry. Of these, mass spectrometry is probably the most informative, but it lacks the ability to give quantitative information. In contrast, the primary value of amino acid analysis today is in the determination of absolute peptide content. It can also provide useful information about peptide composition and stoichiometry. A third application is in monitoring a problem synthesis on-the-fly. Most automated peptide synthesizers have the capability to monitor cycle-to-cycle coupling efficiencies. However, these devices offer a very limited view of how the synthesis is actually proceeding. Questions often arise during the synthesis as to the efficiency of coupling of a particular addition. If a problem occurs early in an extended synthesis protocol, it may be more efficient to terminate the synthesis and remake the peptide rather than to continue the synthesis. The status of the synthesis can be rapidly determined by performing on-resin hydrolysis and subsequent amino acid analysis, all of which can be carried out on a same-day basis. The robust nature and high reproducibility of automated amino acid analysis make it a method of choice for peptide validation in the pharmaceutical industry. It has also found extensive use in the research environment for "calibrating" other quantitation methods, for example, OD_{280}, fluorescence, and Bradford assays.

The basic methodology for quantitative amino acid analysis was first developed in the early 1950s and has changed relatively little since that time.[1] Amino acid analysis comprises three separate but highly interrelated functions: sample preparation and hydrolysis, separation and derivatization, and data analysis and interpretation. The changes that have occurred over time have centered around the derivatization chemistry and the subsequent separation of those derivatives. The original methodology requires the separation of the free amino acids by ion-exchange chromatography and their subsequent visualization by reaction with a chromophore. This method is commonly referred to as postcolumn amino acid analysis. Methods have been developed to derivatize the amino acids directly in the hydrolyzate and subsequently separate the derivatives by reversed-phase HPLC. This method is commonly referred to as precolumn amino acid analysis.

Sample Preparation and Hydrolysis

When absolute peptide content is required, the sample should be dried to constant weight and accurately weighed into a clean hydrolysis tube. If composition alone is required, then only an approximate initial estimate of the quantity of peptide is required. Peptides can be submitted for analysis as either solids or liquids, but they will need to be dried down prior to hydrolysis. If they are in solution, then low salt or water are the preferred solvents. However, they can also be prepared in standard buffers, salts less than 0.2 M, detergents, or a matrix (e.g., bead, membrane, gel). If matrices other than low salt or volatile solvents are used, the postcolumn chemistry is the method of choice. The reason for this is that the amino acids are derivatized postseparation, and, consequently, there is less chance of matrix molecules interfering with the derivatization chemistry. Samples dissolved in ammonium salts or urea are unsuitable for analysis by either chemistry. The large quantities of ammonia that are generated during hydrolysis will react with the derivatization reagent and complicate the chromatographic separation. Quantities of peptide in the low microgram to low milligram range are optimal for analysis, although the precolumn chemistries can readily handle submicrogram amounts of sample.

The standard hydrolysis conditions for amino acid analysis are 6 N hydrochloric acid, 110° for 24 hr, in the absence of oxygen. Other variations include 160° for 45 min or 150° for 90 min. These hydrolysis times can be reduced even further if microwave hydrolysis is employed.[2] The hydrolysis can be performed in the liquid or gaseous phase, and commercially available instrumentation is available for the complete hydrolysis process (see Re-

[1] S. Moore and W. H. Stein, *Methods Enzymol.* **6,** 819 (1963).
[2] L. B. Gilman and C. Woodward, *in* "Current Research in Protein Chemistry" (J. Villafranca, ed.), p. 23. Academic Press, San Diego, 1990.

agents). A detailed discussion of hydrolysis protocols can be found in Davidson.[3]

A special application is the hydrolysis of resin-bound peptides for monitoring problem syntheses. In this case, the 6 N HCl is replaced with 12 N HCl/propionic acid, 1 : 1 (v/v), and the hydrolysis is performed for 90 min at 150° as described by Westfall and Hesser.[4]

Reagents

Hydrolysis tubes: 10 × 75 mm Pyrex, pyrolyzed by heating for 2 hr at 500°

Pico Tag workstation and hydrolysis chamber (Waters, Milford, MA)

6 N HCl in 1-ml vials (Pierce, Rockford, IL)

12 N HCl/propionic acid in 1-ml vials (Pierce)

After the hydrolysis is completed, the chamber is allowed to cool, the individual sample tubes are removed, and the contents evacuated to dryness. The dried hydrolyzate is resuspended in an appropriate sample diluent and either applied directly, or derivatized and applied, to the separation column. The sample diluent contains a known concentration of an internal standard, commonly norleucine (Nle). The resuspension volume must be accurately recorded if absolute peptide content is required.

Separation and Derivatization

In the postcolumn chemistry, the amino acids are separated by ion-exchange chromatography on a sulfonated divinylbenzene/polystyrene resin. They are eluted from the column with a stepwise salt and pH gradient and derivatized with ninhydrin for detection and quantitation. This scheme is embodied in an automated amino acid analyzer that is commercially available (Beckman Instruments, Fullerton, CA). The primary amine derivatives are monitored at 570 nm and the secondary amine derivatives at 440 nm. The average separation time is approximately 60 min, with an injection-to-injection time of approximately 90 min. This chemistry is essentially unaffected by the matrix material that might be present in the original sample. An example of a chromatographic separation of amino acid standards is shown in Fig. 1.

There are several precolumn chemistries that are available for amino acid analysis. They all require the direct derivatization of the amino acids in the hydrolyzate following hydrolysis and the subsequent separation and

---

[3] I. Davidson, *in* "Methods in Molecular Biology, Volume 64: Protein Sequencing Protocols" (B. J. Smith, ed.), p. 119. Humana Press, Totowa, New Jersey, 1996.

[4] F. Westfall and H. Hesser, *Anal. Chem.* **61,** 610 (1974).

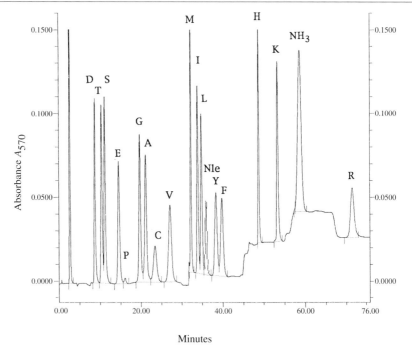

Minutes

FIG. 1. Separation of standard amino acid mixture by ion-exchange chromatography and ninhydrin as the derivatization reagent. Nle, Norleucine; NH₃, ammonia.

identification of the derivatives by reversed-phase HPLC. These methods are inherently more sensitive than the postcolumn method because there is minimal dilution of the sample with the derivatizing agent. In addition, two of the methods employ fluorescence detection, which adds a 5- to 10-fold increase in sensitivity over UV detection methods.

Only one of the methods, derivatization with *o*-phthalaldehyde (OPA), is embodied in an automated instrument that is commercially available (Hewlett Packard, Palo Alto, CA). The fluorescent OPA derivatives are excited at 360 nm, and the emission is monitored at 455 nm, as described by Benson and Hare.[5] The other two derivatization chemistries that are the most popular use essentially custom-built analyzer systems. They consist of a multifunction autosampler for the derivatization process, a binary gradient HPLC system for separation of the derivatized amino acids, and an appropriate detector. The oldest of these two chemistries uses phenyl isothiocyanate as the derivatizing agent and is known as the PTC chemistry

[5] J. R. Benson and P. E. Hare, *Proc. Natl. Acad. Sci. U.S.A.* **72,** 619 (1975).

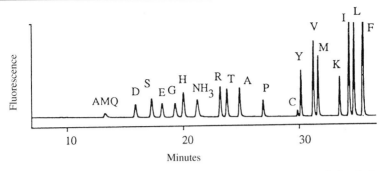

Fɪɢ. 2. Separation of standard amino acid mixture by reversed-phase HPLC after derivatization with AQC. NH₃, ammonia; AMQ, 6-aminoquinoline, a by-product of the reaction.

(see Cohen and Strydom[6] and an excellent review by Molnar-Perl[7]). The PTC derivatives are monitored in the UV at 254 nm. The most recently developed chemistry utilizes aminoquinolylhydroxysuccinimidyl carbamate (AQC), and fluorescence is monitored at 395 nm after excitation of the AQC derivatives at 245 nm. Appropriate details and methodology can be found in Cohen and Michaud[8] and Strydom and Cohen.[9] A typical separation profile of a mixture of amino acid standards using this chemistry is shown in Fig. 2.

All precolumn methods derivatize the amino acids directly in the hydrolyzate mixture. The presence of hydrolyzed matrix material, for example, salts and buffers, can adversely affect the recovery of some amino acids. However, precolumn methods are capable of excellent accuracy and precision when analyzing HPLC-purified peptides.

Data Analysis and Interpretation

The data report from a typical amino acid analysis lists the amino acids in their order of elution and indicates the quantity of each that was present in the aliquot, for example, nanomoles per 50 μl. The variation between repeat analyses of the same hydrolyzate should be better than 1%, and the variation between repeat hydrolyses of the same sample should be better than 5%. An example of a typical amino acid analysis report is shown in Table I. In this example, the peptide sequence is

Ala-Asn-Glu-Arg-Ala-Asp-Leu-Ile-Ala-Tyr-Leu-Lys-Gln-Ala-Thr-Lys

[6] S. A. Cohen and D. J. Strydom, *Anal. Biochem.* **174,** 1 (1988).
[7] I. Molnar-Perl, *J. Chromatogr.* **661,** 45 (1994).
[8] S. A. Cohen and D. P. Michaud, *Anal. Biochem.* **209,** 279 (1993).
[9] D. J. Strydom and S. A. Cohen, *Anal. Biochem.* **222,** 19 (1994).

TABLE I
SAMPLE AMINO ACID ANALYSIS DATA[a]

Amino acid	A (theoretical composition)	B (nmol/50 μl)	C (corrected nmol/50 μl)	D (ng/50 μl)	E (μg/sample; volume 250 μl)	F (unit peptide content/aliquot)
Asx	2	0.72	0.84	93.5	0.47	0.42
Thr	1	0.35	0.41	41.4	0.21	0.41
Glx	2	0.77	0.90	115.0	0.58	0.45
Ala	4	1.40	1.63	115.0	0.58	0.40
Ile	1	0.35	0.41	46.3	0.23	0.41
Leu	2	0.73	0.85	96.0	0.48	0.43
Nle		0.43	0.50			
Tyr	1	0.36	0.42	68.5	0.34	0.42
Lys	2	0.79	0.92	118.0	0.59	0.46
Arg	1	0.33	0.38	60.8	0.30	0.39
Total				754.5	3.77	
Average peptide content per aliquot						0.42
Weight of peptide in sample					3.77 μg	3.78 μg

[a] Column A, from known sequence of peptide; column B, from amino acid analysis data; column C, normalized to 0.50 nmol Nle; column D, calculated by multiplying (C) by molecular weight of each amino acid, less water (18 Da); column E, corrected for volume of resuspended hydrolyzate; and column F, calculated by dividing corrected quantity (C) by the expected composition (A).

and has a molecular mass of 1807.5 as determined by matrix-assisted laser desorption (MALDI) mass spectrometry. Column *B* (Table I) represents the amount of each amino acid that was recovered from the analysis. To obtain stoichiometry, each value is divided by the averaged lowest common denominator (excluding norleucine). This gives the number of moles of each amino acid that is present in the peptide. This can be compared to the expected composition as shown in column *A* (Table I) and will readily verify the fidelity of synthesis. *Note:* Gln and Asn are deamidated during hydrolysis and are quantitated as Glu and Asp.

If absolute quantitation is required, the amino acid analysis data has to be corrected for the amount of sample that was analyzed. The quantities of each amino acid have to be converted to weights for comparison with the weight of the original sample. The Nle concentration in the sample diluent was adjusted such that a 50-μl sample injection onto the analyzer would give 0.50 nmol Nle. The actual recovery of Nle in the analysis was 0.43 nmol (Table I, column *B*), which means that less than 50 μl of sample was analyzed. These data can be corrected by normalizing the recoveries to 0.50 nmol Nle (column *C*, Table I). These values now have to be converted to the actual weights of amino acids. This is achieved by multiplying the values in column *C* (Table I) by the molecular weight of each amino acid, less water (18 Da). The weight of each amino acid (column *D*, Table I) is summed to give the total peptide weight in the aliquot that was analyzed. The total weight of peptide in the sample is obtained by dividing the aliquot volume (50 μl) into the total sample volume (in this example 250 μl) and multiplying the quantities in column *D* (Table I) by that factor (5) to give column *E* (Table I). The sum of column *E* (Table I) is the absolute weight of peptide in the original sample. This method for calculating absolute peptide content can also be used for quantitating proteins.

Unfortunately, not all amino acids are recovered quantitatively under standard hydrolysis conditions. Cys and Trp are totally destroyed and can be quantitated by separate hydrolysis procedures. Cys can be quantitated as cysteic acid after performic acid oxidation of the peptide prior to hydrolysis. Trp can be approximated by hydrolysis with 4 *M* methanesulfonic acid. Ser and Thr can be partially destroyed (up to 10–15%), and some peptide bonds are incompletely hydrolyzed, for example, Ile-Ile and Val-Val (up to 10%). Consequently, the "stoichiometry" method can be used to calculate peptide content. A number of amino acids are stable to these hydrolysis conditions (e.g., Asx, Glx, Gly, Ala, Tyr, Phe, Lys, Arg). The recoveries of these amino acids (column *C*, Table I) are divided by their expected frequency in the peptide (column A, Table I). These lowest common denominator values are averaged to give the unit peptide content in the aliquot. This value is corrected for dilution (5-fold) and multiplied by the molecular weight of the peptide to yield the absolute quantity of the peptide in the original sample. This method is not suitable for determining absolute quantity of a protein. The high copy number for amino acids in the average protein will have to be divided into a small analysis number. This can cause a large fluctuation in the lowest common denominator value.

Summary

Amino acid analysis of synthetic peptides is a robust and highly reliable analytical technique. For HPLC-purified peptides, absolute recovery accuracies of better than 5% are readily attainable. The presence of scavengers and incompletely cleaved protecting groups can have a significant impact on the quality of data that is obtained from crude samples. It should still be possible to obtain a reasonably accurate estimate of the quantity of the synthetic product, but the recovery of some amino acids can be compromised at times. When amino acid analysis is used in conjunction with mass spectrometry and a quantitative separation technique such as HPLC or CE, it can contribute significantly to the complete characterization of a synthetic peptide preparation.

Acknowledgment

    I wish to acknowledge the technical assistance of Adrianne Kishiyama in preparing the figures and table.

[19] Analysis of Synthetic Peptides by High-Performance Liquid Chromatography

By COLIN T. MANT, LESLIE H. KONDEJEWSKI, PAUL J. CACHIA,
OSCAR D. MONERA, and ROBERT S. HODGES

Introduction

High-performance liquid chromatography (HPLC) has proved extremely versatile for the isolation/purification of peptides varying widely in their sources, quantity, and complexity. Indeed, high-performance chromatographic techniques are particularly suited to the purification of a single peptide from the kind of complex peptide mixtures encountered following solid-phase peptide synthesis, where impurities are usually closely related to the peptide of interest (deletion, terminated, or chemically modified peptides), perhaps missing only one amino acid residue, and, hence, may be difficult to separate.

It is not the purpose of this article to present a comprehensive review of HPLC of peptides; there is a host of relevant material readily accessible in the literature, including an extensive article published in the *Methods in Enzymology* series.[1] Several useful articles and reviews on HPLC of peptides can also be found in Refs. 2–4; in addition, Refs. 5 and 6 represent useful resource books in this area. A comprehensive and practical presentation of the topic is supplied by Ref. 7.

Following a brief overview of HPLC of peptides, this article focuses on HPLC applications of particular interest to researchers utilizing solid-phase synthesis methods, for example, problem-solving approaches to difficult peptide separations, HPLC monitoring of formation of desired product, and HPLC as a diagnostic tool to detect unwanted side-chain modification. The use of reversed-phase chromatography (RP-HPLC), the most important mode of HPLC for synthetic peptide purification, is stressed; in addition, the exciting potential of a novel mixed-mode HPLC technique, hydrophilic interaction/cation-exchange chromatography (HILIC/CEC), is demonstrated.

Overview of High-Performance Liquid Chromatography of Peptides

Major Modes of High-Performance Liquid Chromatography Used in Peptide Separations

The major modes of HPLC employed in peptide separations take advantage of differences in peptide size (size-exclusion HPLC, or SEC), net charge (ion-exchange HPLC, or IEC), or hydrophobicity (RP-HPLC). Within these modes, mobile-phase conditions may be manipulated to maximize the separation potential of a particular HPLC column. In addition,

[1] C. T. Mant and R. S. Hodges, *Methods Enzymol.* **271**, 3 (1996).
[2] J. G. Dorsey, J. P. Foley, W. T. Cooper, R. A. Barford, and H. G. Barth, *Anal. Chem.* **64**, 353R (1992).
[3] C. T. Mant, N. E. Zhou, and R. S. Hodges, *in* "Chromatography" (E. Heftmann, ed.), 5th ed., p. B75. Elsevier, Amsterdam, 1992.
[4] C. Schöneich, S. Karina Kwok, G. S. Wilson, S. R. Rabel, J. F. Stobaugh, T. D. Williams, and D. G. Vander Velde, *Anal. Chem.* **65**, 67R (1993).
[5] K. M. Gooding and F. E. Regnier (eds.), "HPLC of Biological Macromolecules: Methods and Applications." Dekker, New York, 1990.
[6] M. T. W. Hearn (ed.), "HPLC of Proteins, Peptides and Polynucleotides: Contemporary Topics and Applications." VCH, New York, 1991.
[7] C. T. Mant and R. S. Hodges (eds.), "HPLC of Peptides and Proteins: Separation, Analysis and Conformation." CRC Press, Boca Raton, Florida, 1991.

a combination of separation techniques (SEC, IEC, and RP-HPLC; multidimensional or multistep HPLC) may be required for efficient peptide purification. The reader is directed to Refs. 8–10 for selected practical examples of approaches to multistep HPLC of peptides, a brief review of which can be found in Ref. 11.

Size-Exclusion Chromatography. Until the early 1990s, SEC has only offered very limited efficiency for peptides, owing to the fact that current size-exclusion packings are designed mainly for protein separations. Thus, the range of required fractionation for peptides (~100–6000 Da) tends to be at the low end of the fractionation range of such columns, and is of little use to the problem of separating the peak of interest from a synthetic peptide crude mixture when the impurities may only be a residue smaller than the desired product. A size-exclusion column designed specifically for the separation of peptides in the molecular weight range 100–7000 (Superdex Peptide, Pharmacia, Piscataway, NJ) has been introduced and promises to raise the profile of this HPLC mode for peptide separations. In addition, size-exclusion columns are of value in the early stages of a multistep peptide purification protocol.[8,11] Nevertheless, SEC remains the least effective major HPLC mode for purification of products arising from solid-phase peptide synthesis.

Ion-Exchange Chromatography. Both anion-exchange (AEC)[1,3,5,7,11,12] and cation-exchange (CEC)[1,3,5,7–14] HPLC have proved useful for fractionation of crude peptide mixtures, particularly where charged species have been deleted from the product of interest. As a general rule, most modern ion-exchange packings are capable of separating charged species differing by only a single net charge,[1,3,7] a not infrequent occurrence in solid-phase peptide synthesis.

If a choice must be made concerning the type of ion-exchange column for general peptide applications, it is recommended that the researcher acquire a strong cation-exchange column. The utility of such a column lies in its ability to retain its negatively charged character (because of the strongly acidic sulfonate functionality characteristic of such packings) in the acidic and neutral pH range. Most peptides are soluble at low pH, where the side-chain carboxyl groups of acidic residues as well as the free C-terminal α-carboxyl group are protonated (i.e., uncharged), thereby

[8] C. T. Mant and R. S. Hodges, *J. Chromatogr.* **326,** 349 (1985).
[9] J. Eng, C.-G. Huang, Y.-C. Pan, J. D. Hulmes, and R. S. Yalow, *Peptides* **8,** 165 (1987).
[10] N. Lundell and K. Markides, *Chromatographia* **34,** 369 (1992).
[11] C. T. Mant and R. S. Hodges, *J. Liq. Chromatogr.* **12,** 139 (1989).
[12] P. C. Andrews, *Pept. Res.* **1,** 93 (1988).
[13] A. J. Alpert and P. C. Andrews, *J. Chromatogr.* **443,** 85 (1988).
[14] T. W. L. Burke, C. T. Mant, J. A. Black, and R. S. Hodges, *J. Chromatogr.* **476,** 377 (1989).

emphasizing any basic, positively charged character of the peptides. Ion-exchange columns are particularly useful in a multistep separation protocol, particularly prior to a final RP-HPLC purification and desalting step.

Reports[15,16] have described a novel approach to the application of IEC, specifically CEC, to peptide purification, where advantage is taken not only of the charged nature of the ion-exchange packing, but also of hydrophilic interactions between peptides, and the hydrophilic ion-exchange matrix. This mixed-mode approach (HILIC/CEC) shows great potential as an HPLC method complementary to RP-HPLC for peptide purifications[15,16] and is featured later in this article.

Reversed-Phase High-Performance Liquid Chromatography. The RP-HPLC mode is overwhelmingly the HPLC mode of choice for peptide separations.[1,2,5,7] It is generally superior to other modes of HPLC with respect to both speed and efficiency; in addition, the availability of volatile mobile phases makes it ideal for both analytical and preparative separations. Although excellent resolution of peptide mixtures may be obtained at acidic or neutral pH, the majority of researchers have carried out RP-HPLC below pH 3.0 to take advantage of acidic volatile mobile phases [particularly aqueous trifluoroacetic acid (TFA)/acetonitrile (CH_3CN) systems] available to the researcher,[1,3,7,17] an important concern when carrying out preparative purification of synthetic peptides or when using RP-HPLC as a final desalting/purification step in a multicolumn protocol. In addition, acidic pH values prevent nonspecific ionic interactions between positively charged residues and any underivatized silanol groups (negatively charged above pH values of about 3–4) on silica-based packings.[18–20] still the most frequently employed packing support for RP-HPLC separations of peptides.

Out of the three major modes of HPLC, RP-HPLC offers the widest scope for manipulation of mobile-phase and stationary-phase characteristics to improve peptide separations. Such manipulations may include varying the ion-pairing reagent (a classic demonstration can be found in Ref. 17), functional groups on the stationary phase, pH, packing support, and temperature; simply varying the gradient rate may also have a profound effect on peptide resolution.[21] As noted previously, there is a considerable volume of literature available to the researcher detailing approaches to peptide

[15] B.-Y. Zhu, C. T. Mant, and R. S. Hodges, *J. Chromatogr.* **548,** 13 (1991).
[16] B.-Y. Zhu, C. T. Mant, and R. S. Hodges, *J. Chromatogr.* **594,** 75 (1992).
[17] D. Guo, C. T. Mant, and R. S. Hodges, *J. Chromatogr.* **386,** 205 (1987).
[18] J. L. Meek, *Proc. Natl. Acad. Sci. U.S.A.* **77,** 1632 (1980).
[19] D. Guo, C. T. Mant, A. K. Taneja, J. M. R. Parker, and R. S. Hodges, *J. Chromatogr.* **359,** 499 (1986).
[20] C. T. Mant and R. S. Hodges, *Chromatographia* **24,** 805 (1987).
[21] T. J. Sereda, C. T. Mant, and R. S. Hodges, *J. Chromatogr.* **695,** 205 (1995).

separations by HPLC, and this is particularly true of RP-HPLC. The next section focuses on specific applications of RP-HPLC for researchers involved in solid-phase peptide synthesis, laying particular stress on problem-solving approaches to separations of synthetic peptides.

Reversed-Phase High-Performance Liquid Chromatography for Purification/Monitoring of Solid-Phase Peptide Synthesis Products

Variation of Functional Groups on Stationary Phase

Although RP-HPLC on stationary phases containing alkyl chains (e.g., C_8, C_{18}) as the functional ligand is still the method of choice for most peptide separations,[3,7] alternative ligands with selectivities different to such traditional stationary phases should not be overlooked by the researcher for specific purification problems arising from peptide synthesis.[22,23] Striking examples of this approach were reported by Zhou et al.[22] in their comparison of a C_8 packing with an acid-stable cyanopropyl (CN) packing.

Contaminants frequently encountered in solid-phase peptide synthesis, including side-chain protecting groups, coupling reagents, cleavage reagents, and scavengers, are often difficult to separate from the desired peptide product during RP-HPLC on C_8 and C_{18} columns. For example, thioanisole is a good scavenger and accelerator of the reaction cleaving the synthesized peptide from the resin support.[24,25] In many cases, this scavenger is eluted with the peptide product of interest from an alkyl-bonded reversed-phase column. An example of this can be seen in Fig. 1A,[22] where thioanisole (T) was not separated from synthetic peptide 35B (Table I) on the C_8 column [dithiothreitol (DTT) was present to prevent interchain disulfide bond formation between peptide molecules]. In contrast, on the CN column (Fig. 1B), whereas the retention time of the peptide was similar to that exhibited on the C_8 (Fig. 1A), thioanisole was now barely retained, thus achieving an easy separation.

Very hydrophobic peptides often pose special problems during RP-HPLC owing to their limited solubility and tendency to aggregate. In addition, they may be adsorbed irreversibly to some reversed-phase sorbents.[26]

[22] N. E. Zhou, C. T. Mant, J. J. Kirkland, and R. S. Hodges, *J. Chromatogr.* **548,** 179 (1991).
[23] B. E. Boyes and D. G. Walker, *J. Chromatogr.* **691,** 337 (1995).
[24] H. Yajima, N. Fujii, S. Funakoshi, T. Watanabe, E. Murayama, and A. Otaka, *Tetrahedron* **44,** 805 (1988).
[25] J. M. Stewart and J. D. Young, "Solid Phase Peptide Synthesis," p. 38. Pierce, Rockford, Illinois, 1985.
[26] C. Edelstein and A. M. Scanu, *in* "Handbook of HPLC for the Separation of Amino Acids, Peptides and Proteins" (W. S. Hancock, ed.), Vol. 2, p. 405. CRC Press, Boca Raton, Florida, 1984.

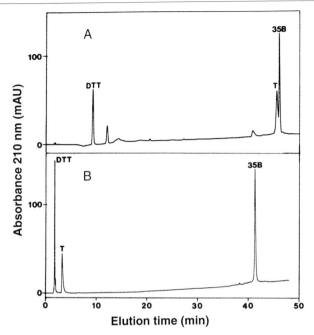

FIG. 1. RP-HPLC of synthetic peptide and thioanisole (T) on (A) C_8 and (B) CN columns. Columns: Zorbax SB-C_8 (150 × 4.6 mm I.D., 5-μm particle size, 94-Å pore size) and Zorbax SB-300CN (150 × 4.6 mm I.D., 6 μm 300 Å; Rockland Technologies, Newport, DE). The HPLC instrument consisted of an HP1090 liquid chromatograph (Hewlett-Packard, Avondale, PA), coupled to an HP1040A detection system, HP9000 Series 300 computer, HP9133 disk drive, HP2225A Thinkjet printer, and HP7440A plotter. Conditions: Linear AB gradient (1% B/min), where eluent A is 0.05% aqueous TFA and eluent B is 0.05% TFA in acetonitrile; flow rate, 1 ml/min; room temperature. The sequence of peptide 35B is shown in Table I. DTT denotes dithiothreitol. (Reprinted from *J. Chromatogr.* **548,** N. E. Zhou, C. T. Mant, J. J. Kirkland, and R. S. Hodges, p. 179, Copyright 1991 with kind permission of Elsevier Science–NL, Sara Burgerhartstraat 25, 1055 KV Amsterdam, The Netherlands.)

Thus, the major problem limiting routine successful purification of very hydrophobic synthetic peptides by RP-HPLC is the excessive strength of hydrophobic interaction between the peptides and alkyl-bonded stationary phases such as C_8 and C_{18} sorbents. Zhou *et al.*[22] demonstrated that a less hydrophobic stationary phase, such as the CN stationary phase, has great potential for this kind of application. Figure 2 compares the elution profile of hydrophobic synthetic peptide P22 (Table I) on C_8 (Fig. 2A) and CN (Fig. 2B) columns. About 12% less acetonitrile in the mobile phase was required to elute P22 from the CN column compared with the C_8 column.

TABLE I

SYNTHETIC PEPTIDES USED IN THIS STUDY

Figure	Peptide sequence[a]	Peptide notation
1	Ac-K-C-A-E-L-E-G-(K-L-E-A-L-E-G)$_n$-amide (where n = 4)	35B
2	Ac-K-K-G-(L)$_{16}$-K-K-A-amide	P22
3	Ac-X-L-G-A-K-G-A-G-V-G-amide, with substitution at position 1 (residue X) denoting peptide analog	N1, M1,[b] etc.
3	Ac-E-A-E-K-A-A-K-E-X-E-K-A-A-K-E-A-E-K-amide, with substitution at position 9 (residue AX9) denoting peptide analog	AN9, AS9, AT9[b]
4	Ac-F-M-H-N-L-G-K-H-L-S-S-M-E-R-V-E-W-L-R-K-K-L-Q-D-V-H-N-F-amide	P2
4	As P2, except NForW[c] instead of W at position 17	P1
5, 6	Ac-E-I-E-A-L-K-A-E-I-E-A-L-K-A-E-I-E-A-L-K-C-E-I-E-A-L-K-A-E-I-E-A-L-K-A-amide	0194
6	Ac-E-I-E-A-L-K-A-E-I-E-A-L-K-C-E-I-E-A-L-K-A-E-I-E-A-L-K-A-amide	0194A
7	Ac-nL-G-G-G-(E-V-S-A-L-E-K)$_5$-amide[c]	P
9	Ac-K-G-C-G-K-E-A-Y-G-amide	1
9	BB-G-A-K-E-A-G-C-G-amide[c]	2
10	Ac-Y-E-C-K-S-L-K-S-E-V-K-S-L-K-S-E-A-K-S-L-K-S-E-V-K-S-L-K-S-E-V-K-S-L-K-S-amide	3
10	Ac-C-G-G-L-E-E-A-E-R-K-S-Q-H-Q-L-E-N-L-E-R-E-Q-R-F-L-K-W-R-L-E-Q-L-amide	4
12, 13	Ac-A-C-K-S-T-Q-D-P-M-F-T-P-K-G-C-D-N	9
14–17	Ac-S-C-A-T-T-V-D-A-K-F-R-P-N-G-C-T-D	KB7
19	Ac-Q-C-G-A-L-Q-K-Q-V-G-A-L-E-K-E-E-G-A-L-E-K-Q-V-G-A-L-Q-K-Q-V-G-A-L-Q-K-amide	011118
20	Ac-K-C-K-S-T-Q-D-E-Q-F-I-P-K-G-C-S-K	03138
21	Ac-K-C-K-S-D-Q-D-P-Q-F-T-P-K-G-C-S-K	03153

[a] Peptide sequences are shown using the one-letter code for amino acid residues.

[b] N1 denotes asparagine substitution at position 1 of this peptide series and M1 denotes methionine substitution at this position, etc.; AN9, AS9, and AT9 denote asparagine, serine, or threonine respectively, at position 9 of peptide sequence.

[c] NForW denotes N-formyltryptophan (peptide P1); nL denotes norleucine (peptide P); BB denotes 4-benzoylbenzyl.

In addition, a greater separation of P22 from impurities (I) was achieved on the CN column.

Effect of Mobile-Phase Salts on Peptide Separations

One novel approach to the manipulation of reversed-phase chromatographic profiles of peptides infrequently considered is the addition of salts to the mobile phase at low pH. Addition of salts (generally 50–100 mM)

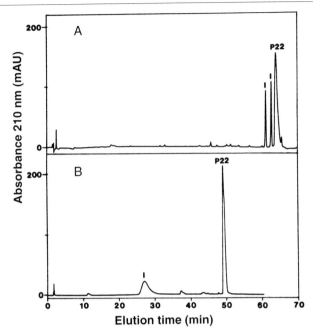

FIG. 2. RP-HPLC of a synthetic hydrophobic peptide on (A) C_8 and (B) CN columns. Columns and conditions as in Fig. 1, except sample was dissolved in 70% eluent A–30% eluent B. The sequence of peptide P22 is shown in Table I. I denotes impurity. (Reprinted from *J. Chromatogr.* **548,** N. E. Zhou, C. T. Mant, J. J. Kirkland, and R. S. Hodges, p. 179, Copyright 1991 with kind permission of Elsevier Science–NL, Sara Burgerhartstraat 25, 1055 KV Amsterdam, The Netherlands.)

to mobile phases over a pH range of approximately 4–7 is traditionally designed to suppress negatively charged silanol interactions with positively charges solutes[18–20]; any selectivity effects tend to be a secondary consideration. However, salt addition does have the potential to offer useful gains in peptide separation selectivity at low pH. The negatively charged ions of the salt reduce the hydrophilicity of positively charged residues by ion-pairing.

To illustrate the potentially beneficial effect of salt on peptide separations at low pH, Fig. 3 compares RP-HPLC separations of selected mixtures of peptides carried out either in an aqueous 0.1% TFA/CH_3CN, pH 2, mobile phase (Fig. 3A,C) or an aqueous 10 mM orthophosphoric acid (H_3PO_4)/CH_3CN, pH 2, mobile phase containing 100 mM sodium perchlorate. The negatively charged perchlorate ion acts as a hydrophilic anionic ion-pairing reagent and, thus, will interact with positively charged groups in peptides; the effect of perchlorate on the retention behavior of a particular

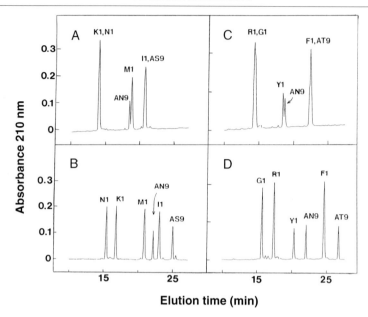

Elution time (min)

FIG. 3. Effect of sodium perchlorate on RP-HPLC of synthetic peptides at pH 2. Column: Zorbax 300-SB C$_8$ (150 × 4.6 mm I.D., 5-μm particle size, 300-Å pore size; Rockland Technologies). Instrumentation: Same as in Fig. 1. Conditions: (A and C) Linear AB gradient (1% B/min), where eluent A is 0.1% aqueous TFA and eluent B is 0.1% TFA in acetonitrile. (B and D) Linear AB gradient (2% B/min, equivalent to 1% acetonitrile/min), where eluent A is 20 m*M* aqueous orthophosphoric acid and eluent B is 20 m*M* orthophosphoric acid in 50% aqueous acetonitrile, both eluents containing 100 m*M* sodium perchlorate. All runs were carried out at a flow rate of 1 ml/min and at room temperature. Peptide sequences and designations are shown in Table I. (From Ref. 30, with permission.)

peptide is then dependent on the number of positive charges the peptide contains. The use of this salt was prompted by a number of considerations: (1) it is a reagent employed in RP-HPLC at both neutral[18–20] and acidic[18,27–29] pH values to improve peak shape; (2) it is UV transparent, allowing peptide bond detection at 210 nm; (3) sodium perchlorate is highly soluble in aqueous acetonitrile eluents, even at relatively high concentrations of this organic modifier[16,30]; (4) it is a strong chaotropic reagent (i.e., perchlorate is an inorganic ion that favors the transfer of nonpolar groups

[27] K. J. Wilson, A. Honegger, and G. J. Hughes, *Biochem. J.* **199**, 43 (1981).
[28] R. L. Emanuel, G. H. Williams, and R. W. Giese, *J. Chromatogr.* **312**, 285 (1984).
[29] H. Gaertner and A. Puigserver, *J. Chromatogr.* **350**, 279 (1985).
[30] T. J. Sereda, C. T. Mant, and R. S. Hodges, *J. Chromatogr.* in press.

to water by altering water structure[18,31] which may be useful in solubilizing hydrophobic peptides). From Fig. 3A, peptides K1 and N1 and peptides I1 and AS9 are coeluted in the TFA system, whereas peptides AN9 and M1 are eluted as a doublet; in contrast, all six peptides are well resolved to baseline in the H_3PO_4/perchlorate mobile phase (Fig. 3B). In a similar manner, from Fig. 3C, peptides R1 and G1 and peptides F1 and AT9 are coeluted, whereas peptides Y1 and AN9 are eluted as a doublet; from Fig. 3D, all six peptides are again baseline resolved by the H_3PO_4/perchlorate mobile phase (see Table I for all peptide sequences). Note that the retention times of all of the peptides increased in the H_3PO_4/perchlorate system (Fig. 3B,D) compared to the TFA system (Fig. 3A,C), this increase being dependent on the magnitude of the positively charged character of a particular peptide.

Thus, peptide mixtures comprising peptides of varying charge (as noted above, a not uncommon situation in crude peptide mixtures) may benefit from the manipulation of selectivity which can be effected through the use of 100 mM sodium perchlorate at pH 2. This is not to suggest that such an approach will replace the traditional aqueous TFA system for most separations, but it may complement such systems for specific applications. In addition, this approach may be useful in situations where it is undesirable to isolate peptides as their trifluoroacetate salts.

Preparative Shallow Gradient Approach to Purification of Closely Related Tryptophan-Containing Analogs

The N-formyl group is the most commonly used protecting group for tryptophan (Trp) during solid-phase peptide synthesis. In addition, N-formyl-Trp is frequently used as an analog of Trp in structure–function studies. Any problems with removal of the formyl group following peptide synthesis, such that a significant proportion of the peptide remains N-formylated, may cause considerable purification problems. For instance, Fig. 4A[32] shows the analytical elution profile of a crude Trp-containing 28-residue peptide (representing a segment of bovine parathyroid growth hormone, bPTH). Generally, by proper selection of the hydrogen fluoride cleavage conditions, all side-chain protecting groups can be removed except for the N-formyl group on Trp. Using some of this peptide, the N-formyl group can then be selectively removed (as both forms of the peptide were required) using an aqueous solution of 1 M ammonium bicarbonate to

---

[31] Y. Hatefi and W.-G. Hanstein, *Proc. Natl. Acad. Sci. U.S.A.* **62,** 1129 (1969).
[32] T. W. L. Burke, J. A. Black, C. T. Mant, and R. S. Hodges, *in* "HPLC of Peptides and Proteins: Separation, Analysis and Conformation" (C. T. Mant and R. S. Hodges, eds.), p. 783. CRC Press, Boca Raton, Florida, 1991.

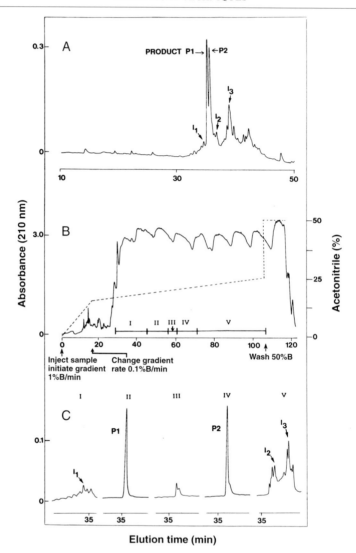

Elution time (min)

Fig. 4. Preparative reversed-phase separation of peptide analogs, P1 and P2, from each other and from various hydrophilic (I_1) and hydrophobic (I_2, I_3) impurities. Column: Aquapore RP300 C_8 (220 × 4.6 mm I.D., 7-μm particle size, 300-Å pore size; Brownlee, Santa Clara, CA). Instrumentation: Same as in Fig. 1, except for a Varian (Walnut Creek, CA) Vista Series 5000 liquid chromatograph. Samples were injected with a 500-μl (analytical runs) or 5-ml (preparative runs) injection loop (Rheodyne, Cotati, CA). (A) Analytical chromatogram of crude peptide mixture. Conditions: Linear AB gradient (1% B/min), where eluent A is 0.05% aqueous TFA and eluent B is 0.05% TFA in acetonitrile. (B) Preparative chromatogram of the same peptide mixture. Conditions: linear AB gradient [100% A/0% B

yield the native peptide (P2, Table I). However, because of the presence of Trp and Met residues, in addition to sequence-specific deprotection problems, it was determined that a significant yield of the N-formylated peptide (P1) could only be obtained by using a special combination of scavengers and temperature during HF cleavage. However, these special cleavage conditions resulted in the cleavage of the N-formyl group from about 50% of the crude peptide product, resulting in about a 1:1 molar ratio of P1 to P2 and, from Fig. 4A, clearly posing a difficult purification problem when both peptides are in the same crude peptide mixture.[32]

Neither SEC nor IEC would be effective separation modes for these peptide analogs, there being no difference in size or net charge between the peptides. Instead, the preparative separation of the analogs represented an ideal application for the shallow gradient approach promoted by this laboratory.[33,34] From Fig. 4A, it can be seen that, despite the very similar hydrophobicities of P1 and P2, the RP-HPLC C_8 column was able to separate the two peptides even with a relatively steep gradient (1% acetonitrile/min), a tribute to the resolving power of this HPLC mode, although baseline resolution was not achieved. Apart from the two peptide analogs, various unknown impurities (e.g., I_1, I_2, I_3) with hydrophobicities similar to that of the two desired peptide products were present, thus complicating the purification problem.

It is clear from Fig. 4A that increasing the sample load of the peptide mixture on the same analytical column and under the same chromatographic conditions would only lead to a worsening in peptide resolution. However, a decrease in the steepness of the gradient during gradient elution of peptide mixtures generally leads to an increase in peptide resolution,[1,11,33] thus allowing for greater sample loads to be applied. Figure 4B shows the elution profile of about 20 mg of the crude peptide mixture on the analytical C_8 column with a 10-fold decrease in gradient steepness (1% acetonitrile/

[33] C. T. Mant, T. W. L. Burke, and R. S. Hodges, *Chromatographia* **24**, 565 (1987).
[34] R. S. Hodges, T. W. L. Burke, C. T. Mant, and S. M. Ngai, *in* "HPLC of Peptides and Proteins: Separation, Analysis and Conformation" (C. T. Mant and R. S. Hodges, eds.), p. 773. CRC Press, Boca Raton, Florida, 1991.

to 84% A/16% B over 16 min (1% B/min), then to 75% A/25% B over a further 90 min (0.1% B/min), followed by an isocratic wash with 50% A/50% B for 10 min prior to reequilibration to 100% A/0% B]. Fractions were collected every minute. Sample load, about 20 mg. (C) Analytical chromatograms of pools I to V from (B). Conditions: Same as in (A). All runs were carried out at a flow rate of 1 ml/min and at room temperature. Peptide sequences are shown in Table I. (Reprinted with permission from Ref. 32. Copyright 1991 CRC Press, Boca Raton, Florida.)

min → 0.1% acetonitrile/min). Despite the unattractive elution profile (due to swamping of the detection system at high peptide concentration) the two peptides were easily obtained in purified form (Fig. 4C), representing about 25% each of the original 20 mg of crude peptide mixture.

The separation shown in Fig. 4 exhibits two key principles of preparative separations of synthetic peptides: (1) large sample loads of crude mixtures may be successfully purified on analytical columns and instrumentation by making efficient use of column packings through the use of shallow gradients, and (2) employing columns under sample overload conditions (i.e., conditions leading to nonlinearity of eluted peaks) causes displacement effects between adjacent solutes,[32–36] so that there is little overlap between adjacent peptide zones, producing high purified peptide yields.

High Temperature Purification of Synthetic Peptides

Peptide or protein retention times in RP-HPLC generally decrease with increasing temperature, owing to increasing solubility of the solute in the mobile phase as the temperature rises.[37–40] In addition, improved peptide resolution, which is due to a more rapid transfer of the solutes between the stationary and mobile phases, generally accompanies a rise in temperature. However, although enhanced peptide resolution is clearly advantageous, there is a relative lack of published data on the usefulness of HPLC at elevated temperature for peptide separations (some selected examples can be found in Refs. 41–44). One reason for this reluctance to carry out high temperature separations in the past is likely due to the increased wear and tear on reversed-phase packings at high temperature. For instance, Guo *et al.*[40] noted the decrease in retentiveness of a C_8 column following a temperature study (26° to 66°) of peptide separations; the high temperatures had

[35] R. S. Hodges, T. W. L. Burke, and C. T. Mant, *J. Chromatogr.* **444**, 349 (1988).
[36] G. B. Cox, *in* "Chromatography in Biotechnology" (C. Horváth and L. S. Ettre, eds.), p. 165. American Chemical Society, Washington, DC, 1993.
[37] K. A. Cohen, K. Schellenger, K. Benedek, B. Grego, and M. T. W. Hearn, *Anal. Biochem.* **140**, 223 (1984).
[38] R. H. Ingraham, S. Y. M. Lau, A. K. Taneja, and R. S. Hodges, *J. Chromatogr.* **327**, 77 (1985).
[39] W. S. Hancock, D. R. Knighton, J. R. Napier, D. R. K. Harding, and R. Venable, *J. Chromatogr.* **367**, 1 (1986).
[40] D. Guo, C. T. Mant, A. K. Taneja, and R. S. Hodges, *J. Chromatogr.* **359**, 519 (1986).
[41] B. A. Marchylo, D. W. Hatcher, J. E. Kruger, and J. J. Kirkland, *Cereal Chemists, Inc.* **69**, 301 (1992).
[42] B. E. Boyes and J. J. Kirkland, *Pept. Res.* **6**, 249 (1993).
[43] W. A. Hancock, R. C. Chloupek, J. J. Kirkland, and L. R. Snyder. *J. Chromatogr.* **686**, 31 (1994).
[44] R. C. Chloupek, W. A. Hancock, B. A. Marchylo, J. J. Kirkland, B. E. Boyes, and L. R. Snyder, *J. Chromatogr.* **686**, 45 (1994).

accelerated the aging of the column. Nonetheless, there are circumstances, such as are described below, where the use of such conditions is unavoidable when purifying specific peptide mixtures. Fortunately, reversed-phase columns with excellent temperature stability are now commercially available,[41–46] thus enhancing the potential of this separation approach.

Figure 5A–C shows the effect of increasing temperature on the elution profile of the crude mixture of peptide 0194 (Table I). This amphipathic α-helical peptide was known to form a highly stable two-stranded α-helical coiled coil.[47] At ambient temperature (Fig. 5A), the poor elution profile may have suggested an unsuccessful synthesis, with the desired product at low yield and in the presence of many contaminants. However, LC mass spectrometry indicated essentially the same peptide mass throughout the major elution profile. As the temperature was raised to 50° (Fig. 5B) and further to 70° (Fig. 5C), there was a distinct improvement in the elution profile; indeed, the profile at 70° indicated a yield of desired product of >90%. It is well known that the reversed-phase HPLC mode is highly denaturing to polypeptide tertiary and quaternary structure[48–54]; thus, the hydrophobic stationary phases characteristic of RP-HPLC, as well as the nonpolar solvents required to elute solutes from such packings, both serve to denature or prevent peptide association or nonspecific aggregation (although, conversely, α-helicity is enhanced by such conditions[55,56]). In addition, temperature elevation also serves to unfold polypeptides.[49,50] From Fig. 5A, it is possible that, at this ambient temperature, this peptide is present in different forms, ranging from a coiled-coil dimer through various unfolded (partially to fully denatured) forms. Indeed, the maintenance of the dimeric form of similar very stable coiled coils has previously been reported by this laboratory.[52,54,57] As the temperature is raised, the various

[45] J. J. Kirkland, J. L. Glajch, and R. D. Farlee, *Anal. Chem.* **61**, 2 (1988).
[46] J. J. Kirkland, C. H. Dilks, Jr., and J. E. Henderson, *LC-GC* **11**, 290 (1993).
[47] J. Y. Su, R. S. Hodges, and C. M. Kay, *Biochemistry* **33**, 15501 (1994).
[48] S. Y. M. Lau, A. K. Taneja, and R. S. Hodges, *J. Chromatogr.* **317**, 129 (1984).
[49] C. T. Mant, N. E. Zhou, and R. S. Hodges, in "The Amphipathic Helix" (R. M. Epand, ed.), p. 39. CRC Press, Boca Raton, Florida, 1993.
[50] C. T. Mant, N. E. Zhou, and R. S. Hodges, *J. Chromatogr.* **476**, 363 (1989).
[51] E. Watson and W.-C. Kenney, *J. Chromatogr.* **606**, 165 (1992).
[52] M. Hanson, K. K. Unger, C. T. Mant, and R. S. Hodges, *J. Chromatogr.* **599**, 77 (1992).
[53] R. Rosenfeld and K. Benedek, *J. Chromatogr.* **632**, 29 (1993).
[54] R. S. Hodges, B.-Y. Zhu, N. E. Zhou, and C. T. Mant, *J. Chromatogr.* **676**, 3 (1994).
[55] N. E. Zhou, C. T. Mant, and R. S. Hodges, *Pept. Res.* **3**, 8 (1990).
[56] V. Steiner, M. Schär, K. O. Barnsen, and M. Mutter, *J. Chromatogr.* **586**, 43 (1991).
[57] R. S. Hodges, P. D. Semchuk, A. K. Taneja, C. M. Kay, J. M. R. Parker, and C. T. Mant, *Pept. Res.* **1**, 19 (1988).

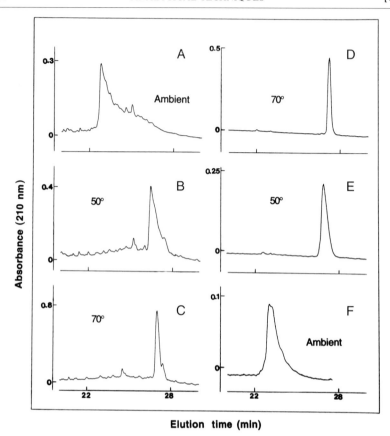

Absorbance (210 nm)

Elution time (min)

Fig. 5. Effect of temperature on RP-HPLC of a synthetic amphipathic α-helical peptide. Column: Zorbax Rx-C$_8$ (150 × 2.1 mm I.D., 5-μm particle size, 300-Å pore size; Rockland Technologies). Instrumentation: Same as in Fig. 1. Conditions: Linear AB gradient (2% B/min), where eluent A is 0.05% aqueous TFA and eluent B is 0.05% TFA in acetonitrile; flow rate 0.25 ml/min. (A–C) Elution profile of crude peptide 0194 at ambient temperature (A), 50° (B), and 70° (C). (D–F) Elution profile of purified peptide at 70° (D), 50° (E), and ambient temperature (F). The sequence of peptide 0194 is shown in Table I.

unfolded forms of the peptide are all progressively denatured to produce the single monomeric species seen in Fig. 5C.

Figure 5D shows the elution profile of the purified (at 70°) peptide and the gradual broadening of the peptide peak as the temperature is lowered to 50° (Fig. 5E) and, finally, to ambient temperature (Fig. 5F). An interesting point to note here is that the retention time of the peak decreases with decreasing temperature. As noted above, the retention times of peptides would normally be expected to decrease with increasing temperature.[37–40]

This laboratory has previously shown that a two-stranded α-helical coiled coil, stable to denaturation by RP-HPLC conditions, is not retained as strongly by a reversed-phase column as the individual strands making up the coiled-coil.[52,54,57] Because the hydrophobic faces of the monomers are shielded when they form a coiled coil, the hydrophobic residues at the monomer–monomer interface are unavailable to interact with the reversed-phase packing; in contrast, these residues are exposed when the peptide is maintained in its monomeric state which is thus more strongly retained by the column. This supports the view that the early eluted species seen in Fig. 5A,F are dimeric and/or partially folded species of the dimeric coiled coil, whereas the single later eluted peak seen in Fig. 5C,D represents the α-helical form of the peptide with an increased retention time due to its preferred binding domain.[52,54,55,57]

An interesting addendum to the results of Fig. 5 can be seen in Fig. 6, which compares the elution profile of peptide 0194 with that of peptide 0194A at ambient temperature and at 70°, the latter peptide differing from the former only in the deletion of the first seven residues (Table I). At ambient temperature (Fig. 6A), a mixture of the two purified peptides shows two major peaks, although there is considerable distortion of the first peak (peptide 0194, as seen previously in Fig. 5F) compared to the second peak (peptide 0194A). However, at 70° (Fig. 6B), both peptides are not only exhibiting excellent peak shape, but they have switched relative elution positions. This observation may be explained by the relative stabilities of the dimeric forms of these peptides. Peptide 0194 is likely principally in a dimeric form at ambient temperature (Fig. 5A,F) albeit with the concomitant presence of partially unfolded species; however, the coiled-coil form of peptide 0194A is less stable than that of peptide 0194 and, hence, is eluted in its monomeric form only (Fig. 6A). As the temperature is raised to 70° (Fig. 6B), both peptides are now eluted in their monomeric forms, with the 35-residue, more hydrophobic peptide 0194 now being eluted later than the 28-residue peptide 0194A. Note the reduced retention time and narrower peak width of peptide 0194A at the higher temperature, both observations characteristic of the effect of elevated temperatures on peptide elution behavior during RP-HPLC.

A somewhat different purification problem where elevated temperatures proved singularly useful is illustrated in Fig. 7 for peptide P (Table I). At ambient temperature (not shown), the elution profile was very poor, with broad, severely tailing peaks; indeed, this situation had not improved appreciably even at 40°, where the location of the desired peptide was still uncertain. However, raising the temperature to 60° and, finally, to 80° produced a much improved elution profile, where peptide P was readily identified and purified. This peptide was readily solubilized only in the

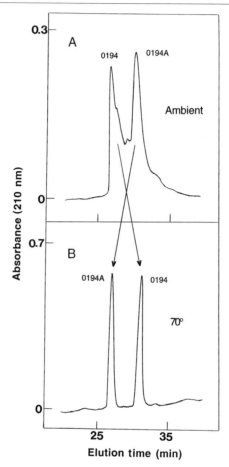

FIG. 6. Effect of temperature on elution profile of a two-peptide mixture. Column: Same as in Fig. 3. The HPLC instrument consisted of an HP1090, Series II, liquid chromatograph, coupled to an HP1040A detection system, HP9000 Series 300 computer, HP7957B disc drive, HP2225A Thinkjet printer, and HP ColorPro plotter. Conditions: Linear AB gradient (2% B/min), where eluent A is 0.05% aqueous TFA and eluent B is 0.05% TFA in acetonitrile; flow rate, 1 ml/min temperature, ambient temperature (A) or 70° (B). The sequences of peptides 0194 and 0194A are shown in Table I.

presence of 6 M guanidine hydrochloride, an indication of a somewhat hydrophobic species with only limited solubility in benign medium. Thus, the appearance of a satisfactory peptide peak only at high temperature suggests either nonspecific aggregation of the peptide on the column at ambient temperature, leading to its nonelution, or perhaps that the peptide is simply too hydrophobic to be eluted without enhancing its solubility in

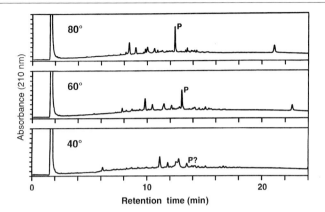

FIG. 7. Effect of temperature on RP-HPLC of a synthetic hydrophobic peptide. Column: Zorbax SB-C$_{18}$ (150 × 4.6 mm I.D., 5-μm particle size, 300-Å pore size; Rockland Technologies). The HPLC instrument was an HP1090, Series II, liquid chromatograph (Hewlett-Packard). Conditions: linear AB gradient (2% B/min, starting from 30% B), where eluent A is 0.1% aqueous TFA and eluent B is 0.09% TFA in acetonitrile; flow rate: 1 ml/min; temperature: 40° (bottom), 60° (middle), and 80° (top). The crude peptide, P, was dissolved in 20 μM Tris buffer, pH 8.0, containing 6 M guanidine hydrochloride. The sequence of peptide P is shown in Table I. (Figure supplied courtesy of Dr. Barry Boyes of Rockland Technologies.)

the mobile phase by elevating the temperature (again note the decrease in peptide retention time with increasing temperature). Whatever the reason for the observed results, Fig. 7 again illustrates the practical benefits of elevated temperatures for RP-HPLC purification of problematic peptides.

Reversed-Phase High-Performance Liquid Chromatography as Tool to Monitor Preferential Formation of Disulfide-Bridged Heterostranded Synthetic Peptides

Cross-linking two different molecules through a disulfide bond is frequently used in synthetic peptide chemistry, protein conjugation, and protein folding studies.[58–60] Although a number of methods to form disulfide bonds have been described (for review, see Ref. 61), the air oxidation

[58] O. D. Monera, N. E. Zhou, C. M. Kay, and R. S. Hodges, *J. Biol. Chem.* **268,** 19218 (1993).
[59] O. D. Monera, C. M. Kay, and R. S. Hodges, *Biochemistry* **33,** 3862 (1994).
[60] O. D. Monera, N. E. Zhou, P. Lavigne, C. M. Kay, and R. S. Hodges, *J. Biol. Chem.* **271,** 3995 (1996).
[61] D. Andreu, F. Albericio, N. A. Sole, M. C. Munson, M. Ferrer, and G. Barany, *in* "Methods in Molecular Biology" (M. W. Pennington and B. M. Dunn, eds.), Vol. 35, p. 91. Humana Press, Totowa, New Jersey, 1994.

method at slightly alkaline pH is the most common.[62-64] A potential problem associated with such techniques, however, is that by random statistical probability, formation of disulfide bonds between two dissimilar peptides is expected to form three disulfide-bridged species in a 1 : 2 : 1 ratio. Thus, the problem addressed in this article is not only how to drive the reaction to form exclusively the heterostranded product (or at least minimize the two undesired homostranded products), but also how to isolate the heterostranded product from the homostranded products. To this end, a comparison is made between simple air oxidation of peptides for disulfide bridge formation versus a thiol activation scheme developed to enhance the formation of asymmetrical disulfide-bridged peptides.

Thiol Activation Scheme for Preferential Formation of Heterostranded Synthetic Peptides. Methods based on the selective activation of one thiol followed by a reaction with a second free thiol to form asymmetric disulfide bonds selectively between proteins or synthetic peptides have been described to alleviate the problem of reduced yields due to symmetric disulfide bond formation.[61] The most versatile of such activating groups are perhaps those based on the pyridine–sulfenyl group,[65-67] with 2,2'-dithiodipyridine (DTDP)[68,69] being employed for this article. The reaction scheme for the formation of heterostranded disulfide-bridged peptides through the use of this reagent is shown in Fig. 8.

Reversed-Phase Monitoring of Disulfide-Linked Peptide Products. DI-SULFIDE-LINKED TWO-STRANDED RANDOM COIL PEPTIDES. Peptides 1 and 2 (Table I) are two random coil peptides used to illustrate the problem of reduced yields of heterostranded products due to the formation of homostranded products during simple air oxidation. From Fig. 9C, it can be seen that air oxidation of peptides 1 and 2 produced the expected three products (the heterostranded peptide and two homostranded peptides), although not in the expected 1 : 2 : 1 ratio of homostranded (peptide 1–peptide 1) : heterostranded (peptide 1-peptide 2) : homostranded (peptide 2-peptide 2). Indeed, peptide 1, in particular, appeared to undergo preferential homostranded disulfide bond formation.

Heterostranded product formation of the peptides was now carried out

[62] A. K. Ahmed, S. W. Schaffer, and D. B. Wetlaufer, *J. Biol. Chem.* **250,** 8477 (1975).
[63] S. M. Lunte and P. T. Kissinger, *J. Liq. Chromatogr.* **8,** 691 (1985).
[64] J. P. Tam, *Int. J. Pept. Protein Res.* **29,** 421 (1987).
[65] M. S. Bernatowicz, R. Matsueda, and G. R. Matsueda, *Int. J. Pept. Protein Res.* **28,** 107 (1986).
[66] F. Rabanal, W. F. DeGrado, and P. L. Dutton, *Tetrahedron Lett.* **37,** 1347 (1996).
[67] K. Akaji, K. Fujino, T. Tatsumi, and Y. Kiso, *J. Am. Chem. Soc.* **115,** 11384 (1993).
[68] J. Carlsson, H. Drevin, and R. Axen, *Biochem. J.* **173,** 723 (1978).
[69] L. H. Kondejewski, J. A. Kralovec, A. H. Blair, and T. Ghose, *Bioconjugate Chem.* **5,** 602 (1994).

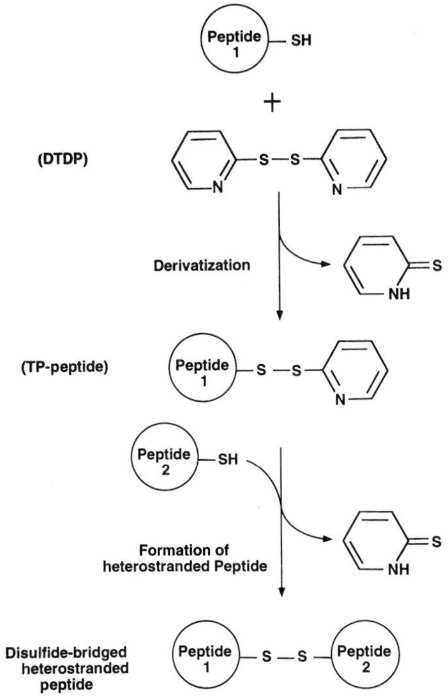

FIG. 8. Reaction scheme for the formation of heterostranded, disulfide-bridged peptides. Peptide 1-SH and peptide 2-SH denote reduced peptides 1 and 2, respectively; DTDP denotes 2,2′-dithiodypyridine; TP-peptide denotes thiopyridine-derivatized peptide. Peptide derivatization: The peptide that forms the disulfide-bridged homostranded peptide at a faster rate is dissolved in ethanol (for increased solubility and ease of solvent removal) at a concentration of 1 mg/ml and a 3-fold molar excess of DTDP added, the solution then allowed to incubate at room temperature for 30 min. On completion of the reaction (confirmed by analytical RP-HPLC and mass spectrometry), ethanol is removed under reduced pressure and the residue is taken up in 0.05% aqueous TFA and extracted three times with diethyl ether to remove unreacted DTDP. Heterostranded disulfide-bridged peptide formation: The lyophilized TP-peptide and the underivatized second peptide are dissolved separately in degassed 50 mM sodium phosphate buffer, pH 7, containing 1 mM EDTA at a concentration of 1 mg/ml. The TP-peptide is slowly titrated with the thiol-containing peptide, typically in 1/20 volume increments, at 3-min intervals, with slow stirring. The progress of the reaction is monitored by RP-HPLC and the identity of the products confirmed by LC/mass spectrometry.

by the reaction scheme shown in Fig. 8. Figure 9D shows the RP-HPLC elution profile of purified DTDP-derivatized peptide 1 prior to titration with the second thiol-containing peptide. The elution profiles of the reaction mixture after titration with either a half-molar equivalent or one-molar equivalent of peptide 2 are shown in Fig. 9E and Fig. 9F, respectively. The

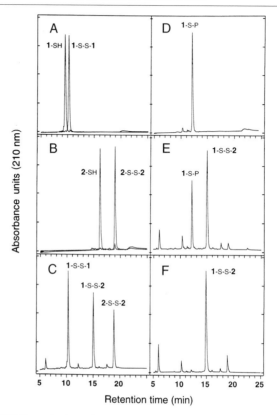

Fig. 9. RP-HPLC monitoring of formation of heterostranded, disulfide-bridged random coil peptides. Column: Zorbax S300-SB C_8 (250 × 4.6 mm I.D., 5-μm particle size, 300-Å pore size; Rockland Technologies). Instrumentation: Same as in Fig. 1. Conditions: Linear AB gradient (1% B/min), where eluent A is 0.05% aqueous TFA and eluent B is 0.05% TFA in acetonitrile; flow rate, 1 ml/min; room temperature. (A) Reduced peptide 1 and air-oxidized, homostranded peptide 1 (overlayed profiles). (B) Reduced peptide 2 and air-oxidized, homostranded peptide 2 (overlayed profiles). (C) Products of air oxidation of peptides 1 and 2. (D) Purified TP-derivatized peptide 1. (E) TP-derivatized peptide 1 after titration with a half-molar equivalent of thiol-containing peptide 2. (F) TP-derivatized peptide 1 after titration with a one-molar equivalent of peptide 2. Air oxidation was carried out by dissolving 1 mg of each of the starting peptides in 2 ml of 100 mM ammonium bicarbonate, pH 8.1, and stirring at room temperature in an open container. TP-peptide denotes thiopyridine-derivatized peptide (reaction scheme shown in Fig. 8). The sequences of peptides 1 and 2 are shown in Table I.

yield of heterostranded product obtained from the 1 : 1 equivalent titration (Fig. 9F) was >80%, a significant increase in yield compared to that achieved by air oxidation (compare Fig. 9C and Fig. 9F).

DISULFIDE-LINKED TWO-STRANDED AMPHIPATHIC α-HELICAL COILED-COIL PEPTIDES. Peptides 3 and 4 (Table I) are amphipathic α-helical peptides designed to form two-stranded parallel coiled coils. Although both peptides readily form homostranded products by air oxidation because of favorable interchain interactions[58,59] (Fig. 10A,B, where the RP-HPLC profiles of the disulfide-bridged peptides are overlayed on the reduced monomeric peptides), they did not produce heterostranded products in significant amounts by air oxidation when mixed together in an equimolar ratio (Fig. 10C). The peptides preferred instead to form the homostranded versions (Fig. 10C) since, as noted above, a purely random oxidation pattern would be expected to give a 1 : 2 : 1 ratio of products (homostranded peptide 3-peptide 3 : heterostranded peptide 3-peptide 4 : homostranded peptide 4-peptide 4). From Fig. 10C, the RP-HPLC retention times of the three products are also similar. Thus, not only is the yield of desired heterostranded product low when air oxidation is employed, but also the resulting three products pose a separation problem.

The reaction scheme outlined in Fig. 8 was now applied to the heterostranded product formation of peptides 3 and 4. Figure 10D shows the RP-HPLC elution profile of purified DTDP-derivatized peptide 3 prior to titration with the second thiol-containing peptide. The RP-HPLC elution profiles of the reaction mixtures obtained by titration of DTDP-derivatized peptide 3 with either a half-molar equivalent or one-molar equivalent of thiol-containing peptide 4 are shown in Fig. 10E and Fig. 10F, respectively. A comparison of Fig. 10C and Fig. 10F clearly shows that the derivatization method (Fig. 10F) resulted in a greatly increased yield (>90%) of the heterostranded product compared to the air oxidation approach (Fig. 10C) and also facilitated the purification of the heterostranded product from trace amounts of the two homostranded products.

Separation of Synthetic Intrachain Disulfide-Bridged Peptides from Reduced Forms by Reversed-Phase High-Performance Liquid Chromatography

Many active fragments of peptides and proteins are synthesized with their native disulfide bonds or the addition of disulfide bonds to confer to the fragments a structure similar to that in the native protein.[70] The main problem encountered in the formation of the intrachain disulfide bridge is

---

[70] M. Z. Atassi, C. S. McDaniel, and T. Manshouri, *J. Protein Chem.* **7**, 655 (1988).

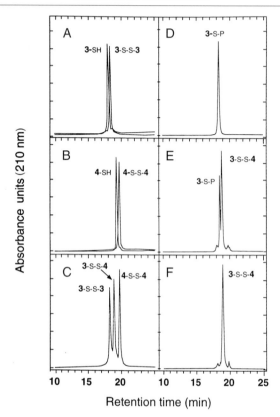

FIG. 10. RP-HPLC monitoring of formation of heterostranded disulfide-bridged amphipathic α-helical peptides. Column, instrumentation, and conditions: Same as in Fig. 9. (A) Reduced peptide 3 and air-oxidized, homostranded peptide 3 (overlayed profiles). (B) Reduced peptide 4 and air-oxidized, homostranded peptide 4 (overlayed profiles). (C) Products of air oxidation of peptides 3 and 4. (D) Purified TP-derivatized peptide 3. (E) TP-derivatized peptide 3 after titration with a half-molar equivalent of thiol-containing peptide 4. (F) TP-derivatized peptide 3 after titration with a one-molar equivalent of thiol-containing peptide 4. The air oxidation process is described in Fig. 9. TP-peptide denotes thiopyridine-derivatized peptide (reaction scheme shown in Fig. 8). The sequences of peptides 3 and 4 are shown in Table I.

the availability of a rapid and simple monitoring method to follow the extent of oxidation of the cysteine residues to cystine (Fig. 11A).[71] In addition, a sensitive and quantitative method is required to resolve the disulfide-bridged (oxidized) peptide from the corresponding cysteine-

[71] K. K. Lee, J. A. Black, and R. S. Hodges, in "HPLC of Peptides and Proteins: Separation, Analysis and Conformation" (C. T. Mant and R. S. Hodges, eds.), p. 389. CRC Press, Boca Raton, Florida, 1991.

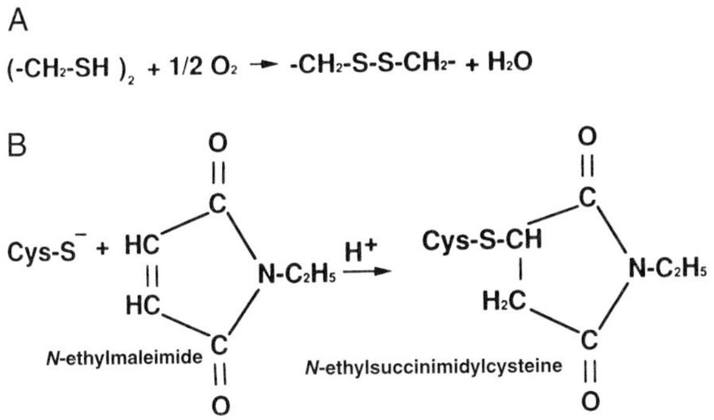

A

(-CH₂-SH)₂ + 1/2 O₂ → -CH₂-S-S-CH₂- + H₂O

B

FIG. 11. (A) Air oxidation of cysteine side chains to form a disulfide bond. The reduced peptide is dissolved in 100 mM ammonium bicarbonate to a concentration of 0.1 mg/ml and the solution stirred in an open container at room temperature. (B) Reaction of cysteine with N-ethylmaleimide (NEM). The extent and/or rate of air oxidation of cysteine side chains is monitored by removing 100-μl samples at various time points and mixing them with 10 μl of an aqueous solution of NEM (1 mg/ml). The mixture is then allowed to sit at room temperature for at least 15 min before injection into a RP-HPLC column. (Reprinted with permission from Ref. 71. Copyright 1991 CRC Press, Boca Raton, Florida.)

containing (reduced) peptide. Assessment of the purity of the oxidized conformer would then be possible by quantitative determination of the undesired reduced conformer. We describe how RP-HPLC, in combination with the use of a free sulfhydryl-specific reagent, may efficiently achieve both of these goals.

N-Ethylmaleimide: A Thiol-Specific Reagent. Figure 11B demonstrates the reaction of cysteine with N-ethylmaleimide (NEM). This modification results in the formation of a stable covalent thioether linkage between the cysteine thiol and NEM; as shown in Fig. 11, the thiolate anion attacks one of the double-bonded carbon atoms, forming N-ethylsuccinimidylcysteine. A covalently modified reduced peptide will thus become more hydrophobic than its reduced form, and should be eluted later from a RP-HPLC column.

Similarly, if one wants to recover the reduced peptide, DTDP (described above) can also be used to derivatize the reduced peptide. In this case, however, a disulfide bridge is formed, allowing recovery of the reduced peptide following reduction of the thiopyridine-derivatized peptide with DTT.

Separation of Reduced and Oxidized Intrachain Disulfide-Bridged Forms of Peptide. From Fig. 12, the reduced and oxidized forms of peptide 9 (Table I) were eluted off the RP-HPLC column at 24.8 min, and a 1 : 1 mixture of the two forms resulted in a single peak (Fig. 12C); clearly, a separation of these two peptide forms appeared extremely difficult. However, when the same peptide mixture was reacted with NEM, it was found

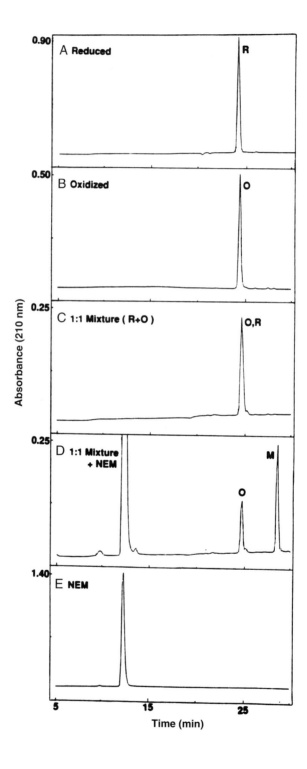

that the modified reduced peptide was eluted from the column at 28.6 min, easily separable from its oxidized form (Fig. 12D). The reagent itself is quite hydrophilic and was eluted from the column at 12 min (Fig. 12E); thus, this reagent has the added advantage of not interfering or coinciding with most peptides, which are generally eluted from a RP-HPLC column with optimum resolution in the 15 to 40% acetonitrile range of the gradient.[72]

Monitoring Formation of Intrachain Disulfide Bridge. Figure 13 illustrates the use of RP-HPLC to monitor the oxidation of a peptide (in this case, peptide 9 from Table I) that shows no separation between the reduced and oxidized conformers. Such a monitoring procedure can be used for any peptides whether they show good or poor resolution. At various time points samples can be removed, and the NEM reaction will stop any further oxidation by covalently modifying the free-SH groups of the reduced peptide, denoted by M in Fig. 13. Monitoring the intrachain disulfide bridge formation by RP-HPLC allows simultaneous quantitation of the two conformers, unlike the commonly employed Ellman procedure which requires spectrophotometric measurements to quantitate free sulfhydryls of the reduced peptides.[73]

Reversed-Phase High-Performance Liquid Chromatography as Tool to Monitor Postsynthesis Modification of Side Chains

The problem of undesirable side-chain modification of synthetic peptides during synthesis and/or during cleavage of the peptide from the resin is well known and has been well documented.[74–76] In addition, the use of

---

[72] M. Hermodson and W. C. Mahoney, *Methods Enzymol.* **91**, 352 (1983).
[73] G. L. Ellman, *Arch. Biochem. Biophys.* **82**, 70 (1989).
[74] M. Bodanszky and J. H. Kwei, *Int. J. Pept. Protein Res.* **12**, 69 (1978).
[75] M. Bodansky, "Peptide Chemistry." Springer-Verlag, New York, 1988.
[76] J. D. Young, R. H. Angeletti, S. A. Carr, D. R. Marshak, A. J. Smith, J. T. Stults, L. C. Williams, K. R. Williams, and G. B. Fields, *in* "Peptides: Chemistry, Structure and Biology," Proceedings of the Thirteenth American Peptide Symposium (R. S. Hodges and J. A. Smith, eds.), p. 1088. Escom, Leiden, The Netherlands, 1994.

FIG. 12. RP-HPLC separation of reduced (R) and oxidized (O) forms of peptide 9 following modification of the reduced peptide with NEM. Column: Same as in Fig. 4. The HPLC instrument consisted of a Spectra-Physics (San Jose, CA) SP8700 solvent delivery system and a SP8750 organizer module combined with a Hewlett-Packard HP1040A detection system, HP3390A integrator, HP85 computer, HP9121 disk drive, and HP7470A plotter. Conditions: Same as in Fig. 1. The NEM-modified peptide is denoted by M. Conditions for air oxidation and chemical modification with NEM are given in Fig. 11. The sequence of Peptide 9 is shown in Table I. (Reprinted with permission from Ref. 71. Copyright 1991 CRC Press, Boca Raton, Florida.)

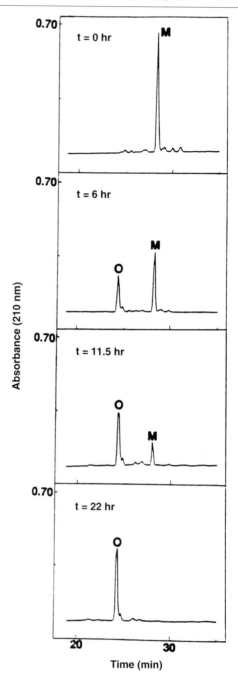

RP-HPLC for monitoring the appearance of the resulting peptide impurities has also been reported.[77] However, as shown below, not all peptide impurities necessarily occur as a result of synthesis and/or cleavage, even if this appears to be the case at first glance; in addition, the verification of the problem may require a quite complex combination of peptide digests, mass spectrometry, amino acid analysis, peptide sequencing, and RP-HPLC.

From Table I, peptide KB7[78] represents the C-terminal region of a strain of *Pseudomonas aeruginosa* pilin and requires an intrachain disulfide bridge for its epithelial cell binding activity. Following synthesis by standard *tert*-butyloxycarbonyl (Boc) chemistry and HF cleavage, the reduced crude peptide was analyzed by RP-HPLC at pH 2 (Fig. 14A). From Fig. 14A, it can be seen that there was a single dominant peak (which was subsequently shown by mass spectrometry to have the expected mass for the desired product), suggesting a successful synthesis. This peptide was then subjected to air oxidation for 16-hr to produce the required intrachain disulfide bridge. Figure 14B,C shows the RP-HPLC elution profiles of the oxidized KB7 at pH 2 and pH 6, respectively. Clearly, in contrast to the reduced peptide, the oxidized KB7 exhibited, in addition to a dominant peak, a major impurity (note the switchover of the two peaks between the two pH values, which is discussed as follows). These two major peaks were purified by RP-HPLC at pH 6 and, following mass spectrometry, were found to have very similar masses of 1826.2 and 1825.5 for the smaller and larger peaks, respectively (the latter figure being the expected mass for the desired product). These very similar mass values certainly ruled out that the impurity was a deletion product during synthesis, a position already supported by the absence of a major impurity in the elution profile of the reduced peptide (Fig. 14A). In addition, this impurity only appeared after air oxidation to produce the internal disulfide bridge. Thus, some kind of modification to the peptide while undergoing oxidation appeared to be the problem.

One candidate for the gain of a single mass unit in a peptide is deamidation of Asn or Gln to Asp or Glu, respectively; thus, replacement of -NH₂

[77] G. Szókán, A. Török, and B. Penke, *J. Chromatogr.* **387**, 267 (1987).
[78] W. Y. Wong, R. I. Irvin, W. Paranchych, and R. S. Hodges, *Protein Sci.* **1**, 1308 (1992).

FIG. 13. RP-HPLC monitoring of the formation of the intrachain disulfide bridge with time during air oxidation of peptide 9 by RP-HPLC. Column, instrumentation, and conditions: Same as in Fig. 12. The reduced peptide was modified with NEM prior to injection onto the column to increase the separation between the oxidized (O) and reduced peptides. The NEM-modified peptide is denoted by M. Conditions for air oxidation and chemical modification with NEM are given in Fig. 11. The sequence of peptide 9 is shown in Table I. (Reprinted with permission from Ref. 71. Copyright 1991 CRC Press, Boca Raton, Florida.)

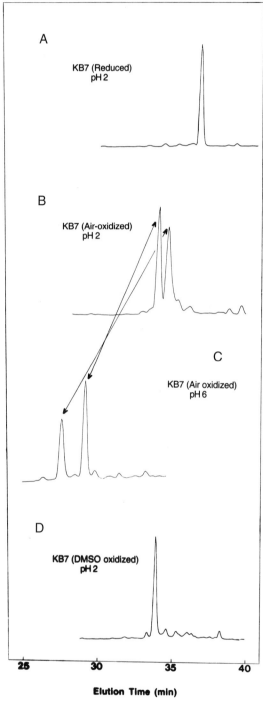

(mass 16) with -OH (mass 17) on the Asn or Gln side chains produces the required gain of a single mass unit. Examination of the sequence of KB7 shows an Asn residue at position 13, more specifically, an Asn residue followed by a Gly residue. Nonenzymatic deamidation of Asn residues represents one of the major chemical degradation pathways of polypeptides both *in vivo* and *in vitro*.[79,80] Furthermore, studies on model peptides have shown that, under neutral to basic pH conditions (note the basic conditions used for air oxidation of KB7; Figure 14B,C), the reaction proceeds via a succinimide derivative, which by spontaneous hydrolysis generates either an α-linked (aspartyl) or a β-linked (isoaspartyl) Asp residue.[80–82] In addition, the rate of deamidation is strongly influenced by the nature of the side chain of the adjacent carboxyl residue, with Asn-Gly being one of the most susceptible sequences.[81,83,84]

Other studies have highlighted that a central problem in the synthesis of aspartic acid-containing peptides is aspartimide formation, followed by a subsequent ring opening of the α-aminosuccinimide ring by aqueous bases providing largely the β-linked isoaspartyl Asp residue, that is, an α- to β-amide bond isomerization.[85,86] In a similar manner to the Asn-Gly sequence noted above, Asp-Gly sequences have been reported to produce extensive imide formation.[85,87,88] If a deamidation process via a cyclic succinimide

[79] H. T. Wright, *Crit. Rev. Biochem. Mol. Biol.* **26**, 1 (1991).
[80] S. Capasso, A. Di Donato, L. Esposito, F. Sica, G. Sorrentino, L. Vitagliano, A. Zagari, and L. Mazzarella, *J. Mol. Biol.* **257**, 492 (1996).
[81] T. Geiger and S. Clarke, *J. Biol. Chem.* **262**, 785 (1987).
[82] K. Patel and R. T. Borchardt, *Pharm. Res.* **7**, 703 (1990).
[83] K. Patel and R. T. Borchardt, *Pharm. Res.* **7**, 787 (1990).
[84] R. Tyler-Cross and V. Schirch, *J. Biol. Chem.* **266**, 22549 (1991).
[85] M. A. Ondetti, A. Deer, J. T. Sheehan, J. Pluscec, and O. Kory, *Biochemistry* **7**, 4069 (1968).
[86] J. P. Tam, M. W. Rieman, and R. B. Merrifield, *Pept. Res.* **1**, 6 (1988).
[87] K. Suzuki and N. Endo, *Chem. Pharm. Bull. Jpn.* **26**, 2269 (1978).
[88] C. C. Yang and R. B. Merrifield, *J. Org. Chem.* **41**, 1032 (1976).

FIG. 14. Monitoring the oxidation of a synthetic peptide by RP-HPLC. Column: Zorbax 300-SB C_8 (150 × 4.6 mm. I. D., 5-μm particle size, 300-Å pore size; Rockland Technologies). Instrumentation: Same as in Fig. 1, except for a Varian Vista Series liquid chromatograph. Conditions: pH 2 (A, B, and D), linear AB gradient (0.5% B/min), where eluent A is 0.05% aqueous TFA and eluent B is 0.05% TFA in acetonitrile; flow rate, 1 ml/min; room temperature. pH 6 (C), linear AB gradient (1% B/min, equivalent to 0.5% acetonitrile/min), where eluent A is 10 mM aqueous ammonium acetate and eluent B is 10 mM ammonium acetate in 50% (v/v) aqueous acetonitrile; flow rate, 1 ml/min; room temperature. Conditions for air oxidation are given in Fig. 9. Dimethyl sulfoxide (DMSO) oxidation was carried out by dissolving the peptide (1 mg/ml) in 10 mM phosphate buffer, pH 6.5, containing 5% (v/v) DMSO and stirring at room temperature for 16 hr. The sequence of peptide KB7 is shown in Table I.

derivative was occurring to the Asn residue of KB7 during air oxidation under basic conditions, it may have been expected to observe a maximum of three components in the elution profile of the oxidized peptide (the desired Asn-substituted product plus α-Asp and β-Asp derivatives; Fig. 15). From Fig. 14B,C, it can be seen that only two components were detected by RP-HPLC at both pH values. Thus, how could one verify that the suggested deamidation process had indeed occurred during oxidation; in addition, which peak (from Fig. 14B,C) contained more than one component (assuming, of course, that the two forms of the Asp-substituted derivative were present)?

Some initial support for deamidation having occurred to KB7 lies in the relative retention times of the two peaks seen in Fig. 14B,C at pH 2 and 6, respectively. At pH 2, a protonated carboxyl group is more hydropho-

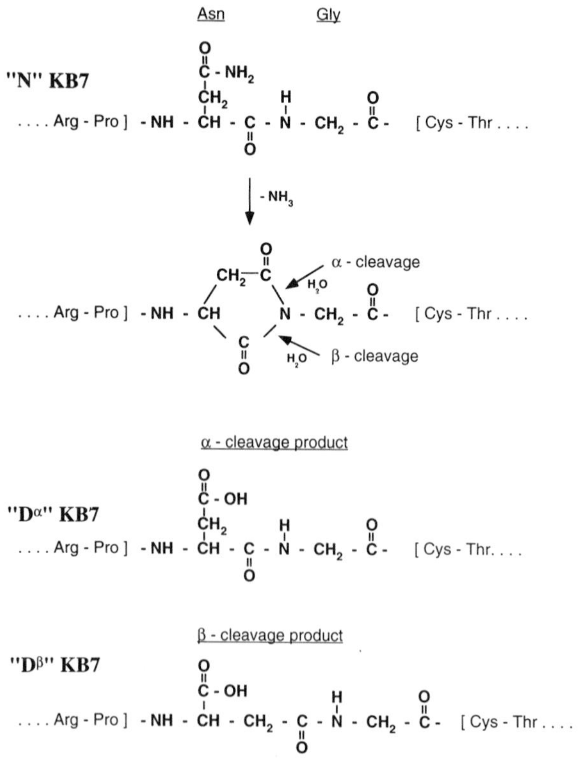

Fig. 15. Proposed process of deamidation of Asn residue of peptide KB7 via a cyclic succinimide intermediate. The sequence of peptide KB7 is shown in Table I.

bic than an amide based on a comparison of hydrophobicity coefficients[19,89] of Asp and Asn, respectively; thus, at pH 2 (Fig. 14B), it would be expected that the smaller, later retained peak would be the Asp derivative(s). At pH 6, the carboxyl group of Asp would be negatively charged, that is, it would have an increase of one negative charge over that of the native Asn-substituted KB7; thus, under these conditions, the observed switch in elution order (Fig. 14C) would not be surprising considering that an ionized carboxyl group is more hydrophilic than an amide.[19,90]

Figure 16 outlines the approach taken to identify absolutely the components of the two peaks shown in Fig. 14B,C. From Fig. 16A, the putative oxidized Asp-substituted KB7, denoted "$D^{\alpha,\beta}$" KB7 (assuming the presence of both the α and β isomers of Asp) was cleaved with trypsin to produce the D^β and D^α tryptic fragments shown. Reduction of the disulfide bridge still present in these fragments should have then produced a common N-terminal fragment (I_F) and either the C-terminal fragment containing the β-isomer (D_F^β) or the α-isomer of Asp (D_F^α). The same approach was applied to Asn-substituted KB7, denoted "N" KB7 (Fig. 16B), with similar results to Fig. 16A, albeit with only one tryptic fragment (N) and one reduced C-terminal fragment containing Asn (N_F); the N-terminal fragment (I_F) is, of course, identical to that from Fig. 16A. It was then proposed to subject all three C-terminal fragments (D_F^α, D_F^β, N_F) to sequence analysis for final peptide identification.

For the above scheme to succeed, efficient purification of peptide fragments was required, and it was hoped that RP-HPLC would be able to achieve this goal. Figure 17 shows RP-HPLC elution profiles at pH 6 of the peptides obtained at each step of the scheme shown in Fig. 16, and the excellent correlation between these figures is quite clear. Thus, from Fig. 17A, the smaller peak produced two peaks following tryptic digest, followed by three separate components (I_F is common to both tryptic fragments) following reduction of these two tryptic fragments (see Fig. 16A). In contrast, the larger peak produced only one peak following tryptic digest, followed by only two components (Fig. 17B). Automated sequence analysis, as well as mass spectrometry, of the three C-terminal fragments made a positive identification of the α-Asp- and Asn-containing fragments (D_F^α and N_F); sequencing of the suspected β-Asp containing fragment (D_F^β) (which was shown to have the expected mass for an Asp-containing fragment) stopped at this residue, but by process of elimination, as well as the demonstration by amino acid analysis of an Asp residue in this fragment, the

[89] T. J. Sereda, C. T. Mant, F. D. Sönnichsen, and R. S. Hodges, *J. Chromatogr.* **676,** 139 (1994).
[90] O. D. Monera, T. J. Sereda, N. E. Zhou, C. M. Kay, and R. S. Hodges, *J. Pept. Sci.* **1,** 319 (1995).

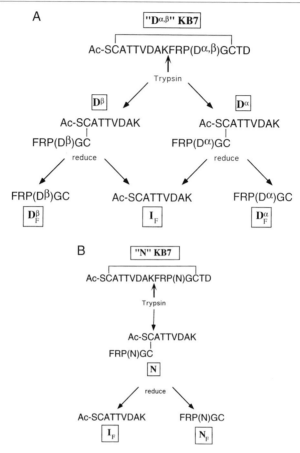

FIG. 16. Protocol for identification of components resulting from deamidation of Asn residue of peptide KB7. (A) Identification of the deamidated peptide components, where the Asn at position 13 has been deamidated via a cyclic succinimide intermediate (Fig. 15) to form a mixture of the aspartyl (α-Asp) and isoaspartyl (β-Asp) derivatives. (B) Identification of the native, Asn-containing, peptide component. The boxed symbols denote the various peptides and tryptic peptide fragments at each stage of the protocol.

denotion of this fragment as D_F^β was reasonable. On the basis of these results, as well as the correct mass identified for I_F, all of the peaks shown in Fig. 17 were denoted as shown. The elution profiles shown in Fig. 17 were very similar to the same runs carried out at pH 2 (except for the switchover of the two original peaks as seen in Fig. 14B,C); however, separations of peptide fragments were generally better at the higher pH, enabling straightforward purification of the fragments at each stage.

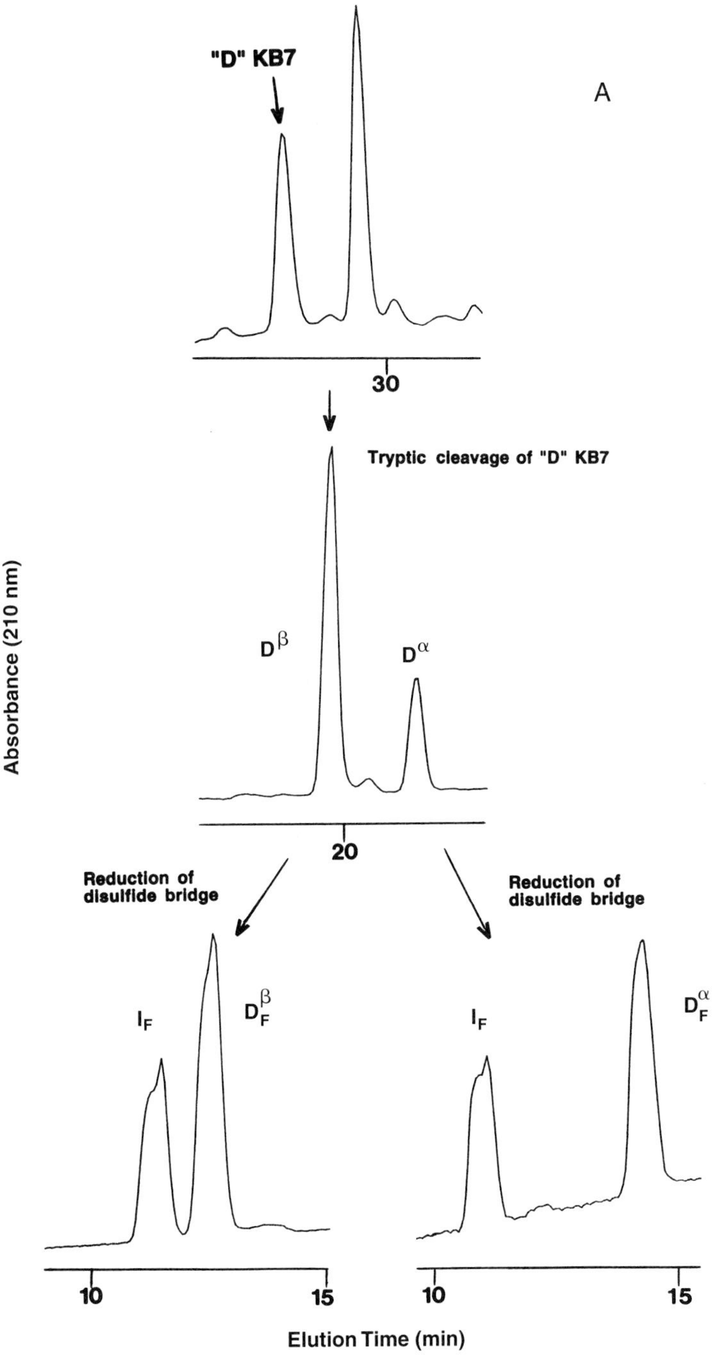

"D" KB7

A

Tryptic cleavage of "D" KB7

D^{β} D^{α}

30

20

Reduction of
disulfide bridge

Reduction of
disulfide bridge

I_F D_F^{β}

I_F D_F^{α}

10   15

10   15

Elution Time (min)

Absorbance (210 nm)

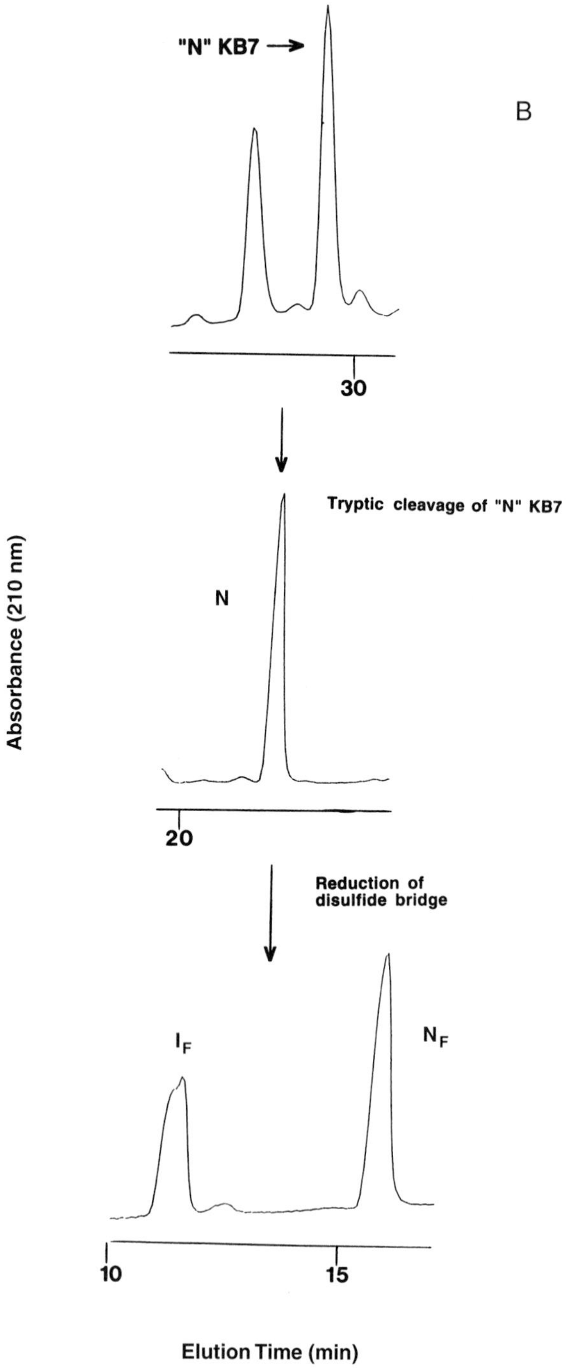

B

"N" KB7 →

30

Tryptic cleavage of "N" KB7

N

20

Reduction of
disulfide bridge

I_F

N_F

10   15

Absorbance (210 nm)

Elution Time (min)

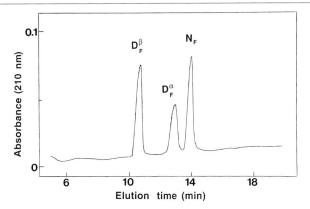

FIG. 18. RP-HPLC at pH 6 of reduced, tryptic fragments of peptide KB7 containing Asn (N_F), α-Asp, (D_F^α), or β-Asp (D_F^β) residues. Column and instrumentation: Same as in Fig. 14. Conditions: Same as pH 6 conditions shown in Fig. 14. See Fig. 16 for sequences of peptide fragments and Fig. 17 for origins of RP-HPLC purified fragments.

An interesting point to note from Fig. 17A is the detection of two peaks (D^β, D^α) following tryptic digestion of "D" KB7, that is, the two isomers of the Asp-peptide fragments were readily separable once the constraints of the internal disulfide bridge had been released by cleavage of the peptide within the disulfide bridge. This ability of RP-HPLC to distinguish between the two isomers of Asp is underscored in Fig. 18, which shows baseline resolution of the reduced C-terminal fragments of the Asp-containing peptides (D_F^β, D_F^α), as well as an excellent separation from the Asn-substituted fragment (N_F).

In summary, this approach of combining several analytical techniques has answered the questions posed earlier, that is, deamidation of Asn had taken place, likely effected by the basic conditions employed in air oxidation, with a concomitant production of α- and β-isomers of Asp. In addition, the three expected components were present, with two components (the two Asp-isomer substituted peptides) being eluted as a single peak (the smaller peak at both pH 2 and pH 6, Fig. 14B,C, respectively).

FIG. 17. RP-HPLC at pH 6 of peptide components obtained at each stage of the protocol outlined in Fig. 16. Column and instrumentation: Same as in Fig. 14. Conditions: Same as pH 6 conditions shown in Fig. 14. (A) Elution profiles correspond to protocol shown in Fig. 16A. (B) Elution profiles correspond to protocol shown in Fig. 16B. Tryptic digestion was carried out by dissolving 1 mg of peptide in 500 μl of 50 mM aqueous ammonium bicarbonate, pH 7.6, containing 1 mM calcium chloride; to this solution was added 1 μg of trypsin, followed by incubation for 24 hr at 37°. The sequence of peptide KB7 is shown in Table I.

Having positively identified the problem, the final step was to employ a different method of oxidation of the peptide. Thus, reduced KB7 was now subjected to oxidation by dimethyl sulfoxide (DMSO)[91] for 16 hr at pH 6.5, a pH value considerably less basic than the pH 8.1 value employed for air oxidation. The resulting single peptide peak observed during RP-HPLC of the oxidized peptide (Fig. 14D), and that exhibited the correct peptide mass by mass spectrometry, represented a satisfying finale to the tracing of a postsynthesis problem and its ultimate solution.

Hydrophilic Interaction/Cation-Exchange Chromatography for Purification/Monitoring of Solid-Phase Peptide Synthesis Products

General Principles

The term hydrophilic interaction chromatography (HILIC) was coined to describe separations based on solute hydrophilicity.[92] Thus, separation by HILIC, in a manner similar to normal-phase chromatography (to which it is related), depends on hydrophilic interactions between the solutes and a hydrophilic stationary phase, that is, solutes are eluted from an HILIC column in order of increasing hydrophilicity (decreasing hydrophobicity). Hydrophilic interaction chromatography is characterized by separations being effected by a linear gradient of decreasing organic modifier concentration, that is, starting from a high concentration of organic modifier (typically, 70–90% aqueous acetonitrile). This high initial organic modifier concentration serves to promote hydrophilic interactions between the solute and the HILIC stationary phase, these interactions decreasing with decreasing organic modifier concentration, with subsequent elution of the peptides.

Although the major process governing peptide retention behavior on ion-exchange columns involves ionic interactions between the column matrix and the peptide solutes, all ion-exchange packings have, in our hands, also exhibited some hydrophobic character leading to long peptide retention times and peak broadening.[14] Most researchers prefer to avoid separations based on such mixed-mode ionic–hydrophobic column behavior. Thus, an organic solvent, such as acetonitrile, is frequently added to the mobile phase buffers to suppress any such hydrophobic packing characteristics.[14] However, we have demonstrated that manipulation of the acetonitrile concentration in the mobile-phase buffers enabled considerable flexibility in the separation of basic (potentially positively charged) peptides on a strong cation-exchange column.[15,16] Important examples of this approach may be

[91] J. P. Tam, C.-R. Wu, W. Liu, and J.-W. Zhang, J. Am. Chem. Soc. 113, 6657 (1991).
[92] A. J. Alpert, J. Chromatogr. 499, 177 (1990).

found in Ref. 16. Thus, it was shown that, as the level of acetonitrile was raised from 10 to 90% (v/v) (in 10% steps), the separation process (linear AB increasing gradients of sodium perchlorate with the same level of acetonitrile in both buffers) became increasingly mixed-mode (hydrophilic and ionic interactions), with HILIC interactions becoming dominant at high acetonitrile concentrations. Indeed, mixed-mode hydrophilic and ionic (in this case, cationic) chromatography (hence the term HILIC/CEC) was shown to combine the most advantageous aspects of two widely different separation mechanisms: a separation based on hydrophilicity/hydrophobicity differences between peptides and the large selectivity advantages of ion-exchange chromatography for the separation of peptides of varying net charge. Sodium perchlorate was the salt chosen for this mixed-mode approach because of its high solubility in aqueous solution even in the presence of high concentrations of organic modifier.[16,30]

Application of Hydrophilic Interaction/Cation-Exchange Chromatography to Purification and Analysis of Solid-Phase Synthesis Products

As noted earlier in this article, RP-HPLC is the method of choice for most preparative separations of peptides; in addition, it is commonly employed analytically to check the purity of a purified product. However, a single peak obtained during RP-HPLC is not necessarily a guarantee of peptide purity. Thus, a complementary HPLC method with different selectivity to RP-HPLC is required for a more accurate assessment of peak purity. HILIC/CEC is an excellent candidate for such a complementary HPLC method. Indeed, as is demonstrated later, such a complementary method may be better suited for specific peptide purifications than even the ubiquitous RP-HPLC mode.

Cysteine Deletion. Figure 19 outlines the purification and analysis of a 35-residue synthetic peptide (peptide 011118 from Table I). The crude peptide elution profile (Fig. 19, top) at first glance indicated a successful synthesis with relatively few impurities, enabling an easy purification of the reduced peptide by RP-HPLC. The analytical RP-HPLC run of the purified peptide (Fig. 19, middle) showed a single, symmetrical peak. The lack of any shoulder on this peak, obtained on a very efficient column, suggested excellent peptide purity. However, mass spectrometry of this peptide showed not only the expected product mass, but a second strong signal exhibiting a peptide mass 103 units less than the expected product, indicating deletion of the Cys residue at position two of the peptide (Table I). Based on side-chain hydrophobicity coefficients derived from model synthetic peptides,[19,89,90] the Cys side chain is relatively hydrophobic, and RP-HPLC on an efficient RP-HPLC column as employed here may have

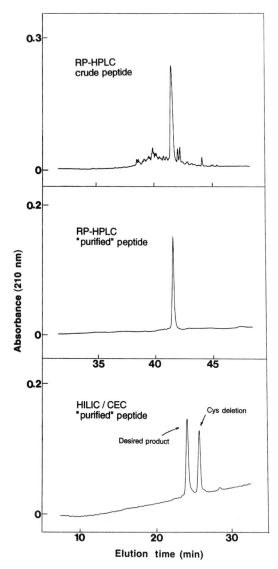

FIG. 19. Analysis and purification of a 35-residue synthetic peptide by RP-HPLC and mixed-mode HILIC/CEC. Columns: RP-HPLC column same as in Fig. 3; the CEC column was a PolySulfoethyl A strong cation-exchange column (200 × 4.6 mm I.D., 5-μm particle size, 300-Å pore size; PolyLC, Columbia, MD). Instrumentation: Same as in Fig. 6. RP-HPLC conditions (top and middle): Same as in Fig. 1. HILIC/CEC conditions (bottom): linear AB gradient (2.5 mM sodium perchlorate/min from 30 mM sodium perchlorate, following 10-min isocratic elution with 30 mM sodium perchlorate), where eluent A is 5 mM aqueous triethylammonium phosphate containing 65% (v/v) acetonitrile and eluent B is eluent A containing 0.4 M sodium perchlorate; flow rate, 1 ml/min; room temperature. The sequence of the desired peptide product (denoted peptide 011118) is shown in Table I.

been expected to detect such a deletion product. However, the contribution which a residue makes to the overall hydrophobicity/hydrophilicity of a peptide diminishes with increasing peptide chain length,[93] and this fact, together with any other conformation and/or nearest neighbor effects specific to the peptide,[89] has served to mask the loss of a Cys residue when applying RP-HPLC as a purity check. Clearly, RP-HPLC is not the mode of choice for separating the desired peptide from its Cys-deletion impurity. In addition, IEC or SEC are also ruled out as potential candidates for purification modes owing to, respectively, the identical net charge and essentially identical size of the two peptides.

Figure 19 (bottom) illustrates the excellent separation of the two peptides when HILIC/CEC was applied to the "purified" peptide shown in the middle RP-HPLC profile. Desalting of these two peptides by RP-HPLC, followed by mass spectrometry, identified two pure peaks with the elution positions denoted as shown. The loss of the Cys residue would make the peptide more hydrophilic (i.e., less hydrophobic), hence the observation that the Cys-deletion impurity is eluted later than the desired peptide product during mixed-mode HILIC/CEC. This result clearly showed that the RP-HPLC-purified peptide contained a major impurity of just less than 50%.

Serine Deletion. Figure 20 outlines the purification and analysis of a 17-residue synthetic peptide containing an intrachain disulfide bridge (peptide 03138 from Table I). Following purification of the crude oxidized peptide mixture (Fig. 20, top) by RP-HPLC, analysis of the purified product showed a single, symmetrical peak (Fig. 20, middle), indicating, as above (Fig. 19), excellent peptide purity. However, mass spectrometry of this single peak again showed, in addition to the expected mass, a second signal; in this case, this second signal exhibited a peptide mass 87 units less than the desired product, indicating deletion of one of the Ser residues. This deletion was almost certainly at position 4 of the sequence (see Table I), because the Ser residue at position 16 was part of a core peptide common to other peptide analogs and this observed Ser deletion was only detected when a second Ser was added during elongation of this core peptide. On the basis of side-chain hydrophobicity coefficients,[19,89,90] the polar Ser side chain contributes little to the retention behavior of a peptide during RP-HPLC and is classed, in RP-HPLC terms, as only a slightly hydrophilic group. Thus, it is not surprising that the deletion product would be difficult to detect, let alone resolve by RP-HPLC, particularly considering the length of the peptide (see above comments concerning peptide chain length effects) and the position of the residue within the intrachain disulfide bridge

[93] C. T. Mant, T. W. L. Burke, J. A. Black, and R. S. Hodges, *J. Chromatogr.* **458,** 193 (1988).

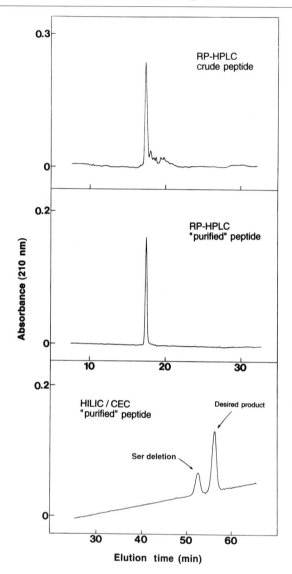

FIG. 20. Analysis and purification of a 17-residue synthetic peptide by RP-HPLC and mixed-mode HILIC/CEC. Columns, instrumentation, and conditions: Same as in Fig. 19. The sequence of the desired peptide product (denoted peptide 03138) is shown in Table I.

(compare to nonresolution by RP-HPLC of the α- and β-isomers of the Asp-substituted, disulfide-bridged peptide shown in Fig. 14). In a manner similar to that of the Cys-deletion product described above, IEC or SEC are again not suitable for purification of the desired peptide product.

Figure 20 (bottom) illustrates the efficacy of the HILIC/CEC approach to resolving two peptides inseparable by RP-HPLC (middle profile of Fig. 20). This mixed-mode approach appeared to enhance the hydrophilic contribution of a Ser residue, as the Ser-deletion impurity is now baseline resolved from the desired product. The Ser-deletion peptide is eluted prior to the desired product (confirmed by mass spectrometry following RP-HPLC of the individual peaks), because loss of this residue has made the peptide less hydrophilic (i.e., more hydrophobic) than the native peptide.

Multiple Deletions. Figure 21 shows an interesting example of a synthesis where, following purification of the crude peptide (Fig. 21, top) by RP-HPLC, multiple deletion products still contaminated the desired peptide product (peptide 03153 in Table I; note that this is an analog of peptide 03138). Although the "purified" peptide clearly still contained more than one component (Fig. 21, middle), the extent of this contamination became clear only when HILIC/CEC was applied to this semipurified product. Three major peaks were detected (their identities confirmed by mass spectrometry following desalting by RP-HPLC), with the desired product making up only about 50% of the "purified" peptide shown in the middle profile of Fig. 21. Loss of the polar Thr residue[19,89,90] at position 12 of the peptide (see Table I) would make this Thr deletion impurity less hydrophilic (i.e., more hydrophobic) than the desired product, hence its elution prior to the native peptide. Mass spectrometry of the earliest eluted peak indicated that it contained two components (a clear shoulder can be seen), one component exhibiting an Asp deletion and the other both an Asp and a Thr deletion. At the pH value of the mobile phase (pH 6.5), the Asp residue would be negatively charged, that is, very hydrophilic (even more so than the polar Thr residue),[19,90] hence a loss of this residue would make the peptide considerably less hydrophilic; thus, its early elution prior to both the desired product and the Thr-deletion impurity was not surprising. It is likely (although not proved) that the shoulder on the leading edge of the first peak is the double-deletion product, adding the effect of losing the hydrophilic contribution of a Thr residue to that of losing an ionized Asp residue. The importance of the HILIC/CEC profile in this case lies in demonstrating that the synthesis of peptide 03153 was relatively unsuccessful (the desired product likely accounting for only about 30–40% of the crude peptide mixture). A decision could then be made whether to purify the peptide by a multistep protocol (e.g., mixed-mode HILIC/CEC followed by RP-HPLC in this case) or (as was subsequently decided) to resynthesize the peptide.

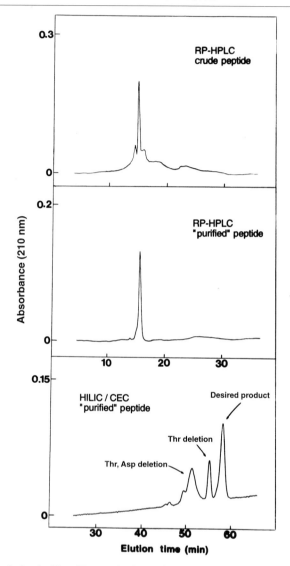

FIG. 21. Analysis of a 17-residue synthetic peptide by RP-HPLC and mixed-mode HILIC/CEC. Columns, instrumentation, and conditions: Same as in Fig. 19. The sequence of the desired peptide product (denoted peptide 03153) is shown in Table I.

Acknowledgments

This work was supported by the Medical Research Council of Canada, the Protein Engineering Network of Centres of Excellence, and the Canadian Bacterial Diseases Network of Centres of Excellence. We thank Dawn Lockwood for typing the manuscript and T. W. Lorne Burke for help with preparing the figures. We also thank our Synthesis Group for technical support: P. D. Semchuk, I. Wilson, and L. Daniels.

[20] Capillary Electrophoresis

By AGUSTIN SANCHEZ and ALAN J. SMITH

Introduction

Capillary electrophoresis (CE) is one of several analytical methods that can be used for the characterization of synthetic peptides. Essentially, it is a high-resolution, high-sensitivity, rapid, and quantitative electrophoretic separation technique. Capillary electrophoresis separates molecules on the basis of charge and, to a lesser extent, size. It can be considered an orthogonal separation system to reversed-phase high-performance liquid chromatography (RP-HPLC), which separates on the basis of hydrophobicity. Capillary electrophoresis has found multiple uses in the field of analytical chemistry, and methods have been developed to analyze small molecules such as drugs and metabolites and larger molecules such as proteins and oligonucleotides. The speed and high resolution capabilities of CE have found ready application in the area of synthetic peptide chemistry. One major role has been to analyze the products of automated peptide synthesis, where both deletion products and incompletely deprotected peptides are commonly present. In fact, CE can be an excellent analytical tool for any synthetic peptide manipulation that results in a change in the net charge or shape of the molecule.

Capillary electrophoresis has also been used to monitor the progress of removal of the protecting groups during the process of cleavage and deprotection. This capability is particularly important when dealing with the deprotection of peptides that contain one or several Arg residues, which is one of the slowest amino acids to deprotect. Because most CE separations of peptides are conducted under low pH conditions, the basic amino acids will be fully protonated at all times. As Arg residues are deprotected the net charge on the molecule increases. This results in the deprotected peptide becoming resolved from its protected counterpart. Consequently, the rate of deprotection can be monitored over time as the appearance of the more positively charged species.

In another case, CE has been used to monitor the oxidation or reduction (i.e., formation or breakage) of disulfide bonds. Typically, an initial CE electropherogram is taken of the peptide dissolved in its oxidative or reductive solution. Over a period of hours or days aliquots are analyzed in order to observe the appearance of any new peaks that would correspond to the peptide undergoing the formation of a disulfide bond or, in the case of reduction, breakage of the disulfide bond. Because no net change in charge results from this manipulation, the separation must be based on a change in shape (cyclization) or size (interchain or intrachain disulfides).

Separation Theory

Separation of peptides by CE occurs as a result of two interacting forces: electrophoretic migration and electroosmotic flow.[1] Electrophoretic migration is the movement of charged molecules in solution toward a positive or negative electrode. However, electroosmotic flow refers to the bulk electrolytic flow caused by the charged inner capillary wall and the applied potential. Silanol groups along the inner capillary wall ionize to form a negative charge when in contact with the electrolytic solution. The negatively charged silanol groups attract the electrolyte cations and create a double layer or region of charge separation along the capillary wall. As the voltage is applied, cations in the electrolyte nearest the capillary wall migrate toward the cathode, pulling electrolyte molecules along with them. This creates a net flow toward the cathode that can then be modulated by changes in voltage, pH, buffer concentration, additives, and capillary wall coatings. Consequently positive, negative, and neutral sample molecules will all be swept toward the cathode, but at different rates with the rate of the positively charged molecules being the fastest.

Materials and Methods

Capillary electrophoresis separations are based on electrophoretic separations within a glass capillary, which keeps diffusion to a minimum and gives rise to its high sensitivity and high resolution capabilities. Under the influence of a high voltage, samples differentially migrate through the capillary, and the components are monitored by an on-line detector.

---

[1] R. M. McCormick, in "Handbook of Capillary Electrophoresis" (J. Landers, ed.), Chap. 12. CRC Press, Boca Raton, Florida, 1994.

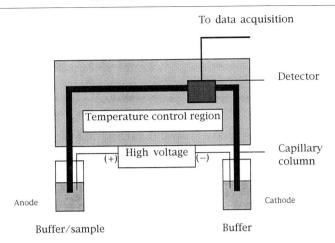

FIG. 1. Schematic of a basic CE instrument.

In its simplest form, a CE instrument consists of two electrode buffers joined by a glass capillary, and a variable wavelength UV detector (see Fig. 1). A third reservoir containing the sample is also required. The sample aliquot is loaded onto the capillary from the sample reservoir, which is temporarily placed as the anode electrolyte. The sample can be loaded either electrokinetically or mechanically (pressure or vacuum). The anode end of the capillary is then removed from the sample reservoir, returned to the anodic buffer reservoir, and electrophoresis is commenced.

The small internal volume of the capillaries require very high voltages (20–30 kV) to achieve electrophoretic separations, and for this reason they are incorporated into dedicated CE instruments. The passage of these high field strengths can result in high temperatures within the capillary. Cooling of the capillary can be accomplished by the use of either forced ambient air (passive) or refrigerated (active) cooling. A number of instruments are commercially available for this purpose.

To improve the tensile strength of the glass capillaries, they are coated with a plastic (polyimide) film. At a point close to the cathode reservoir, a small amount of the polyimide film is burned away, and the exposed capillary serves as the optical window for the detector, while the capillary itself serves as the flow cell.

Standard analytical capillaries have internal diameters of 50–75 μm and lengths of 20–50 cm, which correspond to a capillary volume of 1 to 3 μl. Typically, the sample injection volumes are between 5 and 50 nl, and the peptide starting concentration is between 10 and 20 μg/ml. Ideally the peptide should be dissolved in water; in cases where peptide solubility is limited, however, a minimal amount of trifluoroacetic acid (TFA) or acetonitrile may be added to the solid peptide and water added to attain the final peptide concentration. High salt concentrations should be avoided because they will interfere with the charge separation of the peptides and dramatically reduce resolution.

Capillary electrophoresis separation efficiency is directly related to the solute mobility and the applied voltage. Therefore, the higher the applied voltage, the higher the efficiency of the separations. Consequently, between 100,000 and 200,000 theoretical plates can routinely be achieved for sample molecules with an operating voltage between 20 and 30 kV.

One potential complication with the separation of peptides in open tube capillaries is the irreversible adsorption of some peptides to the silica wall, which leads to peak tailing and reduced separation efficiencies. However, at low pH this has proved not to be a significant problem.

The most widely used buffer for CE analysis of peptides is 50 mM sodium phosphate with a pH between pH 2.00 and 3.00. It should be noted that when using underivatized silica capillaries it is important to precondition them either when new or when changing separation buffers. If precoated capillaries are used, however, they should never be preconditioned, as this will remove their coating, thus ruining the capillary. Buffers can be changed in these capillaries by preequilibration.

A second potential complication is that the absolute mobility of a given peptide will vary over the lifetime of the capillary or the buffer. For this reason, an internal standard should be included with the sample when absolute mobilities are required.

Preconditioning Underivatized Capillaries

1. Place underivatized fused silica capillary in instrument.
2. Flush capillary with 10 column volumes of 1.0 N sodium hydroxide at 0.5 psi.
3. Flush with 10 column volumes water.
4. Flush with 4 column volumes 0.25 M phosphoric acid.
5. Flush with 10 column volumes water.
6. Flush with 4 column volumes 0.25 M sodium phosphate, pH 2.30.
7. Store in same buffer until use.
8. When switching to a new buffer a 4-hr equilibration is advised, and steps 2 and 3 are eliminated.

Analysis of Peptides

1. The peptide concentration should be 10–20 μg/ml. Peptides are usually dissolved in 0.1% TFA (v/v) or water. However, peptides with

a predominance of acidic or hydrophobic residues should be dissolved in a minimal amount of a 50:50 (v/v) mixture of acetonitrile/water and then diluted to the final concentration using 0.1% TFA.

2. Load onto CE sample table that has been equilibrated to 37°.
3. Separate the peptide mixture using the following conditions:

Capillary   75 μm, uncoated, fused silica
Electrolyte   50 mM sodium phosphate, pH 2.3
Sample table temperature   37°
Run temperature   25°
Voltage   25 kV
Sample injection   10 sec at 0.5 psi
Detection   200 or 214 nm
Run time   30 min

4. Operate instrument in accordance with the manufacturer's instructions.

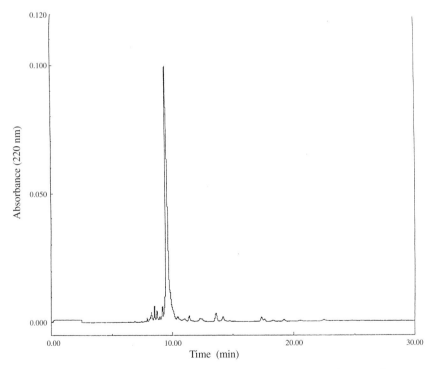

FIG. 2. Electropherogram of a synthetic peptide showing the presence of very small amounts (<2% each) of contaminants.

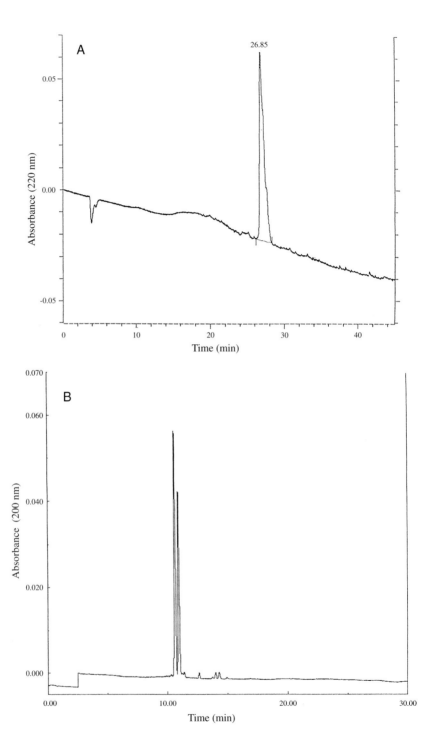

Postrun and Storage

1. Flush capillary with 4 column volumes of 0.1 N sodium hydroxide at 0.5 psi.
2. Flush with 4 column volumes water.
3. Flush with 4 column volumes 0.25 M sodium phosphate, pH 2.30.
4. The capillary may be left on the instrument or removed and placed in a closed container with its ends left uncapped.

Results

The electropherogram in Fig. 2 is a good example of the resolutional capability of CE when used to assess the purity of a peptide. The peptide was synthesized on a Rainin (Emeryville, CA) Symphony multicolumn peptide synthesizer and analyzed using the previously stated CE conditions. The electropherogram shows a large peak, corresponding to the full-length peptide, with many smaller peaks migrating before and after the major peak. These smaller peaks can be attributed to incomplete deprotection of the peptide (e.g., *tert*-butyl groups left on the hydroxyl-containing amino acids or pentamethyldihydrobenzofuransulfonyl groups on Arg) or deletions of amino acid residues along the peptide sequence. None of these small peaks contained more than 2% of the full-length product. The HPLC profile gave a similar representation although the adjacent peaks were considerably broader.

The RP-HPLC profile of a peptide with the sequence

Gly-Asp-Gly-Val-Val-Leu-Gln-Lys-Thr-Gln-Arg-Pro-Ala-Gln-Gly-OH

showed a single major peak with a slight shoulder (Fig. 3A). When a portion of the sample was subjected to matrix-assisted laser desorption–ionization (MALDI) mass spectrometry, two molecular masses were determined, one corresponding to the full-length peptide (1554.22) and one to des-Gly peptide (1498.09). When the same peptide was subjected to CE analysis (Fig. 3B) two peaks were clearly resolved. This is a powerful example of the high resolutional capabilities of CE, as the differences between the two species represents no net change in the charge on the molecule and only a small change in molecular size.

Fɪɢ. 3. (A) A RP-HPLC chromatogram of the synthetic peptide H-Gly-Asp-Gly-Val-Val-Leu-Gln-Lys-Thr-Gln-Arg-Pro-Ala-Gln-Gly-OH. Conditions: Linear gradient of 0–50% acetonitrile/water (0.1% TFA); flow rate, 1.0 ml/min; column, Vydac C_{18}, 0.4 × 25 cm. (B) Electropherogram of the same peptide. The two peaks represent the full-length and des-Gly peptides.

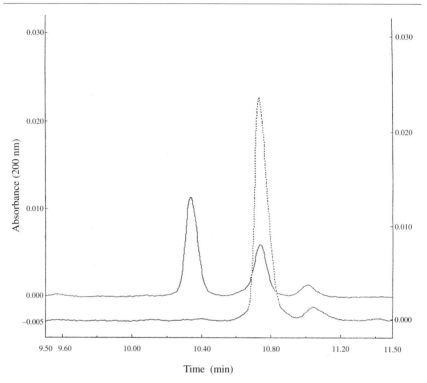

9.50 9.60   10.00   10.40   10.80   11.20   11.50

Time (min)

FIG. 4. Comparative electropherogram of the linear form of a peptide (dotted line) to its cyclized counterpart (solid line). The peptide sequence was Phe-Cys-Phe-Trp-Lys-Thr-Cys-Thr-OH.

Figure 4 shows the results of monitoring the oxidation of a peptide synthesized in response to a 1995 Association of Biomolecular Resource Facilities (ABRF) study.[2] In that study, which included more than 75 participating laboratories, a request had been made to synthesize the following peptide and cyclize it via the oxidation of sulfhydryl groups of the Cys residues:

Phe-Cys-Phe-Trp-Lys-Thr-Cys-Thr-OH.

The conditions specified for the oxidation of the crude peptide were as follows:

[2] R. H. Angeletti, L. Bibbs, L. F. Bonewald, G. B. Fields, J. S. McMurray, W. T. Moore, and J. T. Stults, in "Techniques in Protein Chemistry VII" (D. R. Marshak, ed.), p. 261. Academic Press, San Diego, 1996.

1. Dissolve the peptide at 1 mg/ml in 0.1 M ammonium acetate.
2. Stir for 3 days.
3. Lyophilize.

The peak (Fig. 4, dotted line) at 10.73 min represents the fully reduced, linear form of the peptide. After 3 days of oxidation, a new peak (Fig. 4, solid line) appeared at 10.33 min, corresponding to the cyclized form of the peptide via the disulfide bond. The abundance of the original peak had diminished proportionally. The oxidized species showed an increased mobility over the starting material, presumably reflecting a more compact molecule. The slower migrating material at 11.05 min probably represents interchain oxidized dimer. A similar behavior pattern was observed when the oxidation was followed by RP-HPLC.

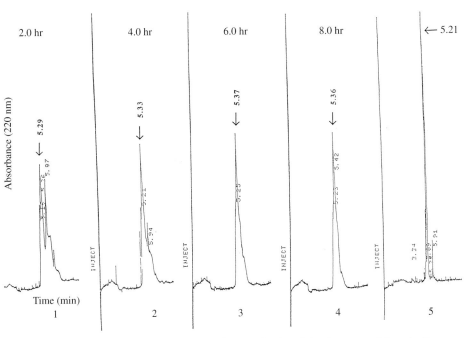

Fig. 5. Frames 1 through 4 show the real-time monitoring of the cleavage and deprotection of a peptide in 2-hr increments for a total of 8 hr. The peptide sequence was Met-Ala-Ala-Val-Ala-Ala-Ala-Ala-Gly-Leu-Tyr-Gly-Leu-Gly-Glu-Asp-Arg-Gln-His-Arg-Lys-Lys-Gln-NH₂. The cleavage and deprotection reagent used was TFA:phenol:ethanedithiol:thioanisole (85:10:2.5:2.5)(v/v). It is clearly seen that two major peaks appear after 2 hr and coalesce into one within 6 hr. Frame 5 shows the electropherogram of the extracted and lyophilized crude peptide.

Capillary electrophoresis was used to monitor the deprotection of a synthetic peptide containing the following sequence:

Met-Ala-Ala-Val-Ala-Ala-Ala-Ala-Gly-Leu-Tyr-Gly-Leu-Gly-
Glu-Asp-Arg-Gln-His-Arg-Lys-Lys-Gln-NH$_2$.

The Arg residues were protected by 4-methoxy-2,3,6-trimethylbenzenesulfonyl (Mtr) groups. As the peptide was undergoing deprotection over several hours, 10-μl aliquots were removed and immediately diluted 10-fold with 10% acetonitrile/water and a CE electropherogram taken (Fig. 5). The series of electropherograms show the appearance of new peak migrating faster than the original peak, that is, having greater positive charge (indicated by the arrow). After 6 hr, the original peak has all but disappeared, and the new peak, representing the fully deprotected peptide, predominates.

Summary

Capillary electrophoresis is a rapid and versatile electrophoretic technique that has found several applications in the field of peptide synthesis. The high resolution and charge-based separation capabilities make it very complementary to RP-HPLC. In addition, both are essentially quantitative, which makes them ideal as purity assessment techniques. Capillary electrophoresis can be used to monitor any charge-related changes in synthetic peptides such as incomplete deprotection or chemical modifications. Capillary electrophoresis can also be very sensitive to changes in the shape or size of a peptide. However, its greatest value for the characterization of synthetic peptides is when it is used in combination with other analytical techniques such as mass spectrometry, HPLC, and amino acid analysis.

[21] Fast Atom Bombardment Mass Spectrometry of Synthetic Peptides

By Šárka Beranová-Giorgianni and Dominic M. Desiderio

Introduction

Mass spectrometry (MS) is a valuable analytical tool that enables researchers to obtain rapid and accurate information about the molecular weight and primary structure (i.e., amino acid sequence) of peptides and other biomaterials. The success of MS in solving problems in the biological sciences is mainly due to the introduction and development of techniques that are capable of directly ionizing involatile, labile biomolecules.

The first widespread ionization method suitable for peptide analysis, devised in the early 1980s by Barber and co-workers, was fast atom bombardment (FAB) ionization.[1-3] Its invention resulted in a major breakthrough in MS and permitted MS to expand into new areas of chemistry and biology. FAB-MS has enjoyed an immense popularity since the mid-1980s. New ionization techniques have been developed, such as matrix-assisted laser desorption–ionization[4] (MALDI) and electrospray ionization[5] (ESI); in several aspects, these two methods are superior to FAB. These new techniques broadened significantly the applicability of MS to the analysis of macromolecules of a biological origin, and vigorous research is underway that is directed toward further improvement of MALDI and ESI. However, FAB still plays a major role, especially in peptide analysis, and will certainly do so for some time to come.

An important application of MS techniques has been in the area of synthetic peptides. The development of automated solid-phase peptide synthesis instruments has allowed the production of many different peptides within a short time, and it has also increased the demand for methods that would provide an efficient and accurate analysis of those synthetic peptides. Mass spectrometry, because of its sensitivity, speed, and high degree of molecular specificity, is a vital part of those analytical protocols that are employed to characterize synthetic peptides.

The objective of this article is to provide an overview of the basic principles of FAB-MS and of FAB tandem mass spectrometry (FAB-MS/MS) and of their use in analyzing synthetic peptides. Examples are given of the application of FAB-MS to determine molecular weights and of FAB-MS/MS to establish amino acid sequences of synthetic peptides.

Clearly, the chemist knows which peptide should have been synthesized. However, the assessment of the homogeneity of the peptide and the unequivocal confirmation of the appropriate molecular weight and of the corresponding amino acid sequence are essential criteria to confirm the identity of the synthetic peptide. Often, synthetic peptides are used in clinical studies or other long-term, expensive, difficult, and important re-

[1] M. Barber, R. S. Bordoli, R. D. Sedgwick, and A. N. Tyler, *J. Chem. Soc., Chem. Commun.*, 325 (1981).

[2] M. Barber, R. S. Bordoli, G. V. Garner, D. B. Gordon, R. D. Sedgwick, L. W. Tetler, and A. N. Tyler, *Biochem. J.* **197**, 401 (1981).

[3] M. Barber, R. S. Bordoli, G. J. Elliot, R. D. Sedgwick, and A. N. Tyler, *Anal. Chem.* **54**, 645A (1982).

[4] M. Karas and F. Hillenkamp, *Anal. Chem.* **60**, 2299 (1988).

[5] M. Yamashita and J. B. Fenn, *J. Phys. Chem.* **88**, 4451 (1984).

search projects. Therefore, it is always imperative to determine the molecular weight, amino acid sequence, and purity of the synthetic peptide, and also to provide certainty that other peptides are not present. This article focuses on these goals.

Certain areas of FAB-MS and its applications are not covered here. Those topics include biological peptides and proteins (proteolytic digests, endogenous peptides in biological extracts, etc.); glycosylated, sulfated, and phosphorylated peptides; peptide quantification; determination of disulfide bonds in peptides; and combinatorial peptide libraries. Other publications[6-9] thoroughly review those other aspects, and the interested reader is referred to those sources for further details. Furthermore, because separate articles in this volume are devoted to the MALDI and ESI of synthetic peptides,[10,11] these ionization techniques are not discussed here.

Experimental Considerations

Basic Principles of Fast Atom Bombardment Ionization

Mass spectrometry deals with gaseous ions that are created and studied in the high vacuum environment that exists inside the mass spectrometer. The production of ions from the sample to be analyzed is the key event that is essential for the success of an MS experiment. As mentioned earlier, peptides and other biomolecules in their native state became amenable to MS analysis only after the invention of suitable ionization techniques. Fast atom bombardment ionization was the first widely used desorption–ionization technique.

During FAB, the sample, which is dissolved in a drop of a nonvolatile liquid matrix, is ionized by the impact of a high energy (kiloelectron volts) particle beam. The particles employed are either fast Xe atoms (in FAB) or Cs^+ ions (in liquid secondary ion mass spectrometry, or LSIMS[12]); similar results are obtained with either a neutral or charged projectile. Throughout this article, FAB is used to describe both variants of the technique. The peptide ions in FAB are believed to arise by the desorption of preformed ions and by gas-phase reactions that occur in the region immediately above

[6] D. M. Desiderio (ed.), "Mass Spectrometry of Peptides." CRC Press, Boca Raton, Florida, 1991.
[7] J. A. McCloskey (ed.), *Methods Enzymol.* **193** (1990).
[8] W. E. Seifert, Jr., and R. M. Caprioli, *Methods Enzymol.* **270**, 453 (1996).
[9] B. L. Gillece-Castro and J. T. Stults, *Methods Enzymol.* **271**, 427 (1996).
[10] W. T. Moore, *Methods Enzymol.* **289**, [23], 1997 (this volume).
[11] D. J. Burdick and J. T. Stults, *Methods Enzymol.* **289**, [22], 1997 (this volume).
[12] W. Aberth, K. M. Straub, and A. L. Burlingame, *Anal. Chem.* **54**, 2029 (1982).

the sample–matrix droplet.[13] Fast atom bombardment ionization of a peptide mainly occurs by the attachment of one H^+ ion to the analyte molecule (M) to produce the singly charged, protonated molecule ion, $(M + H)^+$. Because FAB imparts little excitation into the $(M + H)^+$ ion, a large population of the primarily produced molecular ions remains intact, gives rise to strong, long-lasting currents, and allows an easy determination of the molecular weight of the peptide.

The matrix plays a crucial role in the ionization event because it mediates the transfer of energy to the analyte and it reduces the amount of radiation damage that the analyte would have suffered in the absence of a matrix. The matrix constantly replenishes the population of surface molecules of the peptide, and thus provides a stable ion current during bombardment. Therefore, the choice of an appropriate matrix is an important experimental concern, and several reviews have been devoted to this subject.[14,15] Generally, the matrices that are most commonly used for the FAB analysis in the positive ion mode of synthetic peptides are glycerol, thioglycerol, a mixture of dithioerythritol/dithiothreitol, and nitrobenzyl alcohol. To enhance the protonation of the peptide, acids such as trifluoroacetic acid, acetic acid, and HCl are added to the peptide–matrix mixture.

A significant drawback of FAB is the ion suppression effect. The rules that govern whether a peptide is located on the surface or within the volume of the sample–matrix droplet reflect the relative hydrophobicity of each component in the peptide–matrix mixture (an estimate of the hydrophobicity of a peptide can be obtained using the Bull and Breese index[16,17]). A relatively more hydrophobic material will migrate to the surface of the matrix and will ionize preferentially under FAB; therefore, the ionization of a mixture of peptides is not quantitative. This effect can, in the most unfavorable cases, lead to a complete suppression of the signal of one or more compounds if there is a large difference in the relative hydrophobicity/ hydrophilicity of each individual component. This problem of signal suppression can be overcome by purification [e.g., by high-performance liquid chromatography (HPLC)] of the crude reaction mixture.

After their creation in the ion source, the peptide ions are directed into the analyzer region of the mass spectrometer. The most common types of analyzers that have been coupled with FAB are the magnetic (B), electric (E), and quadrupole (Q) analyzers. Modern mass spectrometers often com-

[13] C. Fenselau and R. J. Cotter, *Chem. Rev.* **87**, 501 (1987).
[14] E. De Pauw, *Mass Spectrom. Rev.* **5**, 191 (1986).
[15] E. De Pauw, A. Agnello, and F. Derwa, *Mass Spectrom. Rev.* **10**, 283 (1991).
[16] H. B. Bull and K. Breese, *Arch. Biochem. Biophys.* **161**, 665 (1974).
[17] S. Naylor, A. F. Findeis, B. W. Gibson, and D. H. Williams, *J. Am. Chem. Soc.* **108**, 6359 (1986).

bine several different types of analyzers; the complexity of the instrumentation needed is dictated by the type of experiment to be performed. A simple configuration is sufficient for FAB-MS measurements, which will produce data about the molecular weight of the analyte. FAB-MS/MS yields amino acid sequence information and involves separation of the ion of interest, its decomposition, and analysis of the decomposition products; therefore, a more sophisticated instrument is required.

Sample Requirements

Amount of Peptide. Very simply, the more peptide that is available for a FAB-MS analysis, the better and more dependable that analysis will be. One can demonstrate the "limit of detection" of a particular experiment,[18] but in a practical sense, it is better to have available for FAB-MS/MS nanomole to picomole amounts of a synthetic peptide to establish a good signal-to-noise (S/N) ratio of the $(M + H)^+$ ion and of the amino acid sequence-determining fragment ions in the product ion spectrum.

Purity of Sample. Purity is a very important analytical objective and is a function of the molecular specificity and sensitivity of the analytical method that is used. In an HPLC analysis, a "homogeneous peak" is usually defined based on the width of the peak corresponding to the UV absorption at 210 nm (absorption of the peptide bond) or at 260 nm (absorption of an aromatic residue). In some cases, a complete UV scan has been determined. Nevertheless, there is a possibility that the HPLC apparatus, which may have been electronically "auto-zeroed," may actually present the analyst with an artificial zero. Electrically and chemically, that "zero" could still be at a high level above true background. Even a purified synthetic peptide preparation could still contain small amounts of failure sequences, oxidized sulfur atoms, residues that have not been completely deblocked, etc., that may escape detection by HPLC. Furthermore, different peptides with very similar hydrophobicities cannot be readily separated with reversed-phase (RP) HPLC. Therefore, HPLC is not always able to provide a rigorous purity assessment of a synthetic peptide.

The FAB-MS and FAB-MS/MS techniques can be successfully used to establish a presence of a synthetic peptide in a mixture. In such a case, more than one $(M + H)^+$ ion, namely, the $(M + H)^+$ ion of each peptide, will be found in the corresponding FAB-MS spectrum. The appropriate precursor ion for the target peptide can be selected in an MS/MS experiment, and the amino acid sequence can be obtained from that selected

---

[18] E. Tolun, C. Dass, and D. M. Desiderio, *Rapid Commun. Mass Spectrom.* **1**, 77 (1987).

precursor ion. In that manner, the presence of the correct synthetic peptide in the mixture can be readily determined by FAB-MS/MS; however, the target peptide must still be purified for its subsequent use. A similar strategy can be employed for the identification of impurities in a synthetic peptide sample.

Although the presence of other components does not usually hinder the FAB-MS analysis of a synthetic peptide, a problem may arise when analyzing very crude mixtures, where significant amounts of contaminants could lead to the suppression of the signal of the target peptide on FAB ionization (see above). Generally, such mixtures should be purified, at least partially, prior to the MS experiment.

Molecular Weight Determination by Fast Atom Bombardment Mass Spectrometry

The molecular weight of a synthetic peptide is a very important piece of analytical information. FAB-MS can determine the molecular weight rapidly and with great mass accuracy, even within a mixture of compounds. The molecular weight of the peptide that can be successfully analyzed depends on the capability of the ionization method and the mass range of the analyzer used. FAB can produce reliable results for peptide ions of up to approximately 6000 Da. The sensitivity of the method and the accuracy of the molecular weight obtained decrease as the mass of the peptide increases.

Figure 1 shows a FAB mass spectrum of the synthetic peptide Tyr-D-Arg-Phe-Lys-NH_2 (DALDA, molecular weight 611.3) at a mass resolution of approximately 1500. The signals in the spectrum derive from several different origins. First, peptides yield, on ionization, an $(M + H)^+$ multiplet due to the naturally occurring isotopes (mainly ^{13}C, ^{15}N, ^{18}O). Modern mass spectrometers can routinely resolve that isotope multiplet and can provide the monoisotopic mass of the peptide with an accuracy of a fraction of a mass unit. Here, FAB ionization of DALDA gives an $(M + H)^+$ ion at m/z 612.3, which dominates the spectrum shown in Fig. 1. The m/z value is in agreement with the calculated monoisotopic molecular weight of DALDA (611.3).

Second, ionization of the matrix produces a series of protonated oligomers. For example, with glycerol (molecular weight 92), ions due to $(n \times 92 + H)^+$, for $n = 1, 2, 3, \ldots$, etc. at m/z 93, 185, 277, etc., respectively, could appear. In the spectrum shown in Fig. 1, the ion at m/z 185 arises from protonated glycerol dimer and is clearly discernible. The matrix signals are known and are readily identifiable, and thus they do not present any obstacle for the spectral interpretation and the detection of the peptide

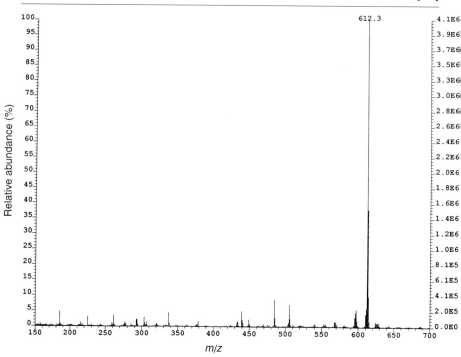

Fig. 1. The FAB mass spectrum of the synthetic tetrapeptide Tyr-D-Arg-Phe-Lys-NH$_2$ (DALDA).

(M + H)$^+$ ion. The FAB spectra of common FAB matrices have been published.[19]

Third, in-source decomposition of an excited precursor ion, typically the (M + H)$^+$ of the peptide, can generate corresponding fragment ions. The FAB-MS spectrum that is shown in Fig. 1 contains a set of signals that may be amino acid sequence-determining fragment ions; for example, at m/z 505, 484, 449, 439, 337, and 303. Note that their abundance relative to the (M + H)$^+$ ion does not exceed 10%; that relatively low value demonstrates that the degree of fragmentation of FAB-generated ions is low. The extent of decomposition also depends on the primary structure of the peptide. Although those fragment ions present in the FAB mass spectrum of DALDA do provide some information about the amino acid sequence of that peptide, such an outcome is rather limited only to those peptides that possess favorable dissociation characteristics. More frequently, the

[19] C. E. Costello, *Methods Enzymol.* **193**, 875 (1990).

FAB spectrum of a synthetic peptide does not unequivocally reveal the entire amino acid sequence of that peptide.

Fourth, ionization of other components of the analyte mixture (e.g., residual reagents, side products), if they are present, can contribute to the ions that are observed in a FAB spectrum. The appearance of such ions provides the molecular weight of each impurity and may help in their identification. It should be pointed out, however, that such information is qualitative in nature, because different mixture components will not be ionized with the same efficiency on FAB.

Fifth, attachment of cations other than H^+ (e.g., metals such as Na^+ and K^+) to the analyte (and matrix) can give, for example, $(M + Na)^+$ adducts. Finally, under certain conditions, chemical reactions between the peptide and the matrix can occur. For example, on bombardment, radiation damage of certain matrices produces radicals that could interact with various reactive centers (e.g., with the side chains of basic amino acids) in the peptide molecule. Such a process would generate adducts at masses higher than the $(M + H)^+$ ion of the peptide.[20] The use of a different matrix and/or an adjustment of the pH of the sample mixture should help to identify those signals that arise from reactions of the analyte with the FAB matrix.

The determination of the molecular weight is a necessary step in the characterization of a synthetic peptide. However, the molecular weight alone is not an unequivocal datum that can be used to confirm the identity of a peptide, even for synthetic peptides. Obviously, the molecular weight and the amino acid sequence of a synthetic peptide are not complete unknowns, because the chemist knows the desired outcome of the synthesis. Therefore, if the molecular weight found by FAB-MS matches the expected value, then such a result provides a favorable indication, but not a solid proof, that the correct peptide has been synthesized. The amino acid sequence of the peptide still must be verified. That task cannot often be accomplished by FAB-MS alone, owing to the lack of an unequivocal link between the $(M + H)^+$ ion and its products, and of structure-specific fragmentation. Such data can readily be obtained from an MS/MS experiment.

Amino Acid Sequence Determination by Tandem Mass Spectrometry

Tandem mass spectrometry can readily determine the amino acid sequence of a peptide. MS/MS encompasses the advantages of MS (i.e., speed, sensitivity, and reliability), and it has (together with the Edman degradation method and X-ray crystallography) the highest degree of molecular speci-

[20] C. Dass and D. M. Desiderio, *Anal. Chem.* **60,** 2723 (1988).

MS-1

precursor ion
selection

precursor ion
fragmentation

MS-2

product ion
analysis

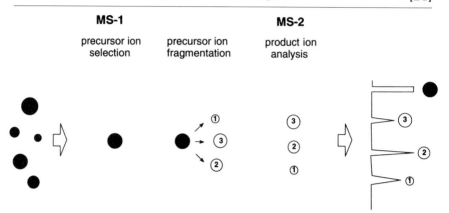

FIG. 2. Basic principles of MS/MS. The filled and numbered circles depict the precursor and product ions, respectively.

ficity of any analytical technique. Another significant asset of MS/MS is its capability to obtain amino acid sequence information from a mixture of components without any extensive purification. As long as the ion current due to the $(M + H)^+$ ion of interest is sufficiently high and long-lasting, the final purification of that ion is achieved in the first stage of the MS/MS experiment.

The basic principles of MS/MS[21] are illustrated in Fig. 2. The FAB-produced $(M + H)^+$ ion of interest is selected as the precursor ion in the first region of the instrument (MS-1). That mass-selected $(M + H)^+$ ion undergoes activating collisions, usually with a gaseous target; as a consequence of these collisions, that precursor ion dissociates. The dissociation products are analyzed in the second mass spectrometer (MS-2) and are recorded in the corresponding MS/MS spectrum (the product ion spectrum). The resulting spectrum contains product ions that originate solely from the mass-selected peptide $(M + H)^+$ ion and that reflect the primary structure of the peptide. Thus, from the MS/MS results, the amino acid sequence of the peptide can be deduced.

Two types of MS/MS experiments are distinguished, based on the magnitude of the kinetic energy (E_{kin}) of the colliding precursor ion. High-energy MS/MS involves precursor ions with a E_{kin} of 5–10 keV and occurs in those instruments that consist of a combination of magnetic and electric analyzers. However, MS/MS in a quadrupole mass spectrometer occurs under low energy conditions, which entail precursor ions with a kinetic energy of less

---

[21] K. L. Busch, G. L. Glish, and S. A. McLuckey, "Mass Spectrometry/Mass Spectrometry: Techniques and Applications of Tandem Mass Spectrometry." VCH, New York, 1988.

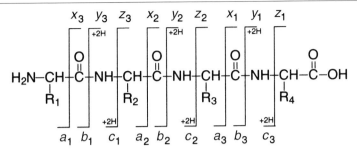

FIG. 3. General fragmentation pattern of protonated peptide ions, and nomenclature of the amino acid sequence-determining fragment ions.

than 100 eV. The nature of the two energy regimes affects the fragmentation characteristics of the precursor ion. That difference must be taken into account when interpreting MS/MS results, and particularly when comparing low- and high-energy MS/MS spectra.

Principles and Nomenclature of Peptide Ion Fragmentation

The details of peptide ion fragmentations have been discussed elsewhere,[22,23] thus a short overview is given here. Briefly, the major amino acid sequence-determining fragment ions of a protonated peptide precursor ion arise from the cleavage of a bond in the backbone of the peptide, with charge retention on either the N or the C terminus. The N-terminal ions are designated a_n, b_n, and c_n, and the C-terminal ions are denoted x_n, y_n, and z_n (Fig. 3). The ions of the same type with adjacent n values differ by the mass of the amino acid residue. Thus, from a complete series of fragment ions, the amino acid sequence can be readily deduced. Alternatively, the primary structure can be established if more than one incomplete but overlapping series are present in the MS/MS spectrum.

Applications of Fast Atom Bombardment Mass Spectrometry and Fast Atom Bombardment Tandem Mass Spectrometry to Study Synthetic Peptides

Because of the favorable characteristics outlined above, FAB-MS and FAB-MS/MS have become an essential part of many of the current analyti-

---

[22] P. Roepstorff and J. Fohlman, *Biomed. Mass Spectrom.* **11**, 601 (1984).
[23] K. Biemann, *Biomed. Environ. Mass Spectrom.* **16**, 99 (1988).

cal strategies that are used for the characterization of synthetic peptides, and numerous examples have appeared in the literature. This section discusses several illustrative applications of the FAB-MS and FAB-MS/MS techniques. The review presented here is by no means inclusive; rather, it is intended to demonstrate to the reader, with selectively chosen data from the literature, the usefulness and potential of the methodology.

Determination of Amino Acid Sequence of Synthetic Opioid Peptides

Several synthetic enkephalin-related peptides have been characterized by low-energy MS/MS.[24] These peptides are intended for use in a long-term research program that focuses on the involvement of opioid peptides in obstetrics; each peptide has been designed to possess specific opioid receptor-binding properties. MS/MS was used to verify the primary structure of these peptides and to select a suitable product ion for their future quantification in biological fluids.

Figure 4 shows the low-energy product ion spectrum of the synthetic hexapeptide Tyr-D-Thr-Gly-Phe-Leu-Thr (DTLET; molecular weight 700.3). The amino acid sequence of DTLET and the product ions that were observed are depicted in Fig. 5. The spectrum of Fig. 4 contains a signal for the $(M + H)^+$ precursor ion at m/z 701 and a number of its dissociation products that derive from both termini. The C-terminal fragment ions contain a complete y series that includes ions at m/z 538 (y_5), 437 (y_4), 380 (y_3), 233 (y_2), and 120 (y_1). The mass differences between the consecutive y product ions, namely, 163 ($MH^+ - y_5$), 101 ($y_5 - y_4$), 57 ($y_4 - y_3$), 147 ($y_3 - y_2$), and 113 ($y_2 - y_1$), correspond to tyrosine, threonine, glycine, phenylalanine, and leucine residues, respectively. Furthermore, the spectrum of Fig. 4 contains C terminus-containing fragment ions of the z series at m/z 521 (z_5), 420 (z_4), and 216 (z_2), and an x_5 ion (m/z 564). The N-terminal product ions include a nearly complete set of b ions (except b_1) and a partial a series (a_1, a_4, and a_5). In summary, the product ion spectrum of DTLET contains a combination of different types of fragment ions that allows one to deduce readily and with confidence the amino acid sequence of this peptide.

The MS/MS analysis of another synthetic opioid peptide analog, N,N-diallyl-Tyr-Aib-Aib-Phe-Leu (ICI 174,864; molecular weight 691.4, where Aib is 2-aminoisobutyric acid), produces the spectrum shown in Fig. 6; the amino acid sequence of ICI 174,864 and the observed fragmentation pattern are depicted in Fig. 7. This peptide has a blocked amino terminus and

[24] J.-L. Tseng, L. Yan, G. H. Fridland, and D. M. Desiderio, *Rapid Commun. Mass Spectrom.* **9**, 264 (1995).

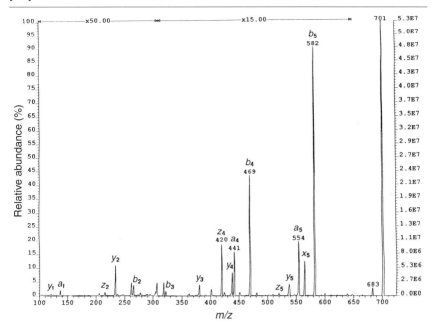

FIG. 4. Low energy MS/MS spectrum of the $(M + H)^+$ ion of Tyr-D-Thr-Gly-Phe-Leu-Thr (DTLET). [From J.-L. Tseng, L. Yan, G. H. Fridland, and D. M. Desiderio, *Rapid Commun. Mass Spectrom.* **9,** 264 (1995). Copyright John Wiley & Sons Limited. Reproduced with permission.]

would, therefore, not be amenable to sequencing by Edman degradation. MS/MS reveals unequivocally the primary structure of this pentapeptide from the presence of the structure-specific fragment ions a_1 (m/z 216), a_2 (m/z 301), a_4 (m/z 533), b_1 (m/z 244), b_2 (m/z 329), b_3 (m/z 414), and b_4 (m/z 561). Note that all of the prominent product ions derive only from the N terminus; this fact results from the stabilizing effect of the two allyl

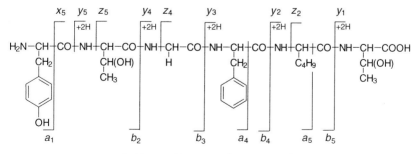

FIG. 5. Observed fragmentation pattern of DTLET.

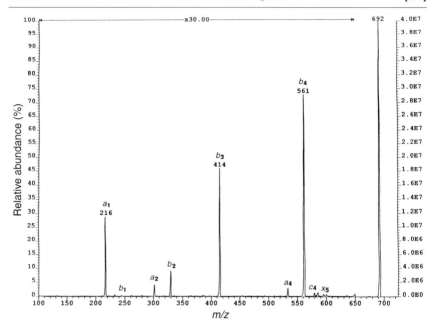

FIG. 6. Low energy MS/MS spectrum of the (M + H)+ ion of N,N-diallyl-Tyr-Aib-Aib-Phe-Leu (ICI 174,864). [From J.-L. Tseng, L. Yan, G. H. Fridland, and D. M. Desiderio, *Rapid Commun. Mass Spectrom.* **9**, 264 (1995). Copyright John Wiley & Sons Limited. Reproduced with permission.]

groups, which supply a high electron density to the amino terminus of this pentapeptide.

Assessment of Homogeneity of Synthetic Peptides by Fast Atom Bombardment Mass Spectrometry

The purification of a synthetic peptide usually involves one or more separation steps, most commonly using RP-HPLC. Although RP-HPLC is a very powerful analytical tool, it alone may not be sufficient for the definitive confirmation of the purity of the peptide. Bösze *et al.*[25] reported on such a finding.

This study involved the preparation and characterization of two synthetic oligopeptides with the amino acid sequences Leu-Arg-Ala-Leu-Arg-Gln-Met (molecular weight 886.5) and Leu-Ala-Pro-Glu-Asp-Pro-

[25] S. Bösze, M. Mák, H. Medzihradszky-Schweiger, and F. Hudecz, *J. Chromatogr. A* **668**, 345 (1994).

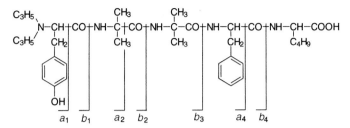

FIG. 7. Observed fragmentation pattern of ICI 174,864.

Glu-Asp-Ser-Ala-Leu-Leu-Glu-Asp-Pro-Val-Gly-NH$_2$ (molecular weight 1764.9). The heptapeptide Leu-Arg-Ala-Leu-Arg-Gln-Met is structurally related to the C-terminal region of the cytokine interleukin-6; the heptadecapeptide amide Leu-Ala-Pro-Glu-Asp-Pro-Glu-Asp-Ser-Ala-Leu-Leu-Glu-Asp-Pro-Val-Gly-NH$_2$ contains a portion of the sequence of a glycoprotein from the herpes simplex virus.

Both peptides were synthesized by the solid-phase technique, and the crude reaction mixtures were separated by gel permeation and were repeatedly purified by RP-HPLC. The molecular weight and primary structure of the HPLC-purified peptides were investigated by amino acid analysis, Edman degradation, and MS. The results showed that repeated HPLC purification of the Leu-Arg-Ala-Leu-Arg-Gln-Met preparation successfully removed all contaminants and yielded the target peptide in a pure form. In contrast, although the final HPLC profile of the heptadecapeptide Leu-Ala-Pro-Glu-Asp-Pro-Glu-Asp-Ser-Ala-Leu-Leu-Glu-Asp-Pro-Val-Gly-NH$_2$ also displayed only a single peak (at ∼12.5 min, Fig. 8a), MS analysis of the collected fraction revealed the presence of two additional components. The main signal in the corresponding FAB-MS spectrum (at m/z 1766.2, Fig. 8b) was in agreement with the calculated molecular weight of Leu-Ala-Pro-Glu-Asp-Pro-Glu-Asp-Ser-Ala-Leu-Leu-Glu-Asp-Pro-Val-Gly-NH$_2$. However, two other ions at m/z 1527.5 and 1254.7, possibly arising from truncated and other modified derivatives, were also found. It is interesting to note that the impurities in the heptadecapeptide sample were also detected by Edman degradation, but not by the amino acid analysis.

Although the precise identity of the impurities was not determined, that study decisively showed that an overinterpretation of the HPLC results could lead to erroneous conclusions concerning the purity of synthetic peptide, and that a combination of MS with other analytical techniques should be used to obtain an accurate characterization of a synthetic peptide.

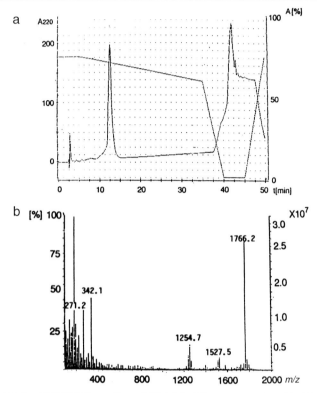

FIG. 8. (a) Final RP-HPLC profile and (b) FAB-MS spectrum of purified Leu-Ala-Pro-Glu-Asp-Pro-Glu-Asp-Ser-Ala-Leu-Leu-Glu-Asp-Pro-Val-Gly-NH_2. (Reprinted from *J. Chromatogr. A.* **668,** S. Bösze, M. Mák, H. Medzihradszky-Schweiger, and F. Hudecz, p. 345, Copyright 1994 with kind permission of Elsevier Science–NL, Sara Burgerhartstraat 25, 1055 KV Amsterdam, The Netherlands.)

Characterization of Side Products in Synthetic Peptides by Fast Atom Bombardment Tandem Mass Spectrometry

The primary structure of impurities in a synthetic peptide sample can be assessed rapidly by FAB-MS/MS, as demonstrated by Mathews *et al.*[26] for the development of the synthesis of the renin substrate decapeptide Pro-His-Pro-Phe-His-Leu-Val-Ile-His-D-Lys (molecular weight 1223.7, Fig. 9). When the decapeptide was prepared with the benzyloxymethyl (Bom) His-protecting group, FAB-MS analysis revealed the presence of several

[26] W. R. Mathews, T. A. Runge, P. E. Haroldsen, and S. J. Gaskell, *Rapid Commun. Mass Spectrom.* **3,** 314 (1989).

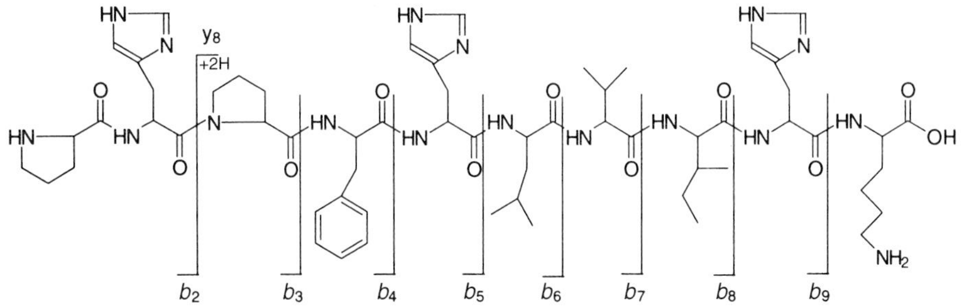

FIG. 9. Amino acid sequence of the synthetic renin substrate decapeptide. The MS/MS fragmentation pattern of the (M + H)$^+$ ion is indicated. [From W. R. Mathews, T. A. Runge, P. E. Haroldsen, and S. J. Gaskell, *Rapid Commun. Mass Spectrom.*, **3**, 314 (1989). Copyright John Wiley & Sons Limited. Adapted with permission.]

by-products, including components with masses 12 and 14 Da higher. MS/MS was used to investigate the structure of those side products. The results of the MS/MS analysis of the (M + H)$^+$ and the (M + 12 + H)$^+$ ions are shown in Fig. 10.

The (M + H)$^+$ ion of the target decapeptide yielded a nearly complete set of amino acid sequence-determining b-type ions, b_3 to b_9 (see Figs. 9 and 10a). In the product ion spectrum of the (M + 12 + H)$^+$ ion (Fig. 10b), two series of b-type fragment ions were observed. In the first series, between $b_2 + 12$ and $b_9 + 12$ (m/z 247, 344, 628, 741, 840, 953, and 1090), all of the product ions displayed the shift of 12 mass units, suggesting that the 12-Da modification is located at the two N-terminal amino acids. At the same time, the appearance of a parallel b-type product ion series, b_5 to b_9 (m/z 616, 729, 828, 941, 1078), which showed no change compared to the fragmentation pattern of the target decapeptide, pointed out a modified C terminus. These data indicated that the (M + 12 + H)$^+$ ion consists of two different isobaric components. Additional support for that conclusion came from the MS/MS analysis of the synthetic peptide mixture after acetylation; those data demonstrated that, for each component, the structural modification blocked a site of N-acetylation. MS/MS data of the (M + 14 + H)$^+$ by-product of the synthesis were consistent with the methylation of the side chain of the C-terminal Lys.

Thus, in this investigation, FAB-MS/MS allowed the determination of the structure of the side products and helped to pinpoint the modification sites. It was postulated that the impurities are either methylene-incorporated (M + 12) or methylated (M + 14) renin decapeptide derivatives (Fig. 11), which arise from the reaction of the peptide with the formaldehyde that is liberated from the Bom group during the deprotection step. On the

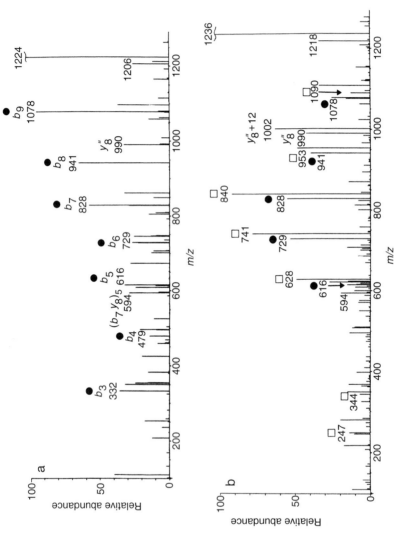

FIG. 10. Tandem mass spectra of (a) the $(M + H)^+$ ion (m/z 1224) and (b) the $(M + 12 + H)^+$ ion (m/z 1236) from the synthetic renin substrate decapeptide. The filled circles and empty squares denote the b and $b + 12$ product ion series, respectively. [From W. R. Mathews, T. A. Runge, P. E. Haroldsen, and S. J. Gaskell, *Rapid Commun. Mass Spectrom.*, **3**, 314 (1989). Copyright John Wiley & Sons Limited. Reproduced with permission.]

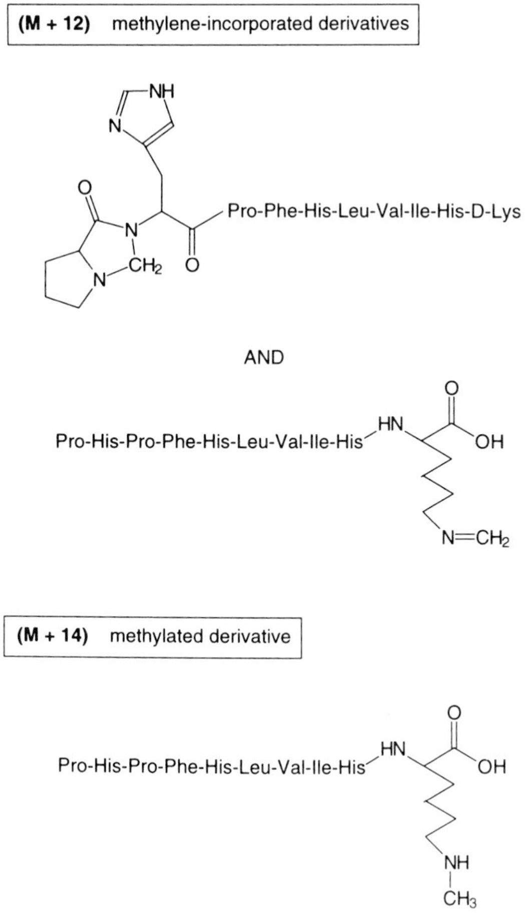

FIG. 11. Proposed structures of the by-products that are observed in the synthetic renin substrate decapeptide preparation. [From W. R. Mathews, T. A. Runge, P. E. Haroldsen, and S. J. Gaskell, *Rapid Commun. Mass Spectrom.*, **3**, 314 (1989). Copyright John Wiley & Sons Limited, adapted with permission; and from M. A. Mitchell, T. A. Runge, W. R. Mathews, A. K. Ichhpurani, N. K. Harn, P. J. Dobrowolski, and F. M. Eckenrode, *Int. J. Pept. Protein Res.* **36**, 350 (1990).]

basis of these findings, changes in the synthesis of Pro-His-Pro-Phe-His-Leu-Val-Ile-His-D-Lys were implemented, and they included the use of a different His-protecting group or the addition of a formaldehyde scavenger in the cleavage reaction.[27]

[27] M. A. Mitchell, T. A. Runge, W. R. Mathews, A. K. Ichhpurani, N. K. Harn, P. J. Dobrowolski, and F. M. Eckenrode, *Int. J. Pept. Protein Res.* **36**, 350 (1990).

Analysis of Synthetic Peptide by Microcolumn High-Performance Liquid Chromatography–Continuous-Flow Fast Atom Bombardment Mass Spectrometry

The principle of the continuous-flow FAB (cf-FAB) technique, developed by Caprioli *et al.*,[28,29] involves a constant delivery of an aqueous sample solution, containing only a few percent of the FAB matrix, into the ion source. The features of cf-FAB enabled its extensive use in the interfacing of mass spectrometers with separation devices such as HPLC. Probably the most significant advantages offered by cf-FAB, as compared to conventional (static) FAB, are the decrease in the matrix-derived chemical noise and the reduction of the suppression effect; these characteristics are of particular value when analyzing complex mixtures.

Continuous-flow FAB-MS in an on-line combination with a microcolumn HPLC (μHPLC) has been employed by McKellop *et al.*[30] to improve the analysis of crude reaction mixtures from an automated solid-phase synthesis. In some cases, static FAB-MS may not be able to distinguish between the elimination of a protecting group as a result of FAB-induced fragmentation and the $(M + H)^+$ ion of the deprotected peptide because the masses of the products from both processes are identical. Figure 12 shows the static FAB mass spectrum of the crude mixture of the protected pentapeptide Abu-Asp(O*t*Bu)-Gln-Pro-Ahx (where Abu is 2-aminobutyric acid, *t*Bu is *tert*-butyl, and Ahx is 2-aminohexanoic acid). The spectrum contains three major signals: the expected $(M + H)^+$ ion at m/z 613.4 for the correct *tert*-butyl-protected peptide, an $(M + H)^+$ ion corresponding to a peptide containing an additional Abu residue (m/z 698.4), and an ion at m/z 557.3. The mass of the m/z 557.3 signal is consistent with the molecular weight of the deprotected (undesired) peptide product. However, the loss of the *tert*-butyl protecting group, which constitutes the major fragmentation pathway of Abu-Asp(O*t*Bu)-Gln-Pro-Ahx, would also produce the ion at m/z 557.3. The exact origin of the m/z 557.3 ion cannot be determined from the static FAB experiment.

Analysis of the same peptide mixture by μHPLC/cf-FAB-MS yields the desired information. The overlap of the selected ion chromatograms of the ion currents that are produced by the m/z 613.4 and 557.3 ions and the full scan spectrum acquired during the elution of the protected peptide (Fig. 13) both indicate that the signal at m/z 557.3 is not due to an impurity, but rather that it arises by an ionization-induced fragmentation of the protected Abu-Asp(O*t*Bu)-Gln-Pro-Ahx.

[28] R. M. Caprioli, T. Fan, and J. S. Cottrell, *Anal. Chem.* **58,** 2949 (1986).
[29] R. M. Caprioli (ed.), "Continuous-Flow Fast Atom Bombardment Mass Spectrometry." Wiley, Chichester, 1990.
[30] K. McKellop, W. Davidson, G. Hansen, D. Freeman, and P. Pallai, *Pept. Res.* **4,** 40 (1991).

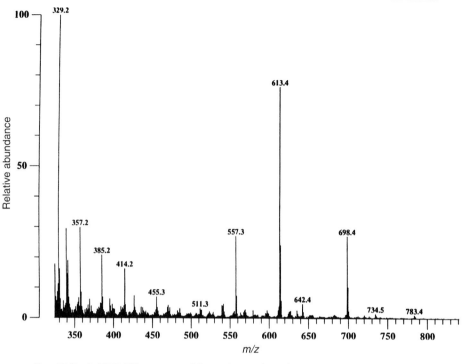

FIG. 12. Static FAB-MS spectrum of the crude reaction mixture obtained from the synthesis of Abu-Asp(OtBu)-Gln-Pro-Ahx. [Reproduced with permission from K. McKellop, W. Davidson, G. Hansen, D. Freeman, and P. Pallai, *Pept. Res.* **4**, 40 (1991).]

Multicenter Evaluation of Peptide Synthesis

The Association of Biomolecular Resource Facilities (ABRF) has established a Committee on Peptide Synthesis and Mass Spectrometry (CPSMS). The ABRF includes over 130 service laboratories that synthesize and analyze peptides and proteins for academic, government, and private institutions. The quality of the various synthetic protocols is periodically evaluated by the CPSMS.[31,32] A model peptide with a specific amino acid sequence that was designed by CPSMS is synthesized by the participating member laboratories using various synthetic methodologies. The resulting peptides are analyzed by an array of analytical techniques such as amino acid analysis,

[31] G. B. Fields, S. A. Carr, D. R. Marshak, A. J. Smith, J. T. Stults, L. C. Williams, K. R. Williams, and J. D. Young, *in* "Techniques in Protein Chemistry IV" (R. H. Angeletti, ed.), p. 229. Academic Press, San Diego, 1993.

[32] G. B. Fields, R. H. Angeletti, S. A. Carr, A. J. Smith, J. T. Stults, L. C. Williams, and J. D. Young, *in* "Techniques in Protein Chemistry V" (J. W. Crabb, ed.), p. 501. Academic Press, San Diego, 1994.

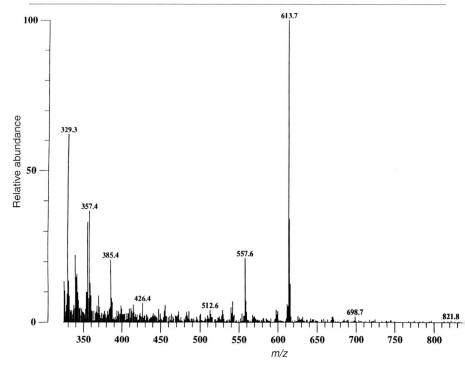

Fig. 13. Full scan cf-FAB mass spectrum obtained during the elution of the μHPLC peak corresponding to the protected peptide Abu-Asp(O*t*Bu)-Gln-Pro-Ahx. [Reproduced with permission with K. McKellop, W. Davidson, G. Hansen, D. Freeman, and P. Pallai, *Pept. Res.* **4**, 40 (1991).]

HPLC, capillary electrophoresis, and MS. That integrated approach permits the critical examination of the successes of individual synthetic steps, and helps to uncover any potential weakness of the various synthetic protocols. The CPSMS also compares the performance and applicability of the analytical techniques that are utilized for synthetic peptide analysis.

Summary

Fast atom bombardment mass spectrometry plays a continuing and effective role in the rapid and efficient analysis of synthetic peptides. In this article, the basic principles of FAB-MS and FAB-MS/MS are reviewed, and the limitations and pitfalls of the method are discussed. The potential of the technique is illustrated by several selected applications.
The molecular weight of a synthetic peptide can be readily and accu-

rately determined by FAB-MS. The sensitivity of the FAB-MS method also makes it extremely useful for the evaluation of the purity of a peptide. FAB-MS/MS allows the elucidation of the primary structure of the target peptide, even in a mixture, and also permits the rapid identification of synthetic side products. Hence, FAB-MS and FAB-MS/MS aid in the unequivocal characterization of synthetic peptides. Furthermore, the information that is gained from the FAB analysis can help to unravel potential problems that may be associated with the synthesis.

Acknowledgments

The authors gratefully acknowledge financial support from the National Institutes of Health (GM 26666).

[22] Analysis of Peptide Synthesis Products by Electrospray Ionization Mass Spectrometry

By Daniel J. Burdick and John T. Stults

Introduction

The postsynthetic analysis of peptides is a crucial step to ensure that the correct molecular structure has been produced. Of the various analytical tools available for structure verification and elucidation, mass spectrometry (MS) provides the most direct and most efficient molecular measurement. A number of mass spectrometric techniques are widely available and have been successfully used for synthetic peptide analysis, including fast atom bombardment (FAB),[1,2] matrix-assisted laser desorption/ionization (MALDI),[3–5] and electrospray ionization (ESI).[6,7]

It is our experience that each of these three techniques provide very similar results for confirmation of correct synthesis and identification of synthetic by-products. Furthermore, in a number of annual studies by the Association of Biomolecular Resource Facilities (ABRF), peptides synthe-

[1] W. E. Seifert and R. M. Caprioli, *Methods Enzymol.* **270**, 453 (1996).
[2] S. Veranová-Giorgianni and D. M. Desiderio, *Methods Enzymol.* **289**, 21, 1997 (this volume).
[3] J. T. Stults, *Curr. Opin. Struct. Biol.* **5**, 691 (1995).
[4] R. C. Beavis and B. T. Chait, *Methods Enzymol.* **270**, 519 (1996).
[5] W. T. Moore. *Methods Enzymol.* **289**, 23, 1997 this volume.
[6] J. F. Banks, Jr., and C. M. Whitehouse, *Methods Enzymol.* **270**, 486 (1996).
[7] M. Mann and M. Wilm, *Trends Biochem. Sci.* **20**, 219 (1995).

sized by member laboratories were analyzed by FAB-, MALDI-, and ESI-MS. These mass spectrometric techniques were compared for their ability to determine the proportion of correct product and the identities of by-products.[8–11] The studies showed that each of the three techniques provided nearly identical results, and the results were in accord with data from high-performance liquid chromatography (HPLC) and N-terminal sequencing for most samples. Of the analytical techniques studied (amino acid analysis, N-terminal peptide sequencing, and HPLC), however, MS was the only one that provided direct evidence for the presence/absence of the expected synthetic product. For example, the chemistries used for amino acid analysis and N-terminal sequencing may cleave residual protecting groups to show good results for a poor synthesis. Confirmation of proper synthesis by HPLC retention time requires a peptide standard for comparison. It is important to remember, however, that molecular mass measurement provides data only on the composition of a peptide; isomeric compounds (e.g., rearrangements, swaps of amino acids due to synthesizer/operator error) are not distinguished. Fortunately, these types of synthetic errors are normally minor. When there is a question on the sequence, tandem MS (see below) can be used to determine the sequence.

An advantage of electrospray ionization (ESI) over the other techniques is its routine interface to HPLC. Indeed, liquid chromatography–mass spectrometry (LC-MS) is an important tool for peptide analysis, as we demonstrate. Furthermore, ESI sources are frequently found on triple quadrupole and ion trap instruments, which can provide peptide fragmentation by tandem MS for peptide sequence determination or location of sites of modification. The routine use of ESI in peptide synthesis has been described extensively,[12,13] and its use for characterizing peptide libraries has been of

[8] G. B. Fields, S. A. Carr, D. R. Marshak, A. J. Smith, J. T. Stults, L. C. Williams, K. R. Williams, and J. D. Young, in "Techniques in Protein Chemistry IV" (R. H. Angeletti, ed.), p. 229. Academic Press, San Diego, 1993.

[9] G. B. Fields, R. H. Angeletti, S. A. Carr, A. J. Smith, J. T. Stults, L. C. Williams, and J. D. Young, in "Techniques in Protein Chemistry V" (J. W. Crabb, ed.), p. 501. Academic Press, San Diego, 1994.

[10] G. B. Fields, R. H. Angeletti, L. F. Bonewald, W. T. Moore, A. J. Smith, J. T. Stults, and L. C. Williams, in "Techniques in Protein Chemistry VI" (J. W. Crabb, ed.), p. 539. Academic Press, San Diego, 1995.

[11] R. H. Angeletti, L. Bibbs, L. F. Bonewald, G. B. Fields, J. S. McMurray, W. T. Moore, and J. T. Stults, in "Techniques in Protein Chemistry VII" (D. R. Marshak, ed.), p. 261. Academic Press, San Diego, 1996.

[12] L. C. Packman, M. Quibell, and T. Johnson, *Pept. Res.* **7**, 125 (1994).

[13] S. S. Smart, T. J. Mason, P. S. Bennell, N. J. Maeij, and H. M. Geysen, *Int. J. Pept. Protein Res.* **47**, 47 (1996).

particular interest lately.[14-17] To demonstrate the use of ESI for peptide synthesis, the characteristics of ESI are described below, followed by examples of the use of electrospray ionization–mass spectrometry (ESI-MS) for improving the efficiency of peptide synthesis and purification, and for establishing the quality of the product. This article emphasizes quadrupole and ion trap instruments because these are currently in widest use.

Characteristics of Electrospray Ionization

Electrospray produces ions from molecules in solution by spraying the solution from a sharp tip in the presence of a high electric field at atmospheric pressure. The spray produces small droplets that are highly charged. Rapid desolvation and charge-induced breakup of the droplets leads to droplets of submicron size. Although the exact mechanism of ion formation remains unclear, ions may be formed by complete evaporation of solvent from droplets that contain a single ion, or ions may be expelled from a droplet by the high electric field on the droplet surface.[18] The ions move from atmospheric pressure to the high vacuum of the mass spectrometer usually through a series of capillaries, nozzles, and/or skimmers. During this transfer process into the high vacuum, the ions experience collisions with residual gas molecules at an intermediate pressure that provides sufficient energetic excitation to remove residual solvent molecules, yet retain the covalent bonds until the ions are detected. Another characteristic of the ions is their propensity to acquire multiple charges. The charges attached are normally protons in the positive ion mode. For peptides the number of added protons (charges) is roughly correlated with the number of basic sites on the molecule and its size, but there is frequently a distribution of charge number that may vary with instrument settings and solution pH. For example, a 12-residue peptide with one basic amino acid (Arg, Lys, His) and a free amino terminus, in a solution of 0.1% trifluoroacetic acid (TFA)/ 30% acetonitrile (v/v), at an orifice voltage of 80 V, would likely show the $[M + 2H]^{2+}$ charge peak as most abundant with a lesser $[M + H]^+$ peak.

The mass spectrometer measures the mass-to-charge ratio (m/z) of the ion. The charge is determined by the isotope spacing [e.g., the ^{12}C and ^{13}C isotopes are 0.5 u (unified atomic mass unit) apart for +2 ions] or by the

[14] J. W. Metzger, K. H. Wiesmueller, V. Gnau, J. Bruenjes, and G. Jung, *Angew. Chem., Int. Ed. Engl.* **105,** 894 (1993).
[15] J. W. Metzger, C. Kempter, K.-H. Wiesmueller, and G. Jung, *Anal. Biochem.* **219,** 261 (1994).
[16] Y. Dunayevskiy, P. Vouros, T. Carell, E. A. Wintner, and J. Rebek, Jr., *Anal. Chem.* **67,** 2906 (1995).
[17] Y. M. Dunayevskiy, P. Vouros, E. A. Wintner, G. W. Shipps, T. Carell, and J. Rebek, Jr., *Proc. Natl. Acad. Sci. U.S.A.* **93,** 6152 (1996).
[18] P. Kebarle and L. Tang, *Anal. Chem.* **65,** 972A (1993).

presence of two or more different charges from the same mass. Most commercial instrument data systems provide software to identify the charge and transform or deconvolute the mass-to-charge axis of a spectrum into a mass axis. One should be cautious in relying solely on transformed spectra, however, because the charge is not determined unambiguously in all cases. The multiple charges allow even large molecules to be analyzed with mass spectrometers of limited m/z range. As a result, the ions formed by ESI can be analyzed in a variety of mass spectrometers (quadrupole, ion trap, magnetic sector, time-of-flight). For peptides, the resulting data from these instruments are nearly identical in terms of ion abundances. Differences arise chiefly in resolution (ability to separate adjacent peaks) and mass accuracy.

Mixtures of peptides are frequently analyzed by ESI-MS. The relative intensities of peaks are often assumed to be indicative of the ratios of peptide abundances. In many cases this is true, especially when the peptides are closely related, as may be the case with a peptide and its synthetic by-products. Indeed, in the ABRF studies mentioned earlier,[8-11] the ratios of ion abundances of various products were surprisingly consistent with the available quantitative data from other techniques. A separate study of 460 crude peptides showed a high correlation between ESI-MS and HPLC for determining the percentage of the expected product.[13] Nonetheless, one should exercise caution when using ion abundances for quantitation. Molecules may have significant differences in ionization efficiencies, and selective suppression of signals is a common occurrence in mixtures. Thus, in the absence of well-characterized internal standards or independent analyses, results should be viewed as qualitative—"semiquantitative" at best—to avoid drawing misleading conclusions.

The mass of a synthetic peptide is an excellent indication of the composition of the peptide. Differences from the expected mass are often due to incomplete removal of protecting groups. Table I gives a list of mass differences commonly observed for by-products of peptide synthesis. The amino acid sequence, however, is not given by the mass alone, so residues that are in the incorrect order (e.g., due to operator error) or for which compensating substitutions have identical mass (e.g., Gly_2 = Asn) are not identified. Tandem mass spectrometry (see below) can be used to provide information on the sequence.

One of the limitations of ESI is its extremely limited tolerance of nonvolatile solution components. Sodium, even in small amounts (>0.01 mM) will also act as a charge carrier and may lead to extra peaks at $+22$ u. At higher concentrations (>10 mM), sodium and other ionic species may totally suppress analyte ionization. Fortunately, convenient desalting methods are available (see below). Samples taken directly from volatile HPLC solvent systems (e.g., water/acetonitrile with TFA, HCl, formic acid, or acetic acid modifier) are ideally suited for ESI. Most detergents, however, even in minute amounts, are detrimental to electrospray. Exceptions include ≤0.1% 3-[(3-Cholamidopropyl)dimethylammonio]-1-propanesulfonate CHAPS or octyl-β-glucoside.[18b]

When compared with the other available mass spectrometric techniques ESI offers the advantages of relatively low cost, convenient automation, and ease of interfacing on-line with chromatography or capillary electrophoresis systems. In addition, the multiple charges on most peptides enhance the ability of MS to provide structure elucidation or confirmation by fragment ion analysis.

Methods and Materials

9-Fluorenylmethyloxycarbonyl Peptide Synthesis

Peptide 1 (see Fig. 2 for structure) is synthesized using standard automatic 9-fluorenylmethyloxycarbonyl (Fmoc) solid-phase methods on Wang (p-alkoxybenzyl alcohol) resin.[19] Piperidine (20%) in 1-methyl-2-pyrrolidinone (NMP) is used to remove the Fmoc groups. Amino acids are activated using (1H-benzotriazol-1-yl)-1,1,3,3-tetramethyluronium hexafluorophosphate (HBTU) and N,N-diisopropylethylamine (DIEA) and double coupled, first for 1 hr and then 30 min. Peptide 2 is synthesized using standard manual Fmoc solid-phase methods on Wang resin. One single 20-min coupling is used; the amino acids are activated with (benzotriazol-1-yloxy)tris (dimethylamino)phoshonium hexafluorophosphate (BOP), and the Fmoc groups are removed with 20% (v/v) piperidine in N,N-dimethylacetamide (DMA).

On completion of the synthesis, the peptides are cleaved from the resin in a TFA solution containing 5% triethylsilane and 5% water (v/v). After evaporating the TFA and scavengers from the resin, the peptide is extracted with a solution of 10% (v/v) acetic acid in water and lyophilized.

tert-Butyloxycarbonyl Peptide Synthesis

Peptide 3 is synthesized using standard automatic *tert*-butyloxycarbonyl (Boc) solid-phase methods on Merrifield resin.[20] The Boc group is removed with a solution of 45% (v/v) TFA in dichloromethane (DCM) for 20 min. Amino acids are activated symmetric anhydrides and double coupled for 1 hr and then 30 min. On completion of the synthesis, the peptide is cleaved

[18b] R. R. Ogorzalek Loo, N. Dales, and P. C. Andrews, *Protein Sci.* **3,** 1975 (1994).
[19] G. B. Fields and R. L. Noble, *Int. J. Pept. Protein Res.* **35,** 161 (1990).
[20] G. Barany and R. B. Merrifield, *in* "The Peptides," (E. Gross and J. Meienhofer, eds.), Vol. 2, p. 1. Academic Press, New York, 1980.

TABLE I

Mass Shifts for Commonly Observed Peptide Modifications

Add-on mass	Formula or abbreviation	Compound name
−18.0	H_2O	Water
14.0	Me or OMe	Methyl or methyl ester
16.0	O	Oxide
28.0	Et or OEt	Ethyl or ethyl ester
28.0	CHO	Formyl
32.0	O_2	Oxide
42.0	Ac	Acetyl
45.0	NO_2	Nitro
56.1	*t*Bu	*tert*-Butyl
71.1	Acm	Acetamidomethyl
76.1	OPh	Phenyl ester
80.1	SO_3H	Sulfonate
82.1	OcHex	Cyclohexyl ester
86.1	Bum	*tert*-Butyloxymethyl
88.2	StBu	*tert*-Butylsulfenyl
90.1	Anisyl	Anisyl
90.1	Bzl	Benzyl
90.1	OBzl	Benzyl ester
93.2	EDT	1,2-Ethanedithiol
96.0	Tfa	Trifluoroacetyl
97.1	OSu	*N*-Hydroxysuccinimide
100.1	Boc	*tert*-Butyloxycarbonyl
104.1	Bz	Benzoyl
104.2	MeBzl	4-Methyl benzyl
106.0	Thioanisyl	Thioanisyl
106.0	Thiocresyl	Thiocresyl
114.1	Aoc	*tert*-Amyloxycarbonyl
117.1	HOBt	Hydroxybenzotriazole ester

(*continued*)

from the dried resin using 5% anisole and 5% thiocresol in HF for 1 hr at 0°. The resin is filtered and washed with ether to remove excess scavengers and protecting groups, and the peptide is extracted with 10% (v/v) acetic acid in water and lyophilized.

Peptide Disulfide Cyclization

Reduced peptides are oxidized prior to lyophilization. A saturated solution of iodine in acetic acid is added dropwise until a persistent yellow color remains. After 5 min, zinc powder is added to the stirring mixture in portions until the solution becomes clear. The excess zinc is then removed by filtration.

TABLE I (continued)

Add-on mass	Formula or abbreviation	Compound name
118.2	diMe-Bzl	Dimethylbenzyl
120.2	Bom	Benzyloxymethyl
120.2	Mob or Mblzl	p-Methoxymbenzyl
121.1	ONp	p-Nitrophenyl
124.6	ClBzl	Chlorobenzyl
134.1	Z, Cbz	Benzyloxycarbonyl
134.2	Ada	Adamantyl
135.1	ONb	p-Nitrobenzyl ester
153.2	Nps	2-Nitrophenylsulfenyl
154.2	Npys	3-Nitro-2-pyridinesulfenyl
154.2	Tos	Tosyl (4-toluenesulfonyl)
157.8	di-Br	3,5-Dibromo
159.0	Dcb	Dichlorobenzyl
166.1	Dnp	2,4-Dinitrophenyl
166.1	OPfp	Pentafluoronophenyl
168.6	Cl-Z	2-Chlorobenzyloxycarbonyl
179.1	4Nz	p-Nitrobenzyloxycarbonyl
180.2	Tmob	2,4,6-Trimethoxybenzyl
180.2	Xan	Xanthyl
182.2	Mts	Mesitylene-2-sulfonyl
212.3	Mtr	4-Methoxy-2,36-trimethylbenzenesulfonyl
213.0	Br-Z	2-Bromobenzyloxycarbonyl
222.2	Fmoc	9-Fluorenylmethyloxycarbonyl
226.3	Mbh	Dimethoxybenzhydryl
233.3	Dns	Dansyl
238.3	Bpoc	2-(p-Biphenyl)isopropyloxycarbonyl
242.3	Trt	Trityl (triphenylmethyl)
251.8	di-I	3,5-Diiodo
266.4	Pmc	2,2,5,7,8-Pentamethylchroman-6-sulfonyl

Analytical Reversed-Phase High-Performance Liquid Chromatography

Peptides are dissolved in a 50/50 mixture of acetonitrile/water containing 0.1% TFA to a concentration of 1 mg/ml, and a 10-μl sample is analyzed on a 2.1 $\times$ 100 mm HPLC column packed with 5 μm Vydac C_{18} (The Separations Group, Hesperia, CA). A linear gradient of 0 to 60% (v/v) acetonitrile/water in 60 min is used. The flow rate is 2 ml/min, and the eluant is monitored by UV detection at 214 nm.

Preparative Reversed-Phase High-Performance Liquid Chromatography

The crude peptide 2 (74 mg) is dissolved in a 1/9 mixture of acetonitrile/water containing 0.1% (v/v) TFA and loaded onto a 1 $\times$ 50 cm HPLC column packed with 15–20 μm Vydac C_{18}. A gradient of 10 to 25% (v/v) acetonitrile/water in 60 min is run at a flow rate of 9 ml/min, and 1-min fractions are collected. The eluant is monitored by UV detection at 214 nm.

Mass Spectrometry

Single infusion spectra are acquired with a Sciex API-I single quadruple mass spectrometer with an IonSpray source (Pe-Sciex, Foster City, CA). Crude samples are normally infused at a concentration of 1 mg/ml via a syringe pump flowing at 5 μl/min. Preparative column fractions are infused directly without further dilution in the same manner as the crude samples. The scans are acquired in multichannel analyzer (MCA) mode with a step size of 0.2 u. Scans are averaged until a peak-top ion count of approximately 1×10^6 is achieved.

Although most samples are analyzed without desalting, when salts or other buffer components are present, the sample may be desalted by reversed phase HPLC, or more simply by a Sep-Pak cartridge (Millipore, Bedford, MA) or other, smaller reversed-phase desalting cartridges (LC Packings, San Francisco, CA; Michrome BioResources, Auburn, CA). The elution may be any volatile buffer system (e.g., 0.1% (v/v) TFA, water, acetonitrile). Other solvents, such as methanol, acetic acid, and formic acid, work well also. We find that 5% formic acid/15% water (v/v) in methanol is a particularly effective elution buffer for step desalting.

On-line High-Performance Liquid Chromatography/Electrospray Ionization Mass Spectrometry

The on-line HPLC/ESI-MS system we use couples a Sciex API-I single quadrupole mass spectrometer and an IonSpray source to a Waters (Milford, MA) dual-pump HPLC system. The RP-HPLC methods we use are described above except that 100 μl of sample is injected onto the column. Eluent from the HPLC is diverted to the mass spectrometer by way of a post UV-detector split with a tee. By careful adjustment of a needle valve placed on the waste line, a flow of 50 μl/min to the mass spectrometer can be maintained. Peaks in the total ion current (TIC) profile are averaged to give the mass of the corresponding peak on the HPLC.

Tandem Mass Spectrometry

Fragment ion spectra are acquired with a Finnigan LCQ quadrupole ion trap mass spectrometer (Finnigan, San Jose, CA). The crude peptide sample is diluted to approximately 5 pmol/μl with 1% acetic acid/50% methanol (v/v), then infused at 0.5 μl/min into a microscale electrospray interface.[21]

[21] M. T. Davis, D. C. Stahl, S. A. Hefta, and T. D. Lee, *Anal. Chem.* **67**, 4549 (1995).

The precursor mass of interest is selected for tandem mass spectrometry (MS/MS) in the Tune Plus program. The automatic gain control (AGC) MS^n target is set to 8×10^7. Singly charged ions are isolated with a 2-u width and fragmented with 20% collision energy. Doubly charged ions are fragmented with 39% collision energy. The low mass in the spectrum is set to the minimum allowed (28% of the precursor m/z). Each spectrum is the sum of 20 scans, each scan being the sum of 3 microscans.

Results and Discussion

Synthesis Process

Figure 1 presents a general flowchart that describes how a peptide request becomes a final product. First, requested peptides are evaluated individually to determine what synthetic strategy should be used to make the compound. Evaluation criteria include the following: the length of the peptide; the presence of problematic amino acid combinations within the sequence; the incorporation of unusual amino acids or side-chain modifications; and whether the peptide is linear or cyclic. The peptide is then made, cleaved, and lyophilized. Second, the crude peptide is analyzed before purification by analytical HPLC and ESI-MS. This allows us to evaluate

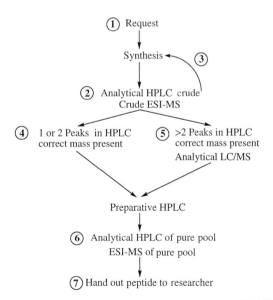

FIG. 1. Flowchart of the peptide synthesis and purification process. ESI-MS is used routinely to verify the identity of the synthetic product and to guide the purification.

the relative purity of the peptide and to ensure that the correct product is in the crude mixture. Third, sometimes cases arise where the correct mass cannot be found in the crude mixture. At this point, the chemist will try a variety of analytical techniques to verify the crude ESI-MS result. Some of these techniques include the desalting of the sample, ESI-MS analysis in the opposite polarity, and analysis of hand-collected analytical HPLC fractions. If these types of analyses show that the correct mass does indeed not exist in the sample, the peptide will have to be remade. However, the ESI-MS provides the chemist with vital information about the synthesis and what can be done to improve it. For instance, if protecting groups remain on the peptide, the cleavage process time may be lengthened; if there are amino acid deletions, the chemist may want to double couple or use a "hotter" activating reagent. In general, the ESI-MS gives sufficient information to achieve a successful second synthesis of the peptide.

Fourth, if there are one or two peaks in the HPLC profile and the correct mass is found in the crude ESI-MS, then the molecule is purified. In the case of two peaks, both are collected and the fractions analyzed by ESI-MS to determine which peak is correct. Fifth, if there are more than two peaks and the correct mass is found in the ESI-MS of the crude material, an LC-MS run will be done to determine which peak is the correct product. Once the proper peak is identified, the appropriate preparative gradient can be set to purify the peptide. Sixth, after the correct peak is collected, an analytical HPLC is run to assure purity and another ESI spectrum is acquired to verify the mass of the final, purified material. Finally, after lyophilization, the peptide is delivered to the requester.

To illustrate three common uses of ESI-MS in peptide synthesis and purification, three peptides (Fig. 2) were synthesized by three different methods. These peptides were chosen because they are known to have

Peptide 1:   Phe-Cys-Phe-Trp-Lys-Thr-Cys-Thr-NH$_2$
                      | |
                      S————————————S

Peptide 2:   Lys- His- Asp- Pro- Cys- Gly- Trp- Asn-
             Gly- Pro- Arg- Pro- Met- Arg- Gly- OH

Peptide 3:   Asp- Ile- Leu- Pro- Ser- Pro- His- Cys- Met- OH

FIG. 2. Sequences of the peptides synthesized in this study. Peptides 1 and 2 were made using an Fmoc strategy, and peptide 3 was made using a Boc strategy.

some synthetic problems, represent cyclic and linear peptides, and have been well characterized in the past.

Electrospray Ionization–Mass Spectrometry in Peptide Synthesis and Cyclization

Disulfide cyclizations are common in peptide synthesis. A variety of methods are used to oxidize cysteine-containing peptides, but there are only a few ways to monitor the oxidation. Traditionally, HPLC is used to look for a change in peak retention time. However, this can be time consuming, and at times, inaccurate. Sequence and conformation of the peptides can affect HPLC separation, and sometimes subtle changes in retention time are difficult to detect. ESI-MS provides an easy, quick method to follow disulfide formation.

Peptide 1 was originally synthesized as part of a multicenter study on disulfide bond formation methods[11] and was synthesized using automatic Fmoc chemistry. A 10-μl aliquot of the crude mixture was diluted with 50 μl of 50% (v/v) acetonitrile/water and then analyzed by direct infusion into the mass spectrometer. Figure 3 shows the spectrum for the reduced peptide. The theoretical monoisotopic mass of the reduced peptide 1 is 1033.5 u. The molecular ions $[M + H]^+ = 1034.6$ and $[M + 2H]^{2+} = 517.8$ are clearly evident and after calculation give the expected formula weight of $M = 1033.6$ u. This quick analysis confirms the presence of the reduced peptide in the crude cleavage mixture.

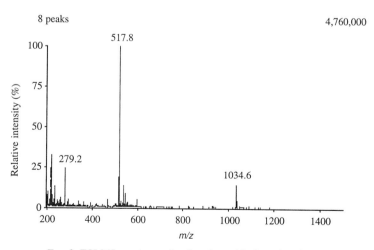

FIG. 3. ESI-MS spectrum of reduced peptide 1 crude mixture.

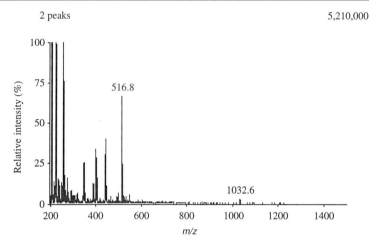

FIG. 4. ESI-MS spectrum of the oxidized peptide 1 crude mixture. The 2 u loss relative to the reduced peptide indicates that the disulfide bond has been formed.

Figure 4 shows the spectrum of the oxidized peptide 1. The shift in the molecular ions $[M + H]^+ = 1032.6$ and $[M + 2H]^{2+} = 516.8$ are clearly evident. After calculation, the formula weight of the oxidized peptide is determined to be 1031.6 u. The theoretical monoisotopic mass of the oxidized peptide is 1031.4, and this loss of 2 u is consistent with the formation of the cystine disulfide bond. The large lower mass peaks are iodine and salt adducts that will be removed from the peptide by preparative HPLC.

Electrospray Ionization–Mass Spectrometry in Peptide Purification: Screening of Chromatography Fractions

Electrospray ionization spectrometry is a powerful and convenient tool in peptide purification. Besides use of ESI-MS to verify the presence of the compound of interest in the crude sample or the purity of the final product, ESI-MS provides quick analysis of fractions during purification. Analysis of fractions by ESI-MS has two advantages over analytical isocratic HPLC. First, ESI-MS is much more rapid than isocratic HPLC for determining the presence of impurities in the fractions. The complete analysis of a preparative peak (multiple fractions) can be done in the time needed to run a single isocratic HPLC. Second, ESI-MS gives quick verification of the mass of the peptide.

Figure 5 shows the analytical HPLC trace of crude peptide 2 after lyophilization. This peptide, synthesized originally as part of a multicenter study on cleavage and deprotection methods,[9] was synthesized using manual

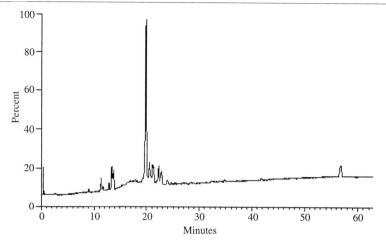

FIG. 5. HPLC chromatogram of peptide 2 crude mixture (UV detection at λ = 214 nm).

Fmoc methods. The main peak in the chromatogram is about 70% of the total area under the curve, with the other 30% in small peaks around the main peak.

Figure 6 shows the ESI-MS spectrum of the crude lyophilized material. The theoretical formula weight for peptide 2 is 1706.8 u. The ESI-MS spectrum shows that the correct molecular ions $[M + H]^+$ = 1708.2, $[M + 2H]^{2+}$ = 854.6, and $[M + 3H]^{3+}$ = 570.2 are present in the crude mixture,

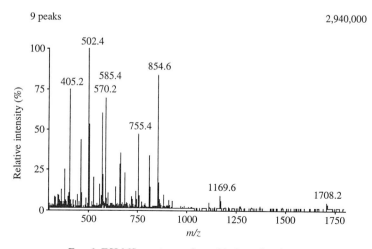

FIG. 6. ESI-MS spectrum of peptide 2 crude mixture.

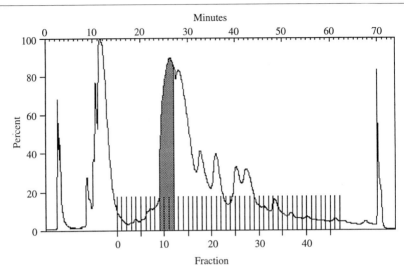

Fig. 7. Preparative HPLC chromatogram of peptide 2. The dark area of the major peak indicates the fractions that contain the pure product.

which after calculation give an expected formula weight of 1707.2. There are also many impurities present as well.

Figure 7 shows the preparative HPLC run. The main peak shows a split at the top, possibly indicating the presence of a shoulder that is not clearly evident in the analytical chromatogram. Figure 8 shows ESI-MS spectra of fractions 8 through 15 across the major peak in the preparative HPLC run. It can be clearly seen that fractions 8 and 9 have no correct material in them, but fractions 10 through 12 are very pure as evidenced by the strong molecular ions of peptide 2, $[M + 2H]^{2+} = 855$ and $[M + 3H]^{3+} = 570$. In fraction 13, an impurity as evidenced by the peak $[M + 2H]^{2+} = 585$, starts to be seen. In fraction 14, the impurity is stronger, and the molecular ion $[M + H]^{+} = 1170$ is also present. These two ion signals become stronger in subsequent fractions while the peaks of interest decrease. This is clear evidence that the peak of the correct material is tailing into the peak of a deletion sequence. The total analysis time was approximately 10 min for all 20 fractions. On the basis of these data, a pure pool of fractions 10 to 12 was taken and lyophilized.

Electrospray Ionization–Mass Spectrometry in Peptide Purification:
Liquid Chromatography–Mass Spectrometry

In peptide synthesis, we often encounter compounds that are not pure, regardless of the care taken in their synthesis. They usually have several

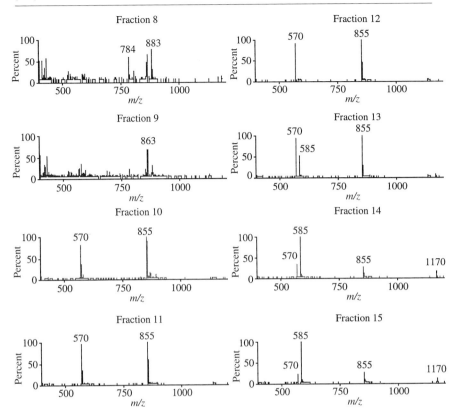

FIG. 8. ESI-MS spectra of fractions 8 through 15 from the preparative HPLC experiment. Fractions 10 to 12 are pure product. Later fractions show ions for a deletion peptide.

major peaks in the analytical chromatogram, and the crude ESI-MS spectrum shows impurities in addition to the correct mass. In cases such as these, we find that a single LC-MS experiment can save many hours of additional analysis and possible resynthesis of the peptide. Peptide 3 represents a good example.

Figure 9 shows the analytical HPLC chromatogram of peptide 3 (see Fig. 2). There are two major peaks with retention times of 21 and 22.5 min. The major peak at 21 min also has a leading edge shoulder, and there are also several minor, later eluting peaks. The crude peptide spectrum is shown in Fig. 10. The theoretical exact mass for peptide 3 is 1011.5 u. The molecular ions $[M + H]^+ = 1012.6$ and $[M + 2H]^{2+} = 506.9$ are clearly seen and after calculation give the expected mass of 1011.7. The spectrum also shows many impurities that probably correspond to the other peaks in the analyti-

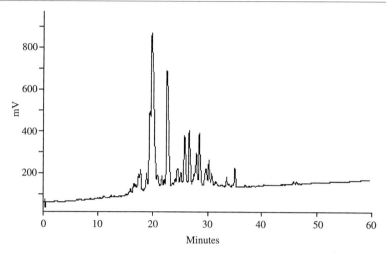

FIG. 9. Crude HPLC chromatogram of peptide 3 crude mixture showing two major peaks and several minor impurity peaks.

cal HPLC chromatogram. By doing an LC-MS experiment, we should be able to identify the product peak and the impurity peaks as well.

Figure 11 shows the total ion current (TIC) profile from the LC-MS experiment and an expanded view around the peaks of interest. The correct peptide product is found in the first peak. However, only the trailing edge of the peak has pure peptide in it. The leading edge of the peak is contami-

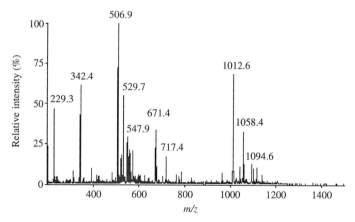

FIG. 10. ESI-MS spectrum of peptide 3 crude mixture. The product moleculer ions are present, as well as several impurities.

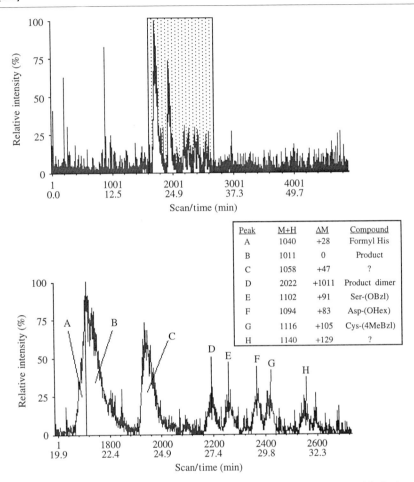

FIG. 11. Total ion current (TIC) profile of the LC-MS experiment for peptide 3. Average compound masses are as follows: (A) 1012 u and 1040 u; (B) 1011 u; (C) 1058 u; (D) 2022 u; (E) 1102 u; (F) 1094 u; (G) 1116 u; and (H) 1140 u.

nated by formylhistidine-containing peptide. The second major peak shows a +46 u shift from the expected mass, and its identify is not immediately apparent. The minor peaks are all peptide product with other side-chain protecting groups still intact, and disulfide-bonded dimer. This information tells us that we must construct a preparative HPLC gradient to optimize collection of the trailing edge of peak one, because we are able to ignore the second peak and the minor peaks in our preparative run, and we can use a very shallow gradient from 17 to 23% (v/v) acetonitrile/water to obtain the best separation of the desired peak. The LC-MS data also give us valuable information about the synthesis of the peptide. Because there are no detectable deletion sequences in the spectra we can safely assume that the synthesis was successful. The presence of several peaks that still have side-chain protection probably indicates that there was not enough HF used in the cleavage.

Sequence Verification by Tandem Mass Spectrometry

When one needs to know the precise location of a modification, or to identify a by-product whose identity is not apparent from its mass alone, fragmentation of the peptide will often provide the needed structural information. To derive fragment ions from a peptide in a mixture, or in the presence of a high background, two stages of mass analysis are required, termed tandem MS, or MS/MS.[22,23] Most often this type of instrument is a triple quadrupole mass spectrometer. The first quadrupole functions as a mass filter to isolate ions of just the peptide of interest. These ions are passed into the second quadrupole, which is operated with a higher pressure of gas, usually argon, to promote fragment-inducing collisions [collision-induced dissociation (CID)]. The second quadrupole is operated in the radiofrequency-only (rf-only) mode, which produces no mass analysis, but focuses the ions that would otherwise be scattered during CID. The fragment ions are analyzed with the third quadrupole.

A more recently introduced instrument is the quadrupole ion trap mass spectrometer.[24] Ions are accumulated and trapped in an rf field imposed on a ring and two end-cap electrodes. By scanning the rf voltage, ions can be sequentially ejected through a hole in one of the end caps to a detector. The resulting mass spectrum is nearly identical to that produced by a quadrupole instrument. Tandem MS experiments are performed by adjusting the fields to trap ions from a single peptide. These ions are excited to undergo CID with background helium gas in the trap. The fragment ions are analyzed by sequential ejection as described above. Relative to triple quadrupole instruments, ion traps are generally lower in cost, smaller in size, capable of higher resolution, and higher in sensitivity for MS/MS.

To illustrate the utility and limitations of MS/MS, one of the by-products from the previous example was analyzed further. The mass spectrum of the peptide 3 crude mixture (see Fig. 10) displayed a singly charged peak,

[22] D. F. Hunt, J. R. Yates, III, J. Shabanowitz, S. Winston, and C. R. Hauer, *Proc. Natl. Acad. Sci. U.S.A.* **83**, 6233 (1986).
[23] A. L. McCormack, J. K. Eng, and J. R. Yates, *Methods (San Diego)* **6**, 274 (1994).
[24] J. C. Schwartz and I. Jardine, *Methods Enzymol.* **270**, 552 (1996).

$[M + H]^+ = 1058.4$. This peak is 46 u higher than the expected peptide. A +46 u mass difference does not correspond to a commonly observed by-product (see Table I). To confirm that this peptide is a modified version of the correct sequence, to show the location of the modification, and to attempt to identify the modification, a fragment ion spectrum was measured. The data were acquired from the crude mixture with a quadrupole ion trap mass spectrometer. Spectra were measured for both the $[M + H]^+ = 1058.4$ and $[M + 2H]^{2+} = 529.7$ precursor ions, but the singly charged precursor gave more useful fragmentation (see Fig. 12). The spectrum shows a series of y ions (charge retained on the C terminus of the peptide) and b ions (charge retained on the N terminus). (Further information on peptide fragmentation can be found elsewhere.[23,25,26]) The y ions observed (y_3–y_8) are all shifted by +46 u, indicating that the modification is localized to the three residues at the C terminus (His-Cys-Met). The b_7 ion shows no mass shift, but the b_8 and b_9 ions are shifted by +46 u, indicating that the modification is found on Cys-8. The fragment ion assignments were confirmed by comparison with the MS/MS spectrum of the unmodified peptide, $[M + H]^+ = 1012$ (data not shown), that displayed the same fragment ion series with no corresponding mass shifts.

The +46 u modification has been shown to reside on the Cys residue in an otherwise unmodified sequence, but the fragment ion spectrum in this example provides no additional information on the identity of the modification. Further experiments would be necessary in this case to identify this modification (e.g., high-resolution exact mass measurement, analysis of the amino acids following hydrolysis) because the modification is not one commonly observed. The presence of the expected peptide as the largest component in the crude mixture suggests that the modification occurred during the cleavage step. In addition, the presence of other by-products with protecting groups remaining (see above) indicates an insufficient amount of HF for the cleavage. To eliminate these by-products, the cleavage step on a subsequent resynthesis could be modified to use more HF and a different scavenger cocktail.

When a triple quadrupole or ion trap instrument is not available, fragment ions may also be generated in the ion source region of the electrospray instrument by operating the skimmer or orifice at a higher than normal potential.[27] As a result, the collisions with background air molecules, mentioned above for desolvation, become more energetic and cause fragmentation of the peptide backbone. These "in source" collisions can provide

[25] K. Biemann, *Methods Enzymol.* **193**, 886 (1990).
[26] B. L. Gillece-Castro and J. T. Stults, *Methods Enzymol.* **271**, 427 (1996).
[27] V. Katta, S. K. Chowdhury, and B. T. Chait, *Anal. Chem.* **63**, 174 (1991).

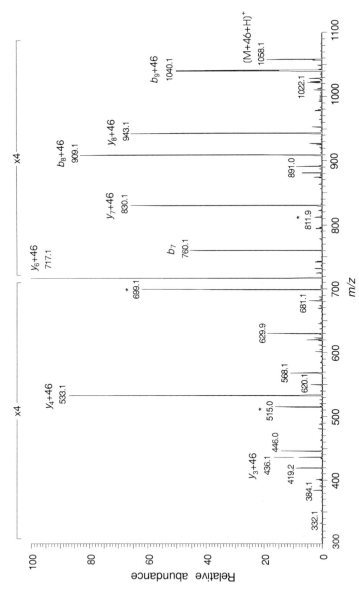

FIG. 12. MS/MS fragment ion spectrum of the $[M + H]^+ = 1058.4$ peak from the peptide 3 crude mixture (Fig. 10). Data were acquired with a quadrupole ion trap mass spectrometer. The b ions have the charge retained on the N terminus. The charge is retained on the C terminus for the y ions. An asterisk indicates a loss of water from the corresponding fragment ion. A mass shift for the fragment from the unmodified peptide is indicated by "$+46$". The lack of the modification for b_7, with its appearance for $b_8 + 46$, indicates that the modification is located at Cys-8.

significant structural information on the peptide. For these fragment ions to be meaningful, however, they must be derived from single peptide species. Thus, this method is useful only for pure peptides. Nevertheless, it has the advantage that it is extremely easy to perform and can be done with a relatively inexpensive single quadrupole instrument.

Summary

Electrospray ionization mass spectrometry is an easy, rapid method for the verification of proper peptide synthesis and for the identification of most synthetic by-products. A synthesis–purification scheme has been described that uses mass analysis to (1) confirm the presence of the proper product in the crude peptide mixture, (2) guide the purification process, and (3) confirm the mass and purity of the final product. Even though many of these steps could be performed just as well with other ionization techniques, the liquid-flow characteristics of electrospray source are clearly an advantage when LC-MS is required. In addition, the ease with which fragment ions can be generated to provide structural information, even with the least sophisticated instruments, is a further advantage of ESI-MS.

Although much of the operation described here was done manually, many of the steps could be automated with little additional effort (e.g., use of an autosampler). Quadrupole and ion trap instruments are widely available at present and provide the chemist with a variety of instruments from which to choose. Electrospray time-of-flight instruments will be commercially have just become available and should also provide similar results.[28] As electrospray instruments continue to evolve, the instruments display greater performance and enhanced user-friendly interfaces, yet are lower in price and smaller in size. These features should lead to even more widespread use for the characterization of synthetic peptides.

---

[28] J. F. Banks, Jr., and T. Dresch, *Anal. Chem.* **68,** 1480 (1996).

[23] Laser Desorption Mass Spectrometry

By WILLIAM T. MOORE

Introduction

A full characterization of a synthetic peptide product requires interrogation of structure at several levels using an array of orthogonal techniques. The more complex the technique array, the more stringent the analysis and the most complete the picture of and confidence in the structure. Synthetic peptide products that have interesting biological properties and that will serve as templates for pharmacological design of mimetics will undergo the most extensive characterizations. An extensive analysis includes amino acid analysis, reversed-phase high-performance liquid chromatography (HPLC) using different solvent systems and column matrices, possibly capillary zone electrophoresis (CZE) involving different buffer systems, elemental analysis, mass spectrometry (MS) using different ionization methods and mass filters, and, at the most structurally informative end of the analysis array, three-dimensional (3D) structural analyses using two-dimensional (2D) nuclear magnetic resonance (NMR) spectroscopy and X-ray crystallography.

This extensive range of analytical techniques requires a heavy investment in both time and resources, often the cooperation of individuals in geographically separated groups, and is reserved and justified for only the most interesting model structures that lead to fundamental understanding of biomedically important structure–function relationship problems. However, synthetic peptides not characterized to this degree are often employed in the early phase of a scientific project and are considered by most biomedical investigators to be readily available from either commercial or local core facilities as routine and reliable tools. There is often a mistaken core facility client "vending machine mentality" that assumes so many dollars in, so much peptide out in about the same time that it requires one to acquire their favorite beverage. At the service side, often peptide synthesis is provided by busy and overworked operators who often necessarily have to approach their automated synthesizers with a "black box mentality" with the hope and faith that the instrument reliably performs as promised by the instrument manufacturer.

In the recent past, both inside and outside the peptide synthesis core facility, nonpeptide chemists have often taken too much for granted concerning the quality of the peptides that are used to initiate their studies. Peptide synthesis analytical surveys performed in the early 1990s under the auspices of the Association of Biomolecular Resource Facilities (ABRF) have demonstrated the need for a serious reality check. The 1991 ABRF peptide synthesis study[1] of a routine synthetic peptide problem indicated that 17% of the volunteered products had no desired product present and that only 28% of crude products submitted had greater than 75% desired product and only 68% of purified products had desired product in excess of 75%. The next survey[2] on a simple synthesis problem showed slight improvement; however, 48% of crude preparations and 21% purified products still contained less than 75% of the desired product. More recent ABRF peptide synthesis surveys[3–6] have indicated steady improvement,[6a] which has been attributed to instrumental design improvements, more reliable chemistries and reagents provided by commercial sources, and, it is hoped, a heightened awareness of quality control in the minds of those associated with peptide synthesis core laboratories.

One positive contributing factor I want to stress is the increased involvement of MS in peptide synthesis core facilities. Either mass spectrometric instrumentation has been acquired by peptide synthesis core facilities or working relationships have been established between peptide synthesis laboratories and MS laboratories. However, although the more recent surveys reveal continuing improvement, the studies indicate the need for constant quality control vigilance. Isolated errors often crop up unexpectedly. Quality control vigilance is aided by MS.

[1] A. J. Smith, J. D. Young, S. A. Carr, D. R. Marshak, L. C. Williams, and K. R. Williams, in "Techniques in Protein Chemistry III" (R. H. Angeletti, ed.), p. 219. Academic Press, San Diego, 1992.

[2] G. B. Fields, S. A. Carr, D. R. Marshak, A. J. Smith, J. T. Stults, L. C. Williams, K. R. Williams, and J. D. Young, in "Techniques in Protein Chemistry IV" (R. H. Angeletti, ed.), p. 227. Academic Press, San Diego, 1993.

[3] G. B. Fields, R. H. Angeletti, S. A. Carr, A. J. Smith, J. T. Stults, L. C. Williams, and J. D. Young, in "Techniques in Protein Chemistry V" (J. W. Crabb, ed.), p. 501. Academic Press, San Diego, 1994.

[4] G. B. Fields, R. H. Angeletti, L. F. Bonewald, G. B. Fields, J. S. McMurray, W. T. Moore, J. T. Stults, and L. C. Williams, in "Techniques in Protein Chemistry VI" (J. W. Crabb, ed.), p. 539. Academic Press, San Diego, 1995.

[5] R. H. Angeletti, L. Bibbs, L. F. Bonewald, G. B. Fields, J. S. McMurray, W. T. Moore, and J. T. Stults, in "Techniques in Protein Chemistry VII" (D. R. Marshak, ed.), p. 261. Academic Press, San Diego, 1996.

[6] R. H. Angeletti, L. Bibbs, L. F. Bonewald, G. B. Fields, J. W. Kelly, J. S. McMurray, W. T. Moore, and S. T. Weintraub, in "Techniques in Protein Chemistry VIII" (D. R. Marshak, ed.), p. 875. Academic Press, San Diego, 1997.

[6a] R. H. Angeletti, L. F. Bonewald, and G. B. Fields, Methods Enzymol. 289, [32], this volume (1997).

The remainder of this article concerns the use of only one type of MS, matrix-assisted laser desorption–ionization mass spectrometry (MALDI-MS). The objective of this article is to demonstrate that MALDI-MS is a valuable analytical tool for monitoring peptide synthesis. I hope to show that this form of MS can be applied successfully even at a minimalist level and that it can be a satisfactory approach to evaluate the progress and validate the outcome of most routine synthetic peptide problems that may be encountered in a typical protein laboratory. Examples of MALDI-MS integration into quality control of automated peptide synthesis are presented, pointing out advantages and pitfalls where necessary. Other MSs using other ionization methods and offering enhanced MS/MS capabilities are presented elsewhere in this volume.[6b,c] I want to caution that there is no single analytical method for characterization of a synthetic peptide preparation and to advise at the outset that MS should be considered a qualitative method and should only be considered semiquantitative at best. No single analytical method substitutes for an all-encompassing orthogonal approach for the highest analytical stringency.

This article illustrates approaches that one can take with a lower cost pioneer instrument lacking costly high-end research instrument refinements to maximize the characterization of a peptide synthesis. Examples of MALDI-MS characterization of crude preparations and purified preparations derived from analytical and preparative HPLC are presented. Practical issues concerning matrix preparation, sample preparation, matrix selection, mass discrimination, and ion suppression problems as a function of peptide concentration are addressed as well as the use of mass shift assays for clarifying chemistries which would normally require high resolution that might not necessarily be available using external mass spectrometric calibrations.

Description of Matrix-Assisted Laser Desorption–Ionization Mass Spectrometry

Matrix-assisted laser desorption–ionization mass spectrometry (MALDI-MS) is an outgrowth of the direct laser desorption mass spectrometry (LD-MS) of small organic molecules that was initially developed in the 1960s and 1970s.[7,8] Karas and Hillenkamp introduced MALDI-MS in

---

[6b] S. Beranová-Giorgianni and D. M. Desiderio, *Methods Enzymol.* **289**, [21], this volume (1997).

[6c] D. J. Burdick and J. T. Stults, *Methods Enzymol.* **289**, [22], this volume (1997).

[7] F. J. Vastola, R. O. Mumma, and A. J. Pinone, *Org. Mass Spectrom.* **3**, 101 (1970).

[8] M. A. Posthumus, P. G. Kistemaker, H. L. C. Meazelaar, and M. C. ten Neuver de Brauw, *Anal. Chem.* **50**, 985 (1978).

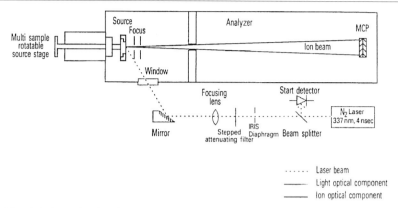

FIG. 1. Schematic of a linear MALDI/TOF mass spectrometer, a Micromass (formerly Fisons Instruments, Beverly, MA) TofSpec.

1987[9,10] when they demonstrated that adding a small molecular weight organic acid matrix to an analyte could overcome molecular photodissociation of the sample ions induced by direct laser irradiation of the sample. The Hillenkamp group at Munster and the Chait group at Rockefeller University pioneered the peptide and protein application of this technique through instrument, matrix, and sample handling development.[11] A major advantage of MALDI-MS is that the mass range extends from the low molecular weight range to the very high molecular weight range up to 200,000 and greater. Time-of-flight (TOF) mass analyzers are suitable for these mass ranges, and several MALDI-MS instruments are commercially available, ranging from the user-friendly benchtop-dimension instruments to stand alone platform research grade instruments with multiple features such as automated sample introduction, multiple lasers, resolution improving reflectron ion mode and "delayed extraction" capabilities, and tandem MS abilities using post source decay analysis to detect fragmentation of a peptide yielding sequence ions permitting primary structure information. A product review of MALDI-MS instruments was published in *Analytical Chemistry* in 1995.[12]

The least complex instrument, a typical linear mode MALDI/TOF-MS system, is shown in Fig. 1. Such a system is suitable for monitoring most problems encountered in solid-phase peptide synthesis laboratories. Sample

[9] M. Karas, D. Bachmann, U. Bahr, and F. Hillenkamp, *Int. J. Mass Spectrom. Ion Proc.* **78**, 53 (1987).
[10] M. Karas and F. Hillenkamp, *Anal. Chem.* **60**, 2299 (1988).
[11] F. Hillenkamp, M. Karas, R. C. Beavis, and B. T. Chait, *Anal. Chem.* **63**, 1193A (1991).
[12] D. Noble, *Anal. Chem.* **67**, 497A (1995).

is allowed to cocrystallize with matrix on a target prior to introduction into the source of the mass spectrometer. Through a set of fixed optics a pulsed laser beam from a nitrogen laser (337 nm) is directed to the target surface for irradiation of the sample. Light energy absorbed mostly by the matrix induces an explosive ejection of a plume of matrix and analyte ions into the vacuum of the source. Ions are extracted and accelerated by a strong electric field (of the order of 25 to 30 KeV) and the ion packets are focused by a set of high-voltage plates (Einzel lenses) into the field-free region of the flight tube. All ions are given the same kinetic energy, and thereby their respective velocities are mass dependent and reach a multichannel plate electron multiplier ion detector at different times measured in microseconds, with smaller molecular weight ions reaching the detector before larger molecular weight ions. Transient signal recordings representing the spectra are then summed and processed into graphical mass spectra by a computerized data system.

The instrument may be calibrated with well-characterized synthetic peptides, known proteins such as cytochrome c, and at the low mass end ions derived from the matrix. The highest mass accuracy is achieved with the use of internal standards; however, the use of this type of calibrant can be problematic owing to ion suppression effects induced by the presence of the calibrant and the time involved to empirically derive proper proportions of calibrant and analyte to minimize these effects and achieve good peak shape for all components, analytes, and calibrants. External calibrations performed at times close to the unknown sample acquisition time in a "pioneer" linear mode instrument such as that depicted in Fig. 1 having resolutions of around 200 yield, in practice, mass accuracy errors of ±0.1 to ±0.5%, which translates to mass error of 1 to 3 amu for most synthetic peptide products. For most synthetic peptide problems where there is foreknowledge of desired product this level of error may be tolerated. Those problems that call for 1 amu accuracy or less, such as evaluation of deamidation or disulfide formation, require the development of a mass shift assay to clarify the small mass change reactive groups or analysis on a higher resolution instrument. An example of a disulfide formation problem is presented later to illustrate the use of a mass shift assay for analytical clarification when using a low-resolution instrument.

Matrix-Assisted Laser Desorption–Ionization Mass Spectrometry
Sample Preparation and Matrix Considerations

Several matrices are available for MALDI-MS, and various ones are more appropriate for the specific type of analyte.[11] Choice of matrix and method of sample deposition remain largely empirical. For synthetic peptide work it is convenient to work with primarily two matrices, α-cyano-4-hydroxycinnamic acid (CHCA)[13] and 2-(4-hydroxyphenylazo)benzoic acid (HABA).[14] CHCA is usually suitable for smaller peptides (<4000) and is the matrix of choice for most synthetic peptide evaluations. HABA is suitable for larger peptides (>4000) and for some peptides that are not well visualized in CHCA. One has to be cautioned, though, against mass discrimination effects of these matrices. HABA will discriminate for larger peptides and against smaller peptides, and CHCA, vice versa. These matrices are also suitable for peptides that are soluble in the solvent 50% acetonitrile and 0.1% trifluoroacetic acid [50% ACN, 0.1% TFA (v/v)].

The CHCA matrix is prepared by making the solvent 50% acetonitrile, 0.1% TFA saturated in CHCA.[13] The HABA matrix is prepared by dissolving 10 mg of HABA per 15 ml of 50% acetonitrile as suggested in the original report describing the use of this matrix.[14] Extremely hydrophobic peptides that are not soluble in 50% acetonitrile and 0.1% TFA will be tested in alternative solvents such as 60% chloroform and 40% methanol, neat methanol, or even tetrahydrofuran or dichloromethane. A matrix prepared in the appropriate solvent chosen for the peptide is used for sample preparation.

It should be noted that in 1996 we received from vendors CHCA batches of lower quality than that originally obtained earlier. The more recent material has a darker brownish color when compared to the original lighter yellow preparations and if used directly results in the acquisition of poor spectra with at least a 10-fold loss in sensitivity. The material may be recrystallized, but we have found it convenient to simply extract–wash it one time with 50% acetonitrile, 0.1% TFA. This can be simply done in a 1.5-ml graduated Eppendorf tube by adding matrix powder to the 0.25-ml mark (~80 to 100 mg), vortexing, suspending in 1.4 ml 50% acetonitrile, 0.1% TFA for 30 sec, and discarding the first supernatant after centrifugation. Vortex resuspension of the wash–extracted pellet in a 1.4-ml fresh aliquot of 50% acetonitrile, 0.1% TFA saturates the fresh solvent aliquot in the remaining wash–extracted CHCA, and, after centrifugation, the supernatant serves an appropriate matrix with a 10- to 30-fold improvement in sensitivity.

Figure 2A shows the differences in the CHCA-related matrix ions for the unextracted and wash–extracted CHCA. Figure 2A (*top*) shows dominant ion signals at masses 165 and 102. Figure 2A (*bottom*) indicates a marked reduction of the 165 and 102 ion signals and the dominance of the ion signals usually observed with higher quality CHCA, especially the 190

[13] R. C. Beavis, T. Chaudhary, and B. T. Chait, *Org. Mass Spectrom.* **27,** 156 (1992).
[14] P. Juhasz, C. E. Costello, and K. Bieman, *J. Am. Soc. Mass Spectrom.* **4,** 399 (1993).

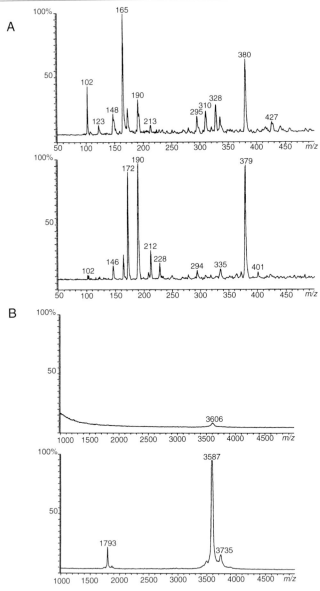

FIG. 2. Restoration of matrix efficiency by simple wash–extraction of the lower quality darker yellow batches of cyano-4-hydroxycinnamic acid (CHCA) that are presently commercially available. (A) *Top:* Matrix background for the lower quality CHCA; *bottom:* matrix background observed after the wash–extraction step. (B) *Top:* MALDI-MS analysis of 2 pmol of a test peptide in the matrix prior to wash–extraction; *bottom:* result obtained for 2 pmol of test peptide following wash–extraction.

and 379 ion signals representing the $(M + H)^+$ for the monomer and dimer of CHCA, respectively. To demonstrate the effect of the CHCA wash–extraction on the restoration of the sensitivity of detection, we compared analyses of a synthetic peptide. Figure 2B shows the MALDI-MS analysis of 2 pmol of the 45-residue synthetic peptide Cys-Cys-His-His-Gly-Gly-Arg-Arg-Gly-Gly-Thr-Thr-Cys-Cys-Asn-Asn-Tyr-Tyr-Tyr-Tyr-Ser-Asn-Ser-Ser-Tyr-Tyr-Ser-Ser-Phe-Phe-Trp-Leu-Ala-Ser-Leu-Asn-Pro-Glu-Arg-Met-Phe-Arg-Lys-Pro-Pro. Figure 2B (*top*) shows MALDI-MS analysis of the 2 pmol of peptide spotted in the poorer quality CHCA matrix directly. Figure 2B (*bottom*) shows the restoration of sensitivity after the simple wash–extraction step. At least a 10- to 30-fold increase in sensitivity is achieved.

A drawback to MALDI-MS is that the quantitation or the relative signal height intensities of the peptide components revealed in a MALDI-MS profile can be affected by peptide concentration.[4] To address this problem in our laboratory and to carefully assess the complexity and quantity of components that we may find in a peptide product, we examine the peptide product at different concentrations. For a typical analysis of a synthetic peptide preparation we prepare synthetic peptide solutions at 1 mg/ml and perform serial dilutions at 1/10, 1/100, and 1/1000 in the matrix and interrogate the peptide at the three dilutions. These dilutions represent peptide amounts ranging from a few picomoles to hundreds of picomoles of product. If the peptide is relatively pure, similar spectra are observed for all three dilutions. Figure 3A shows a MALDI-MS analysis of an HPLC-purified preparation of the 45-residue peptide described above at the 2, 20, and 200 pmol level (*top* to *bottom*, respectively). Essentially the identical ion profile is observed in each spectrum. The only differences in the spectra are the emergences of the doubly charged ions for the peptide at the more dilute concentrations.

If the synthetic peptide product arising from a problematic synthesis turns out to be a complex mixture, it is essential to interrogate at serial dilution to evaluate the quantitative relationships of the components by permitting the release of ion suppression phenomena that are dependent on peptide concentration. Different ion signal intensities are often observed for the same component in the spectra for the diluted samples owing to ion suppression effects. Figure 3B shows the MALDI-MS interrogation at dilution of an attempt to synthesize a 22-residue peptide having the sequence Ala-Pro-Val-Gly-Leu-Val-Ala-Arg-Leu-Ala-Asp-Glu-Ser-Gly-His-Val-Val-Leu-Arg-Trp-Leu-Pro [theoretical $(M + H)^+$ = 2357]. Figure 3B (*top*) represents approximately 500 pmol of the mixture, and Fig. 3B (*middle* and *bottom*) represents 50 and 5 pmol, respectively.

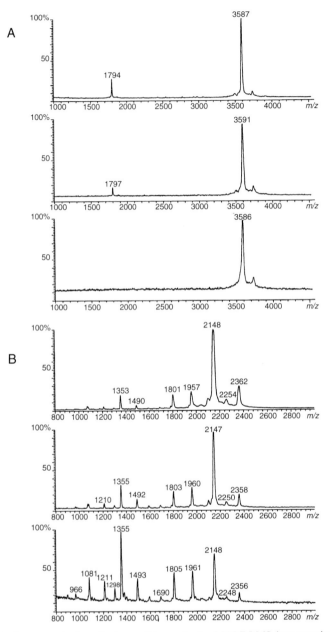

Fig. 3. Assessment of synthetic peptide purity by MALDI-MS interrogation at serial dilution and discernment of potentially misleading ion suppression effects. (A) MALDI mass pectra obtained from a highly purified synthetic peptide at the 2, 20, and 200 pmol level (*top* to *bottom*, respectively). (B) MALDI mass spectra of a highly impure crude synthetic peptide product at the 500, 50, and 5 pmol level (*top* to *bottom*, respectively).

The MALDI-MS analysis (Fig. 3B) indicated that the particular synthesis was extremely problematic. The multitude of ion signals are consistent with a very complex mixture of various amounts of deletion products lacking in various combinations of Pro, Leu, Val, Trp, Arg, His, and Asp residues. The varying ion signal intensities for the respective components depending on the concentration of the peptide mixture also provide a good example for demonstrating the limitations of quantitative estimation from such data. Misleading ion suppression and mass discrimination effects are observed in this set of comparative data. Note in Fig. 3B (*bottom*) showing the MALDI-MS analysis of a 5 pmol amount of the mixture that the 1355 ion signal representing an extensively deleted peptide in 8 to 10 residues appears to be the dominant product. The spectra representing 50 and 500 pmol of product (Fig. 3B, *middle* and *top*, respectively) show a different distribution and are for the most part in agreement with one another. In these two spectra examination of the quantitative relationships suggests that the ion signal having a 2148 mass assignment is the major product. This example is explored further to show the necessity of coupling MALDI-MS with an HPLC analysis to achieve a more realistic clarification of the quantitative relationhips of the various components.

Characterization of a Peptide Synthesis by Coupling Matrix-Assisted Laser Desorption–Ionization Mass Spectrometry and High-Performance Liquid Chromatography Analysis

The example of the problematic synthesis of the 22-residue peptide having the sequence Ala-Pro-Val-Gly-Leu-Val-Ala-Arg-Leu-Ala-Asp-Glu-Ser-Gly-His-Val-Val-Leu-Arg-Trp-Leu-Pro [theoretical $(M + H)^+ = 2357$] introduced above has been exploited to show the definite advantage of coupling MALDI-MS with analytical and preparative reversed-phase (RP) HPLC analysis. In this particular experiment, even though the overall synthesis could be considered a problematic or even a "failed" synthesis, the coupling of MALDI-MS analysis to the HPLC analyses permitted the harvesting of several milligrams of the desired product. Figure 4A shows both the analytical and preparative (inset) HPLC profiles for the complex crude peptide mixture derived from the synthesis having the MALDI-MS analysis presented in Fig. 3B. The conditions for the respective RP-HPLC runs are presented in the legend. Analytical HPLC analysis on a 40-μg load of the mixture indicated three major components, labeled 1, 2, and 3 in the analytical HPLC profile shown in Fig. 4A. As shown in Fig. 4B, MALDI-MS analysis of 2-μl aliquots of peaks, 1, 2, and 3 permitted immediate mass identification of the major products. Peak fractions 1 and 2 were

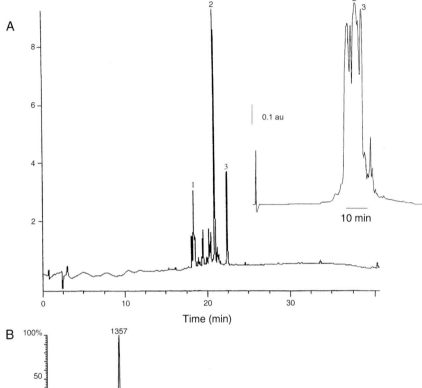

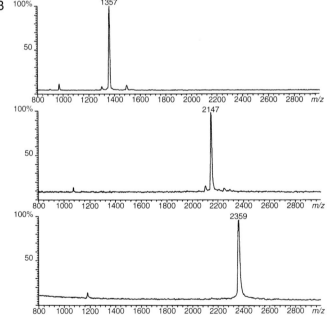

found to be deletion peptides, and peak 3 was found to have a mass consistent with that of the desired product.

For purification of the desired product, 135 mg of the synthetic peptide mixture was subjected to preparative RP-HPLC. The inset of Fig. 4A presents the preparative RP-HPLC profile. Further analytical HPLC and MALDI-MS analyses of the center cuts of the preparative peak fractions 2 and 3 (inset of Fig. 4A) are presented in Figs. 5 and 6, respectively. The analytical HPLC profile shown in Fig. 5A and the MALDI-MS analysis shown in Fig. 5B indicate that the major peak (fraction 2) in the preparative HPLC analysis (inset of Fig. 4A) corresponded to fraction 2 in the analytical HPLC profile of the complex mixture (Fig. 4A). The analytical HPLC profile shown in Fig. 6A and the MALDI-MS analysis shown in Fig. 6B indicate that the fractions labeled 3 in both the analytical and preparative HPLC profiles (Fig. 4A) corresponded to the desired product. Figure 5A is the analytical HPLC profile derived from an aliquot of the eluant from the center cut of the preparative HPLC peak labeled 2 in the Fig. 4A inset. Figure 5B is the MALDI-MS interrogation by serial dilution of this material. Once again notice the marked and misleading ion discrimination effects observed for the set of data presented in Fig. 5B. The MALDI-MS analysis at high concentration (1/10 dilution, Fig. 5B, *top*) suggests that the preparative HPLC fraction 2 is pure, containing a deletion product at mass 2146. However, on further dilution the additional presence of a more extensively deleted product at mass 1805 becomes evident. At the lowest concentration (1/1000 dilution, Fig. 5B, *bottom*) this more extensively deleted product is the major ion signal in the spectrum.

It should be noted that the MALDI-MS analyses of the 1/100 dilutions in both Figs. 3B and 5B appear to be the only spectra that depict realistically semiquantitative information and that are in agreement with the quantitative relationships observed in the respective analytical RP-HPLC profiles (Figs. 4A and 5A). Analytical HPLC and MALDI-MS analyses of the desired product derived from the center cut of fraction 3 of the preparative RP-HPLC run (inset of Fig. 4B) are shown in Fig. 6A,B, respectively. This product is shown to be highly pure by both MALDI-MS interrogation at

FIG. 4. Characterization of a highly impure crude synthetic peptide product by analytical and preparative reversed-phase HPLC and identification of the peaks by MALDI-MS analysis. (A) Analytical HPLC performed on a 4.6 × 200 mm Vydac C_{18} column at a flow rate of 1.5 ml/min. Solvent A, 0.1% TFA; solvent B, 0.1% TFA in acetonitrile; elution with a linear gradient of 5 to 90% B/30 min. *Inset:* Preparative HPLC performed on a 20 × 200 mm Vydac C_{18} column at a flow rate of 10 ml/min. Elution with a linear gradient of 5 to 90% B/40 min. (B) MALDI-MS analysis of fractions derived from the analytical HPLC fractions labeled 1, 2, and 3.

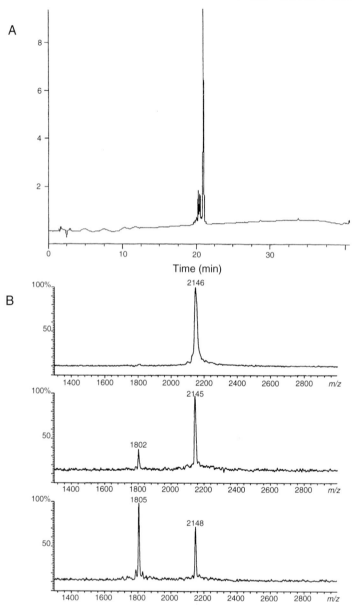

Fig. 5. Characterization of the preparative HPLC fraction 2 (Fig. 4A) by analytical HPLC and MALDI-MS. (A) Analytical HPLC profile of the center cut of fraction 2 derived from the preparative HPLC run presented in the inset of Fig. 4A. Conditions are those described in the legend to Fig. 4. (B) MALDI-MS interrogation by dilution of the preparative HPLC fraction 2.

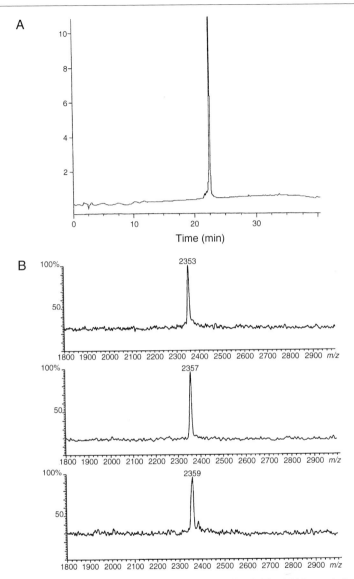

FIG. 6. Characterization of the preparative HPLC fraction 3 (Fig. 4A) by analytical HPLC and MALDI-MS. (A) Analytical HPLC profile of the material obtained from the center cut of preparative fraction 3. (B) MALDI-MS interrogation by dilution of the preparative HPLC fraction 3.

dilution and high resolution analytical RP-HPLC criteria. Note the lack of any emerging ion signals on MALDI-MS interrogation by serial dilution, indicating the high degree of purity.

Monitoring Automated Peptide Synthesis Stepwise by Matrix-Assisted Laser Desorption–Ionization Mass Spectrometry

In the case of troubleshooting problematic syntheses it is advantageous to be able to generate a cycle-by-cycle stepwise historical record of a particular synthesis by coupling microcleavage and deprotection chemistries and mass spectrometric analysis of the generated products. This information would be useful in locating trouble spots and identifying any specific sites of deletion or insertion. This capability was first demonstrated using fast atom bombardment (FAB)-MS and exploiting microcleavage chemistries on peptide–resin aliquots that were automatically removed by a resin sampling feature that is available on some automated peptide synthesizers [Perkin-Elmer Applied Biosystems (PE-ABI, Foster City, CA) Model 430 and 431A peptide synthesizers].[15] Subsequently, the potential of MALDI-MS analysis for this purpose was also demonstrated.[16] Figure 7 shows the utility of this approach for the evaluation of a successful synthesis of a 22-residue peptide having the sequence Asp-Val-Arg-Val-Gln-Val-Leu-Pro-Glu-Val-Arg-Gly-Gln-Leu-Gly-Gly-Thr-Val-Glu-Leu-Pro-Cys. From four TFA microcleavage and deprotection chemistries performed on four resin pools, and MALDI-MS analysis of the product mixtures derived from the respective pools, the entire stepwise assembly record of this particular synthesis was obtained.

Individual peptide–resin sample aliquots were sorted into four pools to minimize any ion suppression effects that may arise if all the peptide–resin samples had been pooled and cleaved in one reaction to generate the historical record. Four pools were chosen to limit the number of microcleavage reactions and thus to minimize effort and time. The first pool consisted of peptide–resins obtained from synthetic cycles 1 through 7, covering assembly of the peptide from the C-terminal Cys to the next six residues (Cys-Pro-Leu-Glu-Val-Thr-Gly) toward the N terminus, the direction of chemical synthesis. The second pool consists of peptide–resins removed after cycles 8 through 12 comprising the next region of extension (Gly-Leu-Gln-Gly-Arg). The third pool of peptide–resin products, from cycles 13 through 17, covers the region Val-Glu-Pro-Leu-Val, and the fourth

[15] W. T. Moore and R. M. Caprioli, in "Techniques in Protein Chemistry II" (J. Villifranca, ed.), p. 511. Academic Press, San Diego, 1991.
[16] B. T. Chait, R. Wang, R. C. Beavis, and S. B. H. Kent, Science 262, 89 (1993).

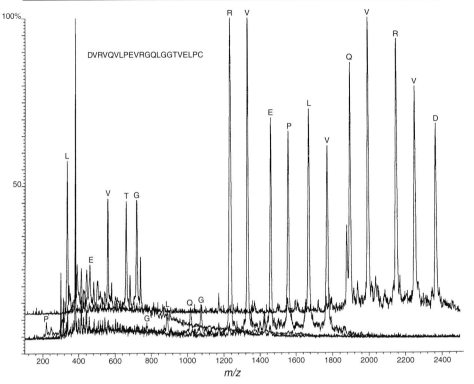

Fig. 7. Stepwise assessment of the automated solid-phase synthesis of a 22-residue peptide by microcleavage/MALDI-MS analysis. Peptide synthesis was performed on a PE-ABI Model 431A automated peptide synthesizer set up and programmed for an Fmoc/*tert*-butyl synthesis strategy. After each addition cycle the automated peptide–resin sampling feature obtained a peptide–resin aliquot. Peptide–resins were pooled into four groups, subjected to microcleavage chemistry, and analyzed by MALDI-MS. The four spectra obtained from each pool are presented as overlapped spectra.

pool covers cycles 18 through 22 comprising the extension of the peptide to the N terminus (Gln-Val-Arg-Val-Asp). The four resin pools were subjected to a microcleavage and deprotection procedure similar to that previously described[15] but adapted for a 9-fluorenylmethyloxycarbonyl (Fmoc) synthesis strategy (a microtization of the Fmoc cleavage reagent K[17]). Cleaved and deprotected peptide products were then interrogated by MALDI-MS analysis on 10-fold serial dilutions in CHCA matrix.

The four spectra derived from those dilutions that yielded maximal information are presented as overlapping spectra in Fig. 7. Note the ion

[17] D. King, C. G. Fields, and G. B. Fields, *Int. J. Pept. Protein Res.* **36,** 255 (1990).

suppression effect that is observed on MALDI-MS analysis for the peptides derived from the microcleavage of the second peptide–resin pool (ion signals for peptides having the assembly sequence Gly-Leu-Gln-Gly-Arg having masses 774 through 1229). The peptide product resulting from the incorporation of the extremely basic Arg residue designated at mass 1229 suppresses the ion signal intensities of the prior assembly products present in this pool. Note the low signal heights for the products labeled in the ion signal sequence G, L, Q, and G falling between masses 774 through 1073 (Fig. 7). It should also be noted that a microcleavage-derived pyroglutamic acid formation was detected for the product labeled Q at mass 1895, as indicated by the −17 amu satellite peak observed for this product. The lack of this −17 amu component in the final crude product (data not shown) indicates that pyroglutamic acid formation occurred as a consequence of the microcleavage chemistry and did not occur during the automated synthesis. Because the analysis of the peptide–resin pool products is stepwise and "time-based," the position of a deletion, insertion, or any modification, if either had occurred, could have been unambiguously and quickly determined. The premise is that any change would be detected as a mass shift in descendants derived only from that cycle where the event occurred.

An interesting variation on this theme has been described exploiting either very acid-labile[18–20] or photolabile linkages[21] to the resin beads. With this approach peptide–resin beads may be exposed to matrix and directly irradiated, with detection of released resin-bound peptides. Either the acidity in the matrix or the laser light promotes a dissociation of a subset of the peptide products from the bead. The postsynthesis cleavage and deprotection steps are thereby eliminated. These modifications lead to the potential for a more direct analysis approaching an on-line peptide synthesis monitoring strategy involving MALDI-MS.

Monitoring Disulfide Bond Formation by Matrix-Assisted Laser Desorption–Ionization Mass Spectrometry-Based Mass Shift Assay

This section covers a MALDI-mass spectrometric assay devised to evaluate a disulfide formation problem. MALDI-MS was performed on a low resolution instrument using external calibrations that resulted in a mass accuracy error that would not permit reliable discernment of the 2 amu

[18] B. J. Egner, G. J. Langley, and M. Bradley, *J. Org. Chem.* **60**, 2652 (1995).
[19] B. J. Egner, M. Cardno, and M. Bradley, *J. Chem. Soc., Chem. Commun.*, 2163 (1995).
[20] G. Talbo, J. D. Wade, N. Dawson, M. Manoussios, and G. W. Tregear, *Lett. Pept. Sci.* **4**, 121 (1997).
[21] M. C. Fitzgerald, K. Harris, C. G. Shevlin, and G. Siuzdak, *Bioorg. Med. Chem. Lett.* **6**, 979 (1995).

mass differences that would be required for distinguishing between the oxidized and reduced forms of the peptide. This assay was developed in response to the challenge presented by the ABRF-1995 Peptide Synthesis study by the Peptide Synthesis Research Committee (PSRC) of the Association of Biomolecular Resource Facilities (ABRF).[5] This study was designed to evaluate four peptide thiol oxidation procedures for the generation of cyclic peptides. The target peptide for this study was cyclo[Cys-Phe-Trp-Lys-Thr-Cys]-Thr-NH$_2$ [(M + H)$^+$ = 1033.2]. The MALDI-MS mass shift assay for free thiols was based on the previous work of Zaluzec et al.[22] demonstrating the use of the organomercurial compound p-hydroxy-mercuribenzoate (PHMB) to probe free thiols in peptides and proteins by MALDI-MS analysis.

The synthetic product was subjected to four different cyclization methods. Methods I and II were postcleavage procedures, and methods III and IV were on-resin procedures.[5] Cyclized product was harvested by lyophilization and dissolved in 50% acetonitrile, 0.1% TFA at 1 mg/ml concentration. This material was diluted 1/10 into a 1.9 mM solution of PHMB dissolved in 0.1% ammonium hydroxide. The reaction was allowed to proceed for 15 min at room temperature. The PHMB peptide reaction mixtures were diluted 1/10 in CHCA matrix and spotted for subsequent MALDI-MS analysis.

Figure 8 is the set of data that we obtained for the peptide product that we submitted for evaluation by the ABRF PSRC. Figure 8A is the PHMB/MALDI-MS analysis result for the linear form of the peptide. The ion signals having observed masses at 1358 and 1680 represent the addition of one (+321) and two molecules of PHMB (+642), respectively, to the available thiols in the linear peptide. Figure 8B is the result for the peptide cyclized by method I involving treatment of the linear peptide in 7 g/l ammonium acetate for 3 days prior to harvest by lyophilization. The MALDI-MS profile in Fig. 8B indicates that some material was oxidized (note the ion signal at mass 1032) and that some material was still in the linear form as demonstrated by the PHMB adducted material (note 1356 and 1676 species). Spectrophotometric evaluation of the free sulfhydryl groups by Ellman's reagent, 5,5'-dithiobis(nitrobenzoic acid),[23] indicated that this product was only 24% cyclized, which correlates well with the PHMB/MALDI-MS analysis. Figure 8C shows the profile derived from the PHMB/MALDI-MS analysis of the method II cyclization procedure involving exposure of the linear form of the peptide to direct oxygenation in

[22] E. J. Zaluzec, D. A. Gage, and J. T. Watson, J. Am. Soc. Mass Spectrom. 5, 359 (1994).
[23] J. M. Stewart and J. D. Young, "Solid Phase Peptide Synthesis," 2nd ed., p. 116. Pierce, Rockford, Illinois, 1984.

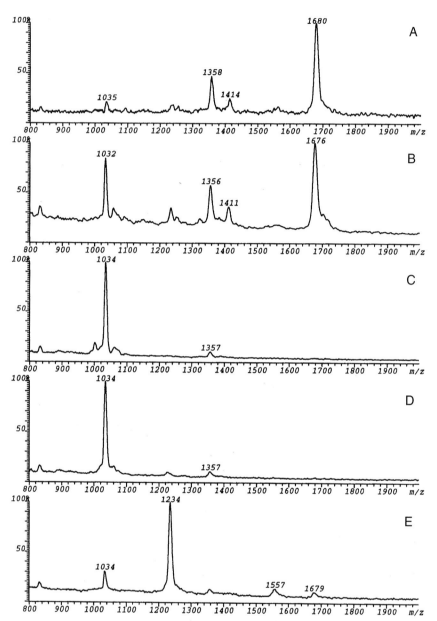

Fig. 8. Assessment of four different peptide cyclization methods through disulfide formation by a MALDI-MS mass shift assay using the thiol adducting reagent PHMB. (A) PHMB/MALDI-MS analysis of the linear thiol-containing form of the peptide. (B) PHMB/MALDI-MS analysis of a peptide subjected to oxidation method I. (C) PHMB/MALDI-MS analysis of a peptide subjected to oxidation method II. (D) PHMB/MALDI-MS analysis of a peptide subjected to oxidation method III. (E) PHMB/MALDI-MS analysis of a peptide subjected to oxidation method IV. PHMB adds a mass of 321 for one thiol and 642 for two thiols.

0.1 M ammonium bicarbonate buffer for 24 hr. Very little PHMB adducted material was observed, indicating an extensive cyclization. Spectrophotometric evaluation using Ellman's reagent indicated 100% cyclization.

Figure 8D,E shows the PHMB/MALDI-MS evaluations for the on-resin cyclizations methods III and IV, respectively. Method III required that the peptide-bound resin be treated with a 1.5 molar excess of 0.4 M thallium trifluoroacetate in dimethylformamide for 1 hr before washing and cleaving. The PHMB/MALDI-MS analysis profile in Fig. 8D indicates an extensively cyclized product that was confirmed by spectrophotometric assay using Ellman's reagent, suggesting 100% cyclization by this method also. The PHMB/MALDI-MS analysis of the product generated by the method IV cyclization procedure clearly indicates a problem with this method. The ion signal at mass 1234 is a Hg ($+200$) adduct of the peptide. Method IV required treating the peptide–resin with a 4-fold molar excess of 0.1 M mercuric acetate in dimethylformamide for 1 hr followed by treatment with a 10-fold molar excess of 2-mercaptoethanol prior to washing and cleavage. The problem of mercury adduction has been previously noted with this procedure. It is of interest to note that spectrophotometric assay using Ellman's reagent suggested 100% cyclization. The product is cyclized but via a mercuric ion insertion. Divalent mercuric ions (Hg^{2+}) are known to have the potential to bridge two thiol groups.[24]

Probing Synthetic Peptide Racemization Problems by Coupling
    Enzymatic Treatment with Matrix-Assisted Laser
    Desorption–Ionization Mass Spectrometry

Enantioselective exopeptidase hydrolysis has been employed to address racemization problems in peptide synthesis.[25] In this type of study, the release of terminal amino acids are quantitatively measured by amino acid analysis. The present example, inspired by the ABRF-1996 PSRC study,[6] employs a similar approach with a test peptide designed to address the racemization that is possible with activated His derivatives.[26] However, for the ABRF-1996 PSRC study,[6] enantioselective enzymatic hydrolysis was coupled with MALDI-MS to investigate any problems encountered in the synthesis of the test peptide Arg-Glu-Arg-His-Ala-Tyr [$(M + H)^+$ = 832].

---

[24] T. M. Jovin, P. T. England, and A. Kornberg, *J. Biol. Chem.* **244,** 3009 (1969).
[25] M. Bodansky and A. Bodansky, "The Practice of Peptide Synthesis," p. 237. Springer-Verlag, New York, 1984.
[26] G. B. Fields, Z. Tian, and G. Barany, *in* "Synthetic Peptides: A User's Guide" (G. A. Grant, ed.) p. 77. Freeman, New York, 1992.

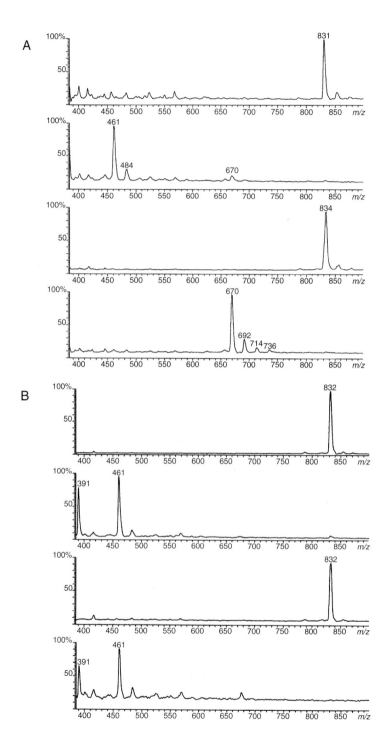

To demonstrate the power of MALDI-MS in this type of racemization study, results obtained on two reference peptides are presented. One reference peptide was made with all-L amino acids (Arg-Glu-Arg-His-Ala-Tyr). The other reference peptide was made with all-L amino acids except for His, which was incorporated as the D-form of the amino acid (Arg-Glu-Arg-D-His-Ala-Tyr). Figure 9A shows the results obtained by applying a coupled carboxypeptidase A/MALDI-MS analysis of the reference peptides. The top two spectra of Fig. 9A are the results obtained from non-enzyme-treated control all-L peptide (*top*) and the carboxypeptidase A-treated all-L peptide (*second from top*). The ion signal at mass 461 seen in the spectrum in Fig. 9A (*second from top*) is for the tripeptide product Arg-Glu-Arg. Carboxypeptidase A removed the three C-terminal residues, Tyr, Ala, and His. The third spectrum from the top in Fig. 9A represents the product containing the D-His residue (non-enzyme-treated control). Treatment of this reference peptide with carboxypeptidase A resulted in removal of solely the C-terminal Tyr as indicated by the observed ion signal at mass 670 (Fig. 9A, *bottom*). The presence of a D-His at the P_2 subsite on the substrate (Schecter and Berger nomenclature for protease subsites[27]) was enough to prevent further processing by the carboxypeptidase A. The theoretical $(M + H)^+$ for the sequence Arg-Glu-Arg-D-His-Ala is 669, and the presence of this ion is taken to be a signature for racemization at the His position in this particular peptide. Note that the slight amount of 670 ion signal in the second spectrum from the the top in Fig. 9A suggests that a slight amount of racemized product was obtained from the synthesis of the all-L peptide. Interestingly, as shown in Fig. 9B, treatment of the reference peptides with trypsin did not show any enantiomeric selectivity of trypsin for these synthetic peptide substrates. In both preparations the identical tryptic products were observed: Arg-Glu-Arg $[(M + H)^+ = 461]$ and His-

---

[27] I. Schecter and A. Berger, *Biochem. Biophys. Res. Commun.* **27**, 157 (1967).

FIG. 9. Racemization analysis by coupling enantioselective enzymatic hydrolysis with MALDI-MS. (A) Carboxypeptidase A treatment of reference peptides. The top two spectra represent the all-L form of the peptide, Arg-Glu-Arg-His-Ala-Tyr. The top most spectrum is the untreated control, and the second spectrum from the top is carboxypeptidase A treated. The bottom two spectra are the fully "racemized" reference peptide Arg-Glu-Arg-D-His-Tyr. The third spectrum from the top is the untreated control, and the bottom most panel is carboxypeptidase A treated. (B) Trypsin treatment of the same reference peptides. The top two spectra represent the all-L form peptide Arg-Glu-His-Ala-Tyr. The top most spectrum is the untreated control, and the second spectrum from the top is trypsin treated. The bottom two spectra represent the fully "racemized" reference peptide Arg-Glu-Arg-D-His-Tyr. The third spectrum from the top is the untreated control, and the bottom most spectrum is trypsin treated.

Ala-Tyr $[(M + H)^+ = 391]$ whether or not D-His was present in the P'_1 position (Schecter and Berger nomenclature for protease substrate sub-sites[27]).

Summary

The examples presented indicate that MALDI-MS is a useful tool for evaluating the progress of peptide synthesis at all the necessary levels: automated assembly, cleavage and deprotection chemistries, RP-HPLC analyses and purifications, and structural validation of the final product. The technique, if judiciously applied, permits the evaluation of complex peptide mixtures and often provides a semiquantitative overview. We have found that the availability of this method has enabled the provision of high-quality peptide reagents for use in the local research environment. The integration of this methodology into our peptide synthesis facility has also enabled and encouraged us to undertake more challenging synthetic problems such as phosphopeptide synthesis, peptide cyclizations, and peptide modification chemistries that would not ordinarily be offered if the laboratory lacked this technology. MALDI-MS is one of the more versatile and readily integrable mass spectrometric methods that can be incorporated into the average peptide synthesis laboratory.

Acknowledgments

The author thanks Dr. John D. Lambris of the Department of Pathology and Laboratory Medicine, The School of Medicine, University of Pennsylvania, Philadelphia, for continual support and encouragement and for realizing the value of mass spectrometry and fostering the inclusion of this technology at the Protein Chemistry Laboratory, and Ms. Lynn A. Spruce for expert and conscientious technical assistance in all the areas of peptide synthesis. The Association of Biomolecular Resource Facilities (ABRF) Peptide Synthesis Research Committee (PSRC) is also acknowledged for inspiring solutions to some well-designed and testable problem cases in solid-phase peptide synthesis.

Section III

Specialized Applications

[24] Protein Signature Analysis: A Practical New Approach for Studying Structure–Activity Relationships in Peptides and Proteins

By Tom W. Muir, Philip E. Dawson, Michael C. Fitzgerald, and Stephen B. H. Kent

Introduction

The emergence of new techniques in structural biology and molecular biology has greatly improved our ability to study the molecular basis of protein function. These approaches are complementary in the type of information they provide: a high-resolution protein structure often suggests the generation of site mutants of the protein to test the validity of structure-based hypotheses. One of the drawbacks to such a synergistic approach, combining structural knowledge with molecular biology, is the time and effort required to generate and study all the individual mutant proteins. To address this, we have introduced a new technique, protein signature analysis (PSA), which provides a more rapid route to the systematic modification (of the covalent structure) and analysis of a peptide or protein.[1-3] This approach takes advantage of our ability to prepare proteins by total chemical synthesis.[4]

The principle of protein signature analysis is illustrated in Fig. 1 and consists of three steps.

1. Total chemical synthesis is used to generate an array of proteins in which an analog unit (e.g., a dipeptide derivative) is systematically substituted throughout the region of interest in the polypeptide chain, such that each member of the array contains a single copy of the analog unit at a unique and defined position. Note that this strategy differs from scanning mutagenesis approaches,[5,6] because all possible positions are explored at once.

---

[1] P. E. Dawson, M. C. Fitzgerald, T. W. Muir, and S. B. H. Kent, *J. Am. Chem. Soc.* in press (1997).
[2] T. W. Muir, P. E. Dawson, M. C. Fitzgerald, and S. B. H. Kent, *Chem. Biol.* **3**, 817 (1996).
[3] P. E. Dawson, T. W. Muir, M. C. Fitzgerald, and S. B. H. Kent, submitted for publication (1996).
[4] T. W. Muir, *Structure* **3**, 649 (1995).
[5] B. C. Cunningham and J. A. Wells, *Science* **244**, 1081 (1989).
[6] J. P. Tam, Y.-Z. Lin, Y.-Z. Wu, Z.-Y. Shen, M. Galantino, W. Liu, and X.-H. Ke, *in* "Peptides: Chemistry, Structure, and Biology, Proceedings of the Eleventh American Peptide Symposium" (J. E. Rivier and G. R. Marshall, eds.), p. 75. ESCOM, Leiden, The Netherlands, 1989.

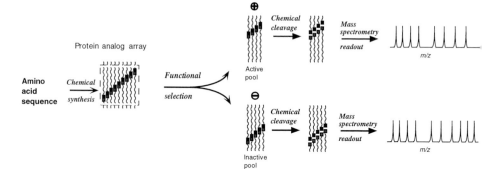

FIG. 1. Principle of protein signature analysis. The approach involves three steps: (1) chemical synthesis of a self-encoded array of proteins in which an analog unit is scanned through the primary sequence of the protein; (2) functional selection of the array of synthetic protein analogs; and (3) single step compositional readout of the resulting active and inactive populations.

2. The array of protein analogs is then subjected to functional selection, resulting in separation into two pools, active and inactive.

3. In the final step, the molecular composition of each pool of protein analogs is determined in a single operation using a built-in chemical decoding system in combination with a mass spectrometric readout. The resulting data provide a qualitative signature relating the effects on activity to substitution of the analog structure throughout the region of interest in the protein molecule.

Protein signature analysis is a chemistry-driven method that provides a functional profile of the effects of structure variation throughout a peptide or protein sequence.[1-3] The technique integrates advances in the total chemical synthesis of proteins[7] and biomolecular mass spectrometry[8] into a practical new tool for studying structure–activity relationships in peptides and small proteins. The concept of PSA has been described in detail elsewhere,[1] as has its application to studying structure–activity relationships in proteins.[2] In this article, important experimental features of the approach are discussed.

Methods

Synthesis of Protein Arrays

The following general procedure has been employed in the synthesis of the Crk-N polypeptide analog arrays.[1,3] A combination of machine-

[7] T. W. Muir and S. B. H. Kent, *Curr. Opin. Biotechnol.* **4,** 420 (1993).
[8] B. T. Chait and S. B. H. Kent, *Science* **257,** 1885 (1992).

assisted synthesis[9] and manual split-resin synthesis[1] is used in the assembly of the protein arrays. Specifically, manual split-resin protocols are employed only in that part of the synthesis where the analog unit is being incorporated; otherwise, standard machine-assisted stepwise SPPS (solid-phase peptide synthesis) is used throughout. *In situ* neutralization/(1*H*-benzotriazol-1-yl)-1,1,3,3-tetramethyluronium hexafluorophosphate (HBTU) activation protocols for *tert*-butyloxycarbonyl (Boc)-based SPPS[9] are used in both the manual and automated portions of the synthesis. Crk-N syntheses are typically performed on a 0.2 mmol scale, and standard side-chain protecting groups[9] are used, except for the indole moiety of tryptophan, which is left unprotected for reasons discussed elsewhere in this article. The dipeptide analogs Boc-Gly-[COS]-β Ala and Boc-Gly-[COS]-Gly, are each prepared in solution as previously described.[10]

The apparatus we used in the split-resin portion of a synthesis comprises two manual SPPS reaction vessels and an intermediate analog attachment vessel (Fig. 2). Standard stepwise Boc-SPPS[9] of the target polypeptide sequence is performed simultaneously in both manual SPPS reaction vessels. At each cycle of split-resin synthesis, the N^α-Boc-deprotected peptide–resin in vessel 1 is suspended in 10 ml dimethylformamide (DMF). An aliquot (~5 μmol) of the suspension is then removed as a fraction by volume and transferred to the analog attachment vessel. The resin aliquot is neutralized with 10% diisopropylethylamine (DIEA) in DMF and then reacted with the appropriate analog unit (25 μmol), preactivated as a 1-hydroxybenzotriazole (HOBt) active ester. Following attachment of the analog unit, the resin aliquot is transferred to manual reaction vessel 2 where stepwise Boc-SPPS is resumed. Note that because dipeptide analog units are used in the Crk-N work, each resin aliquot is transferred to vessel 2 exactly two cycles after it is removed from vessel 1. This ensures that the analog unit replaces the desired dipeptide sequence within the protein.

On completion of the split-resin portion of the synthesis, the resin in vessel 2 is transferred to a modified Applied Biosystems (Foster City, CA) 430A peptide synthesizer[9] and the remainder of the amino acids in the sequence added in stepwise fashion. Following synthesis, the peptide–resin mixture comprising the completed protein array is cleaved from the resin support with simultaneous removal of side-chain protecting groups by treatment with liquid HF (1 hr at 0°) containing 5% (v/v) *p*-cresol as a scavenger. After evaporation of the HF, the crude protein array is precipitated with cold diethyl ether, dissolved in 1 : 1 (v/v) CH_3CN/water, 0.1% trifluoroacetic acid (TFA), and then lyophilized. The composition of each protein array

[9] M. Schnölzer, P. Alewood, A. Jones, D. Alewood, and S. B. H. Kent, *Int. J. Pept. Protein Res.* **40**, 180 (1992).
[10] H. Hojo and S. Aimoto, *Bull. Chem. Soc. Jpn.* **64**, 111 (1991).

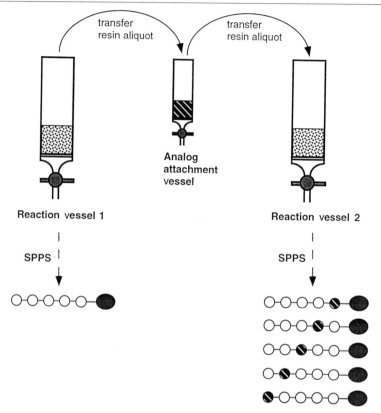

Fig. 2. Solid-phase synthesis of chemically defined protein arrays. In this modified split-resin technique the newly derivatized resin aliquot is transferred to a second reaction vessel rather than back to the first reaction vessel. Use of the second reaction vessel means that the growing peptidyl–resin originally present in vessel 1 is steadily siphoned into vessel 2. On completion of the split-resin synthesis, the complete array of synthetic peptide/protein analogs is present, as a single product mixture, within reaction vessel 2. Furthermore, use of this protocol ensures that each component of the protein array contains only a single copy of the analog unit at a unique and defined position.

is characterized as follows: crude protein array (~1 mg) is dissolved in a cleavage buffer consisting of 1 M NH_2OH, 20 mM NH_4HCO_3, pH 5.5, buffer (1 ml) and stirred for 15 min. The cleaved arrays are then exchanged into a 70% CH_3CN/30% H_2O, 0.1% TFA solvent system using a 1-ml low pressure, disposable C_{18} column (Sep-Pak, Waters, Milford, MA) and immediately analyzed by matrix-assisted laser desorption–ionization (MALDI) mass spectrometry.

Synthesis of C3G-Derived Ligand and Control Peptides

The peptides corresponding to the C3G-derived ligand, Ac-Cys-Trp-Acp-Pro-Pro-Pro-Ala-Leu-Pro-Pro-Lys-Lys-Arg-NH$_2$ (where ACP is ε-aminocaproic acid), and the control, Ac-Cys-Acp-Tyr-Gly-Gly-Phe-Leu-NH$_2$, are synthesized on 4-methylbenzhydrylamine resin (0.93 mmol/g Peninsula Laboratories, Belmont, CA) according to optimized solid-phase methods.[9] Following synthesis the peptides are cleaved from the resin support with simultaneous removal of side-chain protecting groups by treatment for 1 hr at 0° with HF containing 5% (v/v) *p*-cresol as a scavenger. After evaporation of the HF, the crude peptides are precipitated with cold diethyl ether containing 2% (v/v) 2-mercaptoethanol and then lyophilized. The crude peptides are then purified by semipreparative high-performance liquid chromatography (HPLC) using a 25–50% acetonitrile gradient over 30 min. The products are characterized by electrospray mass spectrometry (ESMS). Ac-Cys-Trp-Acp-Pro-Pro-Pro-Ala-Leu-Pro-Pro-Lys-Lys-Arg-NH$_2$: observed mass, 1543 ± 1 Da; calculated mass for $C_{74}H_{118}N_{20}O_{14}S_1$ (average isotope composition), 1543.9 Da. Ac-Cys-Acp-Tyr-Gly-Gly-Phe-Leu-NH$_2$: observed mass, 813.5 ± 0.5 Da; calculated mass for $C_{39}H_{56}N_8O_9S_1$ (average isotope composition), 814.0 Da.

Synthesis of Peptide Affinity Columns

The C3G peptide affinity column is prepared as follows. The peptide Ac-Cys-Trp-Acp-Pro-Pro-Pro-Ala-Leu-Pro-Pro-Lys-Lys-Arg-NH$_2$ is dissolved in 50 mM Tris, 5 mM EDTA, pH 8.0 (10 mg in 2 ml), and shaken with Sulfolink resin (Pierce, Rockford, IL) for 1 hr. Unreacted iodoalkyl groups on the resin are then blocked by treatment with 50 mM cysteamine, 50 mM Tris, and 5 mM EDTA, pH 8.0, buffer. The loading of the column is determined by UV absorbance to be approximately 5 μmol/ml. A similar procedure is used to attach the control peptide Ac-Cys-Acp-Tyr-Gly-Gly-Phe-Leu-NH$_2$ (where Tyr-Gly-Gly-Phe-Leu is leucine enkephalin) to the Sulfolink support.

Affinity Selection of Protein Arrays

The crude protein array (typically around 1.5 mg) is dissolved in 20 mM HEPES, 50 mM NaCl, pH 7.3, buffer (600 μl) and loaded on to a 1 ml C3G peptide affinity column preequilibrated with the same buffer. After 6–8 hr (required for optimal binding) the nonspecifically bound material is eluted from the column by washing with 0.5 M NaCl, 0.1 M sodium phosphate, pH 7.0, buffer (six times, 1 ml each time). Eluted material (typically in the first and second wash) is immediately cleaved by dilution into 1 M NH_2OH, 20 mM NH_4HCO_3, pH 5.5, buffer. Following further column washing with 1 M NaCl, 0.1 M sodium phosphate buffer, pH 7.0, the specifically bound material (active pool) is chemically cleaved and thus simultaneously eluted from the affinity column by washing with 1 M NH_2OH, 20 mM NH_4HCO_3, pH 5.5, buffer (four times, 1 ml). To accommodate the MALDI analysis, both active and inactive pools are desalted and exchanged into a 70% CH_3CN/30% H_2O, 0.1% TFA solvent system using a low pressure, disposable C_{18} column (Sep-Pak, Waters).

Matrix-Assisted Laser Desorption–Ionization Analysis of Peptide Arrays

Following the desalting step, samples are immediately analyzed by MALDI mass spectrometry. All samples are prepared by adding 2 μl of the desalted column fraction to 5 μl of a saturated solution of α-cyano-cinnaminic acid in 50% acetonitrile in water, 0.1% TFA. From this mixture, 2 μl (containing ~1–10 pmol of each peptide component) is added to a stainless steel probe tip and the solvent allowed to evaporate under ambient conditions. Mass spectra are recorded using a prototype laser desorption, linear time-of-flight mass spectrometer from Ciphergen Biosystems (Palo Alto, CA). Samples are desorbed/ionized using 337 nm radiation output from a nitrogen laser (Laser Science, Newton, MA). All spectra are acquired in the positive ion mode and summed over 20–50 laser pulses. Time-to-mass conversion is accomplished by internal calibration using the [M + H]$^+$ signals from the largest and smallest peptide components in each array.

Results and Discussion

In this article, the technical aspects of the protein signature analysis method are illustrated within the context of our work on the N-terminal SH3 domain from the cellular adaptor protein c-Crk[2] (hereafter referred to as Crk-N). Our objective in these studies is to use protein signature analysis to explore the structure–activity relationship within the 58-residue Crk-N protein domain (residues 134–191 of the murine sequence). In the following sections, the important practical aspects, including experimental design, underlying each of the three steps in PSA (synthesis, selection, and readout) are discussed.

Synthesis of a Defined Array of Protein Analogs

Before undertaking the chemical synthesis of an array of protein analogs, a number of factors must be taken into consideration, namely, the experimental strategy used to generate the protein array, the design of the analog unit to be incorporated, and the type of solid-phase chemistry to be used.

Experimental Strategy. If our objective is to obtain a defined array of protein analogs, then clearly two distinct synthetic routes are available. Each of the protein analogs could of course be individually synthesized and the products of these syntheses combined to give the desired array. This option has to be ruled out because of the prohibitively large amount of work involved. Alternatively, the complete array of protein analogs could be constructed simultaneously in the course of one synthesis. This strategy is much more attractive since, in principle, it provides a quicker and more direct route to the desired protein array.

Any approach designed to allow the "one pot" synthesis of an array of protein analogs must ensure that each component of the array contains only a single copy of the analog unit at a unique and defined position. This requirement rules out the use of the "divide and recombine" approach commonly employed in the synthesis of peptide libraries[11]: this would result in multiple copies of the analog unit being incorporated into each peptide chain. Such multiple incorporations of the analog unit can, in fact, only be avoided by physically separating the peptide–resin which has been derivatized with the analog unit from the peptide–resin which has yet to be.

The modified split-resin procedure illustrated in Fig. 2 provides a route to the synthesis of defined arrays of protein analogs.[1] The approach involves the use of two reaction vessels, with identical synthetic manipulations, aimed at the same target sequence, being carried out in each. At appropriate cycles of the synthesis an aliquot of the peptide–resin is removed from the first vessel and the analog unit attached. This derivatized resin aliquot is then transferred to the second reaction vessel and the remainder of the amino acids in the sequence added in stepwise fashion. Continual siphoning of peptide–resin aliquots from vessel 1 into vessel 2 (with analog attachment in between) results in the generation of the complete protein array as a single product mixture. It should be stressed that this procedure allows the entire array of proteins to be chemically synthesized in approximately the same length of time (and for around the same cost) as it would take to construct a single polypeptide using stepwise SPPS.

The split-resin procedure has been successfully used to generate a number of peptide and protein arrays.[1-3] These studies have revealed several important experimental parameters. First, the starting scale of the synthesis must be adjusted so that sufficient resin is present to allow the appropriate number of equimolar aliquots (defined by the desired number of analogs

---

[11] A. Furka, F. Sebestyen, M. Asgedom, and G. Dibo, *Int. J. Pept. Protein Res.* **37,** 487 (1991).

in the final protein array) to be removed from vessel 1 during the course of the split-resin synthesis. Although the size of the resin aliquot being removed is optional, we have found that aliquots in the range 3–10 μmol are both convenient to withdraw and are large enough to ensure that adequate amounts of the final protein array will be generated. Second, it is important that each of the protein components is equally represented (in molar terms) in the final protein analog array. This requirement means that approximately the same molar amount of peptide–resin must be removed from vessel 1 at each withdrawal step. Here three factors must be taken into account: the swelling properties of the growing peptide–resin; the amount of peptide–resin remaining in vessel 1 at any given point in the chain assembly; and the substitution of the peptide–resin during the synthesis.

Variations in the swollen volume of the peptide–resin during synthesis potentially complicate the removal of equimolar aliquots using fractional volume. The swelling properties of the peptide–resin depend both on the type of resin matrix being used and the sequence of the growing peptide chain. The swollen volume of standard cross-linked polystyrene resins can change appreciably during the course of a long synthesis (>5-fold in some cases[12]), whereas other supports such as polyacrylamide or grafted polyethylene glycol–polystyrene-based resins may change volume to a lesser extent.[13] The effect of swelling variations on the accuracy of volumetric transfers can be minimized by withdrawing (as an aliquot) a known fraction of the total volume, from a suspension of the peptide–resin diluted in large amount of solvent (typically at least 10 times the swollen volume of resin in DMF). It is important that such a suspension be nearly homogeneous, and an efficient mixing method such as nitrogen agitation should be employed during the withdrawal step.

Removing equimolar aliquots of resin from vessel 1, means that the total number of moles of peptide–resin remaining in vessel 1 will decrease in a linear fashion during the synthesis. In addition, the substitution of the peptide–resin (as moles of peptide per gram of peptide–resin) decreases significantly during the course of a long synthesis, meaning that the weight of an equimolar aliquot will correspondingly increase. Both these effects (which act in concert) must be suitably compensated for when withdrawing each successive resin aliquot. In practice, approximately equimolar aliquots can be obtained by simply increasing the volume fraction of the resin suspension removed in a linear fashion throughout the course of the experi-

[12] V. K. Sarin, S. B. H. Kent, and R. B. Merrifield, *J. Am. Chem. Soc.* **102,** 5463 (1980).
[13] J. S. Fruchtel and G. Jung, *Angew. Chem., Int. Ed. Engl.* **35,** 17 (1996).

ment. For example, in the case of a ten-membered array, 1/10 of the resin suspension is removed in round one, 1/9 in round two, 1/8 in round three, and so on.

In the synthesis of the Crk-N arrays, the following procedure has been used during the peptide–resin withdrawal step. The peptide–resin in vessel 1 is suspended in DMF to a known volume (typically 10 ml). While gently mixing this suspension (to prevent settling of the resin bed), the calculated fraction of the resin suspension is removed volumetrically from vessel 1 and transferred to an intermediate reaction vessel for analog attachment.

Nature of Analog Unit. The design of the analog unit to be substituted into the peptide or protein sequence is dependent both on the type of structure–activity relationship that is being probed and on the chemical readout strategy being employed to identify each component in the resulting array of protein analogs. The former of these design criteria is case specific, and the analog unit can be synthetically tuned to address a particular structural or functional question. For example, in the Crk-N work we were interested in the functional roles of the individual amino acid side chains in the protein, as well as the polypeptide backbone which connects them. Two different dipeptide analog units, -Gly-[COS]-Gly- and -Gly-[COS]-βAla-, were used in these studies.[2] Replacement of a dipeptide sequence in the protein with the first of these (-Gly-[COS]-Gly-) resulted in the deletion of two adjacent side chains, whereas the second (-Gly-[COS]-βAla-) resulted in the incorporation of an extra methylene unit in the backbone and the deletion of two side chains. Thus, comparison of the results from the two PSA experiments allowed functional effects to be assigned to a particular chemical modification.[2] The ability to iterate with related analog units represents one of the most powerful features of the PSA approach.

Both of the dipeptide analog units described above contain a central thioester linkage. This is incorporated to facilitate single-step compositional readout of the arrays of Crk-N protein analogs. Building a latent chemical cleavage site such as a thioester group into the analog unit means that each protein analog in the final array will contain this chemical marker at a unique position in the polypeptide sequence. Consequently, exposure of the protein array to conditions that selectively cleave at the chemical marker will result in the "unzipping" of the protein analogs with the generation of two sets of daughter peptide fragments. Significantly, each of these peptide fragments will have a unique mass indicative of the position of the analog unit within the sequence of an original protein analog. Each protein analog in an array is thus self-encoded. Chemical groups such as thioesters can be easily incorporated into an analog unit,[10] and they can be selectively cleaved under very mild conditions (e.g., 0.1 M hydroxylamine at neutral pH[14]).

One consequence of using a thioester group as a cryptic cleavage site is that the intrinsic stability of the dipeptide analog must be considered. Analog units that place a nucleophilic group δ or ε to the carbonyl of the thioester should be avoided owing to the risk of acyl rearrangement.[14] For this reason, amino acid residues such as Ser, Thr, Cys, Asn, Gln, and His should not be incorporated into thioester-containing dipeptide analogs. The chemically stable -Gly-[COS]-Gly- and -Gly-[COS]-βAla- units used in the Crk-N studies are each prepared in solution,[10] and then directly incorporated into the protein arrays as HOBt active esters. Note that the risk of racemization via oxazalone formation, which prevents most dipeptides from being directly coupled to peptide chains, is not a concern when using thioester-linked dipeptide analogs, which cannot form the oxazalone.

Compatibility with Solid-Phase Chemistries. The strategy used during solid-phase synthesis of the protein array [i.e., Boc versus 9-fluorenylmethyloxycarbonyl (Fmoc)] is implicitly linked to the chemical structure of the analog unit being substituted into the sequence. The latent cleavage site within the analog unit must be unreactive to the acylation and N^α-deprotection steps of the chain assembly, as well as to the final deprotection/cleavage conditions. Thioester-containing analog units, such as the units in the Crk-N studies, are stable under acid conditions but are rapidly cleaved by exposure to high pH, nucleophiles, and most reducing agents [note that alkyl thioesters are stable to nonnucleophilic reducing agents such as tris(2-carboxyethyl)phosphine[15]]. Thus, the standard N^α-deprotection conditions associated with the Boc strategy (i.e., TFA treatments) are compatible with thioester-containing analogs, whereas the N^α-deprotection conditions commonly used in the Fmoc-strategy (i.e., piperidine treatments) will result in rapid aminolysis of the thioester linkages and thus premature peptide chain cleavage. In general, the Boc-based SPPS strategy can be successfully employed when using nucleophile-sensitive analog units such as those containing thioester linkages.

The acylating power of the thioester unit is well documented in peptide chemistry.[14,16] Acyl-transfer reactions can occur between the thioester group of the analog unit and the α-amino group of the peptide following each successive deprotection step in the synthesis. This undesired acylation

[14] T. Wieland, *in* "The Roots of Modern Biochemistry" (V. D. Kleinkauf, ed.), p. 213. de Gruyter, New York, 1988.
[15] M. Baca, T. W. Muir, M. Schnolzer, and S. B. H. Kent, *J. Am. Chem. Soc.* **117,** 1881 (1995).
[16] P. E. Dawson, T. W. Muir, I. Clark-Lewis, and S. B. H. Kent, *Science* **266,** 766 (1994).

reaction will be most pronounced after deprotection of the residue immediately adjacent to the analog unit, owing to the facile nature of diketopiperazine formation.[17] These side reactions will be especially problematic when the TFA salt of the α-amino group is neutralized as a discrete step prior to addition of the next activated amino acid. However, the use of *in situ* neutralization protocols[9] greatly reduces the likelihood of the free α-amino group attacking the thioester moiety, simply because amine neutralization occurs in the presence of a large molar excess of the more activated Boc-amino acid. The thioester cleaving N-acylation reaction must compete with the kinetically favored amino acid coupling reaction. Thus, use of *in situ* neutralization protocols is preferred for chain assembly of thioester-containing protein analogs.

The majority of the side-chain protecting groups commonly used in the Boc-based SPPS strategy are chemically compatible with thioester-containing analog units.[9] The exceptions are those which are removed by treatment with either nucleophiles (e.g., primary and secondary amines), heavy metal salts [e.g., $Hg(OAc)_2$], or nucleophilic reducing agents (e.g., thiols). Thus, the commonly used amino acid derivatives Boc-His(DNP), Boc-Trp(CHO), and Boc-Cys(Acm) (where DNP is dinitrophenyl, CHO is formyl, and Acm is acetamidomethyl) cannot be used in concert with thioester-containing analog units.[15,18] Note that Boc-His(Bom), Boc-Trp(unprotected), and Boc-Cys(MeBzl) (where Bom is benzyloxymethyl and MeBzl is 4-methylbenzyl) should be used in these instances. Similarly, the choice of scavengers used during the final deprotection/cleavage step is somewhat restricted by the presence of thioester bond within a peptide.[15] Specifically, the use of thiol- and thioester-based scavengers [e.g., thiocresol, dimethyl sulfide (DMS), ethanedithiol (EDT)] will result in unwanted cleavage of the thioester bond within the analog peptide. In contrast, both cresol and anisole can be used with thioester-containing peptides without any appreciable problems.[15,18,19]

Functional Selection of Protein Array

An important implication of the self-encoding strategy used in protein signature analysis is that all postcleavage operations can be performed using an array of protein analogs free in solution. Physical resolution of mixtures of peptides or proteins in solution can be achieved using a range of well-established chromatographic or electrophoretic techniques. In prin-

[17] B. F. Gisin and R. B. Merrifield, *J. Am. Chem. Soc.* **94**, 3102 (1972).
[18] M. Schnölzer and S. B. H. Kent, *Science* **256**, 221 (1992).
[19] L. E. Canne, A. R. Ferre-D'Amare, S. K. Burley, and S. B. H. Kent, *J. Am. Chem. Soc.* **117**, 2998 (1995).

ciple, all of these biochemical separation approaches can be used to select, and in the process resolve, the intact (parent) array of protein analogs based on a particular biological or physicochemical property. For example, affinity chromatography could be used to separate an array of protein analogs on the basis of function, giving rise to active and inactive pools. Alternatively, techniques such as ion-exchange and gel-filtration chromatographies could be used to separate folded from unfolded members of an array of protein analogs. Indeed, the use of two or more orthogonal selection methods may provide valuable complementary information on the structural and functional properties of each member of the array of protein analogs. The ability to use a chromatography-based selection strategy in PSA should be contrasted with combinatorial approaches in which the screening assay is carried out using an immobilized or cell-anchored library. In these cases, selected beads/cells usually have to be visually identified and then mechanically separated on an individual basis.

Of interest in the Crk-N work was the effect on ligand-binding activity of scanning the -Gly-[COS]-Gly- and -Gly-[COS]-βAla-analogs through the Crk-N sequence The most straightforward way to select for ligand binding is to use affinity chromatography. The Crk-N domain specifically recognizes a 10-residue proline-rich sequence from the protein, C3G.[20] A small peptide is synthesized in which an [acetyl-Cys-Trp]- moiety is attached to the N terminus of the C3G sequence (Pro-Pro-Pro-Ala-Leu-Pro-Pro-Lys-Lys-Arg) via an aminocaproic acid spacer (included to separate the binding sequence from the [acetyl-Cys-Trp]- unit). Incorporation of the unique Cys residue allows the peptide to be covalently attached to commercially available iodoalkyl-derivatized agarose beads (Sulfolink, Pierce) via a thioether linkage. The Trp residue allows the loading of the column to be determined by monitoring the decrease in UV absorbance of the supernatant in the coupling reaction. A similar synthetic strategy is used to generate a control affinity column in which a nonspecific peptide sequence, leucine enkephalin (Tyr-Gly-Gly-Phe-Leu), is covalently immobilized on the same type of agarose beads.

It is quite probable that a range of biological activities will be present within an array of protein analogs, with some members possessing near wild-type activity and others having significantly reduced or even no activity. To deal with this complex situation a simple binary system is employed in the selection of the Crk-N arrays. In this strategy, that fraction of the array of protein analogs with binding activity above a chosen threshold is considered active, whereas the remaining fraction with lower binding activity is considered inactive. The threshold level can of course be varied in such

[20] B. S. Knudsen, S. M. Feller, and H. Hanafusa, *J. Biol. Chem.* **269**, 32781 (1994).

a selection system, thereby allowing quite detailed functional information to be extracted through iteration of the selection process.

In the case where affinity chromatography is to be used, the threshold level is determined by the experimental conditions used to wash the affinity column. Stringency can be adjusted by including varying concentrations of salts, detergents, or even competing wild-type protein in the wash buffer. In the Crk-N studies, a wash protocol involving the use of high salt-containing buffers is employed. This procedure was chosen after a series of preliminary studies in which wild-type Crk-N, spiked with various contaminants, was applied to the C3G affinity column. Stringent conditions were developed that allow the contaminants to be completely washed from the column without eluting the Crk-N protein.[3] In an important control experiment, those same wash conditions were found to strip wild-type Crk-N from the leucine enkephalin control column, thereby confirming that the binding to the C3G-column is ligand-specific.

Functional selection of the arrays of Crk-N analogs is performed using the C3G peptide affinity column. As a control against artifactual nonspecific binding, the arrays are also passed over the nonspecific leucine enkephalin column. In a typical experiment, the nonspecifically bound components in the mixture (inactive pool) are washed off the column with high salt-containing buffers and then immediately chemically cleaved by dilution into a cleavage buffer containing 1 M NH$_2$OH. The remaining specifically bound material (active pool) is eluted from the C3G column by washing the affinity column with the same 1 M NH$_2$OH cleavage buffer. Exposure to this buffer system disrupts the specific protein–ligand interactions by chemically cleaving the protein analogs (at the analog unit), resulting in their fragmentation and hence elution from the column. Aminolysis of the thioester bonds within the Crk-N protein analogs was found to occur extremely quickly in the presence of such a high concentration of hydroxylamine (complete cleavage within 15 min).

Mass Spectrometric Readout of Protein Array

The final step in protein signature analysis is to identify each of the protein components in the pools obtained from the selection experiment. In considering how best to achieve this, it should be remembered that the protein analogs in the array differ from one another only in the position of the analog unit in the sequence; otherwise, their sequences are identical. A closely related mixture of this type cannot be fully resolved using even sophisticated separation techniques such as HPLC.[1] Equally, biomolecular mass spectrometry cannot be directly used to unambiguously identify each parent protein analog in an array. This is because many of the protein analogs will have the same molecular mass owing to degeneracies in the masses of the naturally occurring amino acids.

It is for the above reasons that the latent chemical cleavage site is built into the analog unit. As previously discussed, "chemical unzipping" of the array specifically at the analog unit will lead to the generation of two sets of daughter peptide fragments, an N-terminal set and a C-terminal set. Importantly, each peptide fragment in these sets will have a unique length, and thus a unique mass (Fig. 1). Readout of these chemically decoded peptide fragments can be performed, in one operation, using matrix-assisted laser desorption–ionization mass spectrometry (MALDI-MS), a technique well suited to the analysis of peptide mixtures of this type.[8,21] Each component in the resulting MS signature reflects the presence of the corresponding full-length polypeptide chain (containing the analog unit) in that pool of intact protein analogs. Furthermore, the position of the analog unit in the original polypeptide chain is defined by the position of the corresponding signal in the mass spectrometric signature. This self-encoding system is especially powerful because all the information is read out in a single step from a pool of molecules free in solution.

Ideally, the MS signatures of the active and inactive pools from the functional selection experiment should be subsets of the MS signature obtained for the parent array of protein analogs. A great deal of care, however, must be taken when interpreting these MS signature data, and, in particular, one should always be mindful of the dangers associated with extracting quantitative information from the intensity of the peaks in a mass spectrum. Although MALDI-MS does not provide quantitative information, informative qualitative comparisons can be made between data sets provided the mode of sample preparation is the same in each.[22] Several factors are known to contribute to the relative intensity of the peaks in a MALDI mass spectrum of a peptidic mixture. These include the relative abundance of each component in the mixture; the chemical structure of each component (some molecules ionize better or differently than others); and the chemical nature of the saturated matrix solution being used.[22] If the MALDI mass spectra from a series of mixtures are to be usefully compared, it is critical that the same buffer system, ionization matrix, and mode of sample deposition are used throughout.

Typical results from a PSA experiment are shown in Fig. 3. In this experiment every dipeptide sequence throughout the region of Crk-N defined by residues 156–165 was substituted with a -Gly-[COS]-βAla- analog

[21] Y. Zhao, T. W. Muir, S. B. H. Kent, E. Tischer, J. M. Scardina, and B. T. Chait, *Proc. Natl. Acad. Sci. U.S.A.* **93**, 4020 (1996).
[22] S. L. Cohen and B. T. Chait, *Anal. Chem.* **68**, 31 (1996).

unit. Functional selection of the nine-membered array of protein analogs using a C3G-peptide affinity column gave rise to active and inactive pools, which were then chemically decoded using hydroxylamine. As a reference, the parent array of Crk-N protein analogs was also chemically cleaved by treatment with hydroxylamine. Each of the three decoded mixtures were exchanged into a 70% CH_3CN/30% H_2O, 0.1% TFA solvent system using 1-ml disposable reversed-phase columns (Pierce). This step was found to greatly improve the quality of the MALDI mass spectra subsequently obtained. Note that although it is possible to obtain high-quality MALDI mass spectra on peptidic mixtures containing as little as 0.5 pmol of each component,[8,23] significant concentrations of salt (or in this case hydroxylamine) in the sample preparations can interfere with the ionization process.[22] The MS signatures obtained from the parent protein array, active pool, and inactive pool are shown in Fig. 3b. Identical sample preparation techniques were used in each, and thus the significant differences observed can be attributed to actual changes in the composition of the mixtures being analyzed.

Cleavage of the nine-membered parent array of Crk-N analogs will give rise to a total of 18 peptide fragments: 9 C-terminal and 9 N-terminal. Interestingly, only the nine N-terminal peptide fragments were observed with high intensity under the MALDI-MS conditions used (Fig. 3b), even though the C-terminal peptide family was necessarily present in equimolar amounts. Indeed, the same type of selective detection was observed in the MALDI mass spectra of the parent array, as well as active and inactive pools (Fig. 3b). The reason for the preferential MALDI-MS detection of the N-terminal peptide fragments is not fully understood. One possible explanation is that the N-terminal peptide fragments are more readily ionized under the MALDI conditions than the C-terminal peptide fragments, perhaps as a result of certain amino acid sequence differences between the two sets.[22]

According to the readout strategy described earlier, the unambiguous identification of any given member in a pool of protein analogs requires that just one of its unique peptide fragments be identified in the MS signature (as both are equally diagnostic). Thus, the absence of strong MS signals from the C-terminal peptide fragments in no way interferes with the readout process: if anything, it simplifies the interpretation of the mass spectra. Each signal in the parent MS signature is quite characteristic of a particular peptide fragment, and it can be assigned by simply comparing the observed mass with the expected masses of the daughter peptide fragments (Table I). It is thus possible to trace each MS signal back to the substitution of a

[23] B. T. Chait, R. Wang, R. C. Beavis, and S. B. H. Kent, *Science* **262**, 89 (1993).

a

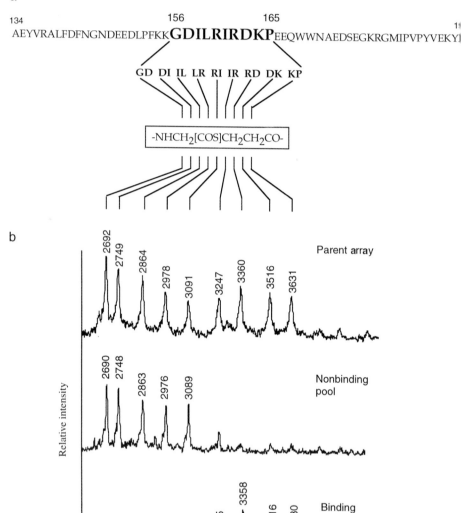

134   156   165   1

AEYVRALFDFNGNDEEDLPFKK**GDILRIRDKP**EEQWWNAEDSEGKRGMIPVPYVEKY

GD DI IL LR RI IR RD DK KP

-NHCH₂[COS]CH₂CH₂CO-

b

Relative intensity

Parent array

2692 2749 2864 2978 3091 3247 3360 3516 3631

Nonbinding pool

2690 2748 2863 2976 3089

Binding pool

3245 3358 3516 3630

2600   3000   3500   4000

m/z

TABLE I
PREDICTED MASSES OF N- AND
C-TERMINAL FRAGMENTS[a]

Residues replaced by analog	Mass (MH⁺) of	
	N-terminal fragment	C-terminal fragment
$G^{156}D^{157}$	2691.9	4279.9
$D^{157}I^{158}$	2749.0	4166.8
$I^{158}L^{159}$	2864.1	4053.6
$L^{159}R^{160}$	2977.2	3897.4
$R^{160}I^{161}$	3090.4	3784.3
$I^{161}R^{162}$	3246.6	3628.1
$R^{162}D^{163}$	3359.7	3513.0
$D^{163}K^{164}$	3515.9	3384.8
$K^{164}P^{165}$	3631.0	3287.7

[a] Obtained by chemically decoding the array of Crk-N protein analogs containing the -Gly-[COS]-βAla- dipeptide analog in the region defined by residues 156–165.

unique dipeptide unit in the Crk-N sequence with the -Gly-[COS]-βAla analog.

Having assigned the MS signature of the parent array, it is then straightforward to read out the composition of the active and inactive pools by simply comparing their MS signatures with that of the parent array (Fig. 3b). This reveals that only four dipeptide sequences ($I^{161}R^{162}$; $R^{162}D^{163}$; $D^{163}K^{164}$; and $K^{164}P^{165}$) within the region being studied could be replaced by Gly-[COS]-βAla without significant loss of binding activity. The remaining five protein analogs were present within the inactive pool. It is striking that the MS signatures obtained for the active and inactive pools

FIG. 3. Protein signature analysis of Crk-N(156–165). (a) The analog unit -Gly-[COS]-βAla- was placed at each of the 9 possible positions within the 10-amino acid stretch of Crk-N defined by residues 156–165 (highlighted region in sequence). (b) MALDI-MS signatures were obtained from the parent array of protein analogs (top), the inactive pool from affinity selection (middle), and the active pool from affinity selection (bottom). Each peak in the three MS signatures represents a full-length analog polypeptide chain. The identify of each decoded peptide fragment can be simply assigned by comparing the observed mass with those of the expected masses of the daughter peptide fragments (Table I). Interestingly, only the N-terminal daughter peptide fragments were observed with high intensity under the MALDI conditions used in this experiment.

are clearly subsets of the original parent array. Combining them would, to the first approximation, regenerate the parent signature.

Interpretation of Protein Signature Analysis Data

Protein signature analysis combines the advantages associated with systematic modification of the covalent structure of a protein (e.g., see Refs. 5, 6, and 7), with the practical convenience of combinatorial techniques,[24] resulting in a unique strategy which permits the synthesis, selection, and characterization of defined, that is to say, nonstochastic, mixtures of protein analogs. In its present form, protein signature analysis provides a wealth of qualitative functional information in the form of mass spectrometric signatures. By simply comparing these MS signatures from before and after a selection step (as in Fig. 3), it is possible to determine the relative functional effects of performing a chemical modification throughout an entire region of a protein sequence. Thus, one is able to quickly construct a qualitative functional profile across the protein sequence. Such a result is usually the outcome of exhaustive structure–activity studies.

An important feature of the protein signature analysis approach is that it is practical (because of the rapidity of the technique) to perform iterations of a PSA experiment in which certain of the key experimental parameters are themselves systematically varied. For example, it may be desirable to subject the same array of protein analogs to a series of selection conditions in which the activity threshold is varied (see earlier), or it may be useful to alter the chemical nature of the analog unit itself. To illustrate how useful iteration can be, we once again turn to our studies on Crk-N.[2] Table II summarizes the results of two rounds of protein signature analysis in which the analog units -Gly-[COS]-βAla- and -Gly-[COS]-Gly- were separately substituted throughout the same region of the Crk-N sequence (residues 156–165).

Under the affinity selection conditions used, only four members of the -Gly-[COS]-βAla-containing array of protein analogs exhibited appreciable binding activity. The lack of activity in the remaining five protein analogs could have been the result of any or all of the following factors: insertion of an extra methylene in the backbone; deletion of the H-bond donor in the central amide; or deletion of pairs of side chains. Information bearing on these possibilities was obtained by generating a second array of protein analogs in the same region of the protein. In this case, -Gly-[COS]-Gly-was used as the analog unit in which the extra backbone methylene group found in the original analog was no longer present. In contrast to the

---

[24] G. Quinkert, H. Bang, and D. Reichert, *Helv. Chim. Acta* **79**, 1260 (1996).

TABLE II
RESULTS FROM TWO ROUNDS OF PROTEIN SIGNATURE ANALYSIS
ON Crk-N[a]

Residues replaced by analog	Analog unit	
	-Gly-[COS]-βAla-	-Gly-[COS]-Gly-
$G^{156}D^{157}$	X	✓
$D^{157}I^{158}$	X	✓
$I^{158}L^{159}$	X	✓
$L^{159}R^{160}$	X	✓
$R^{160}I^{161}$	X	✓
$I^{161}R^{162}$	✓	✓
$R^{162}D^{163}$	✓	✓
$D^{163}K^{164}$	✓	✓
$K^{164}P^{165}$	✓	✓

[a] Each dipeptide sequence in the region 156–165 was replaced by the analog unit shown, and the resulting nine-membered arrays of protein analogs were subjected to functional selection using a C3G-peptide affinity column. Indicated are those protein analogs found to be either active (✓) or inactive (X) under the conditions used.[2]

previous experiment, all nine -Gly-[COS]-Gly-containing protein analogs exhibited appreciable binding activity in an identical affinity selection assay (see Table II).

The results summarized in Table II show that neither the pairwise deletion of side chains nor deletion of the H-bond donor in the central amide moiety was primarily responsible for the lack of binding activity exhibited by the five inactive members of the original -Gly-[COS]-βAla-containing array of protein analogs. Rather, it can be inferred from comparison of the two sets of data that the observed effects on activity had been mainly caused by insertion of the extra methylene in the polypeptide backbone by the original unit. Thus, iteration of the PSA approach revealed that the region defined by residues Crk-N(161–165) is tolerant to backbone engineering, whereas the region defined by residues Crk-N(156–161) is not.[2]

Conclusion

In this article, we have outlined a series of experimental guidelines based on observations from the small number of PSA studies conducted thus far.[1-3] The protein signature analysis approach is, however, still in its infancy, and many of these guidelines may change as the technique continues to evolve. In particular, the development of nucleophile-stable readout chemistries (based on, for example, photochemical or enzymatic cleavage sites) will permit both Boc- and Fmoc-based SPPS to be used in the synthesis of protein arrays. Equally, different readout chemistries may also have to be explored if PSA is to be integrated with the chemical ligation approach to protein synthesis,[4] a prerequisite to extending the PSA approach to the study of larger protein molecules. Other possibilities for future study might include incorporation of panels of analog units at each position in the protein sequence; use of analog units containing secondary structure mimetics, fluorescent probes, or even nuclear magnetic resonance (NMR) probe nuclei; and further development of the binary selection system described earlier.

In conclusion, the protein signature analysis principle is completely new to protein science, although a conceptually similar strategy has been devised for studying nucleic acids.[25-28] The approach combines advances in solid-phase peptide synthesis,[9] chemical ligation,[18] and MALDI-MS,[8,21,23] and it allows a qualitative functional profile to be obtained over an extended region of a peptide or protein sequence. Consequently, protein signature analysis will be of considerable value in exploring the structure–activity relationships of peptides and small proteins.

Acknowledgments

We express thanks to Brian Chait, Julio A. Camarero, and David Cowburn (all at The Rockefeller University) for many valuable discussions during the preparation of the manuscript.

[25] K. C. Hayashibara and G. L. Verdine, *J. Am. Chem. Soc.* **113**, 5104 (1991).
[26] K. C. Hayashibara and G. L. Verdine, *Biochemistry* **31**, 11265 (1992).
[27] J. L. Mascarenas, K. C. Hayashibara, and G. L. Verdine, *J. Am. Chem. Soc.* **115**, 373 (1993).
[28] C. Min, T. D. Cushing, and G. L. Verdine, *J. Am. Chem. Soc.* **118**, 6116 (1996).

[25] *In Vitro* Incorporation of Synthetic Peptides into Cells

By JANELLE L. LAUER and GREGG B. FIELDS

Introduction

The therapeutic use of synthetic peptides in *in vitro* cellular systems or *in vivo* are limited in many cases by cell membrane impermeability. Numerous procedures have been developed to promote peptide transport through cell membranes, including electroporation,[1,2] use of glass beads,[3] attachment of a signal or leader peptide sequence,[4–13] liposome/vesicle delivery,[14–17] and bacterial toxin permeabilization.[18,19]

Many cell permeabilization techniques have undesirable consequences. Electroporation or the use of glass beads to induce cell permeabilization is rather disruptive to the cells, often causing unacceptable levels of cell death. Permeabilization, while being reversible, is also nonspecific.[1] Both methods allow for the transport of antibodies or other large molecules (150–200 kDa) into cells. [2] These procedures may also disrupt postpermeabilization receptor function, and thus an extended recovery period may be necessary. Such characteristics make electroporation or glass bead-induced permeabilization less than optimal for peptide delivery into cells.

Liposome encapsulated peptides, proteins, and adjuvants have been used to introduce a variety of compounds into the intracellular environment.[14–17] Liposomes can be an efficient means of delivery for antigens and

[1] P. Barja, A. Alavi-Nassab, C. W. Turck, and J. Freire-Moar, *Cell. Immunol.* **153**, 28 (1994).

[2] B. Schieffer, W. G. Paxton, Q. Chai, M. B. Marrero, and K. E. Bernstein, *J. Biol. Chem.* **271**, 10329 (1996).

[3] D. F. Fennell, R. E. Whatley, T. M. McIntyre, S. M. Prescott, and G. A. Zimmerman, *Arteriosclerosis Thrombosis* **11**, 97 (1991).

[4] E. P. Loret, E. Vives, P. S. Ho, H. Rochat, J. Van Rietschoten, and W. C. J. Johnson, *Biochemistry* **30**, 6013–6023 (1991).

[5] D. Derossin, A. H. Joliot, G. Chassaing, and A. Prochiantz, *J. Biol. Chem.* **267**, 10444 (1994).

[6] S. Fawell, J. Seery, Y. Daikh, C. Moore, L. L. Chen, B. Pepinsky, and J. Barsoum, *Proc. Natl. Acad. Sci. U.S.A.* **91**, 664 (1994).

[7] M. Mousli, T. E. Hugli, Y. Landry, and C. Bronner, *Immunopharmacology* **27**, 1 (1994).

[8] J. W. Izard, M. B. Doughty, and D. A. Kendall, *Biochemistry* **34**, 9904 (1995).

[9] L. J. Cross, M. Ennis, E. Krause, M. Dathe, D. Lorenz, G. Krause, M. Beyerman, and M. Bienert, *Eur. J. Pharmacol.* **291**, 291 (1995).

[10] Y. Z. Lin, S. Y. Yao, R. A. Yeach, T. R. Torgenson, and J. Hawiger, *J. Biol. Chem.* **270**, 14255 (1995).

[11] L. Theodore, D. Derossi, G. Chassaing, B. Llirbat, M. Kubes, P. Jordan, H. Chneiweiss, P. Godement, and A. Prochiantz, *J. Neurosci.* **15**, 7158 (1995).

[12] M. T. Saleh, J. Ferguson, J. M. Boggs, and J. Gariepy, *Biochemistry* **35**, 9325 (1996).

[13] N. Bodor, L. Prokai, W. M. Wu, H. Harag, S. Jonalagadda, M. Kawamura, and J. Simpkins, *Science* **257**, 1698 (1992).

[14] B. P. Barna, M. J. Thomassen, P. Zhou, J. Pettay, S. Singh-Burgess, and S. D. Deodhar, *J. Leukocyte Biol.* **59**, 397 (1996).

[15] G. Gregoriadis, *Immunomethods* **4**, 210 (1994).

[16] C. R. Alving, *J. Immunol. Methods* **140**, 1 (1991).

[17] A. Nii, T. Utsuge, D. Fan, Y. Denkins, C. Pak, D. Brown, P. van Hoogevest, and I. J. Fidler, *J. Immunother.* **10**, 236 (1991).

[18] B. B. Damaj, S. R. McColl, W. Mahana, M. F. Crouch, and P. H. Naccache, *J. Biol. Chem.* **271**, 12783 (1996).

[19] R. L. Wange, N. Isakov, T. R. Burke, A. Otaka, P. P. Roller J. D. Watts, R. Aebersold, and L. E. Samelson, *J. Biol. Chem.* **270**, 944 (1995).

adjuvants.[15,16] However, delivery of peptides using this method can be inefficient owing to the possibility of liposomes entering the cellular environment in an intact form. Peptides incorporated in this manner still remain inaccessible to the cytoplasm even though they appear to be within the cell, making this assay difficult to monitor.[6]

Bacterial toxin permeabilization of the cellular membrane has been widely used. The benefits of these procedures lie in their specificity based on the toxin chosen for use. Each toxin has been examined extensively with respect to the size and nature of the pore it forms. For example, α-toxin of *Staphylococcus aureus* produces 1- to 1.5-nm pores, which allow for the transport of Ca^{2+} and small nucleotides.[20] Streptolysin O produces much larger pores (≥ 30 nm), which allow for the free passage of proteins.[21] The major drawback of these procedures is their irreversibility. This feature of bacterial toxins makes them difficult to adapt to common cellular assays. Irreversibility also results in the loss of soluble effector molecules that could affect cellular processes, and thus makes interpretation of data difficult.[22]

The attachment of a leader or signal sequence onto peptides has come into favor in the last 2 years. The approach is based on the knowledge that most transported proteins in both prokaryotic and eukaryotic systems are synthesized with an N-terminal extension. This sequence targets the protein for export from the endoplasmic reticulum as well as protects it from protease degradation. Once the peptide crosses the desired membrane, the signal sequence is removed by specific proteases, and the free protein is allowed access to the cytoplasm. General properties of signal sequences include (1) a positively charged N terminus, (2) a hydrophobic core region, and (3) a somewhat polar C terminus.[8] The presently used signal sequences have been taken from the Tat protein of human immunodeficiency virus type-1 (HIV-1),[6] cationic secretagogues of the peptidergic mast cell activation pathway,[7,9] or bacterial toxins.[12] In each case, the signal sequence from these proteins is N-terminally attached to the active peptide. The signal sequence allows for the chimeric molecule to cross the cell membrane, after which the peptide can exert its activity. An interesting adaptation of this design was developed by Bodor et al.,[13] who attached a 1,4-dihydrotrigonellinate redox targetor along with a cholesteryl group to the desired peptide. The targetor specifically delivers the peptide conjugate to the blood–brain barrier (BBB), and the cholesteryl moiety increases the lipophilic nature of the molecule, helping it cross the BBB.[13] Once across the BBB, lipases and esterases remove the targetor and cholesteryl species from the peptide and

[20] S. Bhakdi and J. Tranum-Jensen, *Microbiol. Rev.* **55**, 733 (1991).
[21] S. Bhakdi, U. Weller, I. Walev, E. Martin, D. Jonas, and M. Palmer, *Med. Microbiol. Immunol.* **182**, 167 (1993).
[22] M. A. Alave, K. E. DeBell, A. Conti, T. Hoffman, and E. Bonvini, *Biochem. J.* **284**, 189 (1992).

allow the peptide access to the central nervous system. To date, this concept has only been used to deliver peptides across the BBB. In theory, it could be extended to deliver peptides into specific cellular systems. The benefit of developing such a method would be its applicability in *in vivo* systems.

The present signal sequence approach cannot be widely used owing to several drawbacks. Signal sequences are somewhat lengthy (15–25 amino acids) as well as very hydrophobic in nature, making the chimeric peptide difficult to synthesize and purify. Some of the signal sequences being developed can only be used for mast cells,[7,9] making this approach not a generally applicable delivery mechanism. Also, the signal sequences are not cleaved from the target peptide on translocation into the cytoplasm. This could interfere with the activity of the peptide within the cell. A general, widely applicable, signal sequence has not yet been developed for peptide incorporation into intracellular environments.

To circumvent the various problems described above, Lauer *et al.*[23] have examined a novel permeabilization method. The GIBCO-BRL (Gaithersburg, MD) reversible permeabilization kit, called TRANS-PORT, utilizes a water-soluble lipid derivative to produce pores in cell membranes. The mild permeabilization procedure does not appreciably interfere with receptor function or cell viability.[24] The lipid allows for the intracellular transport of compounds up to 2–3 kDa, sufficient for most peptide uses. Once the cells are permeabilized, and the peptide is in the cytoplasm, the lipid derivative can be bound and removed from the cellular environment by adding a protein solution. Membrane recovery can be monitored visually by assessing trypan blue extrusion from the cytoplasm. Reestablishment of membrane integrity will vary with cell type, but it should take place in 30–60 min. The kit has been used on a variety of cell types including both anchorage-dependent and -independent cells. Levels of permeabilization can vary with the type of cell, culture conditions, and proliferative state.[24] Actively growing cells and those in suspension appear to work better than growth-arrested or attached cells.[24] Some primary cell types such as adipocytes and peripheral T lymphocytes are hypersensitive to the lipid and lyse on contact.[24]

Experimental

The TRANS-PORT kit is used to test the effects of a peptide on melanoma cell adhesion to type IV collagen.[23] The procedure discussed here is altered somewhat from the general protocol given in the product literature so that >90% permeabilization can be achieved with this specific cell type. M14#5 melanoma cells are cultured in modified Eagle's medium (MEM) supple-

[23] J. L. Lauer, L. T. Furcht, and G. B. Fields, *J. Med. Chem.* **40** (1997).
[24] GIBCO-BRL Life Technologies Product Literature, Grand Island, New York (1994).

mented with 10% fetal bovine serum, 1 mM sodium pyruvate, 0.1 mg/ml gentamicin, 0.05 mg/ml streptomycin, and 50 units/ml penicillin. The cells are split and replated on day 1 of the assay so that they will be 80–90% confluent on day 3. On day 2, during the exponential growth phase of the cells, Tran[35]S-label (ICN Pharmaceuticals, Inc., Costa Mesa, CA) is added and the cells are placed back into the incubator overnight. This will allow for the most efficient membrane incorporation of the radioisotope label. On day 3, the cells are released with 5 ml of trypsin–EDTA (140 mM NaCl, 5 mM KCl, 5 mM D-glucose, 7 mM NaHCO$_3$, 7 mM EDTA, 0.05% trypsin). They are diluted up to 10 ml with medium, pelleted, and washed with 10 ml of intracellular buffer (ICB, pH 7.1; 120 mM KCl, 10 mM NaCl, 1 mM KH$_2$PO$_4$, 5 mM NaHCO$_3$, 10 mM HEPES, 1 μM MgCl$_2$, and 0.2 mM EGTA). The cells are pelleted again and redistributed in ICB at a final concentration of 600,000 cells/ml. Cellular concentration is determined for the needs of the adhesion assay, not the requirements of the permeabilization procedure. GIBCO recommends using the cells at or below 2,000,000 cells/ml. This particular type of melanoma cells attain efficient permeabilization at concentrations up to 5,000,000 cells/ml.

Cell suspension (125 μl) is added to each tube (for peptide assays, the tube contains 1–100 μM peptide dissolved in ICB; for control experiments to determine permeabilization levels, the tube contains 25 μl of trypan blue). The water-soluble lipid derivative (15 μl TRANS-PORT reagent) is added to each tube and the tubes are placed at 37° for 10 min. Permeabilization levels are determined by counting the number of cells that incorporate the trypan blue and comparing that number to the total number of cells. A minimum of 90% permeabilization is routinely observed. The protein solution (30 μl, stop reagent) is added, and the cells are diluted to 50,000 cells/ml with adhesion medium (MEM supplemented with 20 mM HEPES and 2 mg/ml ovalbumin). At this point in the procedure GIBCO recommends a mild centrifugation step to remove the lipid from the cells. Adding the stop reagent and diluting the cells in adhesion medium has been found to produce greater cell viability for the assay. The cells are then placed at 37° for 60, 90, or 120 min of recovery to allow membrane integrity to be reestablished. When the recovery step is completed, the cells are plated on 96-well Immulon plates (Dynatech Laboratories, Inc., Chantilly, VA) coated with 10 μg/ml type IV collagen [and blocked with 2 mg/ml ovalbumin in phosphate buffered saline (PBS)], 100 μl of cells per well. The cells are allowed to adhere to the substrate at 37° for 60, 90, or 120 min. Nonadherent cells are removed by washing the plate three times with 200 μl of adhesion medium. Scintillation fluid is added to each well, and radioactivity is determined by a plate counter. Each sample is placed in six wells of the 96-well plate. Three of the wells are washed and three are left unwashed in order to determine the level of adhesion for each sample individually.

TABLE I

POSTPERMEABILIZATION ADHESION OF MELANOMA CELLS TO TYPE
IV COLLAGEN

Membrane recovery (min)	Adhesion time (min)	Percentage of Control Adhesion (%)
Nonpermeabilized	60	100.0 ± 1.3
Nonpermeabilized	120	100.0 ± 0.7
60	60	42.9 ± 1.2
60	90	74.2 ± 2.6
60	120	106.3 ± 1.0
90	60	50.6 ± 0.4
90	90	84.3 ± 0.2
90	120	103.6 ± 1.6
120	60	28.1 ± 0.6
120	90	56.9 ± 0.5
120	120	30.0 ± 0.8

Results and Discussion

The results indicate that cells allowed to recover for 60 min and then allowed to adhere for 120 min have comparable adhesion levels to nonpermeabilized cells plated on type IV collagen for 120 min (Table I). All peptide assays are then run under these conditions using 1–250 μM peptide dissolved in ICB. Each peptide analog is based on a gene discovered by Pullman and Bodmer, designated cell adhesion regulator (CAR). Transfection of the cDNA of this gene into colon cancer cells increases their adhesion to collagen and laminin, without effecting the expression of the integrin receptors that mediate adhesion to those substrates.[25] The proposed protein product of the gene includes a consensus sequence for phosphorylation of Tyr residues. Phosphorylated and nonphosphorylated analogs of this region have been synthesized and tested for their ability to interfere with melanoma adhesion to type IV collagen.[23] For active peptides, Val-Glu-Ile-Leu-Tyr-NH$_2$, and Val-Glu-Ile-Leu-Tyr(PO$_3$H$_2$)-NH$_2$, significant differences in activity can be seen at concentrations as low as 10 μM (Table II). When a negative control peptide is used, Tyr-Leu-Ile-Glu-Val-NH$_2$, concentrations as high as 100 μM show no activity, thus showing the specificity of the CAR sequence used (Table II). Controls are also run by adding identical amounts of peptide to nonpermeabilized cells (Table II). This treatment has no effect on adhesion, demonstrating that the peptide site of action is within the cytoplasm of the cell might be acted on by cellular kinases and/

[25] W. E. Pullman and W. F. Bodmer, *Nature* (*London*) **356**, 529 (1992).

TABLE II

INHIBITION OF MELANOMA CELL ADHESION TO TYPE IV COLLAGEN BY
INTRACELLULAR INCORPORATION OF PEPTIDE ANALOGS OF CELL
ADHESION REGULATOR

Peptide analog	Concentration (μM)	Inhibition of Cell Adhesion (%)
Val-Glu-Ile-Leu-Tyr-NH$_2$	10	0.0 ± 0.0%
	50	2.6 ± 3.8
	100	35.3 ± 9.1
	250	46.0 ± 5.4
Val-Glu-Ile-Leu-Tyr(PO$_3$H$_2$)-NH$_2$	10	30.5 ± 4.2
	50	45.4 ± 5.8
	100	48.1 ± 0.5
	250	51.8 ± 3.1
Tyr-Glu-Leu-Ile-Val-NH$_2$	10	8.2 ± 2.6
	50	0.0 ± 0.0
	100	0.0 ± 0.0
	250	0.0 ± 0.0

or phosphatases, indicating that intracellular function is only minimally disrupted.[23]

Further control experiments are run to determine whether the peptide in question actually enters the cell on permeabilization. An analog of the CAR peptide is produced that contains the active sequence with an N-terminally linked fluorescein label. The label is added as a free acid using double couplings of standard 9-fluorenylmethyloxycarbonyl (Fmoc)-amino acid chemistry.[26,27] The labeled peptide is used in conjunction with fluorescene microscopy to determine that (1) when the cells are not permeabilized the peptide does not enter the cells and (2) when the cells are permeabilized the peptide crosses the cellular membrane and can be visualized in the cytoplasm. Results obtained with the fluorescent peptide confirm the notion that the peptides do enter the cells, and do so only when the cells are permeabilized. These studies have used fluorescein as the fluorescent label. There are numerous other fluorescent groups, as well as means of attachment, that have been utilized to demonstrate peptide incorporation into cells. Fluorescein can be conjugated as an isothiocyanate derivative,[1] or peptides can be detected using an antipeptide antibody system linked to a rhodamine label.[10]

[26] G. B. Fields, Z. Tian, and G. Barany, in "Synthetic Peptides: A User's Guide" (G. A. Grant, ed.), p. 77. Freeman, San Francisco, 1993.
[27] G. B. Fields and R. L. Noble, Int. J. Pep. Protein Res. 35, 161 (1990).

Summary

This article gives a specific example of how a reversible permeabilization kit can be adapted for use with a specific peptide-based protocol. It was relatively easy to design and run the control experiments to determine the necessary adaptations. The basic procedure given with the TRANS-PORT kit is amenable to modification such that is can be used with numerous other *in vitro* procedures.

[26] Construction of Biologically Active Protein Molecular Architecture Using Self-Assembling Peptide–Amphiphiles

By YING-CHING YU, TEIKA PAKALNS, YOAV DORI,
JAMES B. MCCARTHY, MATTHEW TIRRELL, and
GREGG B. FIELDS

Introduction

Many researchers have attempted to create proteinlike assemblies for the purpose of studying protein folding.[1] In addition, functional synthetic proteins have been developed from designed assemblies.[2–14] The *de novo* design of proteins has been based on the knowledge of individual amino acid propensities to form distinct secondary structures, such as α helices

---

[1] K. H. Mayo and G. B. Fields, *in* "Protein Structural Biology in Bio-Medical Research" (N. Allewell and C. Woodward, Eds.), p. 565. JAI Press, Greenwich, Connecticut, 1997.
[2] T. Sasaki and E. T. Kaiser, *J. Am. Chem. Soc.* **111**, 380 (1989).
[3] K. W. Hahn, W. A. Klis, and J. M. Stewart, *Science* **248**, 1544 (1990).
[4] D. E. Robertson, R. S. Farid, C. C. Moser, J. L. Urbauer, S. E. Mulholland, R. Pidikiti, J. D. Lear, A. J. Wand, W. F. DeGrado, and P. L. Dutton, *Nature (London)* **368**, 425 (1994).
[5] P. T. P. Kaumaya, A. M. VanBuskirk, E. Goldberg, and S. K. Pierce, *J. Biol. Chem.* **267**, 6338 (1992).
[6] A. Grove, M. Mutter, J. E. Rivier, and M. Montal, *J. Am. Chem. Soc.* **115**, 5919 (1993).
[7] G. Tuchscherer, B. Dömer, U. Sila, B. Kamber, and M. Mutter, *Tetrahedron* **49**, 3559 (1993).
[8] C. G. Fields, D. J. Mickelson, S. L. Drake, J. B. McCarthy, and G. B. Fields, *J. Biol. Chem.* **268**, 14153 (1993).
[9] A. J. Miles, A. P. N. Skubitz, L. T. Furcht, and G. B. Fields, *J. Biol. Chem.* **269**, 30939 (1994).
[10] B. Grab, A. J. Miles, L. T. Furcht, and G. B. Fields, *J. Biol. Chem.* **271**, 12234 (1996).
[11] L. F. Morton, I. Y. McCulloch, and M. J. Barnes, *Thromb. Res.* **72**, 367 (1993).
[12] G. H. R. Rao, C. G. Fields, J. G. White, and G. B. Fields, *J. Biol. Chem.* **269**, 13899 (1994).
[13] L. F. Morton, P. G. Hargreaves, R. W. Farndale, R. D. Young, and M. J. Barnes, *Biochem J.* **306**, 337 (1995).
[14] T. Tanaka, A. Nishikawa, Y. Tanaka, H. Nakamura, T. Kodama, T. Imanishi, and T. Doi, *Protein Eng.* **9**, 307 (1996).

and β sheets, and modeling of these secondary structures to associate by long-range interactions and create tertiary structures.[15–17] Mutter and colleagues[16] developed the template-assembled synthetic protein (TASP) approach based on the concept that templates would promote secondary structure formation and minimize aggregation of larger complexes. When one considers the contributions to the free energy of folding, a template is predicted to decrease the chain entropy of the unfolded state (due to increased excluded volume effects), while favoring intramolecular interactions (formation of a hydrophobic core) instead of intermolecular interactions (aggregation).[16] Lattice statistical mechanics predicts that surfaces would enhance secondary structure formation by minimizing unfolded states.[18,19] Surfaces have been shown to induce peptide secondary structures, either α-helical or β-sheet structures.[20–22] The template effect has since been documented for synthetic four α-helical bundles, as the α helicity of the individual peptide chains was greatly enhanced on incorporation onto a template containing several other α-helical peptide sequences.[23–25] Thus, one method for creating novel biomaterials with distinct proteinlike structures is to directly synthesize the peptide onto a template that (1) induces secondary and tertiary structures and (2) either serves directly as a biomaterial or is compatible with biomaterial surfaces.

Ideally, one wants to create a system by which synthetic linear peptide chains self-assemble into desirable secondary and tertiary structures. We envisioned the construction of a novel "peptide–amphiphile" to serve this purpose, whereby a peptide "headgroup" has the propensity to form a distinct structural element, while a hydrophobic "tail" serves to align the peptide strands and induce secondary and tertiary structure formation as well as providing a hydrophobic surface for self-association and/or interac-

[15] W. F. DeGrado, *Adv. Protein Chem.* **39**, 51 (1988).

[16] M. Mutter, *Trends Biochem. Sci.* **13**, 260 (1988).

[17] J. S. Richardson, D. C. Richardson, N. B. Tweedy, K. M. Gernert, T. P. Quinn, M. H. Hecht, B. W. Erickson, Y. Yan, R. D. McClain, M. E. Donlan, and M. C. Surles, *Biophys. J.* **63**, 1186 (1992).

[18] M. R. Wattenbarger, H. S. Chan, D. F. Evans, V. A. Bloomfield, and K. A. Dill, *J. Chem. Phys.* **93**, 8343 (1990).

[19] H. S. Chan, M. R. Wattenbarger, D. F. Evans, V. A. Bloomfield, and K. A. Dill, *J. Chem. Phys.* **94**, 8542 (1991).

[20] W. F. DeGrado and J. D. Lear, *J. Am. Chem. Soc.* **107**, 7684 (1985).

[21] G. Ösapay and J. W. Taylor, *J. Am. Chem. Soc.* **112**, 6046 (1990).

[22] G. Ösapay and J. W. Taylor, *J. Am. Chem. Soc.* **114**, 6966 (1992).

[23] M. Mutter, G. G. Tuchscherer, C. Miller, K.-H. Altmann, R. I. Carey, D. F. Wyss, A. M. Labhardt, and J. E. Rivier, *J. Am. Chem. Soc.* **114**, 1463 (1992).

[24] P. E. Dawson and S. B. H. Kent, *J. Am. Chem. Soc.* **115**, 7263 (1993).

[25] S. Vuilleumer and M. Mutter, *Biopolymers* **33**, 389 (1993).

tion with other surfaces. Our initial effort was to use this noncovalent, self-assembly approach for building a collagenlike structural motif by the solid-phase synthesis of peptide–amphiphiles that incorporate a long-chain dialkyl ester lipid tail onto a collagen-model peptide headgroup.[26,27] Design of synthetic lipids requires consideration of four building blocks: tail, linker/connector, spacer, and headgroup.[28] Our design features C_{12}, C_{14}, and C_{16} dialkyl tails, a Glu linker, a -$(CH_2)_2$- spacer, and a collagen–model peptide headgroup. In addition, the peptide headgroup incorporates a cell binding sequence.

Experimental Methods

Synthesis of Lipophilic Tails

Construction of tail compounds[26] first involves the acid-catalyzed condensation of Glu with the appropriate fatty acid alcohol to form the dialkyl ester of Glu (p-toluene sulfonate salt). For example, hexadecanol (44.85 g, 0.185 mol) and Glu (13.6 g, 0.092 mol) are mixed with 21.0 g (0.102 mol) of p-toluene sulfonate in toluene, and the mixture is heated until an equimolar amount of water is recovered in a Dean-Stark trap. The toluene is removed, and the product is recrystallized from acetone in approximately 80% yield. The free amino group of Glu is then treated with succinic anhydride and triethylamine to create a free carboxylic acid. 1′,3′-Dihexadecyl-L-Glu (20 g, 26 mmol) and triethylamine (5.5 ml, 39 mmol) are dissolved in tetrahydrofuran–CHCl₃ (1:1, v/v), and 3.9 g (39 mmol) of succinic anhydride is added under stirring. The mixture is kept for 4 hr at 30°. The product obtained after removal of the solvent is recrystallized from acetone and ethanol in approximately 94% yield. The dialkyl ester tail may then be used directly for solid-phase synthesis or converted to a p-nitrophenyl ester. To synthesize the latter compound, 1′,3′-dihexadecyl-N-sucinyl-L-Glu (6.90 g, 9.9 mmol) and p-nitrophenol (1.65 g, 11.9 mmol) are dissolved in dichloromethane (DCM), and 2.05 g (9.9 mmol) of N,N′-dicyclohexylcarbodiimide as well as a catalytic amount (80 mg) of 4-dimethylaminopyridine are added to the reaction mixture on an ice bath. The reaction is continued for 2 hr on the ice bath and for 24 hr at room temperature. The resultant dicyclohexylurea is filtered off, and the reaction product is precipitated with cold dry ethanol in a yield of about 85%.

[26] P. Berndt, G. B. Fields, and M. Tirrell, *J. Am. Chem. Soc.* **117**, 9515 (1995).
[27] Y.-C. Yu, P. Berndt, M. Tirrell, and G. B. Fields, *J. Am. Chem. Soc.* **118**, 12515 (1996).
[28] T. Kunitake, *Angew. Chem., Int. Ed. Engl.* **31**, 709 (1992).

The two other dialkyl ester tail precursors, 1′,3′-ditetradecyl-N-[O-(4-nitrophenyl)succinyl]-L-Glu [designated (C_{14})$_2$-Glu-C_2-pNp] and 1′,3′-didodecyl-N-[O-(4-nitrophenyl)succinyl]-L-Glu [designated (C_{12})$_2$-Glu-C_2-pNp], are synthesized as described above for 1′,3′-dihexadecyl-N-[O-(4-nitrophenyl)succinyl]-L-Glu [designated (C_{16})$_2$-Glu-C_2-pNp]. The C_{14} and C_{12} tails are prepared using 1-tetradecyl alcohol and 1-dodecyl alcohol, respectively.

Synthesis of Peptide–Amphiphiles

Peptide–resin assembly is performed by Fmoc solid-phase metholology.[29–32] All standard peptide synthesis chemicals and solvents are analytical reagent grade or better. Peptide–resins are characterized by Edman degradation sequence analysis as described previously for "embedded" (noncovalent) sequencing.[32] Peptide–resins are then lipidated[26] with the appropriate (C_n)$_2$-Glu-C_2 tail (Fig. 1). For example, the peptidyl–resin is incubated with 4-fold molar excesses of 1′,3′-dihexadecyl N-[O-(4-nitrophenyl)succinyl]-L-Glu and 1-hydroxybenzotriazole in N,N-dimethylformamide–DCM (1 : 1, v/v) for 4 hr. The reaction is followed by the ninhydrin test.[33] Cleavage and side-chain deprotection of peptide–amphiphile–resins proceeds for 1 hr using either ethanedithiol–thioanisole–phenol–water–trifluoroacetic acid (TFA) (2.5 : 5 : 5 : 5 : 82.5) or water–TFA (1 : 19, v/v).[34,35] Peptide–amphiphile cleavage solutions are not extracted with methyl *tert*-butyl (*t*Bu) ether prior to purification.

Peptide Purification and Analysis

Preparative reversed-phase high-performance liquid chromatography (RP-HPLC) purification is performed on a Rainin (Woburn, MA) Auto-Prep System. Peptide–amphiphile purification is achieved using a Vydac 214TP152022 C_4 column (15–20 μm particle size, 300-Å pore size, 250 × 22 mm) at a flow rate of 10 ml/min. The elution gradient is 55–90% B in 20 min where A is 0.05% TFA in water and B is 0.05% TFA in acetonitrile.

[29] G. B. Fields and R. L. Noble, *Int. J. Pept. Protein Res.* **35**, 161 (1990).
[30] C. G. Fields, D. H. Lloyd, R. L. Macdonald, K. M. Otteson, and R. L. Noble, *Pept. Res.* **4**, 95 (1991).
[31] J. L. Lauer, C. G. Fields, and G. B. Fields, *Lett. Pept. Sci.* **1**, 197 (1995).
[32] C. G. Fields, V. L. VanDrisse, and G. B. Fields, *Pept. Res.* **6**, 39 (1993).
[33] G. B. Fields, Z. Tian, and G. Barany, in "Synthetic Peptides: A User's Guide" (G. A. Grant, ed.), p. 77. Freeman, New York, 1992.
[34] D. S. King, C. G. Fields, and G. B. Fields, *Int. J. Pept. Protein Res.* **36**, 255 (1990).
[35] C. G. Fields and G. B. Fields, *Tetrahedron Lett.* **34**, 6661 (1993).

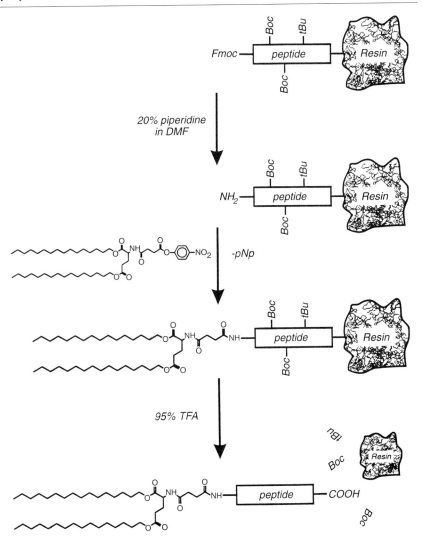

Fig. 1. General scheme for synthesis of peptide–amphiphiles. DMF, Dimethylformamide; pNp, 4-nitrophenol. [Reprinted with permission from *J. Am. Chem. Soc.* **117,** 9517 (1995). Copyright 1995 American Chemical Society.[26]]

Detection is at 229 nm. Alternatively, peptide–amphiphiles may be purified by normal-phase liquid chromatography.[36] Analytical RP-HPLC is performed on a Hewlett Packard (St. Paul, MN) 1090 Liquid Chromatograph equipped with a Hypersil C_{18} column (5 μm particle size, 120-Å pore size, 200 × 2.1 mm) a flow rate of 0.3 ml/min. The elution gradient is 0–60% B in 45 min where A and B are the same as for peptide purification. Diode array detection is at 220, 254, and 280 nm.

Edman degradation sequence analysis is performed on an Applied Biosystems (Foster City, CA) 477 A Protein Sequencer/120A Analyzer. Matrix-assisted laser desorption–ionization mass spectrometry (MALDI-MS) is performed on a Hewlett Packard G2025A laser desorption time-of-flight mass spectrometer using a sinapinic acid matrix. Peptide–amphiphile samples are dissolved in either water or water–acetonitrile containing 0.05% TFA. We have found MALDI-MS to be particularly effective for peptide–amphiphile characterization.[26,27]

Circular Dichroism Spectroscopy. Spectra are recorded on a JASCO J-710 spectropolarimeter using a thermostatted 0.1-mm quartz cell. Thermal transition curves are obtained by recording the molar ellipticity ($[\theta]$) in the range of 10–80° at $\lambda = 225$ nm. The peptide–amphiphile concentration is 0.5 mM in water at 25°.

Nuclear Magnetic Resonance Spectroscopy. Freeze-dried samples for nuclear magnetic resonance (NMR) spectroscopy are dissolved in D_2O or $D_2O–H_2O$ (1 : 9, v/v) at peptide and peptide–amphiphile concentrations of 3–5 mM. The NMR spectra are acquired on a 500 MHz Bruker AMX-500 spectrometer at 10°, 25°, 50°, and 80°. Two-dimensional (2D) total correlation spectroscopy (TOCSY) and nuclear Overhauser effect spectroscopy (NOESY) are performed with 256 $t1$ increment and 1024 complex data points in the $t2$ dimension. TOCSY spectra are obtained at mixing times of 40–150 msec. The NOESY spectra are obtained at mixing times of 60–250 msec. The spectra widths are 6024 Hz in both dimensions.

Atomic Force Microscopy. Atomic force microscopy (AFM) is performed on a Digital Instruments Nanoscope III Multimode Atomic Force Microscope. Contact or tapping modes are used with a tapping mode fluid cell. Methods are as described[37–39] using samples in pure water. Bilayers

[36] T. M. Winger, P. J. Ludovice, and E. L. Chaikof, *J. Liquid Chromatogr.* **18,** 4117 (1995).
[37] J. A. N. Zasadzinski, C. A. Helm, M. L. Longo, A. L. Weisenhorn, S. A. C. Gould, and P. K. Hansma, *Biophys. J.* **59,** 755 (1991).
[38] R. Viswanathan, D. K. Schwartz, J. Garnaes, and J. A. N. Zasadzinski, *Langmuir* **8,** 1603 (1992).
[39] J. B. Huebsch, G. B. Fields, T. G. Triebes, and D. L. Mooradian, *J. Biomed. Mater. Res.* **31,** 555 (1996).

are formed by first layering mica surfaces with 1,2-distearoyl-*sn*-phosphati-dylethanolamine (DSPE), then adding synthetic amphiphile.

Promotion of Cell Spreading

Molar mixtures of the synthetic amphiphiles $(C_{18})_2$-Glu-C_2-COOCH$_3$ and $(C_{16})_2$-Glu-C_2-(Gly-Pro-Hyp)$_4$-[IV-H1] are dissolved in methanol–CHCl$_3$ (1 : 9, v/v) and spread over a pure water subphase. After 15 min, monolayers containing the amphiphiles are compressed laterally to a final surface pressure of 40 mN/m at the air–water interface in a Langmuir trough at 25°. The monolayers are then deposited onto siliconized glass coverslips during downstroke immersion in the trough. After the coverslips are transferred under water to sterile biological assay plates, the medium is changed to Dulbecco's modified Eagle's medium (DMEM) with 10% (v/v) HEPES buffer. K1735 M4 mouse melanoma cells are cultured as described[40] and allowed to adhere for 1 hr at 37°. Adherent cells are fixed with 2% glutaraldehyde then stained with 0.12% (w/v) Diff-Quik Stain Set, which uses methanol to fix cells, eosin Y to stain the cytoplasm, and azure A and methylene blue to stain proteins. Cells are imaged with a video-enhanced microscope, and cell spreading is quantitated by measuring the average surface area occupied by cells using the Metamorph Image Analysis Program.

Characterization of Triple Helical Structure

The triple helix is a supersecondary structure characteristic of colla-gen.[41,42] Collagenlike triple helices are also found in macrophage scavenger receptors types I and II[43–45] and bacteria-binding receptor MARCO,[46] com-plement component C1q,[47] pulmonary surfactant apoprotein,[48] acetylcho-

[40] A. J. Miles, J. R. Knutson, A. P. N. Skubitz, L. T. Furcht, J. B. McCarthy, and G. B. Fields, *J. Biol. Chem.* **270**, 29047 (1995).

[41] G. N. Ramachandran, *Int. J. Pept. Protein Res.* **31**, 1 (1988).

[42] B. Brodsky and N. K. Shah, *FASEB J.* **9**, 1537 (1995).

[43] T. Kodama, M. Freeman, L. Rohrer, J. Zabrecky, P. Matsudaira, and M. Krieger, *Nature (London)* **343**, 531 (1990).

[44] L. Rohrer, M. Freeman, T. Kodama, M. Penman, and M. Krieger, *Nature (London)* **343**, 570 (1990).

[45] J. Ashkenas, M. Penman, E. Vasile, S. Acton, M. Freeman, and M. Krieger, *J. Lipid Res.* **34**, 983 (1993).

[46] O. Elomaa, M. Kangas, C. Sahlberg, J. Tuukkanen, R. Sormunen, A. Liakka, I. Thesleff, G. Kraal, and K. Tryggvason, *Cell (Cambridge, Mass.)* **80**, 603 (1995).

[47] B. Brodsky-Doyle, K. R. Leonard, and K. B. Reid, *Biochem. J.* **159**, 279 (1976).

[48] B. Benson, S. Hawgood, J. Schilling, J. Clements, D. Damm, B. Cordell, and R. T. White, *Proc. Natl. Acad. Sci. U.S.A.* **82**, 6379 (1985).

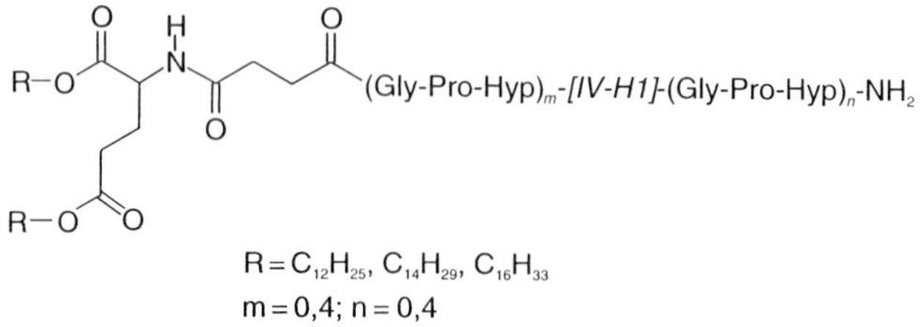

R—O ... N ... (Gly-Pro-Hyp)$_m$-[IV-H1]-(Gly-Pro-Hyp)$_n$-NH$_2$

R—O

R = C$_{12}$H$_{25}$, C$_{14}$H$_{29}$, C$_{16}$H$_{33}$
m = 0,4; n = 0,4

FIG. 2. General structure of the peptide–amphiphiles described here. The sequence of [IV-H1] is Gly-Val-Lys-Gly-Asp-Lys-Gly-Asn-Pro-Gly-Trp-Pro-Gly-Ala-Pro. [Reprinted with permission from *J. Am. Chem. Soc.* **118**, 12517 (1996). Copyright 1996 American Chemical Society.[27]]

linesterase,[49] and mannose binding protein.[50,51] The triple helix consists of three polypeptide chains, each in an extended, left-handed poly-Pro II-like helix, which are staggered by one residue and then supercoiled along a common axis in a right-handed manner.[41,42] Geometric constraints of the triple helical structure require that every third amino acid is Gly, resulting in a Gly-X-Y repeating sequence. For our initial studies, the α1(IV)1263–1277 collagen sequence Gly-Val-Lys-Gly-Asp-Lys-Gly-Asn-Pro-Gly-Trp-Pro-Gly-Ala-Pro ([IV-H1]), which is known to promote melanoma cell adhesion and spreading,[8,52,53] was combined with the lipid to create collagenlike peptide–amphiphiles (Fig. 2).[27] The formation of triple-helical structure within the peptide–amphiphile has been characterized by (1) compression of stable peptide–amphiphile monolayers, (2) circular dichroism spectra and melting curves, and (3) two-dimensional NMR spectra. The [IV-H1] peptide and variants without lipid tails were not surface active, but formation of monolayers at the air–water interface was observed for all investigated collagenlike peptide–amphiphiles. For (C$_{16}$)$_2$-Glu-C$_2$ and (C$_{14}$)$_2$-Glu-C$_2$ derived peptide–amphiphiles, surface pressure (which can be interpreted as a measure of resistance of amphiphile molecules against lateral compression) could be detected at surface areas of 2–3 nm^2/molecule.[27] The surface pressure increased gradually as the monolayer was compressed for peptide–amphiphiles containing both [IV-H1] and Gly-Pro-Hyp repeats. At a surface area of 0.6 nm^2/molecule no further compression was possible,

[49] M. Schumacher, S. Camp, Y. Maulet, M. Newton, K. MacPhee-Quigley, S. S. Taylor, T. Friedmann, and P. Taylor, *Nature* (*London*) **319**, 407 (1986).
[50] K. Drickamer, M. S. Dordal, and L. Reynolds, *J. Biol. Chem.* **261**, 6878 (1986).
[51] S. Oka, N. Itoh, T. Kawasaki, and I. Yamashina, *J. Biochem.* (*Tokyo*) **101**, 135 (1987).
[52] M. K. Chelberg, J. B. McCarthy, A. P. N. Skubitz, L. T. Furcht, and E. C. Tsilibary, *J. Cell Biol.* **111**, 261 (1990).
[53] K. H. Mayo, D. Parra-Diaz, J. B. McCarthy, and M. Chelberg, *Biochemistry* **30**, 8251 (1991).

and the monolayer reached the maximum surface pressure and collapsed. The common value of 0.6 nm^2/molecule for the limiting surface area of the $(C_{16})_2$-Glu-C_2-(Gly-Pro-Hyp)$_4$-[IV-H1] and $(C_{14})_2$-Glu-C_2-[IV-H1]-(Gly-Pro-Hyp)$_4$ peptide–amphiphiles can only be explained assuming a fully stretched, elongated peptide head group.[23] Prior X-ray crystallographic analyses of a triple-helical peptide revealed hexagonal-packed trimers with axis-to-axis distances of 1.4 nm.[54,55] The calculated surface area for this triple-helical peptide would be 1.7 nm^2/trimer, very close to the surface area of 1.8 nm^2/trimer for the peptide–amphiphiles studied here.

Collagens in triple-helical conformation exhibit a circular dichroism (CD) spectrum similar to a poly-Pro II helix, with positive ellipticity from $\lambda = 215$–240 nm.[56] At 25°, (Gly-Pro-Hyp)$_4$-[IV-H1]-(Gly-Pro-Hyp)$_4$ was found to exhibit this characteristic CD spectrum (Fig. 3). For (Gly-Pro-Hyp)$_4$-[IV-H1] and [IV-H1]-(Gly-Pro-Hyp)$_4$ a small magnitude of positive ellipticity at $\lambda = 225$ nm was observed, whereas the [IV-H1] peptide did not show any positive ellipticity at this wavelength. Of the peptide–amphiphiles, $(C_{12})_2$-Glu-C_2-[IV-H1] displayed a CD spectrum similar to that of [IV-H1] (no positive ellipticity at $\lambda = 225$ nm), whereas the other three amphiphiles showed a large magnitude of positive ellipticity at $\lambda > 220$ nm. Most remarkably, the ellipticity per residue for the amphiphilic compounds $(C_{12})_2$-Glu-C_2-(Gly-Pro-Hyp)$_4$-[IV-H1], $(C_{12})_2$-Glu-C_2-[IV-H1]-(Gly-Pro-Hyp)$_4$, and $(C_{12})_2$-Glu-C_2-(Gly-Pro-Hyp)$_4$-[IV-H1]-(Gly-Pro-Hyp)$_4$ was about five times larger than that of (Gly-Pro-Hyp)$_4$-[IV-H1]-(Gly-Pro-Hyp)$_4$ (Fig. 3) and approximately equal to that of the triple-helical peptide (Gly-Pro-Hyp)$_{10}$.[57–59] These ellipticity per residue values indicate a maximal ordered structure for $(C_{12})_2$-Glu-C_2-(Gly-Pro-Hyp)$_4$-[IV-H1], $(C_{12})_2$-Glu-C_2-[IV-H1]-(Gly-Pro-Hyp)$_4$, and $(C_{12})_2$-Glu-C_2-(Gly-Pro-Hyp)$_4$-[IV-H1]-(Gly-Pro-Hyp)$_4$. It appears that all residues in these three peptide–amphiphiles are in triple-helical conformation.

A triple-helical assembly can be distinguished from a simple, nonintercoiled poly-Pro II structure by its thermal denaturation behavior. A triple helix is relatively sensitive to temperature, as it is stabilized by a hydrogen-bonded intra- and interstrand water network.[54,55] Triple-helical melts are highly cooperative.[56] The thermal stability of peptides and peptide–amphiphiles were studied by monitoring ellipticity at $\lambda = 225$ nm as

[54] J. Bella, M. Eaton, B. Brodsky, and H. M. Berman, *Science* **266,** 75 (1994).

[55] J. Bella, B. Brodsky, and H. M. Berman, *Structure* **3,** 893 (1995).

[56] E. Heidemann and W. Roth, *Adv. Polym. Sci.* **43,** 143 (1982).

[57] B. Brodsky, M. Li, C. G. Long, J. Apigo, and J. Baum, *Biopolymers* **32,** 447 (1992).

[58] C. G. Long, E. Braswell, D. Zhu, J. Apigo, J. Baum, and B. Brodsky, *Biochemistry* **32,** 11688 (1993).

[59] M. Li, P. Fan, B. Brodsky, and J. Baum, *Biochemistry* **32,** 7377 (1993).

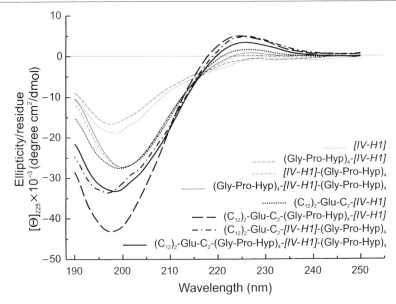

Fig. 3. CD spectra of collagen–model peptides and peptide–amphiphiles. Positive values of ellipticity in the range $\lambda = 215$–245 nm are attributed to an ordered, poly-Pro II-like structure.[38] Among the investigated peptides, only (Gly-Pro-Hyp)$_4$-[IV-H1]-(Gly-Pro-Hyp)$_4$ shows this structure distinctly. However, except for (C$_{12}$)$_2$-Glu-C$_2$-[IV-H1], all peptide–amphiphiles display a positive signal, with the residual ellipticity corresponding to the maximum values reported for triple-helical structures. Solutions of (C$_{12}$)$_2$-Glu-C$_2$-Gly (the lipid tail without a collagenous head group) show little positive or negative ellipticity over the range $\lambda = 190$–250 nm (data not shown). [Reprinted with permission from *J. Am. Chem. Soc.* **118**, 12517 (1996). Copyright 1996 American Chemical Society.[27]]

a function of increasing temperature. Among the peptides, only (Gly-Pro-Hyp)$_4$-[IV-H1]-(Gly-Pro-Hyp)$_4$ gave a typical sigmoidal transition associated with the transformation of triple-helical to single-stranded structure ($T_m = 36°$).[27] [IV-H1]-(Gly-Pro-Hyp)$_4$ showed a small magnitide of positive ellipticity which decreased nearly linearly from 5 to 20° then flattened out, as did (Gly-Pro-Hyp)$_4$-[IV-H1]. The molar elllipticities of the peptide–amphiphiles decreased very gradually between 10 and 30°, then more markedly starting at around 30–40°, with some traces of positive CD detectable up to 80°.[27] The midpoint of the transitions (T_m) was found to be at 50 ± 5°. The melting curve was fully reversible on cooling. Although the change in ellipticity was large, thermal transitions for the peptide–amphiphiles were broad. These observations suggest that the (Gly-Pro-Hyp)$_4$-[IV-H1]-(Gly-Pro-Hyp)$_4$ and (C$_{12}$)$_2$-Glu-C$_2$-(Gly-Pro-Hyp)$_4$-[IV-H1]-(Gly-Pro-Hyp)$_4$ structures consist of packed poly-Pro II-like helices, possibly triple-helical, and that the lipid tail remarkably enhanced the stability of this assembly.

The structures of the collagen-model peptides and peptide–amphiphiles were further investigated by 2D ^{1}H NMR spectroscopy. The Pro and Hyp spin systems in TOCSY were identified by the lack of amide protons and reference to the chemical shifts of the side-chain protons from other collagenlike peptides.[57–59] The chemical shift of the Pro and Hyp side-chain protons is sensitive to their conformation.[57–59] For our peptides and peptide–amphiphiles, there are Pro residues that are found both surrounding and within the [IV-H1] sequence, whereas Hyp residues surround the [IV-H1] sequence. At 10°, the relatively few cross-peaks found in the Pro/Hyp region of the ^{1}H NMR spectra (Fig. 4) indicate that the Pro and Hyp residues of (Gly-Pro-Hyp)$_4$-[IV-H1]-(Gly-Pro-Hyp)$_4$ are in a limited number of conformations, as expected for a compound with an ordered structure. The cross-peaks at 4.6 ppm are comparable to those observed for the triple-helical, template-bound peptide Kemp triacid-[Gly-(Gly-Pro-Hyp)$_3$-NH$_2$]$_3$.[60] The spectra of (Gly-Pro-Hyp)$_4$-[IV-H1]-(Gly-Pro-Hyp)$_4$ at 50° show additional cross-peaks at 4.85 ppm, indicating less ordered conformation at higher temperature. Some of these additional cross-peaks are consistent with the multiple states that exist for the Pro residues within the [IV-H1] sequence when in a non-triple-helical conformation.[53] After the (Gly-Pro-Hyp)$_4$-[IV-H1]-(Gly-Pro-Hyp)$_4$ peptide is lipidated with a C$_{12}$ tail, similar NMR spectra are obtained (Fig. 4). For example, (C$_{12}$)$_2$-Glu-C$_2$-(Gly-Pro-Hyp)$_4$-[IV-H1]-(Gly-Pro-Hyp)$_4$ at 25° shows a few well-defined cross-peaks, indicating ordered conformation of the peptide–amphiphile. Consistent with our CD observations, the NMR spectra of (C$_{12}$)$_2$-Glu-C$_2$-(Gly-Pro-Hyp)$_4$-[IV-H1]-(Gly-Pro-Hyp)$_4$ at 80° indicate more disorder than at 25°. Additional cross-peaks are seen at 4.85 ppm, in similar fashion to (Gly-Pro-Hyp)$_4$-[IV-H1]-(Gly-Pro-Hyp)$_4$ at 50°. Overall, the CD and NMR spectra of the (Gly-Pro-Hyp)$_4$-[IV-H1]-(Gly-Pro-Hyp)$_4$ peptide and the (C$_{12}$)$_2$-Glu-C$_2$-(Gly-Pro-Hyp)$_4$-[IV-H1]-(Gly-Pro-Hyp)$_4$ peptide–amphiphile suggest that both can spontaneously form a well-ordered poly-Pro II-like triple-helical structure. Similar NMR spectra were obtained for the (C$_{12}$)$_2$-Glu-C$_2$-[IV-H1]-(Gly-Pro-Hyp)$_4$ and (C$_{12}$)$_2$-Glu-C$_2$-(Gly-Pro-Hyp)$_4$-[IV-H1] peptide–amphiphiles.

The peptide–amphiphiles appear to self-assemble into highly ordered poly-Pro II-like triple-helical structures when dissolved in aqueous subphases. Evidence for the self-assembly process has been obtained from monolayer observations and CD and NMR spectroscopies. Peptide–amphiphiles had surface areas, CD spectra, and Pro and Hyp side-chain

[60] M. Goodman, Y. Feng, G. Melacini, and J. P. Taulane, *J. Am. Chem. Soc.* **118**, 5156 (1996).

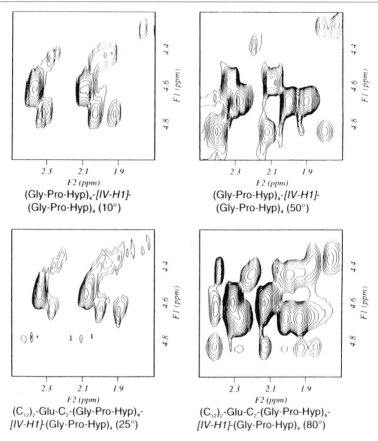

FIG. 4. TOCSY spectra of the Pro/Hyp region for (Gly-Pro-Hyp)$_4$-[IV-H1]-(Gly-Pro-Hyp)$_4$ (*top*) and (C$_{12}$)$_2$-Glu-C$_2$-(Gly-Pro-Hyp)$_4$-[IV-H1]-(Gly-Pro-Hyp)$_4$ (*bottom*) in D$_2$O at peptide and peptide–amphiphile concentrations of 3–5 mM. The 25° spectra of the peptide–amphiphile indicate a highly ordered structure similar to (Gly-Pro-Hyp)$_4$-[IV-H1]-(Gly-Pro-Hyp)$_4$ at 10°. At 80°, the additional cross-peaks at 4.85 ppm for the peptide–amphiphile are similar to those seen for (Gly-Pro-Hyp)$_4$-[IV-H1]-(Gly-Pro-Hyp)$_4$ at 50°. [Reprinted with permission from *J. Am. Chem. Soc.* **118,** 12519 (1996). Copyright 1996 American Chemical Society.[27]]

conformations characteristic of triple helices, and they exhibited large structural transitions as monitored by CD melting curves.

The lipid hydrophobic interactions of the peptide–amphiphiles exert a significant influence on collagen–model structure formation and stabilization. For example, although the [IV-H1]-(Gly-Pro-Hyp)$_4$ sequence has the potential of forming a triple helix, it is realized only in the amphiphilic compound. The triple helix is also exceptionally stable when formed in

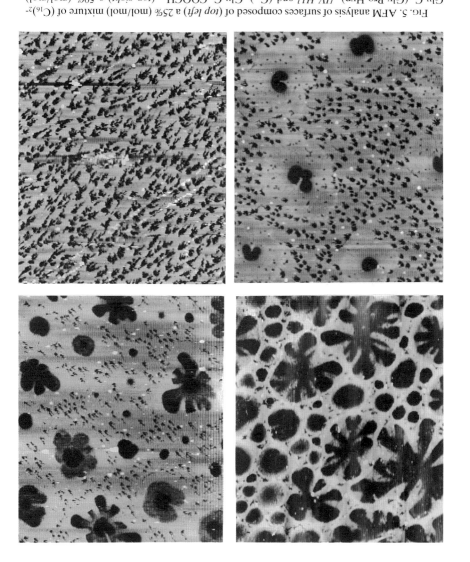

Fig. 5. AFM analysis of surfaces composed of (*top left*) a 25% (mol/mol) mixture of $(C_{16})_2$-Glu-C_2-(Gly-Pro-Hyp)$_4$-*[IV-HI]* and $(C_{18})_2$-Glu-C_2-COOCH$_3$, (*top right*) a 50% (mol/mol) mixture of $(C_{16})_2$-Glu-C_2-(Gly-Pro-Hyp)$_4$-*[IV-HI]* and $(C_{18})_2$-Glu-C_2-COOCH$_3$, (*bottom left*) a 75% (mol/mol) mixture of $(C_{16})_2$-Glu-C_2-(Gly-Pro-Hyp)$_4$-*[IV-HI]* and $(C_{18})_2$-Glu-C_2-COOCH$_3$, and (*bottom right*) 100% $(C_{16})_2$-Glu-C_2-(Gly-Pro-Hyp)$_4$-*[IV-HI]* at 37°. The scan size was 20.00 μm, the scan rate was 1.938–2.977 Hz, and the number of samples was 256.

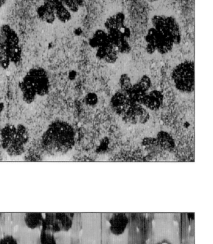

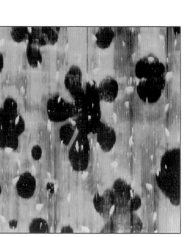

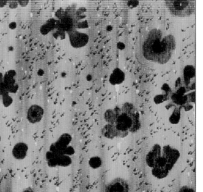

FIG. 6. AFM analysis of (*left*) glass, (*middle*) gold-coated, and (*right*) mica surfaces layered with a 50% (mol/mol) mixture of $(C_{16})_2$-Glu-C_2-(Gly-Pro-Hyp)$_4$-[IV-H1] and $(C_{18})_2$-Glu-C_2-COOCH$_3$ at 37°.

the presence of the lipid modification. The difference in the denaturation temperatures between the structured (Gly-Pro-Hyp)$_4$-[IV-H1]-(Gly-Pro-Hyp)$_4$ peptide and the corresponding C$_{12}$ peptide–amphiphile is about 15–20°. The tight alignment of the N-terminal amino acids achieved through the association of the lipid part of the molecule in a monolayer could be a general tool for initiation of peptide folding.

Surface Association of Peptide–Amphiphiles

The ability of peptide–amphiphiles to form stable and evenly dispersed bilayers on surfaces was studied using AFM. Initial studies used mica surfaces. Layers composed of a 25% (mol/mol) mixture of the peptide–amphiphile (C$_{16}$)$_2$-Glu-C$_2$-(Gly-Pro-Hyp)$_4$-[IV-H1] and the methyl-capped amphiphile (C$_{18}$)$_2$-Glu-C$_2$-COOCH$_3$ showed little mixing of the peptide–amphiphile with the starlike structures formed by the methyl-capped amphiphile (Fig. 5, top left; see color insert). When the mixture was increased to 50%, the peptide–amphiphile dispersed fairly evenly between the amphiphile starlike structures (Fig. 5, top right; see color insert). Layers composed of a 75% (mol/mol) mixture of (C$_{16}$)$_2$-Glu-C$_2$-(Gly-Pro-Hyp)$_4$-[IV-H1] and (C$_{18}$)$_2$-Glu-C$_2$-COOCH$_3$ showed very good dispersion of the peptide–amphiphile, leaving the amphiphile starlike structures isolated (Fig. 5, bottom left; see color insert). Layers composed of pure peptide–amphiphile were fairly evenly covered, with some defects (Fig. 5, bottom right; see color insert).

Studies were expanded to include comparisons of mica surfaces to glass and gold-coated surfaces. A 50% (mol/mol) mixture of (C$_{16}$)$_2$-Glu-C$_2$-(Gly-Pro-Hyp)$_4$-[IV-H1] and (C$_{18}$)$_2$-Glu-C$_2$-COOCH$_3$ was layered onto either glass, gold-coated, or mica surfaces. AFM images showed very similar dispersion patterns on all three surfaces (Fig. 6; see color insert). Thus, the ability of peptide–amphiphiles to form stable, well-dispersed bilayers appears to be applicable to a variety of surfaces.

Biological Activity of Collagenlike Peptide–Amphiphiles

M4 mouse melanoma cell spreading was examined on monolayers composed of either pure (C$_{18}$)$_2$-Glu-C$_2$-COOCH$_3$ or (C$_{16}$)$_2$-Glu-C$_2$-(Gly-Pro-Hyp)$_4$-[IV-H1] or molar mixtures of these synthetic amphiphiles. Little cell spreading and cytoplasmic extrusions occurred on either 100% (C$_{18}$)$_2$-Glu-C$_2$-COOCH$_3$ (Fig. 7) or (C$_{16}$)$_2$-Glu-C$_2$-(Gly-Pro-Hyp)$_4$-[IV-H1] (data not shown), whereas fairly extensive spreading occurred on a 50% molar mixture of the amphiphile and peptide–amphiphile (Fig. 7). The extent of cell spreading can be quantitated by considering the cell shape factor, defined

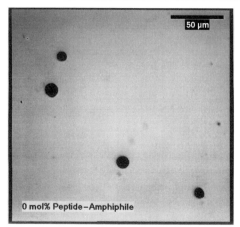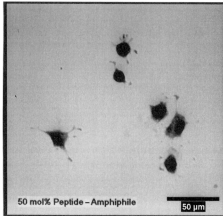

FIG. 7. Video-enhanced microscopic analysis of melanoma cell spreading on Langmuir–Blodgett films composed of (*left*) $(C_{18})_2$-Glu-C_2-COOCH$_3$ and (*right*) 50% (mol/mol) mixture of $(C_{18})_2$-Glu-C_2-COOCH$_3$ and $(C_{16})_2$-Glu-C_2-(Gly-Pro-Hyp)$_4$-[IV-H1].

as $4\pi A/P^2$ (where A is area and P is perimeter). A circle has a value of 1.0, and cell spreading increases as the shape factor decreases. Melanoma cells on either $(C_{18})_2$-Glu-C_2-COOCH$_3$ or $(C_{16})_2$-Glu-C_2-(Gly-Pro-Hyp)$_4$-[IV-H1] are nearly circular (shape factor ≈ 1), whereas cells on a 50% molar mixture of $(C_{18})_2$-Glu-C_2-COOCH$_3$ and $(C_{16})_2$-Glu-C_2-(Gly-Pro-Hyp)$_4$-[IV-H1] have spread considerably (shape factor < 1) (Fig. 8). It appears that the presence of the peptide alone is not sufficient to cause cell spreading to increase. Both the molar concentration and the molecular packing of

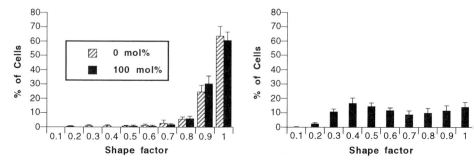

FIG. 8. Shape factor analysis of melanoma cell spreading on Langmuir–Blodgett films composed of (*left*) either 0% $(C_{16})_2$-Glu-C_2-(Gly-Pro-Hyp)$_4$-[IV-H1] [i.e., 100% $(C_{18})_2$-Glu-C_2-COOCH$_3$] or 100% $(C_{16})_2$-Glu-C_2-(Gly-Pro-Hyp)$_4$-[IV-H1] and (*right*) 50% (mol/mol) mixture of $(C_{18})_2$-Glu-C_2-COOCH$_3$ and $(C_{16})_2$-Glu-C_2-(Gly-Pro-Hyp)$_4$-[IV-H1].

the peptide–amphiphile in monolayers play an important role in mediating the cellular response.

Other Syntheses of Peptide–Amphiphiles

Our peptide–amphiphiles have been assembled on the solid-phase utilizing dialkyl ester tail compounds. Other research groups have constructed peptide–amphiphiles using tail compounds such as dialkylamides,[61] phospholipids,[36,62–64] fluorescent phospholipids,[65] and tripalmitoyl-Cys [(Pam)$_3$-Cys].[66] Most research groups have incorporated their lipophilic compounds by solution-phase[61,62,64–67] or chemoselective ligation[36,63] methods. To our knowledge, only one laboratory has reported previously the incorporation of a branched lipid (1,2-dimyristoyl-*sn*-glycerol derivative) onto a peptide by solid-phase methods.[68] Mild solid-phase methods should allow for the assembly of peptide–amphiphiles containing some of these other lipids, further extending the utility of this family of compounds.

Other Applications of Peptide–Amphiphiles

In general, amphiphile systems may form a great variety of structures in solution including micelles and vesicles.[69] Israelachvili devised a dimensionless group, the surfactant number (N_s), to predict the geometry of the final amphiphile aggregate.[70] The surfactant number is defined as $N_s = v/a_o l$, where v is the volume of the hydrocarbon chain tail, a_o is the optimal head group area, and l is the maximal extended length of the hydrocarbon chain tail. For $N_s < 1/3$, spherical micelles are formed; for N_s between 1/3 and 1/2, cylindrical micelles are formed; for N_s between 1/2 and 1, flexible bilayers and vesicles are formed.[70] Because our peptide–amphiphiles have a

[61] T. Shimizu and M. Hato, *Biochim. Biophys. Acta* **1147**, 50 (1993).

[62] F. Macquaire, F. Baleux, E. Giaccobi, T. Huynh-Dinh, J.-M. Neumann, and A. Sanson, *Biochemistry* **31**, 2576 (1992).

[63] T. M. Winger, P. J. Ludovice, and E. L. Chaikof, *Biomaterials* **17**, 437 (1996).

[64] R. K. Jain, C. Gupta, and N. Anand, *Tetrahedron Lett.* **22**, 2317 (1981).

[65] N. L. Thompson, A. A. Brian, and H. M. McConnell, *Biochim. Biophys. Acta* **772**, 10 (1984).

[66] W. Prass, H. Ringsdorf, W. Bessler, K.-H. Wiesmüller, and G. Jung, *Biochim. Biophys Acta* **900**, 116 (1987).

[67] X. Cha, K. Ariga, and T. Kunitake, *Bull. Chem. Soc. Jpn.* **69**, 163 (1996).

[68] H. B. A. de Bont, J. H. van Boom, and R. M. J. Liskamp, *Recl. Trav. Chim. Pays-Bas* **111**, 222 (1992).

[69] D. F. Evans and H. Wennerstrom, "The Colloidal Domain: Where Physics, Chemistry, Biology, and Technology Meet." VCH, New York, 1994.

[70] J. N. Israelachvili, *in* "Intermolecular and Surface Forces" (J. N. Israelachvili, ed.), 2nd Ed., p. 366. Academic Press London, 1992.

large head group and thus a_o is relatively large, it is anticipated that N_s < 1/3 and micellization is probable. However, we have found that one can mix peptide–amphiphiles with vesicle-forming lipids (such as dilauryl phosphatidylcholine) to form stable mixed vesicles with collagen-model, triple-helical peptide head groups. Vesicles featuring collagen coatings have already been shown to be advantageous for targeted drug delivery.[71]

In addition to our results, several different peptide–amphiphiles have been reported that form distinct structures and/or are biologically active. A number of peptide-lipids have been shown to oligomerize and form stable structures.[61,65,66] When the peptide acetyl-Lys-Gly-Arg-Gly-Asp-Gly-amide is attached to dodecylphosphocholine via the N^ε-amino group of Lys, 2D ^{1}H NMR indicated that the peptide–amphiphile forms a turn conformation (type II or I′) while the peptide itself mixed with the phospholipid does not.[62] A Fourier transform infrared (FTIR) reflection absorption spectroscopic study of N-octadecanoyl-(Gly)$_n$ ethyl ester monolayers and Langmuir–Blodgett films, where $n = 1$–5, demonstrated that the peptide–amphiphiles could form poly(Gly) II helices, whereas the oligo(Gly) peptides themselves could not.[67] A series of peptides prepared with an N-terminal (Pam)$_3$Cys moiety were found to activate murine spleen cells.[66] Finally, combining the human thrombin receptor peptide agonist Ser-Phe-Leu-Leu-Arg-Asn-(β-Ala)$_3$-Tyr-NHCH$_2$CH$_2$SH with the phospholipid distearoylphosphatidylethanolamine produced a peptide–amphiphile that promoted platelet aggregation with EC$_{50}$ = 38 + 3 μM, only one order of magnitude worse than the peptide itself.[63]

Summary

The peptide–amphiphiles described here provide a simple approach for building stable protein structural motifs using peptide head groups. One of the most intriguing features of this system is the possible formation of stable lipid films on solid substrates, or the use of the novel amphiphiles in bilayer membrane systems, where the lipid tail serves not only as a peptide structure-inducing agent but also as an anchor of the functional head group in the lipid assembly. The peptide–amphiphile system potentially offers great versatility with regard to head and tail group composition and overall geometries and macromolecular structures. For building materials with molecular and cellular recognition capacity, it is essential to have a wide repertoire of tools to produce characteristic supersecondary structures at surfaces and interfaces.

[71] M. J. Fonseca, M. A. Alsina, and F. Reig, *Biochim Biophys. Acta* **1279,** 259 (1996).

Acknowledgments

We gratefully acknowledge the technical assistance of Cynthia A. Guy. This research is supported by the Center for Interfacial Engineering (an NSF ERC), the Earl E. Bakken Chair for Biomedical Engineering, and National Institutes of Health Grants KD444 94, AR01929, and CA63671.

[27] Chemical Synthesis and Nuclear Magnetic Resonance Characterization of Partially Folded Proteins

By Elisar Barbar, Christopher M. Gross, Clare Woodward, and George Barany

Introduction

Within the past few years, the protein folding paradigm has shifted from the view that there is a unique trajectory of molecules moving through conformational space to the view that there are multiple folding routes involving ensembles of conformations. To help provide an experimental base for this "New View",[1] our research is concerned with the dynamic structure of partially folded proteins that are models for transient species formed during folding.[2,3] Partially folded conformations of proteins are commonly induced by extreme solvent conditions such as low pH and addition of alcohol or denaturants.[4–7] In contrast, we have developed an approach by which appropriately designed protein analogs, accessible only by chemical synthesis, form partially folded conformations at physiological pH without addition of denaturants or alcohol. As a case study to develop the tools for this general approach, our work has focused on the small 58-residue antiparallel β-sheet protein bovine pancreatic trypsin inhibitor

[1] K. Dill and H.-S. Chan, *Nat. Struct. Biol.* **4,** 10 (1997).

[2] P. Jennings and P. Wright, *Science* **262,** 892 (1993).

[3] J. Balbach, V. Forge, N. Vannuland, S. L. Winder, P. J. Hore, and C. M. Dobson, *Nat. Struct. Biol.* **2,** 865 (1995).

[4] A. T. Alexandrescu, P. A. Evans, M. Pitkeathly, J. Baum, and C. M. Dobson, *Biochemistry* **32,** 1707 (1993).

[5] M. Buck, S. E. Radford, and C. M. Dobson, *Biochemistry* **32,** 669 (1993).

[6] C. J. Falzone, M. R. Mayer, E. L. Whiteman, C. D. Moore, and J. T. Lecomte, *Biochemistry* **35,** 6519 (1996).

[7] H. Molinari, L. Ragona, L. Varani, G. Musco, R. Consonni, L. Zetta, and H. L. Monaco, *FEBS Lett.* **381,** 237 (1996).

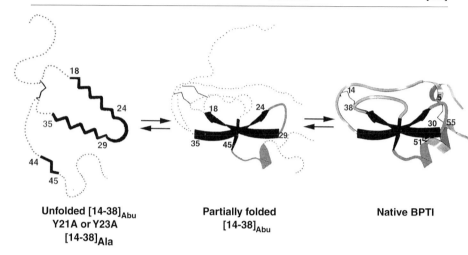

Unfolded [14-38]$_{Abu}$ Partially folded   Native BPTI
Y21A or Y23A   [14-38]$_{Abu}$
[14-38]$_{Ala}$

FIG. 1. Representation of [14–38]$_{Abu}$ NMR-detected structure along with model of unfolded [14–38]$_{Abu}$. Native BPTI is shown for comparison. The most stable region in partially folded [14–38]$_{Abu}$ includes the 18–24, 29–35 antiparallel strands of β sheet (darkest shaded ribbons). Shading indicates segmental stability; more stable regions are shaded darker. The dotted lines correspond to the more flexible part of the protein for which both conformations are nonnative.

(BPTI).[8-14] Partially folded BPTI provides a model for the structure, stability, and dynamics of the ensemble of early folding intermediates in which the hydrophobic core is mostly structured and nativelike, while the rest of the protein fluctuates among nonnative conformations (Fig. 1). Our results provide experimental support for the evolving concepts of protein folding.

To destabilize BPTI, the wild-type form of which contains three disulfide bridges connecting residues 5 to 55, 14 to 38, and 30 to 51, we replace cystine cross-links with pairs of the cysteine isostere α-amino-n-butyric acid (Abu) or alanine (Ala). In addition, key Tyr residues in the hydrophobic core are replaced by Ala. Chemical synthetic methods used to access these molecules, including variants containing stable isotopically labeled residues at strategic positions, are outlined in this article (Table I). The structures of the resultant BPTI analogs, which are either partially folded or completely unfolded, are then characterized by two-dimensional ^{1}H homonuclear and

[8] M. Ferrer, C. Woodward, and G. Barany, *Int. J. Pept. Protein Res.* **40,** 194 (1992).
[9] M. Ferrer, Ph.D. Thesis, University of Minnesota, Minneapolis (1994).
[10] E. Barbar, G. Barany, and C. Woodward, *Biochemistry* **34,** 11423 (1995).
[11] M. Ferrer, G. Barany, and C. Woodward, *Nat. Struct. Biol.* **2,** 211 (1995).
[12] H. Pan, E. Barbar, G. Barany, and C. Woodward, *Biochemistry* **34,** 13974 (1995).
[13] G. Barany, C. M. Gross, M. Ferrer, E. Barbar, H. Pan, and C. Woodward, *in* "Techniques in Biochemistry VII" (D. Marshak, ed.), p. 503. Academic Press, San Diego, 1996.
[14] E. Barbar, G. Barany, and C. Woodward, *Folding Design* **1,** 65 (1996).

TABLE I
SYNTHETIC BOVINE PANCREATIC TRYPSIN INHIBITOR PROTEIN ANALOGS[a]

| BPTI Species | Coupling chemistry | Mass (amu) | |
		Calculated	Found
Naturally occurring[b]	BOP	6512.7	6511.7 ± 1.3
[R]$_{Abu}$[b]	BOP	6409.2	6409.6 ± 1.0
[14–38]$_{Ala}$	TBTU	6386.2	6385.0 ± 1.0
[14–38]$_{Abu}$	HATU	6443.6	6443.2 ± 0.6
[^{15}N]$_8$–[14–38]$_{Abu}$[c]	BOP	6451.6	6450.2 ± 1.4
[^{15}N]$_9$–[14–38]$_{Abu}$[d]	HBTU	6452.6	6452.8 ± 0.4
[14–38]$_{Abu}$Tyr23Ala[e]	BOP	6351.6	6350.0 ± 1.0
[14–38]$_{Abu}$Tyr21Ala[e]	BOP	6351.6	6350.4 ± 1.3
[30–51]$_{Abu}$	HBTU	6443.6	6442.1 ± 0.6

[a] All proteins were assembled on *p*-alkoxybenzyl ester (PAC)–PEG–PS supports with a continuous-flow automated synthesizer as outlined in this article.
[b] As reported in *Int. J. Pept. Protein Res.* **40,** 194 (1992).
[c] ^{15}N labels at Leu-6, Gly-12, Ala-16, Leu-29, Gly-37, Ala-40, Ala-48, and Gly-56, as outlined in *Biochemistry* **34,** 11423 (1995).
[d] ^{15}N labels at Phe-4, Leu-6, Phe-22, Ala-25, Ala-27, Gly-28, Phe-33, Phe-45, and Ala-48, as outlined in *Biophys. Chem.* **64,** 45 (1997).
[e] As reported in *Folding Design* **1,** 65 (1996).

^{15}N heteronuclear nuclear magnetic resonance (NMR) methods. Although it is well established how NMR methods can determine the solution structures of small folded proteins to high resolution, NMR data acquisition, data processing, and analysis of partially folded or denatured proteins is considerably less straightforward. This is due to extensive conformational averaging, spectral overlap in the amide, CαH, and aliphatic regions that correspond to random coil chemical shifts, as well as the presence of either slow chemical exchange giving rise to extra peaks or intermediate chemical exchange causing peak broadening. Nevertheless, it is possible by techniques described herein to extract structural information about component conformations (including identification of minor populations) and to monitor microscopic probes of global folding/unfolding. Other characterization techniques that we have applied, including circular dichroism (CD), fluorescence enhancement on ANS binding, native gel electrophoresis, and biological assay studies, have been described previously[9,11,14] and are not covered in this article.

Most partially folded proteins for which structural analyses are available are α-helical.[2,4,15,16] Our work on the predominantly β-sheet structure of

[15] J. Baum, C. M. Dobson, P. A. Evans, and C. Hanley, *Biochemistry* **28,** 7 (1989).
[16] F. M. Hughson, P. E. Wright, and R. L. Baldwin, *Science* **249,** 1544 (1990).

BPTI is complemented by characterizations of other proteins with this feature.[6,7,17] In addition, structural studies by Shortle on mutants and fragments of staphylococcal nuclease, which serve as models of denatured states under physiological conditions,[18] bear relevant similarities to our approach.

Chemical Synthesis

Over the years, refinements in solid-phase peptide synthesis, including better supports, anchoring chemistries, coupling reagents, and protecting group combinations, have advanced the methods toward preparation of small proteins.[19–22] The total chemical synthesis of BPTI and its analogs is difficult, but it can be achieved on a semiroutine basis by using an automated continuous-flow synthesizer for carrying out under an inert atmosphere the repetitive chain assembly steps.[8,9,13] Carefully optimized protocols of N^{α}-9-fluorenylmethyloxycarbonyl (Fmoc) chemistry are applied (Tables II and III), and a number of precautions outlined herein should be noted. The polymeric supports of choice are polyethylene glycol–polystyrene (PEG–PS) graft resins,[23–26] which have superior performance characteristics in flow-through columns. Overall isolated yields of BPTI analogs, based on C-terminal Ala attached as a p-alkoxybenzyl ester, are in the 3–6% range. The final material is obtained routinely in >99% purity and is suitable for addressing a number of biophysical questions.

[17] N. Schonbrunner, J. Wey, J. Engels, H. Georg, and T. Kiefhaber, J. Mol. Biol. 260, 432 (1996).
[18] D. Shortle, Curr. Opin. Struct. Biol. 1, 24 (1996).
[19] G. Barany, N. K. Cordonier, and D. G. Mullen, Int. J. Pept. Protein Res. 30, 705 (1987).
[20] G. B. Fields, Z. Tian, and G. Barany, in "Synthetic Peptides: A User's Guide" (G. A. Grant, ed.), p. 77. Freeman, New York, 1992.
[21] B. Merrifield, in "Peptides: Synthesis, Structures, and Applications" (B. Gutte, ed.), p. 93. Academic Press, San Diego, 1995.
[22] G. Barany and M. Kempe, in "A Practical Guide to Combinatorial Chemistry" (S. H. DeWitt and A. W. Czarnik, eds.), pp. 3–49. American Chemical Society Books, Washington, DC, 1997.
[23] G. Barany, N. A. Solé, R. J. Van Abel, F. Albericio, and M. E. Selsted, in "Innovation and Perspectives in Solid Phase Synthesis: Peptides, Polypeptides and Oligonucleotides 1992" (R. Epton, ed.), p. 29. Intercept, Andover, UK, 1992.
[24] G. Barany, F. Albericio, N. A. Solé, G. W. Griffin, S. A. Kates, and D. Hudson, in "Peptides 1992: Proceedings of the Twenty-Second European Peptide Symposium" (C. H. Schneider and A. N. Eberle, eds.), p. 267. Escom, Leiden, The Netherlands, 1993.
[25] S. Zalipsky, J. L. Chang, F. Albericio, and G. Barany, React. Polym. 22, 243 (1994).
[26] G. Barany, F. Albericio, S. A. Kates, and M. Kempe, in "Chemistry and Biological Application of Polyethylene Glycol" (J. M. Harris and S. Zalipsky, eds.), in press. American Chemical Society Books, Washington, DC, 1997.

TABLE II
PROTOCOL FOR CONTINUOUS-FLOW Fmoc SOLID-PHASE SYNTHESIS OF $[30–51]_{Abu}$[a]

Cycle No.	Wild-type residue	Derivative	Coupling time (min)	Deprotection protocol	Notes[b]
0	A58	Fmoc-Ala-O-PEG–PS	N/A	N/A	A
1	G57	Fmoc-Gly-OH	60	I	B
2	G56	Fmoc-Gly-OH	30	D	C
3	C55	Fmoc-Abu-OH	60	S	
4	T54	Fmoc-Thr(tBu)-OH	60	S	
5	R53	Fmoc-Arg(Pmc)-OH	90	S	
6	M52	Fmoc-Met-OH	60	S	
7	C51	Fmoc-Cys(Trt)-OPfp	60	S	D
8	D50	Fmoc-Asp(OtBu)-OH	60	S	
9	E49	Fmoc-Glu(OtBu)-OH	90	S	
10	A48	Fmoc-Ala-OH	60	S	
11	S47	Fmoc-Ser(tBu)-OH	60	S	
12	K46	Fmoc-Lys(Boc)-OH	75	E	E
13	F45	Fmoc-Phe-OH	60	S	
14	N44	Fmoc-Asn(Tmob)-OH	75	S	
15	N43	Fmoc-Asn(Tmob)-OH	60	E	
16	R42	Fmoc-Arg(Pmc)-OH	90	E	
17	K41	Fmoc-Lys(Boc)-OH	60	S	
18	A40	Fmoc-Ala-OH	60	S	
19	R39	Fmoc-Arg(Pmc)-OH	90	S	
20	C38	Fmoc-Abu-OH	90	S	
21	G37	Fmoc-Gly-OH	60	S	
22	G36	Fmoc-Gly-OH	75	S	
23	Y35	Fmoc-Tyr(tBu)-OH	75	S	E
24	V34	Fmoc-Val-OH	90	E	
25	F33	Fmoc-Phe-OH	90	E	
26	T32	Fmoc-Thr(tBu)-OH	90	E	
27	Q31	Fmoc-Gln(Tmob)-OH	72	E	
28	C30	Fmoc-Cys(Trt)-OPfp	72	E	D
29	L29	Fmoc-Leu-OH	72	E	
30	G28	Fmoc-Gly-OH	72	E	
31	A27	Fmoc-Ala-OH	75	E	
32	K26	Fmoc-Lys(Boc)-OH	90	E	
33	A25	Fmoc-Ala-OH	72	E	
34	N24	Fmoc-Asn(Tmob)-OH	90	E	
35	Y23	Fmoc-Tyr(tBu)-OH	90	E	
36	F22	Fmoc-Phe-OH	90	E	
37	Y21	Fmoc-Tyr(tBu)-OH	90	E	
38	R20	Fmoc-Arg(Pmc)-OH	120	E	
39	I19	Fmoc-Ile-OH	60	E	
40		Fmoc-Ile-OH	30	N/A	F
41	I18	Fmoc-Ile-OH	60	E	G
42		Fmoc-Ile-OH	30	N/A	F
43	R17	Fmoc-Arg(Pmc)-OH	240	E	
44	A16	Fmoc-Ala-OH	90	E	

(continued)

TABLE II (continued)

Cycle No.	Wild-type residue	Derivative	Coupling time (min)	Deprotection protocol	Notes[b]
45	K15	Fmoc-Lys(Boc)-OH	90	E	
46	C14	Fmoc-Abu-OH	90	E	
47		Fmoc-Abu-OH	90	N/A	F
48	P13	Fmoc-Pro-OH	90	E	
49	G12	Fmoc-Gly-OH	90	E	
50	T11	Fmoc-Thr(tBu)-OH	90	E	
51		Fmoc-Thr(tBu)-OH	72	N/A	F
52	Y10	Fmoc-Tyr(tBu)-OH	60	E	E
53	P9	Fmoc-Pro-OH	72	E	
54		Fmoc-Pro-OH	72	N/A	F
55	P8	Fmoc-Pro-OH	72	E	
56		Fmoc-Pro-OH	72	N/A	F
57	E7	Fmoc-Glu(OtBu)-OH	60	E	
58		Fmoc-Glu(OtBu)-OH	78	N/A	F
59	L9	Fmoc-Leu-OH	72	E	
60		Fmoc-Leu-OH	40	N/A	F
61	C5	Fmoc-Abu-OH	120	E	
62	F4	Fmoc-Phe-OH	120	E	
63	D3	Fmoc-Asp(OtBu)-OH	120	E	
64	P2	Fmoc-Pro-OH	78	E	
65		Fmoc-Pro-OH	72	N/A	F
66	R1	Fmoc-Arg(Pmc)-OH	240	E	
67		Fmoc-Arg(Pmc)-OH	60	N/A	F
68		Fmoc-Arg(Pmc)-OH	60	N/A	H

[a] The synthesis of $[30–51]_{Abu}$ was carried out starting with 0.75 g Fmoc-Ala-O-PAC–PEG–PS (0.24 nmol/g). Couplings were mediated by HBTU/HOBt/NMM (4:4:6), with the activated protected Fmoc-amino acid solution circulating through the column at 8.3 ml/min for the indicated time. Deprotections were according to code I (initial), D (diketopiperazine reduction), S (standard), or E (extended), as described further in Table III. Side-chain protecting groups were as follows: tert-butyl (tBu) ethers and esters for Ser, Thr, Tyr, Asp, and Glu; tert-butyloxycarbonyl (Boc) for Lys; 2,2,5,7,8-pentamethylchroman-6-sulfonyl (Pmc) for Arg; 2,4,6-trimethoxybenzyl (Tmob) for Asn and Gln, and trityl (Trt) for Cys. Other analogs were synthesized by closely related protocols (Table I and text). NA, Not applicable.

[b] Key to notes: A, the first amino acid Ala was already on the support as a PAC ester, through coupling of a preformed handle derivative; B, initial Fmoc deprotection difficult; C, potential diketopiperazine formation minimized by use of short deprotection at a high flow rate, followed by quick coupling; D, coupling of Fmoc-Cys(Trt)-OPfp (4 equivalents) in DMF was carried out in the presence of HOBt (8 equivalents), as well as HBTU [8 equivalents; unnecessary from a chemical point of view, but convenient in terms of the configuration of reagent reservoirs on the PerSeptive (Framingham, MA) 9050 instrument]; E, drop in Fmoc-dibenzofulvene adduct peak height, as explained further in text. F, double coupling of indicated Fmoc-amino acid (i.e., no intervening deprotection before second coupling); G, consecutive Ile-Ile; and H, Arg-1 is notoriously difficult to introduce quantitatively. Despite three couplings, final side-chain protected Fmoc-$[30–51]_{Abu}$–PEG–PS resin remained ninhydrin-positive.

TABLE III

DEPROTECTION PROTOCOLS[a]

Protocol	Code	First wash (min)	Flow rate (ml/min)	Second wash (min)	Flow rate (ml/min)
Initial	*I*	6.0	30.0	2.3	30.0
DKP reduction	*D*	1.5	24.9	2.3	24.9
Standard	*S*	1.0	8.3	5.0	3.0
Extended	*E*	3.0	10.0	7.0	3.0

[a] The deprotection reagent was DBU–piperidine–DMF (1:10:39).

Deprotection/Coupling Cycles

As the BPTI chain grows on the polymeric support, Fmoc removal becomes increasingly difficult, and our current optimized protocol (Tables II and III) takes this into consideration. The problem was revealed by broadening of the Fmoc adduct peaks observed during continuous-flow monitoring, and confirmed by weaker Kaiser ninhydrin tests from separate manual syntheses.[9] Generally, for the first 30 cycles, the peaks are sharp (half-width ≤ 0.5 min); peaks from later synthesis cycles show some broadening and tailing but still return to baseline well within the total time allotted for Fmoc removal. Continuous-flow deprotection is now carried out with 2% 1,8-diazabicyclo[5.4.0]undec-7-ene (DBU) added to the standard piperidine–N,N-dimethylformamide (DMF) (1:4, v/v) mixtures; this reagent mixture is applied in two pulses for a total of 3–10 min depending on where in the sequence the step is applied. We find that our DBU–piperidine deprotection conditions lead to sharp peak shapes throughout the syntheses.

The key to success in BPTI chain assembly is to address properly the consistent "difficult" sequences. The points which require extra attention, namely, residues Gly-57 and Gly-56, Ser-47 to Phe-45, Tyr-35, and Tyr-10, were identified by real-time monitoring, as well as Edman sequential degradation preview analysis of completely assembled BPTI analogs.[8,9] At the very start of the synthesis, procedures are modified (see Tables II and III), due to sluggish deprotection of the starting Fmoc-Ala-O-PAC–PEG-(Nle)–PS and the risk of diketopiperazine formation subsequent to deprotection of the second residue (Gly-57). During the cycles for incorporation of Lys-46, Tyr-35, and Tyr-10, respectively, the Fmoc deprotection peak heights dropped by approximately 50%, 50%, and 25% with respect to the previous cycles. Because heights are not necessarily proportional to area, these monitoring data are just a qualitative gauge of the progress of the synthesis. The "difficult" points are not affected by alterations in Fmoc removal conditions (Table III) or coupling reagents {the N-[(1H-benzotriazol-1-yl)(dimethylamino)methylene]-N-methylmethanaminium hexafluorophosphate/1-hydroxybenzotriazole/N-methylmorpholine (BOP/HOBt/NMM) protocol reported originally[8] is equally effective as protocols involving the aminium salts N-[(1H-benzotriazol-1-yl)(dimethylamino)methylene]-N-methylmethanaminium hexafluorophosphate N-oxide (HBTU) or N-[(dimethylamino)-1H-1,2,3-triazolo[4,5-b]pyridin-1-ylmethylene]-N-methylmethanaminium hexafluorophosphate N-oxide (HATU)}, although use of the aminium salts is operationally easier and less expensive. Incorporation of Cys residues remains best done by using the preformed pentafluorophenyl (Pfp) esters in the presence of HOBt. Studies from our laboratory in model systems have shown that S-protected Cys racemizes extensively on activation with BOP or aminium salts, but the Pfp method is "safe."[27]

Cleavage and Purification

On completion of chain assembly, the dried protected peptide–resin is treated with piperidine–DMF (1 : 1, v/v), 30 min, to remove the N-terminal Fmoc group, followed by freshly prepared Reagent R[28]: trifluoroacetic acid–thioanisole–1,2-ethanedithiol–anisole (90 : 5 : 3 : 2) to cleave acidolyzable side-chain protecting groups and release chains from the support [~88 to 95% yield, as judged by amino acid ratios, with respect to Nle internal reference amino acid (IRAA), of the recovered resin]. Reagent R was originally optimized for removal of Pmc from Arg, and it has the added virtue of providing a strong reducing environment that allows Cys residues to be obtained in the free sulfhydryl form. The crude protein is precipitated with ether at 4°, and further purification occurs by C_4 reversed-phase high-performance liquid chromatography (HPLC). We currently use relatively flat gradients (0.1 to 0.3% CH_3CN per minute), which allow for higher loadings onto the column and better resolution of close peaks. The homogeneity requirement for material to be pooled after such an initial step is not as stringent as for later stages; hence, broader cuts can be tolerated to ultimately increase yields. Oxidation of analogs that require a single disulfide is best achieved by Tam's dimethyl sulfoxide (DMSO) method,[29] in part because of protein insolubilities at the optimal pH values required for other methods.

[27] Y. Han, F. Albericio, and G. Barany, *J. Org. Chem.* **62**, 4307 (1997).
[28] F. Albericio, N. Kneib-Cordonier, S. Biancalana, L. Gera, R. I. Masada, D. Hudson, and G. Barany, *J. Org. Chem.* **55**, 3730 (1990) and references cited therein.
[29] J. P. Tam, C.-R. Wu, W. Liu, and J.-W. Zhang, *J. Am. Chem. Soc.* **113**, 6657 (1991).

Analytical Characterization of Purified Bovine Pancreatic Trypsin Inhibitor Analogs

Stringent characterization of the species synthesized is crucial to ensure that the structural conclusions from the NMR experiments correspond to the intended covalent structure of the synthetic protein, rather than an impurity (including possible disulfide-linked intermolecular dimers). We routinely carry out an Ellman assay of free sulfhydryls before and after oxidation, monitor the purification steps by analytical reversed-phase HPLC, and carry out capillary zone electrophoresis (CZE) on individual fractions during preparative HPLC purification. Masses of all purified proteins are verified by ion electrospray mass spectrometry (ESMS), and compared to calculated values. ESMS gives sets of peaks with mass-to-charge (m/z) ratios derived from a parent mass and variable charges, $z = +1$ to as high as $+7$. That such peaks are convoluted to a single parent mass MH^+ is strong evidence for homogeneity, and the absence of peaks for $2\ MH^+/z$ (odd z) argues against the presence of dimers. Further verification that only intramolecular disulfide bonds have formed is provided by native gel electrophoresis.[30]

Nuclear Magnetic Resonance Methods: Overview and Rationale for Selective Stable Isotope Labeling

With pure chemically synthesized BPTI analogs in hand, characterization proceeds by NMR and other techniques. Covered below are methods for sample preparation, data acquisition, data processing, and interpretation, with an emphasis on aspects unique to the study of partially folded proteins. The major obstacle in characterizing these molecules by two-dimensional homonuclear NMR techniques is the extensive resonance overlap caused by loss of tertiary interactions. That overlap complicates sequential resonance assignments, which in turn makes it difficult to obtain quantitative information. Furthermore, it is common in partially folded conformations for ^1H resonances to be broader, and for there to be more of them, due to chemical exchange that is intermediate or slow on the NMR time scale. In uniformly ^{15}N-labeled partially folded BPTI, the presence of multiple conformations for each amide resonance gives rise to extra peaks and broadened resonances in some cases. Selective labeling, which is accomplished by chemical synthesis incorporating the appropriate labeled residues, has the advantage that spectra that are otherwise crowded by these

---

[30] D. Goldenberg and T. E. Creighton, *Anal. Biochem.* **138,** 1 (1984).

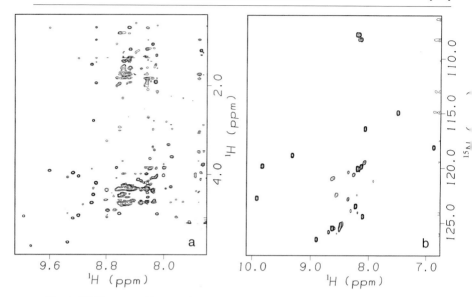

FIG. 2. NMR spectra (500 MHz) of partially folded $[14-38]_{Abu}$ at 1° and pH 5.0 in ^1H_2O. (a) $^1H-^1H$ NOESY spectrum and (b) $^{15}N-^1H$ HSQC spectrum of specifically labeled sample. Overlapping peaks between 8.2 and 8.8 ppm are due primarily to a disordered conformation in chemical exchange with a more ordered conformation. This overlap is much reduced in (b). Also, minor peaks that are not in the homonuclear NOESY are easily observed in (b).

extra peaks become much simpler, and allow extraction of quantitative information from resolved cross-peaks. The relative populations of conformations in slow chemical exchange, their temperature dependence, and their interconversion rates between conformations, can then be obtained from two-dimensional heteronuclear experiments. Another advantage of selective labeling is the capability to quantify intensities of minor conformations that are otherwise masked by more intense, crowded overlapping peaks in homonuclear or uniformly labeled spectra. For example, in partially folded $[14-38]_{Abu}$, specific labeling in strategic positions in the sequence provides ^{15}N-bound ^1H reporters whose cross-peaks are resolved clearly in the heteronuclear NMR spectra. Figure 2 shows a comparison of the $^1H-^1H$ NOESY (nuclear Overhauser effect spectroscopy)[31] spectrum showing the amide proton connectivities to the aliphatic protons with the $^{15}N-^1H$ HSQC (heteronuclear single quantum coherence)[32] spectrum of the same amide region. The overlap in $^{15}N-^1H$ spectrum with only nine

[31] J. Jeener, B. H. Meier, P. Bachmann, and R. R. Ernst, *J. Chem. Phys.* **71,** 4546 (1979).
[32] G. Bodenhausen and D. Ruben, *Chem. Phys. Lett.* **69,** 185 (1980).

residues labeled is much reduced, and the minor peaks of the disordered conformation of Phe-22 and Phe-33 are observed.

Sample Preparation and Initial Analysis

As is well known for native proteins, their solubilities, aggregation states, and stabilities are affected by concentration, temperature, pH, and buffer composition. Partially folded proteins are even more sensitive to changes in these parameters. We are often operating at concentrations that are very near the solubility limit. Hence, it becomes crucial to find an optimal set of conditions that can be used consistently for every sample.

Sample Handling. Spectra are acquired in buffers that are made in 90% $H_2O/10\%$ D_2O or 99.9% D_2O. In the latter case, the amide protons are all exchanged and not visualized by NMR. The lyophilized pure partially folded protein is dissolved at acidic pH (i.e., pH 2–4), and this solution is then slowly brought up to the appropriate pH by dialysis. It is important to note that the pH should not be adjusted directly by addition of base, because the protein aggregates irreversibly at higher pH. Dialysis (molecular weight cutoff 1000) is performed against water without added salt at pH 4.5–5.0 in the presence of 0.02% sodium azide. Deuterated sodium acetate at pH 5.0 is then added to a final concentration of 50 mM. Also, to minimize irreversible aggregation, the sample should never be left at room temperature for extended periods of time, nor shaken vigorously. Analytical HPLC is frequently used before and after several days of NMR acquisition to check for possible sample degradation with time.

Concentration Effects. In partially folded predominantly β-sheet proteins with exposed hydrophobic residues, aggregation is likely to occur due to hydrophobic stacking of the sheets, or because exposed hydrophobic clusters form intermolecular noncovalent bonds. The concentration of partially folded proteins in our NMR studies is $\leq$0.5 mM, less concentrated than solutions generally used for native proteins or for mostly helical partially folded proteins (the latter are generally studied at 2–3 mM). To rule out the possibility of aggregation effects, spectra are also obtained for samples diluted 10-fold; for our systems, line widths, chemical shifts, and nuclear Overhauser effects (NOEs) were found to be unchanged.[10] Pulse field gradient experiments are also performed to check for aggregation at the temperatures and concentrations under which these studies are carried out.[33]

Structure and Stability as Functions of pH. The effect of pH on the structure and stability of partially folded $[14–38]_{Abu}$ has been investigated

[33] H. Pan, G. Barany, and C. Woodward, *Prot. Sci.* **6,** 1 (1977).

by circular dichroism (CD).[11] The protein has some stable structure in the range of pH 4.5 to 7.0. NMR experiments for all BPTI analogs were conducted at pH 4.5–5.0 for two reasons: (1) to increase the solubility, as the protein has a tendency to aggregate above pH 6.5, and (2) to minimize amide proton–solvent exchange, which is faster at higher pH. Rapid exchange with water causes the disappearance or broadening of ^{1}H amide resonances. The pH of the sample is checked both before and after a set of experiments, and is found to be stable. With added buffer, the pH is set to pH 5.0, whereas without added buffer, the sample is adjusted to pH 4.5 with dilute HCl. There is no detected structural difference between these pH values.

Temperature Calibration. When the thermodynamics of folding are being investigated by NMR, it is important to calibrate the temperature under which NMR experiments are conducted, especially when unfolding temperature parameters are to be compared to those determined from other techniques. We have measured the temperature of the protein during NMR experiments in two ways. First, before and/or after the experiment, the protein sample is removed from the spectrometer and replaced by a sealed methanol sample for the temperature range 270–300 K and with ethylene glycol for the temperature range 300–350 K. The temperature of the equilibrated methanol is then determined from the difference between CH_3 and OH proton chemical shifts, which are linearly related to temperature.[34] Second, temperatures of the aqueous protein NMR samples are measured directly by means of a thermistor inserted after the sample was pulsed with the usual sequence for 1 or 5 hr. The sample tube is removed from the probe, and the temperature readings are recorded every 10 sec for 3 min; readings are extrapolated to zero time to give the actual temperature. We carried out this experiment to rule out the effect of spin–lock pulses and decoupling power on sample heating.[35] The temperature has also been calibrated for different levels of air flow. Because an increase in air flow was found to cause a slight decrease in the actual temperature, we are careful to keep this parameter constant in our experiments.

Temperature Effects. Thermal denaturation[11,35] of partially folded [14–38]$_{Abu}$ is discussed later in this article. From the NMR-derived thermal curves, the population of the more folded conformation is stable in the temperature range of 1–9° for most residues. Spectra for assignments and NOE measurements are hence obtained at this same temperature range, that is, 1–9°.

[34] A. L. Van Geet, *Anal. Chem.* **40,** 2227 (1968).
[35] E. Barbar, V. LiCata, G. Barany, and C. Woodward, *Biophys. Chem.* **64,** 45 (1997).

Nuclear Magnetic Resonance Acquisition Techniques as Adapted to Partially Folded Proteins

Solvent Suppression. When spectra are obtained in 90% H_2O/10% D_2O, the water signal is most commonly suppressed by presaturation. Shimming, irradiation frequency, and irradiation power must be adjusted and optimized for every new sample. However, extra care should be taken to suppress water in spectra taken on partially folded proteins. When the solvent resonance is irradiated for saturation, the spectra of denatured conformations may suffer from a loss of amide proton signal intensity due to saturation transfer via exchange with solvent. Denatured conformations lose amide signals to a greater extent than the more folded conformation, hence introducing error in the measurements of the relative intensities of both conformations. Furthermore, weak broad amide peaks corresponding to minor conformations may be lost completely.

For homonuclear NOESY spectra (Fig. 2a), we alternatively use a jump–return sequence.[36] This technique selectively excites the protein resonances, while leaving the equilibrium magnetization of the water protons undisturbed. Hence, saturation transfer from water to protein resonances is minimal. NOESY spectra are also acquired with a pulse field gradient for water suppression using a 3–9–19 pulse sequence.[37]

For heteronuclear HSQC spectra (Fig. 2b), the residual water resonance is suppressed by a high-power ^1H spin–lock purge pulse, applied for 0.7 msec, incorporated in the experiment.[38] Spin–lock purge pulses preferentially dephase the solvent magnetization by independently manipulating the protein and solvent resonances. Alternatively, in the "water flip-back" variation of HSQC, a z-axis field gradient selectively dephases the solvent signal without affecting the protein, or returns the water magnetization to equilibrium prior to acquisition.[39]

When spectra are obtained in D_2O, the residual solvent signal is presaturated using a very weak radio frequency (rf) field during the recycle delay that does not significantly attenuate the α-proton resonances that are close to the water signal.

T_1, T_2 Relaxation Effects. The relaxation rates of the longitudinal magnetization ($1/T_1$) determine the length of the recycle delay needed for acquisition. For unfolded flexible proteins, the rate of tumbling is greater than in

[36] P. J. Hore, *J. Magn. Reson.* **55,** 283 (1983).
[37] M. Piotto, V. Saudek, and V. Sklenar, *J. Biomol. NMR* **2,** 661 (1992).
[38] B. A. Messerle, G. Wider, G. Otting, C. Weber, and K. Wüthrich, *J. Magn. Reson.* **85,** 608 (1989).
[39] S. Grzesiek and A. Bax, *J. Am. Chem. Soc.* **115,** 12593 (1993).

a native protein of the same size. Therefore, a longer recycle delay, that is, 3 sec or longer rather than the usual 1 sec, is used to ensure that the magnetization has completely returned to equilibrium.

The digital resolution in two-dimensional spectra depends on the total acquisition time in the indirect dimensions ($t_{1,max}$). Increasing the resolution in a spectrum requires that t_{max} be increased by recording additional data points. This is productive only if the signals of interest have sufficiently long transverse relaxation times, T_2. However, in partially folded proteins with multiple conformations in chemical exchange, the apparent T_2 is very short. Therefore increasing the number of points in the indirect dimension is counterproductive because increments of t_1 only contribute noise to the spectrum without increasing the resolution between resonance signals. Therefore, 128 data points in t_1 are collected in our heteronuclear experiments, and then linear prediction (explained below) is used to increase the resolution.

Mixing Time. For through-space transfer experiments, the efficiency of mixing depends on the distance between interacting spins. Distance information can then be obtained from the intensity of the NOEs at different mixing times. In the presence of multiple conformations of varying structures that interconvert either fast or slow on the NMR time scale, extraction of distance information for structural determination from NOESY cross-peak intensities is less straightforward. This is because relatively few (out of many) conformations give rise to NOEs; weak NOE cross-peaks do not necessarily mean that the distances between interacting spins are large. To increase the chance of detecting meaningful signals, NOESY mixing times used for less ordered proteins should be longer than in native proteins of similar size. Thus, longer mixing times, for example, 200–400 msec, allow the assignment and identification of resonances that are in proximity but not very populated. Yet another reason for the use of longer mixing time is to build up NOEs between resonances that are either flexible or separated by a distance longer than 5 Å (the usual distance required to give rise to an NOE); in this way information about long-range contacts can be determined.

Decoupling Schemes. Heteronuclear spin decoupling is achieved by applying composite pulses. The power used for a certain length of time should be optimized without causing sample heating, amplifier droop, or other instrumental imperfections. The most commonly used sequences for spin decoupling are WALTZ-16[40] and GARP.[41] GARP decouples a wider spectral width than WALTZ-16 for the same power level. In folding/un-

[40] A. J. Shaka, J. Keeler, T. Frenkiel, and R. Freeman, *J. Magn. Reson.* **52,** 335 (1983).
[41] A. J. Shaka, P. B. Barker, and R. Freeman, *J. Magn. Reson.* **64,** 547 (1985).

folding thermodynamics, it is important to minimize sample heating that might be caused by strong rf pulses.

Nuclear Magnetic Resonance Data Processing

In our work, data are processed and analyzed on a Silicon Graphics work station using the program FELIX 95.0 (Biosym, San Diego) and/or Bruker UXNMR.

Linear Prediction. Linear prediction (LP) is performed only in the $t_1(^{15}N)$ dimension. Forward–backward LP calculation is used to obtain average LP coefficients for suppression of noise and artifacts. Typically, resolution in the indirect dimension (ω_1) was increased to extend the data from 128 or 200 points (originally recorded) to 400 points; 50 coefficients were chosen, and root reflection was used to ensure that calculated frequency components will decay as a function of time. For homonuclear spectra, resolution in ω_1 was increased to extend the data from 380 to 512 points.

Window Functions. For homonuclear spectra, data points are weighted by either a 30° or 60° shifted square sine bell in each dimensions, and zero filled to form $2K \times 2K$ real matrices. With 30° there is more resolution, but this comes at the expense of losing broad and less intense signals. We use a 30° shift to process spectra of species in intermediate chemical exchange, as a 60° shifted square sine bell gives very little resolution in the peaks in the random coil region. Figure 3 shows a comparison of the same data processed using 30° and 75° shifts. For heteronuclear spectra from which quantities were extracted and compared, data points were weighted with a Gaussian resolution enhancement window function with line broadening of -20 Hz and 0.08 degree of Gaussian character in each dimension, and zero filled to form $2K \times 1K$ real matrices. More resolution enhancement resulted in the loss of broad signals. Another window function used is sine bell with a 90° shift in both dimensions and a window size of $1K$ in the ω_1 dimension and $2K$ in the ω_2 dimension.

Baseline Correction. The baseline correction function used is ABL Flatten from FELIX 95.0; this automatically selects noise points and performs a baseline correction for each point. The algorithm is performed only in the ω_2 dimension. The input values for the noise level and the peak size in points are chosen to be 3 and 18, respectively, for a $2K$ data point set in ω_2.

Line Width Measurements. Line widths were obtained by curve fitting to an in-phase Gaussian. Because line widths in an exchanging system vary with temperature, cross-peak volumes that take into consideration changes in line width are more accurate than just peak heights for estimating relative

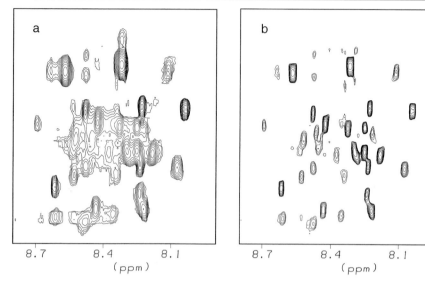

FIG. 3. Effect of data processing on quality of Y23A[14–38]$_{Abu}$ TOCSY spectrum. (a) Data were processed with a squared sine bell shifted 75°. (b) Data were processed exactly the same but shifted 30°.

populations of conformations. Broader lines also lead to a worse signal-to-noise (S/N) ratio (measured as the resonance peak height divided by the root-mean-square baseline noise in the spectrum), which in turn increases the error in peak height measurements.

Nuclear Magnetic Resonance Data Analysis Adapted to Partially Folded Proteins

When partially folded states of proteins are characterized by NMR, only a limited number of structural restraints per residue are measured. NMR parameters such as chemical shifts, NOE intensities, and coupling constants are averaged over an ensemble of conformations that are widely different in character but interconvert rapidly on the NMR time scale. These data lead to information about the structures of conformations preferred in the partially folded states, but an average single structure is not calculated.

Resonance Assignments. Backbone assignments can be determined from a combination of ^1H–^1H two-dimensional NOESY[31] and TOCSY (total correlation spectroscopy)[42] spectra recorded in H_2O, whereas TOCSY spec-

[42] C. Griesinger, K. Otting, K. Wüthrich, and R. R. Ernst, *J. Am. Chem. Soc.* **110,** 7870 (1988).

tra in D_2O are used to get side-chain assignments. Both NOESY and TOCSY spectra of partially folded $[14-38]_{Abu}$ have numerous cross-peaks well removed from the random coil envelope, as well as multiple unresolved peaks in the random coil region. For the resolved cross-peaks, assignments are made from combined analysis of TOCSY spin systems and NOESY primary and secondary structural connectivities, and long-range NOEs, following standard procedures.[43] To assign most of the clustered resonances, the well-resolved cross-peaks are taken as starting points for correlations (due to chemical exchange with the NH resonances) in TOCSY and ROESY (rotating frame Overhauser effect spectroscopy)[44] spectra.

For the unfolded variants, $[R]_{Abu}$, $[14-38]_{Ala}$, Y21A$[14-38]_{Abu}$, and Y23A$[14-38]_{Abu}$, the observed chemical shift dispersion is much less than for native BPTI or partially folded $[14-38]_{Abu}$. Partial assignments were determined from a combination of NOESY and TOCSY spectra acquired with variable temperature and mixing time in H_2O as well as D_2O.[12,14]

Several proteins were resynthesized with site-specific ^{15}N labels (Table I), and [15]N–[1]H HMQC–TOCSY (heteronuclear multiple quantum coherence–total correlation spectroscopy) and NOESY[45] two-dimensional spectra were recorded and interpreted to verify assignments and exchange cross-peaks.

Chemical Shifts. Changes in chemical shifts indicate that nuclei exist in different local magnetic environments. The lower chemical shift dispersion characteristic of partially folded and unfolded proteins indicates loss of stable tertiary structure. Signals in the aliphatic region (-0.5 to 3.5 ppm), downfield-shifted α-proton region (5 to 6.5 ppm), and aromatic region (6.5 to 8 ppm), while of lower intensity than in native BPTI, are nevertheless diagnostic of some nativelike conformations among the many conformations populated. Chemical shifts different from random coil values contain contributions from secondary and tertiary structural elements.[46]

Nuclear Overhauser Effects. The principal use of the NOE effect in NMR analysis of native proteins is to determine distances between pairs of protons. For partially folded or unfolded proteins, the NOEs are also affected by conformational averaging. Besides cross-relaxation, chemical exchange can also lead to cross-peaks in NOESY spectra. A cross-peak is observed between two sites if the exchange rate between the species is of the same order of magnitude as the mixing time. Chemical exchange peaks

[43] K. Wüthrich, "NMR of Proteins and Nucleic Acids." Wiley, New York, 1986.
[44] A. Bax and D. G. Davis, *J. Magn. Reson.* **63,** 207 (1985).
[45] A. Bax, R. H. Griffey, and B. L. Hawkins, *J. Magn. Reson.* **55,** 301 (1983).
[46] D. S. Wishart, B. D. Sykes, and F. M. Richards, *J. Mol. Biol.* **222,** 311 (1991).

and NOE peaks can be distinguished using ROESY spectra, as explained later.

Short-, medium-, and long-range NOEs are assigned in spectra recorded in H_2O. Short-range sequential and medium-range NOEs are indicative of secondary structure. Strong $d_{\alpha N}(i, i + 1)$ NOEs are indicative of extended β-sheet strands. Strong $d_{NN}(i, i + 1)$ in addition to medium-range $d_{\alpha N}(i, i + 3, 4)$ and $d_{\alpha \beta}(i, i + 3, 4)$ NOEs usually indicate an α-helical structure. Long-range backbone NOEs, namely $NH(i)-C\alpha H(j)$, $NH(i)-NH(j)$, and $C\alpha H(i)-C\alpha H(j)$, are indicative of β-sheet structure with residues i and j facing each other (Fig. 4).

Signals from the amide groups are used for backbone resonance assignments and secondary structure analyses, but these are absent in spectra recorded in D_2O. Nevertheless, useful information can be obtained from such experiments, particularly the assignment of α-proton resonances that overlap with the water signal in H_2O spectra. NOESY and TOCSY spectra

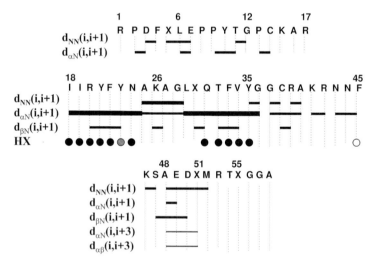

FIG. 4. Summary of sequential assignments for [14–38]$_{Abu}$ at 5°, pH 4.5, for the folded (f) conformation. The bars below the sequence indicate the observed NOE connectivities. Thicker bars show larger intensity. The strong $\alpha N(i, i + 1)$ NOEs, in the absence of $NN(i, i + 1)$, are indicative of β-sheet conformation. The medium-range $\alpha N(i, i + 3)$ and $\alpha \beta(i, i + 3)$ between residues 48 and 51 are indicative of the presence of the first turn of the C-terminal helix. In the bottom row, filled circles mark the amide peaks that exchange the slowest, whereas hatched and blank circles indicate medium and fast rates, respectively. The rest of the amide proton rates could not be measured either because the exchange was too fast or because of overlapping protons. In either case, they exchange much faster than NH's of the core residues. [Reproduced from *Biochemistry* **34**, 11423 (1995).]

are then used to get assignments of the side chains from correlations of α-protons to the rest of the chain. Aromatic/aliphatic NOEs obtained from D_2O NOESY spectra at varying mixing times (150–400 msec) give information related to tertiary structure and hydrophobic contacts. In $[14–38]_{Abu}$, a number of long-range NOEs reflecting tertiary structure are observed.[10] Their intensities reflect both the population and compactness of partially folded conformations.

Conformational Exchange. Conformations detected in partially folded proteins are less stable, and their unfolding is less cooperative compared to native states. Partially folded structures are in equilibrium with less ordered conformations. For most partially folded and denatured proteins studied to date, this equilibrium is fast or intermediate on the NMR time scale. However, $[14–38]_{Abu}$ has shown more complicated equilibria: an ensemble of more ordered states is in slow exchange with another ensemble of less ordered states, and within each ensemble conformers are in intermediate or fast exchange with each other. Slow chemical exchange arises when the same proton has a different chemical shift in each of two or more conformations that interconvert on the chemical shift time scale of 1 msec or longer, thereby giving rise to a distinct signal for each conformation.[47] The NMR chemical shift time scale is defined by the difference between frequencies of two exchanging resonances. Conformational exchange is observed in $[14–38]_{Abu}$ as NH–NH cross-peaks in the low-field region of TOCSY (Fig. 5), NOESY, and ROESY spectra. TOCSY and ROESY spectra are obtained with varying spin–lock mixing times and low spin–lock power; a weak spin–lock field with a 90° pulse of 34 μsec is used in TOCSY experiments. The exchange cross-peaks are assigned by correlation to previously assigned resonances of the more ordered conformation and verified by identification of the respective spin systems in full TOCSY spectra. ROESY spectra can confirm that the cross-peaks are due to chemical exchange; these experiments show exchange cross-peaks having the same sign as diagonal peaks, whereas those cross-peaks due to the NOE effect are of opposite sign.[44]

To verify assignments and exchange cross-peaks, and to identify minor populations, $^{15}N–^1H$ HMQC–NOESY spectra were recorded on site-specific, ^{15}N-labeled protein samples. The presence of substantially more cross-peaks than the number of ^{15}N-labels reflects multiple conformations. A typical $^{15}N–^1H$ HMQC–NOESY spectrum (Fig. 6) contains three types of cross-peaks: (1) NOEs, (2) exchange peaks of the same NH in two conformations (autopeaks; these are the most intense), and (3) cross-correlation peaks of the autopeaks (unlabeled peaks at the vertices of dashed

[47] J. Sandström, "Dynamic NMR." Academic Press, New York, 1982.

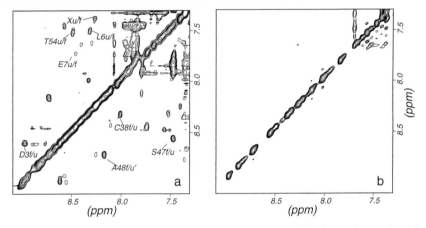

FIG. 5. TOCSY spectra of (a) $[14-38]_{Abu}$ and (b) native BPTI, showing exchange of amide protons between two conformations in the partially folded protein. Spectra were acquired in 90% ^1H_2O/10% ^2H_2O with 60 msec mixing time and low spin–lock power. The cross-peaks shown are due to exchange between more folded (f) and more disordered (u) states of the protein. Some assigned cross-peaks are labeled. These were also observed in ROESY and NOESY spectra. No NH–NH exchange peaks are observed in equivalent spectra of the native protein. [Reproduced from *Biochemistry* **34**, 11423 (1995).]

lines). For example, there are two autopeaks for 40 NH, one for a partially folded (f) conformation (7.5 ppm/117.9 ppm) and a second for a partially disordered or unfolded (u) conformation (8.5 ppm/125 ppm). The dashed lines connecting the 40 NH peaks intersect at their cross-correlation peaks (7.5 ppm/125 ppm and 8.5 ppm/117.9 ppm). The observation of cross-correlation peaks depends on the mixing time during which exchange is allowed to develop. By varying the mixing times, the interconversion rates between conformations can be estimated from intensities of cross-correlation and exchange peaks.[48]

Relative Line Widths. Nuclear magnetic resonance signals corresponding to partially folded states typically have increased line widths as compared to either fully unfolded or native (fully folded) states. The major contributors to line broadening are intrinsic relaxation rate constants as well as the presence of multiple, slowly interconverting conformations. In an exchanging system, line widths are affected by the chemical shift differences between conformations, the relative populations, and the exchange rates between conformations. For example, in $[14-38]_{Abu}$, which is an ensemble of conformations in slow and intermediate chemical exchange, the line widths of NH resonances are considerably broader than for native or fully unfolded

[48] E. Barbar, G. Barany, and C. Woodward, in preparation (1997).

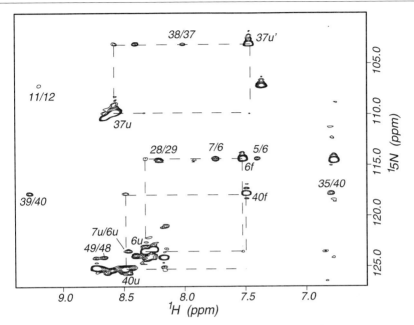

FIG. 6. ^{15}N–^{1}H HMQC–NOESY spectrum of specifically labeled ^{15}N-[14–38]$_{Abu}$. Auto-peaks and exchange cross-peaks of some residues are connected by dashed lines. Cross-peaks labeled with two numbers are due to NH–NH sequential NOEs, except 35/40 which is an NOE between 35CεH at 6.80 ppm and NH of 40 at 7.48 ppm. Peaks labeled with one number are autopeaks. Dashed lines connect autopeaks and cross-correlation peaks. The residues labeled in this sample are different from those shown in Fig. 2b. [Reproduced from *Biochemistry* **34**, 11423 (1995).]

BPTI. Further, in less ordered conformations of [14–38]$_{Abu}$, the line widths among NH resonances vary depending on the position of the residue in the protein sequence. In one-dimensional traces of HSQC spectra (Fig. 7), the line width of the signal corresponding to the disordered conformation of residue 22 (40 Hz) is much broader than that of the disordered conformation of residue 6 (16 Hz); this implies that residue 22 is in a more structured conformation than residue 6.

Hydrogen Exchange Rates. Hydrogen isotope exchange rates identify the residues that are either least accessible to water and/or that are involved in intramolecular hydrogen bonds.[49] In folded native proteins, some of the amide hydrogens that are in the core of the protein exchange extremely slowly, on the order of weeks and months. For partially folded [14–38]$_{Abu}$,

[49] C. Woodward, I. Simon, and E. Tüchsen, *Mol. Cell. Biochem.* **48**, 135 (1982).

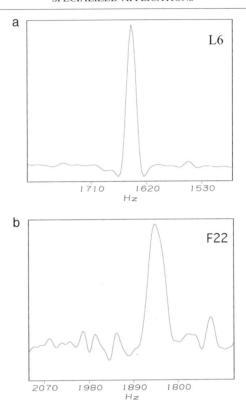

FIG. 7. Line widths of the peaks that correspond to the (u, u′) conformations vary among residues. Using the HSQC spectrum of Fig. 2b, at 9°, one-dimensional slices through the unfolded conformations of (a) L6 and (b) F22 show a sharper signal for L6u (line width of 16 Hz) than for F22u′ (line width of 40 Hz).

amide exchange is much faster, but rates for the residues corresponding to the most stable part of the protein can still be measured. It is difficult to measure the amide exchange rates in individual conformation(s) with stable structure, because the experiments provide an average over the whole protein population.

Rates for $[14–38]_{Abu}$ were obtained by measuring peak intensities (for well-resolved peaks) versus time in a series of one-dimensional spectra at pH 4.5 and 1°.[10] Acquisition of two-dimensional spectra to get information on the unresolved peaks was not possible because of the limited solubility of the protein, as well as the requirement to carry out measurements at pH 4.5 (needed for a folded conformation but well above the pH_{min} of slowest exchange, estimated at pH 3.5). Pseudo-first-order rate constants are obtained from nonlinear least-squares fit of an exponential rate equation to experimental data. The slowly exchanging NH's in [14–38]$_{Abu}$ have rate constants that are smaller by factors of 10–40 than those in equivalent small peptides. In [14–38]$_{Abu}$, 21 NH and 22 NH have the slowest exchange rates, indicating that these are in the most stable part of the protein. However, 23 NH exchanges much more rapidly, suggesting that the native interaction between 23 NH and Asn 43 side chain is greatly diminished (Fig. 4). For the fast-exchanging backbone amide protons that are not resolved in one-dimensional spectra, exchange rates can be obtained in H$_2$O from saturation transfer with the solvent. Two identical spectra should be obtained: one with water saturation and one without.[50] The exchange rates are then estimated from the loss of peak intensity due to exchange with solvent during saturation. There are other techniques described in the literature[51–53] that are more accurate for measuring fast exchanging protons, but they are optimal for concentrated samples and not applicable in our case due to the limited solubility.

Case Study: Conclusions from Studies on [14–38]$_{Abu}$ and Other Bovine Pancreatic Trypsin Inhibitor Variants with One Disulfide Bridge

Using the combination of chemical synthesis and NMR tools described throughout this article, we have studied an analog of BPTI, termed [14–38]$_{Abu}$, in which four of the Cys residues are replaced by Abu while the single disulfide bridge between residues 14 and 38 is retained (near the trypsin binding site, distant from the hydrophobic core). We find that [14–38]$_{Abu}$ has the characteristics of a highly ordered, β-sheet molten globule and undergoes cooperative, temperature-induced denaturation with a midpoint of 19° when monitored by CD and 15° when monitored by NMR (Fig. 8). Thermal denaturation was measured by CD as a decrease in negative molar ellipticity as a function of temperature at 220 nm. With NMR, the disappearance of the ^{15}N–^{1}H cross-peaks corresponding to the more folded conformation was measured as a function of temperature. The difference between thermal unfolding curves reported by NMR and equivalent curves monitored by CD indicates that there is a difference between macroscopic and microscopic reporters and implies deviation from simple two-state behavior.

[50] N. R. Krishna, D. H. Huang, and J. D. Glickson, *Biophys. J.* **26**, 345 (1979).
[51] G. Gemmecker, W. Jahnke, and H. Kessler, *J. Am. Chem. Soc.* **115**, 11620 (1993).
[52] R. W. Kirwacki, R. B. Hill, J. M. Flanagan, J. P. Caradonna, and J. H. Prestegard, *J. Am. Chem. Soc.* **115**, 8907 (1993).
[53] S. Koide, W. Jahnke, and P. E. Wright, *J. Biomol. NMR* **6**, 306 (1995).

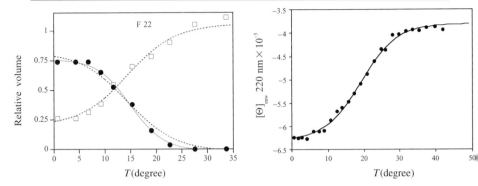

FIG. 8. Temperature unfolding curves for $[14-38]_{Abu}$ as determined by NMR (left) and CD (right). Whereas CD monitors the average temperature dependence of the whole protein, the microscopic NMR probe monitors the temperature dependence of specific residues in specific conformations. For residue F22, the "(f)-curve" is the variation of (f) cross-peak volumes with temperature (circles) and the "(u)-curve" is the variation of (u) peak volumes with temperature (squares). Solid lines are fits to the (f)-curve. Dashed lines represent the simultaneous fit to (f)- and (u)-curves.

In $[14-38]_{Abu}$, each region of the molecule undergoes independent motions that result in an ensemble in which residues 18–24 and 29–35 are in a nativelike antiparallel sheet over 75% of the time, residues 44–48 are in a nativelike conformation 25–40% of the time, and residues 1–17 and 48–57 fluctuate between two nonnative conformations. Gly-37 NH samples three slowly interconverting conformations. The stable structure in partially folded $[14-38]_{Abu}$ primarily involves residues not in the vicinity of the 14–38 disulfide cross-link; this suggests that the first native disulfide does not play a role in enthalpic stabilization of the early intermediates. By arguments summarized elsewhere,[9–12,14] we conclude that the decrease in the chain entropy of extended species on formation of any native disulfide bond leads to rapid, cooperative formation of an interconverting ensemble in which the slow exchange core is more stable while the rest of the molecule is more mobile.

In $[14-38]_{Abu}$, the presence of two conformations in slow chemical exchange makes it possible to characterize the stability and dynamics of both conformations. Thermodynamic analysis of $^{15}N-^{1}H$ exchange peak volumes as a function of temperature in the range 1–35° indicates that partially folded $[14-38]_{Abu}$ undergoes local segmental motions as well as cooperative global unfolding. Segmental motions are monitored by exchange cross-peaks at low temperature (1–9°), where different regions of the protein vary in the extent to which they are disordered before unfolding of the core. The relative abundance of partially folded versus disordered conformations changes throughout the molecule, indicating that various regions of the partially folded protein are disordered to different extents prior to the onset of thermal denaturation. As the temperature is raised, the partially folded ensemble undergoes global unfolding with a T_m around 15°, and the conformations sampled include globally unfolded species.

Nuclear magnetic resonance analysis shows that Y21A[14–38]$_{Abu}$, Y23A[14–38]$_{Abu}$, and [R]$_{Abu}$ are all mixtures of conformations, with the predominant form being unfolded. Another variant, [14–38]$_{Ala}$ in which four of the Cys residues are replaced by Ala, has the major conformation unfolded as well.[54–56] Hence, no T_m is obtained for these species. However, information about some nativelike structure in the minor conformations is obtained from (a) chemical shift of α-protons and (b) NOEs between aromatic protons and downfield-shifted α-protons, consistent with a minor conformation of possible β-sheet structure. All of the above species are equilibrium ensembles of conformations of molecules that are either in slow or intermediate exchange on the NMR time scale.[12,14]

This article illustrates how chemical synthesis coupled with rigorous NMR analysis can provide insights into protein folding. Generalizable advantages of chemical synthesis include the capability to introduce stable isotope labels at specific positions, and to incorporate nongenetically coded amino acid residues, for example, Abu, that are better models of the protein packing. (Note that the addition of four methylene groups, for example, [14–38]$_{Ala}$ versus [14–38]$_{Abu}$, is enough to convert an unfolded structure to a partially folded one.) In addition, partially folded species like the ones we are interested in are degraded rapidly by cellular proteases, with the result that recombinant expression systems will be a less successful method for producing these species.

Acknowledgments

We thank Drs. Yvonne Angell and Michael Hare for helpful comments and critical reading of the manuscript. Our experimental work was supported by National Institutes of Health Grants GM 26242 (C. W.) and GM 51628 (G. B. and C. W.) and an NIH postdoctoral fellowship GM 17341 (E. B.).

[54] M. Dadlez and P. S. Kim, *Nat. Struct. Biol.* **2,** 674 (1995).
[55] M. Dadlez and P. S. Kim, *Biochemistry* **35,** 16153 (1996).
[56] C. Gross, G. Barany, and C. Woodward, unpublished observation (1996).

[28] Multiple Antigen Peptide System

By JAMES P. TAM and JANE C. SPETZLER

Introduction

Multiple antigen peptide (MAP) was introduced in 1988[1] as a multimerization system to present peptides in a clustered, branched dendrimeric format. Although this system was first intended for peptide immunogens and antigens,[1–3] applications have broadened to areas such as inhibitors, artificial proteins, affinity purifications, and intracellular transport.[4]

A central component defining branched architectures is the core matrix that multimerizes peptides and gives the MAP a cascade type of arrangement (Fig. 1). The core matrix contains two or three levels of geometric and asymmetrical lysine (Lys) branches (Fig. 1E) or symmetrical β-Ala-Lys branches (Fig. 1F).

Lysine is the most commonly used amino acid in the core matrix because it has two ends, the α- and ε-amino groups, available for the branching reactions. When Lys is used as a repeating unit in an octameric MAP, the core matrix is asymmetrical, with a long arm consisting of the side chain and a short arm consisting of the α-amino group that varies from 8 to 18 carbon atoms from the first branched C_α atom. We have also designed a symmetrical core matrix (Fig. 1F) consisting of Lys(β-Ala) as a repeating unit.[5] Asymmetrical or symmetrical MAPs are different from conventional polylysyl conjugates because the lysyl core matrices are oligomeric, devoid of cationic side chains, and contain only amide bonds. It is also different from the conventional multimerization approach which gives end-to-end polymers of varying sizes.

The advantages of MAPs are that they consist of nearly pure antigens and are, therefore, immunologically focused. An octameric MAP of a 15-mer peptide would have peptide antigen accounting for >90% by weight while the branching Lys core is small (<10% by weight) and immunologically silent.[2,6] This is in contrast to conventional, large, immunologically

[1] J. P. Tam, *Proc. Natl. Acad. Sci. U.S.A.* **85**, 5409 (1988).
[2] D. N. Posnett, H. McGrath, and J. P. Tam, *J. Biol. Chem.* **263**, 1719 (1988).
[3] J. P. Tam and Y.-A. Lu, *Proc. Natl. Acad. Sci. U.S.A.* **86**, 9084 (1989).
[4] J. P. Tam, *in* "Peptides: Synthesis, Structures and Applications" (B. Gutte, ed.), p. 93. Academic Press, San Diego, 1995.
[5] W. Huang, B. Nardelli, and J. P. Tam, *Mol. Immunol.* **31**, 1191 (1994).
[6] G. Del Giudice, C. Tougne, J. A. Louis, P.-H. Lambert, E. Bianchi, F. Bonelli, L. Chiappinelli, and A. Pessi, *Eur. J. Immunol.* **20**, 1619 (1990).

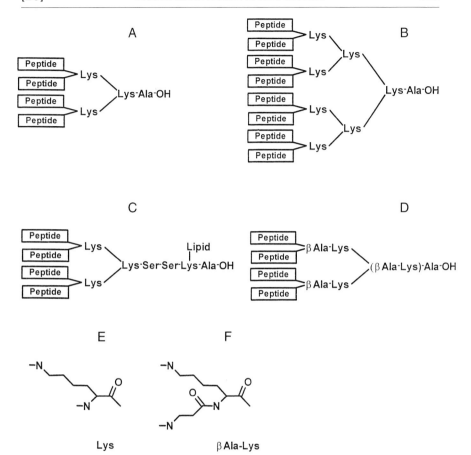

FIG. 1. Different designs of a cascade of peptide dendrimers of MAPs using Lys as the building unit of the core matrix. (A) Tetravalent core matrix, (B) octavalent core matrix, (C) tetravalent MAP with lipid attachment, (D) symmetrical MAP with β-Ala-Lys as a building unit of the core matrix, (E) asymmetrical core matrix, (F) symmetrical core matrix with β-Ala-Lys branches.

active protein carriers whose conjugated peptide antigens account for <20% of their weight and which are randomly distributed on the protein carrier. As a result, multimerization in the MAP design can overcome the lack of effectiveness of linear peptides in raising antibodies.[6–8]

[7] J. P. Tam and F. Zavala, *J. Immunol. Methods* **124,** 52 (1989).
[8] K. Kamo, R. Jordan, H.-T. Hsu, and D. Hudson, *J. Immunol. Methods* **156,** 163 (1992).

TABLE I
METHODS FOR PREPARING MULTIPLE ANTIGEN PEPTIDES WITH DIVERSE ORIENTATIONS
AND STRUCTURES

Bond[a]	Peptide	Polarity[b]	Core	Chemistry	Refs.
Direct approach					
Amide	Linear	C	Asymmetrical	Boc or Fmoc	1, 10
Amide	Two chains	C	Asymmetrical	Boc and Fmoc	3
Amide	Linear	C	Symmetrical	Fmoc	5
Amide	Linear	C	Lipidated	Fmoc	5, 10
Amide	Cyclic	S	Asymmetric	Boc or Fmoc	11
Modular approach					
Thioether	Linear	S	Asymmetrical	Boc or Fmoc	12
Thioether	Linear	S	Lipidated	Fmoc	13
Thioether	Cyclic	S	Asymmetrical	Boc	14
Thiazolidine	Linear	N	Asymmetrical	Boc and Fmoc	15–17
Thiazolidine	Cyclic	C	Asymmetrical	Fmoc	18
Oxime	Linear	N	Asymmetrical	Boc and Fmoc	16
Hydrazone	Linear	N	Asymmetrical	Boc and Fmoc	16, 17

[a] Bond between the peptides and the MAP core matrix.
[b] Attachment of peptide to core matrix in the direction of C → N (C), N → C (N), or side chain (S).

For most studies, a tetra- or octameric MAP is sufficient. However, the number of branches is largely dependent on the number of amino acid residues. With peptides of >15 amino acids, no real advantage exists for using an octameric over a tetrameric MAP.[9] In this article, we describe various methods for the preparation of MAPs containing one (homomeric) and two or more linear peptides (heteromeric MAPs) in various orientations as well as those containing cyclic structure (McAP, multiple cyclic antigen peptide; Table I).[10–18]

[9] M. J. Francis, G. Z. Hastings, F. Brown, J. McDermed, Y.-A. Lu, and J. P. Tam, *Immunology* **73**, 249 (1991).
[10] J. P. Defoort, B. Nardelli, W. Huang, D. D. Ho, and J. P. Tam, *Proc. Natl. Acad. Sci. U.S.A.* **89**, 3879 (1992).
[11] J. C. Spetzler and J. P. Tam, *Pept. Res.* **9**, 290 (1996).
[12] Y.-A. Lu, P. Clavijo, M. Galantino, Z.-Y. Shen, and J. P. Tam, *Mol. Immunol.* **28**, 623 (1991).
[13] J. P. Defoort, B. Nardelli, W. Huang, and J. P. Tam, *Int. J. Pept. Protein Res.* **40**, 214 (1992).
[14] L. Zhang and J. P. Tam, *J. Am. Chem. Soc.* **119**, 2363 (1997).
[15] C. Rao and J. P. Tam, *J. Am. Chem. Soc.* **116**, 6975 (1994).
[16] J. Shao and J. P. Tam, *J. Am. Chem. Soc.* **117**, 3893 (1995).
[17] J. C. Spetzler and J. P. Tam, *Int. J. Pept. Protein Res.* **45**, 78 (1995).
[18] T. D. Pallin and J. P. Tam, *J. Chem. Soc., Chem. Commun.*, 2021 (1995).

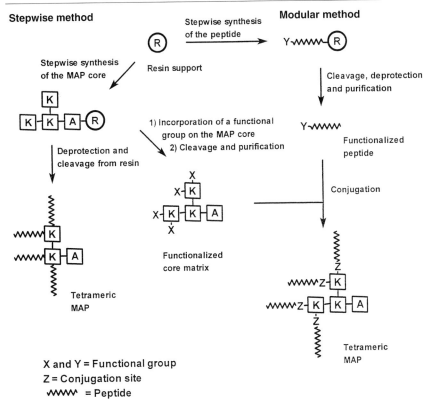

FIG. 2. Stepwise and modular approaches in the preparation of MAPS.

Approaches: Direct Stepwise versus Modular Approaches for Synthesis of Multiple Antigen Peptides

Two general approaches exist (direct versus modular) for preparing MAPs (Fig. 2). Most syntheses are achieved by a direct approach using a stepwise method in which MAPs are prepared sequentially, which is similar in practice to the synthesis of a single-chain peptide on a solid support.[19] Because MAPs are macromolecules and because side products accumulated during the assemblage stage are amplified, thus leading to microheterogenicity, considerable effort is required for purification. For most immunizations, this heterogenicity does not appear to play a significant role as judged from the many successful applications.[4] However, for clinical or other non-immunological applications, chemically unambiguous MAPs are desirable.

[19] R. B. Merrifield, *J. Am. Chem. Soc.* **85**, 2149 (1963).

Thus, we have developed an indirect approach for the modular methods of synthesis for MAPs by ligating purified unprotected peptide segments to a MAP core matrix.[20] Both approaches are described herein (Table I).

Stepwise Method for Preparing Multiple Antigen Peptides

The stepwise preparation of MAPs is convenient and is a repetitive, continuous operation.[1] It starts at the C terminus with a small amino acid, and then continues to build the core with a diprotected Boc-Lys(Boc) in Boc (*tert*-butyloxycarbonyl) chemistry[19] or Fmoc-Lys(Fmoc) in Fmoc (9-fluorenylmethyloxycarbonyl) chemistry[21,22] to reach the desired branching level.[1,3] The selected peptide antigen is then sequentially elongated to the lysinyl core matrix on the resin to form the desired MAPs. This stepwise method produces peptide antigens with a C → N orientation.

tert-Butyloxycarbonyl Chemistry in Stepwise Synthesis of Homomeric Multiple Antigen Peptides

The Boc–benzyl chemistry is a strategy based on the acid-differential lability of the temporary α-amino protecting group (Boc) and the permanent side-chain protecting group (benzyl alcohol derivatives). Thus, the temporary N^α-amino Boc group is subjected to repetitive deprotection cycles, and the side-chain benzyl protecting groups and the peptide linkage to the resin support are removed at the end of the synthesis by a strong acid such as HF[23] and trifluoromethanesulfonic acid (TFMSA).[24] A typical procedure for the Boc chemistry is shown in Table II.

General Comments

1. All stepwise, solid-phase MAP syntheses can be accomplished manually or assisted by an automated synthesizer on a simple amino acid resin such as Boc-Gly-, Boc-Ala-, or β-Ala-OCH$_2$-4-hydroxymethylphenylacetic acid (Pam) resin. *p*-Methylbenzhydrylamine can also substitute for the Pam resin. Boc-β-Ala-OH is often used because it can function as an internal standard to calculate the molar ratio of other amino acids present in the MAP during amino acid analysis.

[20] J. P. Tam and J. C. Spetzler, *Biomed. Pept., Proteins Nucleic Acids* **1**, 123 (1995).

[21] C.-D. Chang and J. Meienhofer, *Int. J. Pept. Protein Res.* **11**, 246 (1978).

[22] E. Atherton, H. Fox, D. Harkiss, C. J. Logan, R. C. Sheppard, and B. J. Williams, *J. Chem. Soc., Chem. Commun.*, 537 (1978).

[23] J. P. Tam, W. F. Health, and R. B. Merrifield, *J. Am. Chem. Soc.* **105**, 6442 (1983).

[24] H. Yajima, N. Fujii, H. Ogawa, and H. Kawatani, *J. Chem. Soc., Chem. Commun.*, 107 (1974).

TABLE II
SCHEDULE FOR *tert*-BUTYLOXYCARBONYL–BENZYL CHEMISTRY

Step	Reagent	Volume (ml)/g resin	Number of operation × min
1	DCM	15	2 × 1
Deprotection			
2	TFA prewash	15	1 × 1
3	TFA deprotection	15	20 × 1
4	DCM	15	5 × 1
Neutralization			
5	DIEA	15	1 × 1
6	DIEA	15	5 × 1
7	DCM wash	15	5 × 1
Coupling[a]			
8	Boc-AA in DCM	10	1 × 1[b]
9	DCC in DCM	1	45 × 1[b]
10	Additional DMF	12	15 × 1[b]
Wash			
11	DMF	15	1 × 1
12	DCM	15	4 × 1
Monitor by ninhydrin test[c]			

[a] Typical coupling procedure for all amino acids except Asn, Gln, and Arg. For those three amino acids, use DCC/HOBt coupling by adding equal molar ratio of DCC and HOBt at step 9. A longer reaction time may be required since the activation step to form the amino acid HOBt ester is slow.

[b] Do not drain. The purpose of adding DMF is to increase the polarity of the solvent in the coupling mixture to change the rate and solvation properties of the resin.

[c] If ninhydrin test shows <99.7% completion, repeat steps 5–12 and use preformed symmetrical anhydride method and DMF as the solvent for coupling.

2. A low loading of the resin, ranging from 0.05 to 0.4 mmol/g, is recommended. This is necessary because the loading is geometrically increased in accordance with the number of branchings. For example, with an initial loading of 0.15 mmol/g, a tetravalent MAP would result in about 0.6 mmol/g and an octavalent MAP about 1.2 mmol/g. Because most commercially available resins have a high level of loading, ranging from 0.5 to 1.0 mmol/g, coupling the second amino acid at 0.15 mmol/g to a highly preloaded Boc-aminoacyl resin such as Boc-Ala-OCH$_2$-Pam resin, followed by permanently blocking the remaining amino groups by acetic anhydride, is desirable. Similarly, the second Boc-amino acid at about 0.15 mmol/g can be loaded to the Ala-OCH$_2$-Pam resin, followed by acetylation. These procedures usually yield a resin with a loading of about 0.1 mmol/g, assuming a 67% coupling yield.

3. Three common, commercially available resin supports used for the synthesis of MAPs in Boc chemistry are (*i*) chloromethyl resin (often referred to as the Merrifield resin), (*ii*) preloaded Boc-aminoacyl-OCH$_2$-Pam resin, and (*iii*) *p*-methylbenzhydrylamine resin. Resins *i* and *ii* give a MAP with a C$^\alpha$-COOH while resin *iii* gives a C$^\alpha$-carboxyamide.

4. The protecting group scheme is as follows. Use the Boc group for the α-NH$_2$ terminus and the following side-chain protection groups for the trifunctional amino acids: Asp(OBzl), Glu(OBzl), Lys(ClZ), Arg(Tos), Ser(Bzl), Thr(Bzl), Tyr(BrZ), Cys(Acm), Trp(For), and His(Dnp). Asn and Gln are used unprotected.

5. Deprotection by trifluoroacetic acid (TFA, 20 min) is preceded by one TFA prewash (2 min) and neutralization by *N*,*N*-diisopropylethylamine (DIEA, 1 min × 3) in *N*,*N*-dimethylformamide (DMF). The coupling is mediated with a 3-fold excess of *N*,*N*-dicyclohexylcarbodiimide (DCC) and 1-hydroxybenzotriazole : 7-aza-1-hydroxybenzotriazole   (HOBt : HOAt, 1 : 4, mol/mol) in DMF (Table II).

6. After completion of the synthesis, a sequence of deprotecting procedures is effected sequentially as follows: (a) when His(Dnp) is present, thiolysis of the MAP–resin is used to release the Dnp protecting group from His(Dnp), either under neutral (1 *M* thiophenol in DMF at 50° for 24 hr) or basic conditions (1 *M* mercaptoethanol in DMF and 5% DIEA thrice for 0.5 hr); (b) when Trp(For) is present, aminolysis by 20% piperidine or hydroxylamine in DMF for 2 hr is used to release the formyl group from Trp(For). However, this procedure is susceptible to aspartimide formation with the following sequences of Asp-X where X is Gly, Ala, or Asn. In general, Trp(For) can be released by low–high HF provided that it is performed properly. The *N*$^\alpha$-Boc is released by 50% TFA containing 1% dimethyl sulfide (DMS) and 0.5% dithioethane, neutralized with 10% DIEA in DMF. The resin is then washed and dried *in vacuo* prior to the HF cleavage. Low HF (HF : *p*-cresol : DMS, 25 : 10 : 65, v/v/v) is performed for 2 hr at 0°. Then HF is evaporated *in vacuo*, and new HF is added to give HF : *p*-cresol (9 : 1, v/v; high HF) for 1 hr at 0°.

7. The low–high HF method is a two-step procedure to remove the MAP from the resin support. However, often laboratories shortcut this procedure to only high HF to save time. Both methods can give the desired MAP, but the low–high HF procedure is highly recommended. The crude peptide is then washed with cold ether/mercaptoethanol (99 : 1, v/v) to remove *p*-thiocresol and *p*-cresol and extracted into 8 *M* urea in 0.1 *M* Tris-HCl buffer, pH 8.0. To remove the remaining aromatic by-products generated in the cleavage step, MAPs are dialyzed (SpectraPor 6, Fisher Scientific, molecular weight cutoff 1000) in 8 *M* urea and then in 0.1 *M* acetic acid twice for 5–6 hr to remove the urea. The MAPs are lyophilized

in water three times to remove the acetic acid. *Note:* It is important to have the base treatment after the HF treatment to revert two major types of side products to the desired products. The first type is the *O*- to *N*-acyl migration of the Ser-X sequence. The second is the hydrolysis of glutamyl esters such as Glu(*O-p*-cresol) or Glu(*S-p*-cresol) to Glu in the presence of H_2O_2, at approximately pH 8 (protection for Cys and Met are needed). The latter side reaction occurs when the *p*-cresol or *p*-thiocresol combines with the glutamyl acylium ion to form the phenyl ester or thioester. When anisole is used as a scavenger, the side product is a ketone and is not reversible. This reaction points to the advantage of *p*-cresol over anisole as a scavenger in the strong acid cleavage reactions.

8. Multiple antigen peptides prepared from the stepwise synthesis are generally purified by gel filtration or C_8 reversed-phase high-performance liquid chromatography (RP-HPLC). Because of the microheterogeneity present in MAPs obtained by this method using either Boc or Fmoc chemistry, broad peaks are often observed. This is in contrast to those MAPs obtained from modular synthesis, which provides sharp peaks with high homogeneity. The usual spectrum of analytical methods for peptides and proteins can be applied to the characterization of MAPs, including amino acid analysis, mass spectrometric analysis, sodium dodecyl sulfate–polyacrylamide gel electrophoresis (SDS–PAGE), and capillary electrophoresis.

9-Fluorenylmethyloxycarbonyl Chemistry in Stepwise Synthesis of Homomeric Multiple Antigen Peptides

Where Boc chemistry is acid-driven, Fmoc chemistry is base-driven. Fmoc chemistry also does not require the use of very strong acid to liberate the side-chain protecting groups and the resin support linkage. As a result, the resins, protected amino acids, repetitive cycles, and cleavage methods are different from Boc chemistry. An advantage of the base-driven chemistry is that it is obviates the need for the neutralization step and simplifies the repetitive cycles. Furthermore, the final cleavage of the peptide from the resin support only requires a mild acid such as TFA for the Fmoc-*tert*-butyl chemistry, whereas the Boc-benzyl chemistry uses HF or TFMSA. For those laboratories that are not equipped with or have no access to a HF apparatus, this procedure may be preferable.

General Comments

1. Comments 1 and 2 from the preceding section regarding the low loading on resin supports and the use of β-Ala as an internal amino acid are applicable for this chemistry.

2. All amino acids are protected as N^α-Fmoc derivatives. The side-chain protecting scheme follows the conventional approach of using *tert*-butyl-based derivatives for Asp, Glu, Ser, Thr, Tyr, Lys, and Trp; trityl for Asn and Gln; and Pmc for Arg.

3. Two common types of TFA-labile resins are being used. To generate a peptide with a free carboxylic acid, the *p*-alkoxybenzyl alcohol resin (Wang resin) is used. To produce a peptide with an α-carboxamide, alkoxy-substituted benzhydrylamine resins such as 4-[(2′,4′-dimethoxyphenyl)-Fmoc-aminomethyl]phenoxy resin (Rink resin) and tris(alkoxy)benzhydry-lamine resin (Breipohl resin) are used. These types of resins are readily available from commercial sources.

4. Repetitive deprotection is effected by two cycles of 20% piperidine in DMF for a total of 20 min (Table III). Because DMF is the primary solvent for this chemistry, a very good grade of DMF with a very low limit of dimethylamine must be used to minimize side reactions. Alternatively, *N*-methylpyrrolidone (NMP), which is a more stable solvent and less prone to autocleavage to give amines, can be used for DMF, particularly for the coupling reaction. The peptide bond is usually mediated by using a 3-fold excess of the base-driven onium coupling reagents such as benzotriazol-1-yloxytris(dimethylamino)phosphonium hexafluorophosphate (BOP) and

TABLE III
SCHEDULE FOR 9-FLUORENYLMETHYLOXY–*tert*-BUTYL CHEMISTRY

Step	Reagent	Volume (ml)/g resin	Number of operation × min
Deprotection			
1	20% Piperidine	15	1 × 2
2	20% Piperidine	15	1 × 8
3	DMF	15	5 × 1
Coupling[a]			
4	Fmoc-AA in DMF	12	1 × 1[b]
5	HOBt in DMF	1	1 × 1[b]
6	DCC in DCM	1	60 × 1[b]
Wash			
7	DMF	15	3 × 1
8	CH₃CN	15	1 × 1
9	DMF	15	3 × 1
Monitor by ninhydrin test[c]			

[a] Typical coupling procedure for all amino acids.
[b] Do not drain.
[c] If ninhydrin test shows <99.7% completion, repeat steps 4–6 and use HBTU method for coupling.

2-(1*H*-benzotriazol-1-yl)-1,1,3,3-tetramethyluronium hexafluorophosphate (HBTU) (2–3 equivalents of DIEA). *Note:* A 3-fold excess is used instead of a 4-fold excess in the Boc chemistry because Fmoc reagents are generally more expensive.

5. After completion of the synthesis, the final cleavage is effected with a mixture of TFA:triisopropylsilane (TIS):thioanisole:H$_2$O (92.5:2.5:2.5:2.5, by volume) for 1.5 hr at ambient temperature. For Arg-rich MAPs, a cleavage time longer than 3 hr is often required to remove the Pmc group. The TFA is concentrated *in vacuo* below 25°, and the MAP is precipitated and washed by cold ether. In most cases, a small-scale deprotection with 30 mg of resin is recommended to establish the optimal deprotection time and to provide an evaluation of the product by HPLC. The appearance of multiple peaks in the HPLC chromatogram suggests alternative scavengers and a change of duration for the deprotection procedure.

6. The optimal volume of TFA deprotection may have to be established. In general 10 ml per gram of peptide–resin may not be the ideal ratio. However, the larger the TFA volume, the greater the difficulty for workup. If a larger volume of TFA is to be used, the workup could contain a step for the reduction of the TFA volume *in vacuo*. *Note:* The TFA solution should never be dried because *tert*-butyl trifluoroacetate is a source of alkylating agent that causes side reactions.

7. For purification and characterizations of MAPs, see comment 8 in the preceding section.

Heteromic Synthesis of Chimeric Multiple Antigen Peptides

A simple approach for the synthesis of chimeric MAPs containing two different kinds of peptides is to link them tandemly in a continuous array. A more challenging approach is to synthesize them on the different arms of the core matrix. This can be achieved using a core matrix bearing two different, orthogonally protected amino groups such as (1) Boc–Fmoc,[3] (2) Fmoc–Aloc,[25] (3) Fmoc–Dde,[26] and (4) Npys–Fmoc.[27] The allyoxycarbonyl (Aloc) group can be removed by Pd^0 in the presence of a nucleophile such as morpholine, whereas 1-(4,4-dimethyl-2,6-dioxocyclohexylidene)ethyl (Dde) is removed by 2% (v/v) hydrazine in DMF; 3-nitro-2pyridine-sulfenyl (Npys) is removed under neutral conditions by a trivalent phosphine or thiol. With these combinations of protecting groups, either arm may be

[25] S. A. Kates, S. Daniels, and F. Albericio, *Anal. Biochem.* **203,** 245 (1993).
[26] B. W. Bycroft, W. C. Chan, S. R. Chabra, P. H. Teesdale-Spittle, and P. M. Hardy, *J. Chem. Soc., Chem. Commun.*, 776 (1993).
[27] N. Ahlborg, *J. Immunol. Methods* **179,** 269 (1995).

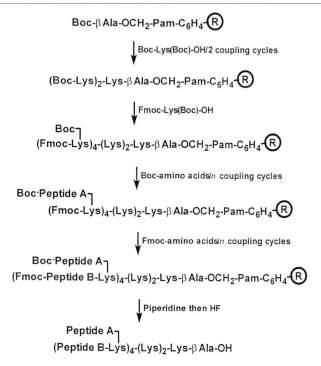

Boc-β Ala-OCH$_2$-Pam-C$_6$H$_4$-Ⓡ

↓ Boc-Lys(Boc)-OH/2 coupling cycles

(Boc-Lys)$_2$-Lys-β Ala-OCH$_2$-Pam-C$_6$H$_4$-Ⓡ

↓ Fmoc-Lys(Boc)-OH

Boc⌐
(Fmoc-Lys)$_4$-(Lys)$_2$-Lys-β Ala-OCH$_2$-Pam-C$_6$H$_4$-Ⓡ

↓ Boc-amino acids/n coupling cycles

Boc-Peptide A⌐
(Fmoc-Lys)$_4$-(Lys)$_2$-Lys-β Ala-OCH$_2$-Pam-C$_6$H$_4$-Ⓡ

↓ Fmoc-amino acids/n coupling cycles

Boc-Peptide A⌐
(Fmoc-Peptide B-Lys)$_4$-(Lys)$_2$-Lys-β Ala-OCH$_2$-Pam-C$_6$H$_4$-Ⓡ

↓ Piperidine then HF

Peptide A⌐
(Peptide B-Lys)$_4$-(Lys)$_2$-Lys-β Ala-OH

FIG. 3. Scheme for the stepwise approach in the preparation of the heteromeric MAP containing two different peptide chains.

used to obtain the diepitope MAP (Fig. 3). Other methods for diepitope MAPS, such as fragment condensation and ligation,[28–30] have been proposed.[31-]

Strategy and Considerations

1. The combination of Boc–Fmoc commonly is used because it employs both the acid- and base-driven deprotecting methods. For synthesis of the core matrix of a MAP containing two different epitopes, Fmoc-Lys(Boc) is used to give selective protection of the Lys branching of the core at the third level to give four each Lys(Boc) and Fmoc-Lys end groups.

2. The order of the chemistry is such that Boc chemistry (see Boc comments 1–8) is used for the synthesis of the first peptide because the

[28] C.-F. Liu and J. P. Tam, *Proc. Natl. Acad. Sci. U.S.A.* **91,** 6584 (1994).
[29] C.-F. Liu and J. P. Tam, *J. Am. Chem. Soc.* **116,** 4149 (1994).
[30] J. P. Tam, Y.-A. Lu, C.-F. Liu, and J. Shao, *Proc. Natl. Acad. Sci. U.S.A.* **92,** 12485 (1995).
[31] G. McLean, A. Gross, M. Munns, and H. Marsden, *J. Immunol. Methods* **155,** 113 (1992).

tert-butyl-based side-chain protection of Fmoc chemistry is not stable to repetitive TFA treatments. During the Boc synthesis, the neutralization time is reduced to 1 min to minimize premature cleavage of the Fmoc group on the core matrix.

3. Synthesis of the second peptide uses Fmoc chemistry (see Fmoc comments 1–7) and starts after the completion of the first peptide chain using Boc chemistry. The final cleavage after assemblage of both peptides requires a strong acid such as HF (see comment 7 on Boc chemistry).

Stepwise Synthesis of Lipidated Multiple Antigen Peptides

An improvement to produce better synthetic peptide immunogens is the introduction of a lipophilic moiety to the MAP carboxyl terminus (lipidated MAP). The lipophilic moiety also plays additional roles such as anchoring the lipid matrix and being a built-in adjuvant. Two types of lipophilic moieties have been used. The first type is a single lipid chain, and the second type is a clustered lipid chain anchored by a Cys such as a tripalmitoyl-derivatized Cys, tripalmitoyl-*S*-glyceryl-Cys (P3C).[32] Both types of lipids are, in turn, attached to the side chain of one or more Lys residues on the COOH terminus of a MAP[5,19] (Fig. 1C).

Strategy and Considerations

1. The protecting group strategy for lipid attachment to the side chain of Lys requires orthogonal protecting groups. These include the use of the following lysyl derivatives of both Boc or Fmoc chemistry.

Lys derivative	Cleavage
Boc-Lys(Fmoc)	20% piperidine
Fmoc-Lys(Dde)	2% hydrazine
Fmoc-Lys(Mtt)	1% TFA
Fmoc-Lys(Aloc)	Pd[0]
Fmoc-Lys(Npys)	R$_3$P in DMF

2. For a single-chain lipid attached to the Lys, it may be desirable to form two or more Lys and palmitoyl-Lys. In such a case, two or more orthogonally protected Lys will be used in tandem.

3. In the example that follows, the synthesis of B2SM-PL$_n$ (n = 1–4) is described.[5] B2 is the linear peptide sequence Lys-Ser-Ile-Arg-Gln-Gly-

[32] K. Deres, H. Schild, K.-H. Wiesmuller, G. Jung, and H. Rammensee, *Nature* (*London*) **342**, 561 (1989).

Pro-Gly-Arg-Ala-Phe-Val-Thr-Ile-Gly-Lys, segment 312–329 from the V3 loop of gp120 of human immunodeficiency virus type-1 (HIV-1), IIIB strain. B2 is in a tetravalent MAP format with a symmetrical core matrix (B2SM) (Fig. 1F). Palmitoyl-Lys (PL) is introduced as a lipid anchor.

Procedure. The B2SM-PL$_n$ MAPs (n = 1–4) are synthesized by solid-phase methodology on Boc-Ala-OCH$_2$-Pam resin (0.10 mmol/g) using a combination of Boc or Fmoc strategies. Removal of the Fmoc group is carried out by 20% piperidine in DMF, whereas the Boc group is removed by 50% TFA in DCM. After removal of the Boc group on the resin (Boc-Ala-OCH$_2$-Pam resin), one or more rounds of Fmoc-Lys(Boc) coupling are carried out sequentially to the alanyl resin. The N^α-Boc groups on Lys are then removed and the palmitic acids (6 equivalents by moles) are coupled by the symmetrical anhydride method using N,N'-dicyclohexylcar-bodiimide (DCC, 3 equivalents by moles) to form Fmoc-[Lys(Pal)]$_n$-Ala-OCH$_2$-Pam resin. The Ser-Ser linker is introduced by two consecutive rounds of Boc-Ser(Bzl) coupling by the HBTU method. After deblocking the Boc group on Ser(Bzl), a tetravalent core matrix and the peptide antigen is coupled stepwise using HBTU as coupling reagent. The B2SM-PL MAPs are cleaved from the resin by stirring in HF/thiocresol/p-cresol (90:3:7, v/v/v) at 0° for 1 hr. After extraction with 8 M urea in 0.1 M Tris-HCl buffer, pH 7.4, and dialysis against 0.1 M Tris-HCl buffer, pH 7.4, with decreasing urea concentrations to 0 M for 24 hr, reasonably pure B2SM-PLs are obtained. Amino acid analyses and laser desorption mass spectrometry showed satisfactory results for all synthetic MAP-PLs.

Stepwise Synthesis of Multiple Cyclic Antigen Peptides

Theory and Design. Constrained peptides are highly desirable for mim-icking the conformation from which the peptide is derived. Multiple cyclic antigen peptides (McAPs) are designed for this purpose.[11] McAPs have a tripartite design (Fig. 4). The first two parts are common regions. Part 1 at the C terminus contains a short peptide followed by a tetralysyl core matrix. Part 2 contains a specific chemical cleavage site that gives a mono-meric cyclic peptide to verify the yield of the intrachain cyclization. This part may not be necessary if verification is not needed. The third part is the variable region and contains a peptide antigen framed by a Lys at the COOH terminus and a Cys(StBu) at the amino terminus. A masked alde-hyde is attached to the side chain of the COOH-terminal Lys to allow an end-to-side chain cyclization via thiazolidine formation with the Cys at the amino terminus (Fig. 4). For model study as well as for their use as standards to verify intrachain cyclization, monomeric peptide precursors correspond-ing to the variable region containing antigen peptides are also prepared.

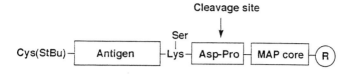

MAP core = Lys$_2$-Lys-

FIG. 4. The three-part design of a McAP precursor to form an end-to-side chain McAP. Part 1 is the MAP core matrix containing a short peptide and a tetralysyl architecture. Part 2 is the specific cleavage site. Part 3 is the peptide antigen, which is flanked by Cys(StBu) at the amino terminus and Lys(Ser) at the C terminus.

The protecting strategy for the two reactive moieties, Cys and aldehyde, is designed to be compatible with Fmoc chemistry and protecting group removals in aqueous conditions. Sulfonyl *tert*-butyl is chosen for Cys as Cys(StBu) can be removed under aqueous conditions by a trialkylphosphine and is stable to the oxidation by periodate.

The masked form of the aldehyde is the 1,2-amino alcohol of an amino-free Ser anchored on the lysyl side chain and which can be converted to an aldehyde by periodate oxidation.[33] The use of trialkylphosphine in our scheme has two advantages. First, it inhibits intermolecular disulfide formation as a side reaction, and second, it allows a "one-pot" reaction of unmasking the thiol protecting group and cyclization.

General Comments

1. This procedure combines both solid- and solution-phase chemistry. First, the precursor of McAP is synthesized by stepwise solid-phase synthesis and cleaved from the resin support. Then, in solution, the aldehyde moiety is generated by sodium periodate oxidation of the Ser 1,2-amino alcohol on the side chain of Lys. Cyclization to form the thiazolidine ring and concurrent removal of the *tert*-butyl sulfonyl protection group on Cys are mediated by an excess of water-soluble triscarboxylethylphosphine (TCEP).

2. For verifying the cyclic structure and quantifying the cyclization yield, the cleavage site Asp-Pro is incorporated at the COOH terminus of the peptide antigen preceding the tetralysyl core matrix. The bond between Asp and Pro is cleaved by formic acid to give the monomeric cyclic peptide. A solution of 70% formic acid (0.5 ml) is added to lyophilized McAP (0.5 mg). The cyclized McAP sample is incorporated at either 37° for 48 hr or

[33] R. Clamp and L. Hough, *Biochem. J.* **94,** 17 (1965).

60° for 24 hr, followed by dilution with water and then lyophilization. Each monomeric sample is compared with the authentic sample (see comment 4) by RP-HPLC, and the intrachain yield is calculated from amino acid hydrolysis (use β-Ala of the core matrix as standard) of the McAP and the cyclic monomer (another authentic amino acid).

3. Serine as a masked aldehyde is coupled to the side chain of Lys using the protecting group scheme Fmoc-Mtt. The 4-methyltrityl (Mtt) group is removable by 1% TFA without affecting the other, more acid-stable, *tert*-butyl protecting group.

4. Intrachain cyclization is favored because of the chain-ring tautomerization (open chain versus ring) of the amino-aldehyde peptides.[14] For verifying the intrachain cyclization of the McAP, the cyclic monomeric peptide obtained after formic acid treatment of the McAP is synthesized. The linear peptide precursor is synthesized by the stepwise solid-phase method. A masked aldehyde as a protected Ser is placed on the side chain of a Lys, and Cys(StBu) is placed at the amino terminus of the peptide antigen. The unprotected precursor is liberated from the resin support and cyclized as described.

5. The peptide Gly-Pro-Gly-Arg-Ala-Phe-Tyr-Thr-Thr-Lys-Asn-Ile-Gly-Gly, referred as CP-18, is derived from the V3 loop of the surface protein gp120 of HIV-1.[11]

Procedure. The linear precursor of McAP is synthesized using Fmoc chemistry (see above). The synthesis starts from Fmoc-β-Ala-Wang resin (0.1 mmol/g) and all couplings are accomplished using BOP/DIEA in DMF. After removing the Fmoc group by 20% piperidine, two cycles are performed of single coupling of Fmoc-Ser(tBu) (3 equivalents each), followed by Fmoc-Lys(Fmoc) (3 equivalents). The tetrabranching [Fmoc-Lys(Fmoc)]$_2$-Lys-Ser-Ser-β-Ala-Wang resin is achieved using 6 equivalents of Fmoc-Lys(Fmoc). For the cleavage site sequence Asp-Pro, and all subsequent coupling reactions, 12 equivalents (0.6 mmol) of amino acids are used. The order of assembly is Fmoc-Lys(Mtt), peptide antigen sequences, and Fmoc-Cys(StBu). The Mtt group of Lys(Mtt) is then selectively removed by 1% TFA, 5% TIS in DCM, and Fmoc-Ser(tBu) is incorporated at the Lys side chain. Deprotection/cleavage and cyclization protocols are as follows: (1) deprotection/cleavage, see Fmoc chemistry, general comments 5–7 and (2) cyclization. The lyophilized McAP precursor is dissolved in 10 mM sodium phosphate (pH 6.8), and Ser at the side chain of Lys is oxidized by NaIO$_4$ followed by removal of the tBu sulfenyl protecting group on Cys and concomitant cyclization to form intrachains on the McAP. A typical procedure is as follows. (1) Sodium periodate (0.75 mg, 8 equivalents) in water (27 μl) is added to a solution of crude precursor of McAP-

CP-18 (4 mg, 0.44 μmol) in 10 mM sodium phosphate buffer (pH 6.8, 2 ml). The mixture is stirred for 2 min, and the McAP is purified by semipreparative HPLC and lyophilized. (2) Tris(2-carboxyethyl)phosphine (1.5 mg, 40 equivalents) in 10 mM sodium acetate buffer (pH 4.2, 47 μl) is added to a solution of oxidized McAP-CP-18 (1.5 mg, 0.17 μmol) in sodium acetate buffer (1.5 ml). The mixture is stirred at 22° for 48 hr, and the progress of the cyclization is monitored by Ellman's reagent until all the thiols are consumed. The product is again purified by semipreparative HPLC.

Modular Approach

One approach for obtaining highly purified MAPs is through orthogonal ligation, which joins purified unprotected peptide segments to the core matrix. Modular synthesis by orthogonal ligation has the advantage of the flexibility of incorporating several types of epitopes to form di- or triepitope MAPs and the option of choosing their orientation to be attached to the core matrix. In addition, MAP products are easily purified by conventional methods. Two general methods (Tables IV and V) based on thiol and

TABLE IV

CHEMOSELECTIVE LIGATION BY THIOL CHEMISTRY VIA THIOALKYLATION, THIOL ADDITION, AND THIOL–DISULFIDE EXCHANGE

Thiol nucleophile	Electrophile	Reaction pH	Product	Remarks
$\diagup\diagdown$SH	$X-CH_2-\overset{O}{\overset{\|}{C}}-$	6–8	$\diagup\diagdown S-CH_2-\overset{O}{\overset{\|}{C}}-$	X = Cl or Br
$\overset{O}{\overset{\|}{-C}}-SH$	$Br-CH_2-\overset{O}{\overset{\|}{C}}-$	4–5	$\overset{O}{\overset{\|}{-C}}-S-CH_2-\overset{O}{\overset{\|}{C}}-$	
$\diagup\diagdown$SH	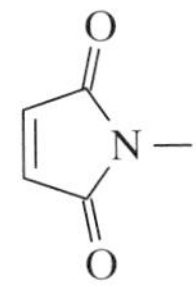 maleimide	7	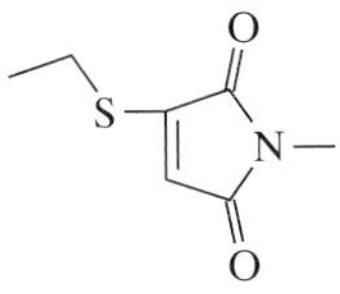 thioether-maleimide adduct	
$\diagup\diagdown$SH	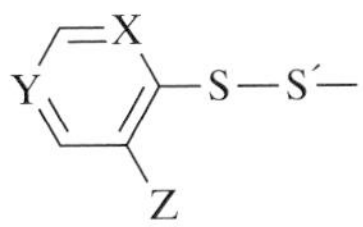 $Y\diagup\diagdown^{X}_{Z}-S-S'-$	8	$\diagup\diagdown S-S'-$	X = N, Y = C, and Z = NO$_2$ or X = C, Y = N, and Z = H

TABLE V
CHEMOSELECTIVE LIGATION BY WEAK BASE–ALDEHYDE CHEMISTRY

Weak base	Reaction pH	Product	Remark
HX— NH$_2$—⟨structure with C=O⟩	3–5	⟨thiazolidine/imidazolidine ring, N–H, C=O⟩	X = S or N
NH$_2$—OCH$_2$C(=O)—	5	—CH=N—OCH$_2$—C(=O)—	
NH$_2$—NH—C(=O)—	5	—CH=N—NH—C(=O)—	
NH$_2$—NH—⟨C$_6$H$_4$⟩—C(=O)—	5	—CH$_2$—N=N—⟨C$_6$H$_4$⟩—C(=O)—	

carbonyl chemistries have been developed to ligate unprotected peptides to form MAPs with nonamide bonds. New ligation methods have been developed to form an amide bond that can be applicable to the synthesis of peptide dendrimers.[28–30,34] In both thiol and carbonyl chemistries, a reactive pair consisting of a nucleophile and an electrophile is placed, respectively, on the purified synthetic peptide monomer and the core matrix during solid-phase synthesis. Usually, a weak base that is significantly more nucleophilic than α- or ε-amines is used for the ligation selectively in aqueous buffer at or below pH 7. Applicable weak bases include alkyl thiol, acyl thiol, 1,2-aminothiol (N-terminal Cys), 1,2-aminoethanol (Thr), hydroxylamine, acylhydrazine, and arylhydrazine. The other reactive component is usually an activated electrophile such as haloacetyl, activated unsymmetrical disulfide, maleimide, or aldehyde. Orthogonality is achieved when these mutually reactive groups are brought together under aqueous conditions with the weak base as the sole nucleophile to react with the electrophile so that protecting other functional groups on the peptide is unnecessary.

[34] P. E. Dawson, T. W. Muir, I. Clark-Lewis, and S. B. H. Kent, *Science* **226,** 776 (1994).

Thiol Chemistry

Thiol chemistry exploits the selective reactivity of sulfhydryls in alkylation with halocarbonyl, sulfur–sulfur exchange with disulfide, and addition to conjugated olefins (Table IV). The application of thioalkylation of unprotected peptide segments on MAPs was first demonstrated by Lu *et al.*[12] and subsequently by Defoort *et al.*[13] In both cases, a chloroacetyl group is incorporated on the Lys core matrix and coupled to a purified, unprotected peptide with an N-terminal Cys to yield a MAP with unambiguous structure as determined by mass spectrometric analysis. The reverse placement with thiol on the core matrix can be achieved by using the *S*-acetyl group attached to the lysinyl core matrix and haloacetyl groups on the peptide.[35] In this approach, a cysteinyl moiety is activated as thiopyridyl or nitropyridylsulfenyl (Npys). Drijhout and Bloemhoff[35] have successfully shown that this chemistry is effective for the synthesis of MAPs in a one-pot reaction performed with a thiolated MAP core matrix and an activated *S*-(Npys)-cysteinyl peptide.

Thioethers can also be formed by adding a thiol of Cys to an activated double bond of a maleimide group. This method is convenient because *N*-alkyl- or *N*-arylmaleimide groups are available either as free carboxylic acid or as an active ester such as *N*-hydroxysuccinimide,[36–38] which can be incorporated as a premade unit in solid-phase synthesis. The maleimido group on Lys and Phe has been shown to be stable to 100% TFA for 3 hr[37] and is fully compatible with the Fmoc chemistry in peptide synthesis when the maleimido group is added last to the peptide sequence. Three examples (examples 1–3) are now presented to illustrate the applicability of thiol chemistry. Examples 1 and 2 show attachment of the peptide through the amino terminus while Example 3 shows attachment of the peptide through the side chain.

Example 1: Coupling via Thioether Formation

GENERAL COMMENTS

1. A tetra- or octavalent (Lys$_2$-Lys-Ala or Lys$_4$-Lys$_2$-Lys-Ala) MAP core is synthesized by the stepwise solid-phase synthesis method using either Boc chemistry or Fmoc chemistry.[12] Normal resin loading, ranging from 0.3 to 0.9 mmol/g, is recommended because only four (for four branches) or five (for eight branches) couplings steps are required.

[35] J. W. Drijfhout and W. Bloemhoff, *Int. J. Pept. Protein Res.* **37,** 27 (1991).
[36] J. Carlsson, H. Drevin, and R. Axen, *Biochem. J.* **173,** 723 (1978).
[37] O. Keller and J. Rudinger, *Helv. Chim. Acta* **85,** 531 (1975).
[38] T. P. King, Y. Li, and L. Kochoumian, *Biochemistry* **171,** 499 (1978).

2. After removing the N-terminal protection group (Boc or Fmoc), a 4 molar excess of chloroacetic acid is coupled to the free amino groups of the core matrix. Bromoacetic acid also can be used.

3. The chloroacetylated MAP core is cleaved from the resin using either HF (Boc chemistry, comment 7) or TFA (Fmoc chemistry, comment 5).

4. The crude chloroacetylated MAP core is usually quite pure (>95%) as judged by analytical RP-HPLC and can be used immediately without further purification.

5. The peptide–antigen is synthesized by stepwise solid-phase synthesis using either Boc or Fmoc chemistry (see above). For conjugation to the chloroacetylated MAP core, an additional acetylated Cys is coupled to either the C termini (peptide–antigens derived from internal or N-terminal region) or the N termini (peptide–antigens derived from the C-terminal region) of the peptide–antigen. Acetylation of amino-terminal Cys minimizes dimerization.

6. Before the conjugation reaction, the disulfide bonds of the peptide–antigen are reduced to sulfhydryl groups using water-soluble tris(2-carboxyethyl)phosphine (TCEP). Dithiothreitol (DTT) cannot be used because it may react with the chloroacetylated MAP core.

7. Purification and characterization of the peptide are generally similar to that suggested in comment 8 in the Boc chemistry section.

PROCEDURE. A typical procedure is as follows: The peptide (28 μmol) is dissolved in 6 M guanidine hydrochloride (5 ml) containing EDTA (50 mg EDTA/ml) and 3 equivalents of TCEP. The solution is kept under nitrogen and adjusted to pH 6–6.5 with 1 M Tris-HCl. The solution is flushed with nitrogen for 30 min and then sealed for 2 hr. The tetravalent chloroacetylated MAP core (4 μmol) is added to the above solution, and the pH is adjusted to pH 8.9–9. The mixture is stirred for 45 min at 18°. To remove excess peptide–antigen, the MAP conjugate is dialyzed (SpectraPor, M_r cutoff 1000) against 0.1 M NH$_4$HCO$_3$ or 10% acetic acid, followed by deionized water, and then lyophilized to dryness.

Example 2: Coupling of Peptides to Lipidated Multiple Antigen Peptides via Thioether Formation

GENERAL COMMENTS

1. Chloroacetic acid is coupled[13] via the symmetrical anhydride method to the Lys$_2$-Lys-Ser-Ser-Lys(P3C)-Ala-resin (see Example for preparation detail).

2. The MAP core is cleaved from the resin with 96% TFA, 2% DMS, and 2% anisole for 2 hr. It is not possible to analyze the product by RP-HPLC because of its very hydrophobic properties. Therefore, the crude

product is purified by repeated precipitation in cold ether, redissolved in glacial acetic acid and lyophilized.

Example 3: Coupling of Free, End-to-End Cyclic Peptides to Form Multiple Cyclic Antigen Peptide via S- → N-Acyl Transfer Reaction and Thioether Formation

COMMENTS

1. The peptide, which contains a Cys residue at the N terminus and a thioester at the C terminus, is synthesized by the stepwise solid-phase method using the Boc/benzyl strategy.[14] The peptide thioester is formed by converting the C-terminal Boc-amino acid to its 3-thiopropionic acid ester and attaching it to the 4-methylbenzhydrylamine (MBHA) resin (Fig. 5).

2. After side-chain deprotection, cleavage from the resin, and purification, the unprotected cysteinyl peptide thioester is end-to-end cyclized via an S- to N-acyl transfer reaction at pH 5.2. Water-soluble TCEP is added to prevent disulfide formation. The resulting end-to-end cyclic peptide contains a thiol side chain (from Cys) that is then used for attachment to the oligomeric lysyl core matrix containing halocarbonyl moties (see Example 1).

PROCEDURE. A typical protocol for preparing cyclic peptides is as follows (Fig. 5). The thioester peptide (analog of enkephalin) is synthesized by Boc chemistry using the BOP coupling protocol (see section on Boc chemis-

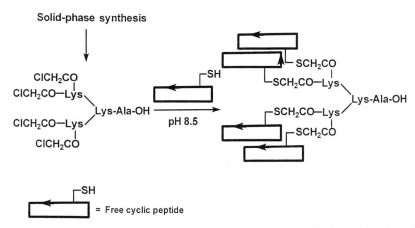

FIG. 5. Synthetic scheme for McAP using the modular approach of attaching through cysteinyl moiety of free cyclic peptides.

try). The thioester resin is prepared according to Hojo and Aimoto[39] by converting the C-terminal Boc-amino acid to its 3-thiopropionic acid ester and then coupling it to a MBHA resin (0.3–0.4 g for each synthesis) by N,N′-diisopropylcarbodiimide (DIPCDI)/HOBt. After the synthesis, the peptide thioester is cleaved from the resin by HF (9:1, v/v) at 4° and purified (see section on Boc chemistry). The linear peptide thioester (6 mg, 3.5 mmol) is dissolved in 20 mM Na$_2$HPO$_4$–10 mM citric acid buffer (2 ml, pH 7.5) containing TCEP (1.04 mg, 3.5 mmol). The solution is stirred for 6 hr, and the progress of the intramolecular cyclization is monitored by RP-HPLC. Conjugation of the free cyclic peptide through the pendant thiol moiety to the chloroacetyl MAP core matrix is as follows. The cyclic peptide (2.8 × 10^{-2} mmol) is dissolved in 0.2 M phosphate buffer containing 10 mM EDTA (1.25 ml, pH 7.4). The solution is purged with argon for 10 min, and the tetravalent chloroacetyllysinyl core matrix (1.1 mg, 1.4 10^{-3} mmol) is added. The mixture is stirred for 24 hr at 20°, and the product is purified by RP-HPLC, followed by characterization by matrix-assisted laser desorption–ionization mass spectrometry (MALDI-MS).

Carbonyl Chemistry

Another approach along the same principles of thiol chemistry is to exploit the selectivity of other weak bases, particularly in the condensation reaction with aldehydes (Table IV). There are two types of weak bases for this purpose. The first type consists of conjugated amines whose basicities are lowered by neighboring electron-withdrawing groups such as hydroxyl-amine[40] and substituted hydrazines.[41] The second type contains the 1,2-disubstituted patterns such as 1,2-aminoethanethiol derivatives of N-terminal Cys and 1,2-aminoethanol derivatives of N-terminal Thr. They react with aldehydes to form cyclic compounds that are Pro-like rings such as thiazolidine from Cys and oxazolidine from Thr.[42] An aldehyde moiety on a peptide or MAP core matrix can be obtained by NaIO$_4$ oxidation of N-terminal Ser, Thr, or Cys under neutral conditions to give an oxoacyl group.[33] These N-terminal amino acids serve a dual purpose, as precursors for 1,2-amino weak bases and as aldehydes. The weak base–aldehyde chemistry has also been used to ligate constrained peptides to a MAP core matrix in the preparation of McAP via carbonyl chemistry.[18]

Five examples (Examples 4–8) are described to show the versatility of carbonyl chemistry through thiazolidine, oxime, and hydrazone linkages.

[39] H. Hojo and S. Aimoto, *Bull. Chem. Soc. Jpn.* **64,** 111 (1991).
[40] K. Rose, *J. Am. Chem. Soc.* **116,** 30 (1994).
[41] T. P. King, S. W. Zhao, and T. Lam, *Biochemistry* **25,** 5774 (1986).
[42] J. P. Tam, C. Rao, J. Shao, and C.-F. Liu, *Int. J. Pept. Protein Res.* **45,** 209 (1994).

Again, there is flexibility in attaching the peptide to the core matrix. In Examples 5–7, hydrazide could be placed at the COOH terminus or the side chain to give a different orientation than those illustrated here. Similarly, hydroxyamino acid or Cys anchored on a lysyl side chain can also be used in these examples to give side-chain attachment.

Example 4: Coupling through Thiazolidine

GENERAL COMMENTS

1. A tetra- or octavalent (Lys$_2$-Lys-Ala or Lys$_4$-Lys$_2$-Lys-Ala) MAP core is synthesized as described in Example 1. After Boc or Fmoc deprotection, Ser is coupled to the MAP core using DCC. Then Ser$_4$-Lys$_2$-Lys-Ala or Ser$_8$-Lys$_4$-Lys$_2$-Lys-Ala is cleaved from the resin by HF (Boc chemistry, comment 7) or TFA (Fmoc chemistry, comment 5) and is lyophilized for immediate use without purification.[15–17]

2. Conversion of the Ser-MAP to glyoxyl-MAP is achieved by oxidation with a 2 molar excess of sodium periodate in 0.1 M sodium phosphate buffer, pH 7. Oxidation of Ser is accomplished within a few minutes, and the reaction is then quenched by adding a 2 molar excess of ethylene glycol.

3. Glyoxyl-MAP is purified by RP-HPLC because formaldehyde is generated during the quenching as well as from the oxidation reaction. It is used immediately for the conjugation reaction.

4. The peptide Cys-Asn-Tyr-Asn-Lys-Arg-Lys-Ile-His-Ile-Pro-Gly-Arg-Pro-Arg-Ala is synthesized by stepwise solid-phase synthesis using either Boc or Fmoc chemistry (see above). For conjugation to the aldehydic MAP core, an additional Cys is coupled to the amino terminus of the peptide–antigen.

5. The rate of thiazolidine ring formation for peptide dendrimers depends on several factors such as the size and excess of the peptide–antigen, the solvent, temperature, pH, and number of branches on the MAP core.[16] In this example, a tetrameric MAP is prepared with a peptide that is derived from the principal neutralizing determinant of the surface coat protein gp120 of HIV-1, MN strain.

PROCEDURE. A typical procedure is as follows. The peptide antigen (18 μmol) is dissolved in 20 mM sodium acetate buffer containing EDTA, pH 5 (2.8 ml, 0.8 mM), and the tetravalent aldehydic MAP core solution (1.13 μmol) collected from HPLC is added. The solution is adjusted to pH 5 with pyridine. The deaerated solution is kept under argon and in the dark for 10 hr at 40°. The conjugation reaction is monitored by analytical C$_8$ RP-HPLC. The MAP conjugate is purified by semipreparative HPLC and lyophilized to dryness.

Example 5: Coupling through Oxime

GENERAL COMMENTS

1. A glyoxy-MAP is prepared as previously described in Example 4.[16]

2. The peptide X-Val-Met-Glu-Tyr-Lys-Ala-Arg-Arg-Lys-Arg-Ala-Ala-Ile-His-Val-Met-Leu-Ala-Leu-Ala, where X is NH_2OCH_2CO, is synthesized by the solid-phase method using Fmoc chemistry. For the purpose of conjugating to the aldehyde MAP core, $Boc-NHOCH_2CO_2H$ is coupled via BOP/DIEA to the N terminus of the peptide–antigen.

3. In this example, the peptide–antigen is derived from the surface immunodeficiency virus and consists of 20 amino acid residues. The MAP core contains four branches.

PROCEDURE. A typical procedure is as follows: 0.4 M Na/acetic acid buffer (400 μl) and a glyoxy-MAP stock solution (40 μl, 0.2 μmol) are added to a peptide stock solution (400 μl, 2 μmol). Then dimethyl sulfoxide (DMSO, 800 μl) is added. The solution is adjusted to pH 5.7, and the deaerated solution is incubated 10 hr at 37°. The conjugation reaction is complete within 12 hr as monitored by analytical C_8 RP-HPLC. The MAP conjugate is purified by semipreparative HPLC and lyophilized to dryness.

Example 6: Coupling through Hydrazone Using Hydrazide Succinyl Group

GENERAL COMMENTS

1. The peptide X-Ser-Ser-Gln-Phe-Gln-Ile-His-Gly-Pro-Arg, X = $NH_2NH-COCH_2CH_2CO$- is synthesized by the stepwise solid-phase method using Fmoc chemistry.[16] For the purpose of conjugation to the aldehyde MAP core, $4-Boc-NHNHCOCH_2CH_2CO_2H$ is coupled via BOP/DIEA to the N terminus of the peptide–antigen.

2. The MAP conjugate is prepared similarly to the procedure for coupling through oxime (Example 5).

Example 7: Coupling through Hydrazone Using 4-Hydrazinobenzoyl Group

GENERAL COMMENTS. The peptide (X-Ser-Ser-Gln-Phe-Gln-Ile-His-Gly-Pro-Arg, X = $NH_2NH-C_6H_4-CO$) is prepared by stepwise solid-phase synthesis using Boc chemistry (see Boc chemistry).[17] For conjugation to the aldehyde containing the MAP core (Example 4), $4-Boc-NHNH-C_6H_4-CO_2H$ (Boc-Hob) is coupled via DCC/HOBt to the amino terminus of the peptide–antigen.

PROCEDURE. The tetravalent aldehydic MAP core (1.65 μmol) collected from HPLC is mixed with 4-hydrazinobenzoyl-peptide (Hob-peptide) (13.2 μmol), and the solution is adjusted to pH 5 with 0.2 M sodium acetate buffer, pH 5. The deaerated solution is kept under argon and in the dark for 1 hr at room temperature. The progress of the reaction is complete within 1 hr as determined by analytical RP-HPLC. The MAP conjugate is purified by semipreparative HPLC and lyophilized to dryness.

Example 8: Coupling of Cyclic Peptides through Thiazolidine Linkage to Form Multiple Cyclic Antigen Peptide

STRATEGY AND CONSIDERATIONS

1. A linear unprotected peptide precursor contains three Lys residues for attaching three functional groups [hydroxylamine, Ser, and Cys(S*t*Bu)] through its side chains, two of which are used to form a cyclic structure; the third is used for attachment to the aldehyde-containing MAP core matrix (Fig. 6).[18]

2. An *O*-alkylhydroxylamine group is attached to the side chain of the first Lys and a Ser moiety to the side chain of the second Lys as a masked aldehyde that is converted to an aldehyde by NaIO$_4$ oxidation. A cyclic peptide is created by intramolecular oxime formation from the reaction between *O*-alkylhydroxylamine and the aldehyde.

3. A Cys(S*t*Bu) residue on the side chain of the third Lys moiety is used to ligate the cyclic peptides to the aldehyde moieties of the MAP core (Example 4) without affecting the oxime linkage (Fig. 6).

PROCEDURE. A typical protocol for preparing the cyclic peptide is as follows. The peptide, Ile-Gly-Pro-Gly-Arg-Ala-Phe, is synthesized by Fmoc chemistry with all couplings performed by HBTU/DIEA. A Lys residue is introduced as Fmoc-Lys(Mtt). The Mtt protecting group is removed with 1% TFA, 5% TIS in DCM, and Boc-Ser(*t*Bu) is coupled to the side chain of Lys. The peptide sequence is synthesized followed by incorporation of a Lys residue as the Fmoc-Lys(Mtt) derivative. Boc-(aminooxy)acetic acid is coupled to its side chain. Lys is coupled as the Boc-Lys(Fmoc) derivative, and, after removing the Fmoc group by 20% piperidine, Fmoc-Cys(S*t*Bu) is coupled to the side chain. The peptide is cleaved and purified as described in the Fmoc section (comments 5–7).

After purification, the peptide is dissolved in 10 mM sodium acetate buffer (pH 7) to give a final concentration of 0.8 M. The 1,2-amino alcohol moiety is then oxidized by addition of a 2 molar excess of NaIO$_4$ for 2 min at room temperature to give the glyoxyl derivative, followed by purification using semipreparative HPLC to remove the formaldehyde by-product. Oxime cyclization is effected when the solution collected from semipreparative HPLC is adjusted to pH 5.5. The reaction is completed with 12 hr, and progress of the intramolecular cyclization is monitored by RP-HPLC. The cyclic peptide is characterized by mass spectrometry.

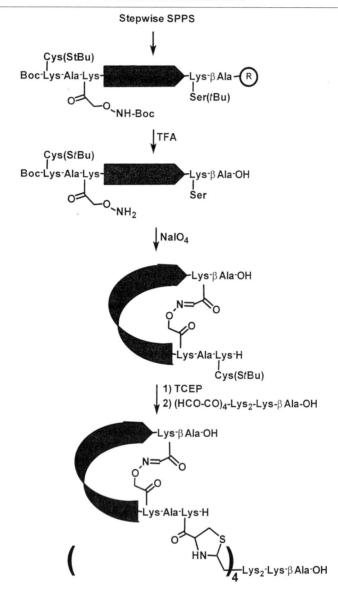

FIG. 6. Synthetic scheme for McAP using the modular approach of attaching via a thiazolidine linkage of free cyclic peptides.

Cysteinyl deprotection and conjugation protocols are as follows. (1) For Cys deprotection, TCEP (1.5 μmol) is added to a solution of cyclic peptide (0.3 μmol) in 50 mM sodium acetate buffer (pH 6, 0.5 ml). The mixture is stirred for 1 hr at room temperature. (2) For conjugation, the tetravalent MAP core containing aldehyde groups (0.036 μmol) is added, and the mixture is adjusted to pH 5 using acetic acid. The solution is stirred for 36 hr at 18°, and the product is purified by preparative HPLC.

Acknowledgment

This work was supported in part by U.S. Public Health Services Grants CA 36544 and AI 37965.

[29] Relaxin

By JOHN D. WADE and GEOFFREY W. TREGEAR

Introduction

In the early 1920s, Frederick Hisaw observed that injection of the serum of pregnant guinea pigs and rabbits into virgin guinea pigs shortly after estrus caused a dilation and relaxation of the pubic symphyses.[1] A similar result was obtained when aqueous extracts of porcine corpora lutea were used.[2] The active component was named relaxin. In the years that followed, much physiological evidence was acquired that showed that relaxin was a protein that is produced in, and acts on, the tissues of the reproductive tract in many mammalian species to facilitate parturition.[3] The principal actions of relaxin were considered to be a lengthening of the pubic ligaments, widening of the pelvis, dampening of uterine contractions, and softening and dilating of the cervix. More recent studies point to a far wider physiological role for the hormone. Relaxin is now recognized to be capable of causing changes in fluid balance,[4] to have potent actions on the heart,[5] and to occur in the male seminal plasma.[6]

[1] F. L. Hisaw, *Proc. Soc. Exp. Biol. Med.* **23,** 661 (1926).
[2] H. L. Fevold, F. L. Hisaw, and R. K. Meyer, *J. Am. Chem. Soc.* **52,** 3340 (1930).
[3] O. D. Sherwood, *in* "The Physiology of Reproduction" (E. Knobil and J. D. Neill, eds.), p. 861. Raven, New York, 1994.
[4] R. S. Weisenger, P. Burns, L. W. Eddie, and E. M. Wintour, *Endocrinology* **137,** 505 (1993).
[5] H. Kakouris, L. W. Eddie, and R. J. Summers, *Lancet* **339,** 1076 (1995).
[6] J. W. Winslow, A. Shih, J. H. Bourell, G. Weiss, B. Reid, J. T. Stults, and L. T. Goldsmith, *Endocrinology* **130,** 2660 (1992).

A-chain

```
        1         5                      20        25
H1    R P Y V A L F E K C C L I G C T K R S L A K Y C
H2    Z L Y S A L A N K C C H V G C T K R S L A R F C
Hi        G I V E Q C C T S I C S L Y Q L E N Y C N
BII       G I V D E C C L R P C S V D V L L S Y C


BII   Z Q P Q A V H T Y C G R H L A R T L A D L C W E A G V D
Hi      F V N Q H L C G S H L V E A L Y L V C G E R G F F Y T P K T
H2    D S W M E E V I K L C G R E L V R A Q I A I C G M S T W S
H1    K W K D D V I K L C G R E L V R A Q I A I C G M S T W S
      1         5         10        15        20        25        30
```

B-chain

FIG. 1. Primary structures of Genes 1 (H1) and 2 (H2) human relaxins, human insulin (Hi), and silkworm Type II bombyxin (BII). Disulfide pairings are shown.

It took 50 years after Hisaw's original discovery before sufficient quantities of purified porcine relaxin were obtained to enable its structural determination. Relaxin was shown to possess a two-chain, three disulfide-bonded peptide structure analogous to that of insulin.[7] The disposition of the invariant Cys residues is identical between the two hormones,[8] although the remaining primary structure differences are as great as 60%. At this time, the primary structure of the relaxin from 22 species is known. In the human, there are two genes, known as Genes 1 and 2, which each code for a functional relaxin (Fig. 1). Gene 2 relaxin is the principal expression product *in vivo*.[9] The great apes also have two relaxin genes. However, only the chimpanzee is capable of expressing two functional relaxin peptides.[10] All other mammalian species have only one relaxin gene. Like insulin, relaxin is made *in vivo* on the ribosome as preprohormone that undergoes proteolytic processing to yield the mature two-chain form. In contrast, two other members of the relaxin/insulin family, insulin-like growth factor I and II, remain intact.

[7] C. Schwabe, J. K. McDonald, and B. G. Steinetz, *Biochem. Biophys. Res. Commun.* **70,** 397 (1976).
[8] E. Canova-Davis, T. J. Kessler, P.-J. Lee, D. T. W. Fei, P. Griffin, J. T. Stults, J. D. Wade, and E. Rinderknecht, *Biochemistry* **30,** 6006 (1991).
[9] P. Hudson, M. John, R. Crawford, J. Haralambidis, D. Scanlon, J. Gorman, G. Tregear, J. Shine, and H. Niall, *EMBO J.* **3,** 2333 (1988).
[10] B. A. Evans, P. Fu, and G. W. Tregear, *Endocr. J.* **2,** 81 (1994).

To date, only one other native peptide with structural features similar to relaxin and insulin has been reported. Bombyxin, a brain secretory peptide of the silk moth *Bombyx mori,* was found to stimulate the prothoracicotropic activity.[11] Five molecular species of the bombyxins have been identified; all consist of two separate peptide chains and six Cys residues, which form three disulfide bonds in a manner identical to that of insulin (Fig. 1). Bombyxin genes also have the same domain organization as insulin genes, with the result that a preprobombyxin is produced.

The insulin-like heterodimeric structure of relaxin necessitates special considerations insofar as its chemical synthesis is concerned. Much of the early and original experience of synthesizing the peptide was based on methods developed in the 1960s by two independent groups for the assembly of insulin.[12,13] Both groups relied on the combination of the two constituent synthetic chains in solution at high pH. Higher yields of insulin were obtained when chain combinations were carried out using S-sulfonated forms of the B-chain and the *S*-sulfhydryl form of the A-chain.[14,15] As the number of possible disulfide conformers is large, these results showed that the individual A- and B-chains of insulin contain sufficient structural information to form insulin, a structure that must also be the most thermodynamically stable. A parallel combination approach was used by the Howard Florey Institute group to yield the first successful chemical preparation of relaxin.[16,17] In contrast to insulin, however, both the *S*-sulfhydryl and S-sulfonated forms of the relaxin B-chain are poorly soluble. This has necessitated the development of modified combination procedures to yield reasonable levels of relaxin.[18] An additional difficulty that arises is that modified relaxins, either chain-shortened or amino acid substituted, are difficult to prepare, presumably because of resulting secondary structural changes, which decrease the efficiency of combination. Similar observations

---

[11] K. Maruyama, H. Hietter, H. Nagasawa, A. Isogai, S. Tamura, A. Suzuki, and H. Ishizaki, *Agric. Biol. Chem.* **52**, 3035 (1988).

[12] Y. T. Kung, Y. C. Du, W. T. Huang, C. C. Chen, L. T. Ke, S. C. Hu, R. Q. Jiang, S. Q. Chu, C. I. Niu, J. Z. Hsu, W. C. Chang, L. L. Cheng, H. S. Li, Y. Wang, T. P. Loh, A. H. Chi, C. H. Li, P. T. Shi, Y. H. Tieh, K. L. Tang, C. Y. Hsing, *Sci. Sin.* **14**, 1710 (1965).

[13] H. Zahn, *Naturwissenschaften* **52**, 99 (1965).

[14] P. G. Katsoyannis, A. C. Trakatellis, S. Johnson, C. Zalut, and G. Schwarz, *Biochemistry* **6**, 2642 (1967).

[15] J.-G. Tang, C.-C. Wang, and C.-L. Tsou, *Biochem. J.* **255**, 451 (1988).

[16] G. Tregear, Y.-C. Du, C. Fagan, H. Reynolds, D. Scanlon, P. Jones, B. Kemp, and H. Niall, *in* "Peptides—Structure, Synthesis and Function" (D. H. Rich and E. Gross, eds.), p. 249. Pierce, Rockford, Illinois, 1981.

[17] P. Hudson, J. Haley, M. John, M. Cronk, R. Crawford, J. Haralambidis, G. Tregear, J. Shine, and H. Niall, *Nature (London),* **301**, 628 (1983).

[18] J. Burnier and P. D. Johnston, European Patent 251,615 (1988).

have been reported for the synthesis of insulin analogs.[19] Despite these difficulties, several syntheses of relaxin from different species and of analogs have been successfully achieved and the resulting products shown to be biologically active.[20-23]

The regioselective disulfide assembly of such two-chain, three-disulfide bonded peptides has become feasible following the development both of novel Cys thiol-protecting groups possessing differential reactivities and of new chemical methods for specific removal of the groups. Such developments are a considerable advance on the procedures originally developed for the synthesis of insulin via an elegant but complex stepwise fragment assembly and subsequent stepwise iodine oxidation of each of the three disulfide bonds.[24] The first reported regioselective disulfide synthesis of relaxin employed selectively S-protected A- and B-chains made by 9-fluorenylmethyloxycarbonyl (Fmoc) and tert-butyloxycarbonyl (Boc) solid-phase synthesis, respectively. The three disulfide bonds were formed by air oxidation, thiolysis, and iodine oxidation.[25] This approach has been employed to make a variety of mammalian relaxins and analogs.[26-28] In contrast, the development of silyl chloride as an efficient means of removal of certain S-protecting groups has broadened the scope for regioselective disulfide bond formation.[29] Using this method, an efficient solid-phase synthesis of insulin was reported in which the two Fmoc-solid phase assembled chains bearing three orthogonal thiol-protecting groups were used. The three disulfide bonds were individually formed by thiolysis, iodine oxidation, and the silyl chloride treatment.[30] A disadvantage of this procedure is that its application to relaxin would require the use of Boc-assembled B-chain in which the indole of the two Trp residues was temporarily pro-

[19] Y.-C. Chu, R.-Y. Yang, G. T. Burke, J. D. Chanley, and P. G. Katsoyannis, *Biochemistry* **26**, 6966 (1987).

[20] E. Canova-Davis, I. P. Baldonado, and G. M. Teshima, *J. Chromatogr.* **508**, 81 (1990).

[21] J. D. Wade, S. S. Layden, P. F. Lambert, H. Kakouris and G. W. Tregear, *J. Protein Chem.* **13**, 315 (1994).

[22] J. D. Wade, F. Lin, Y.-Y. Tan, and G. W. Tregear, *in* "Peptides 1994" (H. L. S. Maia, ed.), p. 319. ESCOM, Leiden, The Netherlands, 1995.

[23] J. D. Wade, F. Lin, D. Salvatore, L. Otvos, Jr., and G. W. Tregear, *Biomed. Pept. Protein Nucleic Acids* **2**, 27 (1996).

[24] P. Seiber, B. Kamber, A. Hartmann, A. Jöhl, B. Riniker, and W. Rittel, *Helv. Chim. Acta* **60**, 27 (1977).

[25] E. E. Bullesbach and C. Schwabe, *J. Biol. Chem.* **266**, 10754 (1991).

[26] B. Rembesia, R. Bracey, E. E. Bullesbach, and C. Schwabe, *Endocr. J.* **1**, 263 (1993).

[27] E. E. Bullesbach, B. G. Steinetz, and C. Schwabe, *Endocr. J.* **2**, 1115 (1994).

[28] E. E. Bullesbach and C. Schwabe, *J. Biol. Chem.* **269**, 13124 (1994).

[29] K. Akaji, T. Tatsumi, M. Yoshida, T. Kimura, Y. Fujiwara, and Y. Kiso, *J. Chem. Soc., Chem. Commun.*, 167 (1991).

[30] K. Akaji, K. Fujino, T. Tatsumi, and Y. Kiso, *J. Am. Chem. Soc.* **115**, 11384 (1993).

tected with the formyl group to prevent known modification by silyl chloride.[29] Removal of the N^{in}-protection after disulfide bond formation can then be achieved by short treatment of the peptide with dilute aqueous NaOH solution.[25] An additional consideration is that S-protected B-chains are poorly soluble, with the result that overall yields may be no better than that achieved by the random chain combination approach. Nevertheless, this strategy affords a potentially effective means of producing relaxin analogs that may not be readily obtainable by other means.

An additional method for the acquisition of relaxin has been reported.[31] Although it involves the use of recombinant DNA-derived peptide, the general protocol should also be readily applied to chemical synthesis. It involves the use of a single-chain relaxin in which the C-peptide segment is replaced with a short 13-residue "mini"-C sequence bearing appropriately selected residues at the B–C and C–A junctions. After expression of the "mini"-C prorelaxin in *Escherichia coli* and partial purification, the peptide is correctly folded in high yield in solution in the presence of oxidized and reduced glutathione. Removal of the "mini"-C peptide is accomplished by sequential treatment of the refolded protein with Asp N and Arg C enzymes. Following purification, the relaxin is obtained in an overall yield severalfold higher than that obtained from an optimized two-chain process. The one-chain process has been used to obtain several Ala-substituted relaxin analogs for biological study.

The availability of synthetic relaxin has enabled a number of important observations about the peptide to be made. Because the predominant source of relaxin in the human is the ovary during pregnancy, it is not possible to obtain adequate amounts of the hormone for biological evaluation. This is particularly so for the Gene 1 relaxin, which appears to function as a local hormone and has yet to be isolated from human tissue. Thus chemical peptide synthesis offers the principal practical source of human relaxin for study. The tertiary conformation of synthetic human Gene 2 relaxin has been determined by X-ray crystallography and shown to possess a dimeric structure very similar to that of insulin.[32] However, the dimer contacts of the two hormones differ greatly as do the spatial orientation of the receptor binding sites. Autoradiography using ^{32}P-labeled synthetic relaxin clearly identified specific relaxin receptors in only three tissues: uterus, brain, and heart.[33] Within both the male and female brain, the

[31] R. Vandlen, J. Winslow, B. Moffat, and E. Rinderknecht, *in* "Recent Progress in Relaxin Research" (A. H. MacLennan, G. W. Tregear, and G. D. Bryant-Greenwood, eds.), p. 59. Global, Singapore, 1995.
[32] C. Eigenbrot, M. Randal, C. Quan, J. Burnier, L. O'Connell, E. Rinderknecht, and A. Kossiakoff, *J. Mol. Biol.* **221**, 15 (1991).
[33] P. Osheroff and W.-H. Ho, *J. Biol. Chem.* **268**, 15193 (1993).

principal relaxin binding regions were in two circumventricular organs, both of which lack a blood–brain barrier and are known to be involved in the control of cardiovascular functions such as blood pressure and fluid and electrolyte homeostasis. A likely neurophysiological role for the hormone was further demonstrated by the effect of the peptide on the plasma osmolality–arginine vasopressin relationship.[4] Subsequent physiological analysis using synthetic human relaxin showed the peptide to possess direct inotropic and chronotropic activity in the isolated rat heart.[5] The peptide is highly active with pD_2 values in the nanomolar range and is more potent than either endothelin or angiotensin II, suggesting a possible use of relaxin for treatment of cardiovascular disorders.[34] Synthetic relaxin was also shown to be a potent inhibitor of collagen deposition,[35] and this has led to its current clinical evaluation as a treatment for connective tissue disorders such as scleroderma.

Methods

The general procedure described below has been developed for the preparation of human Gene 1 relaxin[23] by the random chain combination approach and can be successfully used for most native relaxins and of analogs of the peptide.

Solid-Phase Synthesis

Both the A- and B-chains can be assembled without difficulty by either the Boc- or Fmoc-based solid-phase methods using standard protocols. It is recommended if using the latter method for the synthesis of the A-chain that N^α-Fmoc deprotection be carried out with 1% (v/v) 1,8-diazabicyclo[5.4.0]undec-7-ene (DBU) in dimethylformamide (DMF) to minimize base-catalyzed epimerization of the C-terminal Cys residue.[36] As free Cys thiol groups are required following synthesis and cleavage, S-protecting groups that are labile to the final acid cleavage are employed. In the case of Fmoc assembly the S-trityl group (S-Trt) is preferred, and for Boc assemblies the S-methylbenzyl group (S-MeBzl) is used. Some relaxin A-chains have an N-terminal Pca, which is best coupled to the growing

---

[34] R. J. Summers, Y. Y. Tan, H. Kakouris, and L. W. Eddie, in "Recent Progress in Relaxin Research" (A. H. MacLennan, G. W. Tregear, and G. D. Bryant-Greenwood, eds.), p. 487. Global, Singapore, 1995.

[35] E. N. Unemori and E. P. Amento, in "Recent Progress in Relaxin Research" (A. H. MacLennan, G. W. Tregear, and G. D. Bryant-Greenwood, eds.), p. 481. Global, Singapore, 1995.

[36] J. D. Wade, J. Bedford, R. C. Sheppard, and G. W. Tregear, Pept. Res. 4, 194 (1991).

peptide chain via its 2-dimethylaminoisopropyl chloride hydrochloride (DIC)-activated species.

After cleavage and side-chain deprotection, again using standard procedures, the crude A-chain peptides are extracted into 0.1% (v/v) aqueous trifluoroacetic acid (TFA) and freeze-dried. In contrast, prior freeze-drying of the crude S-thiol B-chain results in a product that cannot be satisfactorily redissolved into solution. For this reason and because of its poor solubility, the crude peptide is extracted into 0.25 mM Tris/8 M urea, pH 8.0, diluted with 0.1% aqueous TFA, clarified by both centrifugation at 3000 rpm for 5 min and filtration through 0.2-μm Acrodiscs (Gelman Sciences, Ann Arbor, MI), and then applied directly onto a reversed-phase high-performance liquid chromatography (RP-HPLC) column for preparative purification (see below).

Purification

Both peptides can be satisfactorily purified by standard preparative RP-HPLC on Vydac C_4 supports using 0.1% (v/v) aqueous TFA as buffer A and 0.1% (v/v) TFA in acetonitrile as buffer B and a gradient of 20 to 50% buffer B over 40 min. Neither peptide requires prior S-reduction unless they have been stored for some time. It is not recommended that crude peptides be used for chain combination as it has been found that even minor impurities can markedly diminish the efficiency of oxidation. Yields of purified S-reduced B-chain are consistently lower than for S-reduced A-chain because of limited solubility. Following purification, the peptides are comprehensively characterized by analytical RP-HPLC and mass spectrometry. The peptides may be stored at 4° for extended periods without polymerization or decomposition. However, there are reports of light-initiated conversion of Met residues to the oxidized form.[37]

Chain Combination

Combination is carried out using a ratio of A- and B-chains of 2–4:1 (w/w). Because of the greater ease of obtaining A- than B-chain, a larger excess of the former is employed. The use of a pH greater than pH 10 is crucial for successful combination; below this level, no reaction occurs. Following combination, the excess unreacted A-chain may be collected by RP-HPLC and rereduced with an excess of dithiothreitol for later reuse. Stock solutions of the following are prepared freshly and thoroughly deaerated by sparging for 10 min with dry nitrogen immediately prior to use.

[37] D. C. Cipolla and S. J. Shire, *in* "Techniques in Protein Chemistry II" (J. J. Villafranca, ed.), p. 543. Academic Press, San Diego, 1991.

A-chain buffer: Either 30 mM dithiothreitol in deionized water alone or B-chain buffer as described below

B-chain buffer: 30 mM dithiothreitol/25 mM Tris/6 M urea/30 mM urea, pH 8.0

Reaction buffer: 2.60 ml of 0.5 M sodium glycinate (pH 10.5), 1.32 ml of acetonitrile, 0.4 ml of n-propanol, 3.08 ml of 6 M urea, 3.16 ml water

For the reaction, 10.0 mg of the A-chain peptide is dissolved in 1.0 ml A-chain buffer, and 5.0 mg of the B-chain is dissolved in 1.0 ml of B-chain buffer. Add the B-chain slowly in small portions to ensure dissolution. Both solutions are added in one portion in this order to 8.0 ml of reaction buffer in an open beaker. The whole is stirred briskly at room temperature. Aliquots are removed immediately and then at 2-hr intervals for monitoring by analytical RP-HPLC on Vydac C_4. The presence of S-reducing agent (dithiothreitol) slows down the rate of combination and allows the chains to align in the correct conformation. In the absence of reductant, a very high level of mismatched relaxin products is generated. Combination times vary greatly but have usually commenced within 3 to 5 hr and are complete within 24 hr. The first molecular event to occur is a rearrangement of the S-reduced A-chain to a family of disulfide isomers[38] and finally a principal product, believed to be peptide with the intramolecular disulfide bond formed, which elutes earlier on RP-HPLC (Fig. 2). Once this so-called oxidized A-chain has formed, the S-reduced B-chains react with it rapidly to generate relaxin. Often, intramolecular disulfide-bonded B-chain forms too and at varying levels. This by-product elutes closely to the relaxin and can make preparative RP-HPLC separation difficult if aqueous TFA-based buffers are used.

When combination is complete as assessed by the absence of starting B-chain, the solution, often turbid at this stage, is acidified to approximately pH 2 by addition of TFA. The mixture is clarified by centrifugation and the supernatant subjected to RP-HPLC preferably using 0.25 N triethylaminephosphate (TEAP), pH 2.5, as buffer A and 40% A/60% acetonitrile as buffer B[39] to ensure optimal separation of the relaxin from any contaminating oxidized B-chain. The purified relaxin is desalted by RP-HPLC in aqueous TFA-based buffers. Combination yields are usually between 20 and 50% depending on the relaxin under study, but overall yields can be

---

[38] P. D. Johnston, J. Burnier, S. Chen, D. Davis, H. Morehead, M. Remington, M. Struble, G. Tregear, and H. Niall, in "Peptides: Structure and Function" (C. M. Deber, V. J. Hruby, and K. D. Kopple, eds.), p. 683. Pierce, Rockford, Illinois, 1985.

[39] J. Rivier and R. McClintock, J. Chromatogr. **268,** 112 (1983).

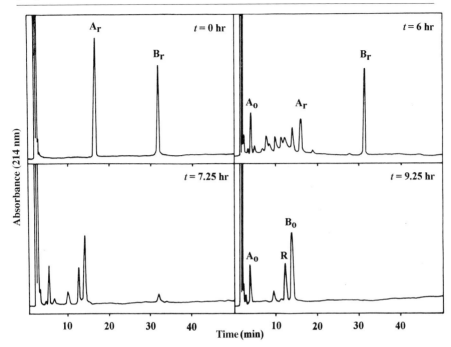

FIG. 2. Synthetic human Gene 1 relaxin chain combination monitoring by RP-HPLC. Conditions: Column, Vydac C_4 (analytical); buffer A, 0.1% aqueous TFA; buffer B, 0.1% TFA in acetonitrile; gradient, 25–40% B in 50 min; flow rate, 1.5 ml/min; wavelength, 214 nm. A_r, S-Reduced A-chain; B_r, S-reduced B-chain; A_o, oxidized A-chain; B_o, oxidized B-chain; R, relaxin.

as low as 0.5% relative to starting synthetic A-chain resin because of the generally poor recovery of B-chains.

Purified synthetic relaxin is characterized by a number of standard criteria[20] including by mass spectrometry. The presence of two chains can also be determined by simple S-reduction of an aliquot and analytical RP-HPLC. Confirmation of the parallel alignment of the two chains is achieved by tryptic digestion of an aliquot of peptide followed by mass spectrometric identification of the appropriate intermolecular disulfide-linked fragments.[8] Synthetic relaxin is stored as a lyophilized powder at 4°. Peptide content is measured by amino acid analysis.

Acknowledgments

Many researchers at the Howard Florey Institute have contributed significantly to the development of synthetic relaxin protocols. We also acknowledge the valuable support and input provided by Drs. Paul Johnston, Richard Vandlen, and John Burnier, all of Genentech, Inc., South San Francisco, CA. The work carried out at the Howard Florey Institute was supported by an Institute Block grant from the National Health and Medical Research Council of Australia.

[30] Solution Nuclear Magnetic Resonance Characterization of Peptide Folding

By KEVIN H. MAYO

Introduction

The advent of solid-phase peptide synthesis has opened the door to various fields in biochemistry: understanding protein folding and folding pathways; protein fragment activity studies; *de novo* design of novel peptides; peptide ligand–receptor interactions; and drug discovery. To more fully understand how any peptide behaves or functions, conformational analyses are required. Two spectroscopic methods, nuclear magnetic resonance (NMR) and circular dichroism (CD), have been called on most frequently to assess the type and degree of peptide folding. This article, while focused primarily on solution NMR characterization of peptide folding, cannot exclude information from CD studies, which are most often presented along with complementary NMR experiments. For this reason, some reference will be made to various CD studies on peptides as related to NMR-based investigations. The usually convoluted CD trace represents a weighted conformational average over all backbone positions, whereas NMR has the capability of reporting weighted averages at specific backbone positions. Although this article is not meant to be a reference compendium for all NMR studies that have been done on peptides, pertinent examples from the literature are discussed. The following sections are presented to give the reader a brief overview of the use of NMR in analyzing conformational populations of peptides in solution.

Peptides in Water

The term structured protein or peptide refers to that overall conformational state which exists as a single, compact fold or as a family of very closely related folds. The "folded" state exists because the polypeptide chain has sufficient thermodynamic "stability" from various internal interactions to maintain a "tight knit" conformation. In aqueous solution, however, any short linear peptide (without sufficient "poly" in it, i.e., less than

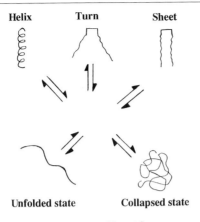

FIG. 1. Simple scheme of peptide folding equilibria showing a few possible backbone conformations as discussed in the text.

about 40 to 50 residues) and lacking sufficient structurally stabilizing forces, exists as a highly fluctuating distribution of conformational states as depicted schematically in Fig. 1. Here, the terms "unfolded," "collapsed," "helix," "turn," and "β sheet" have been used to indicate some of the possible "folding" contributions in this conformational ensemble. Whereas the later three are self-explanatory, the former two are least clear. The "collapsed" state is perhaps best described as a hydrophobically mediated association of side chains resulting in or not resulting in some clear structural elements; at one extreme would lie the "molten globule" state[1] which has a collapsed hydrophobic core and considerable secondary structure but lacks a defined, fixed tertiary structure. A true, fully "unfolded" state would be one where all residues are maximally exposed to solvent; this state probably does not actually exist in water since some folding, however transient, would tend to screen hydrophobic residues from water (i.e., some hydrophobic collapse would occur.[2] This of course depends on the amino acid sequence, composition, and length. In general, equilibria are shifted toward what is most often refered to in the literature as the "random coil" state (shown in Fig. 1 as some combination of unfolded and collapsed state), such that the population of any one "folded" state is usually small.

[1] O. B. Ptitsyn and G. V. Misotnov, in "Conformations and Forces in Protein Folding" (B. T. Nall and K. A. Dill, eds.). American Association for the Advancement of Science, Washington, D.C., 1991.
[2] V. A. Daragan, E. E. Ilyina, C. G. Fields, G. B. Fields, and K. H. Mayo, Protein Sci. 6, 355 (1997).

Early attempts to identify secondary structure formation within linear peptides were generally unsuccessful (see, e.g., Epand and Scheraga[3]). This lack of evidence led to the premature conclusion that long-range interactions occurring within proteins were absolutely essential for the formation of secondary structure and that short linear peptides were largely unstructured in aqueous solution. A major reason for the dearth of observations of secondary structure in short peptides in aqueous solution was the absence of a sensitive probe for small populations of structured conformers within a conformational ensemble. However, one exception did exist early on: the C-peptide of ribonuclease A (residues 1–13) was found to populate helical conformations to approximately 30% (as assayed by using CD) in aqueous solution at 1.7°.[4] Since that time, many peptides have been found by using NMR and CD to populate a variety of "folded" conformations in aqueous solution, in dynamic equilibrium with "random coil" forms (reviewed by Dyson and Wright[5]). The presence of small fractions of ordered structure in equilibrium with predominantly random coil conformations, however, is not only difficult to detect but also difficult to define. Does the detection of, for example, 5% helical structure mean that 5% of the population is in a fully helical conformation, or are there regions of fluctuating structure in different regions which average to that value?

Because populations of "folded" structure are usually very small and the lifetimes of individual states within this conformational ensemble are short on the NMR chemical shift time scale, NMR resonances average, with the most populous "random coil" state being weighted the most. This gives rise to the so-called random coil NMR spectrum of a peptide as exemplified in Fig. 2 (bottom). Here, a 33-residue peptide has been dissolved in 90% H_2O/10% D_2O, pH 6.3, 40°. [1]H resonances are little dispersed from those of very short peptides as defined in tetrapeptides for all 20 common amino acid residues.[6,7] Backbone NH resonances fall between 7.5 and 8.7 ppm; no αH resonances are observed downfield of the water resonance, and resonances arising from side-chain methyls all overlap at around 1 ppm. This spectrum is in contrast to the ^{1}H NMR spectrum of another 33-residue peptide (βpep-4 from Mayo et al.[8]) shown in Fig. 2 (top). There, many backbone NH and αH resonances are downfield shifted

[3] R. M. Epand and H. A. Scheraga, *Biochemistry* **7,** 2864 (1968).

[4] J. E. Brown and W. A. Klee, *Biochemistry* **10,** 470 (1971).

[5] H. J. Dyson and P. E. Wright, *Annu. Rev. Biophys. Biophys. Chem.* **20,** 519 (1991).

[6] A. Bundi and K. Wüthrich, *Biopolymers* **18,** 285 (1979).

[7] K. Wüthrich, "NMR of Proteins and Nucleic Acids." Wiley (Interscience), New York, 1986.

[8] K. H. Mayo, E. Ilyina, and H. Park, *Protein Sci.* **5,** 1301 (1996).

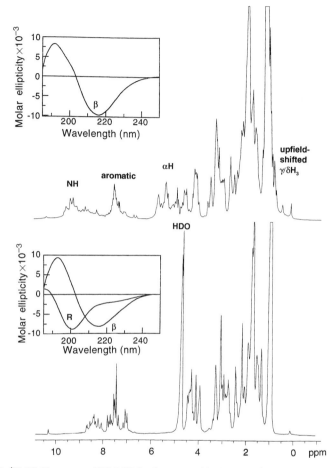

FIG. 2. ^{1}H NMR spectra (600 MHz) of two peptide 33-mers (5 mM peptide in 10 mM potassium phosphate, pH 6.3, 313 K). NMR data were acquired on a Bruker AMX-600 spectrometer. Insets show CD traces of three peptide 33-mers as discussed in the text; CD data were acquired on a Jasco JA-710 spectropolarimeter.

by up to about 1 ppm from their random coil positions, and methyl reso- nances are upfield ring current shifted resulting from proximity to aro- matic side chains. On first inspection, this relatively good chemical dis- persion suggests compact folding. Furthermore, downfield NH and αH resonance shifting is one indication of the specifics of folding, namely, β-sheet folding.

Wishart *et al.*[9] have established a simple formalism for deriving helix and β-sheet populations on the basis of the deviation of observed ^{1}H chemical shifts from a "random coil" reference set. When viewed over a number of residues in a given sequence, downfield shifts indicate β-sheet formation while upfield shifts indicate helix formation. Because an observed chemical shift represents a linear combination of random coil and more folded states, the degree of deviation can be used to discern relative "folded" populations. In the present case (Fig. 2, top), greater than 90% sheet formation is indicated.

More specific structural details result from a complete analysis of various multidimensional NMR data sets.[7] For this peptide 33-mer, as for most peptides studied by NMR, the usual approach consists of acquiring ^{1}H homonuclear two-dimensional (2D) NMR COSY, TOCSY, and NOESY data. COSY (correlated spectroscopy) and TOCSY (total correlated spectroscopy) experiments (and variations thereof) correlate resonances whose protons are through-bond coupled, whereas NOESY (nuclear Overhauser effect spectroscopy) experiments provide information on which protons are through-space dipole–dipole coupled. The power of the nuclear Overhauser effect (NOE) in terms of structural elucidation lies in the fact that an NOE is inversely proportional to the sixth power of the internuclear distance. For short linear peptides which usually exist in solution as an equilibrium of highly fluctuating conformational states, the internuclear distance is considerably more variable than in a compactly folded protein, and derivation of accurate internuclear distances is complicated by the fact that the observed NOE represents some unknown distance-of-closest-approach weighted average. Moreover, what is often less appreciated is that the NOE is also directly proportional to a function which is dependent on internuclear motions. In this respect, interpretation of NOE data, particularly for peptides, should be approached with caution. For peptide conformational analyses, however, NOE data may provide a qualitative picture of preferred conformational populations.

As with any peptide/protein conformational analysis, sequence-specific resonance assignments must be done first. Here, COSY and TOCSY experiments are used initially to help identify or to group spin-coupled resonances into a particular amino acid residue type. This is often referred to as "spin system identification." For example, a residue having a short side chain often can be distinguished from one having a long side chain, or a polar residue from a nonpolar one. Ser, Thr, Gly, and Ala have a more easily distinguishable spin system, whereas Tyr, His, Trp, Phe, Asp, and Asn residues and Glu, Gln, and Met residues usually show characteristic αH–

[9] D. S. Wishart, B. D. Sykes, and F. M. Richards, *Biochemistry* **31**, 1647 (1992).

βH_2 and $\alpha H - \beta H_2 - \gamma H_2$ patterns, respectively. Longer chain aliphatic and Lys spin systems are often problematic to discern. Spin system identification is exemplified in Fig. 3, which shows a "fingerprint" region from a TOCSY spectrum of a 27-residue peptide, $\gamma 27$, derived from the C terminus of the human fibrinogen γ chain.[10] The sequence runs from residues 385 to 411 as Lys-Ile-Ile-Pro-Phe-Asn-Arg-Leu-Thr-Ile-Gly-Glu-Gly-Gln-Gln-His-His-Leu-Gly-Gly-Ala-Lys-Gln-Ala-Gly-Asp-Val. For orientation in this TOCSY contour plot, the NH resonance domain runs horizontally and the αH and side-chain proton resonance domain runs vertically. For any 2D-NMR experiment, the diagonal (not shown) represents the normal NMR spectrum, and off-diagonal cross-peaks shown here identify which resonances are coupled. In this TOCSY experiment, therefore, cross-peaks identify which NH's are through-bond coupled to which αH's and side-chain proton resonances. Although the intent of this section is not to provide an exhaustive spectral analysis, some key points can be made. Notice that the five Gly αH_2 resonances all fall around 4 ppm, and residues with an $\alpha H - \beta H_2$ fragment have their βH_2 resonances falling between 2.5 and 3.5 ppm, with Asn and Asp resonances usually lying more upfield.[7] The $\beta H_2 - \gamma H_2$ protons of the four Glu/Gln spin systems resonate with a characteristic pattern between 1.9 and 2.5 ppm. The two Ala spin systems are identified by a relatively intense βH_3 resonance coupled directly to the respective αH resonance, and even the single Thr-393 can be picked out by the highly downfield shifted βH (4.2 ppm) being coupled to an upfield βH_3. Ile and Val spin systems often may be identified through their more upfield shifted methyl resonances and βH resonances around 1.8 to 2.2 ppm; Val βH resonances are usually the more downfield ones.

Sequence-specific resonance assignments labeled in Fig. 3 resulted from subsequent analysis of a NOESY spectrum (Fig. 4) which was used to identify sequential spin systems, usually by the presence of an NOE between the αH of one residue and the NH of a sequential neighbor. Spin systems are thereby placed in a sequence that can then be compared to the known chemical sequence for sequence-specific resonance assignments. In this example (Fig. 4), TOCSY $\alpha H - NH$ cross-peaks are boxed in, and lines connect $(i, i + 1)$ $\alpha H - NH$ NOEs. A good starting point is usually one of the unambiguous spin systems (e.g., Ala, Gly, Thr). For example, starting with either Ala spin system, one can unambiguously connect six spin systems as Gly-Ala-Xxx-(Glu/Gln)-Ala-Gly, where Xxx is probably a long side-chain spin system and Glu/Gln cannot be distinguished. When compared with the known amino acid sequence given above, this hexapeptide fragment can be unambiguously assigned as Gly^{404}-Ala^{405}-Lys^{406}-Gln^{407}-Ala^{408}-

[10] K. H. Mayo, C. Burke, J. N. Lindon, and M. Kloczewiak, *Biochemistry* **29**, 3277 (1990).

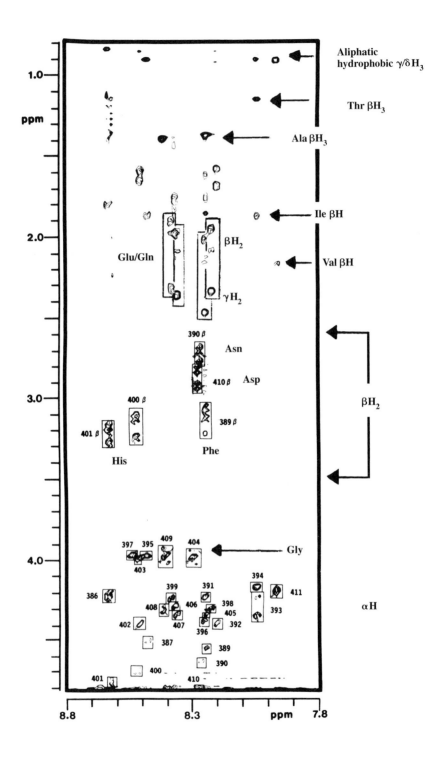

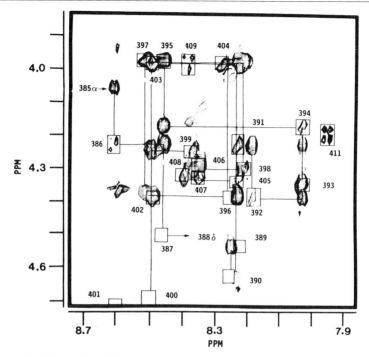

FIG. 4. Backbone NH–αH fingerprint region from a NOESY contour plot. TOCSY cross-peaks have been boxed in, and lines connect spin systems in the sequence. Sample conditions are the same as in legend for Fig. 3.

Gly[409]. Other spin system fragments were likewise identified and assigned. The unique Thr spin system had to be Thr-393 and was used to identify the sequence Pro-388 αH to Gly-395 NH. Because all Gly αH$_2$ (but not NH) resonances overlap somewhat, ambiguity arises and the sequence must start again from a different point. The Ile-387 to Pro-388 connection is made between the Ile-387 αH and the Pro-388 δH$_2$ (upfield, not shown) since Pro has no NH. In this way, the complete amino acid sequence can be assigned and traced as shown in Fig. 4.

Having completed sequence-specific resonance assignments, preferred conformational populations, if present, may be identified from further analysis of NOESY data as well as from analyses of coupling constants, back-

FIG. 3. 2D NMR TOCSY contour plot of the NH–αH and upfield region for a 27-residue peptide derived from γ chain of human fibrinogen, γ27. The peptide concentration was 5 mM in 20 mM potassium phosphate in 90% H$_2$O/10% D$_2$O.

bone NH temperature factors, and the presence of long-lived NHs. In a compactly folded protein, identification of numerous long-range NOEs, that is, greater than $i, i + 3,4,5$, usually can be made; coupling constants range from about 4 Hz (helix) to 9 Hz (sheet); and NH's form hydrogen bonds to yield low NH temperature factors and long-lived NHs. For peptides in water alone, however, the situation is very different. Long-range NOEs are usually not observed; instead, mostly $i, i + 1$ and possibly $i, i + 2$ NOEs are evident; coupling constants most often average to between 6 and 8 Hz (conformationally uninformative), and hydrogen bonds are weak, if present at all, resulting in higher NH temperature factors and few long-lived NH's. $\gamma 27$ peptide was no exception, and a few weak $i, i + 2$ αH–NH NOEs provided some evidence for turn formation, while a series of mostly weak $i, i + 1$ NH–NH NOEs (Fig. 5) suggested transient helix formation from Ile-394 to Leu-402. For further information in performing conforma-

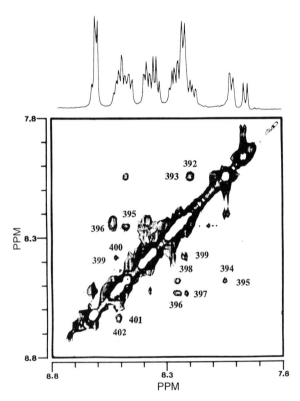

Fig. 5. NH–NH region from a NOESY contour plot. Sample conditions are the same as in legend for Fig. 3.

tional analyses using NMR, the reader is referred to texts by Wüthrich,[7] Clore and Gronenborn,[11] and Evans.[12]

Of further interest to understanding folding of peptides is a comparison of NMR and CD data. For example, both peptide 33-mers discussed above give CD traces (insets to Fig. 2) indicative of significant β-sheet conformation (labeled with β),[13] whereas NMR spectra allow one to differentiate between compact (Fig. 2, top) and noncompact (Fig. 2, bottom) folding. The CD trace for the NMR "random coil" peptide (Fig. 2, bottom), however, does indicate significant β-strand secondary structure (labeled with β). In this case, any apparent β-sheet structure must be highly transient. Moreover, another peptide 33-mer gives both a "random coil" CD trace (labeled with "R" in Fig. 2, inset) and NMR spectrum (not shown; essentially the same as in Fig. 2, bottom); here, no preferred secondary or tertiary structure can be deduced. What makes peptides like these 33-mers fold or not fold is an area of intense investigation in the peptide folding field.

Peptides have been used as models for initial folding events. Fragments derived from a structured protein may mimic the "unfolded" state and allow analysis of local interactions while eliminating the complication of tertiary interactions.[5,14–24] Initial folding of a complete protein sequence does not necessarily give rise to structures that are retained in the final folded form of the protein; examples exist of structures found in peptides that are not seen in the intact protein.[25] However, it does appear that the protein primary sequence codes for secondary structure found in the native

[11] G. M. Clore and A. M. Gronenborn, "NMR of Proteins." CRC Press, Boca Raton, Florida, 1993.

[12] J. N. S. Evans, "Biomolecular NMR Spectroscopy." Oxford Univ. Press, Oxford, 1995.

[13] G. Fasman, ed., "Circular Dichroism and the Conformational Analysis of Biomolecules." Plenum, New York, 1996.

[14] P. E. Wright, H. J. Dyson, and R. A. Lerner, *Biochemistry* **27,** 7167 (1988).

[15] D. Shortle and A. K. Meeker, *Biochemistry* **28,** 936 (1989).

[16] P. S. Kim and R. L. Baldwin, *Annu. Rev. Biochem.* **59,** 631 (1990).

[17] H. J. Dyson, G. Merutka, J. P. Waltho, R. A. Lerner, and P. E. Wright, *J. Mol. Biol.* **226,** 795 (1992).

[18] H. J. Dyson, J. R. Sayre, G. Merutka, H.-C. Shin, R. A. Lerner, and P. E. Wright, *J. Mol. Biol.* **226,** 819 (1993).

[19] J. Sancho and A. R. Fersht, *J. Mol. Biol.* **224,** 741 (1992).

[20] Y. Kuroda, T. Nakai, and T. Ohkubo, *J. Mol. Biol.* **236,** 862 (1994).

[21] H. Shin, G. Merutka, J. Waltho, P. Wright, and H. Dyson, *Biochemistry* **32,** 6348 (1993).

[22] H. Shin, G. Merutka, J. Waltho, L. Tennant, J. Dyson, and P. Wright, *Biochemistry* **32,** 6356 (1993).

[23] J. Waltho, V. Feher, G. Merutka, H. Dyson, and P. Wright, *Biochemistry* **32,** 6337 (1993).

[24] A. D. Kippen, J. Sancho, and A. Fersht, *Biochemistry* **33,** 3778 (1994).

[25] H. J. Dyson, K. J. Cross, R. A. Houghten, I. A. Wilson, P. E. Wright, and R. A. Lerner, *Nature (London)* **318,** 480 (1985).

folded protein; this was demonstrated with NMR by using a series of peptide fragments derived from two different protein folds: the peptides derived from the β-sheet protein plastocyanin showed conformational preferences predominantly for extended β structure in solution,[18] whereas those derived from the four-helix bundle protein myohemerythrin showed considerable propensity for helical structures.[17]

Problems with conformational equilibria always arise. Most often these tend to be associated with the lack of observable NMR parameters. Because of rapid conformational averaging and the shift in equilibrium to the "random coil" state (Fig. 1), chemical shifts are structurally uninformative, J coupling constants average to 6–8 Hz, and NOEs are significantly weakened. In addition, observed NOEs represent some weight average over a number of structural states, complicating conformational analysis. Moreover, NH temperature factors and hydrogen/deuteron exchange data, often used to deduce hydrogen bonding in a structured protein/peptide, are less useful and more difficult to interpret with "unstructured" peptides. A further problem arising from conformational equilibria may occur when various folding interactions are quasi-stable and lifetimes of some conformations fall into an intermediate or slow exchange regime on the NMR chemical shift time scale. This gives rise to line broadening and/or to individual resonances for each long-lived exchange state. In general, the latter is not a problem with short linear peptides which remain monomeric; however, the presence of aggregation can induce any number of unwanted species. Furthermore, cis–$trans$-proline isomerization, which occurs slowly on the NMR chemical shift time scale,[7] can present serious problems particularly in peptides where conformational stability is low compared to a structured protein. This conformational "freedom" tends to shift the cis–trans equilibrium toward the cis state at the expense of the normally more populated trans state. Two sets of resonances are then observed for each isomer; chemical shift separation is greatest for residues sequentially flanking the proline.

It should also be mentioned that even though numerous NMR structural studies have been done using [15]N- and ^{13}C-labeled proteins, most peptide NMR investigations have been limited to ^{1}H NMR techniques. The main reasons for this are (1) the usual small size of the peptide and (2) the cost of [15]N- and ^{13}C-labeled and derivatized amino acid residues required for synthesis. In future studies on peptides, more use probably will be made of selective isotopic enrichment in peptides; some advantages would be increased sensitivity to relatively small conformational populations in a highly overlapped ^{1}H NMR spectrum and possible characterization in the "unfolded" state.[2,26] ^{1}H NMR characterization of peptide folding has been

[26] T. M. Logan, E. T. Olejniczak, R. X. Xu, and S. W. Fesik, *J. Biomol. NMR* **3**, 225 (1993).

done on peptides with conformational populations from turns, helices, coiled coils, β hairpins, β sheets, to collagen triple helices. The following sections exemplify these various structural types.

Helices and Turns

Reverse turns are determined by short-range interactions and limit the conformational space available to a polypeptide chain. It has long been recognized that certain amino acid sequences have a high probability of being part of a turn conformation in proteins,[27,28] and this has more recently been shown by NMR and CD to be true also for peptides in aqueous solution.[29] In some of these short linear peptides, characteristic β-turn NOEs have been observed: NH–NH between turn positions 3 and 4, and αH–NH between turn positions 2 and 4. The weakness of these NOEs indicates partial folding. Several NMR studies done on cell adhesion-promoting Arg-Gly-Asp-containing hexapeptides found evidence for "nested" β turns (see Mayo *et al.*[30] and references therein). Furthermore, Stradley *et al.*[31] found that when turn-type NOEs are observed in short peptides, they usually represent an equilibrium among various types of turns, particularly type-I (III) and type-II.

Turns have been implicated in the folding of helices: the structure known as "nascent helix" formed in short peptides in aqueous solution[32] consists of a series of interlinked turnlike conformations that readily form helix in the presence of solvents such as trifluoroethanol or in the intact protein. A "nascent helix" is characterized by a continuous series of NH–NH NOEs and weak or absent $i, i + 2,3,4$ NOEs. If a sequence forms turnlike structures in solution as an initiating event in the folding process, then it is likely that the structure will persist in longer fragments containing the sequence. By investigating short linear peptides derived from specific structural units in the native protein, that is, helix, turn, or β sheet/turn regions, conformational preferences in early folding intermediates may be postulated under equilibrium conditions. When solvent conditions are the same as native folding conditions, it may be argued that relatively "stable" conformations probably do exist early on in the protein folding conformational ensemble.

[27] G. D. Rose, L. M. Gierasch, and J. A. Smith, *Adv. Protein Chem.* **37**, 1 (1985).
[28] P. Y. Chou and G. D. Fasman, *Adv. Enzymol.* **47**, 45 (1978).
[29] H. J. Dyson, M. Rance, R. A. Houghten, R. A. Lerner, and P. E. Wright, *J. Mol. Biol.* **201**, 161 (1988).
[30] K. H. Mayo, F. Fan, M. P. Beavers, A. Eckardt, P. Keane, W. J. Hoekstra, and P. Andrade-Gordon, *Biochim. Biophys. Acta* **1296**, 95 (1996).
[31] S. J. Stradley, J. Rizo, M. D. Bruch, A. N. Stroup, and L. M. Gierasch, *Biopolymers* **29**, 263 (1990).
[32] H. J. Dyson, M. Rance, R. A. Houghten, P. E. Wright, and R. A. Lerner, *J. Mol. Biol.* **201**, 201 (1988).

That being the case, it may be possible to identify structures which either form first or at least are stable intermediates. These potential "intermediates" may or may not remain as part of the native protein structure. The approach of using synthetic peptides derived from structural units in native proteins has been increasingly exploited. Kemmink and Creighton[33] studied the conformational properties of seven overlapping peptides, 9 to 16 residues long, that comprise the entire primary structure of bovine pancreatic trypsin inhibitor (BPTI). The peptides are largely disordered but have somewhat different average conformational propensities, in line with various secondary structure predictions. Reduced BPTI appears to be approximately the sum of its individual peptides. Several local interactions involving aromatic rings of side chains interacting with groups nearby in the primary structure have been identified and verified by replacing the responsible side chains. Some of these interactions were found to play a role in folding of reduced BPTI. Wright and co-workers[21-23] have undertaken an extensive NMR study of peptide fragments corresponding to regions in myoglobin that have been implicated in early folding events. Waltho *et al.*[23] and Shin *et al.*[21] studied the conformational preferences of peptides corresponding to the G and H helices and of the intervening sequence, termed the G–H turn. Optical methods failed to detect helical structure,[3,34] but small conformational preferences for helical structure were found by NMR.[35,36] Waltho *et al.*[23] and Shin *et al.*[21] demonstrated that a peptide corresponding to the entire H helix contains a significant population of helical structure in solution as a consequence of the removal of nonhelical residues at the C terminus and the additional residues at the N terminus that form part of the H helix in the native protein. This study indicates that formation of stable structure depends to some extent on where the peptide is "clipped." Related to this, Munoz *et al.*[37] showed that inclusion of a "hydrophobic-staple" at the end of a helix promotes increased stability; in an 18-mer, they noted numerous longer-range helix NOEs in about half of the proposed helix. NMR also has been used to establish the existence of N- and C-terminal capping interactions in peptides.[38,39] These primarily

[33] J. Kemmink and T. E. Creighton, *J. Mol. Biol.* **234,** 861 (1993).
[34] J. Hermans and D. Puett, *Biopolymers* **10,** 895 (1971).
[35] J. P. Waltho, V. A. Feher, R. A. Lerner, and P. E. Wright, *FEBS Lett.* **250,** 400 (1989).
[36] J. P. Waltho, V. A. Feher, and P. E. Wright, *in* "Current Research in Protein Chemistry" (J. J. Villafranca, ed.), p. 283. Academic Press, San Diego, 1990.
[37] V. Munoz, F. J. Blanco, and L. Serrano, *Nat. Struct. Biol.* **2,** 380 (1995).
[38] P. C. Lyu, D. E. Wemmer, H. X. Zhou, R. J. Pinker, and N. R. Kallenbach, *Biochemistry* **32,** 421 (1993).
[39] H. X. Zhou, P. C. Lyu, D. E. Wemmer, and N. R. Kallenbach, *J. Am. Chem. Soc.* **116,** 1139 (1994).

arise from hydrogen-bonding interactions of a polar side chain (e.g., Asn, Gln, Ser) with a backbone carbonyl or amine group. Such interactions also appear to play an important role in helix stabilization in short linear peptides. Although native capping interactions have been identified in short linear peptides, nonnative capping interactions have also been observed.[40,41]

β Sheets

In terms of studying short peptides with β-sheet folding tendencies, the situation differs as attempts to observe isolated β sheets have usually been hampered by the strong tendency of these peptides to aggregate and by a general lack of requisite longer range interactions that dictate a more stable β-sheet fold. The smallest β-sheet unit is one in which two antiparallel β strands are joined by a β turn. This is often referred to as a β hairpin. Generally, these too are not stable in water and often aggregate. Some exceptions, however, have been reported. Blanco et al.[42] studied a short linear peptide derived from the B1 domain of protein G and found that it folds to some extent (40%) into a β-hairpin conformation in water. By replacing a turn in a β-hairpin peptide derived from Tendamistat with the sequence NPDG (highest predicted score to adopt a type I β turn), Blanco et al.[43] showed that that peptide also folds into a β hairpin in water. These same turn residues have also been found to stabilize a nonnative hairpin conformation in a peptide derived from the N-terminal β-hairpin of ubiquitin.[44] More recently, Ramirez-Alvarado et al.[45] designed a novel dodecapeptide which was found by NMR to adopt a 30% populated β-hairpin structure in water. Although interstrand NOEs consistent with β-sheet structure were observed (4–9 αH–αH; 3–10 NH–NH; 5–8 NH–NH), a series of relatively strong NH–NH NOEs also were observed from residue 5 to residue 12, suggesting the additional presence of nascent helix conformation[32] in relatively fast exchange with the β-hairpin population. This raises an interesting point that when dealing with short peptides, NOEs representing multiple conformations may be observed, and caution should be exercised when interpreting data.

Two well-characterized examples of β-sheet formation in somewhat longer peptides have been found in a disulfide-bridged heterodimeric model

[40] E. Ilyina, R. Milius, and K. Mayo, Biochemistry 33, 13436 (1994).
[41] B. Odaert, F. Baleux, T. Huynh-Dinh, J. M. Neumann, and A. Sanson, Biochemistry 34, 12820 (1995).
[42] F. Blanco, G. Rivas, and L. Serrano, Nat. Struct. Biol. 1, 584 (1994).
[43] F. J. Blanco, G. Rivas, and L. Serrano, J. Am. Chem. Soc. 115, 5887 (1993).
[44] M. S. Searle, D. H. Williams, and L. C. Packman, Nat. Struct. Biol. 2, 999 (1995).
[45] M. Ramirez-Alvarado, F. J. Blanco, and L. Serrano, Nat. Struct. Biol. 3, 604 (1996).

of a protein folding domain[46] and a zinc-finger peptide where a two-stranded amphipathic β sheet is stabilized by the folding an amphipathic helix onto it.[47] In both cases, the presence of a short length of β sheet was established by interstrand NOEs. Several other peptides have been proposed to form β sheets on the basis of NMR data,[48–50] but for none of these are the data completely convincing because no cross-sheet NOEs were observed. In a protein dissection study, one peptide 20-mer derived from the β sheet of platelet factor-4 (PF4) was shown by NMR to have a nativelike turn conformation that is stabilized by hydrophobic side-chain interactions from flanking β-strand sequences.[40] Moreover, some nonnative helixlike conformational populations were noted in one of the β strands. In related studies, peptide 33-mers comprising the full β-sheet domain from PF4[51] and from interleukin-8 (IL-8)[8] showed native-like, compact β-sheet sandwich folding.

Collagen Model Peptides

Although most NMR peptide conformational studies have been done on peptides derived from globular proteins containing more traditional α-helical, β-sheet, and turn motifs, triple helices do represent another major structural motif. The triple helix is the main structural unit found in all collagens and in a variety of other proteins and is generally composed of the amino acid sequence repeat $(X-Y-Gly)_n$. In fibrillar collagens, this repeat can extend to about 1000 residues long. Nonglobular, fibrillar proteins like most collagens are difficult to work with from a structural perspective. For example, most do not crystallize, obviating X-ray diffraction analysis, and size, solubility, and gelation generally preclude solution NMR studies. Alternatively, one can work with considerably shorter fragments of the triple helix, for example, peptides generated by synthetic techniques. To date, various triple-helical model peptides such as $(Pro-Pro-Gly)_n$ and $(Pro-Hyp-Gly)_n$ have been studied by using NMR.

Mayo *et al.*[52] studied the ^{1}H NMR solution conformation of a nonconsensus collagen peptide Gly-Val-Lys-Gly-Asp-Lys-Gly-Asn-Pro-Gly-Trp-Pro-Gly-Ala-Pro-Tyr (called peptide IV-H1), derived from the protein sequence of human collagen type IV, triple helix domain residues 1263–1277. The NMR data on peptide IV-H1 indicated significant β-turn populations cen-

[46] T. G. Oas and P. S. Kim, *Nature (London)* **336**, 42 (1988).
[47] M. S. Lee, G. Gippert, K. Y. Soman, D. A. Case, and P. E. Wright, *Science* **245**, 635 (1989).
[48] R. Saffrichm, H. R. Kalbitzer, H. Bodenmuller, P. Muhn, R. Pipkorn, and H. C. Schaller, *Biochim. Biophys. Acta* **997**, 144 (1985).
[49] S. C. J. Sumner and J. A. Ferretti, *FEBS Lett.* **253**, 117 (1989).
[50] J. B. Vaughn and J. W. Taylor, *Biochim. Biophys. Acta* **999**, 135 (1989).
[51] E. Ilyina and K. H. Mayo, *Biochem. J.* **306**, 407 (1995).
[52] K. H. Mayo, D. Parra-Diaz, J. B. McCarthy, and M. Chelberg, *Biochemistry* **30**, 8251 (1991).

tered at Lys^3-Gly^4, Lys^5-Gly^6, Pro^9-Gly^{10}, and Pro^{12}-Gly^{13}, and a C-terminal γ-turn within the Ala^{14}-Pro^{15}-Tyr^{16} sequence. Although this multiple turn conformation may or may not be found in native collagen, it is one structural possibility for a nonconsensus collagen sequence, many of which are found in native collagen. By using NMR, CD, and cross-linking, Hoppe et al.[53] showed that the structure of a 35-residue collagen-type sequence from lung surfactant protein D consists of a triple-stranded parallel α-helical bundle in a nonstaggered, and extremely strong, noncovalent association. This type of association between three polypeptide chains may represent a common structural feature immediately following the C-terminal end of the triple-helical region of collagenous proteins.

Li et al.[54] have used ^1H and ^{15}N NMR to structurally characterize two consensus collagen triple-helical peptides: (Pro-Hyp-Gly)$_{10}$, which is considered to be the most stable triple helix model peptide, and (Pro-Hyp-Gly)$_3$Ile-Thr-Gly-Ala-Arg-Gly-Leu-Ala-Gly-Pro-Hyp-Gly (Pro-Hyp-Gly)$_3$ (called T3-785), designed to model an imino acid poor region of collagen. ^{15}N enrichment has been selective at specific Ala and Gly positions. Both peptides associated as trimers with melting temperatures of 60° for (Pro-Hyp-Gly)$_{10}$ and 25° for T3-785. In T3-785, with nonrepeating Xxx-Yyy-Gly units incorporated in the sequence, the three chains of the homotrimer were distinguishable in NMR spectra. The solution conformations of (Pro-Hyp-Gly)$_{10}$ and T3-785 were similar to the model derived from X-ray fiber diffraction data, with the closely packed central residues of the three chains showing a one-residue stagger of the three parallel chains.

In a related NMR study, Long et al.[55] designed a set of four (Pro-Hyp-Gly)$_{10}$ peptides to model the effect of (Gly-X-Y)$_n$ repeat interruptions on triple-helix formation, stability, and folding. An interruption in the middle of this stable triple-helical peptide still allowed formation of trimers in solution, but with decreased stability which depended on the type of interruption. A Gly substitution or an Ala insertion were least disruptive, whereas a Gly deletion was the most destabilizing and a Hyp deletion was intermediate. Their results suggested that such interruptions in the collagen repeat pattern affected both triple-helix and monomer conformations. These studies provided a general approach to analyzing the percent triple helix composition in a collagen peptide. By measuring the loss of monomer resonance intensity and the increase in trimer resonance intensity, Liu et

[53] H. J. Hoppe, P. N. Barlow, and K. B. Reid, FEBS Lett. 344, 191 (1994).
[54] M. H. Li, P. Fan, B. Brodsky, and J. Baum, Biochemisry 32, 7377 (1994).
[55] C. G. Long, E. Braswell, D. Zhu, J. Apigo, J. Baum, and B. Brodsky, Biochemistry 32, 11688 (1993).

al.[56] were able to directly measure by NMR the folding kinetics of the ^{15}N-enriched collagenlike triple-helical peptide (Pro-Hyp-Gly)$_3$Ile-Thr-Gly-Ala-Arg-Gly-Leu-Ala-Gly(Pro-Hyp-Gly)$_4$.

Several problems still exist in studying collagen-based peptides. The repetitiveness of the amino acid sequence presents serious problems with resonance overlap. Moreover, because this block copolymer would have the same folded structure, resonance dispersion due to differences in chemical/magnetic environment is minimal. One way to help overcome this problem is to use selective ^{15}N and/or ^{13}C isotopic enrichment. The problem of strand alignment presents another concern because one relies on an equilibrium among three homostrands to associate the same in all cases. This cannot always be controlled, and various populations with slightly different alignments may arise. Moreover, measurements generally need to be done at low temperature 5° or 10° in order to shift the equilibrium as far as possible toward a more structured state. An approach which covalently attaches the three collagen strands to a chemical scaffold[57] appears to be able to circumvent this problem. This novel synthetic approach may also solve the chemical degeneracy problem, as a heterotrimer may be synthesized.

A number of NMR conformational analyses also have been performed on collagen-derived telopeptides, that is, nonhelical peptides (20–30 residues) found at the N- and C-terminal ends of collagen.[58–63] In general, NMR data suggest mostly extended type structures with the presence of some turn populations. For this work, mixed solvents were required for solubility and conformational stabilization. This serves to introduce the next section for discussion.

Toward Improved Conformational Stability and Better Nuclear Magnetic Resonance Structures

As should be evident from the above discussion, most peptides in water generally have limited conformational stability, even when the temperature

[56] X. Liu, D. L. Siegel, P. Fan, B. Brodsky, and J. Baum, *Biochemistry* **35,** 4306 (1996).
[57] C. G. Fields, C. M. Lovdahl, A. J. Miles, V. L. Matthias Hagen, and G. B. Fields, *Biopolymers* **33,** 1695 (1995).
[58] X. H. Liu, P. G. Scott, A. Otter, and G. Kotovych, *J. Biomol. Struct. Dyn.* **8,** 63 (1990).
[59] A. Otter, P. G. Scott, and G. Kotovych, *Biochemistry* **27,** 3560 (1988).
[60] A. Otter, P. G. Scott, X. H. Liu, and G. Kotovych, *J. Biomol. Struct. Dyn.* **7,** 455 (1989).
[61] A. Otter, G. Kotovych, and P. G. Scott, *Biochemistry* **28,** 8003 (1989).
[62] A. Otter, P. G. Scott, and G. Kotovych, *Biopolymers* **33,** 1443 (1993).
[63] X. Liu, A. Otter, P. G. Scott, J. R. Cann, and G. Kotovych, *J. Biomol. Struct. Dyn.* **11,** 541 (1993).

is lowered. Usually, only short-range NOEs are observed that suggest the presence of turns and/or nascent helix and, as discussed above, few semistable β-hairpin peptides. Numerous studies, therefore, have been directed toward increasing conformational stability, thereby improving the quantity and quality of NMR observables like J coupling constants and NOEs. Aside from using covalent bonds, that is, -S-S- or peptide cyclization, the four most frequently employed approaches toward improving peptide conformational stability are listed below.

Cosolvents

The most exploited cosolvent known to "stabilize" various conformations in aqueous solution is trifluoroethanol (TFE). Other halogenated aliphatic alcohols, however, such as heptafluorobutyric acid (HFB), may also be used.[64] These solvents generally lower the solution dielectric, help overcome solubility problems, and, more important for this discussion, increase conformational stability. In general, titrations are initially done with the cosolvent in the peptide–water solution in order to optimize percent composition. Usually, cosolvent (i.e., TFE) concentrations used are between 20 and 50%, but this varies from study to study. Other cosolvents may be as active at substantially lower concentrations.[64]

The detection of α helices in many peptides has been aided by using TFE, which induces formation of helical structure in peptides that are mainly or partly unstructured in aqueous solution but have a helical propensity.[65] The exact mechanism by which TFE stabilizes secondary structure is unknown, but some studies suggest that TFE interacts preferentially with the helical conformation of a peptide to shift the structural equilibrium toward this state.[21,22,65,66] It has been shown that the stabilization of helical conformations in protein fragments by TFE tends to be restricted to those regions that are helical in the native protein, although this is not always the case.[67] For the fibrinogen γ-chain-derived γ27 peptide discussed earlier, NH–NH NOEs are few and weak in the absence of TFE (see Figs. 4 and 5), whereas in the presence of 40% TFE (Fig. 6), NOEs (particularly NH–NH) are more plentiful and stronger, consistent with the presence of more stable "structure." Helix formation is supported by the presence of helix-constraining $i, i + 3$ and $i, i + 4$ NOEs.[68]

[64] Y. Yang, S. Barker, M.-J. Chen, and K. H. Mayo, *J. Biol. Chem.* **268**, 9223 (1993).
[65] F. D. Sonnichsen, J. E. Van Eyk, R. S. Hodges, and B. D. Sykes, *Biochemistry* **32**, 8790 (1992).
[66] A. Jasanoff and A. R. Fersht, *Biochemistry* **33**, 2129 (1994).
[67] Z. P. Liu, J. Rizo, and L. M. Gierasch, *Biochemistry* **33**, 134 (1994).
[68] F. Fan and K. H. Mayo, *J. Biol. Chem.* **270**, 24693 (1995).

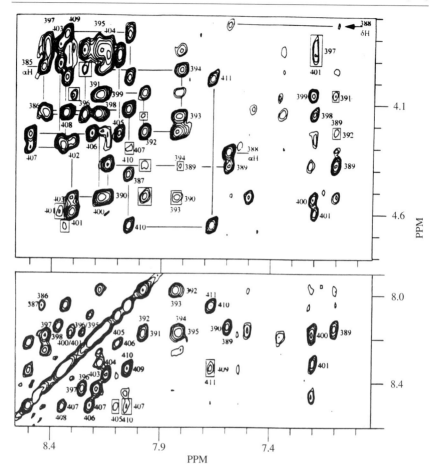

FIG. 6. NH–NH region from a NOESY contour plot. Sample conditions are the same as in legend for Fig. 3 except for the addition of 40% (v/v) trifluorethanol.

Other examples of TFE-induced stabilization of α-helix conformation include studies on the ribonuclease S-peptide[69,70]; peptides derived from bovine pancreatic trypsin inhibitor,[71] a peptide 23-mer from the plasma protein transthyretin,[72] a peptide fragment from α-lactalbumin,[73] and the

[69] J. W. Nelson and N. R. Kallenbach, *Proteins: Struct. Funct. Genet.* **1**, 211 (1986).
[70] J. W. Nelson and N. R. Kallenbach, *Biochemistry* **28**, 5256 (1989).
[71] J. Kemmink and T. E. Creighton, *Biochemistry* **34**, 12630 (1995).
[72] J. Jarvis, S. Munro, and D. Craik, *Biochemistry* **33**, 33 (1994).
[73] L. J. Smith, A. T. Alexandrescu, M. Pitkeathly, and C. M. Dobson, *Structure* **2**, 703 (1994).

phospholipase A_2 fragment residues 38–59.[74] Although helical conformations appear to develop in those regions with a high helical-forming propensity or preexisting helical structure, the stabilizing effects of TFE are not limited to α-helical peptides. Blanco et al.,[75] for example, have observed that TFE also induces nativelike β-hairpin formation in an N-terminal fragment derived from protein G B1 domain. A TFE back-extrapolation procedure[76] has been applied to two barnase-derived synthetic peptides that correspond to sequences 5–21 and 1–36, which contain its first (residues 6–18) and second (26–34) helices. From this analysis, these peptides were found to form 3 and 6% helical structure, respectively, in water at pH 6.3, 25°.[19,76–78] The first helix has been predicted to act as a nucleation site, initiating the folding of barnase[79] and is largely formed in the folding intermediate.[80–82]

With the use of any cosolvent, caution must be exercised in interpreting data because the solution environment can and often does affect conformational distributions in different ways. For example, the transthyretin peptide 23-mer[72] in deuterated dimethyl sulfoxide (DMSO-d_6) shows minor populations of turnlike character, whereas nascent helix between residues 5 and 12 was observed in 20% TFE and water.

Micelles/Vesicles

Many peptides have been observed to interact with micelles, vesicles, and membrane bilayers. Using solution NMR techniques, however, resonance broadening due to the long motional correlation times may preclude conformational analysis. In this respect, solid-state NMR techniques[82a] may be more appropriate to these types of investigations. Nonetheless, in smaller detergent micelles and lipid vesicles, solution NMR has been used success-

[74] M. A. Jimenez, C. Carreno, D. Andreu, F. J. Blanco, J. Herranz, M. Rico, and J. L. Nieto, *Biopolymers* **34**, 647 (1994).

[75] F. J. Blanco, M. A. Jimenez, A. Pineda, M. Rico, J. Santoro, and J. L. Nieto, *Biochemistry* **33**, 600 (1994).

[76] J. Sancho, J. L. Neira, and A. R. Fersht, *J. Mol. Biol.* **224**, 749 (1992).

[77] J. Sancho, L. Serrano, and A. R. Fersht, *Biochemistry* **31**, 2253 (1992).

[78] A. D. Kippen, V. L. Arcus, and A. R. Fersht, *Biochemistry* **33**, 10013 (1994).

[79] L. Serrano, J. T. Jr. Kellis, P. Cann, A. Matouschek, and A. R. Fersht, *J. Mol. Biol.* **224**, 783 (1992).

[80] A. Matouschek, J. T. Kellis, Jr., L. Serrano, M. Bycroft, and A. R. Fersht, *Nature* (*London*) **346**, 440 (1990).

[81] A. Matouschek, L. Serrano, and A. R. Fersht, *J. Mol. Biol.* **224**, 819 (1992).

[82] A. R. Fersht, *FEBS Lett.* **325**, 5 (1993).

[82a] T. A. Cross, *Methods Enzymol.* **289**, [31], 1997 (this volume).

fully. Chupin *et al.*,[83] for example, have studied the PhoE signal peptide 21-mer in a micellar environment and showed that it formed N- and C-terminal α-helical conformations which were separated by a kink at a central glycine. Depending on the size of the macromolecular aggregate and the lifetime of the "bound" peptide, use of the transferred NOE experiment to derive structural information may be appropriate; this is discussed later in this article.

Peptide–Peptide Interactions

In protein folding in general conformational units associate to form a more stable fold. This idea has been carried through to peptide conformational studies by mixing in solution two peptides that would normally be brought together in the native protein. Wu et al.,[84] for example, investigated a peptide model for an early cytochrome *c* folding intermediate consisting of a noncovalent complex between a heme-containing N-terminal fragment (residues 1–38) and a synthetic peptide corresponding to the C-terminal helix (residues 87–104). Those studies indicated that a partially folded intermediate with interacting N- and C-terminal helices is formed at an early stage of folding when most of the chain is still disordered. NMR and CD indicate that the isolated peptides are largely disordered, but, when combined, they form a flexible, yet tightly bound complex with enhanced helical structure. These results emphasize the importance of interactions between marginally stable elements of secondary structure in forming tertiary subdomains in protein folding.

Self-association of peptides also clearly can stabilize conformational populations. For example, the 33-amino acid β-sheet peptide derived from PF4 (residues 23–55)[51] at low concentration exists in aqueous solution in a "random coil" distribution of highly flexible conformational states. Some preferred conformation, however, is observed, particularly within a relatively stable chain reversal from Lys-45 to Arg-49. As the peptide 33-mer concentration and/or temperature is increased, nativelike antiparallel β-sheet folding results. Sheet formation is thermodynamically linked with tetramerization of the peptide, which occurs on the slow chemical shift exchange time scale. Other β-sheet-forming peptides also have been recognized to self-associate and exhibit considerably more stable conformations.[8] Terzi *et al.*[85] had similar findings with the aggregation of fragment 25–35

[83] V. Chupin, J. A. Killian, J. Breg, H. H. de Jongh, R. Boelens, R. Kaptein, and B. Kruijff, *Biochemistry* **34**, 11617 (1995).

[84] L. Wu, P. Laub, G. Elove, J. Carey, and H. Roder, *Biochemistry* **32**, 10271 (1993).

[85] E. Terzi, G. Holzemann, and J. Seelig, *Biochemistry* **33**, 1345 (1994).

of the β-amyloid protein [βAP(25–35)OH] studied under a variety of conditions.

De Novo Design Approach

Most of what has been presented above concerns peptides with sequences derived from native proteins. An alternative approach aimed at improving conformational stability uses basic folding principles to design peptides *de novo*. A number of these *de novo* designed peptides have been designed to promote α-helical and/or turn properties. Regan and DeGrado[86] designed a relatively stable helical peptide. Making use of the hydrophobic effect to engineer a stable helix fold, Hecht *et al.*[87] was one of the first to design a four-helix bundle. Fairman *et al.*[88] designed and characterized another four-chain coiled coil and found that increasing the monomer peptide length in the helix from 21 to 28 to 35 residues resulted in tetramer stabilities in terms of peptide monomer concentration that ranged from 0.18 mM to 51 nM to 0.28 pM, respectively.

Munoz *et al.*[37] have identified a recurrent, stabilizing local structural motif found at the N termini of α helices that they refer to as the hydrophobic-staple; a more stable helix-turn structure is formed when hydrophobic residues from the N-terminal side of a turn interact with hydrophobic residues from the helix. On the basis of such hydrophobic stabilization effects, others have designed various α/α and α/β peptides.[89–91] Whereas most of these peptides form marginally stable helix structures at best, Struthers *et al.*[91] have designed an α/β peptide 23-mer based on a zinc finger motif, which in the absence of coordinating zinc or disulfide bonds forms a compact structure with an N-terminal amphipathic α-helix folded onto a two-stranded amphipathic β sheet.

Although *de novo* design of α helix, helix bundle, and α/β peptides has been successful, designing water-soluble, purely β-sheet-containing peptides has met with limited success. Designing β-sheet peptides has proved more complicated primarily because of their limited solubility via aggregation in water and because of the nature of β-sheet folding, which is dictated by long-range interactions. A small peptide 12-mer, however, has been designed which is water soluble, remains monomeric, and does form some

[86] L. Regan and W. DeGrado, *Science* **241**, 976 (1988).
[87] M. Hecht, J. Richardson, D. Richardson, and R. Ogden, *Science* **249**, 884 (1990).
[88] R. Fairman, H. G. Chao, L. Mueller, T. B. Lavoie, L. Shen, J. Novotny, and G. R. Matsueda, *Protein Sci.* **4**, 1457 (1995).
[89] Y. Fezoui, D. L. Weaver, and J. J. Osterhout, *Protein Sci.* **4**, 286 (1995).
[90] D. J. Butcher, M. D. Bruch, and G. R. Moe, *Biopolymers* **36**, 109 (1995).
[91] M. D. Struthers, R. P. Cheng, and B. Imperiali, *Science* **271**, 342 (1996).

highly transient β-hairpin-like structure.[92] The beta-bellin series[93] and beta-doublet[94] peptides exemplify initial attempts at designing larger β-sheet sandwich peptides; these peptides, usually about 30 residues in length, show limited solubility in water primarily at lower pH values and noncompact β-sheet conformational properties as monitored by CD and NMR. Owing to limited water solubility of beta-bellin 12, for example, NMR assignments could be made only in dimethyl sulfoxide, and no long-range NOEs were observed, indicating the absence of stable folded structure.[93] Beta-Doublet[94] which has the same predicted antiparallel β-sheet motif as beta-bellin, was water-soluble, albeit at low pH; gave a typical β-sheet CD trace, but failed to show any compact folding as monitored by NMR, that is, it gave a ^{1}H NMR "random coil" spectrum.[7] Beta-Bellin 14D[95] is the best in that series and shows good solubility at millimolar concentrations up to pH 5.5. However, it does not give a good compact β-sheet fold even with a covalent disulfide bond between two sandwiched monomers.

Other amphipathic β-sheet forming peptides have been derived from proteins in the α-chemokine family. Studies on β-sheet domain peptides derived from IL-8,[8] Gro-α (growth-related protein-α),[8] and PF4,[51] as well as those on beta-bellins[93,95] and beta-doublet,[94] led to some general guidelines for designing water-soluble, self-association-induced β-sheet-forming peptides.[8] NMR conformational analyses show that formation of β-sheet structure is thermodynamically linked to peptide–peptide associations as discussed in the previous section. One of these β-sheet sandwich peptide 33-mers was found to fold compactly as a tetramer composed of two hetero-dimers.[96] Although the three-stranded antiparallel β-sheet folds in monomer subunits were generally the same, two types of dimers formed by continuing monomeric β sheets into a six-stranded sheet, each with differently aligned interfacial antiparallel β strands. Heterodimers then associated by sandwiching their amphipathic hydrophobic surfaces.

In the future, it is expected that *de novo* designs of various desired peptide conformations will continue with the use of noncommon[91] and D-isomer[97] amino acids.

[92] V. Sieber and G. R. Moe, *Biochemistry* **35,** 181 (1996).
[93] J. S. Richardson, D. C. Richardson, N. B. Tweedy, K. M. Gernert, T. P. Quinn, M. H. Hecht, B. W. Erickson, Y. Yang, R. D. McClain, M. E. Donlan, and M. C. Surles, *Biophys. J.* **63,** 1185 (1992).
[94] T. P. Quinn, N. B. Tweedy, R. W. Williams, J. S. Richardson, and D. C. Richardson, *Proc. Natl. Acad. Sci. U.S.A.* **91,** 8747 (1994).
[95] Y. Yan and B. W. Erickson, *Protein Sci.* **3,** 1069 (1994).
[96] E. Ilyina, V. Roongta, and K. H. Mayo, *Biochemistry* **36,** 5245 (1997).
[97] V. Bobde, Y. U. Sasidhar, and S. Durani, *Int. J. Pept. Protein Res.* **43,** 209 (1994).

Transferred Nuclear Overhauser Effect:
Peptide Ligand–Receptor Interactions

Since the late 1980s, use of the transferred NOE (TRNOE) has made possible conformational studies of small peptides in the "receptor" bound state. The basic experiment employed is the usual NOESY. What differentiates a TRNOE experiment from a standard NOE experiment is the presence of "ligand" and "receptor" components in the experimental system. Figure 7 depicts the basic situation for the TRNOESY experiment.

In the presence of an appropriate receptor, a smaller peptide ligand can exist either free in solution or bound to its receptor. This "receptor" could be an enzyme, a membrane receptor, plasma protein, etc. Although the "receptor" molecular mass could be anything greater than about 25,000 Da, a molecular mass of 200,000 Da will be used for the purpose of this discussion. A peptide of about 12 residues (1400 Da) is often used. In solution, the larger receptor is considerably less mobile than the peptide. Overall tumbling correlation times, τ_R and τ_F, are approximated to be 1×10^{-7} and 1×10^{-9} sec, respectively, and the correlation time for the complex is about that of the receptor itself. Therefore, the effective correlation time, τ_{eff}, in terms of the peptide ligand, is the population-weighted average of the bound and free ligand states:

$$\tau_{eff} = \chi_B \tau_B + \chi_F \tau_F \tag{1}$$

where χ is the mole fraction of ligand in the bound (B) and free (F) states. In this example, 1% peptide bound will contribute nearly equally to τ_{eff} as 99% free. For this reason, usually 5% or less of bound ligand is experimentally required. Furthermore, τ_{eff} is optimized by adjusting the mole fraction

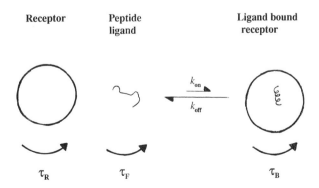

FIG. 7. Schematic representation of a peptide ligand interacting with a "receptor."

of bound and free ligand such that observed ligand resonances for the free state dominate the spectrum; those of the receptor are often either too broad to be observed or can be filtered experimentally.

The NOE can be expressed in the following proportionality as:

$$\text{NOE} \propto f(\tau_{\text{eff}})/r_{ij}^6 \tag{2}$$

where r_{ij} is the interproton distance. The numerator, $f(\tau_{\text{eff}})$, indicates that the NOE is a function of τ_{eff}. From Eq. (2) it is evident that increasing τ_{eff} with some fraction of the peptide bound to receptor can significantly increase the NOE magnitude. Moreover, this experiment is most easily interpreted when the NOE for the free ligand is either positive or near zero, because in the large receptor complex the NOE will be negative; corrections for NOEs arising from the larger free state population are thereby obviated. In this case, the absence of free ligand (peptide) NOEs becomes an advantage.

One further consideration is the interplay of the dipolar cross-relaxation rate, which dictates how fast the NOE will build up and the lifetime of ligand–receptor complex, that is, receptor off-rate, k_{off}. In the receptor-bound state, NOEs build up considerably faster than in the ligand-free state, primarily owing to the much larger value of τ_B. If the ligand remains on the receptor for a time longer than the inverse of the NOE buildup rate, various indirect cross-relaxation processes would occur and no TRNOEs would be observed. The key to adjusting experimental parameters for an effective TRNOE lies in making the ligand off-rate from the receptor complex fast relative to the NOE buildup or cross-relaxation rate. In this case, bound-state NOEs develop in the receptor complex and the ligand is rapidly released; in the ligand-free state, relaxation processes occur much more slowly, and ligand NOEs which developed in the bound state remain for a time and can be detected in the usual NOESY experiment. In general, the equilibrium dissociation constant should be no smaller than about 10×10^{-6} M. With a diffusion-limited ligand on-rate of 5×10^7 M^{-1} sec^{-1}, the ligand–receptor off-rate would be 500 sec^{-1}. For a 200,000-Da receptor complex and an interproton distance of 2.5 Å, the cross-relaxation rate would be about 25 sec^{-1}.

Another useful observable in this experiment is the chemical shift, which also presents itself as a weighted average of free and bound ligand states:

$$\delta_{\text{obs}} = \chi_B \delta_B + \chi_F \delta_F. \tag{3}$$

By analyzing chemical shift changes between free and bound ligand or by varying the ratio of free to bound ligand, it is sometimes possible to argue that resonances most shifted (and broadened) are more directly involved in receptor binding.

This brief overview merely considers some of the salient aspects of the TRNOE experiment. A more thorough analysis of TRNOE data involves, for example, considering various internal motions and implementing a full relaxation matrix approach in which a chemically exchanging multispin system is analyzed. More explicit details of the theory behind the TRNOE can be found in several reviews.[98-100]

Use of the TRNOE to investigate receptor-bound peptide conformation can be exemplified by several 2D NMR studies where "receptor" takes on the role of enzymes: thrombin,[101-103] phosphoglycerate kinase,[104] cyclophilin,[105] the plasma protein fibrinogen,[106] muscle proteins,[107] micelles,[30] lipid vesicles,[108] and a membrane glycoprotein, integrin GPIIb/IIIa.[109] Most of these TRNOE studies were done using synthetic peptides less than 20 residues in length. With thrombin as the "receptor," for example, Ni *et al.*[101] found that TRNOEs of the 14-residue C-terminal fragment (residues 52–65) of hirudin (a 65-residue protein) are consistent with an α-helical structure involving residues 61 to 65 and a hydrophobic cluster involving Phe-56, Ile-59, Try-63 and Leu-64, with all charged residues lying on the opposite face. Thrombin-induced resonance broadening of peptide side chains 55 to 65 argues that this sequence is the predominant binding site of hirudin fragments with thrombin.

Conclusion

Much has been learned concerning basic principles of protein folding from studies on synthetic peptides derived from native protein sequences. This information has been almost immediately incorporated into the *de novo* design of novel peptides with a desired fold. As the state of the art

[98] R. E. London, M. E. Perlman, and D. G. Davis, *J. Magn. Reson.* **97**, 79 (1992).

[99] A. P. Campbell and B. D. Sykes, *Annu. Rev. Biophys. Biomol. Struct.* **22**, 99 (1993).

[100] F. Ni, *Prog. NMR Spectrosc.* **26**, 517 (1994).

[101] F. Ni, Y. Konishi, and H. A. Scheraga, *Biochemistry* **29**, 4479 (1990).

[102] F. Ni, D. R. Ripoll, and E. O. Purisima, *Biochemistry* **31**, 2545 (1992).

[103] J. Srinivasan, S. Hu, R. Hrabal, Y. Zhu, E. A. Komives, and F. Ni, *Biochemistry* **33**, 13553 (1994).

[104] M. Andrieux, E. Leroy, E. Guittet, M. Ritco-Vonsovici, B. Mouratou, P. Minard, M. Desmadril, and J. M. Yon, *Biochemistry* **34**, 842 (1995).

[105] L. T. Kakalis and I. Armitage, *Biochemistry* **33**, 1495 (1994).

[106] L. J. Yao and K. H. Mayo, *Biochem. J.* **315**, 161 (1996).

[107] A. P. Campbell and B. D. Sykes, *J. Mol. Biol.* **222**, 405 (1991).

[108] Z. Wang, J. D. Jones, J. Rizo, and L. M. Gierasch, *Biochemistry* **32**, 13991 (1993).

[109] K. H. Mayo, F. Fan, M. P. Beavers, A. Eckardt, P. Keane, W. J. Hoekstra, and P. Andrade-Gordon, *Biochemistry* **35**, 4434 (1996).

improves, more compactly folded structures are being made. NMR is perhaps the only technique which can provide such detailed information regarding peptide conformational distributions and preferred structures in solution.

[31] Solid-State Nuclear Magnetic Resonance Characterization of Gramicidin Channel Structure

By T. A. Cross

Introduction

Solid-state nuclear magnetic resonance (NMR) is a powerful technique for studying molecular structure in complex environments. Lipid bilayers represent such a complex environment. Two-dimensional or three-dimensional crystallization of membrane proteins in bilayer environments is possible but very challenging.[1,2] Similarly the study of membrane proteins by solution NMR is substantially complicated by the presence of lysolipids or detergents, thereby limiting this technique to very small molecular weight proteins. Consequently, there have been relatively few membrane protein structures solved. Similarly, the structure of fibrous proteins and large protein–protein or protein–nucleic acid complexes have been rarely solved even at modest resolution. Solid-state NMR neither requires crystallization nor isotropic environments; in fact, the more rigid, the better.[3,4] Hence this spectroscopic approach has a great deal of potential for studies of these proteins. This article describes the current state of the development of solid-state NMR structural methods for these environments with special attention to the first high resolution structure of a membrane-bound polypeptide by solid-state NMR, gramicidin A.[5,6]

Unlike an aqueous environment, the lipid environment is heterogeneous as a result of the low dielectric lipid bilayer interstices, the bulk aqueous environment, and the considerable interface region.[7] Furthermore, lipids

---

[1] T. Tsukihara, H. Aoyama, E. Yamashita, T. Tomizaki, H. Yamaguchi, K. Shinzwa-Itoh, R. Nakashima, R. Yaono, and S. Yoshikawa, *Science* **272**, 1136 (1996).
[2] W. Kohlbrandt, *Q. Rev. Biophys.* **25**, 1 (1992).
[3] S. J. Opella, *Methods Enzymol.* **131**, 327 (1986).
[4] T. A. Cross and S. J. Opella, *Curr. Opin. Struct. Biol.* **4**, 574 (1994).
[5] R. R. Ketcham, K.-C. Lee, S. Huo, and T. A. Cross, *J. Biomol. NMR* **8**, 1 (1996).
[6] R. R. Ketcham, B. Roux, and T. A. Cross, submitted.
[7] S. H. White and W. C. Wimbley, *Curr. Opin. Struct. Biol.* **4**, 79 (1994).

are chiral, unlike water, and in biological membranes the bilayers are asymmetric. This environment has been shown to kinetically trap metastable molecular conformations.[8] This stems from two properties of the bilayer interior. First, electrostatic interactions are stabilized by the low dielectric of this environment such that it is very difficult to break hydrogen bonds for the process of searching conformational space. Second, water or other protic solvents that can act as a catalyst for conformational rearrangements are rare.[9] Protic solvents can act to lower the ΔG of activation by destabilizing intramolecular hydrogen bonds and by stabilizing a conformational intermediate in which the polypeptide does not have its amide groups optimally satisfied with intramolecular hydrogen bonds.

There are two fundamentally different types of structural constraints that can be obtained from solid-state NMR experiments. Distance constraints can be obtained from Rotational Echo Double Resonance (REDOR)[10] and rotational resonance[11] experiments and their more recent experimental offspring. Unlike nuclear Overhauser effect (NOE) constraints in solution NMR, the distance constraints obtained in solid-state NMR are typically far more quantitative, and the distances observed can be 10 Å or even greater depending on the gyromagnetic ratios of the interacting nuclei. Such experiments do not require one-, two-, or three-dimensional order because the samples are spun in a magic angle spinner; however, conformational heterogeneity broadens the spectral lines and reduces the quality of the final structural characterization. Therefore, either microcrystalline or carefully lyophilized[12] samples are often used in the sample rotors.

A second approach for structural constraints is to obtain orientational constraints by solid-state NMR. For this approach the samples need to be aligned with respect to the magnetic field direction, B_0, of the NMR spectrometer.[13] Then the orientation-dependent magnitude of numerous nuclear spin interactions, such as the anisotropic chemical shift and dipolar and quadrupolar interactions, can be observed.[14,15] These interactions are represented in the molecular frame of the molecule by a coordinate system or tensor that is fixed with respect to the covalent bonds of the molecular

[8] S. Arumugam, S. Pascal, C. L. North, W. Hu, K.-C. Lee, M. Cotten, R. R. Ketchem, F. Xu, M. Brenneman, F. Kovacs, F. Tian, A. Wang, S. Huo, and T. A. Cross, *Proc. Natl. Acad. Sci. U.S.A.* **93**, 5872 (1996).

[9] F. Xu, A. Wang, J. B. Vaughn, and T. A. Cross, *J. Am. Chem. Soc.* **118**, 9176 (1996).

[10] T. Gullion and J. Schaefer, *J. Magn. Reson.* **81**, 196 (1989).

[11] D. P. Raleigh, M. H. Levitt, and R. G. Griffin, *Chem. Phys. Lett.* **146**, 71 (1988).

[12] A. Christensen and J. Schaefer, *Biochemistry* **32**, 2868 (1993).

[13] T. A. Cross and S. J. Opella, *J. Am. Chem. Soc.* **105**, 306 (1983).

[14] S. J. Opella, P. L. Stewart, and K. G. Valentine, *Q. Rev. Biophys.* **19**, 7 (1987).

[15] T. A. Cross, *Annu. Rep. NMR Spectrosc.* **29**, 123 (1994).

frame. Consequently, by observing the orientation-dependent nuclear spin interactions, not only is the nuclear spin tensor constrained with respect to B_0, but also the molecular frame is constrained with respect to B_0. Furthermore, in a uniformly aligned system a single molecular axis, in the case of gramicidin, the channel axis, is aligned parallel to B_0, and therefore the molecular frame of a local site in the macromolecule is orientationally constrained with respect to a unique global axis of the macromolecule.

Gramicidin A is a linear polypeptide of 15 amino acid residues having the following sequence: formyl-Val-Gly-Ala-DLeu-Ala-DVal-Val-DVal-Trp-DLeu-Trp-DLeu-Trp-DLeu-Trp-ethanolamine. Both end groups are blocked so that no formal charges exist. As an amino terminus to amino terminus symmetric dimer, this polypeptide forms a monovalent cation-selective channel across lipid bilayers. Urry[16] proposed a β-type of helical structure for the channel conformation. Because of the alternating amino acid stereochemistry a β strand would have all side chains on one side of the strand causing the strand to bend into a helix, hence a β helix. The gramicidin A conformation in organic solvents or in crystals prepared from organic solvents is quite different in that the strands are intertwined, but still with a fundamental β-strand conformation. In sodium dodecyl sulfate (SDS) micelles the single-stranded conformation has been characterized by solution NMR methods.[17] The hydrogen bonding and molecular fold are identical to the structure presented here in lipid bilayers, but the side-chain dynamics and rotameric states are very different, most likely reflecting differences in the hydrophobic/hydrophilic interface of these two environments.

Solid-phase peptide synthesis has played a critical role in the development of these solid-state NMR structural methods. By using single-site ^{15}N, ^{13}C, and ^{2}H isotopic labels in the polypeptide, the spectral assignment of the resonances is trivial.[18,19] Furthermore, resolution from other resonances is also optimized, but not eliminated because natural abundance signals can interfere and signals from other conformers or from unoriented material can also interfere. Because of the single site labeling it has been possible during the course of this technique development to reduce the line widths of the resonances from oriented samples, and schemes have been developed for making spectral assignments when multiple resonances are present. Consequently, it is now possible to consider the analysis of spectra from

[16] D. W. Urry, *Proc. Natl. Acad. Sci. U.S.A.* **68,** 672 (1971).
[17] A. L. Lomize, V. Orekov, and A. S. Arseniev, *Bioorg. Khim.* **18,** 182 (1992).
[18] G. B. Fields, C. G. Fields, J. Petefish, H. E. Van Wart, and T. A. Cross, *Proc. Natl. Acad. Sci. U.S.A.* **85,** 1384 (1988).
[19] C. G. Fields, G. B. Fields, R. Noble, and T. A. Cross, *Int. J. Pept. Protein Res.* **33,** 298 (1989).

biosynthetic preparations where amino acid-specific labeling or uniform labeling has been achieved. To get to that point, however, solid-phase synthesis has been essential, and because the polypeptide sequence was synthesized repeatedly with only isotopic substitutions there was an opportunity to optimize the synthetic conditions. Consequently, 80% overall yield and 98% purity on cleavage from the resin was typical for these gramicidin syntheses.[19]

Solid-phase peptide synthesis has also been important in characterizing the nuclear spin interaction tensors, work that has been most accurately achieved through analysis of powder pattern spectra, that is, spectra where a random distribution of molecular orientations are present. To interpret the observed nuclear spin interactions in uniformly aligned samples the magnitude of the tensor elements must be known, as well as the orientation of each tensor element with respect to the molecular frame (i.e., the covalent bonds surrounding the nuclear site). The tensor element magnitudes can be obtained directly from the powder pattern spectra of single site labeled polypeptides, and there is presently no way to accurately achieve this data from multiply labeled samples. The tensor element orientations can be accurately defined by determining the relative orientation of two tensors such as a dipolar interaction tensor and a chemical shift interaction tensor in the molecule of interest.[20,21] Dipolar interactions between covalently attached nuclei have the unique axis of their axially symmetric tensor aligned with the internuclear vector. Therefore, the chemical shift tensor orientation of an amide ^{15}N site can be constrained to the molecular frame by determining the orientation of this tensor to the ^{15}N–^{13}C$_1$ dipolar interaction tensor. Such experiments require the specific double labeling of the peptide bond. Furthermore, in oriented samples these double labels, which are difficult to achieve by biosynthetic means but readily achieved by solid-phase methods, can be used as orientational constraints. These characterizations will not be possible in most biosynthetically labeled preparations where significant assumptions will have to be made about the tensor element magnitudes and orientations.

Preparation of Oriented Samples

The quality of orientational constraints is dependent on how well the samples are aligned with respect to B_0. Such samples are used in a wide range of spectroscopic techniques from neutron scattering to infrared linear dichroism and many more. Equally varied are the number of approaches

[20] Q. Teng and T. A. Cross, *J. Magn. Reson.* **85,** 435 (1989).
[21] C. Wang, Q. Teng, and T. A. Cross, *Biophys. J.* **61,** 1550 (1992).

for generating aligned samples from Langmuir–Blodgett films to magnetic and electric field alignment to shear between glass plates and more. Since these methods were briefly reviewed[22] several methods have advanced quickly. The mixed lipid system that generates discoidal micelles, known as "bicelles," aligns with the disc normal perpendicular to the magnetic field.[23,24] Vold and co-workers added lanthanide shift reagents to these bicellar preparations and changed the orientation of the disc normal from perpendicular to parallel.[25] The orientation of these samples is the result of the weak interactions between B_0 and the diamagnetic susceptibility of the lipids being dominated by the stronger interaction between B_0 and the paramagnetic susceptibility (of the lanthanides). Such samples avoid the laborious task described below of aligning samples between glass plates. However, these samples permit additional dynamics for a bound protein or peptide, and the lanthanides may bind to the protein or polypeptide directly. Therefore, it is not yet clear how useful these samples will be for obtaining the structural constraints described here.

The gramicidin and lipid samples used here have been uniformly aligned with respect to B_0 by first cosolubilizing the peptide and dimyristoylphosphatidylcholine (DMPC; 1 : 8 molar ratio) in an organic solvent. This ensures that there is uniform mixing of the peptide and lipid. If this can be achieved without using an organic solvent or dialysis exhange of detergent for lipid, then some of the concerns described below are avoided. For cosolubilization the peptide and lipid need to be highly soluble in this solvent so that the process of layering the solution on glass plates is not too tedious. Furthermore, the solvent should wet the surface of the glass so that a film is generated covering most of the slide. It is also important for the peptide conformation in the organic solvent to be one that when introduced to a lipid bilayer it can readily convert to the minimum energy conformation in the bilayer experiment. By choosing methanol for gramicidin and DMPC it is possible to kinetically trap nonminimum energy conformations of the peptide in the bilayer.[8] However, the conformer population in 95% benzene/ 5% ethanol (v/v) is such that the minimum energy conformer is readily achieved in the lipid environment on hydration. This dependence on the choice of organic solvent led to the characterization of a solvent history dependence well before it was known which conformers were being trapped.[26–29]

[22] F. Moll III and T. A. Cross, *Biophys. J.* **57,** 351 (1990).
[23] C. R. Sanders, B. J. Hare, K. P. Howard, and J. H. Prestegard, *Prog. Nucl. Magn. Reson. Spectrosc.* **26,** 421 (1994).
[24] C. R. Sanders II and G. C. Landis, *Biochemistry* **34,** 4030 (1995).
[25] R. S. Prosser, S. A. Hunt, J. A. DiNatale, and R. R. Vold, *J. Am. Chem. Soc.* **118,** 269 (1996).
[26] P. V. LoGrasso, F. Moll III, and T. A. Cross, *Biophys. J.* **54,** 259 (1988).
[27] J. A. Killian, K. U. Prasad, D. Hains, and D. W. Urry, *Biochemistry* **27,** 4848 (1988).

The choice of lipid is also important from the standpoint of having a match between the lengths of the hydrophobic regions of the peptide and lipid bilayer. In gramicidin these result in having the monomer/dimer equilibrium strongly favor the dimer. More generally, however, a hydrophobic mismatch will lead to induced curvature in the bilayer resulting in liposome and vesicle formation. Any peptide that is not in a lipid environment with a bilayer normal aligned parallel with B_0 will result in less well-aligned peptide and protein preparations. In the preparations used here the lipid bilayer domains have dimensions of millimeters, that is, they have very extensive structures with a very small fraction of the bilayer domain normal tilted away from parallel to B_0.

The choice of the orienting surface is dictated by a desire to have relatively rigid surfaces that can be arranged and maintained in a sample container with the surface normally parallel to the magnetic field direction. A German company (Marienfeld Glassware, Bad Mergentheim, Germany) makes 70 μm thick glass plates to order, but Prosser, Vold, and co-workers have demonstrated that these plates can be etched to a reproducible average thickness of 21 μm.[30] This considerably reduces the amount of glass in the samples and improves the filling factor substantially. The shape of the glass plates should be chosen so as to minimize the ratio of edge to surface area. It is inevitable that sample seeps from between the glass plates, generating powder patterns in an otherwise sharp line spectrum. By avoiding long thin rectangular shapes this problem is minimized. Following on the concept from Opella's group of using a flat coil it is possible to mold the coil around a stack of glass plates, thereby increasing the filling factor even more.[31] After the film on the glass plates is dried, the plates are stacked (typically 20 or more) in the sample tube and hydrated slowly by exposure to a water-saturated atmosphere in an incubator at 45°, approximately 15° to 20° above the phase-transition temperature. After 2 weeks the samples have nearly doubled their dry weight and are typically clear. The sample tube is then sealed with epoxy and the sample weight is monitored carefully to ensure that the sample is completely sealed and does not dry out.

The quality of oriented samples can be characterized by observing NMR spectra. Preliminary characterization can be made by acquiring the ^{31}P spectra of the phospholipids, but the important characterization is not how well the lipid is aligned but how well the peptide is aligned. By observing the NMR spectra, a time-averaged assessment of orientation is obtained

[28] M. C. Bano, L. Braco, and C. Abad, *FEBS Lett.* **250**, 67 (1989).
[29] K. J. Cox, C. Ho, J. V. Lombardi, and C. D. Stubbs, *Biochemistry* **31**, 1112 (1992).
[30] R. S. Prosser, S. A. Hunt, and R. R. Vold, *J. Magn. Reson.* **109B**, 109 (1995).
[31] B. Bechinger and S. J. Opella, *J. Magn. Reson.* **95**, 585 (1991).

that will conceal dynamic fluctuations about the director axis and other axes. In such spectra two features are important, the amount of powder pattern and the asymmetry of the resonance representing the aligned sample.[22] The powder pattern simply reflects the sample that seeped from between the glass plates. The asymmetry results from a mosaic spread of orientation for a given site with respect to B_0. Figure 1 shows experimental

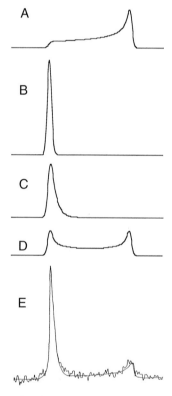

FIG. 1. [31]P line shapes associated with phospholipid bilayer phases and defect structures. (A) Simulated axially symmetric line shape associated with a randomly oriented bilayer sample. (B) Simulated gaussian shape for a bilayer uniformly aligned such that the bilayer normal is parallel to the magnetic field direction. (C) As in (B) with a gaussian distribution of bilayer normal orientations and where the average orientation is parallel to the magnetic field direction. Such a distribution could result from "parabolic focal conic" defects in the bilayer structure. (D) A line shape associated with a cylindrical distribution of orientations where the cylinder axis is perpendicular to the field, such as would be induced by "oily streak" defects. (E) Experimental and observed spectra in which 62% of the signal intensity arises from uniformly oriented sample and 38% from an axially symmetric powder pattern as in (A). [Adapted from Moll and Cross (1990).[22]]

and simulated ^{31}P NMR spectra of both oriented and unoriented samples. In well-oriented samples, the powder pattern is barely detectable and the mosaic spread may be less than $1°$.[32]

Tensor Characterization

To interpret the nuclear spin interaction observables from aligned samples it is essential that the spin interaction tensors be well characterized. This means that both the magnitude of the tensor elements and their orientation with respect to the molecular frame needs to be defined. For dipolar interactions between covalently attached nuclei the unique axis of the axially symmetric tensor is aligned parallel with the internuclear bond vector. The magnitudes of the tensor elements can be calculated from physical constants (gyromagnetic ratios γ and Planck's constant h) along with an accurate value of the bond length, r:

$$\nu_\| = \gamma_I \gamma_S h r^{-3}.$$

[2]H quadrupolar interactions for carbon-bound deuterons are essentially axially symmetric, and in this study they are assumed to be symmetric. Like the dipolar interactions the unique axis of the electric field gradient tensor is parallel to the C–[2]H bond. The magnitude of these tensor elements in studies of the side chains have been achieved from model compounds. The error in using model compounds for ^{2}H tensor characterization is not large compared to the estimated influence of the local dynamic averaging of the tensor. More will be stated about the latter problem below.

The chemical shift interaction tensors are axially asymmetric, and the tensor orientation needs to be experimentally characterized for high resolution structural characterization. The tensor element magnitudes can be determined from static chemical shift powder pattern spectra of an unoriented sample. Because the chemical shift tensors are sensitive to the electron density surrounding the nuclear site it is important to characterize these tensors for the site of interest in the environment of interest. To observe the static chemical shift tensor for a polypeptide in a hydrated lipid bilayer environment it is necessary to eliminate any large amplitude motions that could average the spin interaction tensor. Because distortions are induced on lowering the temperature through the gel to liquid-crystalline phase transition, a flash freezing technique has been adapted from electron microscopy freeze-fracture methods. By plunging thin films of hydrated bilayers into a cryogenic solvent such as liquid propane, the dynamics can be elimi-

---

[32] T. A. Cross, R. R. Ketchem, W. Hu, K.-C. Lee, N. D. Lazo, and C. L. North, *Bull. Magn. Reson.* **14**, 96 (1992).

nated and the structure is undistorted.[33,34] Measurements of the powder pattern below 200 K where most torsional motions cease result in a static powder pattern, the discontinuities of which yield an experimental characterization of the chemical shift tensor elements. For ^{13}C sites next to ^{14}N there is an additional complication. ^{14}N is a quadrupolar nucleus that is dipolar coupled to adjacent ^{13}C sites, resulting in a superposition of chemical shift and dipolar powder patterns. For ^{13}C carbonyl sites and field strengths above 6 T the result of the relatively weak dipolar interaction is a broadening and only modest distortion of the chemical shift powder pattern. Because of the quadrupolar nature of ^{14}N it is rare to have enough data to accurately deconvolute the dipolar and chemical shift patterns.[35] However, the replacement of ^{14}N with ^{15}N results in a much simplified spectrum with spin 1/2–spin 1/2 dipolar interactions. The dipolar and chemical shift patterns can then be readily deconvoluted and accurate ^{13}C chemical shift tensor values determined.

The ^{15}N and ^{13}C chemical shift tensor orientations relative to the molecular frame can be determined from single crystal studies of model compounds.[36] However, the tensor characterization is highly dependent on the electronic environment and hence potentially on the crystal form.[37] Nevertheless, the determination of tensor element orientations from such model compound studies is valuable and the specific orientations can be determined by observing the relative orientation of a dipolar and chemical shift tensor. By observing the powder pattern described above (Fig. 2) that included the ^{15}N–^{13}C dipolar interaction it is possible to determine the orientation of the chemical shift tensor with respect to this dipolar interaction, if it is assumed that one of the tensor elements is in the peptide plane. Observation of both ^{13}C and ^{15}N resonances in such a sample results in the orientation of both tensors relative to the C_1–N bond.[20,21] The results from characterizing many of the chemical shift tensors in gramicidin is that the tensor element magnitudes vary significantly (7%) even among the same amino acid; however, the ^{15}N tensor orientation appears to be very similar from site to site with the exception of Gly, which has a tensor orientation that differs by 6° from the other amino acids.[38] The uniformity of the tensor orientations may reflect the relatively uniform secondary structure

[33] J. N. S. Evans, R. J. Appleyard, and W. A. Shuttleworth, *J. Am. Chem. Soc.* **115**, 1588 (1993).
[34] N. D. Lazo, W. Hu, and T. A. Cross, *J. Chem. Soc., Chem. Commun.*, 1529 (1992).
[35] Q. Teng, M. Iqbal, and T. A. Cross, *J. Am. Chem. Soc.* **114**, 5312 (1992).
[36] G. S. Harbison, L. W. Jelinski, R. E. Stark, D. A. Torchia, J. Herzfeld, and R. G. Griffin, *J. Magn. Reson.* **60**, 79 (1984).
[37] Y. Hiyama, C.-H. Niu, J. V. Silverton, A. Bavoso, and D. A. Torchia, *J. Am. Chem. Soc.* **110**, 2378 (1988).
[38] W. Mai, W. Hu, C. Wang, and T. A. Cross, *Protein Sci.* **2**, 532 (1993).

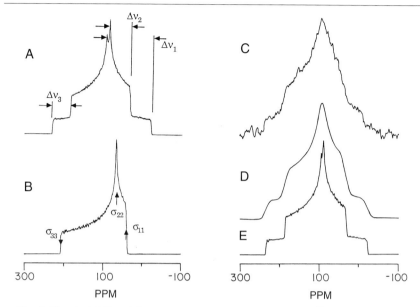

FIG. 2. Chemical shift and dipolar powder pattern spectra of amide ^{15}N sites. (B) Simulated amide ^{15}N chemical shift powder pattern characterized by the magnitudes of three tensor elements, σ_{ii}. (A) In the presence of a doubly labeled, ^{15}N–^{13}C$_1$, peptide bond, the ^{15}N–^{13}C dipolar interaction is superimposed on the chemical shift spectrum. The dipolar splitting of each chemical shift tensor element is shown. Because the dipolar interaction is fixed in the molecular plane along the internuclear vector, the dipolar interaction observed at each tensor element defines the orientation of the chemical shift tensor in the molecular frame. (C) Experimental ^{15}N spectrum of ^{13}C$_1$–Gly^2-^{15}N-Ala^3-gramicidin A. (E) Simulated line shape with σ_{22} perpendicular to the peptide plane and σ_{33} making an angle (β_D) with respect to the C$_1$–N bond of 106°. (D) Simulated spectrum as in (E) with 360 Hz of gaussian broadening. [Adapted from Teng and Cross (1989).[20]]

of gramicidin, and so this may not be a general finding for amide sites in the polypeptide backbone; future studies will soon resolve this question.

Data Collection

The use of isotopic labels results in a considerable improvement in sensitivity. For ^{15}N the low natural abundance (0.3%) and the relatively small number of nitrogen sites versus carbon sites in protein and lipid samples results in single site ^{15}N spectra that are uncomplicated by signals from other sites. ^{13}C carbonyl spectra of single-site ^{13}C$_1$-labeled samples yields just 50% of the observed signal from a labeled site in gramicidin even with a very high mole fraction of peptide. Consequently, much of the solid-state NMR structural approach presented here is based on ^{15}N spectroscopy.

Cross-polarization adds considerably to the experimental sensitivity (maximally the ratio of the gyromagnetic ratios). Moreover, the ^{1}H T_1 relaxation is often much shorter than the ^{15}N or ^{13}C T_1, and consequently the recycle delay can be substantially reduced when using cross-polarization, further enhancing sensitivity. Typically, the rare (^{15}N) and abundant (^{1}H) nuclei are allowed to mix so that all of the rare nuclei are equally or near equally polarized. However, if there is a directly attached proton to the rare nucleus it may be to the experimentalists' advantage to limit the polarization transfer time for just long enough to polarize the rare nucleus from this one proton. In so doing the polarization transfer is more selective, and it is possible to more nearly achieve the maximal sensitivity.[39,40]

Chemical shift spectra are obtained with cross-polarization and high power proton decoupling. Chemical shift spectra that include the ^{15}N–^{13}C or ^{15}N–^{2}H dipolar interaction are obtained using the same experiments without modification. ^{15}N–^{1}H dipolar couplings have typically been obtained with a separated local field experiment,[41] but the spectral resolution is far less than in the ^{15}N–^{2}H dipolar interaction spectra, which provide the same structural information.[42] A new experiment has been developed by Opella, Ramamoorthy, and co-workers, called PISEMA[43] that improves spectral resolution for ^{15}N–^{1}H dipolar observation by an order of magnitude. ^{2}H spectra have been obtained using a quadrupolar echo sequence; in this way signal detection is separated from the radio frequency (RF) pulses by a delay that allows the electronics to quiet down (from acoustic ringing) following the RF pulse. If the ^{2}H signal has rapid T_{2e} relaxation the ^{2}H magnetization will be lost before the signals can be recorded. An advantage for high field ^{2}H NMR spectroscopy is that acoustic ringing is frequency dependent and at high field and high frequencies the echo delay can be shortened considerably.[44]

Before the data can be interpreted as structural constraints the influence of dynamics and structure on the NMR observables must be resolved. A description of the dynamics that will affect the NMR spectra can be broken down into three categories: global motion, large amplitude local motions, and small amplitude librational motions. For membrane-bound peptides that have rapid axial motions parallel to the bilayer normal, these motions

[39] L. Muller, A. Kumar, T. Baumann, and R. R. Ernst, *Phys. Rev. Lett.* **32**, 1402 (1974).
[40] F. Tian and T. A. Cross, *J. Magn. Reson.* **125**, 220 (1997).
[41] J. S. Waugh, *Proc. Natl. Acad. Sci. U.S.A.* **73**, 1394 (1976).
[42] R. R. Ketchem, W. Hu, K.-C. Lee, and T. A. Cross, *Structure* **2**, 699 (1994).
[43] C. H. Wu, A. Ramamoorthy, and S. J. Opella, *J. Magn. Reson.* **109A**, 270 (1994).
[44] V. Soghomonian, M. Cotten, R. Rosanske, and T. A. Cross, *J. Magn. Reson.* **125**, 212 (1997).

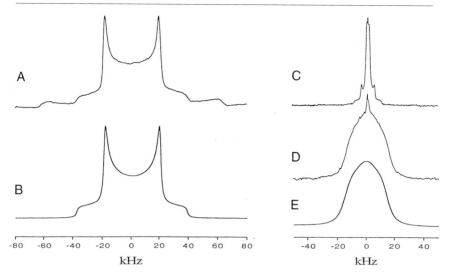

FIG. 3. Experimental and simulated ^{2}H NMR powder pattern spectra of d_8-Val. (A) Experimental spectrum of the polycrystalline amino acid. The spectrum is dominated by the signals from the two d_3-methyl groups. (B) Simulation of the methyl resonances motionally averaged by three-site rotational jumps about the C_β–Me axis. (C) Experimental spectrum of d_8-Val^1-gramicidin A in DMPC bilayers hydrated with 40% by weight deuterium-depleted water, obtained at 36° (phase transition is 28°). (D) Same sample as in (C) but with the temperature reduced to 5°. The global motion of the channel has been quenched. (E) Simulated spectrum of the methyl resonances using a three-state jump model having a residence time of 1.5 μsec and occupancy ratios of 75:15:10 in the three χ_1 states of 184°, 304°, and 64°, respectively. [Adapted from Lee and Cross (1994).[45]]

can be quenched by lowering the sample temperature through the gel to liquid-crystalline phase transition. However, such motions do not average the Z component of the nuclear spin interaction tensors when the bilayer normal is parallel with respect to the magnetic field direction.[45] Consequently, for the interpretation of spectra from oriented samples where the bilayer normal is parallel to the magnetic field direction, the global motions can be ignored, but in powder pattern studies the global motions must be considered and preferably quenched (Fig. 3).

Large amplitude motions such as the three site jump about the χ_1 axis of Val averages the ^{2}H quadrupolar interactions for the C_β–^{2}H and methyl deuterons (Fig. 3). Powder pattern spectra show characteristic line shapes for various motional modes. In fact, it is possible to determine the relative populations of the three states for Val χ_1 motion. However, to accurately

[45] K.-C. Lee and T. A. Cross, *Biophys. J.* **66**, 1380 (1994).

define such parameters it is necessary to quench the global motions which also influence the overall powder pattern line shape.

In addition local librational motions scale the tensor elements. A full description of such motions includes a definition of the axis about which the motion occurs, a description of the motional mode, such as jump or diffusional motions, and a description of the motional frequency. The methyl deuterons of Ala in gramicidin undergo a rapid tetrahedral averaging about the χ_1 axis, and in addition the quadrupolar interaction is scaled by approximately 10% due to motion about the C_α–C_β axis. Such librational amplitudes become larger for sites further removed from the polypeptide backbone. Sometimes these librational motions can be accurately modeled as anisotropic motions. Previously, the preparation of fast-frozen samples was described for the characterization of static chemical shift tensors. By recording spectra of such samples as a function of temperature between 200 K and the gel to liquid-crystalline phase transition temperature, the specific influence on the chemical shift powder pattern line shape can be observed (Fig. 4) and a model of the local motions derived. In this way it has been shown that the predominant librational motion in the polypeptide backbone is a libration about the C_α–C_α axis. Such motions characterized

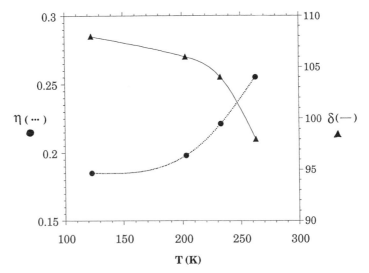

FIG. 4. Asymmetry (η) and anisotropy (δ) of the ^{15}N chemical shift tensor for the ^{15}N–Ala[3] site in gramicidin A solubilized in a fast frozen lipid bilayer preparation. Below 200 K librational motions essentially cease. Librational amplitudes above 200 K average the chemical shift tensor in an anisotropic fashion, demonstrating preferential motion about a single axis which can be shown to be parallel with the C_α–C_α axis. The averaging is consistent with an amplitude of approximately 20° about this axis.

in this way have amplitudes in gramicidin of ±20°.[46] If the motions are characterized by field dependent ^{15}N T_1 relaxation times the amplitudes are observed to be only ±6°.[47] The difference in these characterization methods is that the powder pattern analysis is essentially frequency independent (above 10^5 Hz), whereas the relaxation measurements are frequency dependent, being most sensitive to frequencies near the Larmor frequencies for ^{15}N and ^{1}H. The combination of powder pattern and relaxation analysis results in a full characterization of the librational motion.

With the above characterization of molecular motions and averaged nuclear spin interaction tensors it is possible to interpret the orientational constraints as purely structural constraints. In this way the orientation of the motionally averaged tensor is defined with respect to the magnetic field. Because the spin interaction tensor orientation is known with respect to the molecular frame the orientational constraints restrain the molecular frame to B_0, and because the channel axis and bilayer normal are parallel to B_0 the local molecular frame of the spin interaction site is constrained with respect to a global molecular axis, the channel axis.

Initial Structure

Orientational constraints can be used to define all of the torsion angles in a polypeptide or protein provided that each structural element (separated by adjacent torsion angles) is constrained. In gramicidin, there are 30 ϕ, ψ backbone torsion angles to be solved if the torsion angles in the ethanolamine blocking group are ignored. For the side chains, if methyl torsion angles are ignored because of rapid motions, then there are 4 Val, 4 leu, and 4 Trp χ_1 angles as well as 4 Leu and 4 Trp χ_2 angles for a total of 50 torsion angles to be solved for the near complete structure of the gramicidin channel.

Each peptide plane can be oriented with respect to the channel axis through the interpretation of two dipolar observations, the ^{15}N–^{1}H and ^{15}N–^{13}C$_1$ spin interactions. Each dipolar constraint is interpreted with the following equation:

$$\Delta\nu_{\text{obs}} = \nu_\parallel(3\cos^2\theta - 1)$$

where $\Delta\nu_{\text{obs}}$ is the observed dipolar splitting, $\nu_\parallel$ is the motionally averaged interaction magnitude, and θ is the angle between the internuclear vector and B_0. If $\Delta\nu_{\text{obs}} > \nu_\parallel$ then the sign of $\Delta\nu_{\text{obs}}$ is known (+) and there exist only two possible solutions for θ, an angle with respect to $+B_0$ and an

[46] N. D. Lazo, W. Hu, and T. A. Cross, *J. Magn. Reson.* **107B**, 43 (1995).
[47] C. L. North and T. A. Cross, *Biochemistry* **34**, 5883 (1995).

identical angle with respect to $-B_0$. If $\Delta\nu_{obs} < \nu_\parallel$ then the sign of $\Delta\nu_{obs}$ is not defined and four possible solutions for θ exist. Two such dipolar constraints could generate numerous peptide plane orientations; however, many of the possible solutions are not consistent with the covalent geometry of the peptide plane, that is, the bond angles in the peptide plane. Furthermore, the observed ^{15}N chemical shift can be used as a structural filter to eliminate some of the structural possibilities on the basis of the dipolar data alone. In this way a unique set of peptide plane bond orientations is defined with respect to B_0.[48] An ambiguity in the sign of the cosine of the peptide plane normal to B_0 remains. This ambiguity is referred to as a chirality.[49] Similar results are obtained for each peptide plane. In contrast, with the additional $C_\alpha-^2H$ orientational constraint the $N-C_\alpha-C_1$ plane which joins adjacent peptide planes is uniquely oriented with respect to B_0. The relative orientation of the two peptide planes and the unique $N-C_\alpha-C_1$ plane leads to two possible ϕ and two possible ψ angles. Despite these ambiguities all ϕ, ψ combinations are in the β region of the Ramachandran diagram, and the $C_\alpha-C_\beta$ axis is approximately perpendicular to the channel axis, minimizing steric hindrance between turns of the helix.[5,50]

Despite having several discrete solutions for each amino acid residue a unique molecular fold is generated (Fig. 5). This is caused by the unique properties of the orientational constraints and the similarity in the ϕ, ψ torsion angles for various solutions. First, only certain combinations of ϕ, ψ can be joined from adjacent residues. The ϕ, ψ angles are defined by specific chirality solutions for the two peptide planes joined by the C_α carbon. One of these planes is shared with each adjacent residue, and only identical chirality solutions for the shared plane can be considered a viable structure. In other words, the shared plane must have the same orientation for residue i of a given ϕ, ψ solution as for residue $i + 1$ of a given ϕ, ψ solution. This reduces the number of structural solutions by a factor of 2.

The remaining structural possibilities result entirely from peptide plane chirality ambiguities described as "$+$" and "$-$" for each plane.[49] We have illustrated this range of structures with four extremes: two alternating patterns: $+/-$ and $-/+$ for all of the peptide planes, and two homogeneous patterns: $+/+$ and $-/-$ for all peptide planes.[5] Each of these structures represents the same folding motif, a right-handed helix with approximately 6.6 residues per turn and an average helical pitch of 4.8 Å. The hydrogen bonding pattern is identical in the four structures, and the C_α carbon traces of the backbone are virtually superimposable (Fig. 6). However, the struc-

[48] Q. Teng, L. K. Nicholson, and T. A. Cross, *J. Mol. Biol.* **218,** 607 (1991).
[49] J. Quine, M. Brenneman, and T. A. Cross, *Biophys. J.* **72,** 2342 (1997).
[50] R. R. Ketchem, W. Hu, and T. A. Cross, *Science* **261,** 1457 (1993).

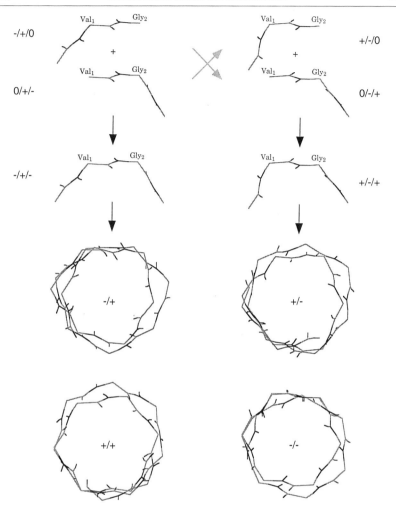

Fig. 5. Dipolar interactions are used to define the orientation of each peptide plane with respect to a global molecular axis; here, this axis is the gramicidin channel axis which is, in turn, aligned parallel to the magnetic field direction. This peptide plane orientation has a single ambiguity known as a chirality; the carbonyl is oriented so that it points either slightly in toward the channel axis (+) or slightly away (−) from the channel axis. Data from adjacent peptide planes are used to define the ϕ, ψ torsion angles. Only combinations of these diplanes with the same chirality for the overlapping peptide plane can be combined. Four regular chirality repeat patterns are shown at the bottom. These four patterns out of the many possible chirality combinations span the range of ambiguities in the structure. Even so the helix sense is uniquely defined, as are the hydrogen bonds and hence the polypeptide fold. [Adapted with permission from Ketchem, R. R., Lee, K.-C., Huo, S. and Cross, T. A. (1996). *J. Biomol. NMR* **8**, 1–14. Copyright 1996 ESCOM Science Publishers B.V.]

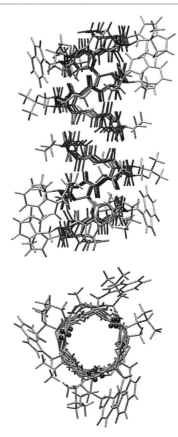

FIG. 6. Superposition of the four structures shown in Figure 5 illustrates the similarity of the backbone structure. For the Leu-10, Leu-12, and Leu-14 carbonyls the oxygen atoms are shown as balls in the end view. Attached to one of the backbone conformers are the side-chain structures constrained by a set of orientational constraints. [Adapted with permission from Ketchem, R. R., Lee, K.-C., Huo, S. and Cross, T. A. (1996). *J. Biomol. NMR* **8,** 1–14. Copyright 1996 ESCOM Science Publishers B.V.]

tures do not represent perfect geometry; the root mean square deviation (rmsd) from average β-strand hydrogen bond distances is approximately 0.5 Å. Considering the local nature of the structural constraints used in this study, it may be surprising that the hydrogen bond distances are so near to ideal values. The orientational constraints are not only precise and quite accurate, but each constraint constrains the molecular frame to a laboratory fixed axis system. We refer to these constraints as absolute constraints, as opposed to relative constraints such as NOE, REDOR, or rotational resonance-derived distance constraints, all of which constrain one part of the macromolecule to another part of the molecule. Consequently, as the structure is assembled from orientational constraints one residue at a time, the errors in each peptide plane are just as likely to cancel as to add to the errors in the adjacent plane. The result is a unique molecular fold that is aligned with respect to the laboratory axis and the anisotropic molecular environment.[50]

The backbone structure defines the orientation of the C_α–H and C_α–C_β bonds, and consequently the side chain structure of the Gly and Ala residues are defined. Besides these residues there are 4 Val, 4 Leu, and 4 Trp residues. We have previously described the importance of characterizing the dynamics of a specific site before making the structural interpretation. Two of the Val residues, Val-1 and Val-7, have been shown to have large amplitude dynamics about the C_α–C_β axis (χ_1 motion).[51] The other two Val, Val-6 and Val-8, are fixed in χ_1 rotameric states. d_8-Val was used to generate the orientational constraints for each of the Val residues. The six methyl deuterons give rise to just two quadrupolar splittings due to rapid three-site jumps about the χ_2 axes. The C_α deuteron resonances were assigned by predicting the splitting from the backbone structure. The C_β deuteron is the only remaining large splitting in the observed spectrum, as the methyl resonances are highly averaged. The χ_1 solutions are determined by calculating the rmsd between observed and calculated quadrupole splittings as a function of χ_1 and the C_α–C_β orientation in the vicinity of the backbone defined angles (Fig. 7). The C_β–^{2}H splitting is used as a filter for the four possible solutions in the rmsd plots. Two possibilities remain: one solution is 5° from the 60° rotameric state, and the other is 25° or 30° from the rotameric state for Val-6 and Val-8, respectively; therefore, these latter χ_1 values are considered less probable. The data analysis for Val-1 and Val-7 is somewhat more complicated, but it appears that the relative populations in the various rotameric states observed at 5° is unchanged above the phase transition temperature. What does appear to change is the residence time of the side chain in each state, which decreases from the microsecond to nanosecond time scale on increasing the temperature from 5° to 35°. The populations determined from low temperature powder pattern data represent an additional constraint, and a unique χ_1 value is defined for the dominant conformer of both Val-1 and Val-7 (177° and −145°, respectively).

For the Leu side chains two different isotopic preparations were used: d_3-(α, β) Leu and d_7-(γ, δ) Leu.[5] An example of the data from oriented d_3-Leu-labeled gramicidin A is shown in Fig. 8. The sharp lines document a very small range of orientational disorder in the samples and hence the

[51] K.-C. Lee, S. Huo, and T. A. Cross, *Biochemistry* **34**, 857 (1995).

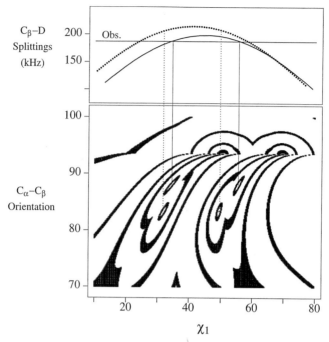

C_β–D Splittings (kHz)

C_α–C_β Orientation

χ_1

FIG. 7. The quadrupole splittings from uniformly oriented d_8-Val[6]-labeled gramicidin A are used to define the χ_1 torsion angle. Unlike Val-1, Val-6 does not have rotational jump motions about this torsional axis. The backbone structure restricts the C_α–C_β axis orientation to $88 \pm 5°$. Four minima are observed in these contour plots which represent the rmsd values between the experimentally observed quadrupolar splittings and the splittings predicted for each conformational locus on the plot. In the upper portion of the figure the predicted C_β–D splitting is predicted as a function of the χ_1 torsion angle, and the horizontal line represents the observed splitting. Only two of the minima are consistent with the C_β–D splitting, and only one of these minima is close to a χ_1 rotameric state ($60°$); therefore, a most probable conformational state is identified. (Reprinted with permission from Ref. 51. Copyright 1995 American Chemical Society.)

precise nature of the constraints. Powder pattern spectra at $5°$ showed that these sites were not undergoing large amplitude motions, but that significant librational amplitudes within the rotameric state exist. The side-chain structure was solved by first constraining χ_1 and then searching χ_2 space. Two rmsd plots were used. The first was identical to the Val plot described earlier in which a limited range of C_α–C_β angles was plotted versus χ_1 and contours were generated by comparing calculated and observed splitting from the d_3-labeled Leu. In this way a unique solution was found for the C_α–C_β orientation, and like the Val sites two values for χ_1 were found, one of which was much closer to a rotameric state than the other. The second

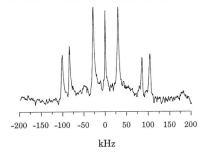

-200 -150 -100 -50   0   50   100 150 200

khz

FIG. 8. Orientational constraints are obtained from samples that are uniformly oriented with respect to the magnetic field direction of the NMR spectrometer. Here, a ^{2}H NMR spectrum of $d_3(\alpha, \beta_1\beta_2)$-Leu^{12}-labeled gramicidin A in planar bilayers aligned between thin glass plates is recorded. The narrow resonances illustrate how well the samples are aligned. A mosaic spread of channel axis orientations can be calculated from the resonance line width, and values of less than 0.3° have been observed.[32] Because the backbone structure defines the C_α–D orientation the 207 kHz splitting is readily assigned. β_1 and β_2 are assigned only after an analysis similar to that shown in Fig. 7. It has been shown that the β_1 splitting is 171 kHz and that the β_2 splitting is 49 kHz.[5]

rmsd plot is of χ_1 versus χ_2, and the contours represent data from both d_3- and d_7-labeled Leu using the defined C_α–C_β orientation. Again, for the most likely χ_1 value (the one closest to a rotameric state) two values of χ_2 are predicted, one much closer to a rotameric state than the other. In this way each of the Leu side-chain conformations were determined.

The four Trp conformations were determined with ^{15}N$_{\varepsilon 1}$-indole labels and d_5-indole labels. Again, the dynamics for the side chains have been characterized.[52] Here ^{15}N powder pattern spectra as a function of temperature using flash frozen preparations show that the indole side chains are fixed in rotameric states and that librational amplitudes about the χ_2 axis of 19° to 29° were observed for the various indoles. Because of the anisotropic motion, the averaged nuclear spin interaction tensor can be calculated for each deuteron, for the ^{15}N$_{\varepsilon 1}$ chemical shift, and for the ^{15}N–^{1}H dipolar interaction. Again an rmsd plot was generated using all of the orientational constraints in a plot of χ_1 versus χ_2. Two possible χ_1 solutions were generated for each indole and for each χ_1 two possible χ_2 angles for a total of four solutions. Each of these solutions have the same orientations of the indole ring with respect to the magnetic field. The NMR data do not further reduce this structural solution set. However, two of the solutions have the indoles radiating directly away from the channel instead of being packed close to the backbone and other side chains. Knowing that the structure is

[52] W. Hu, N. D. Lazo, and T. A. Cross, *Biochemistry* **34**, 14138 (1995).

fairly rigid it is unlikely that they radiate outward. Furthermore, these conformers have a radial component of their dipole moment that is oriented so as to destabilize the cation in the binding site. The indoles are known to stabilize cations in the binding site.[53] Of the four remaining conformer combinations of Trp-9 and Trp-15, two significantly overlap in space, and only one is oriented so as to provide a considerable stacking energy. Finally, only one of the conformers for each of the other indoles, Trp-11 and Trp-13, has the same tangential orientation for the dipole moment. It has been suggested for years that cations may pass through the channel off axis on a helical path. Experimental evidence strongly supports this concept,[54] and with the cations on a helical path the tangential components will affect conductance significantly. Therefore, the channel will operate most efficiently if the tangential components are all oriented in the same direction. On the basis of this fundamental knowledge a most probable set of indole conformers can be established.[55]

With the side-chain structural solutions an initial set of structures for the gramicidin channel is solved. A unique set of hydrogen bonds is identified. The absolute symmetry of the dimer is demonstrated by single chemical shift values and dipolar and quadrupolar splittings for each observed interaction. With this set of structures a unique molecular fold is defined. However, in achieving this molecular fold the data have not been equally weighted; for instance, the dipolar interactions are used quantitatively to define peptide plane orientations, whereas the anisotropic chemical shifts in the backbone are used qualitatively to eliminate structural possibilities. When the side-chain structures are assembled onto the backbone structure, there are several significant van der Waals contacts representing significant flaws in the structure. In addition, the hydrogen bonding geometry is outside the normal range of β-sheet hydrogen bonding geometry. Moreover, there are several assumptions in generating this initial structure which could be alleviated in a refinement procedure, such as the fixed covalent geometry and the planarity of the peptide planes. Therefore, a refinement protocol has been developed to take full advantage of the data, to improve on the structure, and to eliminate some of the assumptions.

Structural Refinement

To deal with the wide range of refinement goals it was necessary to develop a penalty function to refine against, that included contributions

[53] M. D. Becker, R. E. Koeppe II, and O. S. Andersen, *Biophys. J.* **62**, 25 (1992).
[54] F. Tian, K.-C. Lee, W. Hu, and T. A. Cross, *Biochemistry* **35**, 11959 (1996).
[55] W. Hu and T. A. Cross, *Biochemistry* **35**, 14147 (1995).

from both experimental data and global energy.[56] The penalty contributions from the experimental data were developed by calculating the NMR observables from the molecular structure, finding the difference with the observed results, and scaling the difference by the experimental error to generate a unitless contribution to the penalty. The energy contribution to the penalty was calculated with the all-atom PARAM 22 version of the CHARMM force field.[57] The IMAGE facility of CHARMM was used to impose dimer symmetry. For this simulated annealing refinement process three different types of molecular modifications were used.[6] Atom moves were used to relax covalent geometry and other subtle structural modifications by using a diffusion parameter of 0.0005 Å and a very modest starting temperature. Compensating torsional moves were achieved by randomly modifying a ψ_i and ϕ_{i+1} torsional pair within the range of $\pm 3°$ by an equal amplitude but opposite sign. In this way a considerable conformational space near a given chirality solution is searched without greatly distorting the helical parameters and incurring large penalties. If the peptide plane is nearly parallel to the bilayer normal and the channel axis, then both chiralities for the specific peptide plane can be conformationally searched. However, for those peptide planes where the plane makes a significant angle with the lipid bilayer normal, the energy barrier is too great for these torsional moves to transit from one chirality solution to the other. A third type of conformational perturbation was introduced, that of tunneling moves. Here compensating torsional moves were used and the magnitude was chosen so as to maintain the magnitude of cosine θ_n, the angle of the plane normal to the field, while changing the sign of this cosine. In other words, this move models roughly a change in the chirality for this site with a single structural move. With a mix of these three types of conformational moves it is possible to search the complete conformational space in the immediate vicinity of the structural solution set. There is no need to search additional conformational space.

Each of the four initial backbone structures as defined previously with a set of experimentally defined side-chain structures were refined independently. To reduce the relatively large penalty from the van der Waals overlap between the side-chain structures, the Leu-10, Leu-12, Leu-14, and ethanolamine groups were energy minimized using 250 steps of Adopted

[56] R. R. Ketchem, B. Roux, and T. A. Cross, *in* "Biological Membranes: A Molecular Perspective from Computation to Experiment" (K. M. Merz and B. Roux, eds.), p. 299. Birkhaeuser, Boston, 1996.

[57] A. D. Mackerell, Jr., D. Bashford, M. Bellot, R. L. Dunbrack, M. J. Field, S. J. G. Fischer, H. Guo, S. Ha, D. Joseph, L. Kuchnir, K. Kuczera, F. T. K. Lau, C. Mattos, S. Michnick, D. T. Nguyen, T. Ngo, B. Prodhom, B. Roux, M. Schlenkrich, J. Smith, R. Stote, J. Straub, J. Wiorkiewicz-Kuczera, and M. Karplus, *Biophys. J.* **61**, A143 (1992).

Basis Newton–Raphson[58] minimization within CHARMM. The resulting modifications in the torsion angles were relatively small and did not generate a greatly inflated structural penalty.

For the penalty function, 120 orientational constraints from ^{15}N and ^{13}C chemical shifts, $^{15}N-^1H$ or $^{15}N-^2H$ and $^{15}N-^{13}C$ dipolar interactions, and ^2H quadrupolar interactions were used per monomer. In addition, because the hydrogen bonding pattern was uniquely defined by the initial structure 10 N–O and 10 H–O hydrogen bond distances (2.91 $\pm$ 0.3 Å and 1.96 $\pm$ 0.3 Å, respectively)[59] were used as constraints. Finally the energy was used as a constraint. The refinement of the initial four structures was performed 10 times for each initial structure to form an ensemble, each of which has a high degree of self-consistency (atomic rmsd values within each ensemble range from 0.11 to 0.17 Å). This self-consistency is despite significant changes in chirality; in fact, the chirality solutions are highly consistent between refined ensembles despite specific different starting configurations for the chiralities. The all-atom rmsd was 0.48 Å for the 40 structures.[6] A single refined structure was achieved by averaging the ensembles (all refinements were used, no structure was eliminated). On averaging, the covalent geometry is distorted, and a final refinement with only atom moves was performed to correct these minor distortions.

The refinement has met all of the goals that were established earlier. The experimental constraints are weighted equally throughout the structure. The van der Waals overlaps have been eliminated. The covalent geometry and the planarity of the peptide planes have been effectively relaxed. Most importantly, the CHARMM force field and the experimental constraints have been utilized simultaneously to generate the final structure.

The chiralities along the backbone are not represented by any regular pattern, but where cations initially bind in this structure near the carboxyl termini the carbonyls all orient in toward the channel axis. These oxygens help to solvate the cations while they are in the channel. Surprisingly, this structure shows virtually no change upon cation binding, thereby supporting the concept that cations pass through the channel off axis and that small cations such as Na^+ are solvated by only two carbonyl oxygens at any one time (Fig. 9).[54] As seen in Fig. 10 (see color insert), the carbonyl oxygens form a helical array for the cationic path. The helical pitch is 4.9 Å and

[58] B. R. Brooks, R. E. Bruccoleri, B. D. Olafson, D. J. States, S. Swaminathon, and M. Karplus, *J. Comp. Chem.* **4**, 187 (1983).
[59] G. A. Jeffrey and W. Saenger, "Hydrogen Bonding in Biological Systems." Springer-Verlag, Berlin, 1994.

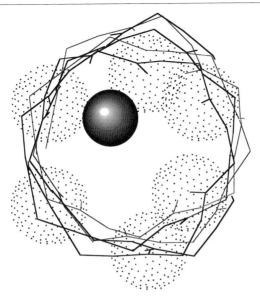

Fig. 9. There is little structural change on addition of Na$^+$ to samples of gramicidin, even for sites near the cation binding site.[54] The carbonyl oxygens of the first turn (C terminus, near the bilayer surface) are shown as stippled surfaces. Na$^+$ is shown as a solid sphere.

the residues per turn is 6.5, slightly different from the proposed model of 6.3 residues per turn.[16]

The indoles are almost ideally oriented to hydrogen bond to the bilayer surface and to direct the indole dipole moment so as to minimize the potential energy barrier for cation transit at the bilayer center. These indoles play an important role in orienting the channel with respect to its anisotropic environment (Fig. 11; see color insert) and in stabilizing the channel conformation which in other low dielectric environments is not the minimum energy conformation.

Summary

The method of using orientational constraints derived from solid-state NMR for structural characterization of polypeptides in heterogeneous environments has now been demonstrated. A very high resolution structure has been achieved that has led to greater functional understanding of this channel. Much can be done to improve this structural technique to make it more efficient and more generally applicable. Others as well as ourselves are applying this approach to membrane proteins.[60-62] Although solid-phase synthesis and specific site isotopic labeling has been essential for the development described here, one of the primary challenges is to be able to use amino acid-specific and uniform labeling of peptides and proteins by biosynthetic means for isotopic incorporation. This will allow for the study of many more proteins and significantly large proteins. Unlike solution NMR structural methods, there are no intrinsic molecular weight limitations. In fact, as the molecular weight increases the molecular motion will become less and the spectroscopic properties will improve. The major limitation will be sensitivity: as the molecular weight increases the number of moles will decrease in the samples, causing sensitivity to decrease. Advances in field strength and NMR technology help to address this problem. With larger molecules and more isotopically labeled sites resolution could also be a problem; however, the two- and three-dimensional methods demonstrated by Opella and co-workers clearly show the potential for enormous resolving power.[63] In the ^{15}N dimension alone it is shown that the resolution is greater than in solution NMR. Although challenges such as spectral assignments have yet to be completely solved, several approaches have been described, and the prospects are excellent for solving this and other problems facing the development of this novel approach for structural elucidation.

Although there is an attempt to get away from solid-phase synthesis to solve larger molecular weight structures, peptide synthesis will continue to be important for generating single- and double-site labeled model compounds for characterizations of spin interaction tensors. Such characterizations will continue to be a very important aspect of this structural approach.

Acknowledgments

The author extends special thanks to G. Fields and C. Fields for the first solid-phase peptide syntheses in my laboratory, to the staff of the Florida State University NMR Facility: J. Vaughn, R. Rosanske, and T. Gedris, and to the staff of the Bioanalytical Synthesis and Services Facility: H. Henricks and U. Goli. This work has been primarily supported by grants from the National Institutes of Health (AI-23007) and National Science Foundation (MCB-9603935).

[60] R. Smith, F. Separovic, T. J. Milne, A. Whittaker, F. M. Bennett, B. A. Cornell, and A. Makriyannis, J. Mol. Biol. **241,** 456 (1994).

[61] A. S. Ulrich, A. Watts, I. Wallet, and M. P. Heyn, Biochemistry **33,** 5370 (1994).

[62] B. Bechinger, L. M. Gierasch, M. Montal, M. Zasloff, and S. J. Opella, Solid State NMR **7,** 185 (1996).

[63] A. Ramamoorthy, L. M. Gierasch, and S. J. Opella, J. Magn. Reson. **111B,** 81 (1996).

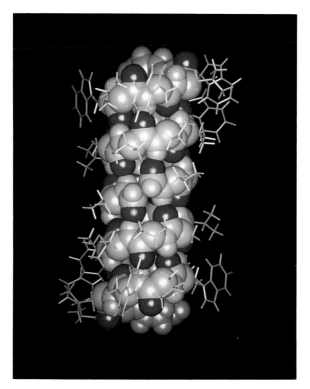

Fig. 10. The carbonyl oxygens shown in red form a helical array for solvation of the cation passing through the gramicidin channel. For the Na^+ ions to be adequately solvated the cation must pass through the channel off axis following the helical array of carbonyl oxygens.

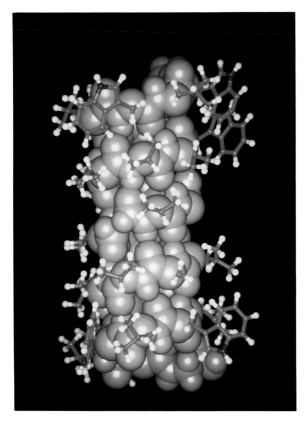

Fig. 11. Side view of the gramicidin channel with the backbone atoms shown as space filling and the side chains (green) in ball and stick. The Trp residues have several very important functional roles in the gramicidin channel. One role is that the Trp side chains are essential for properly orienting the molecule with respect to its anisotropic environment. Unlike water-soluble proteins membrane proteins must be oriented with respect to the surrounding environment. The indole NH groups (blue) are all oriented toward the bilayer surface, presumably so that they can hydrogen bond to the surface or the interfacial region. The top monomer shows the indoles of residues 13, 15, and 9 (left to right, respectively), whereas the symmetric bottom monomer shows the indoles of residues 13 and 11 (left to right, respectively). When the Trp residues are replaced with Phe the gramicidin conformation is primarily double helical, like gramicidin A in low dielectric organic solvents.

[32] Six-Year Study of Peptide Synthesis

By RUTH HOGUE ANGELETTI, LYNDA F. BONEWALD,
and GREGG B. FIELDS

The present volume of *Methods in Enzymology* is devoted to methodology for and utility of solid-phase peptide synthesis. The vast number of diverse products that can be created by the solid-phase approach has made the need for highly efficient synthetic methods especially critical. The Peptide Synthesis Research Committee (PSRC) of the Association of Biomolecular Resource Facilities (ABRF) was formed to evaluate the quality of the synthetic methods utilized in its member laboratories for peptide synthesis. Peptide synthesis, as defined by this committee, includes the chemistries used for peptide assembly and cleavage and the methods used for characterization of the final product.

The PSRC designed a series of studies over 1991–1996 to examine synthetic methods and analytical techniques. The first two studies, ABRF-91 and ABRF-92, focused on assembly and cleavage of synthetic peptides.[1,2] For both studies, peptide sequences were designed that incorporated amino acid residues believed to present challenges in either the peptide–resin assembly or cleavage step. Results from these two studies suggested that the respective cleavage conditions of Fmoc (9-fluorenylmethyloxycarbonyl) and Boc (*tert*-butyloxycarbonyl) solid-phase peptide synthesis were the primary source of synthetic difficulties, as most nondesired products were the result of covalent adducts, not deletions. Studies three and four, ABRF-93 and ABRF-94, thus focused specifically on side reactions during peptide–resin cleavage and side-chain deprotection.[3,4] In 1993, ABRF member laboratories were supplied with a peptide–resin that was assembled by the PSRC. Even with the preassembled peptide–resin, 20% of the crude samples

[1] A. J. Smith, J. D. Young, S. A. Carr, D. R. Marshak, L. C. Williams, and K. R. Williams, *in* "Techniques in Protein Chemistry III" (R. H. Angeletti, ed.), p. 219, Academic Press, Orlando, Florida, 1992.

[2] G. B. Fields, S. A. Carr, D. R. Marshak, A. J. Smith, J. T. Stults, L. C. Williams, K. R. Williams, and J. D. Young, *in* "Techniques in Protein Chemistry IV" (R. H. Angeletti, ed.), p. 227, Academic Press, San Diego, 1993.

[3] G. B. Fields, R. H. Angeletti, S. A. Carr, A. J. Smith, J. T. Stults, L. C. Williams, and J. D. Young, *in* "Techniques in Protein Chemistry V" (J. W. Crabb, ed.), p. 501, Academic Press, San Diego, 1994.

[4] G. B. Fields, R. H. Angeletti, L. F. Bonewald, W. T. Moore, A. J. Smith, J. T. Stults, and L. C. Williams, *in* "Techniques in Protein Chemistry VI" (J. W. Crabb, ed.), p. 539, Academic Press, San Diego, 1995.

did not contain any of the correct product,[3] further emphasizing problems in cleavage conditions. The 1994 study used the same peptide sequence as that used for the 1991 study,[1] based on the potential for side reactions during side-chain deprotection and cleavage. Of the products supplied for the 1994 study, 93% contained ≥25% of the desired product.[4] This represented a reasonable improvement over the 1991 study, where 78% of the crude products contained ≥25% of the desired material.[1] Core facilities had clearly developed improved methodology for minimizing side reactions during peptide–resin cleavage.

Following the 1994 PSRC study, it was felt that more specific potential problems in peptide chemistry needed to be addressed. The fifth study (ABRF-95) was designed to examine disulfide bond formation in the presence of residues susceptible to modification.[5] The sixth study (ABRF-96) examined His racemization during peptide assembly.[6] Overall, the PSRC studies have proved to be very useful for defining (i) both general and specific problems in solid-phase peptide synthesis and (ii) the analytical techniques useful for characterizing synthetic peptides. This article focuses mainly on the common side reactions found during the first 6 years of PSRC studies and the combination of analytical techniques that most effectively detect synthetic successes and failures.

I. Aims of ABRF Studies

The first PSRC study (ABRF-91) sought to determine the potential difficulties in peptide assembly and/or cleavage and side-chain deprotection reactions. Participating ABRF laboratories were asked to synthesize the following peptide by the methodology most commonly used in their facility:

Val-Lys-Lys-Arg-Cys-Ser-Met-Trp-Ile-Ile-Pro-Thr-Asp-Asp-
Glu-Ala-OH   [ABRF-91]

The designed peptide sequence presented several synthetic challenges. The hydrophobic Trp-Ile-Ile region could be difficult to assemble, while the Trp, Cys, and Met residues were sources of potential modifications during the cleavage step. Analysis of products would be by analytical reversed-phase high-performance liquid chromatography (RP-HPLC), amino acid

[5] R. H. Angeletti, L. Bibbs, L. F. Bonewald, G. B. Fields, J. S. McMurray, W. T. Moore, and J. T. Stults, in "Techniques in Protein Chemistry VII" (D. R. Marshak, ed.), p. 261, Academic Press, San Diego, 1996.

[6] R. H. Angeletti, L. Bibbs, L. F. Bonewald, G. B. Fields, J. W. Kelly, J. S. McMurray, W. T. Moore, and S. T. Weintraub, in "Techniques in Protein Chemistry VIII" (D. R. Marshak, ed.), p. 875, Academic Press, San Diego, 1997.

analysis (AAA), capillary electrophoresis (CE), electrospray mass spectrometry (ESMS), and plasma desorption mass spectrometry (PDMS).

In similar fashion to ABRF-91, the peptide designed for ABRF-92 contained both potential assembly problems and residues subject to modification. Facilities were asked to synthesize the following peptide:

Gly-Val-Arg-Gly-Asp-Lys-Gly-Asn-Pro-Gly-Trp-Pro-Gly-
Ala-Pro-Tyr-OH   [ABRF-92]

Analysis of ABRF-92 products would be by RP-HPLC, AAA, CE, Edman degradation sequence analysis, ESMS, PDMS, fast atom bombardment mass spectrometry (FABMS), and matrix-assisted laser desorption–ionization mass spectrometry (MALDI-MS).

The ABRF-91 and ABRF-92 studies ultimately demonstrated that side reactions during peptide–resin cleavage and deprotection were a much greater source of problems than peptide assembly.[1,2] Thus, the ABRF-93 study specifically addressed cleavage problems. The PSRC assembled peptide–resins by either Boc or Fmoc chemistries, and participating ABRF laboratories were requested to cleave and side-chain deprotect the following peptide by the methodology most commonly used in their facility:

Lys-His-Asp-Pro-Cys-Gly-Trp-Asn-Gly-Pro-Arg-Pro-Met-
Arg-Gly-OH   [ABRF-93]

The peptide designed for the ABRF-93 study incorporated Cys, Trp, Met, and multiple Arg residues, all susceptible to difficulties during peptide–resin cleavage. Analysis of products would be by RP-HPLC, AAA, Edman degradation sequence analysis, ESMS, and MALDI-MS.

The results of ABRF-93 suggested that core facilities had improved their techniques for cleaving peptide–resins. To further test if this conclusion applied to more than one peptide, the ABRF-94 study used the same peptide sequence as ABRF-91 and asked participating laboratories to synthesize the peptide by the methodology most commonly used in their facility:

Val-Lys-Lys-Arg-Cys-Ser-Met-Trp-Ile-Ile-Pro-Thr-Asp-Asp-
Glu-Ala-OH   [ABRF-94]

The goal of ABRF-94 was to see if the participating laboratories had truly improved their methods compared with ABRF-91, and hence the overall quality of the product would be improved. Analysis of products would be by RP-HPLC, AAA, Edman degradation sequence analysis, ESMS, and MALDI-MS.

For the fifth PSRC study, the focus was shifted from general problems of peptide synthesis to specific applications. The peptide for ABRF-95 was designed to examine disulfide bond formation chemistries:

Phe-cyclo[Cys-Phe-Trp-Lys-Thr-Cys]-Thr-NH₂   [ABRF-95]

For extra difficulty, a Trp residue was included to test both the efficiency of disulfide bond formation and the modification of susceptible residues by the respective chemistries. The small nature of this peptide allowed for mass spectrometric identification of both the reduced and oxidized peptide products. Analysis of products would be by RP-HPLC, AAA, Edman degradation sequence analysis, ESMS, MALDI-MS, and spectrophotometric analysis following treatment with Ellman's reagent.

The sixth PSRC study approached a specific, but somewhat forgotten, potential problem of peptide assembly, that of His racemization.[7] Peptide ABRF-96 was thus designed to study the extent of His racemization during peptide assembly.

Arg-Glu-Arg-His-Ala-Tyr-OH   [ABRF-96]

Analysis of products would be by RP-HPLC, RP-HPLC following treatment with Marfey's reagent, AAA, ESMS, and MALDI-MS.

II. Methods

Analytical Reversed-Phase High-Performance Liquid Chromatography

Samples are dissolved in 0.1% aqueous TFA and approximately 50 μg analyzed on a Perkin-Elmer (Norwalk, CT) Series 4 HPLC using a Vydac C_{18} column (300 Å pore size, 4.6 × 250 mm)[4,5] or a Bio-Rad (Richmond, CA) rp304 C_4 column (300 Å pore size, 4.6 × 250 mm).[1-3] The linear gradient extending from 0.1% aqueous trifluoroacetic acid (TFA) to 70% acetonitrile (containing 0.09% TFA) over either 33 min[1-4] or 60 min is used.[5] Additional conditions are either (a) a flow rate of 2 ml/min and absorbance monitored at 214 nm using a Perkin-Elmer LC 95 detector[1-4] or (b) a flow rate of 1 ml/min and absorbance monitored at 214 nm using a photodiode array detector.[5] Samples are injected by a Perkin-Elmer SS 100 autosampler. Quantitation is by a Nelson Model 1020 Data System[1-4] or the Waters (Milford, MA) 820 software program.[5]

---

[7] G. B. Fields, Z. Tian, and G. Barany, *in* "Synthetic Peptides: A User's Guide" (G. A. Grant, ed.), p. 77, Freeman, New York, 1992.

Capillary Electrophoresis

Capillary electrophoresis is performed on an Applied Biosystems (Foster City, CA) 270HT using the 600 Data Analysis System. The buffer is either 20 mM sodium phosphate, pH 2.5,[1] or 50 mM sodium biphosphate, pH 4.4.[2] Voltage is 20 kV at 30°, the program is 27 min, and electrophoresis is monitored at 220 nm after a 1 to 5 sec sample injection.

Amino Acid Analysis

Samples (~0.5 μg) are hydrolyzed for 24 hr at 112° in 100 μl 6 N HCl, 0.2% phenol. For ABRF-95, Cys is analyzed as cysteic acid after performic acid oxidation and hydrolysis in 6 N HCl for 24 hr at 120°, whereas Trp is analyzed after a rapid hydrolysis for 25 min at 166° in 6 N HCl containing 4% phenol.[5] Analysis is performed on a Beckman (Fullerton, CA) 6300 with a sulfonated polystyrene cation-exchange column (0.4 × 25 cm). Quantitation is by a Beckman 7300 or a Nelson Analytical Model 4400 Data System.

Edman Degradation Sequence Analysis

Edman degradation sequence analysis of selected samples (dissolved in 0.1% TFA–20% acetonitrile) is performed on an Applied Biosystems (Foster City, CA) 477 A Protein Sequencer/120A Analyzer using BioBrene Plus as described.[8] To identify deprotection products and deletions, 800–900 pmol of sample is sequenced. Protocols for solid-phase sequencing of Boc-synthesized peptide–resins have been described.[9] Protocols for sequencing Fmoc-synthesized peptide–resins are identical to standard protein sequencing protocols.[9]

Mass Spectrometry

Over the course of the ABRF studies, four different mass spectrometric techniques have been used: PDMS, FABMS, ESMS, and MALDI-MS. PDMS is performed on a BioIon 20 (Applied Biosystems) plasma desorption, time-of-flight mass spectrometer using a ^{252}Cf fission fragment source in the positive ion mode at an accelerating potential of 16 kV for 1 hr at 8000-nsec intervals.[1,2] Aliquots (1–10 μl) of the peptide solution are applied

---

[8] Applied Biosystems, Inc., Applied Biosystems Model 477A Protein–Peptide Sequencing System Users Manual, Foster City, CA, 1989.

[9] M. L. Kochersperger, R. Blacher, P. Kelly, L. Pierce, and D. H. Hawke, *Am. Biotechnol. Lab.* **7**(3), 26 (1989).

to an aluminized Mylar target preelectrosprayed with nitrocellulose. Targets are dried under a stream of N_2 and not washed. Initial PDMS analysis indicated that some Fmoc samples contained a +28 Da peak which was not present on reanalysis. As a result, only the latter PDMS analyses are used in calculating the percent desired product.

Fast atom bombardment mass spectrometry is carried out on MS-1 of a JEOL HX110HF tandem double-focusing mass spectrometer.[2] Samples are prepared by addition of a 1-μl aliquot of the peptide solution to 1 μl of m-nitrobenzyl alcohol matrix on the sample target. Ions are generated by FAB with 6 keV Xe atoms. Resolution is set at 3000, and the mass axis is scanned from 375 to 2500 Da in 20 sec.

Electrospray mass spectrometry is performed either on a Sciex API-III triple-quadrupole mass spectrometer[1-3,5] or a Fisons VG Quattro outfitted with a Fisons Electrospray Source.[4] Samples are prepared and analyzed by either (a) dissolving in 0.05% TFA at 0.5 μ/μl and introducing 2 μl into the mass spectrometer via a Beckman Model 507 autosampler[1-3] or (b) dissolving in 1.0 ml of 50% methanol–1% acetic acid, diluting 1 : 10 with 50% acetonitrile–1.0 mM ammonium acetate to give 25 pmol/μl, and injecting a 10-μl aliquot into a 10 ml/min stream of 50% acetonitrile–1.0 mM ammonium acetate.[4] For liquid chromatography (LC)–ESMS, products are separated using a Vydac C_{18} column (300 Å pore size, 2.1 × 250 mm) with a linear gradient of 0–60% B over 45 min, where A is 0.1% aqueous TFA and B is 90% acetonitrile–0.08% aqueous TFA.[2,3]

Matrix-assisted laser desorption mass spectrometry is accomplished initially with a Vestec LaserTech ResearcH laser desorption time-of-flight mass spectrometer, operated at 20 keV.[2,3] Samples are prepared by addition of 1 μl of the diluted peptide solution (~10 pmol) to 1 μl of matrix solution [50 mM 2,5-dihydroxybenzoic acid (DHBA) or sinapinic acid (SA) in water] on the sample target, which are then allowed to dry at ambient temperature. The N_2 laser (337 nm) is triggered at 20 Hz. Spectra are required at 5 nsec resolution and are a summation of ion intensity for 64 laser pulses. Subsequently, MALDI-MS is performed with a Vestec LaserTech ResearcH time-of-flight,[3] Vestec Benchtop IIt linear time-of-flight,[4] or Fisons VG Analytical TofSpec mass spectrometer,[5] operated in the linear mode with an N_2 laser (337 nm). Samples are dissolved in 1.0 ml of 25% acetonitrile–0.1% TFA, then diluted 3 : 100 to give 5–10 pmol/μl. A 0.5-μl aliquot of each sample solution is added to 0.5 μl of matrix (α-cyano-4-hydroxycinnamic acid, saturated solution in 50% acetonitrile–2% TFA). Samples are dried at ambient temperature and pressure. Each spectrum is the sum of ion intensity from 10–50 laser pulses. The mass axis is calibrated externally.

TABLE I
CHEMISTRY UTILIZED BY CORE FACILITIES FOR
ABRF TEST PEPTIDES

| | Fmoc | | Boc | |
Year	Samples submitted	%	Samples submitted	%
1991	18	50	18	50
1992	42	72	16	28
1993	34	74	12	26
1994	80	98	2	2
1995	58	97	2	3
1996	51	96	2	4

Spectrophotometric Detection of Free Sulfhydryl Groups

The percentage of free sulfhydryl groups is determined spectrophotmetrically at 412 nm after reaction of the peptide samples with Ellman's reagent, 5,5'-dithiobis(nitrobenzoic acid).[10] An extinction coefficient of 13,600 M^{-1} cm^{-1} is used.

III. Defining General Problems in Peptide Synthesis

When these studies were initiated, the prevalent dogma was that "standard" peptide synthesis protocols were optimal. Over the course of the 6-year period, it became apparent that many caveats existed to the belief that "automated" peptide synthesis always yielded the correct peptide. Interestingly, we observed a strong shift in the chemistry utilized in core facilities during this time period. The more senior Boc methodology was replaced by Fmoc chemistry (Table I). Study results were clearly affected by this change.

ABRF-91

The first ABRF study approached peptide synthesis from a broad perspective, in that the designed peptide was used to examine both potential assembly and postassembly problems. One conclusion drawn from this study was that peptide assembly did not present a major problem for most facilities. Mass spectrometry indicated that impurities were primarily due

---

[10] J. M. Stewart and J. D. Young, "Solid Phase Peptide Synthesis," 2nd Ed., p. 116. Pierce, Rockford Illinois, 1984.

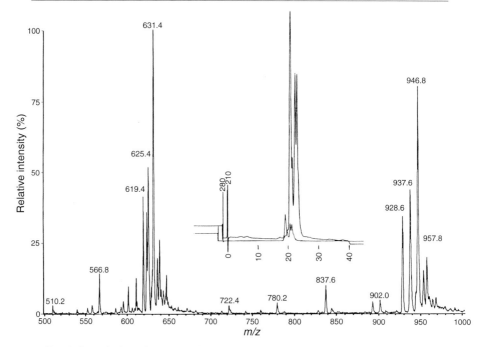

FIG. 1. Quantitation of an ABRF-91 sample by RP-HPLC (inset) and ESMS. The relative abundance of the desired product was 43% by RP-HPLC, 46% by CE (data not shown), and 52% by ESMS ([M + 3H]$^{3+}$ = 631.4 and [M + 2H]$^{2+}$ = 946.8). Other species present in this sample include the desired peptide with a single dehydration ([M + 3H]$^{3+}$ = 625.4 and [M + 2H]$^{2+}$ = 937.6 by ESMS) and a double dehydration ([M + 3H]$^{3+}$ = 619.4 and [M + 2H]$^{2+}$ = 928.6 by ESMS).

to side reactions that occurred during peptide–resin cleavage and side-chain deprotection.[1] For example, ESMS analysis of one peptide synthesized by Boc chemistry showed that, in addition to the desired peptide, products were present with m/z values of 1873 and 1855 Da (Fig. 1). These masses correspond to single and double dehydrations of the desired peptide, which most likely occurred during strong acid cleavage of the peptide–resin.[11] ABRF-91 provided the initial data to suggest that peptide modification during the cleavage step could be significant, and that certain by-products were fairly common (Table II). In addition, reasonable agreement was found between chromatographic and mass spectrometric analyses of samples (Fig. 1).

[11] G. Barany and R. B. Merrifield, in "The Peptides" (E. Gross and J. Meienhofer, eds.), Vol. 2, p. 1. Academic Press, New York, 1979.

TABLE II
COMMON BY-PRODUCTS OBSERVED BY MASS SPECTROMETRY

Mass difference from desired peptide (Da)	Possible product[a]
−36	−2H$_2$O
−18	−H$_2$O
+16	Oxidation
+28	For
+32	Double oxidation
+42	Acetyl
+56	tBu
+67	β-Piperidide
+76	Phenyl ester
+96	Trifluoroacetyl
+100	Boc
+106	[Anisyl + O] or thioanisyl or thiocresyl
+109	Unknown
+122	[Anisyl + 2O] or [thioanisyl + O] or [thiocresyl + O]
+152	Unknown
+172	Adduct from EDT + TFA
+201	Adduct from HMP + TFA
+212	Mtr
+266	Pmc

[a] For, Formyl; tBu, $tert$-butyl; EDT, ethanedithiol; HMP, hydroxybenzyl; Mtr, 4-methoxy-2,3,6-trimethylbenzenesulfonyl;   Pmc,   2,2,5,7,8-pentamethyl-chroman-6-sulfonyl.

ABRF-92

The second ABRF study was developed to further expand on the conclusions of ABRF-91. Again, a peptide was designed that presented both potential assembly and cleavage problems. AAA showed 88% of the crude products and 94% of the purified products submitted for this study to be compositionally correct (Trp was not quantitated). Examples of peptide deletions included 1 des(Asn), 2 des(Pro), 1 partial des(Asn,Gly) and 1 des(2Gly) crude products and 2 des(Pro) purified products. Two crude products with improper compositions showed correct compositions for the subsequent purified products. On the basis of mass spectrometric results, one of these crude products was indeed a complex mixture with no detectable desired peptide, whereas the other crude peptide contained >80% of the desired peptide.

Reversed-phase HPLC analyses indicated a high percentage of successful syntheses, as 88% of the crude products had >25% of the apparent desired peptide and 85% of the purified products had >75% of the apparent desired peptide. In general, CE indicated a higher degree of purity (85%) of the submitted samples than RP-HPLC (75%). However, with 41 samples run consecutively by CE, the main peak migration times advanced progressively by more than 4 min. This variability in CE migration time made it impossible to determine the relative percent content of the correct peptide, whether by comparison to a known standard or by trying to achieve reproducibility of the individual peptide itself, and thus probably accounted for the higher estimated degree of purity as compared to RP-HPLC.

Assessment of product purity by ESMS, FABMS, and PDMS showed *semiquantitative* agreement with RP-HPLC analyses. The desired species molecular ions were $[M + 3H]^{3+} = 543.4$ and $[M + 2H]^{2+} = 814.6$ by ESMS, $[M + H]^{+} = 1627.6$ by FABMS, and $[M + H]^{+} = 1627.9$ by PDMS. By comparison of the relative abundance of these desired molecular ions for all samples, the combined mass spectrometric techniques assigned 12% of the Fmoc-synthesized crude products and 12.5% of the Boc-based crude products as "poor" quality (<25% desired product). One of these Fmoc-synthesized products had a molecular mass 114 Da below that of the desired peptide, which could have been a deletion of 1 Asn or 2 Gly residues. Comparison with AAA and preview sequence analysis indicated an Asn deletion. Interestingly, the RP-HPLC elution time of this des(Asn^8) peptide was the same as the desired peptide. Another Fmoc synthesized crude product was a mixture of peptides 57, 171, and 228 Da below the desired peptide. Comparison with AAA and preview sequence analysis indicated deletions at Gly^1, Gly^7, and Asn^8. Two Fmoc-synthesized crude products (and their subsequent purified products) had molecular masses indicative of a Pro deletion. The final "poor" quality crude product synthesized using Fmoc chemistry contained numerous peptides of higher molecular masses than the desired peptide, most predominantly a +42 Da species. The source of this modification is unknown. Of the two "poor" quality Boc-synthesized crude products, one was primarily a deletion peptide. 300 Da lower than the desired peptide. AAA and preview sequencing showed this to be a des(Gly^1,Gly^4,Trp^{11}) deletion peptide. The other Boc-synthesized poor product was 28 Da higher than the desired product, indicative of a Trp(For) peptide. In this particular case the crude formylated product was the desired product, as the For group was removed in a separate step postcleavage. Overall, incomplete (<90%) deprotection of Trp(For) was detected in Boc-synthesized products 4 times (out of a possible 12) as determined by the combined mass spectrometric analyses. For Fmoc-based syntheses, the predominant side reaction resulted in products of mass 201 Da higher than the desired peptide (Table II), and was generated during peptide-resin cleavage of HMP-based linkers. This side product, which was noted elsewhere,[12] was ultimately assigned to a trifluoroacetyl-4-hydroxybenzyl adduct.[13] Peptide modified by the trifluoroacetyl-4-hydroxybenzyl adduct was readily separated from the desired product by RP-HPLC and easily detected by mass spectrometry (Fig. 2).

Syntheses incorporating Fmoc-Asn without side-chain protection appeared problematic. Of the 4 Fmoc-synthesized crude peptides that contained >10% des(Asn), 2 were assembled without Asn side-chain protection. The only Fmoc-synthesized peptide for which all three mass spectrometric techniques detected >10% dehydration was assembled without Asn side-chain protection by DCC/HOBt (N,N'-dicyclohexylcarbodiimide : 1-hydroxybenzotriazole). This dehydration was probably the result of a side reaction (nitrile formation) attributable to the use of unprotected Asn during coupling.[7] Thus, of the 7 peptides synthesized using Fmoc-Asn, 43% had >10% product of a side reaction directly attributable to Asn incorporation.

Impurities in Boc-synthesized peptides were mostly due to cleavage problems. Of the 12 crude samples synthesized with Trp(For), 4 had >10% of the For group still attached. Deprotection of Trp(For) was complete by treatment of the peptide–resin with approximately 10% piperidine–dimethylformamide (DMF) for 2 hr prior to HF cleavage. Dehydration and *tert*-butylation were also significant problems in Boc-synthesized peptides, as was observed in the ABRF-91 study.[1]

ABRF-93

ABRF-93 was an unusual study, in that participating laboratories were asked only to cleave and side-chain deprotect a peptide–resin standard. The peptide sequence was designed specifically to study cleavage-based side reactions. For Fmoc methods, the predominant nondesired products were of 16, 106, 152, and/or 266 mass units higher than the desired peptide. The +16 mass unit product is the result of oxidation, most probably of Met or Cys but not Trp, generated during peptide–resin cleavage (Table II). Oxidation was significant in ABRF-91 for a test peptide containing both Cys and Met,[1] but not in ABRF-92 when Cys and Met were absent from the test peptide.[2] The +106 mass unit product is the desired peptide plus either a thioanisyl adduct or the combination of oxidation and an anisyl adduct (Table II). Boc-based methods had problems with His side-

[12] C. G. Fields and G. B. Fields, *Tetrahedron Lett.* **34**, 6661 (1993).
[13] C. G. Fields and G. B. Fields, *in* "Innovation and Perspectives in Solid Phase Synthesis, Biological and Biomedical Applications 1994" (R. Epton, ed.), p. 251. Mayflower Worldwide, Birmingham, U.K., 1994.

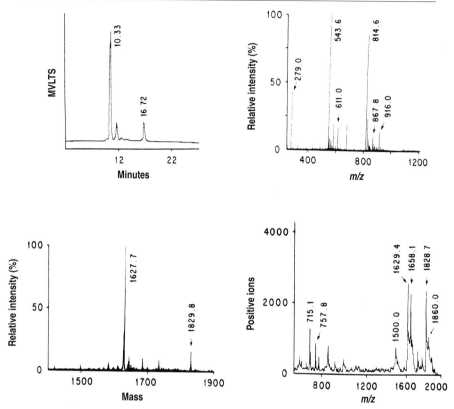

Fig. 2. Quantitation of an ABRF-92 sample by RP-HPLC (*top left*), ESMS (*top right*), FABMS (*bottom left*), and PDMS (*bottom right*). The relative abundance of the desired product was 68% by RP-HPLC, 69% by ESMS ([M + 3H]$^{3+}$ = 543.4 and [M + 2H]$^{2+}$ = 814.6), 75% by FABMS ([M + H]$^{+}$ = 1627.7), and 56% by PDMS ([M + H]$^{+}$ = 1629.4). The relative abundance of the +202 Da trifluoroacetyl-4-hydroxybenzyl adduct was 12% by ESMS ([M + 3H]$^{3+}$ = 611.0 and [M + 2H]$^{2+}$ = 916.0), 11% by FABMS ([M + H]$^{+}$ = 1829.8), and 30% by PDMS ([M + H]$^{+}$ = 1828.7). (Reproduced by permission from Ref. 2, pp. 233–234.)

chain deprotection. Eleven laboratories elected to deprotect His(Dnp) prior to HF peptide–resin cleavage, 7 using thiophenol, 3 using 2-mercaptoethanol, and 1 using piperidine. The removal of the dinitrophenyl (Dnp) by thiophenol or 2-mercaptoethanol prior to peptide–resin cleavage by HF was efficient for this test peptide (Fig. 3). Overall, no correlation was found between the common side reactions observed in ABRF-93 and specific cleavage methods.

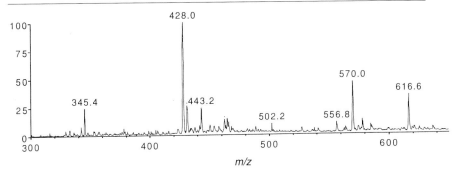

F<small>IG</small>. 3. ESMS analysis of ABRF-93 sample in which the Dnp group was removed from His by thiophenol prior to HF treatment. The desired product had ($[M + 4H]^{4+} = 427.7$ and $[M + 3H]^{3+} = 569.9$).

ABRF-94

To investigate the mechanisms of cleavage-based side reactions, the ABRF-94 study limited the cleavage methods to be used by participating laboratories while requesting synthesis of the same peptide as ABRF-91. Of the crude peptide samples supplied, 2% were synthesized by Boc chemistry and 98% by Fmoc chemistry (Table I). The peptide–resins assembled by Fmoc chemistry were cleaved by either reagent K [ethanedithiol (EDT)–thioanisole–water–phenol–TFA, $1:2:2:2:33$][14] (61%), reagent B (triisopropylsilane–phenol–water–TFA, $2:5:5:88$)[15] (28%), or other cleavage cocktails (11%).

In similar fashion to ABRF-92, assessment of product purity by ESMS and MALDI-MS showed semiquantitative agreement with RP-HPLC analyses. The desired species molecular ions were $[M + 3H]^{3+} = 631.3$ Da and $[M + 2H]^{2+} = 946.3$ Da by ESMS and $[M + H]^+ = 1893.2$ Da by MALDI-MS. For ESMS, samples were run at atypically high concentrations so that minor components were detectable. Further dilution of samples yielded identical results for those cases examined. In contrast, the relative abundances of peaks in a mixture detected by MALDI-MS varied with different dilutions of the sample, along with different spots on the sample target and different laser power settings. In general, the Fmoc-synthesized poor quality products contained several species with masses both below (-36 Da, -18 Da) and above ($+16$ Da, $+56$ Da) the desired peptide. These species are indicative of double and single dehydrations, single oxidations, and tBu adducts (Table II). Of the 29 Fmoc-synthesized peptides for which both

[14] D. S. King, C. G. Fields, and G. B. Fields, *Int. J. Pept. Protein Res.* **36**, 255 (1990).
[15] N. A. Solé, and G. Barany, *J. Org. Chem.* **57**, 5399 (1992).

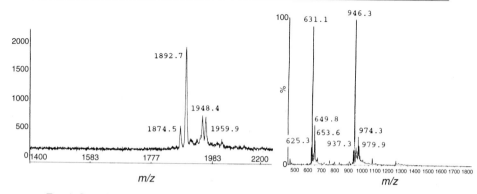

Fɪɢ. 4. Quantitation of an ABRF-94 sample by MALDI-MS (*left*) and ESMS (*right*). The relative abundance of the desired product was 50% by MALDI-MS ([M + H]$^+$ = 1892.7) and 58% by ESMS ([M + 3H]$^{3+}$ = 631.1 and [M + 2H]$^{2+}$ = 946.3). Other species present in this sample include the desired peptide with a single dehydration ([M + 3H]$^{3+}$ = 625.3 and [M + 2H]$^{2+}$ = 937.3 by ESMS, [M + H]$^+$ = 1874.5 by MALDI-MS), *t*Bu adduct ([M + 3H]$^{3+}$ = 649.8 and [M + 2H]$^{2+}$ = 974.3 by ESMS, [M + H]$^+$ = 1984.4 by MALDI-MS), and β-piperidide adduct ([M + 3H]$^{3+}$ = 653.6 and [M + 2H]$^{2+}$ = 979.9 by ESMS, [M + H]$^+$ 1959.9 by MALDI-MS). (Reproduced by permission from Ref. 4, p. 531.)

mass spectrometric techniques detected >5% dehydration, 15 contained a +67 Da species (Fig. 4 and Table II). The +67 Da species was probably the desired peptide modified by a β-piperidide.[16,17] Base treatment used to remove the Fmoc group can result in aspartimide formation (dehydration) from Asp(O*t*Bu) residues; the cyclic aspartimide residue can then incorporate a β-piperidide.[16,17] Thus, peptides containing the +67 Da adduct were not modified during cleavage. Aspartimide formation from Asp(O*t*Bu) residues can be inhibited by adding HOBt or 2,4-dinitrophenol to the piperidine solution.[17] The presence of a β-piperidide adduct in 19% of the Fmoc-synthesized products indicates that dehydration of Asp(O*t*Bu) residues during Fmoc removal can be a significant problem during synthesis.[17] The other problem detected in Fmoc syntheses was the generation of *tert*-butyl (*t*Bu) adducts during peptide–resin cleavage. The modification of peptides by *t*Bu groups (usually via the indole side chain of Trp) can be dependent on the workup of the crude peptide following cleavage.

The overall quality of the peptides synthesized improved greatly from ABRF-91 to ABRF-94. Percentage wise, more laboratories were using Fmoc chemistry since the start of the studies (see Table I). Trp modification,

[16] R. Dölling, M. Beyermann, J. Haenel, F. Kernchen, E. Krause, P. Franke, M. Brudel, and M. Bienert, *J. Chem. Soc. Chem. Commun.*, 853 (1994).
[17] J. L. Lauer, C. G. Fields, and G. B. Fields, *Lett. Pept. Sci.* **1**, 197 (1995).

one of the major problems that occurs during peptide–resin cleavage, was certainly reduced by the introduction of Boc side-chain protection of the Trp indole.[12,18,19] Thus, possible reasons for the improved results from 1991 to 1994 are any combination of (i) the greater percentage of peptides synthesized by Fmoc chemistry, where cleavage conditions are less harsh, (ii) the use of different side-chain protecting group strategies [i.e., Pmc (2,2,5,7,9-pentamethylchroman-6-sulfonyl) instead of Mtr (4-methoxy-2,3,6-trimethylbenzene sulfoyl) for Arg, Boc for Trp] that help reduce side reactions during cleavage, (iii) the use of cleavage protocols designed to minimize side reactions, and (iv) more rigor and care in laboratory techniques.

IV. Specific Problems in Peptide Synthesis

ABRF-95

ABRF-95 was designed to study an analog of octreotide, an 8-residue peptide which is readily oxidized to its cyclical form. The presence of a Trp residue provided a site susceptible to modification during disulfide bond formation. The low m/z of the reduced (1034.2 Da) and oxidized (1032.2 Da) peptides would permit the determination of the relative proportions of linear and cyclized forms in the same sample by analysis of the isotopic distribution in ESMS. The participating laboratories were asked to oxidize the peptide by one or all of four frequently used protocols, two off-resin and two on-resin. Laboratories were instructed that the Acm (acetamidomethyl) group should be used for side chain protection of Cys residues if on-resin protocols were followed. Because Trp would be modified during on-resin oxidation procedures, it was also suggested that the Boc group should be used for Trp protection if Fmoc chemistry were employed, or the For group for Boc chemistry. In protocol I, the cleaved peptide was dissolved in 7 g/liter ammonium acetate, and stirred for 3 days before lyophilization. Protocol II used direct oxygenation for 24 hr in an ammonium bicarbonate buffer at pH 8.5. In protocol III, the resin-bound peptide was treated with a 1.5 molar excess of 0.4 M thallium trifluoroacetate in DMF for 1 hr before washing, and cleavage. For protocol IV, the peptide–resin was treated with a 4-fold molar excess of 0.1 M mercuric acetate in DMF for 1 hr, after which the resin was filtered, treated with a 10-fold molar excess of 2-mercaptoethanol in DMF for 1 hr, filtered and washed with solvents, and cleaved from the resin with H_2O–TFA (1:19, v/v).

[18] P. White, in "Peptides: Chemistry and Biology" (J. A. Smith and J. E. Rivier, eds.), p. 537. Escom, Leiden, The Netherlands, 1992.
[19] H. Choi and J. V. Aldrich, *Int. J. Pept. Protein Res.* **42,** 58 (1993).

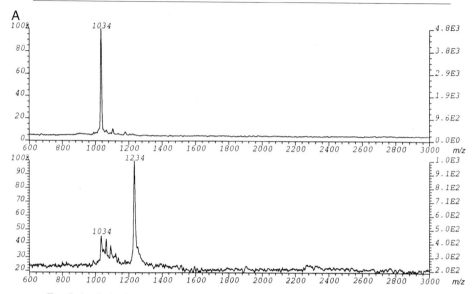

FIG. 5. Analysis of two ABRF-95 samples by (A) MALDI-MS and (B) ESMS. Cyclization of the peptide was performed on-resin by either thallium trifluoroacetate (*top*) or mercuric acetate (*bottom*). The desired peptide has [M + H]$^+$ = 1034 by MALDI-MS and [M + 2H]$^{2+}$ = 516.8 and [M + H]$^+$ = 1032.4 by ESMS. The peptide containing the mercury adduct has [M + H]$^+$ = 1234 by MALDI-MS and [M + 2H]$^{2+}$ = 617.8 and [M + H]$^+$ = 1234.4 by ESMS. (Reproduced by permission from Ref. 5, pp. 266–271.)

Participants were warned that mercury adducts had been observed in peptides oxidized with this procedure. Three laboratories submitted five samples oxidized by a postcleavage potassium ferricyanide method. Another three laboratories used the Ekathiox resin.

The majority of the 60 submitted cyclized peptide samples were prepared by protocol I, which did not require the purchase of specialty reagents. Both postcleavage methods I and II produced the correct cyclized product. However, both procedures, particularly method I, yielded dimers in the majority of samples, which reached as high as 70% of the product in one of the samples. The on-resin cyclization with thallium trifluoroacetate (protocol III) produced high yields of cyclized peptide in all samples submitted, with little or no dimer formation (Fig. 5). The samples for which either Ekathiox resin treatment or potassium ferricyanide oxidation were utilized also appeared to produce the cyclized peptide in good yield, although the sample size was too small to assess. Only a small amount of dimer formation, if any, was observed in these samples. The mercuric acetate method did

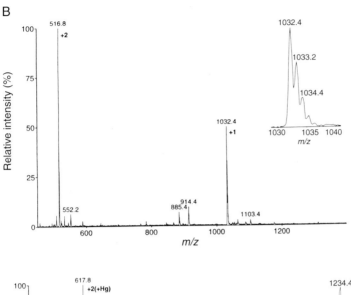

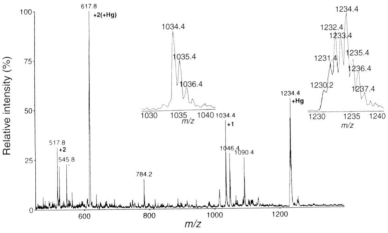

FIG. 5. (*continued*)

not yield the correct peptide product but produced mercury–peptide adduct which is stable by MS (Fig. 5).

Results of ABRF-95 indicated that producing cyclized peptides with the correct structure can be achieved readily by either on-resin or postcleavage techniques. Postcleavage techniques are less expensive and provide reasonable yields of the desired product. In contrast, on-resin techniques produce greater yields of the final product but are more expensive to perform.

ABRF-96

Although the biological importance of synthesizing a specific stereoisomeric form of a peptide is well recognized, little concern is often given to this aspect of purity. ABRF-96 was designed to assess the extent to which racemization occurs in routine peptide synthesis. His was included as a known site of racemization. The other amino acids were chosen partly because their derivatives were well resolved by amino acid analysis following reaction with Marfey's reagent. The Arg adjacent to the His also provided a tryptic cleavage site, and the C-terminal residues were sensitive to digestion with carboxypeptidase A. These choices allowed for introduction of novel methods for racemization analysis, including enzymatic digestion followed by MALDI-MS of the peptide products. The peptide sequence was also chosen because no problems with peptide assembly or cleavage were anticipated.

In analyzing the reference peptides (Arg-Glu-Arg-His-Ala-Tyr-OH and Arg-Glu-Arg-D-His-Ala-Tyr-OH), it was found that trypsin did not discriminate between either peptide. In contrast, carboxypeptidase A removed only the C-terminal Tyr from the peptide containing D-His, instead of both Ala and Tyr. AAA following derivitization with Marfey's reagent was the most sensitive indicator of racemization. Baseline resolution of the reference peptides was also obtained with two conventional RP-HPLC protocols (Fig. 6), but not with a chiral chromatography column. However, one cannot reliably predict whether the stereoisomers of any particular peptide sequence will be separated by RP-HPLC.

Fifty-three peptide samples were submitted by 48 laboratories. Previous studies by this committee have demonstrated the need for multiple analytical methods for the assessment of purity. Therefore the peptides in this study were analyzed by AAA, RP-HPLC, ESMS, and MALDI-MS to determine purity. Two peptide samples had less than 50% of the desired product, and three other samples had less than 70% of the desired product, as judged by their mass spectra, amino acid composition, and RP-HPLC retention time. Overall, the peptides were of excellent quality.

By the above criteria, racemization during peptide assembly was not found to be a serious problem in the majority of the participating laboratories. The minimum racemization observed was 1.9%. Interestingly, of the sets of samples with an unacceptable level of racemization (5–8%), some were among the samples with the highest chemical purity. For example, RP-HPLC profiles of two high purity samples exhibited 2.26 and 7.15% racemization, respectively (Fig. 6). The racemized peptide containing D-His was well resolved from the desired peptide by RP-HPLC. From examination of information provided by the laboratories, no correlation

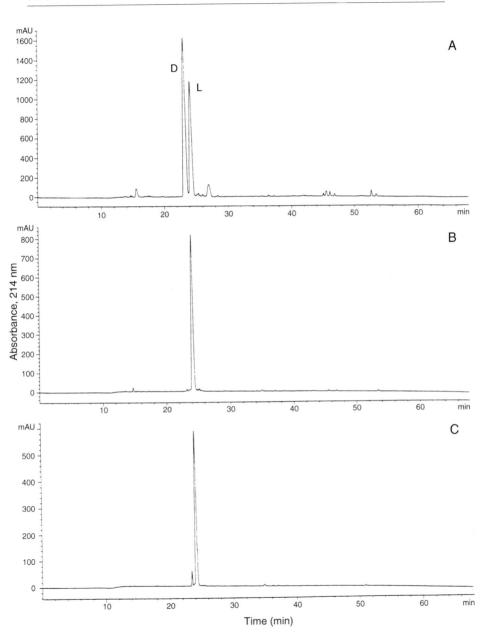

FIG. 6. RP-HPLC analysis of (A) a mixture of Arg-Glu-Arg-D-His-Ala-Tyr-OH (designated "D") and Arg-Glu-Arg-His-Ala-Tyr-OH (designated "L"), (B) an ABRF-96 sample containing 2.26% racemization, and (C) an ABRF-96 sample containing 7.15% racemization.

was found of the extent of racemization with coupling times, type of base in the coupling cocktail, nor the vendor or lot number of the Fmoc-His derivative starting material.

V. Understanding Strengths and Weaknesses of Analytical Techniques Used for Characterizing Synthetic Peptides

One of the benefits of the PSRC studies is that the committee had the opportunity to conduct side-by-side comparison of a variety of analytical techniques. Each technique had its own benefits and drawbacks. Amino acid analysis was the best technique for absolute amino acid quantitation, but it was not helpful for detecting modifications of amino acid residues and was susceptible to inaccuracies, probably due to the presence of scavengers in crude peptide mixtures. For example, in ABRF-94, AAA showed 80% of the crude products to be compositionally correct. Only one product had a complete deletion, corresponding to a des(Val,Lys,Arg) peptide synthesized by Boc chemistry. Of the other samples that were not compositionally correct, 87% showed at least a low Lys value, often accompanied by a low Ser, Thr, and/or Arg value. On the basis of sequence analysis and mass spectrometric results, neither the low Lys nor other low amino acid values could be demonstrated, indicating that AAA of crude peptides may sometimes suffer interferences. A similar effect was seen in the 1993 PSRC study.[3] In contrast, Edman degradation sequence analysis could determine residue position and appeared to be less susceptible than AAA to inaccuracies due to interferences.

Capillary electrophoresis allowed for excellent resolution of desired products from contaminating species but suffered from lack of reproducibility. Both the ABRF-91 and ABRF-92 studies found shifting retention times during CE, even though buffers of different pH values were used in an attempt to improve reproducibility. RP-HPLC using a C_{18} column appeared to provide an accurate estimate of the complexity of the peptide samples. RP-HPLC could overestimate the percentage of nondesired product due to the high UV absorbance of scavengers and side-chain protecting group adducts, and thus requries an authentic standard. Also, deletion peptides may have the same RP-HPLC retention time as the desired material.[2]

The MS techniques examined here allowed for identification of peptide modifications, including residual protecting groups. In one synthesis from the 1994 study, a peptide of one mass unit lower than desired was the result of using an Fmoc-Gln derivative instead of Fmoc-Glu(O*t*Bu). A similar error was found in the 1992 study, where one laboratory used the wrong resin, resulting in a peptide amide instead of the desired peptide acid.[2] The analysis of multiple samples allowed for the easy identification of products that had 1 Da deviation from the desired peptide; such distinctions may not be made when only isolated samples are analyzed. Overall, assessment of product purity by MS showed *semiquantitative* agreement with RP-HPLC analyses for the ABRF-91 to ABRF-96 peptides. MS as used in these studies did not distinguish between products of the same mass, nor did it allow for assignment of the positions of residue deletions and/or modifications.

The specific MS techniques changed over the course of the ABRF studies. Initially, PDMS, FABMS, and ESMS were utilized. FABMS and ESMS worked equally well for the molecular weight range of peptides used in the ABRF studies. PDMS tended to underestimate the quantity of desired product, as ionization of peptides containing side-chain protecting groups was enhanced compared to deprotected peptides.[1] Over time, PDMS became less popular and MALDI-MS rose to the forefront. MALDI-MS is useful for mass analysis over a broad molecular weight range of species. The only drawback to MALDI-MS that we found was that peptide concentration affected MALDI-MS quantitation.[4] Although only used for selected samples, LC-ESMS was particularly useful for characterizing complex mixtures of peptides, enabling direct assignment of RP-HPLC peaks as desired peptide or side products based on the observed molecular weights. For example, from one sample in ABRF-92, LC-ESMS indicated that the RP-HPLC peaks eluting at 10.3, 11.6, and 16.7 min correspond to the desired peptide, the desired peptide plus 106 Da, and the desired peptide plus 202 Da, respectively.

Our experience in the PSRC studies suggested that efficient characterization of synthetic peptides is best obtained by a combination of RP-HPLC and MS, with sequencing by either Edman degradation or tandem MS being used to identify the positions of modifications and deletions. Proper peptide characterization by multiple techniques was essential. A number of common side reactions were documented in the course of these studies. One puzzlement that remains is the lack of correlation between cleavage reagents or workup protocols with product quality. It appears that the experience and accuracy of personnel who carry out the procedures is as important or more important than the instrument or method.

Acknowledgments

We thank all of the ABRF core facilities that participated in these studies and the National Science Foundation for financial support (Grant DIR 9003100), and gratefully acknowledge the invaluable contributions of the PSRC members: R. H. Angeletti, L. Bibbs, L. F. Bonewald, S. A. Carr, G. B. Fields, S. A. Kates, J. W. Kelly, A. Khatri, D. R. Marshak, J. S. McMurray, K. Medzihradsky, W. T. Moore, A. J. Smith, J. T. Stults, S. T. Weintraub, K. R. Williams, L. C. Williams, and J. D. Young.

Author Index

Numbers in parentheses are footnote reference numbers and indicate that an author's work is referred to although the name is not cited in the text.

Pipkorn, R., 660
Pirrung, M. C., 337, 357(5)
Pirsch, M., 90
Pitkeathly, M., 587, 589(4), 664
Pivonka, D. E., 375
Place, G. A., 163
Plattner, J. J., 241
Plaué, S., 179, 195(11)
Pless, J., 32
Pluscec, J., 455
Pluskal, M., 116
Pohl, J., 399, 402(6), 410(6)
Point, J. J., 95
Pokorny, V., 342, 356(60), 358
Polakowski, J. S., 160
Polt, R., 221, 226(6), 241(6), 242(6)
Pombo-Villar, E., 261
Ponasti, B., 199
Pons, M., 85, 91(23), 209, 214(37), 215(37), 318
Ponsati, B., 205(r), 206, 218
Pop, I., 339
Porco, J. A. J., 341
Porreca, F., 221, 226(6), 241(6), 242(6)
Porter, J., 166(c), 169(n), 170–171, 368
Posnett, D. N., 417, 612
Posthumus, M. A., 522
Poteur, L., 262
Potts, J. T., 147
Powell, M. F., 241
Powers, D. B., 302
Powers, J. C., 158–159, 159(91, 92), 162, 162(91, 92, 95)
Powers, T. S., 355, 356(133)
Pozdnev, V. F., 163
Prasad, K. U., 164, 676
Prass, W., 585, 586(66)
Prescott, S. M., 565
Prestegard, J. H., 609, 676
Previero, A., 71
Prezegalinski, E., 161, 162(134)
Priestly, G. P., 50
Prochiantz, A., 565
Prodhom, B., 693
Prokai, L., 565, 566(13)
Prosser, R. S., 676–677
Proussios, K., 47
Ptitsyn, O. B., 647
Puett, D., 658
Pugh, K. C., 84, 85(6), 90(6), 119

Puigserver, A., 434
Pullman, W. E., 569
Pungercar, J., 208, 210(29)
Purisima, E. O., 671

Q

Qasim, M. A., 275, 277, 284(30, 33)
Quan, C., 298, 299(1), 303(1), 641
Quartara, L., 180, 321
Quibell, M., 120, 222, 237(19), 261, 264(59), 316, 318, 330, 330(9), 331(74), 332(19), 416, 500
Quillan, J. M., 345, 350(91)
Quine, J., 686
Quinkert, G., 562
Quinn, T. P., 572, 668

R

Rabanal, F., 135(p), 138, 172, 182, 322, 327, 331(40), 334(40), 444
Rabel, S. R., 427
Rabinovich, A. K., 119
Raboy, B., 15
Radford, S. E., 587
Ragona, L., 587, 590(7)
Rajasekharan Pillai, V. N., 47
Rakhit, S., 322(47), 323
Raleigh, D. P., 673
Ramachandran, G. N., 577
Ramachandran, N. S., 202
Ramage, R., 48, 70, 71(18), 76, 76(18), 77(18), 136(r), 138, 152(j), 153, 157, 166(e), 171
Ramakrishnan, C., 198
Ramalingam, K., 246
Ramamoorthy, A., 682, 696
Ramirez-Alvarado, M., 659
Rammensee, H., 623
Rampold, G., 162
Rance, M., 657, 659(32)
Randal, M., 641
Rao, C., 614, 632, 633(15)
Rao, J. K. M., 270
Rao, V. S. V., 76
Raphy, G., 157, 166(e), 171
Rapp, E., 193
Rapp, W., 47, 83, 87, 90, 92, 92(39), 162, 340
Raschdorf, F., 123
Rauwald, W., 238

Subject Index

A

U